ORDER NUMBER EA-421

AVIATION METEOROLOGY UNSCRAMBLED:

For VFR and IFR Operations/ Certificates and Ratings

By Kenneth B. McCool

International Standard Book Number 0-89100-421-1
For sale by: IAP, Inc., A Hawks Industries Company
Mail To: P.O. Box 10000, Casper, WY 82602-1000
Ship To: 7383 6WN Road, Casper, WY 82604-1835
(800) 443-9250 ❖ (307) 266-3838 ❖ FAX: (307) 472-5106
HBC1192 Printed in the USA

IAP, Inc.
7383 6WN Road, Casper, WY 82604-1835

Fifth Edition
© 1992 by Kenneth B. McCool
All Rights Reserved

Except as permitted under the United States Copyright Act of 1976, no part of this publication may be reproduced or distributed in any form or by any means, or stored in a database or retrieval system, without the prior written permission of the publisher.

Printed in the United States of America

Preface to the fifth edition . . .

The major goal of the fourth edition was to include AVIATION WEATHER SERVICES and related appendices, and to expand the instructor supplements (now over 150 pages). Future changes in AWS will be made in Appendix A; such changes should be mostly minor. The primary effort in the fifth edition was to make the text more readable, and at the same time update the text as well; many sections and/or paragraphs have been rewritten. However, no meteorology text is completely easy to read, unless such a book totally ignores the reasons why the atmosphere behaves as it does. There are many pieces to the puzzle; therefore, one must complete a meteorology text before the pieces appear to fit together.

More information has been included on stability and soaring, as well as a new optional chapter (seventeen – Appendix E) on IFR weather alternate requirements, supplementary information on the AIM (inflight weather advisory broadcasts), and on Arctic and Tropical flying weather. Also included are the answers to frequently asked questions that are difficult to answer. In fact, Appendix D is devoted entirely to the relationship between several important atmospheric variables.

Appendix C has been revised, and now contains questions from all FAA Question Books, those pertaining to the seventeen chapters being grouped by the chapter that contains the answers to those questions. The groupings at the end of each chapter in the text as PRI (Private), COM (Commercial), etc., are not as valid as they once were, for many of the questions have been shifted up or down between FAA Question Books. Actually, more questions have been shifted down than up; the exams have been getting tougher ever since the exam books have been published. Most of the chapter questions, about three-fourths, are still on some exams; the other one-fourth are very close to being the same as ones on current exams, and those that are not will likely reappear in the future.

Appendix C also contains all of the FAA questions on AVIATION WEATHER SERVICES, private through ATP. Appendix C and/or instructor supplements will be updated to reflect significant changes in FAA exam content. FAA private, commercial, and instrument exams are released in even years and instructor and ATP exams in odd years. Appendix C also contains a complete copy of AC 00-30A and AC 00-50A; most of this information is in the text, but the complete copy is contained in Appendix C for easy reference.

And finally, a word about organization. It is true that some content appears in more than one place. This "spiral approach", in which some concepts are reviewed and extended as the book progresses, may be overdone; however, a total reorganization of the text would not be feasible. Indeed, wind shear in one form or another is involved in practically every chapter, and that is one reason why wind shear is such as ominous flying hazard.

About the author. . .

EDUCATION: U.S. Air Force Certificate in Meteorology. B.S., M.S., Mathematics. Ph.D., College Teaching.

AVIATION CERTIFICATES: Commercial Pilot, Instrument Rating. Certified Flight Instructor. Ground Instructor, Advanced and Instrument.

EXPERIENCE: Military Weather Officer (Captain, USAF) – briefer, wing weather officer, detachment commander; Civil Weather Forecaster – television and radio weathercaster (American Meteorological Society Seal of Approval for Television Weathercasting, 1976); Educator, in Aviation, Meteorology, Environmental Science, Physical Science, Mathematics, and Computer Science – college chief ground instructor, taught aviation weather and aviation education courses (including CFI revalidation courses); Meteorological Consultant (approved as a Certified Consulting Meteorologist, CCM, by the American Meteorological Society) – provide industrial meteorological services (President, Meteorological Associates).

PROFESSIONAL AFFILIATIONS: American Meteorological Society, Aircraft Owners and Pilots Association, University Aviation Association.

Book designed to be used. . .

1) as information for aviation personnel at all levels of experience who wish to develop a greater understanding of meteorological principles and their effects on aviation, and

2) as a textbook for aviation students working toward certificates and/or ratings. The text is suitable for the formal classroom setting or for self-study; the material presented provides the background necessary to answer questions asked by the FAA for private through airline transport certificates, and for instructor certificates. In fact, almost three hundred questions from FAA Written Study Guides (private through ATP and flight instructor) are included at the end of the first fifteen chapters, and over three hundred additional questions are in Appendix C (from FAA question books).

Aviation Meteorology Unscrambled and *Aviation Weather Services* (Appendix A), with some topics omitted, provide enough material for a full course of study (approximately forty-eight hours), and when treated very thoroughly, provide enough material for two courses in aviation weather, particularly if laboratory sessions are held for data and chart analysis and application. The author knows that many individuals and aviation schools use *Aviation Weather* in conjunction with *Aviation Weather Services* for a basic course in aviation weather. In fact, this author used that combination for several years. *Aviation Weather* is good information; it, like *Aviation Weather Services,* is an advisory circular and is published by the FAA and NOAA. The author wrote *Aviation Meteorology Unscrambled* partially as an alternative to *Aviation Weather* for weather study, feeling that objectives, questions from written exams, more depth and coverage, etc., were desirable. In these regards, the author believes that *Aviation Meteorology Unscrambled* is far superior – individual readers/instructors will need to decide for themselves. Of course, in addition, the text contains a reprint of *Aviation Weather Services* (future changes should be slight, but Appendix A will be updated to reflect any significant changes). The text also contains many of the features of a workbook due to the objectives and questions in both the body of the text and the appendices. Thus the text is essentially three publications, all in one book.

The reader of the text has a good comprehension of the material when he can answer all of the questions at the beginning and end of each chapter, and in Appendices B and C. The author has prepared other multiple-choice questions for examination purposes that will test the reader's comprehension of the material, both for the text and for *Aviation Weather Services.* As was mentioned in the preface, other instructor supplements are also available (answers to the objectives in the text and Appendix B, and to the multiple-choice questions from FAA Question Books in Appendix C).

INTRODUCTION

To be a successful pilot one must develop good flying habits. For example, a pilot on final approach to a runway who encounters a sudden wind gust as he is about to touch down must react, for there is no time to think through the problem. However, I believe that it is extremely important for pilots to develop proper flying habits through understanding.

When a pilot understands why a habit is necessary and how it resolves a problem, then proper habits will stick with him longer and he will react more quickly when encountering the problem. It is impossible, in the time usually devoted to pilot training, for a pilot to understand the why and how of everything he does in aviation at a given instant. However, that is no excuse not to try; the burden is on both the student and the instructor in such an effort. It is much more efficient (and less expensive) to understand ''why and how'' on the ground than it is to learn by ''trial and error'' in the air. And more importantly, the why and how might save lives at a later date.

This text is designed for all aviation personnel who wish to understand why weather behaves as it does. I have emphasized the COVERED concept throughout — COmprehend, Visualize, Evaluate, Respond, re-Evaluate, and Determine. It is my opinion that too much emphasis has been placed on Respond. The response will be more appropriate and effective when the pilot, in his learning, has first COmprehended, Visualized, and Evaluated the flying situation. Then after a pilot Responds, he is better prepared to re-Evaluate and, if necessary, Determine a new course of action.

Before I discuss the content of the book, I would like to emphasize that I myself am at best an average pilot, in terms of ability. Despite my training and experience, I have been humbled on numerous occasions. Each person who reads this book has certain talents and abilities which far exceed mine; in all probability you will be a much better pilot than me. Thus, please forget that I have advanced degrees and read and study the material as information from me to you in the sincere hope that you will find flying more enjoyable. I hope that you will become more knowledgeable and, hence, more confident, and at the same time increase your safety margin.

My approach has been to ''start at the beginning'' in my presentation of atmospheric principles. The first five chapters are somewhat technical, compared to most aviation books of this nature. However, no background in physics or mathematics is necessary. Of more importance is tenacity, the discipline to read and re-read certain areas many times. The technical level is no more than that for liberal arts (non-science) college students who take an elementary course in meteorology as a science elective. The more experienced pilot will spend less time gaining an understanding of the principles discussed in the early chapters. The remaining chapters are more straightforward. However, the first five chapters are essential for understanding and will prepare the student for further studies in aerophysics, engine performance, etc.

You will find that an understanding of weather principles will help you to visualize or form a mental picture of weather systems, leading you to make the practical decision ''Do I fly?'' or ''Do I stay on the ground?'' What many pilots view as poor weather when they look outside may in fact be good flying weather or may change into good flying weather in just a short period of time. And, of course, the reverse is also true. What may appear to be beautiful flying weather may in fact be poor flying weather or may quickly become so. Having the confidence that comes with exercising judgment concerning weather factors is one of the keys that contribute to the enjoyment of flying.

Now while knowing where smooth air is on a particular flight will increase your enjoyment, certainly of much more importance is flight safety, that is, the prevention of accidents. In a study conducted by the National Transportation Safety Board, one out of every ten accidents resulted in fatalities. One out of every six accidents was weather-related. Even more alarming is the fact that three out of every ten people who perished in an aircraft accident died as a result of a weather-related factor. Out of the three people who perished due to weather (again, this three is out of each ten aircraft fatalities), two of the three died due to continued flight into adverse weather. The following are some of the weather factors or causes relative to weather accidents: unfavorable wind, low ceiling, high density altitude, fog, updrafts and downdrafts, rain, thunderstorm activity, induction icing, snow, turbulence with clouds and thunderstorms, structural icing, and obstructions to vision. By understanding weather principles you will be better able to visualize weather patterns in order to exercise the pilot judgment necessary to prevent accidents.

I have drawn upon many years of training and experience in an effort to provide as complete a text on ''understanding and applying meteorology'' as possible for aviation personnel. In order to ''unscramble'' aviation weather for you, I may have to temporarily scramble it for you. Be patient. Read and study the material until you begin to see this presentation as a unified whole. Each chapter represents a piece of a puzzle. Only after you have completed the entire text will you realize why the material was presented as it was. Just as you must master the airplane in order to become a competent pilot, I hope that this book will enable you to become a more knowledgeable and confident pilot where weather is concerned.

In addition to drawing on my own experience and training, I have tried to include thoughts and topics from various government publications, such as Air Force Manual 51-12, the Airman's Information Manual, Exam-O-Grams, FAA Written Study Guides and Question Books, and Advisory Circulars. Some of the material would be treated differently by the meteorology purist, such as the ''steady-state'' thunderstorm. However, my purpose is to try to enhance understanding, not to try to change the language of aviation meteorology.

I consider the emphasis on upper levels of the atmosphere to be one of the major features of this book. Another major area of concentration has been altimetry; the book also emphasizes aircraft performance. The VFR visibility and cloud minimums have been included. There is a chapter on flight safety, including a comparison of legal flying and use of common sense. And metric units have been included — some of the conversions, in parentheses, are exact and some approximate.

Perhaps the most significant feature of the book is that it encourages the reader to visualize weather. When you receive weather information, your comprehension of weather factors should provide you with a mental image of weather along your route of flight plus potential changes. Only when you can visualize weather in your ''mind's eye'' are you truly a competent pilot with regard to weather considerations. Despite the inclusion of many illustrations in the book, you should develop the ability to visualize as much as possible. In fact, reference to illustrations is accomplished parenthetically, indicating that such figures or tables are usually supplementary to the paragraph containing the reference. For continuity and reinforcement, you should usually complete the entire paragraph before referring to the illustration(s). There is no firm rule as to where in a paragraph a reference is made to an illustration; in general, such illustrations correlate with the entire paragraph.

I would like to thank Dr. Eugene Wilkins, CCM, Jim Branch and Curt Callaway for their suggestions relative to the content and wording of the original transcript. I would also like to thank Ed Curtis of Mountain View College for his suggestions relative to the second edition. However, I accept full responsibility for what has been said and the manner in which it has been stated.

Kenneth McCool, President
Meteorological Associates
September 1, 1989

TABLE OF CONTENTS

LIST OF ILLUSTRATIONS

LIST OF TABLES

Table *Title* *Page*

INTRODUCTION

The rapid expansion of air transportation makes necessary a move toward mass briefings to meet aviation demands. As a results, you, the pilot, must become increasingly self-reliant in getting your weather information. On occasion, you may need to rely entirely on self-briefing.

This advisory circular, AC 00-45C, explains weather service in general and the details of interpreting and using coded weather reports, forecasts, and weather charts, both observed and prognostic. Many charts and tables apply directly to flight planning and in-flight decisions.

This advisory circular is an excellent source of study for pilot certification examinations. Its 14 sections contain information needed by all pilots, from the student pilot to the airline transport pilot.

AC 00-45 is updated periodically to reflect changes brought about by the latest service demands, techniques, and capabilities. The purchase of an updated copy is a wise investment for any active pilot.

Comments and suggestions for improving this publication are encouraged and should be directed to:

National Weather Service Coordinator, AAC-909
Federal Aviation Administration
Mike Monroney Aeronautical Center
P.O. Box 25082
Oklahoma City, OK 73125

Advisory Circular, AC 00-45C, supersedes AC 00-45B, Aviation Weather Service, revised 1979.

NOTE: The following pages have been partially or totally changed by the author of this textbook, due to information obtained from other government documents: 4-8 through 4-11; 4-13; 4-17,18; 8-7,8; 11-1,2; 12-3; and 14-4.

It is quite likely that some FAA Question Books still use some of the old values on a few questions. For example, "WND" in the categorical outlook section of an FA used to refer to sustained or gusty surface winds of 30 knots or more; that value is now 20 knots. If a question on this issue (or any other issue) is encountered on an FAA exam and both 20 and 30 knots are listed as possible answers, then the question should be "written up" by the person taking the test. That is, it should be written that 30 is the correct value from AC 00-45C but that 20 is being used in current FAs. Of course, this is the procedure that I (the author) would use if I were taking an FAA exam and encountered such a question; however, I cannot predict whether or not credit will result from following such a procedure.

REFERENCES:

Holley, John, AVIATION WEATHER SERVICES (Revised by IAP, Inc.).
National Weather Service, OPERATIONS MANUAL (Part D, Chapters 20 and 21).

CHAPTER I
PRESSURE

The reader should demonstrate an understanding of pressure to the extent that he can answer the following questions (as stated below, or as multiple-choice, etc.):

1. How many planets are in our solar system? In which galaxy is our solar system contained? About how much of the earth's surface is covered by water?
2. Define "atmosphere"; "troposphere"; "stratosphere".
3. State the average mixture of dry gases in our atmosphere. State the variability of water vapor in our atmosphere.
4. Name the "states of matter".
5. Of what two basic kinds of raw material is matter composed? What special name is given to this collection of raw material?
6. Define "density".
7. Why is the concept of mass important?
8. Define "velocity; acceleration".
9. State Newton's first law.
10. Define "momentum".
11. State Newton's second law, both verbally and symbolically.
12. An imbalance in force causes a change in velocity and, hence, a change in ______________________________.
13. State Newton's law of universal gravitation. What is the approximate value of the acceleration due to the earth's attraction for an object on its surface?
14. When the acceleration due to the earth's gravitational attraction is used in the equation F = ma, then F is usually called ______________________________.

15. Use the equation m = F/a to determine the mass of an object weighing 19,200 lbs (8.7 kg).
16. Starting with the assumption that the atmosphere is composed of a mixture of gases, explain why the atmosphere has weight.
17. Define "pressure". What are the units of pressure?
18. At what approximate altitude is atmospheric pressure reduced to half its value at mean sea level?
19. Why does density decrease with height? What would be the effect of higher flight altitudes on aircraft performance?
20. Define "isobar"; "pressure gradient"; "high"; "low"; "ridge"; "trough"; "col".
21. Pressure varies more (horizontally, vertically) than it does (horizontally, vertically).
22. Define "pressure altitude".

CHAPTER I

PRESSURE

A. Earth's Atmosphere

We begin by putting our planet in perspective, as merely one of ten planets (includes a recently discovered planet) in our solar system, which is contained within the Milky Way galaxy, one of billions of galaxies in our universe. One of the unique features of our planet is that 70% of its surface is comprised of water. Our atmosphere is a mixture of gases that surrounds the earth and travels through space with it. There is no definite boundary to our atmosphere; it just becomes thinner and thinner as it blends with interplanetary gases

Most weather and flying take place in the troposphere, which extends vertically up to an average height of about 7 miles (11 kilometers) above the surface. Some flights, mostly test and supersonic, take place in the stratosphere, which extends from approximately 7 to 30 miles (11 to 48 km) above the earth's surface (Figure 1-1).

We mentioned that the troposphere is a mixture of gases. On the average, "dry air" is about 78% nitrogen, 21% oxygen, and 1% other gases. The amounts of dry gases are reduced proportionately when water vapor is present. Water vapor usually ranges from near 0 to 5% by volume.

B. Matter

We will be concerned here with three states of matter: solid, liquid and gas. These forms of matter are composed of certain raw material, atoms and molecules (combinations of atoms), particles which are too small to be seen. We will call this raw material mass. Are we drifting too far from aviation? Not at all, for the definition of density is mass per unit volume (.0024 slugs per cubic foot – 1.24 kilograms per cubic meter – is the value for air at standard sea level). The mass of air encountered by the propeller and wing is crucial to their performance.

C. Velocity and Acceleration

Velocity means speed and direction. An airplane travelling at a speed of 100 knots (115 miles per hour) will cover 100 nautical miles in an hour, for speed is the ratio of distance to time. Travelling 50 nautical miles in 30 minutes also represents an average speed of 100 knots. Speeds of 100 knots to the north and south represent different velocities since the directions are different. According to Sir Isaac Newton's first law, or the law of inertia, a body at rest will remain at rest and a body in motion will continue in motion in a straight line at a constant speed unless either body is acted upon by an imbalance in force (net force). For example, in the absence of the frictional forces that create drag, a landing aircraft will continue to roll at the speed at which it lands in a power off landing. However, in the presence of frictional forces and the absence of power to create thrust, the landing aircraft will eventually come to rest – the net retarding force of drag has caused the aircraft to decelerate, for acceleration and deceleration (negative acceleration) are changes in velocity. If the forward acting force of thrust and the rearward acting force of drag are the same value, then the forces are balanced and there is no change in speed.

In the above discussion we focused most of our attention on speed changes rather than on changes in direction. However, we are all familiar with the fact that a banked track will cause a car

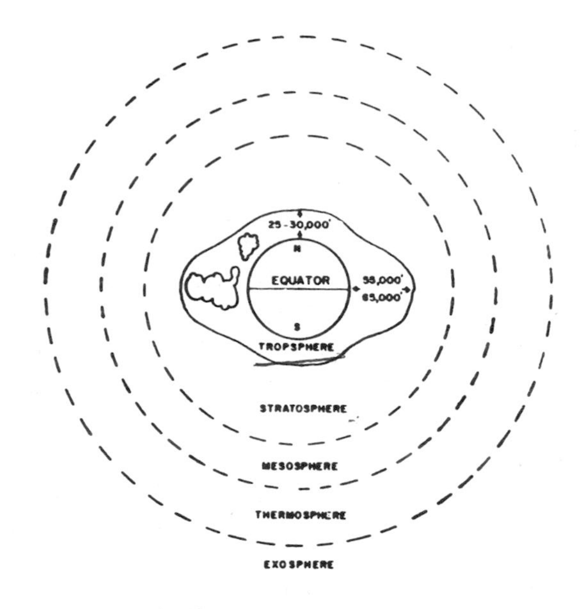

Figure 1-1. Vertical structure of the atmosphere.

to change its direction, and as a pilot you either have or will practice a constant radius of turn at a constant speed in an airplane. Changes in direction will be of special importance in our later study of wind shear (a sudden change in wind speed and/or direction – wind velocity).

D. Momentum and Newton's Second Law

Momentum is defined as the product of the mass of an object and its velocity. Momentum will be important when we discuss maneuvering speeds (Chapter III). In our discussion we will keep mass constant, so changes in momentum can only arise from changes in velocity. And since acceleration means a change in velocity, acceleration causes a change in momentum. Acceleration can be calculated from Newton's second law, F = ma (thus, a = F/m), where F stands for net force, m for mass, and a for acceleration. Thus, a net force (imbalance in the sum of all forces acting on an object) produces acceleration, a change in velocity, and hence a change in momentum. Thus, velocity, acceleration, momentum, and Newton's first and second laws are all "tied together" by the notion of force (nature's pushes and pulls).

E. Calculation of Mass

One of nature's forces is weight, which exists due to the often misunderstood notion of "gravity". If you drop a rock out of your hand, it falls (accelerates) toward the earth. What causes earth's gravitational pull? Well, it is an application of Newton's law of universal gravitation, namely, that in our universe any two given objects attract each other, the force of attraction between them being dependent primarily on the masses of the two objects and the square of the distance separating them (the larger the masses the greater the attraction and the larger the distance the less the attraction). The rock and the earth are accelerated toward each other, but the rock is the only one that can be seen to move when dropped since the earth is so much more massive. Therefore, we disregard the gravitational attraction of the rock for the earth and speak only of earth's gravitation (the gravitational attraction of the earth for the rock).

As the rock falls, gravity is a net force acting on the rock (disregarding air friction), causing it to accelerate at the rate of 32 feet per sec per sec (9.8 meters per sec per sec) near the earth's surface; the rate would be less at a higher altitude in the atmosphere since the rock and the earth would be farther apart. If it were not for the surface of the earth, the rock would continue to pick up speed in its descent – at the end of one second, its speed would be 32 feet per second, at the end of two seconds its speed would be 64 feet per second, etc. An airplane on the surface is not accelerating downward since the weight of the airplane is being opposed by the earth's surface. This is a reflection of Newton's third law – for every force that acts on one body there is a second force equal in magnitude but opposite in direction that acts upon the other body. In straight and level flight lift equals the weight of an airplane, with lift acting upward and weight downward, so the airplane does not accelerate upward or downward.

Assume that we have an airplane weighing 6400 lbs. Since gravity is responsible for weight, we use its value to calculate the mass of the airplane. Now we have the equation 6400 lbs equals mass times the acceleration due to gravity, that is, 6400 lbs (2900 kg) equals mass times 32 feet per sec per sec (9.8 meters per sec per sec). Since 200 times 32 is 6400, we have a mass of 200 units, and for lbs, feet and seconds, this will be the unit called slug. Thus, the raw material of our airplane is 200 slugs (2932 kg). While a slug is not easy to visualize, its importance is in its use in relative measurement; that is, 400 slugs is twice as much mass as 200 slugs and, since force equals mass times acceleration, doubling the mass doubles the force.

Now back to the gases of the atmosphere. Since gas is a form of matter and matter is composed of mass, and since weight is mass times the acceleration due to gravity, we conclude that our atmosphere has weight. Now you may say you would have been convinced of this fact without all the physics. But not only have we shown that the atmosphere possesses weight (rather than assume so), we have presented principles which will be important in your understanding of other aspects of flying, namely, aerodynamics, engine performance, etc.

F. Pressure Defined

Our next step in this discussion is the definition of pressure, as force per unit area; we will use units of pounds for force and square inches for area (we will also use millibars, the metric equivalent). The pressure we are interested in at the moment is that of the atmosphere. Since gases have weight, a force, they exert pressure. To determine this pressure all we must do is place a one square inch surface parallel to the earth's surface at any given height in the troposphere and the pressure will be the weight of the air on the one square inch surface, namely, so many pounds per square inch. Now since the atmosphere is composed of gases and gases possess mass and gases are compressible, we would expect the greater amount of mass to be concentrated toward the surface of the earth. This is reflected in the pressure observed at different levels in the troposphere (Figure 1-2).

Near mean sea level on an average day, pressure is 14.7 lbs per square inch (1013.2 millibars). At 18,000 feet (5.5 km) the pressure is 7.32 lbs per square inch (504.6 mb). Thus, roughly half the mass of the atmosphere is contained in the first 18,000 feet of the atmosphere and, hence, the density of the atmosphere increases closer to the surface (recall that density is mass per unit volume). Therefore, aircraft performance will usually be greater nearer the earth's surface; for example, the propeller will move a greater mass of air rearward, the opposite and equal force due to Newton's third law being forward and called thrust in aviation.

G. Inches of Mercury

The most common indication of pressure is inches of mercury. We emphasize the word indication because 30 inches (76.2 centimeters) of mercury is not in units of lbs per square inch. 29.92 inches (76 cm) is the height that a column of mercury would be forced to attain in an inverted tube placed in a container of mercury near the earth's surface at mean sea level on a standard day. The height reached is a result of the pressure of the atmosphere on the mercury, namely, 14.7 lbs per square inch. If the container is placed at 18,000 feet, the column would rise to only about 15 inches (38.1 cm), once again confirming that about half the weight of the atmosphere is contained in the first 18,000 feet.

H. Pressure Systems

If we choose a fixed height in the atmosphere, say, 10,000 feet (3 km), we can analyze a horizontal surface at that height for pressure differences. There are many reasons for pressure differences, one of these being temperature, which we will describe in our next chapter. But assuming intuitively that there are pressure differences as a result of the vertical distribution of

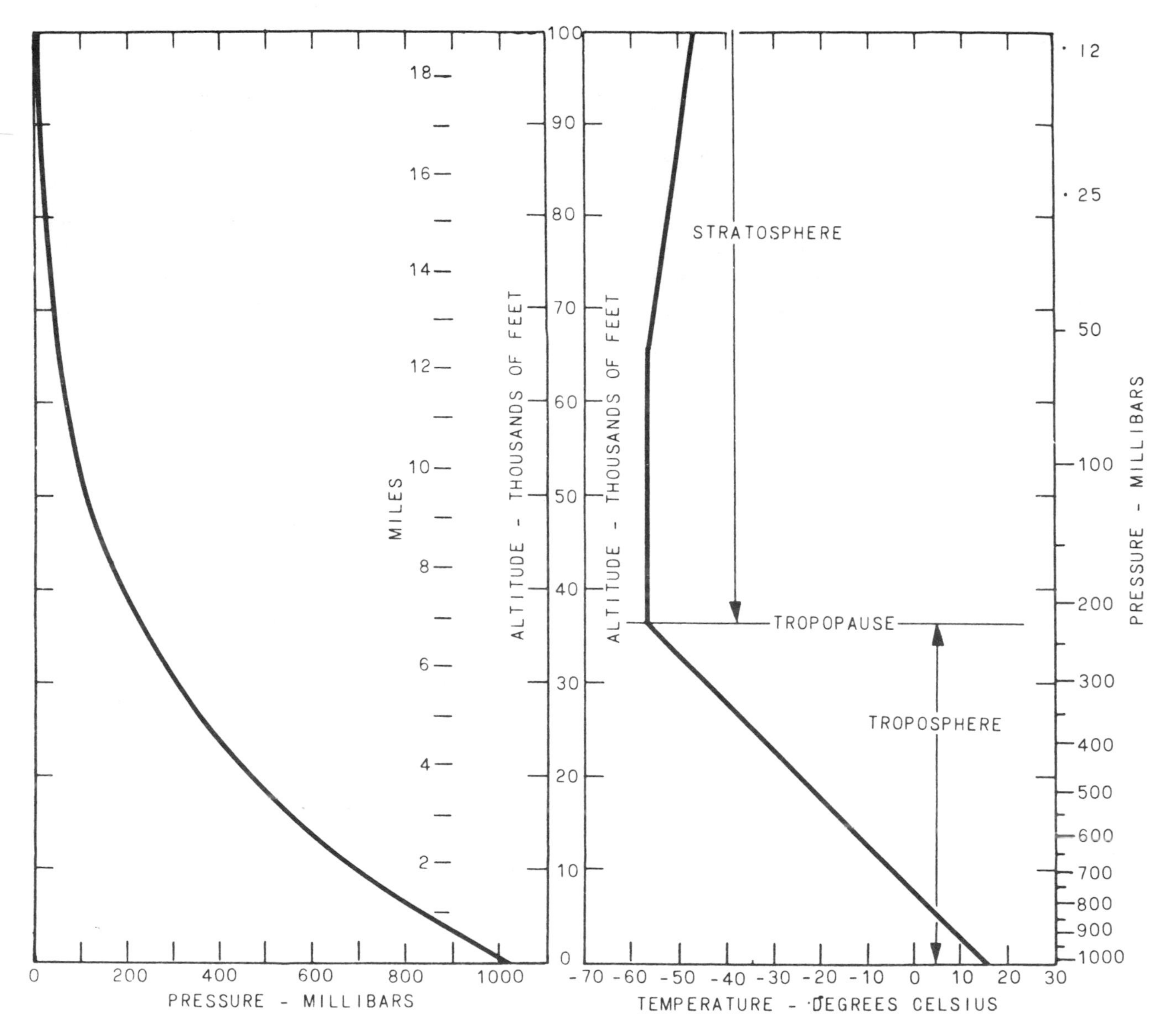

Figure 1-2. Changes of pressure (left) and temperature (right) with height from the surface to 100,000 feet in the standard atmosphere. Note the vertical scale in three units: miles, thousands of feet, and pressure in millibars. Note on the left, about half the atmosphere (the lower 500 millibars) is below 18,000 feet.

mass on a horizontal surface at a given height, we can draw isobars, lines of constant or equal pressure (later, you will see that in the upper levels contours are used rather than isobars). Suppose that at some spot on our horizontal surface the pressure is 700 millibars and that at a fixed circular distance from that point the pressure is 690 millibars. Then we have discovered a so-called high pressure area as a result of this horizontal pressure gradient — a change in pressure in a given distance. The formal definition of a high pressure area (system) is a center of pressure surrounded on all sides by lower pressure. Likewise, a center of pressure surrounded on all sides by higher pressure is called a low pressure area.

Now an isobar does not have to be circular. There may be an elongation or outward stretching in an otherwise circular isobar, for example. The area of elongation is called a ridge if associated with a high pressure area and a trough if associated with a low pressure area. Another

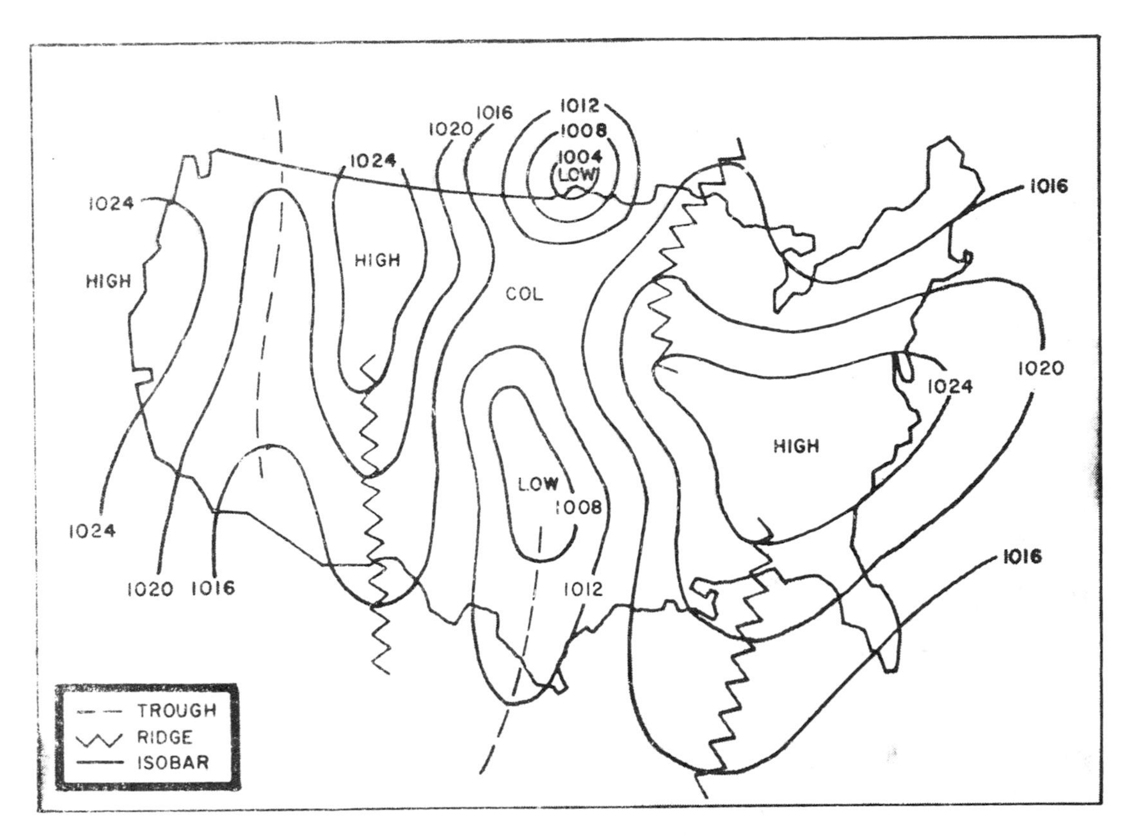

Figure 1-3. Sea level pressure patterns showing isobars, low pressure centers, high pressure centers, troughs, ridges and a col.

pressure area is the col, merely a zone between two highs and two lows, which is of minor importance (Figure 1-3). You are probably already anticipating the behavior or movement of air in a high pressure area. Certainly, air will move outward from a high pressure area, that is, air will move from higher to lower pressure; however, this movement will be seen to be influenced by other factors as well when we discuss wind. As a result of the outward motion in a high, air will be descending from above; the reverse occurs for a low. What do you speculate will be the nature of descent or ascent (straight vertical, spiral, etc.)? — answered in Chapter IV.

I. Pressure Changes

Now pressure does not vary nearly as much in the horizontal as it does vertically; changes in the vertical are reflected by a change in the height of a column of mercury on the order of about one inch per 1000 feet (10 mb per 100 m) in the lower troposphere. The small changes which do occur horizontally are extremely important in determining where the highs and lows are, which have a great influence on our weather, again to be seen later.

J. Pressure Altitude and Altimetry

We look at one final application of pressure in this chapter, namely, pressure altitude, which is defined as the altitude in the standard, or average, atmosphere where you would expect to find the same pressure as where you are (pressure altitude will be discussed in much greater depth in the chapter on altimetry). For example, suppose you are at 10,000 feet and you measure pressure with a barometer to be 19 inches of mercury. Since 20 inches is the approximate standard value found at that height, we have a case of high pressure altitude. This is because if 20 inches is the average value at 10,000 feet, then 19 inches would usually be found at about 11,000 feet, one thousand feet higher (remember, pressure decreases with altitude). Since altimeters, which indicate altitude, are calibrated for standard or average conditions, then other things being equal our altimeter will indicate 11,000 feet, which is 1,000 feet higher than we really are. Since we are at an altitude where pressure is lower than usual, pilots speak of this as "from high to low, look out below". Nineteen is lower than twenty and as a result we are lower than our altimeter indicates. This example has been vastly oversimplified, as you will discover in the chapter on altimetry, but it does at least indicate the importance of pressure altitude.

PRESSURE
Review Questions

CFI

1. By referring to isobars on a weather map, what can a person determine?
   a. Pressure gradient.
   b. Temperature changes.
   c. Areas of magnetic variation.
   d. Areas of precipitation.
2. Which of the following statements is true with respect to either high or low pressure systems?
   a. A low pressure area or trough is an area of descending air.
   b. A high pressure area or ridge is an area of descending air.
   c. Both high and low pressure areas are characterized by descending air.
   d. A high pressure area or ridge is an area of rising air.
3. With respect to pressure systems, which of the following statements is true:
   a. A low pressure area or trough is an area of rising air.
   b. A high pressure area or trough is an area of rising air.
   c. A low pressure area or trough is an area of descending air.
   d. Both high and low pressure areas are characterized by rising air.

CHAPTER II
TEMPERATURE

The reader should demonstrate an understanding of temperature to the extent that he can answer the following questions (as stated below, or as multiple-choice, etc.):

1. Define "energy"; "power"; "work".
2. Define "kinetic energy". What would be the effect on the amount of kinetic energy if mass is doubled? If speed is reduced from 200 to 100 mph (174 kts to 87 kts)?
3. How much of the sun's radiation that encounters earth's atmosphere actually reaches the earth's surface, on the average? What happens to the remaining portion?
4. State the two basic types of heat energy.
5. Define "temperature". State the freezing and boiling points on the Fahrenheit and Centigrade scales.
6. Write the formula which relates the Fahrenheit and Centigrade scales. What is the Fahrenheit temperature when the Centigrade temperature is 25°? What is the Centigrade temperature when the Fahrenheit temperature is 50°?
7. Define "adiabatic".
8. Define "horsepower".
9. State the dry adiabatic lapse rate; the standard temperature lapse rate.
10. If a parcel of air has a temperature of 50°F (10°C) at the surface, then what will be the temperature of the parcel when lifted dry adiabatically to a height of 6,000 feet (1.8 km) above the surface?
11. What are the three primary methods of heat transfer? Which one of these is the chief means of distributing heat in our atmosphere?
12. How do land and water compare relative to temperature changes?
13. What is the primary cause of daily variations in temperature?
14. At what time of day is the highest temperature usually reached? Why?
15. Which month of the year is the warmest? Why?
16. Define "density altitude". If an airplane is operating at an altitude of 10,000 feet (3 km) and the density altitude is 9,000 feet (2.7 km), how will that aircraft perform? Why?
17. Why does colder air aloft not sink when there is a standard lapse rate in the troposphere?
18. State the relationship between pressure, temperature, and density.
19. What is the effect of much colder temperatures aloft on density, and what will be the action of air in such conditions?

CHAPTER II

TEMPERATURE

This chapter will be extremely important, as we explore the key concept of temperature. But just as pressure was related to matter, mass and force, so temperature is related to heat and energy. In fact, temperature will turn out to be a measure of the direction of heat transfer, heat itself being just one form of energy. We will see how heat energy is changed to kinetic energy, that is, energy of motion, and how energy can accomplish work. In fact, energy is the capacity to do work, that is, the equivalent of a force acting through a distance, and power is the rate at which work is accomplished. And does power make a difference in aircraft performance? Well, it certainly does as we shall soon see. Thus, one result of this section will be a preliminary view of aircraft engine performance. We are way ahead of ourselves, so let's go back and develop these concepts clearly and concisely.

A. Kinetic Energy

We begin by examining the concept of energy a little more closely. We have spoken of energy as the capacity to do work and we defined work as a force acting through a given distance. Let's consider kinetic energy, the ability of mass in motion to do work. Suppose your car runs into a tree; this is not desirable, of course. Now, since your car possesses mass and it is in motion, it possesses kinetic energy, and when the car encounters the tree, it will do work on the tree – a great amount of force, the weight of your car, acting through a short distance will greatly alter the shape of the tree. And not only have you performed work on the tree, it has performed work on your car (you may not care to look, but the fact remains), and the work done in either case follows from Newton's third law – for every force exerted by one object on a second object there is an equal and opposite force exerted by the second object on the first object.

One additional note at this point on kinetic energy. The amount of this energy is $\frac{1}{2}mv^2$, that is, one-half times mass times the square of the speed of the object concerned. If speed is doubled, then kinetic energy is quadrupled. For example, if the speed was originally 30 mph (26 kts) and then doubled to 60 mph (52 kts), then 30 squared would be 900 (676) and 60 squared would be 3600 (2704), four times as great. In general, if v is the original speed, then 2v is the new speed; the new kinetic energy is ½ times mass times the square of 2v and this is ½ times mass times $4v^2$ (squaring 2v, that is, 2v times 2v, gives $4v^2$). Before we doubled the speed, our kinetic energy was ½ times mass times v^2, and afterwards, ½ times mass times $4v^2$; thus, the kinetic energy has been quadrupled. What would happen if we tripled the speed? Then since 3 squared is nine, the kinetic energy would be increased by a factor of 9. Incidentally, the aerodynamic forces of lift and drag acting on an airplane also vary as the square of the speed of the airplane.

B. Heat

Now with some understanding of one form of energy, let's go back and analyze heat to see why it can be classified as a form of energy. Our primary source of energy for the earth is the sun, which radiates electromagnetic energy. In fact, one law of physics states that all matter radiates electromagnetic energy. But because of the mass and extremely high temperature of the sun, it obviously radiates more than, say, you or me. One theory of radiation states that this radiation takes the form of waves traveling through space (Figure 2-1). Such waves come in many different lengths, together constituting the so-called electromagnetic spectrum. These waves give us visible light; however, much of the total radiation is not visible.

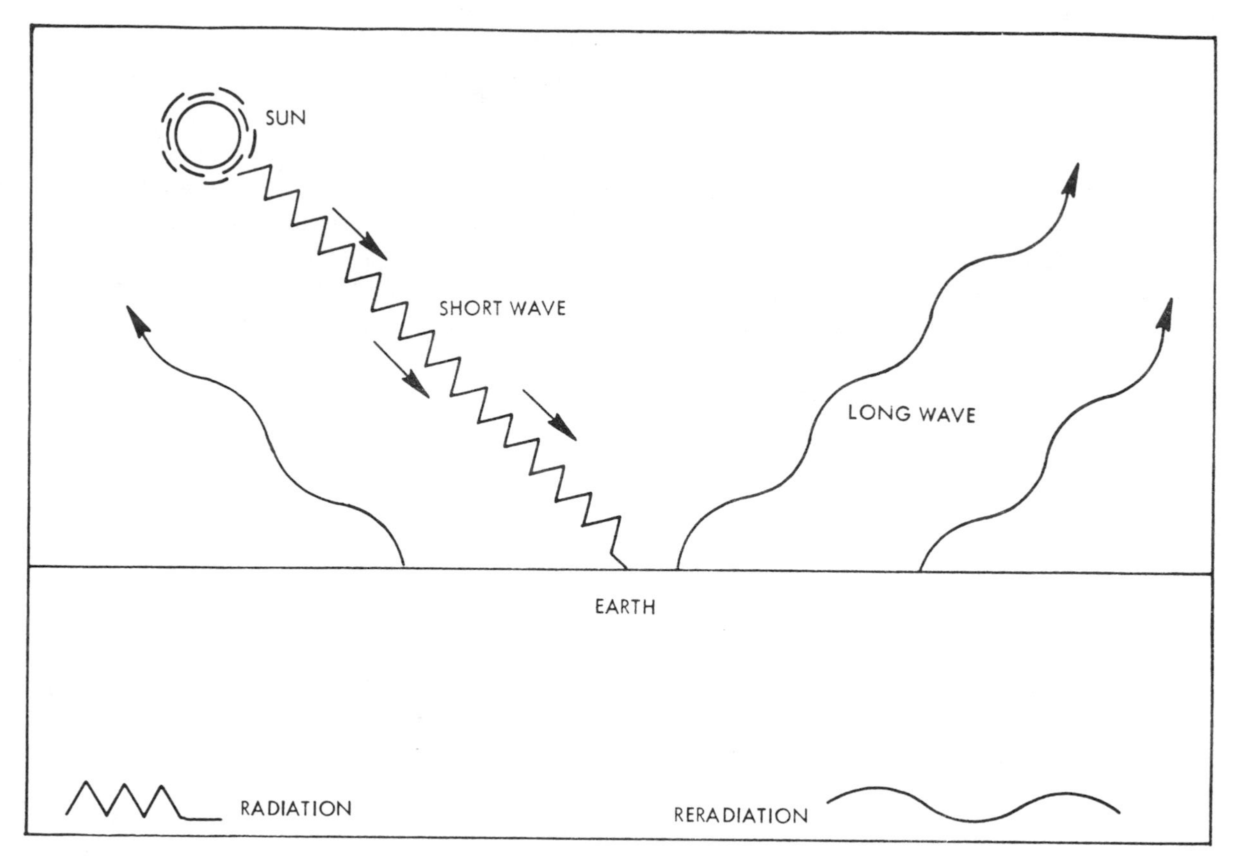

Figure 2-1. The sun radiates its maximum as short waves. Solar radiation reaching the earth is insolation. The earth absorbs the short waves heating the surface. The earth in return reradiates the energy as long wave radiation.

Approximately 43% of the solar radiation that enters earth's atmosphere reaches the surface. The other 57% is either directly absorbed (about 22%, mostly by ozone in the stratosphere) or reflected back to space (Figure 2-2). Short waves absorbed by the earth's surface heat land and water surfaces. Water vapor in the atmosphere absorbs long wave radiation (dry air is largely transparent to both short and long wave radiation; moist air is largely transparent only to short wave radiation). Water vapor converts long waves to heat, which is reradiated in all directions.

We will classify heat into two types, sensible and latent. We know when sensible heat is present by the motion of molecules. While water in a beaker may appear to be motionless, water molecules are moving or oscillating on a scale too small to be seen. Thus, sensible heat is kinetic energy of an internal nature. This internal motion in gases is in the form of high-speed travel of free molecules with frequent collisions with other molecules. Latent heat will be discussed in the chapter on moisture — we will find that it is "hidden" in the processes of evaporation and sublimation — we are concerned primarily with sensible heat here. We now state that sensible heat is a form of energy which molecules of matter possess because of their motion, an internal kind of kinetic energy.

C. Principle of Temperature

It has long been recognized that when two objects are placed in contact that there is usually a flow of heat from one to the other; that is, either the molecules of object A cause greater movement of the molecules of object B or vice versa. In the first case, A is said to have a

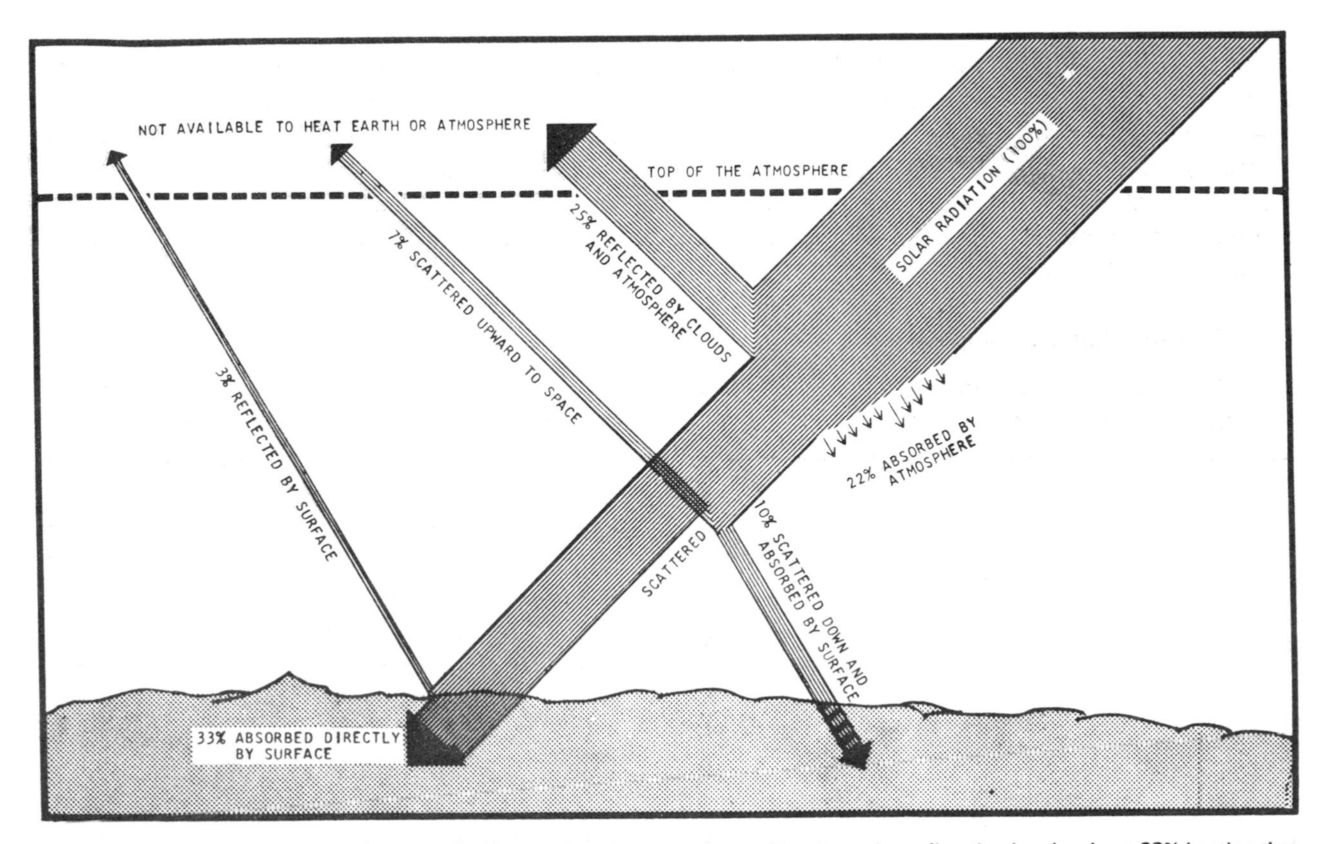

Figure 2-2. Average disposition of solar radiation entering the atmosphere. The atmosphere directly absorbs about 22% heating the air. 35% returns to space and is lost to both earth and atmosphere – 25% reflected by the clouds and atmosphere, 7% lost by scattering, and 3% reflected by the earth's surface. 43% heats the earth's surface, 33% by direct rays and 10% by scattering.

higher temperature than B. Thus, temperature is a measure of the direction of heat transfer. And, incidentally, this type of heat transfer is called conduction. When one places a calibrated tube of mercury in a beaker of water, either the mercury level rises, falls, or remains the same. For example, if the flow of heat is from the water to the thermometer, that is, if the molecular activity in the beaker of water causes a greater molecular activity in the mercury thermometer, then the mercury will expand more than the glass bulb and rise in the tube. Two primary reference points are on the Fahrenheit thermometer, the freezing and the boiling points of water, namely, 32° and 212°; on the Centigrade scale these points are 0 and 100, respectively (Figure 2-3). The formula relating these two scales is $F = (9/5)C + 32$. Standard sea level temperature is 59°F (15°C) (Figure 1-2). Note also that $C = (5/9)(F - 32)$.

D. A Comparison of Heat and Temperature

Let's compare the concepts of heat and temperature. Suppose we have a small and a large beaker of water, each at a temperature of 104°F (40°C). Of course, this means that they have the same degree of molecular activity, each causing the mercury in the thermometer to reach the 104° calibration point. But while both beakers are at the same temperature of 104°, the amount of heat energy present is different. In which beaker do you believe there is more heat? Well, it is the larger beaker. How can we convince you of this? Since energy is the ability to do work, let's see which one performs the most work by pouring the contents of each beaker onto the same size block of ice. I believe you can visualize the result. The water in the large beaker will melt more ice than will the smaller beaker, that is, the larger beaker of water is responsible for more work on the ice than the smaller one. Stated another way, there are more molecules available in the larger beaker to melt the ice. Therefore, while temperature is a measure of the degree or intensity of molecular motion, heat is the total motion (as a consequence of the presence of mass) available to do work (Figure 2-4). Thus, just because two objects have the same temperature, they do not usually possess the same amount of heat, even if they are of the same substance, such as water.

E. The Adiabatic Principle

There is a process in which the temperature of a substance can change without there being a change of heat in the substance. Such a process is called adiabatic; this is a vital concept. Here is an example of that process. On a clear summer afternoon the earth receives a large amount of solar radiation that in turn is reradiated by the earth's surface, causing the sensible heat of the air near the earth's surface to increase. Bubbles of air will start to rise, similar to the rising of bubbles in a heated beaker of water.

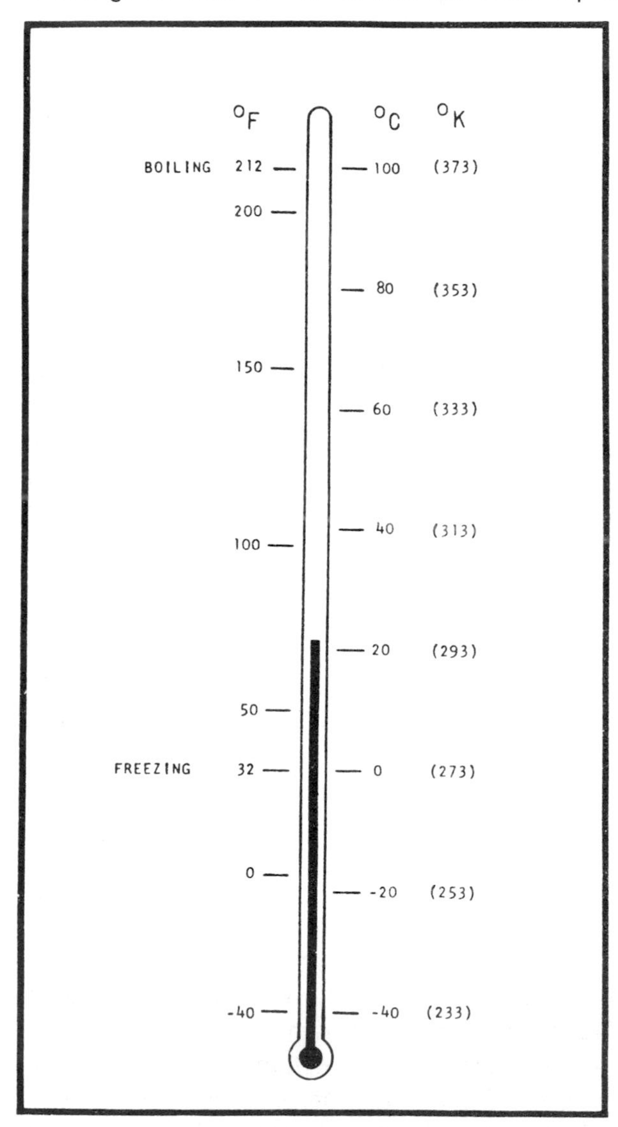

Figure 2-3. Comparison of three temperature scales. Fahrenheit (F) and Celsius (C) are the two scales commonly used. The Kelvin scale (K) is used in some scientific computations of the behavior of gasses. One degree Celsius equals 9/5 degrees Fahrenheit; the two scales read the same at −40°. One degree Kelvin is equal to one degree Celsius. 0°K is theoretically the coldest temperature possible and equals approximately −273°C.

The ascension of bubbles of air is in accordance with the buoyancy principle of fluids. For example, a balloon carrying a gas which is less dense than the density of the air the balloon and its contents is replacing is buoyed, or forced, or accelerated upward by a force equal to the weight of the air displaced. As the balloon ascends, external pressure decreases and the balloon expands; however, the density of the air that is being replaced by the balloon and its contents is rapidly decreasing, that is, the upward buoyant force is being decreased, so that the balloon will rise until the weight of air displaced in the new position is equal to the total weight of the balloon and its contents.

Now back to the rising bubbles of air. As did the balloon, these bubbles of air will expand. In order to expand, the air bubble had to perform work on the surrounding air, namely, molecular collisions provided the force which was exerted outward a given distance until balanced by an equal and opposing force, that is, until there was a balance in density, inward and outward. Now energy is the capacity to do work; through expansion, our bubble produced work on the surrounding air as a result of the motion of its molecules. The resulting reduction in the motion of molecules in our expanded bubble is indicated by a lower temperature.

It is very important to note that no heat was taken away from the bubble (the number of molecules remains the same); what has happened is that a part of the sensible heat (recall that sensible heat is internal kinetic energy as a result of molecular motion) has been used to make the bubble larger in volume. If we now reverse the process, that is, bring the bubble to a lower altitude, the bubble will become smaller, the motion of the molecules will increase, and a temperature increase will thus be indicated. Again, note that the same amount of heat was present, before and after expansion. When the bubble expanded, part of its internal molecular motion was transformed to outward molecular motion, that is, expansion. When the bubble was compressed, it regained its original size and its original level of internal molecular motion and, hence, its original temperature. We had a temperature change in the act of expansion and compression, but heat was not added nor taken away, an adiabatic process.

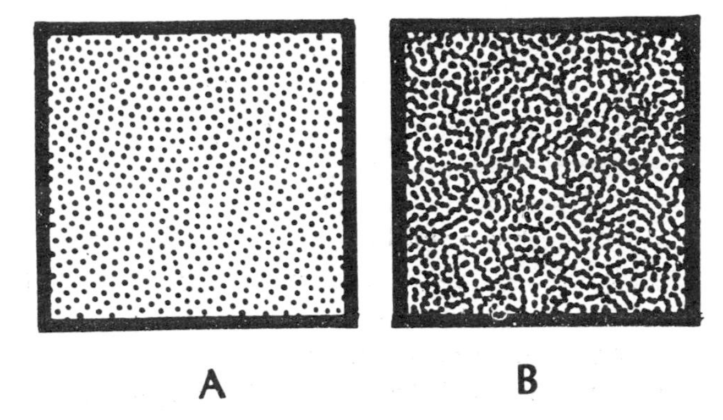

Figure 2-4. Temperature is the average energy of motion of molecules. Temperature can be the same in both A and B. Heat is the total energy of the molecules. B has more molecules than A and has more energy or heat than A because of the greater number of molecules.

The adiabatic principle is of importance in aircraft performance as well, particularly the engine. Adiabatic expansion generates work, that is, moves the piston a fixed distance with a force that is proportional to the mass and molecular motion of the compressed gas after burning. Note the word mass again; the less the mass of the fuel-air mixture, the less the force generated, even with the same degree of molecular activity. So as the airplane ascends in the atmosphere, the density of the air, mass per unit volume, is reduced and, hence, there will be less force on the piston. And this will cause a loss in power, for power is the rate at which work is accomplished. When you learn more about the aircraft engine you will find that a device called a supercharger will reduce this power loss.

One horsepower is defined as 550 ft-lbs per sec (745.7 watts), the ft-lb being the unit of work, a one lb force acting through a distance of one foot, and the second being of course the unit of time. Since power is work per unit time, that is, work divided by time, we see that 550 ft-lbs divided by two seconds gives us 275 ft-lbs per second, which is one-half of one horsepower. Thus, pushing 550 lbs one foot in two seconds represents one-half as much power as pushing that same weight, 550 lbs, the same distance, one foot, in one second. So again, power is the time rate of doing work. Back to our piston. As an airplane climbs, density decreases; thus, the mass of air in the cylinder decreases, the force on the piston decreases, the amount of work decreases and, hence, power decreases since power is work divided by time. The power generated turns the crankshaft which turns the propeller. Because of the force of friction, the power generated in the cylinders is reduced at the crankshaft and the propeller. The concept of power is important in describing such atmospheric phenomena as thunderstorms.

F. The Dry Adiabatic Lapse Rate

Now, back purely to the adiabatic concept. The dry adiabatic lapse rate is approximately 5.5°F (3°C) per 1000 feet (10°C per km). A lapse rate is a decrease in temperature with height.

You will see how the dry adiabatic lapse rate compares to the moist value when we cover the chapter on moisture. The so-called standard lapse rate is 3.5°F (2°C) per 1000 feet (6.5°C per km). This standard lapse rate is merely an atmospheric average value in the troposphere. The primary reason for the temperature increase with height in the stratosphere is the absorption of ultraviolet radiation by ozone (Figure 1-2).

A bubble or parcel of air that rises or descends will have a temperature change of 5.5°F per 1000 feet for reasons we saw earlier. Temperature decreases as the parcel ascends and increases by the same amount as the parcel descends, all without a gain or loss of heat. If the parcel was at a temperature of 75°F at the surface, it would be at a temperature of 20°F (75° − 10 X 5.5) at 10,000 feet and back to 75°F (20° + 10 X 5.5) if brought back to the surface. And this all occurred without a change in the amount of heat. Now certainly temperature varies horizontally as well as vertically. This will be obvious in the definition of fronts, which we will study later.

G. Methods of Heat Transfer

In the material just presented we have touched on three methods of heat transfer, radiation, conduction, and convection. Since the atmosphere is a collection of gases and since gases are poor conductors of heat, we may disregard conduction as being of major importance in the distribution of heat in the atmosphere. Convection is very important for it involves the transfer of heat in our atmosphere by means of mass movement of air parcels, thought of loosely as air bubbles in the vertical sense. However, convection can be either horizontal or vertical. Actually, horizontal convection (usually referred to as advection) is typically on a larger scale, involving the slow transfer of heat over vast distances, such as from equator to pole. In fact, on the whole, advection is the chief means of distributing heat in our atmosphere; latent heat and ocean currents account for important but lesser amounts of heat distribution. The importance of radiation is obvious, the receipt of which by the earth provides the heat for the processes of vertical convection and advection.

H. Temperature Variations

We have seen how temperature varies with altitude. It also varies with regard to topography, time of day, season, and latitude. Concerning topography, primarily land versus water, we observe that water absorbs and radiates heat with less temperature change than does land. This is true for the same mass of each. These influences are both diurnal, that is daily, and seasonal, and will be partially responsible for certain air circulations we will study in the chapter on wind.

Diurnal variations are caused primarily by the rotation of the earth on its axis. During the day the earth receives an excess supply of radiation. At night, solar radiation is nil but the earth's surface radiates heat and, hence, cools markedly. The cooling continues after sunrise until the earth once again receives more radiation than it emits. And, incidentally, the hottest part of the day is not local noon when the sun is most directly overhead, but around three to four pm. Even though the radiation received is greatest at noon, it takes time for this radiation to be absorbed, re-emitted, and transported to effect a temperature increase in the atmosphere.

Not only does the earth rotate on its axis, it also revolves around the sun. The earth's axis is tilted 23½°, and in the path of the earth around the sun the axis is either pointed toward the sun (summer in the Northern Hemisphere), away from the sun (winter) or in between (spring and fall) (Figure 2-5). June 21 is the longest day of the year (the sun is visible for the longest period of time on that day) and, hence, we receive the greatest amount of solar radiation on that day. But the hottest day is not until July, because of a lag comparable to the daily lag. That is, it takes a period of time for this maximum of radiation to result in the distribution of heat

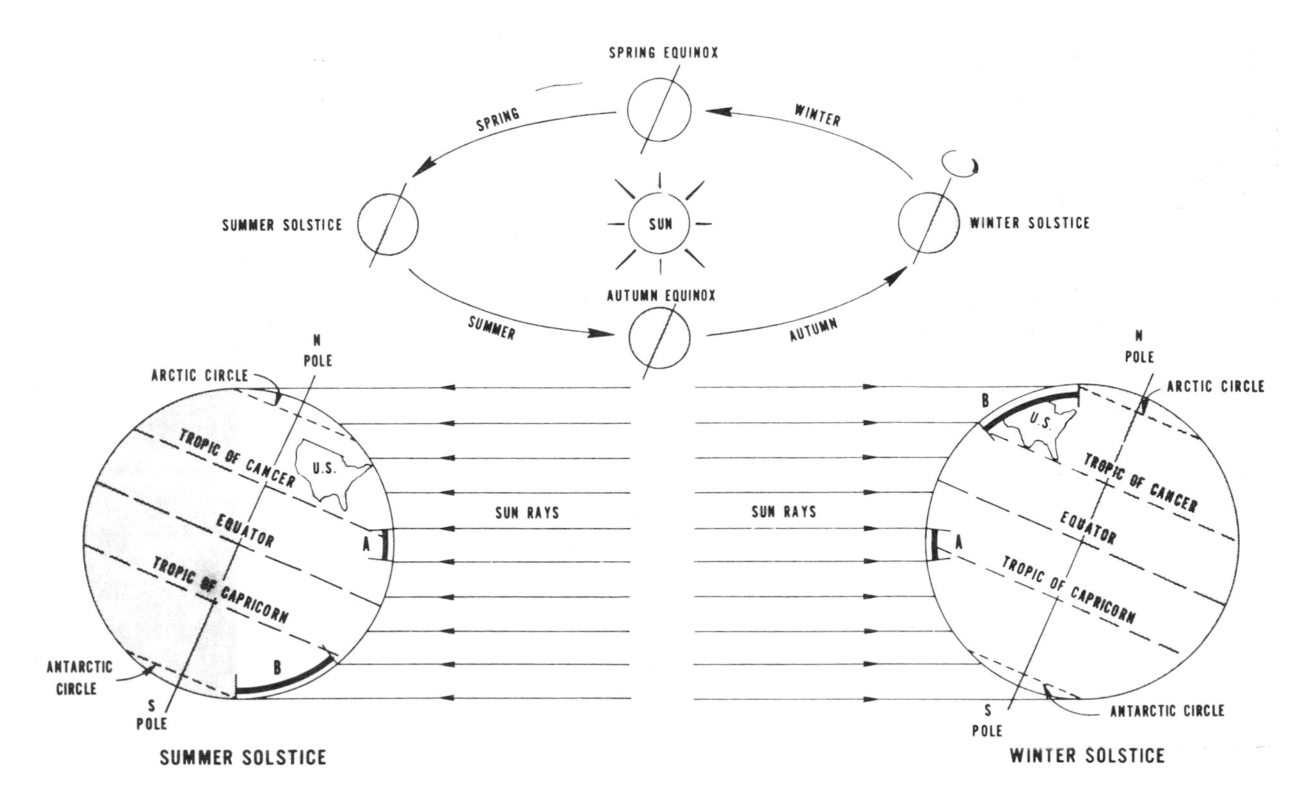

Figure 2-5. The seasons. At the summer solstice (left), the United States passes near the region of maximum heating at A. At the winter solstice, it passes through the region of minimum heating, B.

necessary to indicate the highest daily temperature. The same is true for winter, the shortest day being in December but the coldest not until January.

Finally, we consider the variation due to latitude (Figures 2-6 and 2-7). Since the earth is essentially spherical and since its axis is tilted no more than 23½° toward or away from the sun, the sun is more nearly overhead in equatorial regions than at higher latitudes. Actually, the location of points on the surface where the sun does reach directly overhead at noon ranges from 23½° south to 23½° north of the equator.

I. Principle of Density Altitude

Now we discuss the concept of density altitude, defined as the altitude in the standard atmosphere where the density is the same as where you are. Density altitude is strictly a performance criterion and can be determined primarily by considering temperature and pressure. Charts are available to convert pressure altitude and temperature to density altitude; this would actually be an approximate density altitude, for the moisture in the air will also affect density altitude. But since the amount of water vapor varies only from near 0 to 5% by volume, its influence is not as great as temperature and pressure considerations. We will consider the effect of moisture in that chapter. We are concerned here not with the calculation of density altitude but rather with what it is and something about its effect on aircraft performance. We will calculate density altitude in the chapter on altimetry, performance, and airspeed.

Suppose your airplane is at an altitude of 5000 feet (1525 m) and that density altitude was computed to be 6000 feet (1830 m). Then the aircraft will be trying to perform at 5000 feet in air that is normally found at 6000 feet, and air that is usually at 6000 feet is less dense than air at 5000 feet. This means there is less mass of air for the propeller to turn through (recall that density is mass per unit volume), so there is less thrust and hence less forward movement. In other words, the airplane does not perform as well (Figure 2-8). This has been an example of

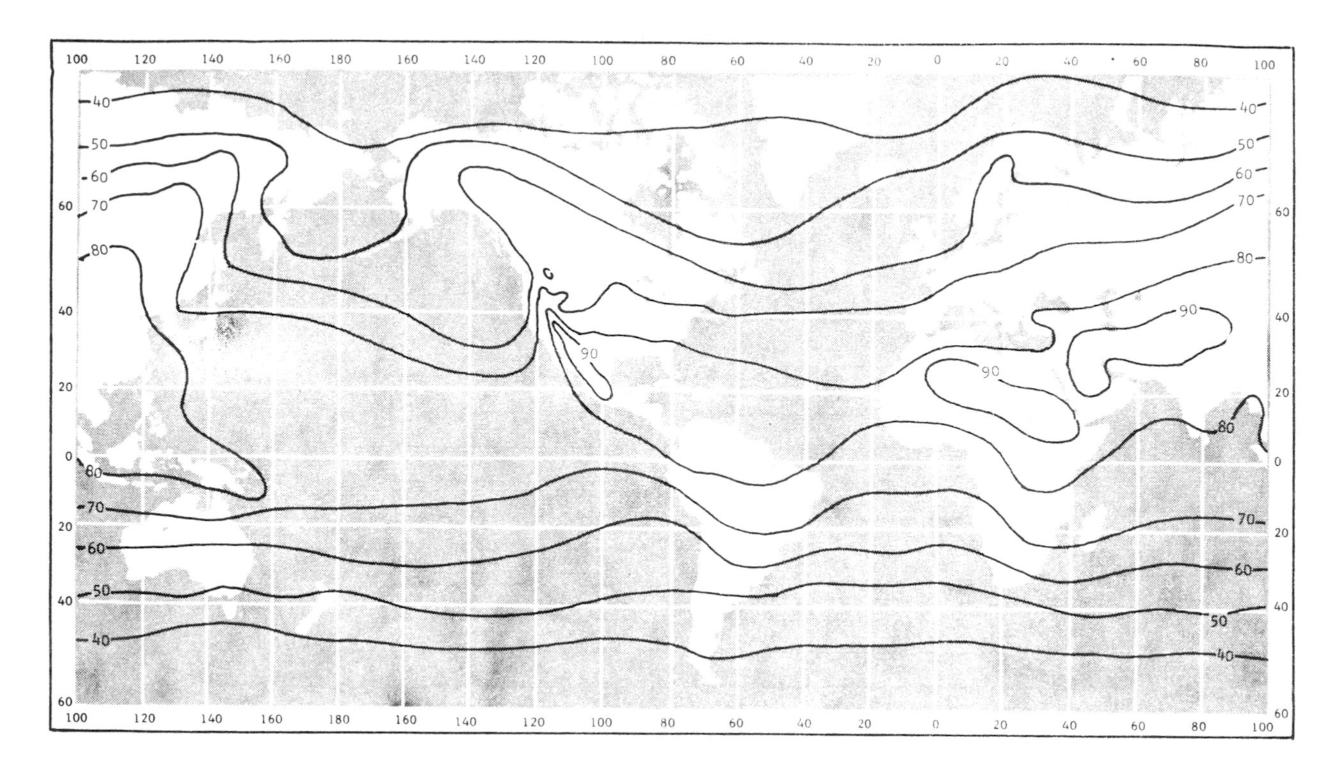

Figure 2-6. Average temperature distribution over the earth in July.

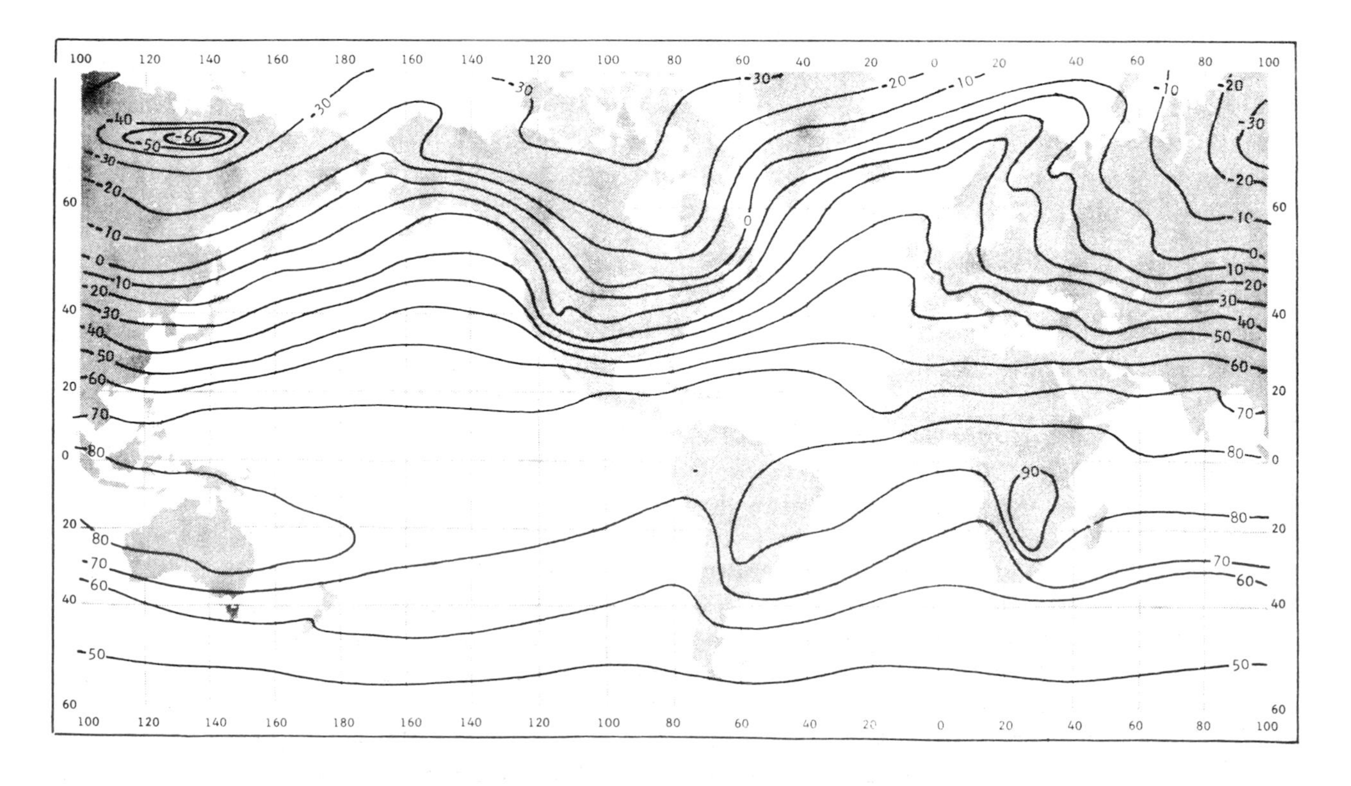

Figure 2-7. Average temperature distribution over the earth in January.

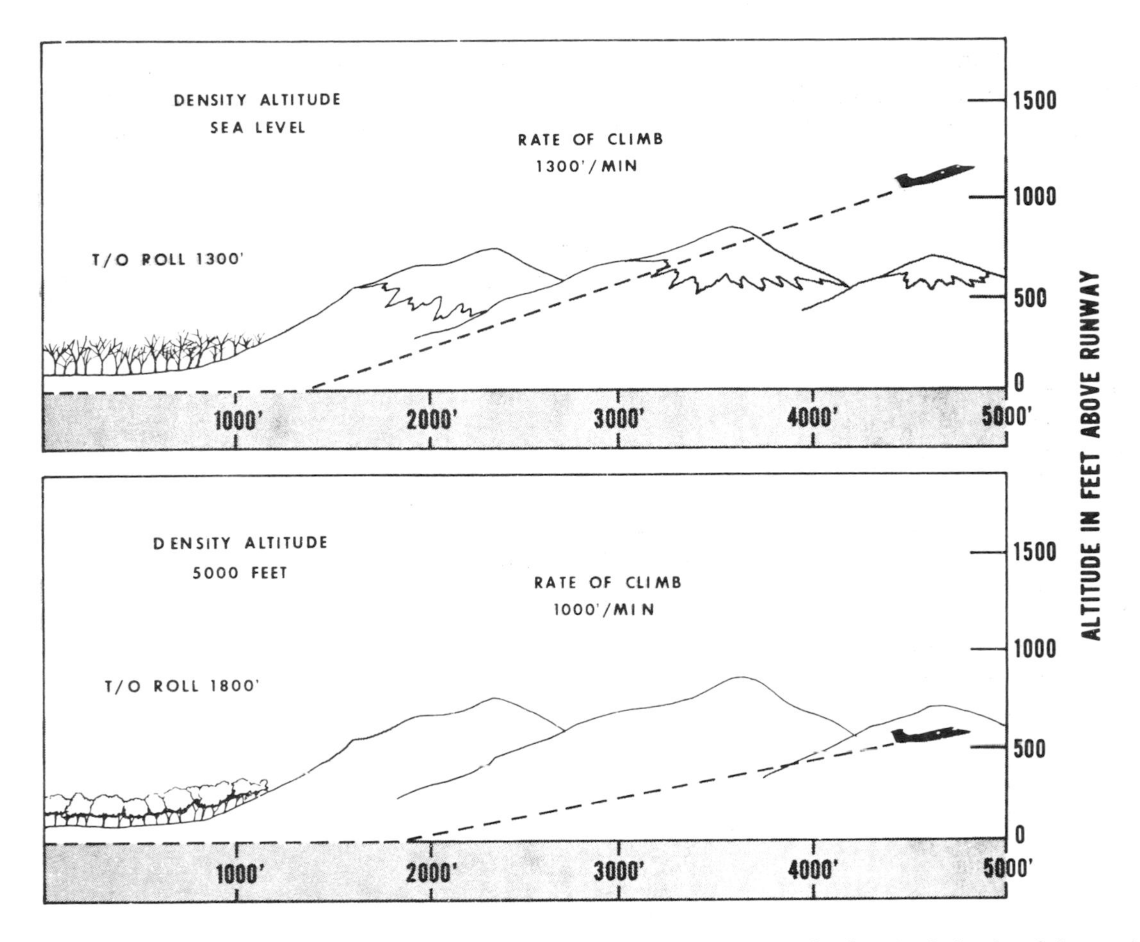

Figure 2-8. Effects of density altitude on takeoff roll and rate of climb. At a density altitude of sea level, the aircraft has an altitude of 1000 feet when reaching a distance of 4500 feet from beginning of takeoff roll. At a density altitude of 5000 feet, the same aircraft is only 500 feet above runway elevation after covering the same distance from beginning its takeoff roll. A density altitude variation of 5000 feet or more is not uncommon at an airport in mountainous areas because of differences in surface temperature and pressure. Density altitude may become so critical that a loaded aircraft cannot become airborne.

high density altitude. Later, we will see many other effects of density altitude, on airspeeds, for example. Concerning altimetry, change in temperature is the chief cause of error in the pressure altimeter, both at the surface and aloft. We will see more of this relationship between pressure and temperature in the chapter on altimetry.

J. Why Cold Air Does Not Always Sink

We conclude this section with a discussion of why colder air aloft does not always sink. First, pressure always decreases with height. However, temperature and density do not always decrease with height, although with a standard lapse rate in the troposphere they do. Thus, with a standard lapse rate the colder air is less dense and will not sink. However, the less dense air will cause pilots to have to use supplemental oxygen at high altitudes.

One of the gas laws from physics states that pressure is directly dependent on the product of temperature and density – specifically, $P = T \times \rho \times k$, where P is pressure, T is temperature, ρ is density, and k is a constant. Consider air at two levels A and B, A at a lower altitude in the troposphere than B. If we do not change the pressure at B but we do decrease the temperature at B, then since pressure depends on the product of temperature and density, the density must increase at B. The density must increase by an even greater amount when the pressure at B increases as the air cools at B; however, the pressure at B is determined primarily by the temperature distribution above B – rather than at B). In any event, the air at B can be so cold that it is more dense than the air at A and will, therefore, descend toward level A. Thus, an increase in density with height can occur when there is much colder air present than is normally found there (or when there is much warmer air below).

TEMPERATURE
Review Questions

CFI

1. What are the standard temperature and pressure values for sea level?
   - a. 59°C and 1013.2 millibars.
   - b. 15°C and 29.92 inches of mercury.
   - c. 68°F and 29.92 millibars.
   - d. 59°C and 1013.2 inches of mercury.

2. If the air temperature is +16°C at an elevation of 3,000 feet MSL and a standard (average) temperature lapse rate exists, what will be the approximate freezing level?
   - a. 14,000 feet MSL.
   - b. 11,000 feet MSL.
   - c. 9,000 feet MSL.
   - d. 7,000 feet MSL.

6,000
16
–
5

CHAPTER III
BASIC AERODYNAMICS

The reader should demonstrate an understanding of basic aerodynamics to the extent that he can answer the following questions (as stated below, or as multiple-choice, etc.):

1. What are the four basic forces acting on an airplane in flight?
2. Define "relative wind".
3. Weight always acts in which direction?
4. Define "angle of incidence".
5. What would be the effects of a layer of ice on an aircraft? Why?
6. Define "dynamic pressure". State Bernoulli's principle.
7. State the equation which defines the amount of lift.
8. What would be the effect on lift if the wing area is reduced; if the density of the air is increased; if the speed is decreased?
9. Define "angle of attack".
10. What two factors determine the coefficient of lift?
11. Describe the effects on C_l of low angles of attack; high angles of attack.
12. What is the stalling angle of attack for most general aviation aircraft?
13. An aircraft stalls when it exceeds its critical ____________________. How does an aircraft recover from a stall?
14. Can an aircraft stall in different attitudes? at different airspeeds? Give an example of each, different from that in the text.
15. Define "load factor".

CHAPTER III

BASIC AERODYNAMICS

This chapter will consist of a discussion of how a pressure difference makes an airplane fly. If you comprehend our discussion of how one walks along a floor, you will be well on your way to understanding much of what makes an airplane fly. This chapter will establish a foundation for future studies on aerodynamics, weather, and engine performance. We will be discussing the four forces of lift, thrust, drag, and weight in straight and level flight. We will see how an aircraft stalls and we will also briefly discuss the concept of load factor.

A. The Four Forces in a Hall

In order to establish a few terms, let's assume that we are walking along a hallway in still air. We observe first that we have created a relative wind from the direction in which we are walking. Even if the air is circulating in various directions and speeds, this does not affect the relative wind, for again it is an artificial wind created by our direction of motion as we walk. For simplicity, we again assume that the air in the hall is still. We are being propelled along our walking path with a force generated by a heat machine known as our body. This force is called thrust. Operating opposite to thrust along our path is a force we call drag. This force exists because our body is composed of mass and we must overcome the resistance of our body to forward movement.

We know from Newton's first law, the law of inertia, that a body at rest will stay at rest unless acted upon by an outside force. In our hall case it is thrust which overcomes the body's inertia, that is, its tendency to remain still. This resistance to forward movement is called drag. From Newton's second law, we will walk along at a constant speed as long as thrust and drag are equal. This fact comes from acceleration, $a = F/m$, where m stands for mass and F for net force acting along the path of motion. As long as there is a balance between thrust and drag, remember they represent opposite forces, then there is a net force of zero acting along the walking path and thus an acceleration of zero, which means we are walking at a constant speed. As long as thrust is greater than drag we gain speed, that is, we accelerate, but when we reach a constant speed, thrust and drag must be equal for acceleration to be zero, as we have seen.

Our weight always acts toward the center of the earth because it is the acceleration due to the earth's attraction for us that accounts for our weight. Without the intervening floor we would be accelerated toward the center of the earth. Note that the floor pushes upward on our weight with a force we might call lift. This lift acts perpendicular to our walking path, hence to our relative wind, and is sufficient to support our weight. The lift that supports an airplane when suspended in the air is different from the "lift" exerted by the floor, as we shall soon see.

B. The Four Forces in the Air

We now pursue the above ideas, with a reminder that we will be considering only straight and level flight in this book. Other flight configurations force us to consider components of lift which are very important to further studies on aerodynamics, but not to our goal here. Thus, we only have one component of lift, acting perpendicular to the relative wind and equal and opposite to weight for straight and level flight (Figure 3-1).

Visualize an aircraft wing suspended freely in still air. Assume for now that the wing is shaped like a very thin flat plate. The pressure at some point on top of the wing is a result of the weight of the air above the point. The pressure at a point on the underside of the wing is the same, just like the air pressure on your chest is the same as it is on your back (pressure in a

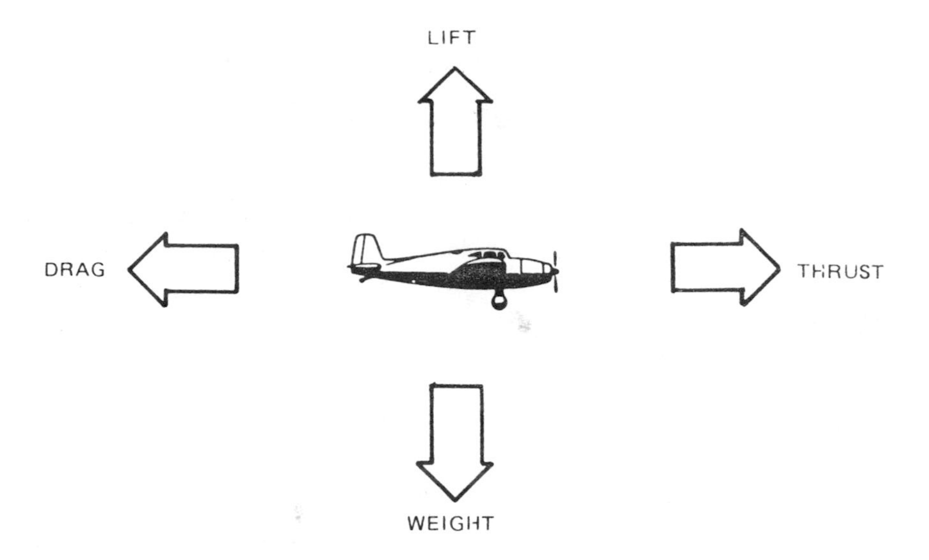

Figure 3-1. Relationship of forces in flight.

gas is independent of direction). If we think of the wing as having zero weight, it would indeed be freely suspended and just float. But the wing does have weight, so it will tend to descend; the wing is stalled, there is nothing to hold it up. The wing will fall to the earth's surface as a result of the force of its own weight.

Suppose we incline the wing at some angle. But since the air and wing are relatively still, the wing again falls. What force would keep the wing up? Note that there is no floor to hold the wing up as did the floor in the hall relative to our body. The wing is suspended in air. We move the wing through the air with the leading edge of the wing angled upward. And how can we move the wing forward? Well, you might guess that we would attach a propeller to a horizontal fuselage to which the wing is attached, slanted upward, at a fixed angle. The angle at which the wing is attached to this fuselage we will call the angle of incidence; note that we designed this of our own choice but for a purpose. Now as the propeller physically pulls the wing forward with a force called thrust, we note that if the propeller is not pulling the wing forward fast enough descent will result. If the speed is above a certain value, the wing will ascend. But when the speed is the right amount, the wing stays level. Enough force, called lift, is being developed to overcome the weight of the total airplane.

In this level flight configuration, lift is acting opposite and is equal to weight. If these forces were not in balance, then with one force stronger than the other, Newton's second law would have caused wing acceleration, namely, with lift more than weight, the wing would be accelerated upward, and the reverse, that is, with lift less than weight the wing would be accelerated downward. This conclusion is derived from $a = F/m$, for if there is no change in force, acceleration is zero; if there is a change in force, that is, lift more, or less than weight, then acceleration results and the wing either goes up or comes down, that is, climbs or descends. Again, lift does not always act directly opposite to weight, but in straight and level flight it does.

We are not trying to confuse you with this discussion, but rather introduce you to those factors which will be mentioned later. For example, we would hope at this point that you understand that a factor such as a layer of ice on an aircraft will increase both weight and drag

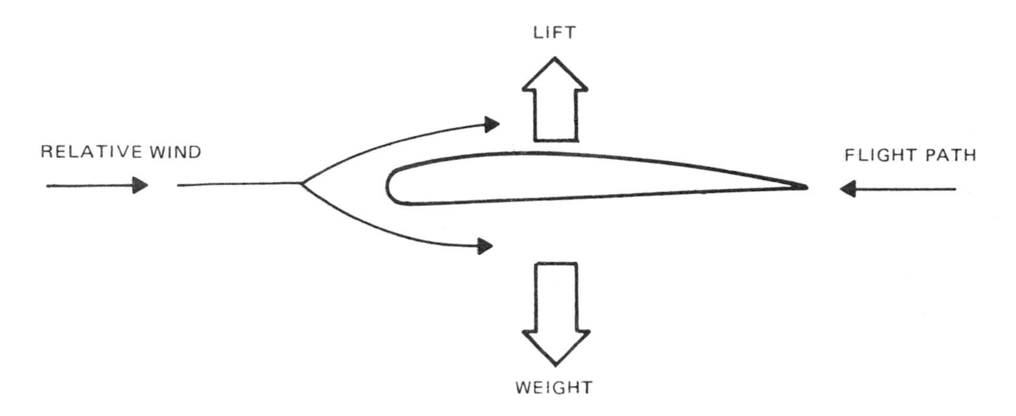

Figure 3-2. The wing, relative wind, and lift.

because mass is increased and, hence, more lift and thrust are needed to stay level at the same speed. You will see how icing affects stall speed shortly.

C. Bernoulli's Principle

To this point we have been considering our lifting force as being developed only from the impact of air on the lower surface of the wing, that is, from the pressure of air on the bottom of the wing from our relative wind. Remember, in our discussion the air is still and the wing is being propelled forward through the air and any wind motion created is relative to the flight path; hence, a relative wind is created, just as when we walked down a hall in still air.

We now place our wing flat in the air by lowering the leading edge of the wing through control useage. Let's picture a curved surface on the top of the wing, rounded at the leading edge, curving up slightly and then curving down to and becoming a point at the trailing edge. Assume the bottom of the wing is flat (Figure 3-2). Let's again propel this wing forward. What will happen? Well, with sufficient speed the flight path of the wing will remain horizontally straight, no climb or descent, because as we will soon see, sufficient lift is being generated to offset weight. Had we not propelled the wing through the air with enough speed, the wing would have again fallen out of control or descended gradually. How does the wing fly? Our wing flies because of dynamic pressure, the pressure which results from the flow of air past the wing. Lift occurred with our inclined wing also, but that was due to impact pressure and here we have no incline to our wing.

How is dynamic pressure generated with our wing parallel to the flight path? The secret lies in the curvature of our wing. As air impacts the curved surface, it is forced upward and backward along the surface, whereas beneath the wing air flows straight backward, opposite the flight path. Now the curvature effect of the wing will only extend so high, that is, there will be a straight-line flow backward at some fixed distance above the wing, an upper boundary. Thus, the curved surface on the top side of the wing and the boundary have created a restriction in the air flow, similar to a pipe with two wide ends of the same size and a restricted passageway in the center.

Assume a pipe is flat along the top and curved inward at the center (Figure 3-3). As water flows through the pipe, the speed of the flow will be the same at the two ends; the flow through the narrow portion will have to be at a higher speed to accomplish this result, a physical fact which can be proven but which we will accept here. The air relative to our wing will behave in the same manner. In order for the speed of the relative wind to be the same after it passes the

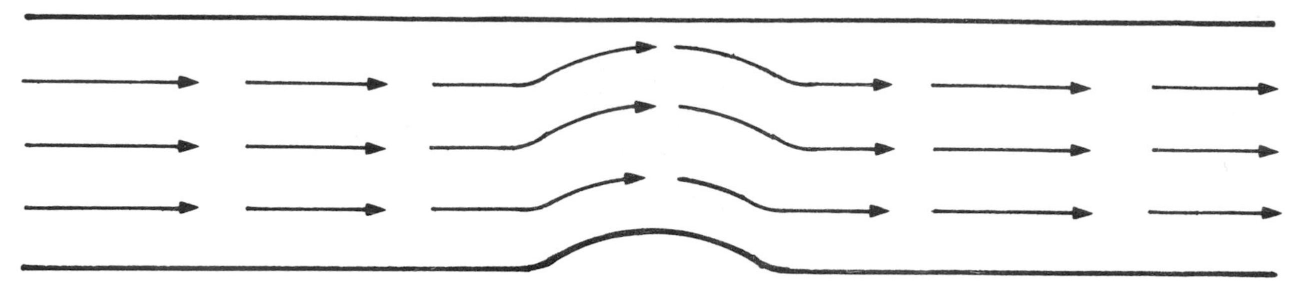

Figure 3-3. Constriction in flow through a pipe. Speed of flow is the same at each end, but greater through the constriction.

wing as it was before encountering the wing, the speed of the air must increase because of the constriction (while air can certainly be compressed, the amount of compression in this aerodynamic process is slight and we disregard it; thus, the density of the air is constant throughout the flow and a faster speed through the constriction is necessary — again, this is a demonstrable fact that we will accept). Dynamic pressure, defined as one-half times air density times the square of the speed, has increased above the wing; that is, the force of the air acting rearward is greater due to the greater speed of the air, just as you may be blown backward by a sudden gust on a windy day. (Since density is mass per unit volume, then dynamic pressure, $\frac{1}{2}\rho V^2$, is $\frac{1}{2}mV^2$ per cubic foot, that is, dynamic pressure may be interpreted as the amount of kinetic energy produced by the motion of a cubic foot of air; note that this is external kinetic energy, as compared to the internal kinetic energy theory of sensible heat.) An increase in dynamic pressure acting rearward is accompanied by a decrease in static pressure acting downward; this statement is a form of the Bernoulli principle.

We now describe an oversimplified example of Bernoulli's principle (Figure 3-4). As mentioned above, the speed of the air must increase through the constriction. Since the air is accelerating, there must be an imbalance in force in accordance with Newton's second law. There are two primary forces involved, one acting rearward contributing to dynamic pressure and one downward contributing to static pressure. Since speed is increasing through the constriction, dynamic pressure is increasing; thus, static pressure must be less — in fact, reduced by the same amount that dynamic pressure is increased. Note that the relationship between dynamic and static pressure is indicated by our three barometers in Figure 3-4. Thus, for our wing, static pressure is reduced on top of the wing, yet the static pressure below the wing is just the atmospheric pressure at that altitude. With relatively lower pressure along the upper surface, the wing now lifts or accelerates upward as a result of air pushing up on the wing in its attempt to flow from higher to lower pressure.

D. Equation of Lift

If you understand the above discussion, you know more about what makes an airplane fly than ninety-nine percent of certificated pilots. But back to our wing. With sufficient speed of the air over the wing, it generates enough lift to overcome weight and the wing flies straight, that is, holds altitude. Note that all of this was accomplished with our wing parallel to the flight path,

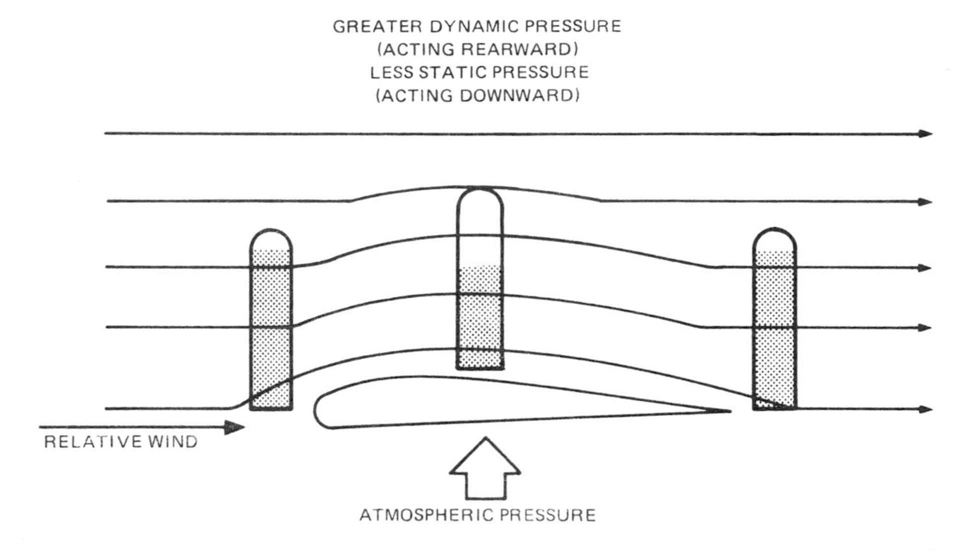

Figure 3-4. Example of the Bernoulli effect.

the pressure difference being created by the curvature of the upper surface. It turns out that the amount of lift is given by the equation lift = $C_l qS$, where C_l is the coefficient of lift – we will look at this factor further in a moment; S is the surface area of the wing – note that if one doubles the area of the wing, lift is doubled; q is the dynamic pressure. Recall that the greater the dynamic pressure, the less the static pressure and the greater the lift. Note the dependence of dynamic pressure on the density and speed of the air (q equals ½ times density times the square of the speed). The more dense the air, the greater the dynamic pressure. If you were to be hit in the face with a given volume of meringue from a pie and then an equal volume of pie filling, the dynamic pressure on your face would be greater from the filling than the meringue because of the greater density of the filling. An increase in speed will greatly increase the dynamic pressure since we must square the speed. This results in a large increase in lift. Thus, the amount of lift depends on the coefficient of lift, density of the air, airspeed, and size of the wing.

E. Coefficient of Lift

We now further discuss the coefficient of lift, C_l. In order to do so we need to first briefly discuss angle of attack. Actually, we have already done so, we just did not call it that. Recall that we had our wing attached to the fuselage at a fixed angle, two degrees is common. In order to make our wing horizontal to the flight path we lowered the nose, which lowered the leading edge of the wing about two degrees. This changed the angle of our wing to the flight path to zero. Note that the angle of incidence did not change, since the wing was built onto the fuselage at a fixed angle. The angle of attack is the angle between the wing and the relative wind (Figures 3-5 and 3-6).

The coefficient of lift is determined by the shape of the wing and its angle of attack. We have already considered other factors (density, airspeed, and wing area). Certainly, the shape of the wing affects lift as a result of its curvature. But angle of attack is also a part of C_l. In a prior discussion, the angle of attack had been zero and C_l had been dependent on only the

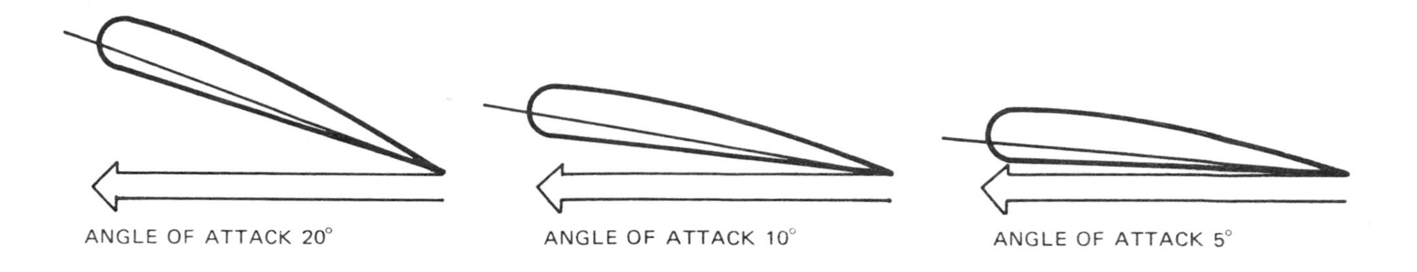

Figure 3-5. Angle of attack is the angle between the wing and the flight path.

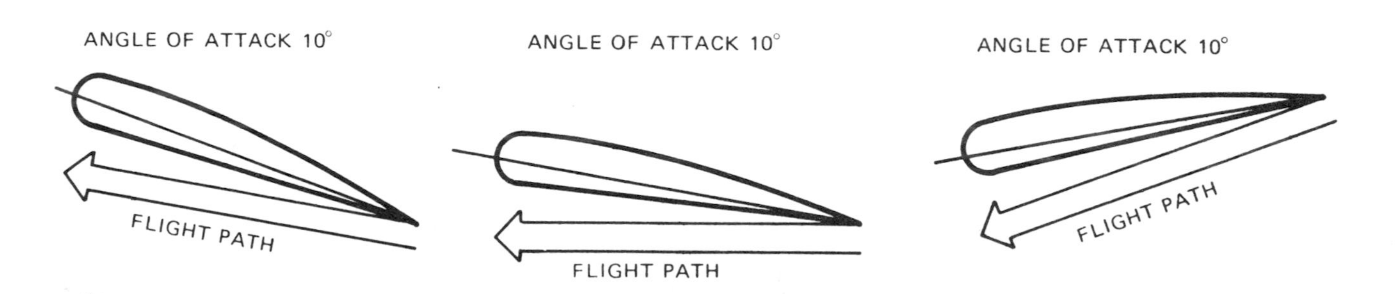

Figure 3-6. Angle of attack is always based on the flight path, not the ground.

shape of the wing. The wing can fly in such a situation as we have seen. However, C_l will increase greatly as angle of attack is increased. As C_l increases, lift increases. To see this, visualize our wing flying straight ahead at a given angle of attack, say, 10 degrees. Now we have the two effects working together that we discussed separately earlier, that is, the wing is inclined to the flight path and the top surface of the wing is curved. We have impact air on the bottom. We also have the airstream flowing along the curved surface generating lower pressure on top of the wing. Thus, C_l depends both on the shape of the wing and its angle of attack. At low angles of attack, impact air is slight, so practically all lift resulting from C_l is resulting from the shape of the wing. At high angles of attack, impact pressure from air hitting the bottom of the wing accounts for about one-third of the value of C_l. Therefore, you can appreciate the value of wing design.

F. Stalls

You might suspect that the angle of attack cannot be increased indefinitely. If the angle of attack is 90 degrees, the wing being perpendicular to the relative wind, then all our lift would be converted to drag, similar to a parachute being horizontal instead of vertical, and you would fall. A parachute does not have to be completely horizontal for you to fall, however. For most wings the stalling angle of attack is 16 to 20 degrees. This may surprise you. Picture a wing at an angle of 20 degrees to its relative wind and recall that the relative wind is opposite the flight path of the aircraft. The aircraft will sink, the new flight path being straight down. How does a wing recover from a stall? Only by a reduction in the angle of attack so that sufficient lift is generated to support the weight of the aircraft. (Add power, if available, to increase the speed of air over the wing and, hence, lift — effective *only* if the angle of attack is reduced.)

1. *In Any Attitude*

A wing can be stalled along any flight path by exceeding its critical angle of attack. For example, if an aircraft flies straight ahead and the nose is pulled up abruptly, the motion of the aircraft is still forward due to the momentum of the aircraft, but if the critical angle of attack has been exceeded, the wing stalls. Recovery is effected by reducing the angle of attack by lowering the nose. If you encounter a sudden downdraft, then as you are forced downward your relative wind is upward, and with a high angle of attack you may stall. Again, to recover, the angle of attack must be reduced by lowering the nose, which will cause a further loss of altitude, until sufficient lift has been re-established to support the weight of the aircraft. Thus, turbulent air near the ground can be very hazardous, for it may change your relative wind and, hence, may cause you to stall.

2. *At Any Airspeed*

An aircraft's stall speed is not constant, not in the above cases nor many others. An example: suppose aircrafts A and B are flying straight and level, aircraft B heavier than aircraft A. Assume they both have the same wing area, the density of the air is constant, their speeds are the same and the shape of the wing is the same. Where will we get the extra lift to offset the extra weight of B so that it will maintain altitude? Well, the only factor left which affects lift is angle of attack. Thus, we increase the angle of attack for B. With more thrust we can bring B back to the same speed as A, but with a higher angle of attack. Picture these two aircraft, A and B, at the same speed, B heavier than A and B with a higher angle of attack. As we increase the angle of attack by raising the nose, I think you can anticipate that B will stall first. B will stall first because it was at a higher initial angle of attack to support its extra weight.

As the angle of attack is increased on both aircraft, more drag is acquired and both aircraft lose speed. At low airspeeds, more thrust will be required to hold altitude. B will reach the critical angle of attack first, but A will still be flying. As A continues to increase its angle of attack, it will continue to lose speed, and then stall when it exceeds the critical angle of attack. Both aircraft stalled when they exceeded the critical angle of attack, but A did so at a slower speed. An increase in weight caused a higher stalling speed. Stated another way, anytime more lift is required to maintain altitude, stall speed goes up. Note that an aircraft stalls when the wing exceeds the critical angle of attack. However, that angle can be exceeded along any flight path and at any speed. These considerations will be particularly important when maneuvering the aircraft and, where weather is concerned, in flying in turbulence and icy conditions. We have considered only enough information on stalls for weather considerations – there is much more to be said about lift and stalls.

G. Load Factor

We briefly mention one other concept, load factor, which is defined as the actual load being supported by the wing divided by the weight of the aircraft. A load factor of 2 means that the wing is supporting twice the weight of the aircraft, or carrying 2 Gs, as is often described. Maneuvering loads provide good examples of this; in a pull-up maneuver, an extra load on the wing is produced. Turbulent air can cause changing loads on the wing even as an aircraft flies essentially straight. At speeds beneath a certain value (termed "maneuvering speed") an aircraft's momentum is reduced, thus the aircraft will stall before structural damage is incurred; at higher speeds the reverse is true – momentum is so great that structural damage will occur before the aircraft can stall.

In presenting this discussion on how the wing flies, we have taken many shortcuts. We have used some words loosely, such as speed, and have left out some terms altogether, such as camber and chord. We treat the concept of speed more fully in the chapter on altimetry and airspeeds. We have tried not to be bogged down by more terminology than absolutely necessary. This has not been a complete treatment, but merely an introduction. We have considered only straight and level flight. This chapter probably raised more questions than it answered, but it will answer a lot of later questions and will give you a much better understanding of weather, engine performance, and other instructional material related to lift and thrust, drag, weight, stalls, and loads.

BASIC AERODYNAMICS
Review Questions

PRI

1. As you maneuver an airplane in the traffic pattern, you should realize that an airplane can be stalled:
   a. Only when the nose is high and the airspeed is low.
   b. Only when the airspeed decreases to the published stalling speed.
   c. At any airspeed and in any flight attitude.
   d. Only when the nose is too high in relation to the horizon.

2. Refer to the illustration on the right. The acute angle "A" is the angle of:
   a. Dihedral.
   b. Attack.
   c. Camber.
   d. Incidence.

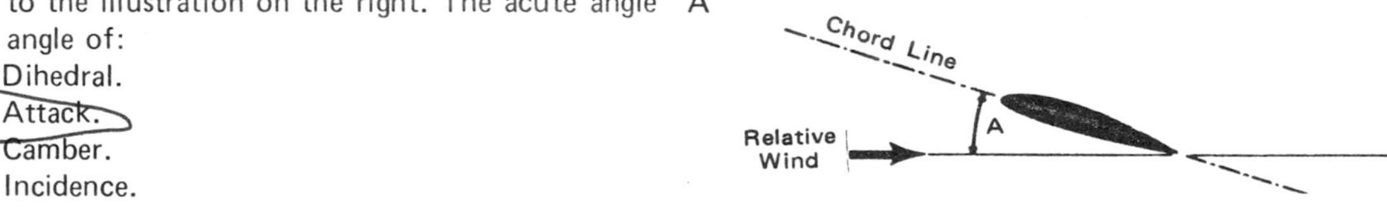

3. The term "angle of attack" is defined as the:
   a. Angle between the wing chord line and the direction of the relative wind.
   b. Angle between the airplane's climb angle and the horizon.
   c. Angle formed by the longitudinal axis of the airplane and the chord line of the wing.
   d. Specific angle at which the ratio between lift and drag is the highest.

COM

4. The angle between the chord line of the wing and the longitudinal axis of the airplane is known as the angle of:
   a. Dihedral.
   b. Incidence.
   c. Attack.
   d. Relative wind.

5. When considering the forces acting upon an airplane in straight and level flight at constant airspeed, which statement is correct?
   a. Drag always acts rearward parallel to relative wind and is less than thrust.
   b. Thrust always acts forward parallel to the relative wind and is greater than drag.
   c. Lift always acts perpendicular to the longitudinal axis of the wing and is greater than weight.
   d. Weight always acts vertically toward the center of the earth.

6. Both lift and drag of an airfoil are:
   a. Proportional to the square of the velocity (V^2) of the relative wind.
   b. Proportional to increases and decreases in the velocity of the relative wind.
   c. Inversely proportional to the air density.
   d. Inversely proportional to the area of the wing.

7. Load factor is the actual weight supported by the wings at any given moment
   a. Divided by the total weight of the airplane.
   b. Multiplied by the total weight of the airplane.
   c. Added to the total weight of the airplane.
   d. Subtracted from the total weight of the airplane.

8. The angle of attack at which an airplane wing stalls will
   a. Change with an increase in gross weight.
   b. Remain the same regardless of gross weight.
   c. Decrease if the center of gravity is moved aft.
   d. Increase if the center of gravity is moved forward.

9. What determines the angle of attack at which an airplane stalls?
   a. Design of the wing.
   b. Load factor.
   c. True airspeed.
   d. Airplane gross weight.

10. In airplanes all stalls are caused by
    a. Exceeding the critical angle of attack.
    b. A loss of airspeed.
    c. Exceeding the critical angle of pitch.
    d. Misuse of the elevators.

11. Which statement is true relating to the factors which produce stalls?
    a. The stalling angle of attack depends upon the speed of the airflow over the wings.
    b. The critical angle of attack is a function of the degree of bank.
    c. To accelerate a stall will always produce a spin.
    d. The stalling angle of attack is independent of the speed of airflow over the wings.

CHAPTER IV
WIND

The reader should demonstrate an understanding of wind to the extent that he can answer the following questions (as stated below, or as multiple-choice, etc.):

1. Define "wind".
2. Explain the effect of the spacing of isobars on the speed of the wind.
3. Air is deflected (right, left) in the Northern Hemisphere due to the rotation of the earth on its axis. What is this deflecting force called?
4. Air flows (clockwise, counterclockwise), (outward, inward) and (upward, downward) in a high.
5. The effect of friction is to deflect the wind to the (left, right), causing the wind to cross isobars from (low to high, high to low) pressure.
6. Other common names for high and low are __________ and __________, respectively.
7. Explain the reasons for the northeast trade winds.
8. Define "wind direction".
9. What is the effect of the tilt of the earth's axis on pressure belts?
10. Explain the development of the land breeze.
11. Explain the development of a summer monsoon.
12. What is a rawinsonde?
13. What formula is used to relate change in pressure to change in height?
14. What happens to true altitude as you fly from high to low pressure at a given indicated altitude?
15. Explain how you could gain true altitude yet be lower than your altimeter indicates.
16. Which instruments operate only off the static port?
17. Explain the effect of flying a "valley" on vertical velocity.
18. At what altitude does the effect of friction become negligible?
19. Suppose the wind at 1,000 feet (300 m) is due north. Frictional effects will cause the wind to be (northwest, northeast) at the surface. Why?
20. On the U.S. surface map pressure is analyzed by studying lines called __________. On constant pressure charts, pressure is analyzed by studying lines called__________.
21. Pressure usually changes more in the (vertical, horizontal). (Vertical, Horizontal) changes are important in locating highs and lows.
22. State the most common pressure levels and their corresponding altitudes.
23. Describe the flow of air when there is high pressure as determined by the 850 mb chart and low pressure as determined by the 700 mb chart.
24. State four ways in which divergence occurs.
25. What are the horizontal and vertical effects on an aircraft crossing a trough? A ridge?
26. How does flying across troughs and ridges compare to flying across highs and lows relative to indicated/true altitude?
27. Define vorticity.
28. Compare absolute and relative vorticity.
29. For relative vorticity, counterlockwise motion is (positive, negative).
30. What is the effect of increasing positive vorticity?
31. State typical vertical speeds. For what phenomena are vertical speeds very large?
32. Describe warm and cold core highs; warm and cold core lows.
33. Define katabatic.
34. What does "crab" mean?
35. Define "weathervaning".
36. Compute the crosswind component for a 35 knot wind at an angle of 50° to the nose.

CHAPTER IV
WIND

We define wind as air in motion. We defined pressure earlier to be force per unit area; symbolically, $P = F/A$, where P stands for pressure, F stands for force, and A stands for area. The force we were concerned with was the weight of air, which is its mass times earth's gravitational pull. Thus, the more mass of air, the greater the weight, or force, and the greater the pressure. Therefore, difference in mass in two adjacent columns of air results in different pressures on a horizontal surface at a given height. When we compare these two columns of air on a horizontal surface, one column represents relatively higher pressure than the other. The result, all other things being equal, would be a flow of air on the given surface directly from the column with higher pressure to the one with lower pressure. However, all other things are not equal.

A. Coriolis Force

Earlier, we observed that the earth rotates on its axis. Consider a record turning counter-clockwise on a turntable. Now while the record is turning, picture yourself taking a piece of chalk and, while standing over the turntable, move the chalk straight outward from the center of the record to the edge. Now relative to you the chalk moved in a straight line. But look at the record. You will see a curved path to the right (Figure 4-1). The movement of the chalk took time and during this period the record was turning counterclockwise. Such is the case on the earth's surface. In fact, regardless of the direction of the flow of air, as long as it has a north or south component of motion, there will be an apparent turning motion to the right in the Northern Hemisphere, left in the Southern Hemisphere.

As you stood above the record, you drew a straight line with the chalk relative to space, so the curved line on the surface is merely an effect of the earth's rotation and does not represent a turning force relative to space. Since we live on the earth, the effect is real to us and we consider this effect to be a force. Well, the result is the same, but you will sometimes hear this phenomenon described as a force and sometimes as an effect, the real difference being only whether it is considered relative to the earth or to space. And, incidentally, the degree of turn or curvature in motion depends on the rotational speed of the earth, latitude, and wind speed; the force increases as latitude and/or wind speed increase.

B. Pressure Gradient and Wind Speed

Now, back to the flow of air, wind, which moves from high to low pressure relative to two columns of air on a horizontal surface at some height. Assume that in our higher pressure column the highest pressure is at the center and that pressure decreases uniformly outward. Then as we saw in the chapter on pressure, we have circular isobars around this center, representing a fixed decrease in pressure from the center (four millibars, for example, which is used in analyzing the United States surface map). Recall that an isobar is a line of constant pressure. So if the central pressure is 1000 millibars, the first circular isobar could represent a value of 996 millibars, that is, that isobar connects points on the horizontal surface we are considering where the weight of the air on that surface is the same. The closer the isobars, the faster the speed of the wind. For example, if a pressure of 996 millibars occurs at 100 miles (160 km) and 992 millibars at 200 miles (320 km) from the center, then the horizontal pressure gradient is 4 millibars per 100 miles. If 996 millibars occurs at 50 miles (80 km) and 992 millibars at 100 miles from the center, then this is a change of 4 millibars per 50 miles (equivalent to 8 millibars per 100 miles); so other things being equal, the speed of the wind is twice as great for the larger

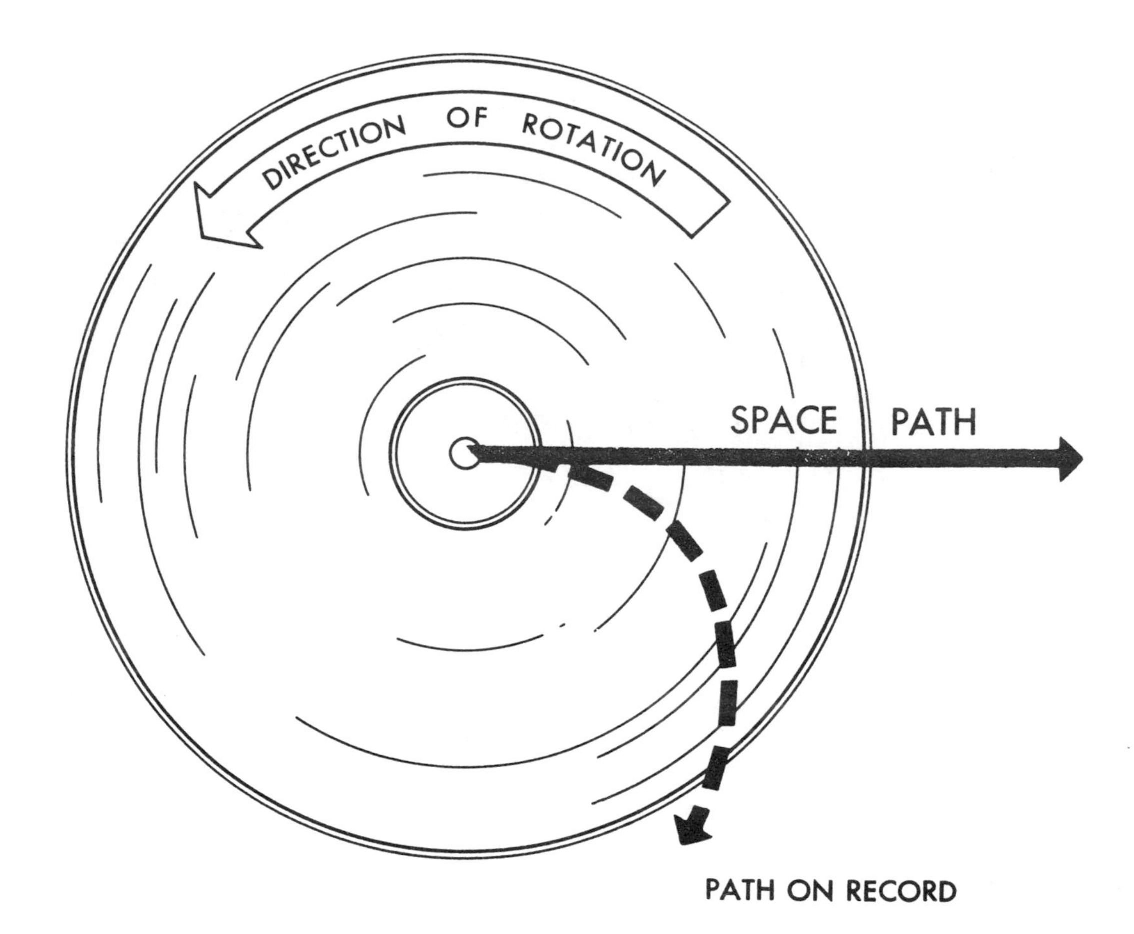

Figure 4-1. Apparent deflective force due to rotation of a horizontal platform. The "space path" is the path taken by a piece of chalk. The "path on the record" is the line traced on the rotating record. Relative to the record, the chalk appeared to curve; in

gradient. Conversely, the greater the spacing of the isobars, the smaller the gradient and the less the speed of the wind.

In the next section we will be discussing pressure gradient and Coriolis forces above the friction level, relative to contour lines. For now, treat contours as isobars, that is, the closer the spacing the greater the wind speed. We will develop the nature of contours in section F of this chapter. You will also discover later that isobars and/or contours are usually closer together in lows than highs, with resultant higher winds in lows than highs.

C. Pressure Gradient and Coriolis Forces (Above the Friction Level – 2,000 Feet)

In Figure 4-2, the pressure gradient force is directed outward. As air accelerates it is deflected to the right by the Coriolis force (as wind speed increases, the Coriolis force also increases) until the wind becomes parallel to the contours. Note that the pressure gradient force determines the wind speed and that the Coriolis force eventually makes the wind direction parallel the straight line contours. Wind flowing at a constant speed along straight line contours is called a geostropic wind.

What if the contours are curved rather than oriented in a straight line? Well, if you were in an airplane and making a turn you know (or will learn) that the horizontal component of lift provides the net force (called centripetal force) that turns the airplane. For right turning contours (clockwise motion), the centripetal force is provided by a greater Coriolis force than the pressure gradient

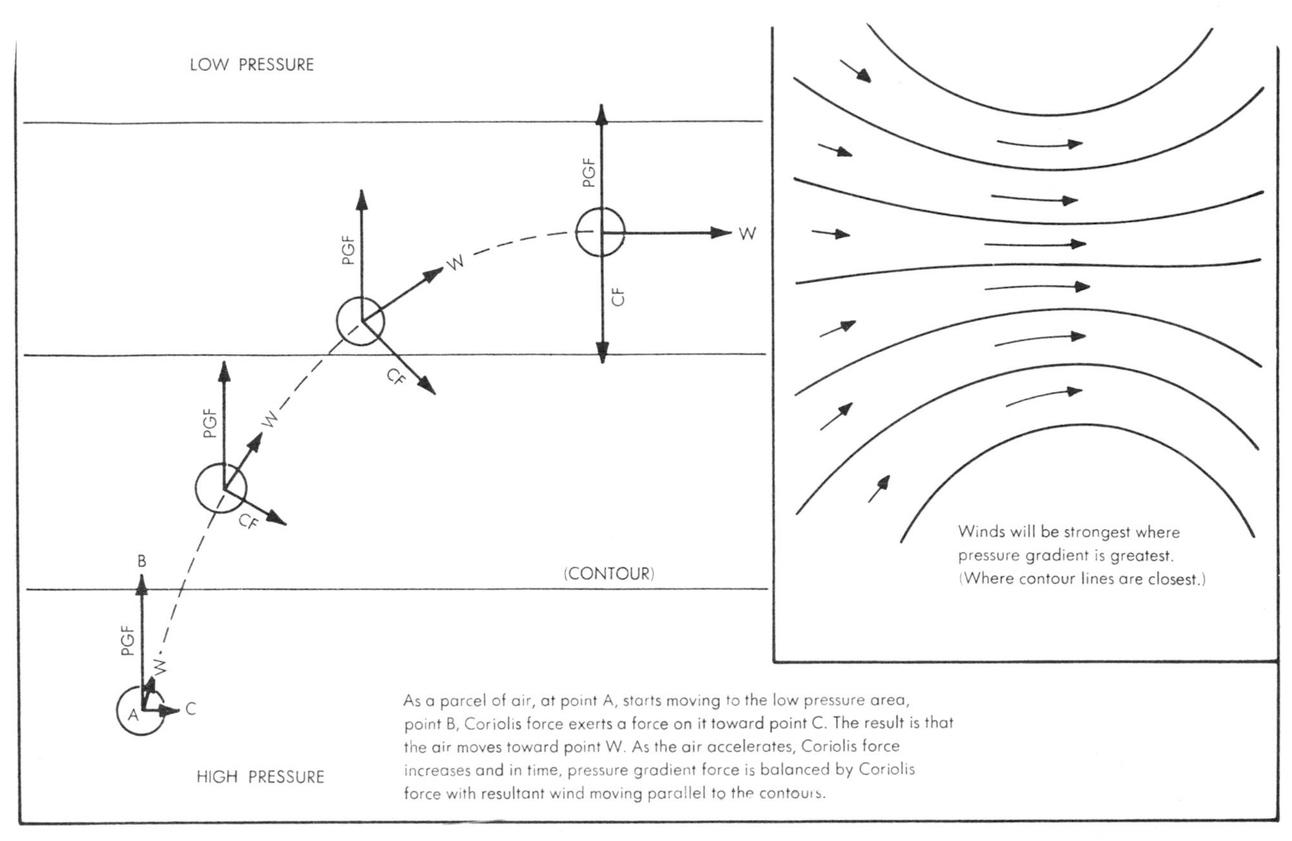

Figure 4-2. Turning of the wind by the deflective (Coriolis) force. On the left, the wind begins parallel to the pressure gradient force, but the pressure gradient and deflective forces are not in balance. The deflective force turns the wind to the right until it is parallel to the isobars as shown on the right, and the two forces are in balance.

force; the centripetal force then is the Coriolis force minus the pressure gradient force. For left turning contours (counterclockwise motion), the centripetal force is provided by a smaller Coriolis force than the pressure gradient force; the centripetal force then is the pressure gradient force minus the Coriolis force. As we shall see later, wind shear in the upper levels is largely responsible for the centripetal force and subsequent turning. Figure 4-2 also shows flow of the wind around curved contours. Note that clockwise rotation is associated with highs and ridges and counter-clockwise rotation is associated with lows and troughs, as expected. Wind flowing at a constant speed around curved contours is called a gradient wind.

D. Pressure Gradient, Coriolis and Friction Forces (Below 2,000 Feet)

While 3,000 feet is probably a more accurate top of the friction layer, 2,000 feet is used in most aviation literature, and in FAA Question Books. Keep in mind that centripetal force is involved in going from Figure 4-3 (straight line isobars) to Figure 4-4 (curved isobars). Without centripetal force, the wind in Figure 4-3 will continue in a straight line across the straight line isobars toward lower pressure. Centripetal force is necessary for the curved flow across curved isobars. In either case, friction causes the flow to be from high to low pressure at an average angle of 30° to the isobars. Thus, air spirals inward toward the center of a low pressure system in a counterclockwise fashion and then ascends vertically (Figure 4-4).

In the Southern Hemisphere the direction of deflection of Coriolis and friction forces are reversed and so is the circulation, that is, air motion will be counterclockwise in a high and

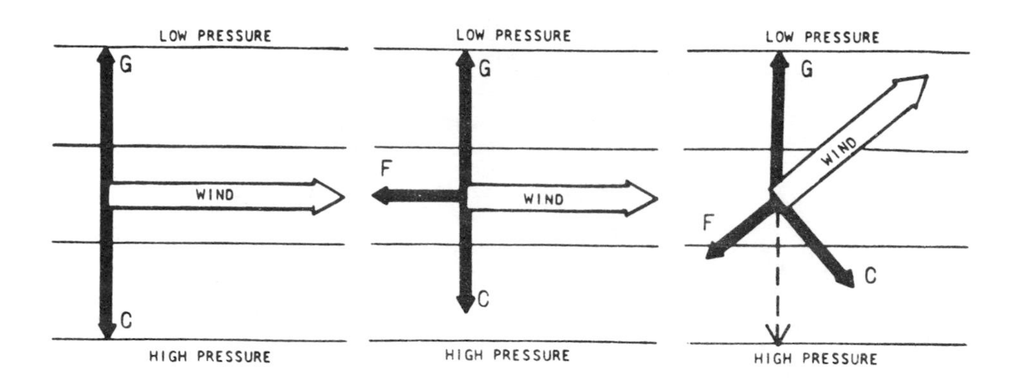

Figure 4-3. On the left, pressure gradient force (G) and Coriolis force (C) are in balance – the same as they were shown on the right in Figure 4-2. In the center, frictional force (F) has slowed the wind. As the wind slowed, Coriolis force decreased and the forces are no longer in balance. The greater pressure gradient force turns wind toward lower pressure as on the right. Frictional and Coriolis forces rotate with the wind. Finally, frictional and Coriolis forces combine (dashed arrow on the right) to exactly balance pressure gradient force. In the real atmosphere, these three forces come into play simultaneously. The forces reach a balance before the wind actually becomes parallel to isobars.

clockwise in a low. However, air will still spiral downward and outward from a high and inward and upward in a low. Other common names for high and low are anticyclone and cyclone, respectively.

Finally, observe that if one stands with one's back to the wind in the upper levels, low pressure is 90° to the left and high pressure is 90° to the right. At the surface the rule has to be adjusted *approximately* 30° due to friction, that is, low pressure will be 60° to the left and high pressure will be 120° to the right. In the Southern Hemisphere, low pressure will be 60° to the right and high pressure will be 120° to the left.

E. A Global Wind Model

We will now investigate the large-scale system of winds. What follows will be merely one theory of global winds. The theory is itself just one way of modeling atmospheric winds. The model presented here has flaws, not in what is described but rather in what is not mentioned. There are at least four other models of atmospheric wind circulation. The values of this model are two-fold. One is the stepwise description of the structure of winds, for they are essentially as described. The other value is in the reasons given for this structure, for we will be reviewing and applying concepts we have previously learned. Again, the weaknesses in the model come from what is not presented rather than what is. Other models are difficult to comprehend and our purpose here is not to compare all the theories but to visualize the general wind structure.

1. A Non-Rotating Earth

We begin our model by considering a non-rotating earth. We know that maximum solar radiation will be received near the equator, in fact, exactly at the equator assuming no tilt to the earth's axis, which we will assume for the time being. Now the sun's radiation is reradiated by the earth's surface as sensible heat. This heating causes the air above the equator to expand vertically, transferring mass to a higher level; the center of mass is also known as the center of gravity – you will determine that value when you compute aircraft weight and balance. Since weight is mass times the acceleration due to gravity, the center of mass, or gravity, is that location where all mass can be effectively placed insofar as balance is concerned. Two people of

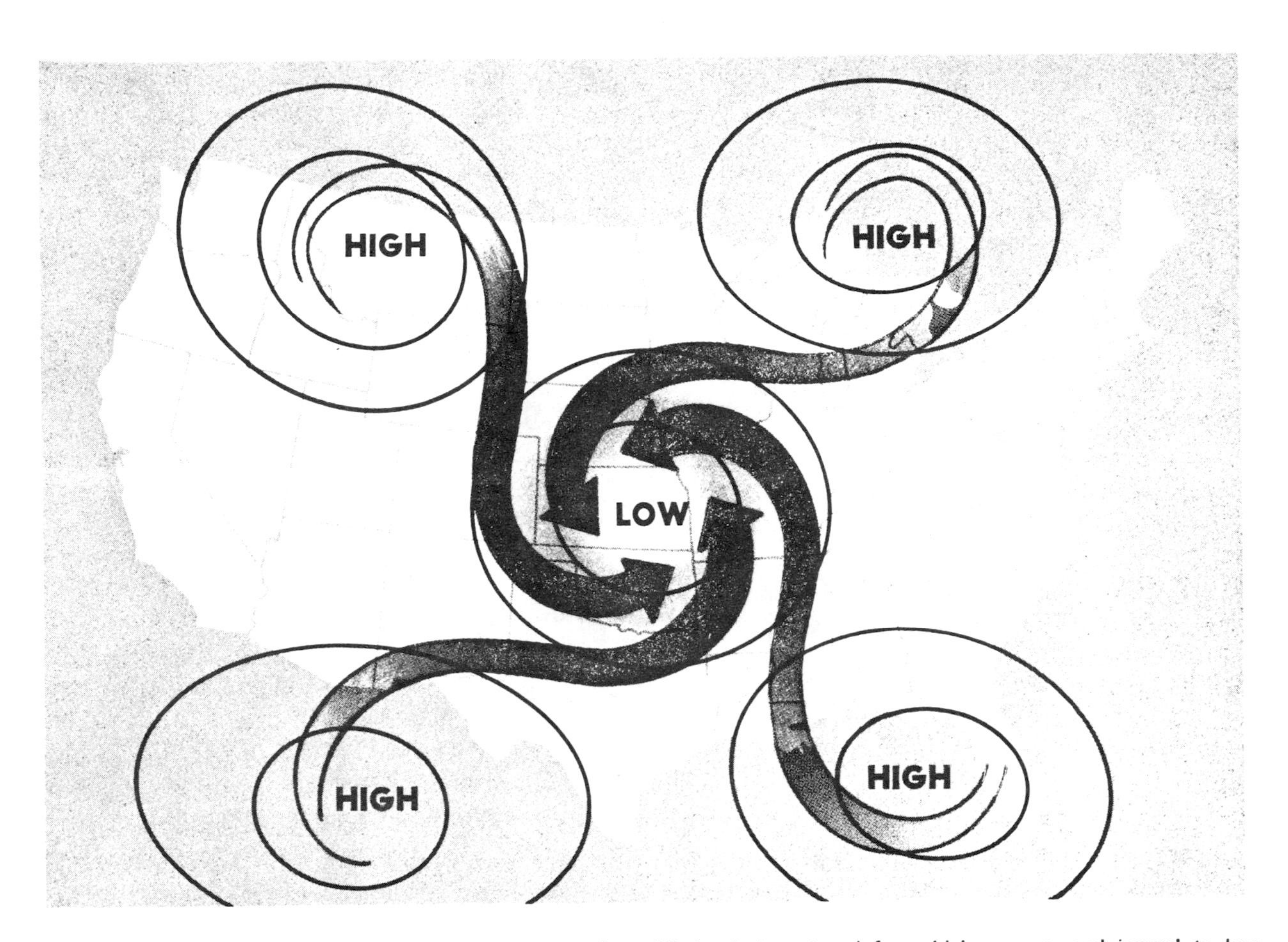

Figure 4-4. Circulation around pressure systems at the surface. Wind spirals outward from high pressure and inward to low pressure, crossing isobars at an angle.

different weights will balance on a teeter-totter at some point closer to the larger person. That would be the point at which both weights could be placed considering only the idea of balance or effective location of weight. Now back to our discussion of air.

The center of mass of the air in our vertically expanded column at the equator is higher than at the poles (we will describe the effects only in the Northern Hemisphere, the results in the Southern Hemisphere being merely a mirror image of the Northern). Let's now take an imaginary surface at the same height above the earth's surface, say, at 20,000 feet (6.1 km) from the equator to the pole. Form a mental picture of this. Since the center of mass is higher at the equator than at the pole, there will be more mass above the 20,000 foot surface at the equator; hence, there is more mass, there is more weight, more force, and more pressure on the 20,000 foot surface above the equator. Thus, air in flowing from high to low pressure will move from the equator toward the pole. As air in the high levels moves toward the pole, air in the lower levels will move toward the equator, setting in motion one large cell of air circulation (Figure 4-5). We do not obtain just this one cell, however, as we shall now see.

2. A Rotating Earth

As air moves from equator to pole in the upper levels, it is deflected to the right by the Coriolis force. Thus, we are now letting the earth rotate on its axis, counterclockwise. By the time air reaches approximately 30° North latitude, it is moving almost due west to east; the

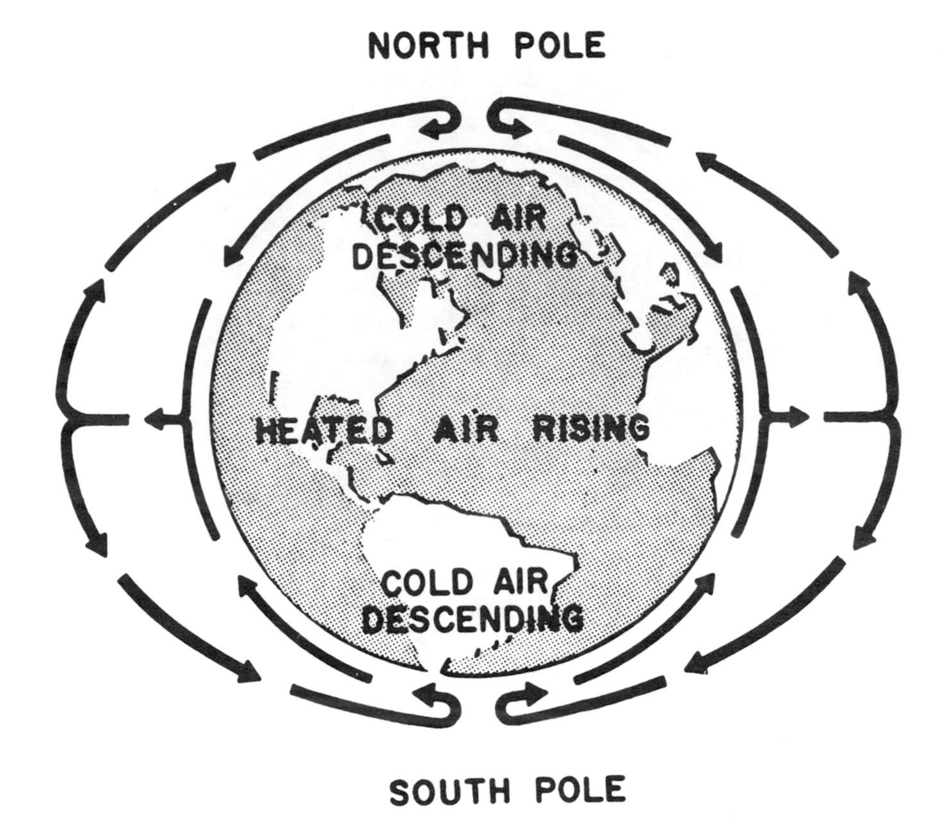

Figure 4-5. Circulation as it would be on a non-rotating globe. Intense heating at the equator lowers the density. More dense air at the poles flows toward the equator forcing the less dense air upward where aloft it flows toward the poles.

effect of friction is gone and pressure and Coriolis forces are close to balancing each other. But the important thing is really that this air motion results in a piling up of mass at that latitude. Thus, there has been a transfer of mass from the equator to 30° North and since the pressure at a given level is a result of the weight of the air above that level, we have relatively higher pressure near the surface of the earth at 30° North than at the equator. Hence, air flows down and spirals outward in a clockwise manner at the surface of the earth at 30° North latitude. The air that flows back to the equator is known as the northeast trade winds, the northeasterly direction resulting from the right-deflecting Coriolis force as air moves from 30° North to the equator.

Of the air that spills out at 30° North, some travels northward toward the pole, being deflected again to the right by the Coriolis force to become the prevailing southwesterlies. This is all occurring at the earth's surface in the vicinity of 30° North latitude. Note that the high at 30° North and the low at the equator feed each other, that is, air sinks down and spirals outward in a clockwise manner in the high, and then as it approaches the low pressure area it begins to flow counterclockwise and spirals inward and upward toward the center of the low. As air reaches high levels, it moves northward and descends in the high. You should now be able to picture this simple circulation in your mind. Simple? Yes, compared to what occurs when we consider other factors.

Let's now consider the behavior of air at the pole. As cold air moves from the pole southward at the surface it is deflected to the right, becoming the polar northeasterlies. This air converges or meets the prevailing southwesterlies near latitude 60° North. The mass of converging air has no place to go but up, expanding adiabatically as it does. When the air reaches

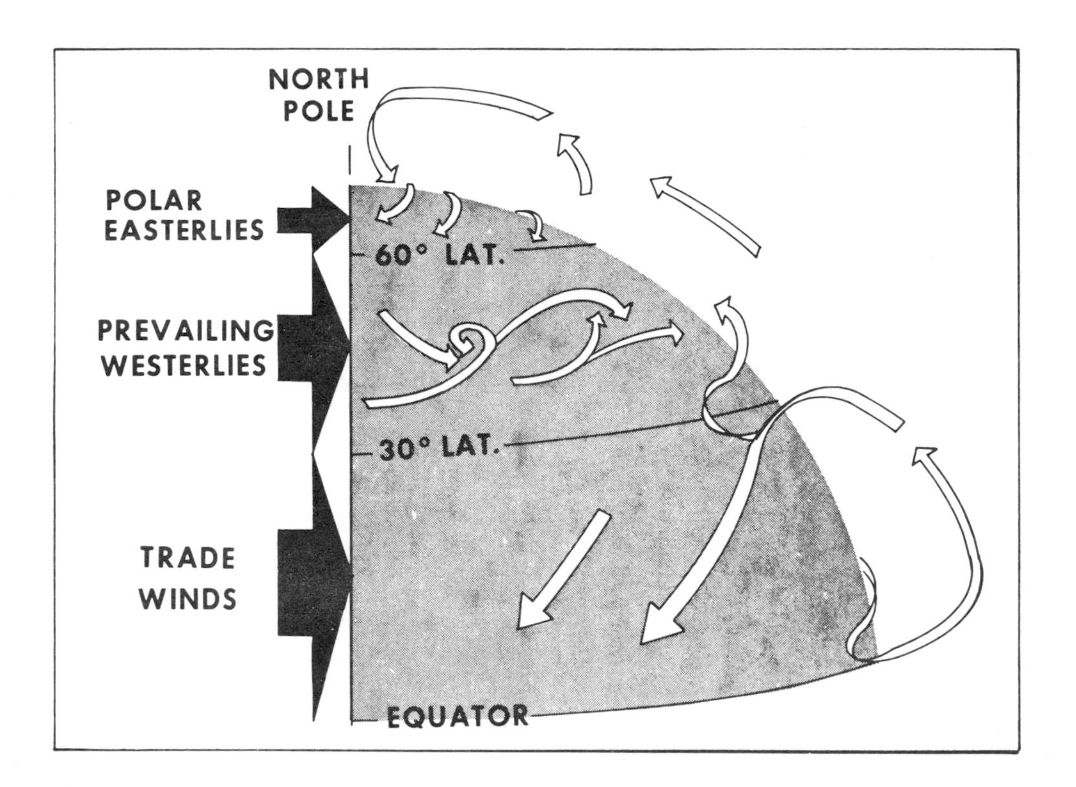

Figure 4-6. Pressure belts as a result of the earth's rotation. Huge masses of cold air frequently break out of the polar high pressure belt forcing their way southward through middle latitudes toward the tropics. Warm air is then forced northward toward the poles. The result is a mixing zone in middle latitudes with ever changing winds and migratory storms. This illustration is an attempt to average the general circulation resulting from uneven heating and the Coriolis force. During any given season, about one-half the surface is under the trade winds; about one-sixth, in the relatively calm polar regions; and about one-third, in the prevailing westerlies with migratory storms.

the upper levels, some of it moves toward the pole and some of it toward the equator. The air moving toward the pole is again deflected to the right becoming a high-level prevailing southwesterly zone of winds, air moving equatorward becoming a high-level northeast trade wind. Air moving to the pole in the upper levels from 60° North descends at the pole; air moving to the south from 60° North converges with air coming from the equator at 30° North, thus descending at 30° North since we are at high levels and the path of least resistance is downward (Figure 4-6). The rotation of the earth on its axis certainly complicates matters, the Coriolis force being the chief factor.

Note that if air could be perfectly still relative to space, the counterclockwise rotation of the earth on its axis would give an east to west wind, for the earth would rotate underneath its atmosphere, resulting in this relative wind. But as we have already seen, the air is not still, for temperature differences usually lead to pressure differences which lead to air motion, wind. Air which moves from the equator to latitude 30° North is deflected to the right becoming a west to east wind; we might as well call this a west wind, for wind direction is defined as the direction from which the wind is coming. The speed of the wind increases as air moves from the equator, overcoming the speed of rotation of the earth on its axis and, hence, becoming a true west wind relative to the earth's surface.

The reason wind speed becomes greater with an increase in latitude is the following example of the law of conservation of angular momentum from physics: suppose you had a rock tied on a string, attached to a stick; now as you rotate the rock around in a circle and let the string wind around the stick, you will observe an increase in the speed of the rock; it goes faster and faster the closer the rock gets to the stick. The same phenomenon is occurring with air as it moves from the equator toward the pole. When air is at the equator as compared to 30° North we are assuming that the air in its northward movement is at the same height above the earth's surface. But a point on the earth's surface at 30° North is closer to the earth's axis, the imaginary line running North to South through the geographical poles. Thus, the horizontal speed of the wind will increase as did the speed of the rock, so that the speed of the west wind increases as air approaches latitude 30° North. Incidentally, this is one reason why we can infer the existence of at least one jet stream, a zone or current of maximum wind speed; it does not occur, however, at the pole where the distance to the axis of rotation is the least.

3. *Tilt of the Earth's Axis*

Let's look again at our model. We have low pressure at the surface at the equator, high pressure at 30° North, low pressure at 60° North, and high pressure at the pole, with air flowing from high to low pressure. The ascending air near latitude 60° North was caused by the convergence of the prevailing southwesterlies and prevailing northeasterlies. The cold air from the pole meeting the warm air from the south forms a more-or-less continuous line around the globe near 60° North known as the polar front. This is a prime factor in causing changes in weather in mid-latitudes. Now we assumed when we began this discussion that the earth's axis was not tilted, but since it is (Figure 2-5), we need to investigate its effect on pressure.

In the summer in the Northern Hemisphere, the 23½° tilt of the earth's axis is toward the sun, placing the heat equator as far north as 23½° North of the geographical equator. The heat equator is just an imaginary circle around the earth's surface on which the sun is directly overhead. The result of all this is a shift in the pressure zones or belts we have mentioned. For example, on June 21, the longest day of the year in the Northern Hemisphere, the heat equator is located at 23½° North. Thus, all the pressure zones are shifted northward 23½°; hence, the polar front is not as far south in summer as it would be in winter, where the reverse is true. Therefore, in summer, cold air does not push as far south as in winter. Now in winter, the heat equator is south of the geographical equator – picture this in your mind; the earth's axis is tilted 23½° away from the sun on the shortest day in the Northern Hemisphere. Thus, all the zones of pressure move southward, including the polar front. Hence, cold air reaches much further south in winter in the Northern Hemisphere. Fronts will be discussed in greater detail in a later chapter.

4. *Land and Sea Contrasts*

We consider another factor yet to be taken into account, the contrast between land and sea. Frictional effects are greater over land areas and it was observed in the chapter on temperature that water absorbs and radiates heat with less temperature change than does land. We know that temperature changes usually yield pressure differences, resulting in wind. Partly because of the contrast of land and water in mid-latitudes, we do not have just one large zone of high pressure around the earth near latitude 30° North, nor just one large zone of low pressure around the earth near latitude 60° North. Rather, these zones of pressure are broken down into several whirlpools of high and low pressure.

Between so-called domes of high pressure there are smaller low pressure pools of air. In fact, since a high pressure system is a center of pressure surrounded on all sides by lower pressure, there must be some kind of low pressure pool or circulation between two highs. However, the overriding feature is high pressure due to the descent of air at approximately 30° North. Similarly, near 60° North several separate low pressure circulations are in evidence, usually on a larger scale than the small-scale lows near 30° North (Figures 4-7 and 4-8). There is an exception to this – the hurricane, a massive low, which develops near the equator and moves northward. So not all lows near the equator are small-scale. The zone of larger-scale lows near

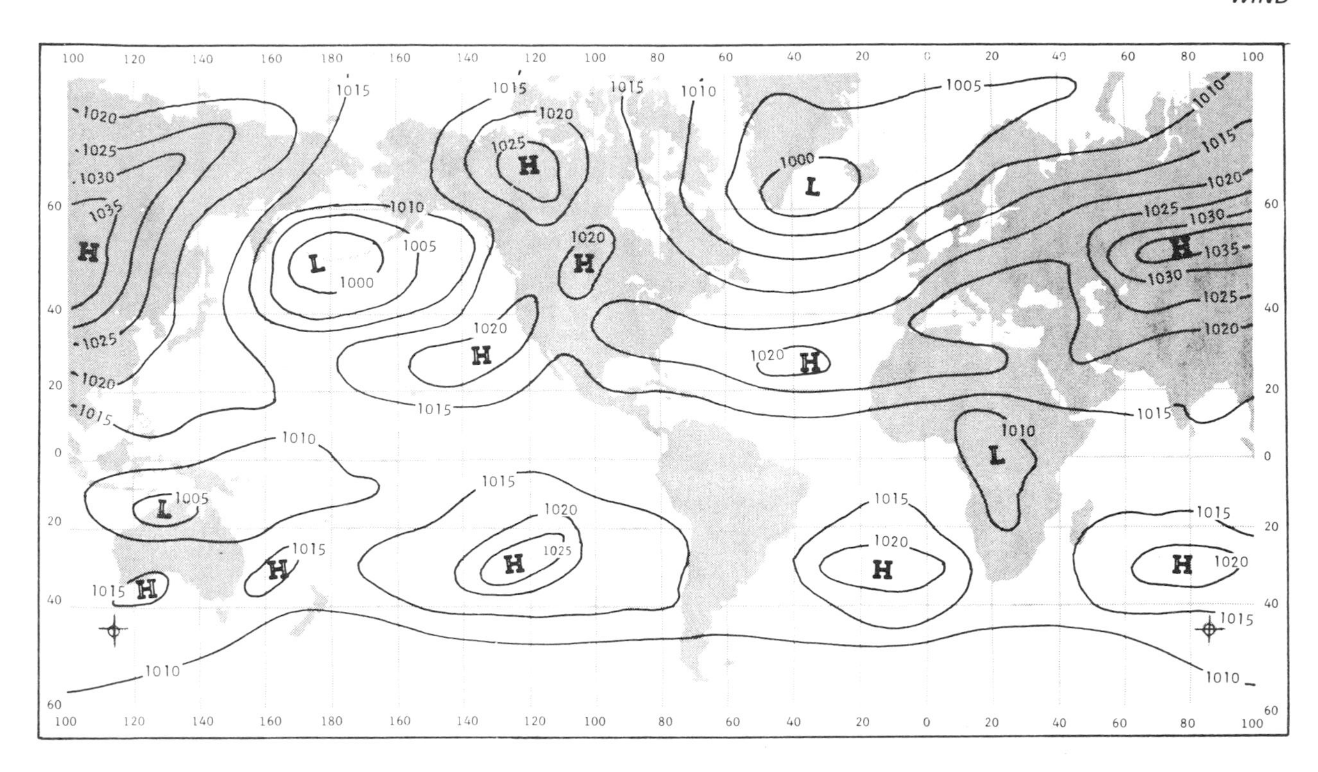

Figure 4-7. Mean world-wide pressure distribution in January. In the cold Northern Hemisphere, cold continental areas are predominantly areas of high pressure while warm oceans tend to be low pressure areas. In the warm Southern Hemisphere, land areas tend to have low pressure, and oceans high pressure. The subtropical high pressure belts are clearly evident in both hemispheres.

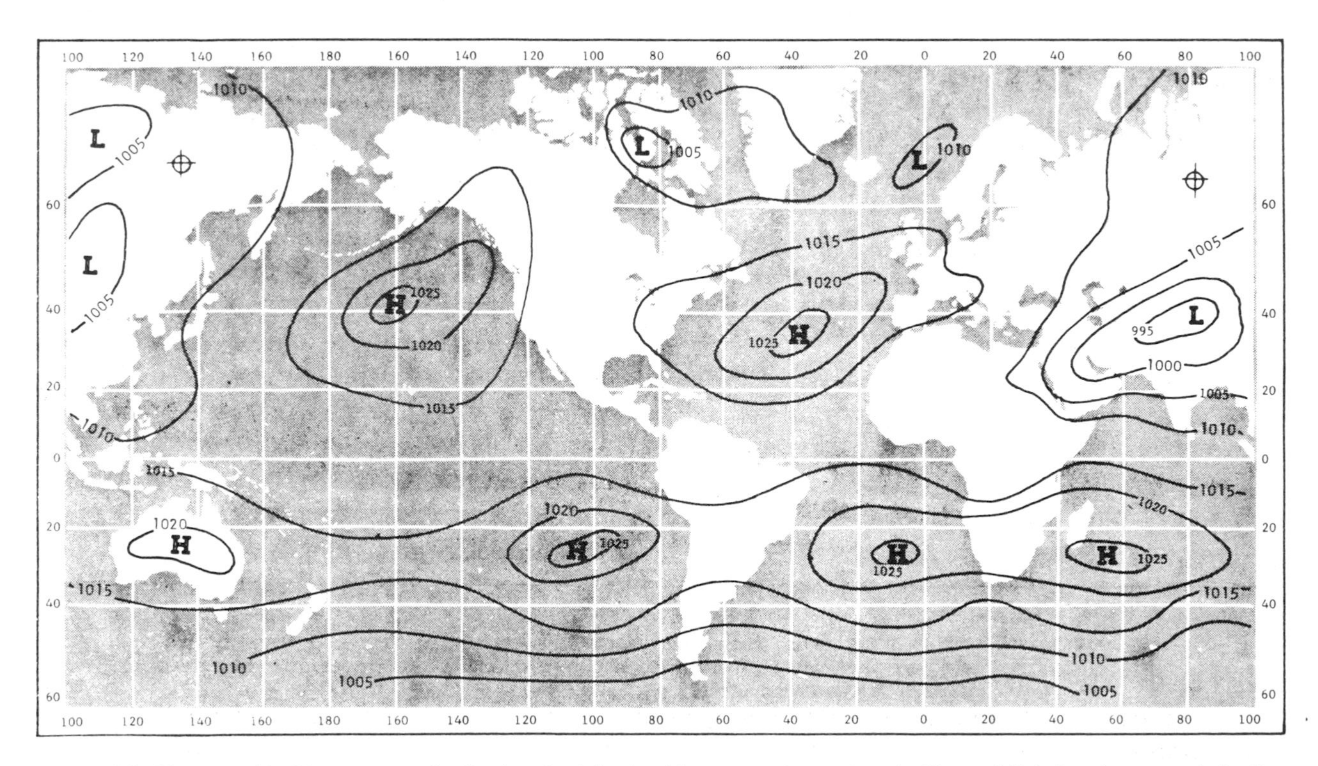

Figure 4-8. Mean world-wide pressure distribution in July. In this season, the pattern in Figure 4-7 is largely reversed. In the warm Northern Hemisphere, land areas tend to have low pressure and cool oceans, high pressure. In the cold Southern Hemisphere, land areas tend to have high pressure and ocean areas, low pressure. Compare with Figure 4-7 and note the seasonal shift in latitude of the subtropical high pressure belts — about 12 degrees in the Northern Hemisphere and 5 degrees in the Southern Hemisphere.

60° North is invaded by pools or cold domes of high pressure from polar regions. Thus, we have a large battleground for position between these giants. As far as movement is concerned, high pressure dominates, for heavier air will displace lighter air. For sheer excitement, however, the low wins hands down, for when moisture enters the picture, an upcoming chapter, the low just literally eats it up, as water vapor, and returns much of it to us as what we call precipitation.

We now examine land-water differences on a small scale. Picture an island surrounded by water. During the day, land heats up more than water, resulting in an expansion of air above the land as the air receives the heat. This results in less pressure over land while we have relatively higher pressure over water. Now we already know how air circulates from high pressure to low pressure. We are interested in the result, namely, that this wind is from water to land giving us the so-called sea breeze. The reverse of this is the land breeze, a nighttime phenomenon. Again, these are local winds, caused by daily variations in temperature (Figure 4-9).

On a larger scale we consider the monsoon, a seasonal wind. For example, Siberia is a huge land mass, and in winter air is extremely cold due to the effect of snow cover — so cold, in fact, that the maximum daily temperature is less than the temperature over oceans at the same latitude. The cold air, representing higher pressure, results in a flow of air from land to sea both day and night. This resultant seasonal wind, the winter monsoon, a seasonal land breeze, completely overwhelms what would ordinarily be a sea breeze in the daytime for a smaller land mass. The reverse of the above would yield the summer monsoon.

F. Constant Pressure Surfaces

The discussion we have just finished will give us an even better picture of what may be called large and small highs and lows. How high up do highs and lows extend? Well, they can extend up to 50,000 feet (15 km) and higher, but this is not the usual case. In order to better understand upper level winds, we go back to the concept of pressure. When we talked about pressure, it was relative to a horizontal surface at some fixed height. The United States surface map is a pressure analysis where all pressures are relative to sea level. Pressure is measured at the earth's surface and then converted to sea level by formula. This can be done since we know the height of the surface relative to sea level. With such a reference level, we can analyze the pressure distribution, drawing isobars in order to locate highs and lows. As we shall see in the chapter on air masses and fronts, the location of these highs and lows will be very important.

We encounter a complication when considering upper level charts. Our sea level conversion formula will suffer for at least two reasons; we are at a greater distance above the surface and we have no way of knowing exactly how high we are. Therefore, what are we to do about the problem? Well, strangely enough, we reverse the process. Instead of fixing height and determining pressure, we will fix the pressure and determine heights, the analysis of height or contour lines being analagous to analyzing isobars, yielding the position of highs and lows. What follows is a description of the process.

1. Determining Hills and Valleys

Let's first fix pressure, say 700 millibars, which occurs at approximately 10,000 feet (3 km) in the standard atmosphere. We now send a rawinsonde up in the atmosphere, a balloon with weather equipment and a miniature radio. The equipment will measure wind direction and speed, pressure, temperature, and humidity (which is indicative of moisture content). All of the information is used in an equation called the hydrostatic equation, which relates the change in pressure to the height through which the change in pressure takes place. Since we know the pressure at the surface, we know the change in pressure. We know the height of the surface above sea level, so we can now compute the height at which 700 millibars of pressure occurs; we actually accomplish this first at 850 millibars, as an intermediate step.

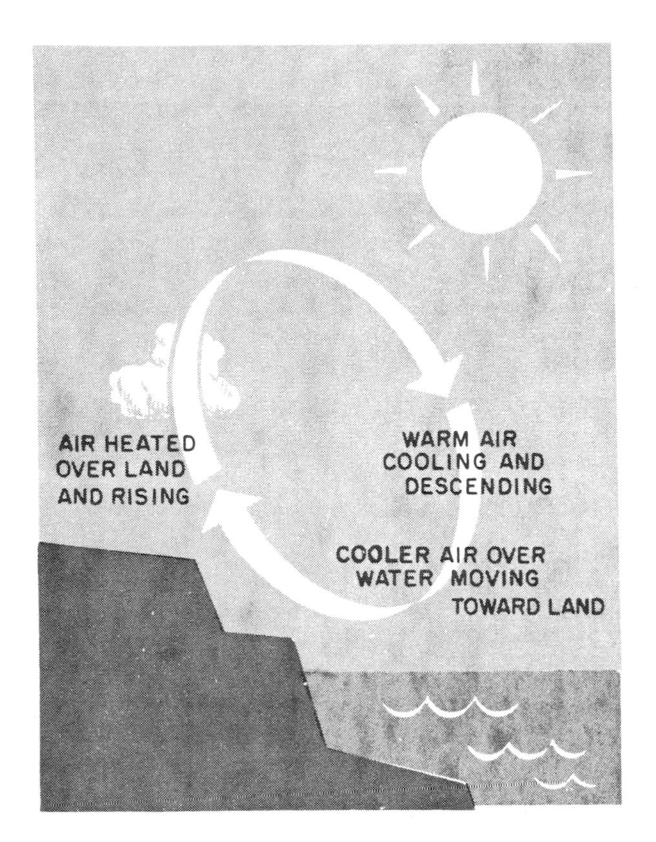

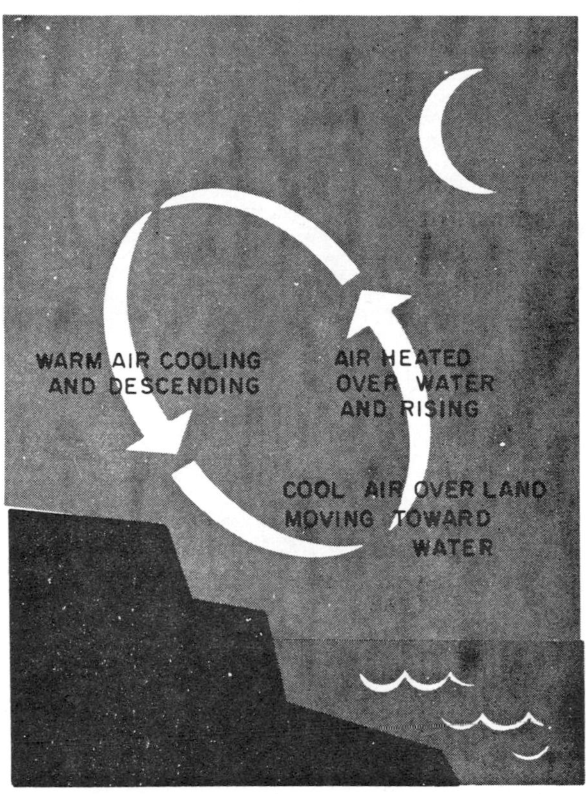

Figure 4-9. Sea breeze and land breeze. These small scale convection currents result from differences in diurnal temperature change between land and water surfaces.

The use of the formula also involves temperature and moisture, as well as two other factors. The bottom line is that all the information is available to calculate the height at which 700 millibars of pressure occurs. If we do this across the continental United States we have a pressure surface (Figure 4-10). This 700 millibar surface is not exactly 10,000 feet above sea level, but varies in height, close to 10,000 feet. You should be picturing in your mind a surface at about 10,000 feet with hills and valleys (Figure 4-11; TA means true altitude). On a standard day the hills and valleys would not be present.

2. *Hills and Valleys as Pressure Systems*

It turns out that hills will be locations of higher pressure, domes of high pressure, and valleys will be locations of lower pressure. We determine this as follows. First, recall that the 700 mb chart is a constant pressure surface. Let's draw lines of constant height on a U.S. map on a table to represent this. For example, draw a line on the map connecting points on the surface of the earth which have a pressure of 700 millibars at a height of, say, 10,100 feet (3080 m) above sea level. Now draw a line for 9,900 feet (3020 m). We have 700 millibars of pressure at two different heights. Let's compare pressure at two points chosen side-by-side on the two lines by drawing a line as perpendicular as we can to the two lines and choosing our points at the intersections. "A" is the point on the 10,100 foot line and "B" is the point on the 9,900 foot line (Figure 4-12).

We cannot compare pressure horizontally unless we fix the height, so we will choose the altitude of 10,100 feet. Now, how about the pressure at a spot 200 feet (60 m) higher than our point B at which we had 700 millibars of pressure at 9,900 feet? Well, if we have 700 millibars of

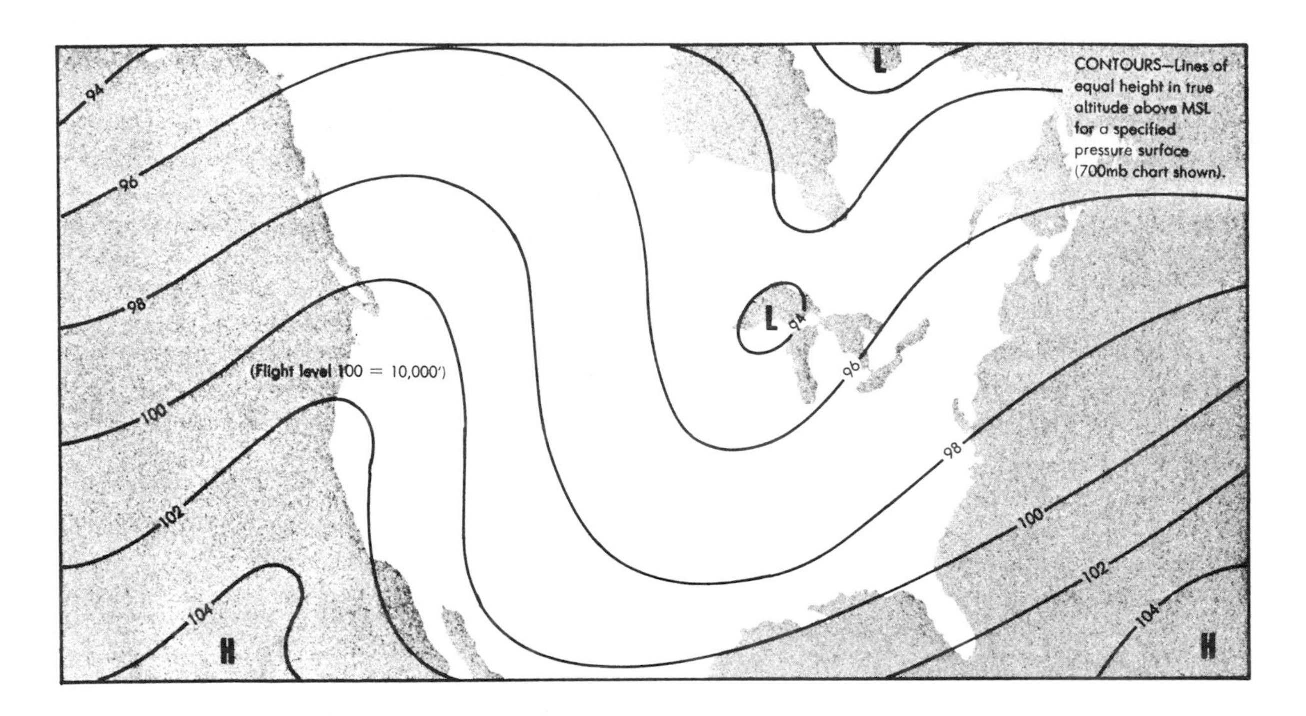

Figure 4-10. Upper Air Chart (700 mb pressure surface).

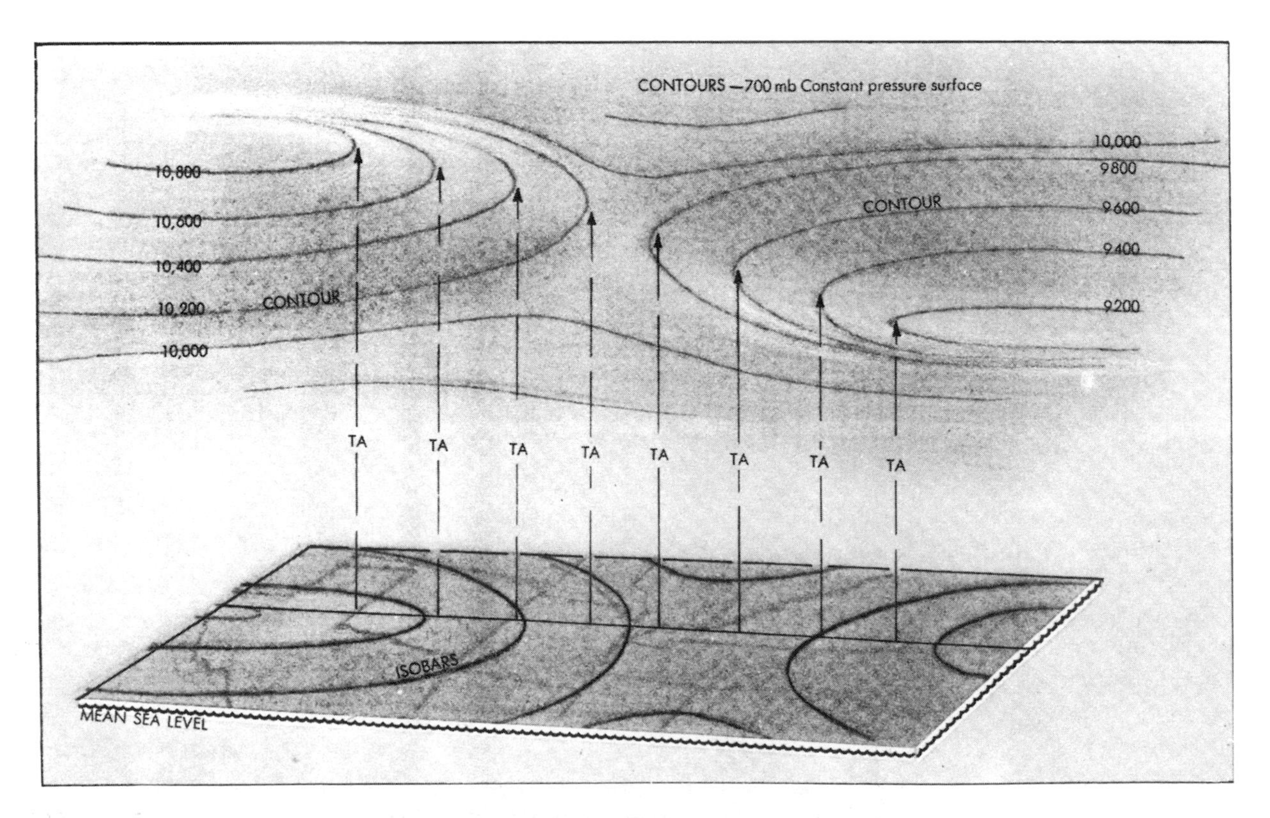

Figure 4-11. Comparison of Isobars and Contours.

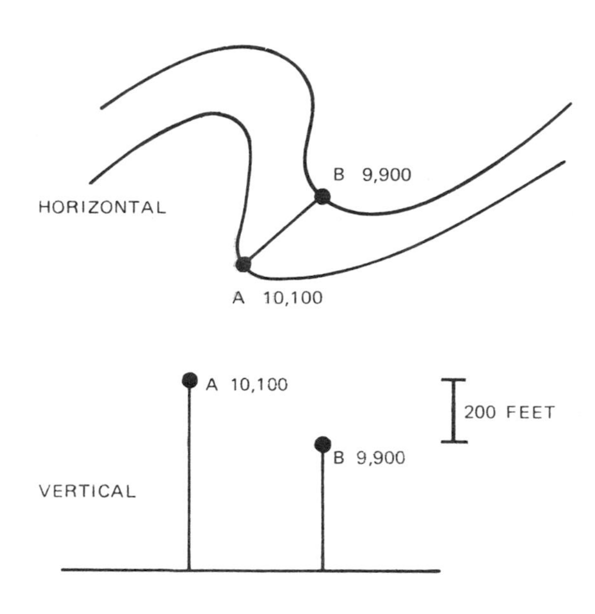

Figure 4-12. Horizontal and Vertical Cross-Sections of a 700-mb Chart.

pressure at 9,900 feet at B, then at 10,100 feet we are throwing away 200 feet of air. Therefore, the pressure at 200 feet above point B must be less than 700 millibars. Thus, comparing pressure at a height of 10,100 feet, we have higher pressure at point A, which is at 10,100 feet, than we do at a spot 200 feet above point B. So comparing pressure at 10,100 feet, the pressure is higher at point A than at the spot 200 feet higher than point B. Thus, hills represent relatively higher pressure and valleys lower pressure.

Since we are drawing lines of constant height for intervals of 200 feet on a U.S. chart on a table top, the hills and valleys disappear, but their height lines are there. Think of it this way – picture 10,800 and 10,600 foot lines around the hill. The 10,800 line will be above the 10,600 line on the hill, representing higher pressure (recall that the 10,100 foot line represented higher pressure than the 9,900 foot line). Now let's lay the pressure surface flat on a table top and press down on the hill. The hill flattens out, that is, disappears, but assuming the height lines were wet paint, they would leave circles on the map on the table (we will assume the hills were perfectly rounded). The 10,800 foot line will lie inside the 10,600 foot line, again representing higher pressure. Thus, we are left with a flat U.S. chart called the 700 millibar chart, analyzed for height.

3. Flying Hills and Valleys

If you were going to fly cross-country at an indicated altitude of 10,000 feet, you could look at the 700 mb chart to see where the highs and lows are relative to your route of flight. Wherever there are highs you could expect downdrafts and expect updrafts wherever there are lows. Suppose we have just taken off and we could inflate our surface with just enough air to pop out the hills and valleys. As you will see when you study the unit on alitmetry, your aircraft altimeter is a barometer which is calibrated to indicate height rather than pressure. Since the altimeter is a barometer, it measures pressure and then indicates an altitude where you would expect to find that pressure in a standard atmosphere (with one correction you will see later). That is, as long as you wish to fly the altitude indicated by your altimeter, you will have to be at an altitude at which the pressure is what your barometer now measures (assuming the "altimeter setting" – to be defined later – remains unchanged). This means that you are following the hills and valleys in order for the barometer part of your altimeter to sense the same pressure in order to give you a constant indicated altitude above sea level. Picture yourself riding along these hills and valleys at a constant indicated altitude, but at a changing true altitude, your actual altitude above sea level (Figure 4-11).

Note that as you fly from hill to valley you are flying from high to low pressure (considering only the horizontal component of motion) and that in the process you are losing true altitude (due to the vertical component of motion). The combined components of motion (horizontal and vertical) preserve the pressure the barometer part of your altimeter senses which gives your aircraft its constant indicated altitude, say, 10,000 feet. If some of the height lines on the hill are higher than 10,000, then, initially, your true altitude is more than 10,000 feet. Hence, the old phrase "from high to low, look out below" does mean you are losing true altitude, all other things being equal, but this does not mean we are necessarily lower than our altimeter indicates.

If we were at a true altitude of 10,000 feet initially and then flew into one of the valleys, we would again be flying from high to low and we would be losing true altitude and we would also be lower than our altimeter indicates. So considering pressure alone, we can say that from high to low we are losing true altitude, but we are not necessarily lower than our altimeter indicates. In general, however, the statement "from high to low, look out below" and thus you are losing true altitude, and you are lower than your altimeter indicates, is a true statement. The rule will work for high to low temperature also (to be seen in the chapter on altimetry), and the rule is just reversed for low to high. Whether or not the rule is always correct, you should assume it is for safety reasons – always make sure you have enough altitude. Once again, you will see in the chapter on altimetry that the above discussion has been oversimplified; this is the reason for the phrase "all other things being equal". Altimetry is a fairly difficult topic, until you finally "see" it. We will complete our discussion on altimetry when we come to that chapter, so do not be overly concerned if you are still "fuzzy" on it right now; this second early introduction to altimetry should be helpful at that time.

Another instrument we might consider for a moment is the vertical velocity, or rate-of-climb indicator. The instrument works off a so-called static port; when pressure differences are noted, this is registered in so many feet per minute. The altimeter system also uses the static port. The barometer part of the altimeter measures pressure at the port, located away from impact air, on the fuselage, for example. Small altitude changes are recommended to be made at no more than 200 feet (60 m) per minute.

When we climb one of the hills or descend one of the valleys we are undergoing a change in vertical velocity not registered on that instrument. This is because we are flying along a constant pressure surface – no change in pressure, so no indication. But when we fly up one of the hills, for example, we are encountering both a real and an artificial downdraft. Even if the air is not descending we would be acquiring an extra load on our aircraft wing as we climb to maintain the same pressure to maintain the same indicated altitude, all accomplished without a change in vertical velocity. Add to this the fact that the air is descending – remember, this hill represents a high pressure system – and you will need even more nose-high attitude to climb and have even a greater load on the wing, all the time having no indicated change in vertical velocity. We will study the effect of the above considerations on aircraft speeds in that chapter. Suffice it to say here that there will be influences since indicated airspeed works off the static port and impact air as well.

4. Contours and Wind Direction

You probably are convinced by now that wind direction and speed are quite variable, depending largely on latitude and altitude alone. We now try to complete the true picture of the atmospheric wind structure. We start at the surface and work upward. As we ascend, the effect of friction gradually disappears, and is virtually nil at an altitude of 2,000 feet (610 m) above the surface. Recall that the effect of friction is to deflect the wind to the left, so as the air gets closer and closer to the surface, below 2,000 feet, the greater this left deflection. The reverse is also true; as you ascend, the effect of friction is less and, hence, the wind has an apparent deflection to the right as the effect of friction decreases. This means that a south wind at the surface will become more southwesterly as height increases and vice versa, other factors being equal – recall that wind direction is the direction from which the wind is coming and leftness and rightness must be measured from that direction. Thus, above 2,000 feet above the surface, the effect of friction has virtually disappeared and the winds in the upper levels closely follow contour lines (Figure 4-13).

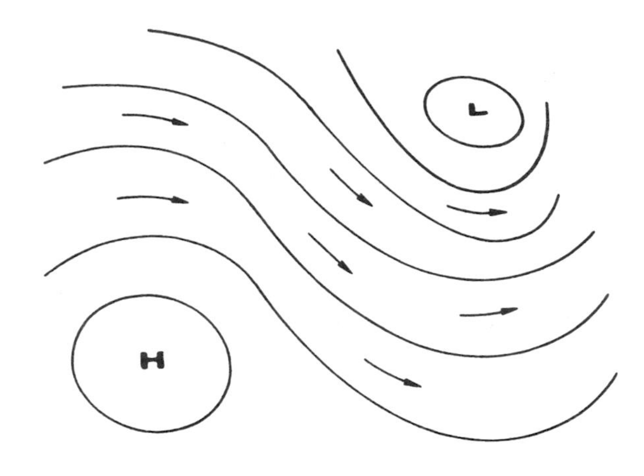

Figure 4-13. Wind flow around a high and a low pressure area (anticyclone and cyclone) relative to the contours above the friction layer. The arrows are winds and solid lines are contours.

5. *Contour Pressure Analysis*

Now again, we emphasize that the U.S. surface chart is a pressure analysis at a fixed height, because pressure values are made at local stations and then converted to mean sea level. So the U.S. surface map is really a sea level chart and the lines drawn connect points at sea level at which the pressure is the same; again, these lines are called isobars. As we have seen, in the upper levels pressure is fixed and we draw lines of constant height, called contours. We treat contours the same as isobars for analysis purposes; however, contours with higher numbers represent higher pressure horizontally. Pressure does not change as much horizontally as it does vertically, but horizontal changes are important in finding where highs and lows are. Use the 700 mb chart for cross-country flying at indicated altitudes near 10,000 feet for, as we have seen, you will very nearly be flying the hills and valleys of the chart – remember to "inflate" the chart as we did before so that the hills and valleys show. How near the 10,000 foot level would the chart be useful? Well, indicated altitudes of 8,000 to 12,000 would usually be very good since the height of this surface would seldom vary more than 2,000 feet from standard from one side of the U.S. to the other (you will see why in the chapter on altimetry).

6. *Common Pressure Levels*

The usual pressure levels for constant pressure charts are 850 mbs, 700 mbs, 500 mbs, 300 mbs, and 200 mbs, corresponding to standard altitudes of approximately 5,000 feet (1.5 km), 10,000 feet (3 km), 18,000 feet (5.5 km), 30,000 feet (9.1 km), and 40,000 feet (12.2 km), respectively. You would not want to use a 700 mb chart for indicated altitudes near 5,000 feet. Why? Because while the 850 mb chart may show a high pressure area at 5,000 feet, the 700 millibar chart might show a low pressure system at 10,000 feet over the same region. These two systems are vastly different and will require different flight planning. For flights between 5,000 and 10,000 feet, you would need to average the features of the two charts. Did I say that there could be a high pressure system on the 850 mb chart and a low pressure system above the same point on the 700 mb chart? Well, yes, that can occur and, in fact, can produce very severe weather.

Picture high pressure as determined on the 850 mb surface and low pressure as determined on the 700 mb surface, at the approximate heights of 5,000 and 10,000 feet, respectively. Air is therefore descending at 5,000 feet and ascending at 10,000 feet. Now the west wind feeds the high and the low, and pushes on them. The west wind will feed the high and low in the sense that it joins their circulation pattern (much of the west wind is diverted around these systems to continue its eastward flow). The west wind also pushes on these systems because of the west to east force, actual movement depending on how strong the respective west wind and vertical currents are.

7. Vertical Structure of Highs

We have described the nature of upper level winds, basically a west to east flow with occasional spiraling columns of ascending or descending air known as lows and highs, respectively. The vertical extent of highs and lows is quite variable, usually less than 20,000 feet (6 km) except in well-defined systems, such as a thunderstorm, which may extend to 60,000 feet (18 km) or higher. How would a high pressure system that extends to 20,000 feet (6 km) appear on these upper level charts? Well, let's visualize the pressure surfaces at their standard levels. A high pressure system would show up on each of these charts by at least one closed contour inside of which there is a higher height value, at the center of a circular contour, for example. If each center is over the same spot on the earth's surface at each pressure level, then the axis of this high would be straight up and down – at this point we will disregard the fact that pressure systems usually have a varying degree of tilt in the vertical. Assuming that the axis of our high is straight up and down and we inflate our constant pressure charts to show the hills and valleys, then the high pressure system is indicated by a series of smaller and smaller hills with height, one directly above the other at each pressure level (Figure 4-11).

Picture our high as a cone with a circular base at the earth's surface and coming to a point at 20,000 feet. As cold air is fed in at the top, it descends and spreads out at all levels. When the air in its descent has reached the surface at the center, we stop for just a moment and look at what has happened. We have a column of air along the centerline with a certain amount of mass, hence exerting a certain pressure at the surface. One second later, assuming the same amount of mass input at the top of the cone, we have the same amount of mass along the centerline, part of which is new mass or old mass in a new location. The new mass makes up for the mass which spiralled away from the centerline during the second. And part of the old mass has descended a little along the centerline, ready to spiral outward during the next second. In this manner a constant high pressure is maintained along the centerline. As we move away from the centerline, the spiralling out of the air has resulted in a spread of mass and, hence, less pressure at the surface.

There is one paradox here, however. We stated in the chapter on temperature that air which descends is compressed and warms adiabatically. How can air descend and both spiral outward and be compressed? Well, the circulation pattern dominates for reasons we will not give here. The maximum compression that occurs will be along the axis of the high, thereby generating a warm-core high through adiabatic heating. Note that we built this high. In that sense it can be said to be dynamic. The cold-core high tends to be less in vertical extent and spreads downward and outward due to its own weight, much as a large drop of molasses on a table would spread. We have shown that a high pressure system can be pictured as a dome of air, somewhat like a cone, with air spiralling downward and outward in a clockwise manner inside the cone.

8. *Divergence*

We have used the word convergence previously. Divergence is the opposite of convergence and occurs when horizontal outflow exceeds horizontal inflow. Divergence occurs when (1) air moves in diverging directions, (2) when downstream winds exceed upstream winds, (3) when air motion is anticyclonic (recall the high – air spreads down and out), and (4) when the wind flows from the north (visualize the air as moving from the north pole, spreading out as do lines of longitude – this explanation is only a simplified approximation) (Figure 4-14). The opposite of these statements describes the ways in which convergence occurs.

In the upper levels, the word confluence is often used in place of convergence (diffluence for divergence). At the surface, converging air can only rise, but in the upper levels with convergence (confluence) the air can be squeezed upward into a low, downward into a high (the deep high of section 7, for example), or into both the low and high. If confluence aloft is greater than divergence at the surface, the high will intensify (becomes stronger, that is, pressure increases).

9. *Troughs and Ridges*

We now look at another feature of upper level charts, troughs and ridges. These features are defined in terms of low and high pressure, as we saw in the chapter on pressure. A ridge, for example, is an outward stretching or elongation of isobars or contours, where the wind flows generally along these lines in a clockwise manner. The same description fits the trough, except the wind flows counterclockwise. Picture an S lying on its side. As air moves along the curve – we have counterclockwise rotation around the left half of the S and clockwise rotation around the right half, assuming that the wind is from the left. Note that this contour line is not closed. When a contour line is closed, we have a high when the circulation is clockwise and a low when the circulation is counterclockwise. Remember, a high is a center of pressure surrounded on all sides by lower pressure, conversely for a low. In either case, "on all sides" forces at least one closed contour line, not necessarily completely circular. Air spirals downward and outward in a high as we have already seen. Now back to the S. The left half represents a counterclockwise circulation and the right half clockwise. Thus, the trough, the left half, is only "part low" and the ridge "part high" since the contour is not closed.

The main observation we wish to make is that while horizontal movement is essentially along the curvature, there is vertical movement of the air, but to a lesser degree than in a high and a low. Picture an imaginary line from left to right through the middle of the S (Figure 4-15). In flying across the trough, you receive both a horizontal and vertical tearing effect; the horizontal effect is due to the wind shift from right to left as you cross the trough. The vertical effect comes from the fact that the air is rising vertically as it moves horizontally; this is due to the fact that a trough is a type of low and, hence, inherits its ascending air characteristics. Thus, you have an updraft as you cross the trough. This is not a low, so air does not spiral upward, instead maintaining its left to right horizontal movement with some vertical motion. And in order to maintain constant indicated altitude you will lose some true altitude – this is the same as it was when you flew the hills and valleys on the inflated contour chart we described earlier, in order to maintain indicated altitude. We can also inflate these ridges and troughs, but they will not have the nice uniform shape of our idealized hills and valleys for our highs and lows. In fact, the inflated shape cannot be a complete hill or valley, for this would force us to draw a closed contour on the constant pressure chart.

10. *Vorticity*

Vorticity refers to turning in the atmosphere; sufficient turning may create what is commonly called a vortex or a rotary motion on a small or large scale. For a parcel of air, relative vorticity (atmospheric turning relative to a fixed point on the earth) is by definition either positive (for counterclockwise, that is cyclonic motion), zero, or negative (for clockwise, that is, anticyclonic

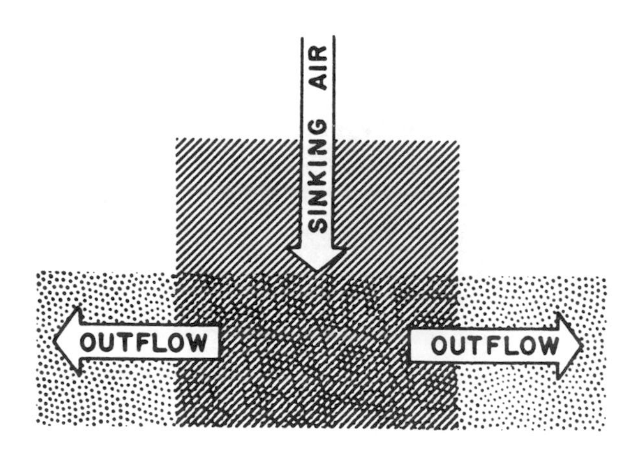

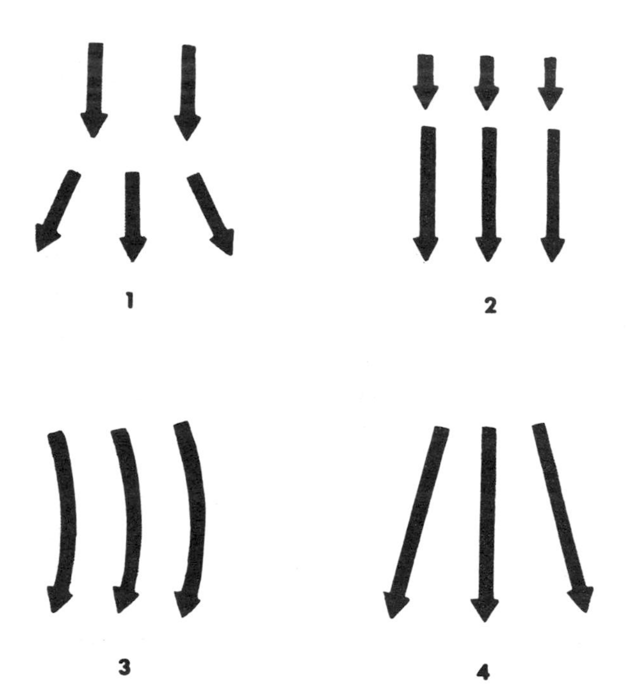

Figure 4-14. Schematic of divergence. Above, the hatched area represents a volume of air before divergence. The dotted area is the same volume after divergence. Below, (1) diverging directions; (2) down stream winds exceed up stream winds; (3) anticyclonic wind flow produces divergence; (4) northerly flow results in divergence.

motion). Spin or turn at a point on the earth's surface is relative to the vertical axis directly above the point.

We must also consider that the earth rotates. If you stood at the north pole you would turn (spin) once each 24 hours. At the equator you will move as the earth rotates, but you will not spin relative to the vertical axis above you. Thus, the earth's vorticity increases from zero at the equator to a maximum at the poles. Since the earth rotates counterclockwise, the earth's vorticity is always positive and turns out to usually be greater than relative vorticity.

Absolute vorticity is the sum of the earth's vorticity and the relative vorticity, that is, the total vorticity as viewed from space. In this chapter we are only concerned with the relative vorticity. In chapter XVI we will be concerned with absolute vorticity.

Since the atmosphere is composed of many different circulations, vorticity is very difficult to measure. But increasing positive vorticity induces upward motion (lows and troughs) and larger negative values of negative vorticity induce downward motion (highs and ridges). The converse is also true – upward motion (lows or troughs) generates positive vorticity and downward motion (highs or ridges) decreases it. Typical vertical speeds are on the order of two to four inches per second (five to ten centimeters per second). Even in large scale highs and lows, vertical speeds are not great; however, they are very large in thunderstorms, tornadoes, and localized turbulence.

Wind shear will be discussed in more detail in the chapter on turbulence. However, the following illustrations will indicate the relationship between vorticity and wind speed and direction (Figures 4-16 and 4-17).

11. Summary of Vertical Pressure Structure

For the above reasons, as well as for reasons we have not yet covered, we may classify warm core highs as deep systems while cold core highs tend to be shallow systems, changing to lower pressure aloft. Warm core lows tend to be shallow, changing to higher pressure aloft, and cold core lows tend to be deep systems. As we will see later, warm core highs are tilted toward warm air aloft (keep in mind that –20°C is warmer than –25°C) and cold core lows are tilted toward cold air aloft. Many pilots believe that "high" means good weather and "low" means bad weather. While this statement is true in general, we will see exceptions to this rule as we go along.

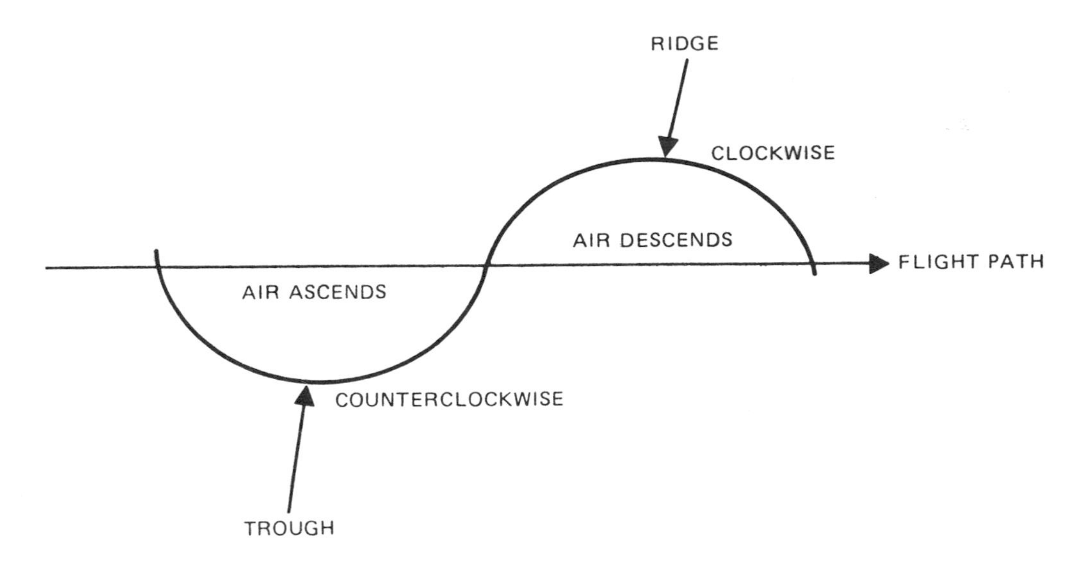

Figure 4-15. Flight Path Across a Trough and Ridge.

G. Katabatic Winds

We now present a brief discussion of the katabatic wind, any wind flowing down an incline. Picture air flowing down a mountain slope (Figure 4-18). This air is warmed adiabatically, for it is compressed under a greater weight of air as it descends. The increase in temperature will be about 5.5°F per 1,000 feet as we saw earlier. We will look at this concept more in the chapter on moisture.

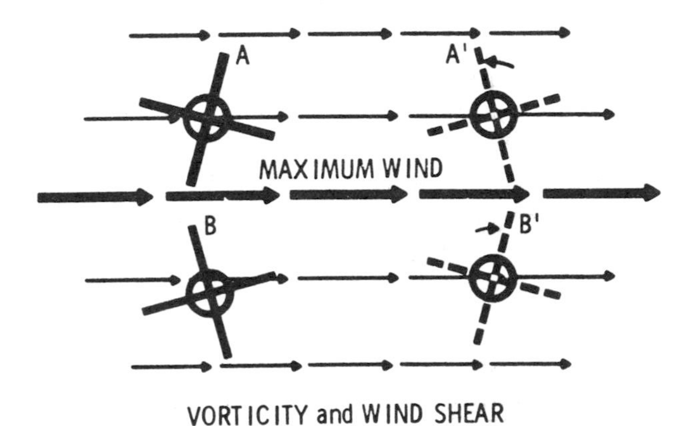

Figure 4-16. Wind speed contributes vorticity. Note the maximum wind in the center with decreasing winds on either side creating shear. The paddle wheel to the left of the maximum (above) will rotate cyclonically due to the difference in wind speeds. This is cyclonic shear with positive vorticity. To the right of the maximum (below) the paddle wheel rotates anticyclonically. This is anticyclonic shear with negative vorticity. Thus a wind maximum (jet) has anticyclonic shear with negative vorticity to the right and cyclonic shear with positive vorticity to the left.

H. Some Effects of Wind on Airplanes

We conclude this section with some of the effects of wind on an airplane, from FAA VFR Exam-O-Gram No. 27. Other effects of wind will be considered in the chapter on airspeed.

Does an airplane in flight travel in the direction it is headed? Not always! The airplane moves forward because of engine thrust pulling in the direction it is headed. However, if the mass of air surrounding the airplane is also moving (wind), then the airplane, in addition to its forward movement, is carried in the same direction and at the same speed as the air mass. Thus, we have two directional forces acting on the airplane — the thrust component and the wind component. If thrust is moving the airplane forward toward the east and the wind is moving it sideward toward the south, then the resultant path over the ground will be east-southeasterly. This sideward movement of the airplane caused by the wind is called "drift" (Figure 4-19).

How can we compensate for drift in order to make good a desired course over the ground? We must head the airplane into the wind at an angle at which the direction of the thrust

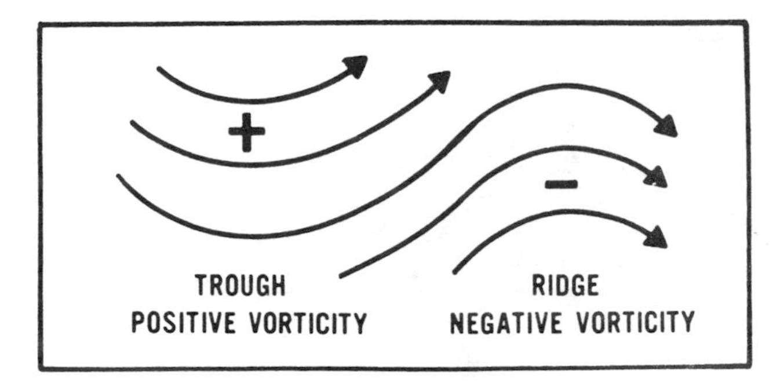

Figure 4-17. Curved flow contributes vorticity. A trough contains positive vorticity and a ridge, negative.

component will compensate for the wind component. This correction angle, or "crab", should be sufficient to make the resultant path over the ground (ground track) coincide with the desired course over the ground (Figure 4-20). The necessary heading can be determined by trial and error, or by wind triangle computations based on true airspeed, true course, and wind direction and speed.

Does wind affect an airplane on the ground the same as in the air? In certain respects, no. In addition to being moved forward through the air by its own power, an airplane in flight is carried in the same direction and at the same speed as the movement of the air mass surrounding it. Since it is free to move with the air mass, the airplane in flight does not "feel" this movement of the mass of air (except when wind shear, or sudden lulls or gusts are encountered). Therefore, after the proper correction for drift is established, control pressure need not be maintained for directional control. However, during ground operation, the friction of the airplane's wheels in contact with the ground resists drifting, creating a pivot point at the main wheels. Since a greater portion of the airplane's surface is presented to the crosswind aft of the wheels than is presented forward of the wheels, the airplane tends to "weathervane" or turn into any crosswind (Figure 4-21). In this case corrective control pressures must be applied and maintained for directional control on the ground. This weathervaning occurs even in tricycle (nose wheel) gear airplanes, unless the wheels are located well aft in relation to the side surface of the airplane.

What effect do crosswinds have on takeoff and landing? While the airplane is free of ground, the wind has the same effect as explained in preceding paragraphs for an airplane in flight. However, on takeoff and landings, an airplane should never be allowed to contact the ground while drifting or while headed in a direction other than that in which it is moving over

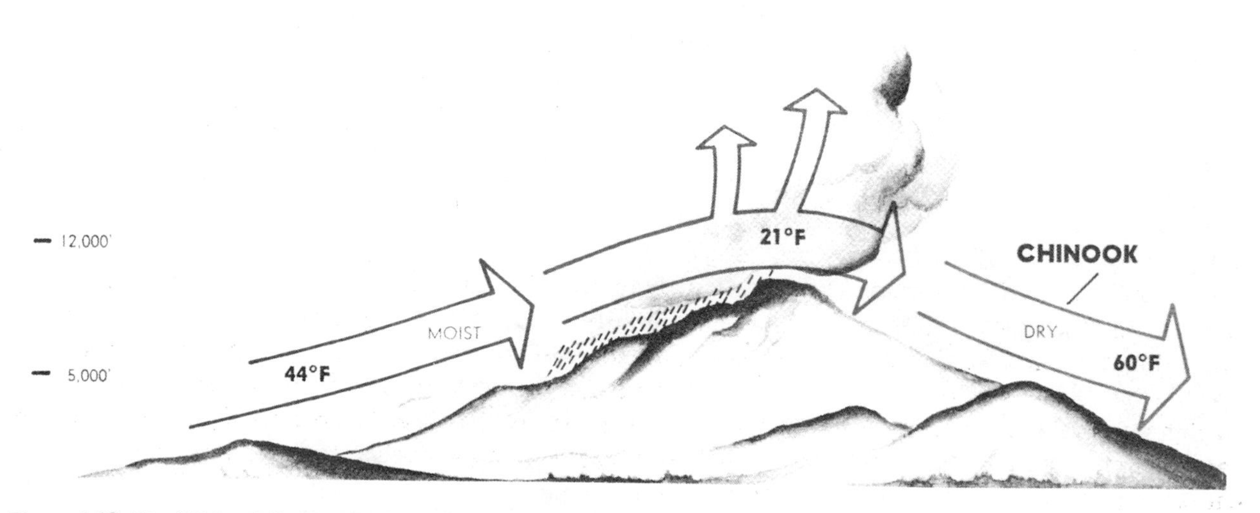

Figure 4-18. The "Chinook". Air cools as it moves upslope and warms as it blows downslope. The Chinook occasionally produces dramatic warming over the plains just east of the Rocky Mountains.

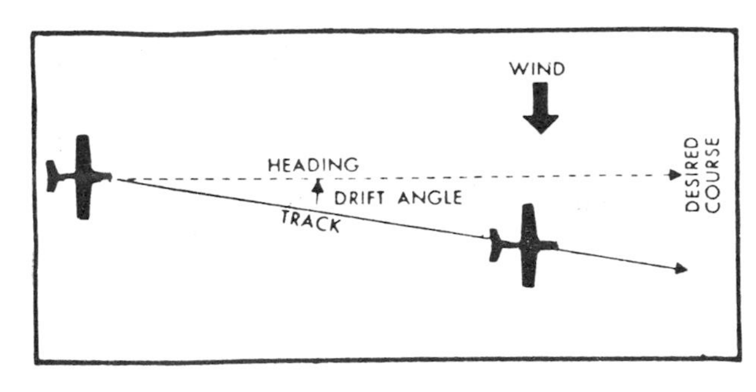

Figure 4-19. Drift Caused by Wind.

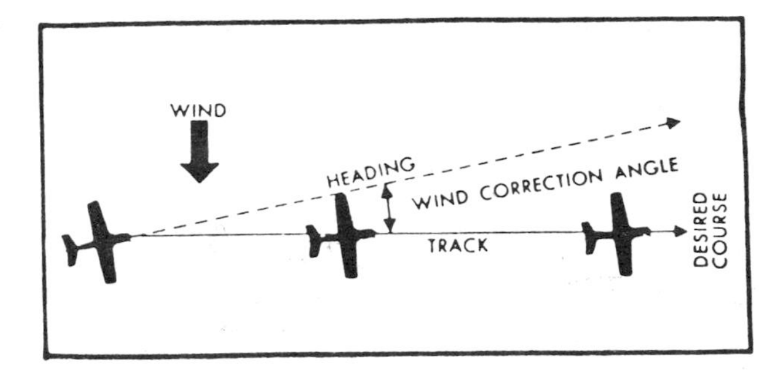

Figure 4-20. Crab to Correct Drift.

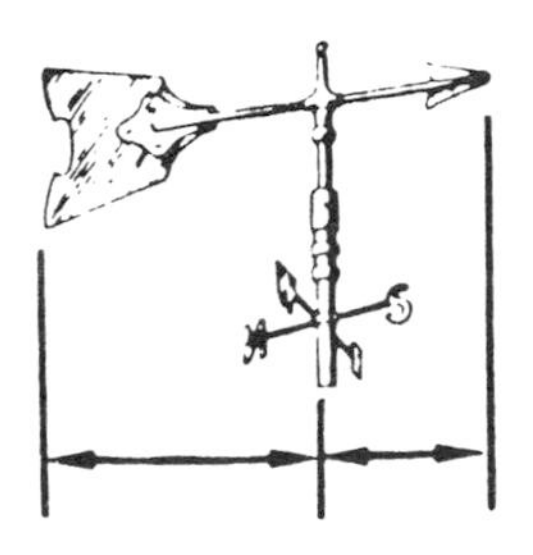

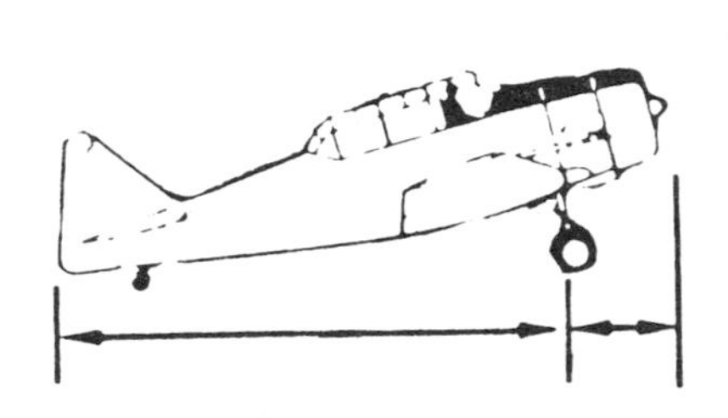

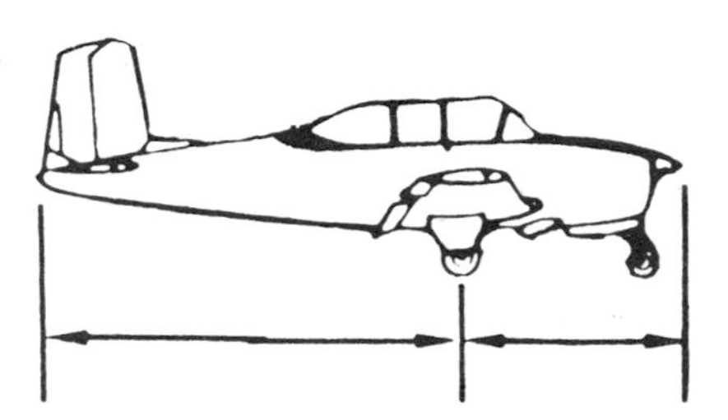

Figure 4-21. Weathervaning.

the ground. Unless proper action is taken to prevent this from occurring, severe side stresses will be imposed on the landing gear, and a sudden swerve or ground loop may occur. When this develops, we have an almost uncontrollable situation, and consequently a serious accident potential.

Can takeoffs and landings be safely made in all crosswind conditions? Not always! Takeoffs and landings in certain crosswind conditions are inadvisable, or even dangerous. If the crosswind is great enough to warrant an extreme drift correction, a hazardous landing condition may result. Therefore, always consider the takeoff or landing capabilities with respect to the reported surface wind conditions and the available landing directions. The absence of proper crosswind techniques, or the disregard for adequate consideration of the airplane's characteristics and capabilities with respect to crosswind conditions, are reflected by the continual rise in accidents involving ground control.

What is the maximum safe crosswind condition? Before an airplane is type certificated by the FAA, it must be flight tested to meet certain requirements. Among these is the demonstration of being satisfactorily controllable with no exceptional degree of skill or alertness on the part of the pilot in 90° crosswinds up to a velocity equal to 0.2 V_{SO}. This means a windspeed of two-tenths of the airplane's stalling speed with power off and gear and flaps down. (If the stalling speed is 60 mph, then the airplane must be capable of being landed in a 12 mph 90° crosswind.) To inform the pilot of the airplane's capability, regulations require that the demonstrated crosswind velocity be made available. Certain Airplane Owner's Manuals provide a chart for determining the maximum safe wind velocities for various degrees of crosswind for that particular airplane (Figure 4-22).

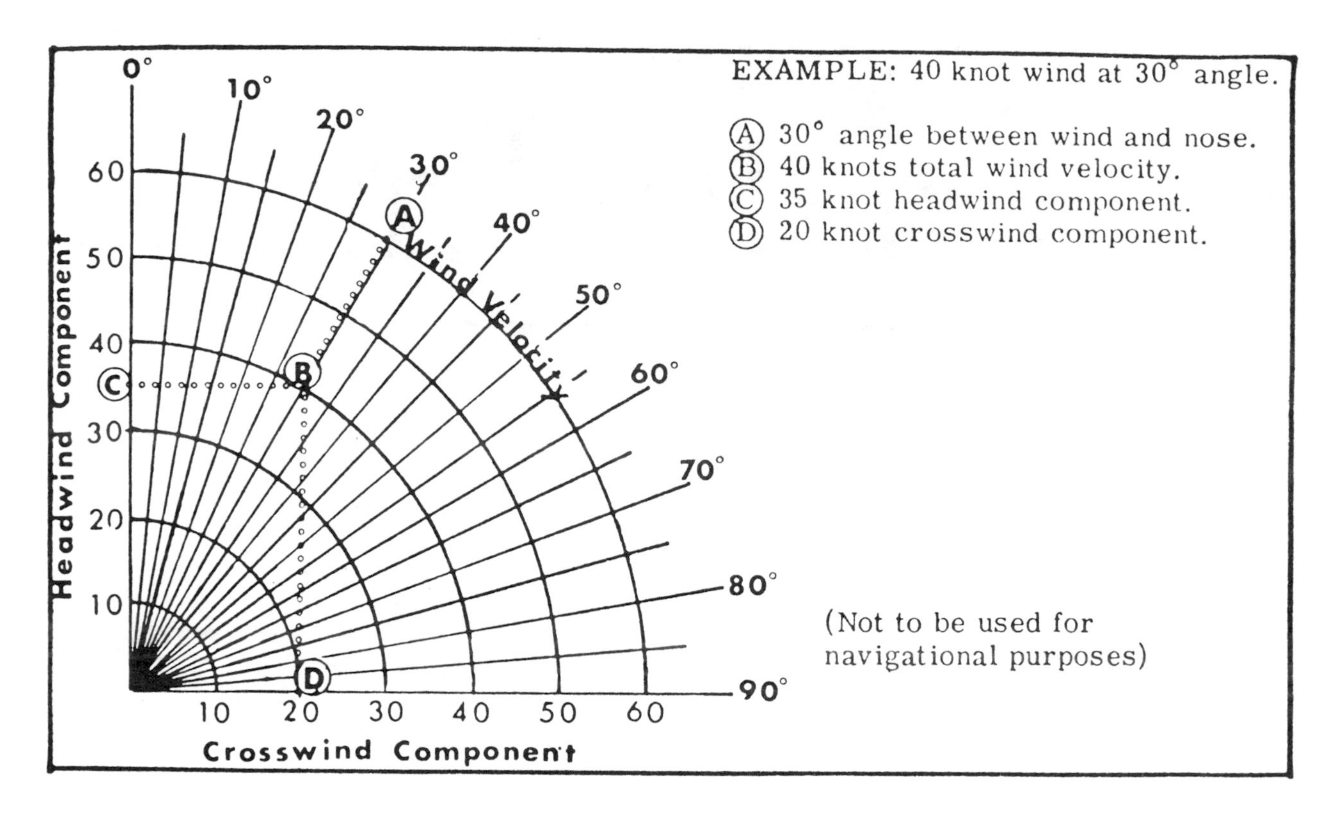

Figure 4-22. Chart for Determining Crosswind Component. (Example only.)

WIND
Review Questions

PRI

1. Select the true statement concerning wind circulation associated with pressure systems in the northern hemisphere.
   a. Wind circulates counterclockwise around high pressure areas and clockwise around low pressure areas.
   b. Wind circulates clockwise around high pressure areas and counterclockwise around low pressure areas.
   c. Wind circulates counterclockwise around both high pressure and low pressure areas.
   d. Wind circulates clockwise around both low pressure and high pressure areas.

2. Concerning wind circulation associated with high and low pressure areas at the surface, select the true statement.
   a. Wind flows across both high and low pressure areas paralleling isobars.
   b. Wind flows outward from high pressure areas and inward to low pressure areas, crossing isobars at an angle.
   c. Wind flows outward in both high and low pressure areas, crossing isobars at an angle.
   d. Wind flows inward to high pressure areas and outward from low pressure areas, crossing isobars at an angle.

3. Select the true statement concerning isobars and windflow patterns around high and low pressure systems.
   a. When the isobars are far apart, crests of standing waves may be marked by stationary lenticular clouds.
   b. Isobars connect contour lines of equal temperature.
   c. When the isobars are close together, the pressure gradient force is greater and wind velocities are stronger.
   d. Surface winds flow perpendicular to the isobars.

4. A pilot planning a long distance flight from west to east in the conterminous United States would most likely find favorable winds associated with high and low pressure systems by planning to fly a course which is:
   a. North of a high.
   b. North of a low.
   c. South of a high.
   d. South of both highs and lows.

5. In the Northern Hemisphere, a pilot making a long distance flight from east to west would most likely find favorable winds associated with high and low pressure systems by flying:
   a. To the north of a high and to the south of a low.
   b. To the south of a high and to the north of a low.
   c. Through the center of highs and lows.
   d. To the north of a high.

6. What causes surface winds to flow across isobars at an angle rather than parallel to isobars?
   a. Heat radiation from the surface.
   b. The difference between air temperature and dewpoint temperature.
   c. Surface friction and windflow toward lower pressure.
   d. Coriolis force.

7. The pitot-static system is a source of pressure for which of the instruments listed below?
   a. Airspeed indicator only.
   b. Airspeed indicator, altimeter, vertical-speed indicator, turn-and-slip indicator, and heading indicator.
   c. Altimeter, vertical-speed indicator, and heading indicator.
   d. Altimeter, vertical-speed indicator, and airspeed indicator.

INS

8. Winds at 5,000 feet AGL on a particular flight are southwesterly while most of the surface winds are southerly. This difference in direction is primarily due to:
   a. A stronger pressure gradient at higher altitudes.
   b. Stronger Coriolis force at the surface.
   c. Friction between the wind and the surface.
   d. The influence of pressure systems at the lower altitudes.

CFI

9. What causes the counterclockwise flow of air around a low pressure in the Northern Hemisphere?
   a. Centrifugal force.
   b. Surface friction.
   c. Pressure gradient.
   d. Coriolis force.

10. In the Northern Hemisphere, what causes the wind to be deflected to the right?
    a. Centrifugal force.
    b. Coriolis force.
    c. The pressure gradient force.
    d. Surface friction.

11. In the Northern Hemisphere, which of the following is a true statement with regard to the flow of air within a low pressure center?
    a. Air flows outward, upward, and counterclockwise.
    b. Air flows inward, downward, and counterclockwise.
    c. Air flows outward, downward, and counterclockwise.
    d. Air flows inward, upward, and counterclockwise.

12. Which statement is true with regard to the general circulation of air associated with a high pressure area in the Northern Hemisphere?
    a. Air flows outward, upward, and counterclockwise.
    b. Air flows outward, downward, and clockwise.
    c. Air flows inward, upward, and clockwise.
    d. Air flows inward, downward, and counterclockwise.

13. To maintain the desired ground track when flying directly toward a low pressure center (in the Northern Hemisphere), what heading correction, if any, would most likely be required?
    a. A correction to the left would be required.
    b. No correction would be required.
    c. No correction is necessary, because Coriolis force will tend to keep the airplane on the desired track.
    d. A correction to the right would be required.

14. In the Northern Hemisphere, the outward flow of air from a high pressure area or the inward flow to a low pressure area is deflected to the:
    a. Right in a high pressure area; left in a low pressure area.
    b. Right in both a high and low pressure area.
    c. Left in both a high and low pressure area.
    d. Left in a high pressure area; right in a low pressure area.

ATP

15. Why does the wind have a tendency to follow the isobars above the friction level?
    a. The Coriolis force tends to counterbalance the horizontal pressure gradient.
    b. The Coriolis force acts perpendicular to a line connecting the highs and lows.
    c. The friction of the air with the earth deflects the air perpendicular to the pressure gradient.
    d. Isobars are lines connecting points of equal wind direction aloft.

CHAPTER V

ALTIMETRY, PERFORMANCE, AND AIRSPEED

The reader should demonstrate an understanding of altimetry, performance, and airspeed to the extent that he can answer the following questions (as stated below, or as multiple-choice, etc.):

1. Define "indicated altitude"; "altimeter setting".

2. Standard sea level pressure is indicated by __________ inches of mercury.

3. Suppose the altimeter is set to 30.00 inches (76.2 cm) – we are not saying that 30.00 is the actual local setting – and the barometer part of the altimeter senses a pressure equivalent to 30 inches of mercury. What will be the indicated altitude?

4. As you fly from an area of warm temperature to an area of cold temperature, what usually happens to true altitude? Where is such an occurrence most important and what should you calculate to compensate?

5. With a local altimeter setting of 29.92 (76 cm) what will be the relation between pressure altitude, indicated altitude, and true altitude when at field elevation? With this same altimeter setting, what will be the relation between these altitudes when flying at an indicated altitude above field elevation?

6. Assume one inch (2.5 cm) of mercury corresponds to 1,000 feet (300 m) of altitude, that our field elevation is 2,000 feet (600 m), and that the pressure at our airport is equivalent to 27.80 inches (70.61 cm) of mercury. What will be our altimeter setting? At the airport when we set in the altimeter setting, what will be our indicated and pressure altitudes?

7. In question 6, what will be your indicated and pressure altitudes if after you take off you are flying at an altitude where the pressure is equivalent to 22.80 inches (57.91 cm) of mercury (assume the same altimeter setting)?

8. In what two ways can pressure altitude be determined?

9. Find density altitude at a field elevation of 650 feet when the temperature is 95°F and the altimeter setting is 29.75. Is this density altitude high or low (compared to field elevation)? What does this suggest about aircraft performance?

10. Suppose you are planning a flight at 5,000 feet. If the temperature is forecast to be 10°C at your flight level and the altimeter setting is 30.15, what is the density altitude for the planned indicated altitude of 5,000 feet? What does this suggest about aircraft performance?

11. Name three kinds of altimeter errors. How much may the altimeter be in error for low altitude operations? High altitude operations?

12. Due to the above errors, what is the value of maintaining the correct altimeter setting when at cruise altitudes?

13. If you do not have a radio, how do you set your altimeter?

14. What is the requirement for altimeter settings when you do have a radio?

15. What information is conveyed about high and low pressure systems from altimeter settings?

16. Define "indicated airspeed"; "calibrated airspeed". Compare these two airspeeds for low-speed operations; for high-speed operations. What are some reasons for these differences? Which of these two speeds is used on airspeed indicators?

17. Define "true airspeed".

18. Suppose you are flying at a constant power setting into a 20 knot headwind and then you turn downwind with a 20 knot tailwind. Assuming no change in the density of the air in the downwind and upwind directions, what will be the effect of this change in direction on indicated airspeed? True airspeed? Groundspeed?

19. Suppose our takeoff speed is 60 knots indicated airspeed for standard mean sea level. Then we should take off at (a lower, the same, a higher) indicated airspeed at an airport located at 5,000 feet? Why?

20. For an indicated airspeed of 100 knots at an altitude of 8,000 feet (2.4 km) estimate your true airspeed. If you are flying with a 12 knot tailwind what will be your approximate groundspeed?

CHAPTER V

ALTIMETRY, PERFORMANCE, AND AIRSPEED

A. Altimetry (See Appendix F for Altimeter setting procedures from the AIM)

The altimeter consists of three parts. One is the Kollsman window, where the altimeter setting is entered. A part you do not see is an aneroid barometer that consists of an evacuated disk that expands or contracts, depending on whether pressure is decreasing or increasing, respectively; this pressure is measured at the static port (Figure 5-1). The altimeter is calibrated so that the difference between what the aneroid senses and the value of the altimeter setting, which will be equivalent to so many inches of mercury, will be indicated on an altitude scale in corresponding feet based on the standard atmosphere.

1. Altimeter Setting

Indicated altitude is relative to mean sea level and is the height we read on the height scale when the Kollsman window is set to the local altimeter setting. This indicated altitude may or may not be your true altitude, that is, your actual altitude above mean sea level. We now define altimeter setting. It is the value we set in the Kollsman window so that the altimeter reads true altitude at field elevation. Thus, indicated and true altitude are the same at field elevation. We shall now see why.

Suppose we are at mean sea level at this moment. Recall that standard sea level pressure is equivalent to 29.92 inches (76 cm) of mercury. In order to determine the altimeter setting, which is always done on the ground, we first determine an adjustment factor for field elevation,

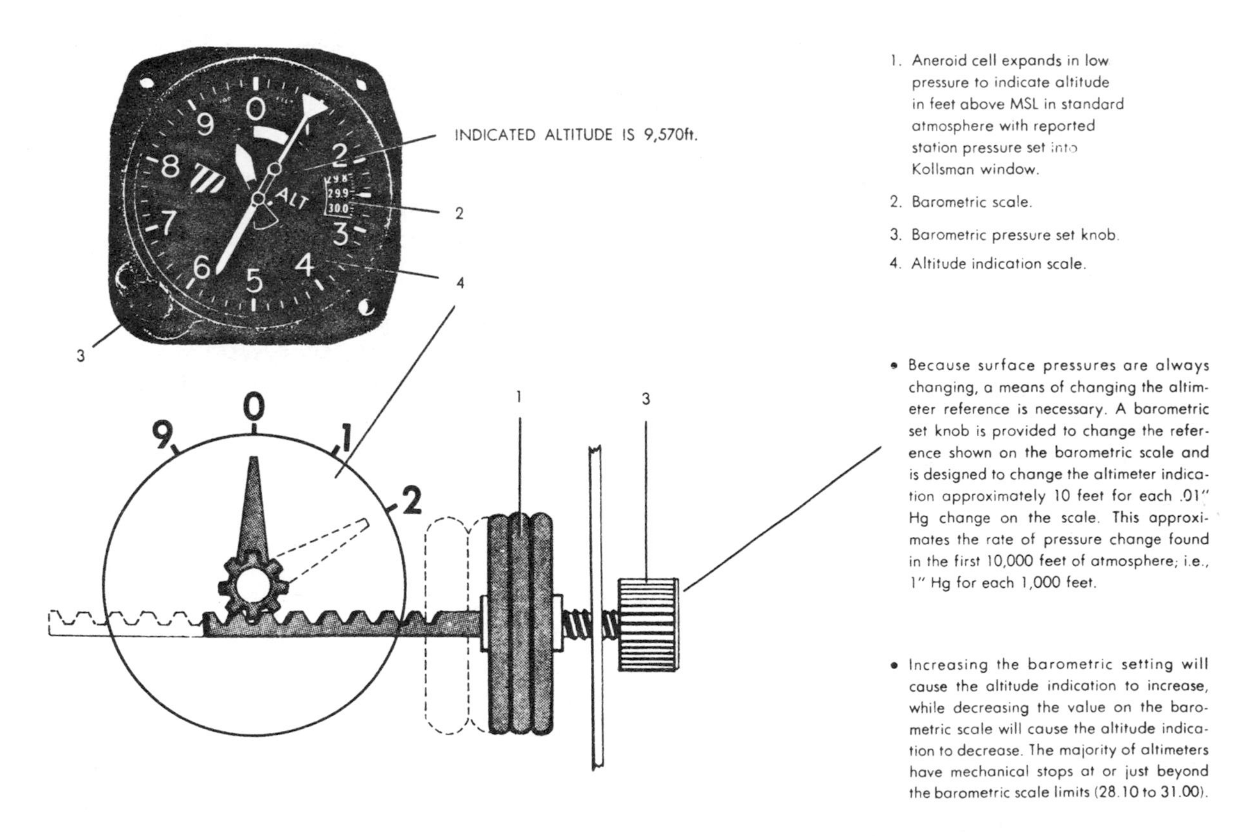

INDICATED ALTITUDE IS 9,570ft.
1
2
3
4
1. Aneroid cell expands in low pressure to indicate altitude in feet above MSL in standard atmosphere with reported station pressure set into Kollsman window.
2. Barometric scale.
3. Barometric pressure set knob.
4. Altitude indication scale.
• Because surface pressures are always changing, a means of changing the altimeter reference is necessary. A barometric set knob is provided to change the reference shown on the barometric scale and is designed to change the altimeter indication approximately 10 feet for each .01" Hg change on the scale. This approximates the rate of pressure change found in the first 10,000 feet of atmosphere; i.e., 1" Hg for each 1,000 feet.
• Increasing the barometric setting will cause the altitude indication to increase, while decreasing the value on the barometric scale will cause the altitude indication to decrease. The majority of altimeters have mechanical stops at or just beyond the barometric scale limits (28.10 to 31.00).

Figure 5-1. Aneroid Altimeter.

which is always 29.92 minus the standard value at field elevation. But since standard pressure is 29.92 at sea level, we have no adjustment factor since 29.92 – 29.92 is zero inches of mercury. Suppose the actual pressure at sea level on this day is 29.50 (74.93 cm). Our adjustment factor is still zero. The second part in determining our altimeter setting is to add our actual pressure and our adjustment factor. In this example our setting will be 29.50 plus zero, which is just 29.50.

Now the barometer part of the altimeter measures 29.50 on the ground at sea level in the above example. The altimeter takes the difference between the setting value of 29.50 and the sensed pressure of 29.50. But that is zero inches of mercury. Remember that one inch of mercury corresponds to about 1,000 feet of altitude change. A change of zero inches of mercury would therefore correspond to zero feet altitude change from mean sea level and, hence, the height scale shows us at an indicated altitude of 0 feet, which is also our true altitude. Notice in this example that our altimeter setting is just the pressure at sea level, regardless of whether it may be high or low, and that since we were at sea level our adjustment factor is zero for field elevation. Thus, we get an indicated altitude of zero, regardless of the actual pressure at sea level.

Suppose that we take off from sea level with an altimeter setting of 29.50. When the barometer part of the altimeter senses a pressure of, say, 28.50 (72.39 cm), this is a change of 1 inch (2.54 cm) of mercury and our altitude scale will show approximately 1,000 feet (305 m). Is 1,000 feet our true altitude? Well, not unless the rate of decrease of pressure with height on that day is standard, that is, approximately 1 inch per 1,000 feet. What if the air had been colder than standard on this particular day near the surface? Picture two columns of air of the same volume based at sea level, both exerting a pressure of 29.50 inches of mercury at sea level (Figure 5-2). Suppose further that one column has a standard rate of change of pressure with height during the first 1,000 feet. Suppose in the other column the air is cold enough so that the first 800 feet (244 m) of air weighs equivalent to 1 inch of mercury. Could the weight of these two entire columns still both be 29.50 at the bottom? How? Well, the air above 800 feet in the second column would have to be warmer than the air above 1,000 feet in the other column.

The two columns are of the same volume and contain the same mass measured at the bottom. But due to the difference in temperature distribution, the altitude at which they indicate a change in pressure is different. The rate of change of pressure in the cold air is 1 inch in 800 feet while for the other column it is 1 inch in 1,000 feet. Thus, with the same altimeter setting of 29.50 and an indicated altitude of 1,000 feet, we find that we are losing true altitude in the colder air. Thus, even if the altimeter setting does not change enroute, if the air you are flying into is colder beneath your airplane than where you have just come, you will be losing true altitude (Figure 5-3 – "QNH" is the indicated altitude and the D-value is the difference between true and indicated; the altimeter setting is assumed to be 29.92).

The above example is especially important for mountain flying. We showed that you could be lower than your altimeter indicates. Now while there are cases in which we could lose some true altitude without actually being lower than our altimeter indicates (see WARMER THAN STANDARD side of Figure 5-3), in many cases a loss of true altitude will mean we are lower than our altimeter indicates. You could find yourself as much as 2,000 feet (610 m) lower than your altimeter indicates (this is one reason why the minimum obstruction clearance altitude – MOCA – is 2,000 feet in mountainous terrain for pilots on instrument flight plans). You have no means of adjusting your altimeter for the temperature structure of the air column beneath you. If you knew this, flight computers could be designed to use this consideration in computing true altitude. You do have, however, the temperature at your particular altitude. If the departure of that temperature from standard is representative of the average departure from standard beneath your aircraft, then you can obtain your approximate true altitude, which is a substantial improvement and should be close enough to be used for terrain clearance. Most flight computers

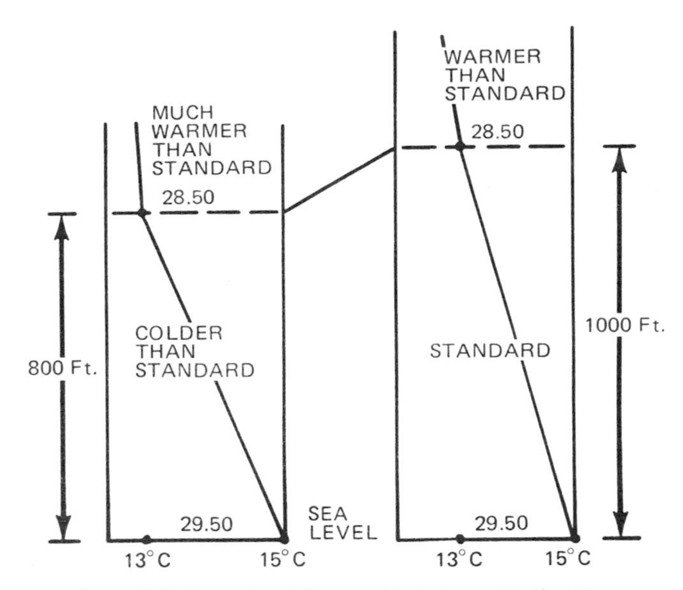

Figure 5-2. Example of Pressure Variation with Height.

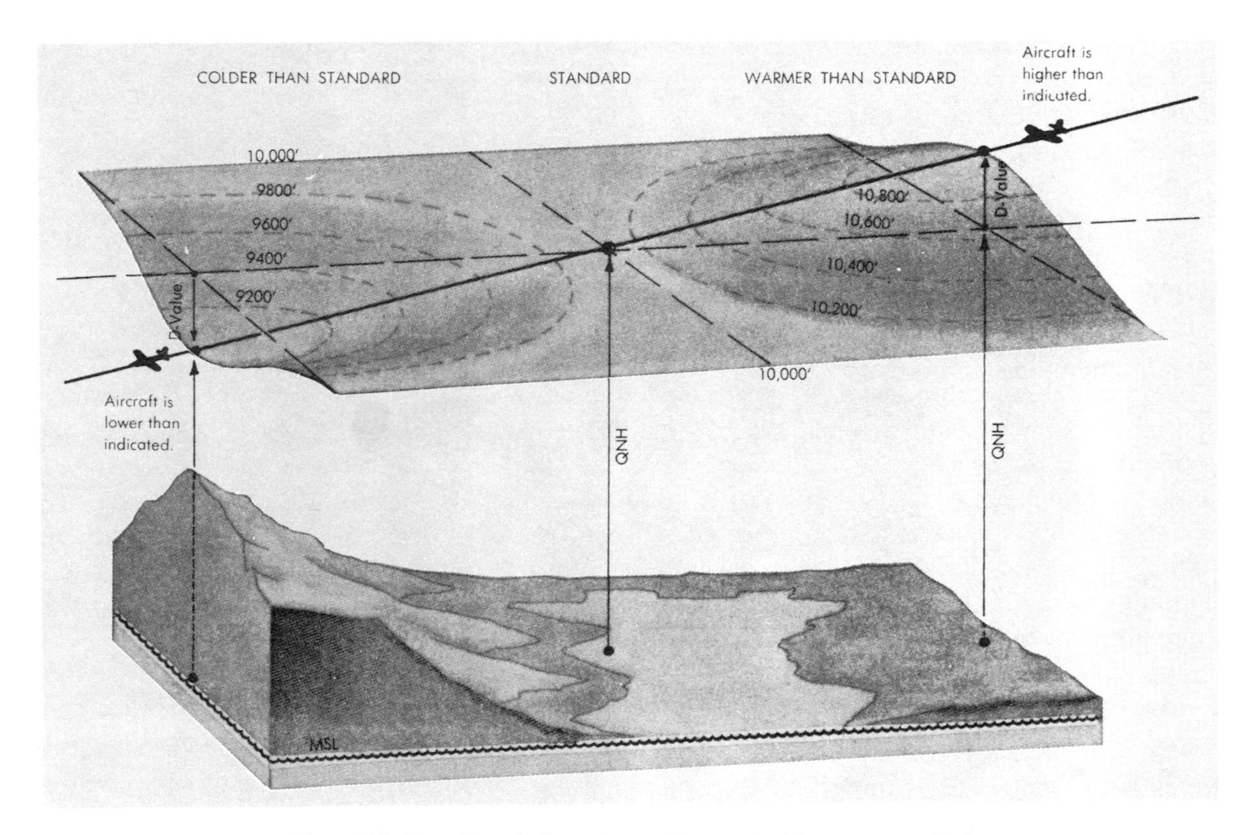

Figure 5-3. True Altitude Error Due to NonstandardTemperatures Aloft.

will have the capability for calculating approximate true altitude. It will be labeled true altitude on the flight computer, but keep in mind that it will really be your approximate true altitude. You will never know for sure your true altitude, except when at field elevation, or when flying on a completely standard day.

Suppose that instead of a station being at sea level we have an airport at 1,000 feet. Now such a field on a standard day would have a pressure of approximately 28.92 inches (73.46 cm) of mercury. Another way of saying this is that the pressure altitude for 28.92 is 1,000 feet. In any event our correction factor for field elevation is 29.92 – 28.92 = 1 inch. Suppose on this particular day the actual station pressure had been 28.92. We then add 28.92 and the 1 inch to get an altimeter setting of 29.92. What will be our indicated altitude while on the ground? Well, again, the altimeter takes the difference between the setting and the pressure it senses, 29.92 – 28.92 = 1 inch, and shows an indicated altitude of approximately 1,000 feet, which is our true altitude. In this example, pressure altitude, indicated altitude, and true altitude are the same, 1,000 feet. Note that if temperature had been standard, then density altitude would also have been 1,000 feet. All of this would have occurred as a result of having standard temperature and pressure at field elevation. However, our absolute altitude, our altitude above the terrain, is zero.

Notice the effect of an altimeter setting of 29.92. Even after we leave the ground, our indicated altitude will be our pressure altitude, for the altimeter will subtract whatever pressure its barometer is sensing from 29.92, and the altitude scale will show where that pressure difference occurs in the standard atmosphere. So, if the barometer sensed 13.03 inches, the indicated altitude is 15,000 since that is where 29.92 – 13.03 = 16.89 occurs in a standard atmosphere. Our true and density altitudes will likely be different due to a non-standard vertical temperature structure.

We have answered the question of how we may find pressure altitude, whether on the ground or in the air (merely set 29.92 on the altimeter and read the altitude scale). Combining outside temperature with pressure altitude, one can determine density altitude, which is an index of aircraft performance as we have already seen.

In our first example we assumed our field was at mean sea level. In the second example we assumed our field was above sea level, at 1,000 feet, and pressure at the field just happened to be 28.92, which gave us an altimeter setting of 29.92. We now look at a third example, one in which we are above sea level and we have an altimeter setting other than 29.92.

We keep our field elevation at 1,000 feet as in our last example. Our correction factor is still 1 inch because 1,000 feet corresponds to an approximate pressure change of 1 inch. We now assume that our actual pressure at the field is 29.32. Then 29.32 + 1 = 30.32, our altimeter setting. Now at field elevation our aneroid barometer senses 29.32, subtracts 29.32 from the setting of 30.32 to give 1 inch, and shows us an indicated and true altitude of 1,000 feet. As in our first example, suppose after taking off our barometer senses 27.32 inches of mercury. Then 30.32 – 27.32 is 3 inches and our indicated altitude will be shown to be about 3,000 feet. But as in our first example, this is not likely to be our true altitude because of a non-standard temperature distribution above field elevation. We will now show how the altimeter determines pressure altitude when the setting is 30.32 (as an example).

2. *How the Altimeter Determines Pressure Altitude*

Suppose we wish to determine pressure altitude as a first step in determining density altitude. As we have already seen, all we need do is set in 29.92 in our Kollsman window and read our indicated altitude. In one of our examples we just happened to have a setting of 29.92 already and, hence, did not have to change the setting. Here, our setting is assumed to be 30.32 and we change to 29.92. Now since the altimeter measures the difference between the setting and the pressure its barometer senses, there will be a .4 inch difference measured relative to 29.92 than as to 30.32. Since .4 corresponds to about 400 feet, in turning the setting from 30.32 to 29.92 the pressure altitude will be 400 feet less than the indicated altitude. So if the indicated altitude had been 8,500, the pressure altitude would have been 8,100 (Figure 5-4). Be sure and change your altimeter back to the correct setting after determining pressure altitude.

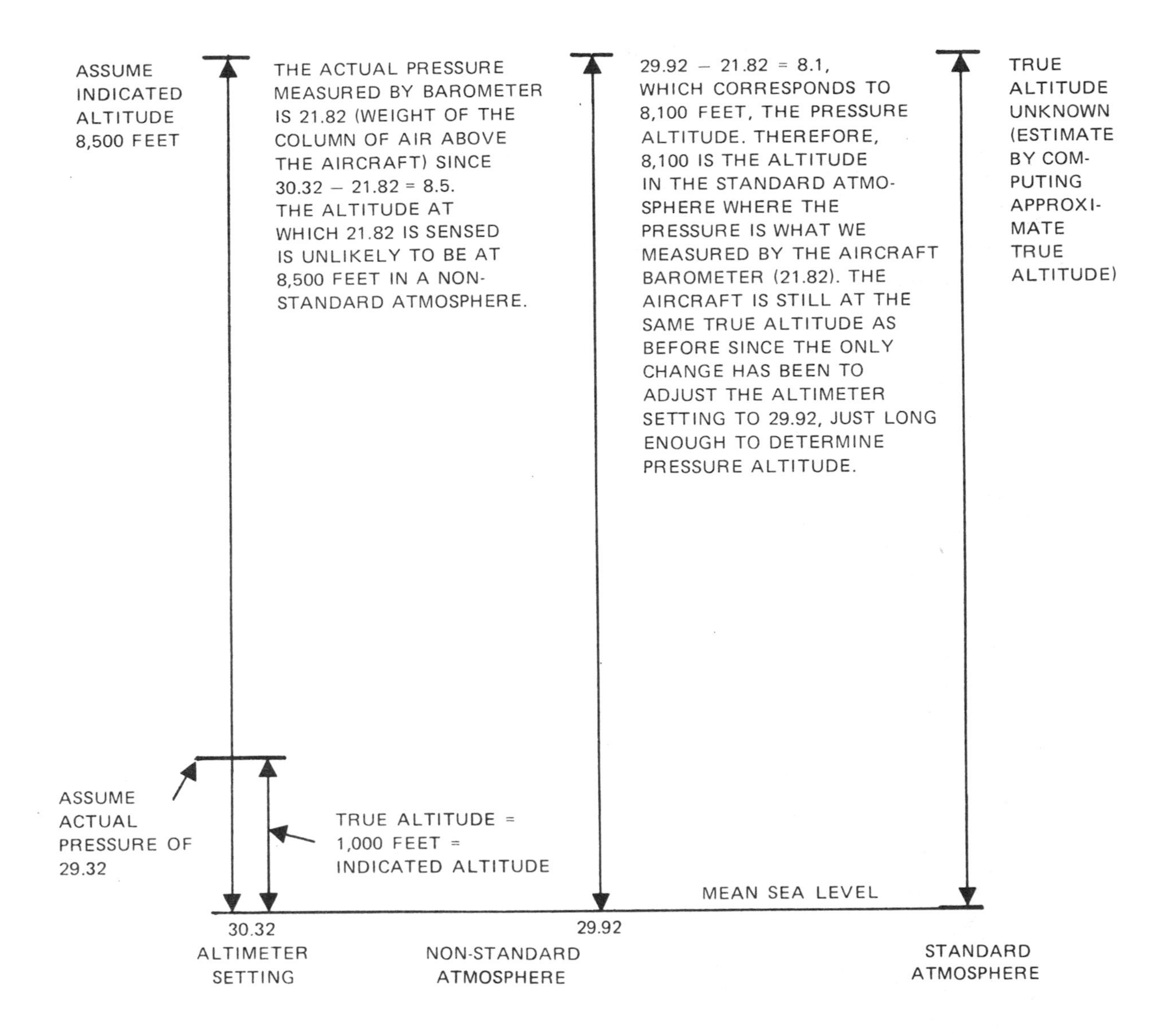

Figure 5-4. Example of Determination of Pressure Altitude.

One way to remember whether to add or subtract is to use the phrase "from high to low, look out below" that we used before. This phrase was used concerning true altitude, but it works here for determining pressure altitude also. 30.32 to 29.92 is high to low, so look out below, subtract. Had the setting been 29.52 we would have had low to high and would add 400 feet to get the pressure altitude. We mention at this point that pilots fly pressure altitude above flight level 18,000.

3. *Chart for Determining Pressure Altitude*

To bring out the relationship of the above factors a little better, we discuss the pressure altitude and density chart (Figure 5-5). If one were doing flight planning on the ground, he would use such a chart to find pressure altitude by using the altimeter setting to obtain the correction factor. We have just determined above that an altimeter setting of 30.32 would give us

Set Altimeter to 29.92 In. Hg.
When Reading Pressure Altitude

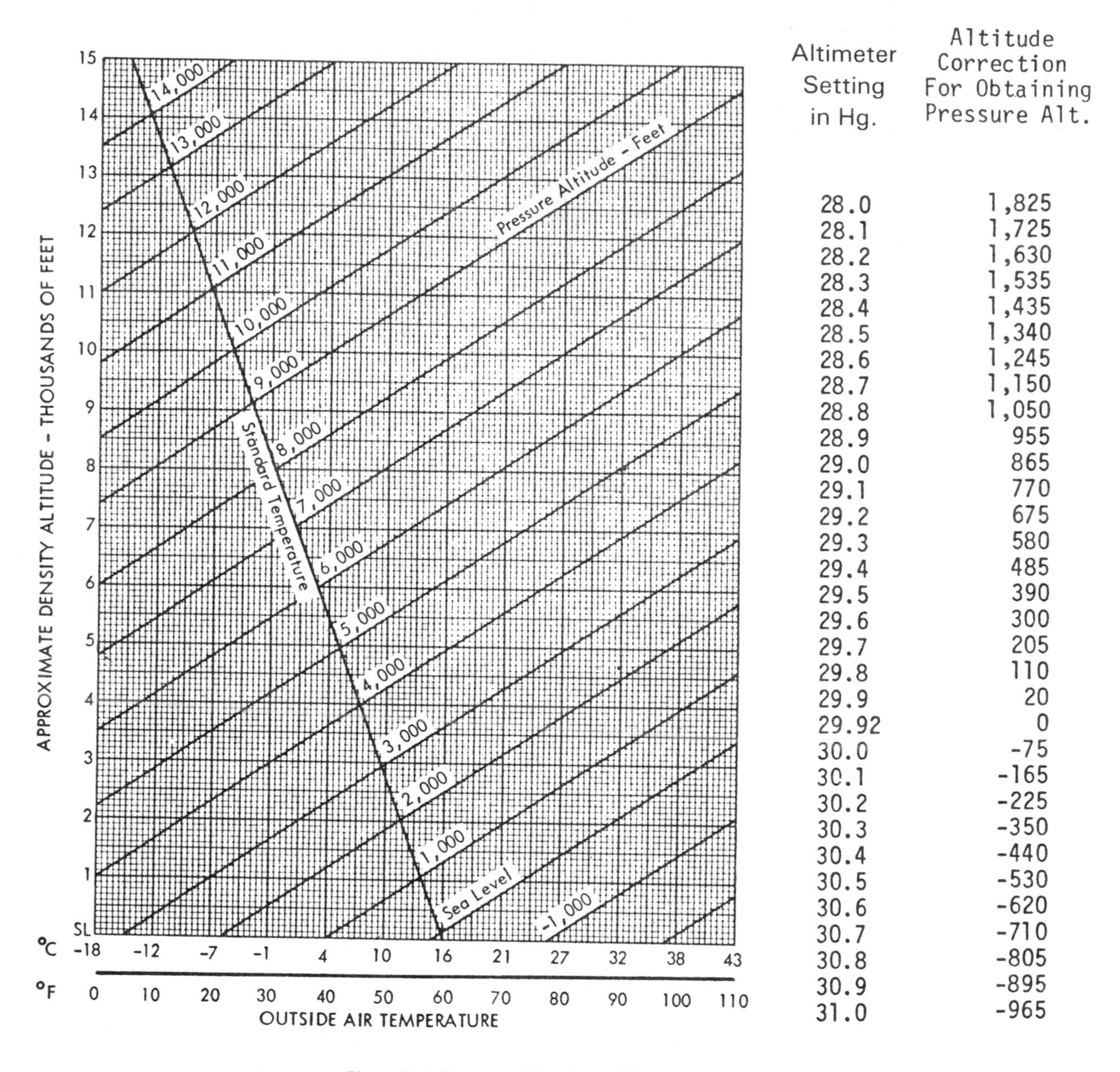

Altimeter Setting in Hg.	Altitude Correction For Obtaining Pressure Alt.
28.0	1,825
28.1	1,725
28.2	1,630
28.3	1,535
28.4	1,435
28.5	1,340
28.6	1,245
28.7	1,150
28.8	1,050
28.9	955
29.0	865
29.1	770
29.2	675
29.3	580
29.4	485
29.5	390
29.6	300
29.7	205
29.8	110
29.9	20
29.92	0
30.0	-75
30.1	-165
30.2	-225
30.3	-350
30.4	-440
30.5	-530
30.6	-620
30.7	-710
30.8	-805
30.9	-895
31.0	-965

Figure 5-5. Pressure Altitude and Density Chart.

an approximate 400 feet that we would need to subtract from our indicated altitude in order to determine the pressure altitude. Note that this chart shows that 30.32 would actually correspond to a correction of 366 feet, interpolating between –348 and –440. Our indicated altitude at field elevation with the correct setting is field elevation. So, if we are at the airport we apply these corrections to field elevation. If field elevation had been 1,000 feet, pressure altitude would have been 1,000 – 366 = 634 feet. You would have determined the same value, 634 feet, by walking out to the airplane, setting 29.92 in the Kollsman window, and reading the indicated altitude; after doing this, if you rotate the setting back to the correct setting value of 30.32, you would read field elevation, 1,000 feet.

The chart saves one from having to walk out to the airplane to determine pressure altitude for flight planning purposes. Combining temperature with pressure altitude one can now determine density altitude for takeoff and landing performance, such as length of runway needed and rate of climb.

4. Density Altitude

One particular value that is needed for cruise flight is true airspeed, for flight planning purposes. Without true airspeed, which we will discuss further shortly, we would not be able to determine our groundspeed and, hence, the time our trip would take us and, therefore, our fuel requirements. We want to determine this information before we leave the ground. You will fly with horsepower ratings or RPM settings recommended for your aircraft by the manufacturer. Those recommendations combined with density altitude will determine a true airspeed, obtained from the performance charts for that airplane.

But we are talking primarily about our cruise true airspeed. How can we know the necessary density altitude at our flight altitude if we have not even taken off yet? Well, if we had pressure altitude and temperature we could determine density altitude and proceed (under standard conditions pressure altitude will equal density altitude). Suppose we were planning on flying an indicated altitude of 8,500 feet as mentioned before, with an altimeter setting of 30.32. Then had we been in the aircraft at this cruise altitude, we could have set 29.92 in the window and we would have gotten a pressure altitude of approximately 8,100 (actually 8,134) by reading our indicated altitude. Or if we had our little density altitude chart along we would have determined the correction factor of 366 feet without even touching our altimeter. We subtract 366 from 8,500 to obtain 8,134 feet. We could then obtain our temperature from our outside air temperature gauge, and then determine density altitude from our chart or our flight computer.

But again, we are not in the airplane, so how do we obtain the temperature and pressure altitude? Well, we obtain our approximate temperature from the aviation arm of the FAA, the Flight Service Station. The pressure altitude is 8,134, which we can get by applying the 366 correction to the planned cruise altitude of 8,500 feet on the ground using our chart. We then combine this with the temperature we obtained from the Flight Service Station. With the density altitude so determined, we can determine everything else we need.

Many flight instructors tell their students to use the planned cruise altitude as their pressure altitude. The usual range of altimeter settings is between about 29.40 and 30.40, which would give us a correction of about plus or minus 500 feet. This is because, as we observed in the chapter on wind, pressure does not vary as much horizontally as it does vertically. As our correction table shows, this variation could be more on the order of plus or minus 1,000 feet, but only in extreme cases.

Noting that the spacing of the pressure altitude lines is about the same as the density altitude lines, a 500 foot (150 m) change occurs in density altitude by not applying a 500 foot pressure correction value. A 10°F (5.5°C) temperature change gives about the same amount of error in density altitude. The effect on true airspeed of a 500 foot error in density altitude is about .5 mph (.4 kts) on most light, single-engine aircraft. Note that while our correction value rarely exceeds 500 feet in determining pressure altitude, that large temperature differences could easily triple the correction value error effect in determining density altitude. While such errors do not result in large true airspeed changes, they are very important in other related areas of performance; for example, a 25°F higher than normal temperature will increase the necessary runway length by 10% for light, single-engine aircraft.

5. *Altimeter Errors*

The type of altimetry we have been studying is called QNH and it is standard throughout most of the world. We have stated that when we have the local altimeter setting set in the Kollsman window that the indicated altitude will be our true altitude at field elevation. This will be the case if there are no errors involved. One source of error is calibration error. Calibrated altitude is indicated altitude corrected for instrument error. Another source of error is the location on the field where atmospheric pressure is taken, which is usually the highest point in the landing area. Still another source of error is the use of flaps and gear, aerodynamic effects causing pressure changes that are reflected on the altimeter and which cause errors of 100 feet (30 m) or more. Thus, in coming in to land, while our indicated altitude is getting closer and closer to our true altitude, we can still have significant error, to a VFR pilot at night, or to an instrument-rated pilot who may be in the clouds. For a VFR daytime pilot these errors are not as significant due to his ability to maintain good visual reference to the ground. But for low altitude operations, be aware of the possibility that your altimeter could be 100 feet or more in error and plan accordingly. The errors we have just discussed are "instrument" errors, as opposed to the atmospheric-induced errors discussed earlier and reviewed in the next few sections.

6. *Vertical Separation*

At cruise altitudes, we know that changes in temperature (which account directly or indirectly for horizontal pressure changes) can cause as much error as 2,000 feet (600 m) or more, the higher we get the greater the possibility of more error – review Figure 5-2 and its discussion. If there is this much error in indicated altitude from true altitude, then what is the value of the altimeter at cruise? Well, as long as you and other pilots in the vicinity have the local setting in the window, vertical separation will be assured. For example, if you are flying at an indicated altitude of 8,500 feet (2.6 km) and you hear on the radio that someone else is at 6,500 feet (2 km), then if you both have the same setting in the window, there is 2,000 feet of separation. If your true altitude is really 8,000 feet (2.4 km), then his will be 6,000 feet (1.8 km); in any event you do have 2,000 feet of separation.

7. *Setting the Altimeter*

If you do not have a radio, you set your altimeter to read field elevation before you take off. By definition, the Kollsman window will show the *local* altimeter setting at that *time.* If you have a radio you are required to keep your altimeter set to a setting within 100 nautical miles of your route of flight.

8. *Altimeter Settings and Pressure Systems*

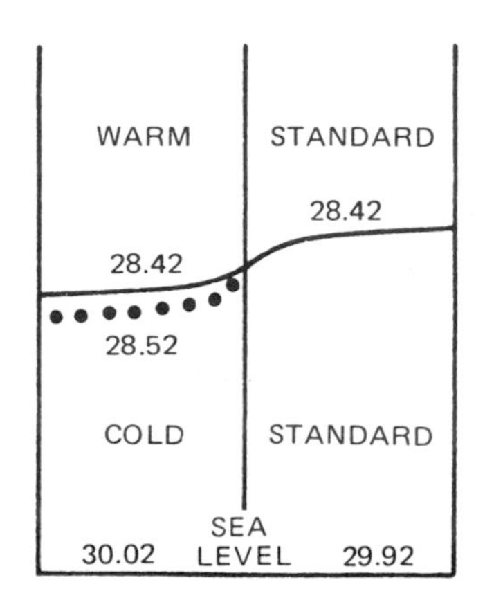

Consider the diagram to the left. Cold and warm are, again, relative to standard. Assume we wish to fly an indicated altitude of 1,500 feet. The altimeter will show about 1,500 feet on the right (29.92 – 28.42 = 1.5). On the left, with colder than standard temperatures below, the altimeter setting might well be 30.02, reflecting higher pressure at the surface. In order to indicate 1,500 feet, the altimeter will have to sense 28.52 inches of mercury (30.02 – 28.52 = 1.5). The 28.52 constant pressure surface is lower than the 28.42 surface. (Not resetting an altimeter causes the aircraft to follow a single constant pressure surface; each change in setting results in a different constant pressure surface to be followed.) Thus, the aircraft will be about 100 feet lower (28.52 – 28.42 = .1) in order to start following the 28.52 constant pressure surface. Combined with the fact that our pressure levels are lowering enroute (from right to left), the higher altimeter setting causes our aircraft to lose even more true altitude. Thus, flying from a high to a low *aloft* but low to high at the *surface* causes a combined error *aloft*. In this example, there would actually have been less error at cruise altitude had the altimeter setting not been changed to 30.02; but, of course, not setting the altimeter would have resulted in improper readings for aircraft separation at cruise and also improper readings for field elevation when landing.

A factor relating to altimeter settings and pressure systems is the diurnal change in pressure. The gravitational pull of the sun and moon result in small changes in surface pressure – high departures occur at 10 a.m. and 10 p.m., low departures at 4 a.m. and 4 p.m., on the order of one to two tenths of an inch. Thus, a setting of 30.00 at noon changing to 29.90 at 4 p.m. would be normal and should not be interpreted as a "low pressure system moving in."

B. Airspeeds

There are several types of speed with which we will be concerned. Indicated airspeed is determined by the difference between ram air pressure and static pressure, as was noted in the chapter on pressure. Calibrated airspeed is indicated airspeed corrected for instrument and position error. At cruise airspeeds indicated and calibrated airspeeds are usually quite close, within three to four knots. At lower airspeeds there may be a departure of 20 knots due to high angles of attack which point the pitot tube at an angle to the relative wind, and due to aerodynamic pressure changes caused by the use of gear and flaps. The speeds on the typical airspeed indicator are calibrated airspeeds. The compressibility of air becomes a problem at speeds above 250 knots. The small opening on the pitot tube cannot accommodate the increased air flow. Thus, there is a need for high speed, high altitude aircraft to calculate equivalent airspeed.

True airspeed is calibrated airspeed corrected for changes in the density of air. We will also be concerned with the groundspeed of an airplane. We will now try to help you better understand these speeds by looking at a boat in water; this example will also help clarify a couple of misconceptions many pilots have concerning these speeds.

C. Comparison of Airspeeds and Waterspeeds

Picture a boat in still water with a power setting sufficient to give a waterspeed of 10 knots (Figure 5-6). Then with no density consideration, our true and our indicated waterspeeds will be 10 knots. Since the water is still, our groundspeed is also 10 knots; think of the boat moving relative to a pier for groundspeed. We now let the water move at 10 knots in the direction opposite to that in which the boat is travelling and we maintain the same power setting as before as we face this current. Now our waterspeed will again be 10 knots; the boat is encountering the same mass of water as when the water was calm. Note in this case that our groundspeed is 0, 10 knots less than when the water was still. Thus, with a constant speed of the current we get the same indicated and true waterspeeds whether the water is coming at us at 10 knots or is still. The only change is in groundspeed, for we have 10 more knots of groundspeed with still water than when the water is against us.

We now make a turn in our boat from upstream to downstream. With our motor off we would be carried downstream with a groundspeed of 10 knots, but with a zero waterspeed since we are moving with the water not through it. But if we turn our motor on and we set in the same power setting as before, we will be travelling 10 knots through water that is itself moving at 10 knots. Thus, our groundspeed is 20 knots (Figure 5-7).

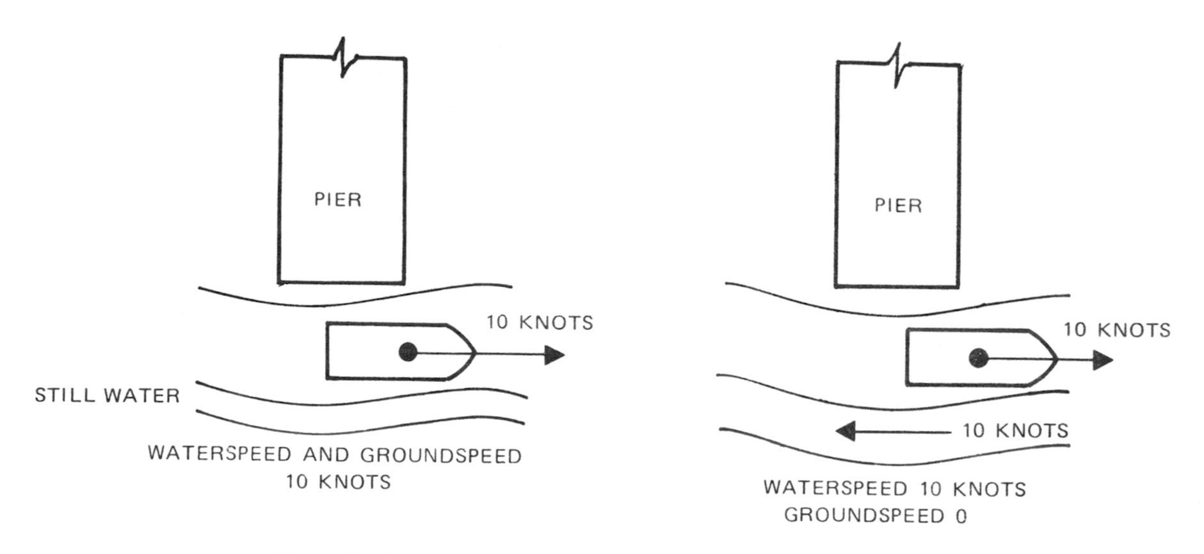

Figure 5-6. Comparison of Speeds in Still Water and Against Current.

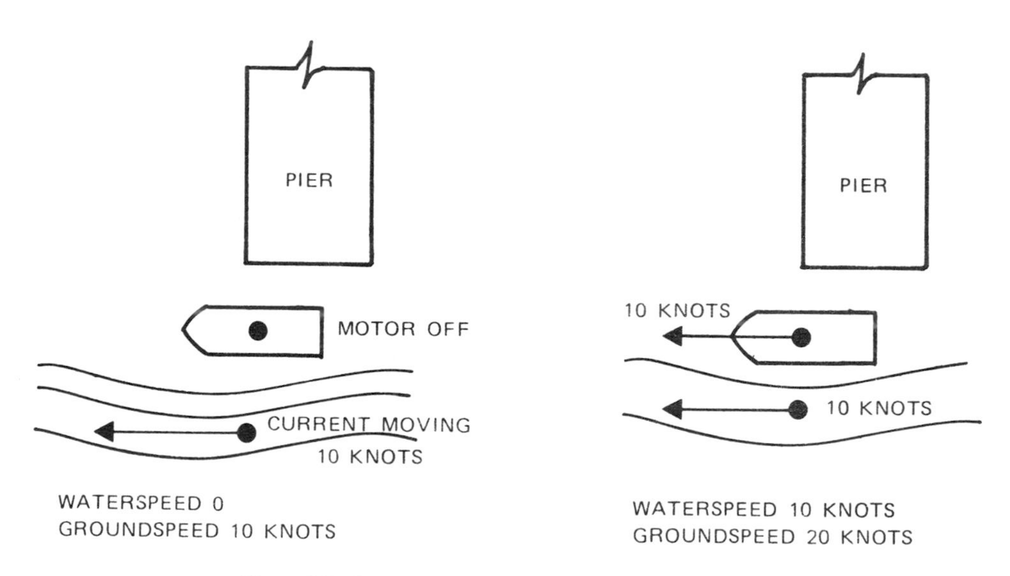

Figure 5-7. Comparison of Speeds in the Direction of Current.

Whether we travelled upstream or downstream our waterspeed was the same for our given power setting. All that changed was our groundspeed. In the air we must consider one other factor, namely, the density of air. On a standard day at sea level, indicated and true airspeed are the same, as in the boat example. Now let's see what would happen at an airport at a 5,000 foot (1.5 km) elevation. Assume the air is still, initially; also assume that we wish to maintain the same indicated airspeed for takeoff at 5,000 feet as we did at sea level, say, 50 knots. Why would we wish to use the same indicated airspeed? Because our aircraft performance depends on the mass of air we encounter. We need to impact the same mass of air in the same time as we did at sea level in order to have the airplane lift off the runway as at sea level. Since air is less dense, that is, thinner at 5,000 feet, we will need to travel at a faster true airspeed in order to impact that same mass of air. This will also increase our groundspeed, so that while we take off at the same indicated airspeed and in the same amount of time, our groundspeed is higher at 5,000 feet than at sea level; thus, we need more runway.

We can approximate our true airspeed in the above example by increasing our indicated airspeed by 2% per 1,000 feet. Two percent times 50 is 1, multiplied by 5 is 5, so our true airspeed is approximately 55 knots. Since the air is still, our groundspeed is also 55 knots.

Now suppose the air is moving at a speed of 10 knots and we are flying directly upwind at an indicated airspeed of 100 knots at 5,000 feet. Then our true airspeed is approximately 100 + 2% X 100 X 5 = 100 + 10 = 110 knots. The speed of 110 is relative to still air. Since the air is moving toward us at 10 knots, our groundspeed is 100 knots, 110 – 10. If we turn our airplane downwind and leave our same power setting in, we will have the same indicated and true airspeeds as when travelling upwind (recall the boat example). But now our groundspeed is 110 + 10 = 120. We conclude that groundspeed is true airspeed plus or minus a tail or headwind component, respectively.

ALTIMETRY, PERFORMANCE, AND AIRSPEED

Review Questions

PRI

1. The pitot system provides impact pressure for only the
   a. Airspeed indicator, vertical-speed indicator, and altimeter.
   b. Altimeter and vertical-speed indicator.
   c. Vertical-speed indicator
   d. Airspeed indicator.

2. Which statement is true in regard to the effects of atmospheric conditions on the indication of a pressure altimeter? When flying in air that is
   a. COLDER than standard temperature the aircraft will be higher than the altimeter indicates.
   b. WARMER than standard temperature the aircraft will be at the altitude indicated on the altimeter.
   c. COLDER than standard temperature the aircraft will be lower than the altimeter indicates.
   d. WARMER than standard temperature the aircraft will be lower than the altimeter indicates.

3. Pressure altitude can be determined by which one of the following methods?
   a. Adjust the altimeter setting window to 29.92 and read pressure altitude directly from the altimeter.
   b. Adjust the altimeter to the airport elevation and read pressure altitude.
   c. Pressure altitude can be determined only by the use of a computer.
   d. Set the altimeter to the current altimeter setting and read pressure altitude directly from the altimeter.

4. Assuming an airport elevation of 3,165 feet, an outside air temperature of 93°F, and an altimeter setting of 30.10" Hg, what is the DENSITY ALTITUDE? NOTE: See Figure 5-5.
   a. 3,000 feet.
   b. 3,850 feet.
   c. 6,800 feet.
   d. 5,800 feet.

5. If the elevation of an airport is 5,480 feet, the altimeter setting is 29.90" Hg, and the outside air temperature is 80°F, what is the DENSITY ALTITUDE at that airport? NOTE: See Figure 5-5.
   a. 2,800 feet.
   b. 5,400 feet.
   c. 8,100 feet.
   d. 9,200 feet.

6. While on the ground at an airport, you can determine the pressure altitude by
   a. Setting the altimeter to the field elevation and reading the value in the altimeter setting window.
   b. Setting the altimeter to zero and reading the value in the altimeter setting window.
   c. Setting 29.92 in the airplane's altimeter setting window and reading the indicated altitude.
   d. Setting the field elevation in the altimeter setting window and reading the indicated altitude.

7. If you plan to land at an airport where the elevation is 7,500 feet, the indicated approach airspeed should be
   a. Higher than that used for a sea level airport, and some power should be used until touchdown.
   b. The same as that used at a sea level airport.
   c. Lower than that used at a sea level airport.
   d. Higher than that used at a sea level airport.

TAKE-OFF DATA

TAKE-OFF DISTANCE FROM HARD SURFACE RUNWAY WITH FLAPS UP

Gross Weight Pounds	IAS At 50 MPH	Head Wind Knots	At Sea Level & 59° F Ground Run	At Sea Level & 59° F Total To Clear 50 Ft. Obs.	At 2500 Ft. & 50° F Ground Run	At 2500 Ft. & 50° F Total To Clear 50 Ft. Obs.	At 5000 Ft. & 41° F Ground Run	At 5000 Ft. & 41° F Total To Clear 50 Ft. Obs.	At 7500 Ft. & 32° F Ground Run	At 7500 Ft. & 32° F Total To Clear 50 Ft. Obs.
2300	68	0	865	1525	1040	1910	1255	2480	1565	3855
		10	615	1170	750	1485	920	1955	1160	3110
		20	405	850	505	1100	630	1480	810	2425
2000	63	0	630	1095	755	1325	905	1625	1120	2155
		10	435	820	530	1005	645	1250	810	1685
		20	275	580	340	720	425	910	595	1255
1700	58	0	435	780	520	920	625	1095	765	1370
		10	290	570	355	680	430	820	535	1040
		20	175	385	215	470	270	575	345	745

NOTES: 1. Increase distance 10% for each 25° F above standard temperature for particular altitude.
2. For operation on a dry, grass runway, increase distances (both "ground run" and "total to clear 50 ft. obstacle") by 7% of the "total to clear 50 ft. obstacle" figure.

8. Existing situation:
   Gross weight . . . . 1,700 lbs.
   Outside temperature . . . . 66°F
   Pressure altitude . . . . 5,000 feet
   Wind (Headwind) . . . . 20 knots
   The TOTAL TAKEOFF DISTANCE required to clear a 50-foot obstacle is (NOTE: Use chart above.)
   a. 575 feet.
   b. 633 feet.
   c. 518 feet.
   d. 930 feet.

9. Given data:
   Gross weight . . . . 2,000 lbs.
   Pressure altitude . . . . 2,500 feet
   Outside temperature . . . . 100°F
   Wind (Headwind) . . . . 10 knots
   Applying the given data to the chart above, the total takeoff distance required to clear a 50-foot obstacle is
   a. 1,005 feet.
   b. 804 feet.
   c. 1,206 feet.
   d. 1,842 feet.

CRUISE AND RANGE PERFORMANCE

Gross Weight – 2300 Lbs. **Standard Conditions** **Zero Wind** **Lean Mixture**

NOTE: Maximum cruise is normally limited to 75% power.

					38 Gal. (No Reserve)		48 Gal. (No Reserve)	
Alt.	RPM	% BHP	TAS MPH	Gal/ Hour	Endr. Hours	Range Miles	Endr. Hours	Range Miles
2500	2700	86	134	9.7	3.9	525	4.9	660
	2600	79	129	8.6	4.4	570	5.6	720
	2500	72	123	7.8	4.9	600	6.2	760
	2400	65	117	7.2	5.3	620	6.7	780
	2300	58	111	6.7	5.7	630	7.2	795
	2200	52	103	6.3	6.1	625	7.7	790
5000	2700	82	134	9.0	4.2	565	5.3	710
	2600	75	128	8.1	4.7	600	5.9	760
	2500	68	122	7.4	5.1	625	6.4	790
	2400	61	116	6.9	5.5	635	6.9	805
	2300	55	108	6.5	5.9	635	7.4	805
	2200	49	100	6.0	6.3	630	7.9	795
7500	2700	78	133	8.4	4.5	600	5.7	755
	2600	71	127	7.7	4.9	625	6.2	790
	2500	64	121	7.1	5.3	645	6.7	810
	2400	58	113	6.7	5.7	645	7.2	820
	2300	52	105	6.2	6.1	640	7.7	810
10,000	2650	70	129	7.6	5.0	640	6.3	810
	2600	67	125	7.3	5.2	650	6.5	820
	2500	61	118	6.9	5.5	655	7.0	830
	2400	55	110	6.4	5.9	650	7.5	825
	2300	49	100	6.0	6.3	635	8.0	800

10. Refer to the above chart. If the cruise altitude is 7,500 feet, using 58% power at 2400 RPM, what would be the range with 48 gallons of usable fuel?
    a. 635 miles.
    b. 810 miles.
    c. 645 miles.
    d. 820 miles.

11. With the conditions shown on the chart above, what would be the approximate true airspeed and fuel consumption per hour at an altitude of 7,500 feet, using 52% power?
    a. 105 MPH TAS, 6.2 gals/hour.
    b. 100 MPH TAS, 7.4 gals/hour.
    c. 118 MPH TAS, 6.9 gals/hour.
    d. 113 MPH TAS, 6.7 gals/hour.

12. With the conditions shown on the above chart, what would be the flight hours' endurance at an altitude of 2,500 feet, using 58% power? NOTE: With 38 gals. fuel-no reserve.
    a. 5.7 hours.
    b. 6.0 hours.
    c. 5.3 hours.
    d. 7.2 hours.

INS

13. Under what condition will true altitude be lower than indicated altitude with an altimeter setting of 29.92 even with an accurate altimeter?
    a. In colder than standard air temperature.
    b. In warmer than standard air temperature.
    c. When density altitude is higher than indicated altitude.
    d. Under higher than standard pressure at standard air temperature.

14. Under which condition(s) will pressure altitude be equal to true altitude?
    a. When the atmospheric pressure is 29.92" of Hg.
    b. When standard atmospheric conditions exist.
    c. When indicated altitude is equal to the pressure altitude.
    d. When the outside air temperature (OAT) is standard for that altitude.

15. What type altitude does a pilot maintain at flight level 210?
    a. Indicated
    b. Pressure
    c. Density
    d. Corrected (approximately true)

16. Altimeter setting is the value to which the scale of the pressure altimeter is set so the altimeter indicates:
    a. Density altitude at sea level.
    b. Pressure altitude at sea level.
    c. True altitude at field elevation.
    d. Pressure altitude at field elevation.

17. Under what condition is pressure altitude and density altitude the same value?
    a. At sea level, when the temperature is 0°F.
    b. When the altimeter has no installation error.
    c. When the altimeter setting is 29.92.
    d. At standard temperature.

18. What is the effect of a temperature increase from 25°F to 50°F on the density altitude if the pressure altitude remains at 5,000 feet?
    a. 1,650-foot increase.
    b. 1,400-foot increase.
    c. 1,200-foot increase.
    d. 1,000-foot increase.

19. What is the effect of a pressure change from 3,000 to 5,000 feet pressure altitude on the density altitude if the temperature remains standard?
    a. 1,800-foot increase.
    b. 2,000-foot increase.
    c. 1,800-foot decrease.
    d. 2,000-foot decrease.

COM

20. The reason altimeters should be adjusted to the same altimeter setting for a specific area is
    a. The cancellation of altimeter error due to position of static source.
    b. The elimination of a need to make in-flight calculations of true altitude.
    c. More accurate terrain clearance in mountainous areas.
    d. To provide better vertical separation of aircraft.

21. If cruising into a 15 MPH headwind and a 180° turn is made so the wind is from directly behind the airplane, the indicated airspeed would:
    a. Be the same and the groundspeed would increase 30 MPH.
    b. Be the same and the groundspeed would increase 15 MPH.
    c. Decrease 15 MPH and the groundspeed would increase 15 MPH.
    d. Increase 30 MPH and the groundspeed would remain the same.

22. If the pitot-static pressure tubes are broken inside a pressurized cabin during a high altitude flight, the altimeter would probably indicate: (use "low to high" rule)
    a. Lower than actual flight altitude.
    b. A fluctuating altitude.
    c. Sea level.
    d. Higher than actual flight altitude.

23. Which of the following will occur if the indicated airspeed is constant and the density altitude increases?
    a. True airspeed will decrease, and groundspeed will increase.
    b. True airspeed will decrease, and groundspeed will decrease.
    c. True airspeed will increase, and groundspeed will decrease.
    d. True airspeed will increase, and groundspeed will increase.

24. How does high density altitude affect the takeoff performance of an airplane?
    a. Increased drag will require more power for acceleration.
    b. Reduced engine and propeller efficiency will decrease acceleration.
    c. Reduced drag will increase the rate of acceleration.
    d. A higher indicated airspeed is required to produce necessary lift.

25. What would occur if the density altitude is 5,000 feet at an airport where the field elevation is 2,000 feet?
    a. Takeoff and landing performance would not be affected.
    b. The altimeter would indicate 5,000 feet when the airplane is on the ground.
    c. Takeoff and landing performance would be the same as an airport with an elevation of 5,000 feet.
    d. The indicated takeoff and landing airspeed should be higher than on a standard day.

26. Assume that an airplane is flying at a constant power setting and at a constant indicated altitude. If the outside air temperature increases, the true airspeed will
    a. Decrease; the true altitude will increase.
    b. Increase; the true altitude will decrease.
    c. Increase; the true altitude will increase.
    d. Decrease; the true altitude will decrease.

27. For a given indicated airspeed, a high density altitude will always result in
    a. An increase in equivalent airspeed.
    b. An increase in true airspeed.
    c. A decrease in true airspeed.
    d. An increase in calibrated airspeed.

28. Assume comparable conditions relative to temperature, wind, and airplane weight. The groundspeed at touchdown at high elevation airports will be
    a. Higher than at sea level.
    b. Lower than at sea level.
    c. The same as at sea level.
    d. Either higher or lower than at sea level, depending on airspeed corrections applied.

29. If 80 MPH indicated airspeed has been used on final approach at an airport at sea level, the indicated airspeed on final approach to an airport where the field elevation is 4,800 feet MSL should be
    a. Lower because the true airspeed is higher.
    b. Higher because the stalling speed is higher.
    c. Lower because the air density is lower.
    d. The same as at sea level fields.

30. Assume a calm wind. During approach and landing at a high elevation airport and using the same indicated airspeed as that used at a sea level airport, the
    a. Groundspeed will be higher and the landing distance will be greater at the higher elevation airport.
    b. Groundspeed will be the same and the landing distance will be the same at each of the airports.
    c. True airspeed will be the same and the landing distance will be the same at both airports.
    d. True airspeed will be lower and the landing distance will be less at the higher elevation airport.

31. As air density decreases, density altitude
    a. Increases when the temperature decreases.
    b. Decreases when the temperature increases.
    c. Decreases.
    d. Increases.

32. If, without adjusting the altimeter setting, a flight is made from an area of high temperature into an area of low temperature and a constant altitude is maintained, the actual altitude of the airplane would be
    a. Lower than the altimeter indicates.
    b. At a level below the standard datum plane.
    c. At the same level as the altimeter indicates.
    d. Higher than the altimeter indicates.

33. If, without adjusting the altimeter setting, a flight is made from an area of low temperature into an area of high temperature and a constant altitude is maintained, the actual altitude of the airplane would be
    a. At a level below the standard datum plane.
    b. At the same level as the altimeter indicates.
    c. Higher than the altimeter indicates.
    d. Lower than the altimeter indicates.

34. If a constant indicated altitude and altimeter setting are maintained and the temperature increases, what would be the effect on the true altitude and pressure altitude?
    a. Both true altitude and pressure altitude decrease.
    b. True altitude remains the same while pressure altitude increases.
    c. Both true altitude and pressure altitude increase.
    d. True altitude increases while pressure altitude remains the same.

35. Assume an altimeter is set to 29.84" Hg and the correct altimeter setting is 30.00" Hg. If under these conditions a landing is made at an airport where the field elevation is 772 feet, the altimeter would indicate approximately
    a. 160 feet.
    b. 612 feet.
    c. 772 feet.
    d. 932 feet.

36. If, while on the ground, a sensitive altimeter is set to 29.92" Hg and the ambient pressure is 29.92" Hg, the altimeter will indicate
    a. Density altitude.
    b. Zero.
    c. Field elevation.
    d. True altitude.

37. Which statement is true regarding a sensitive altimeter?
    a. The altimeter will assure safe terrain clearance if adjusted to the proper altimeter setting.
    b. All aircraft flying at the same indicated altitude with identical altimeter settings will always be at the same true altitude.
    c. If corrections are made for nonstandard temperature and pressure, the altimeter will give an accurate indication relative to terrain clearance.
    d. The altimeter will indicate accurate altitude above terrain only when operating over flat terrain.

38. On a warmer than standard day the pressure level where the altimeter will indicate 4,000 feet would be
    a. Higher than it would under standard conditions.
    b. The same as it would under standard conditions.
    c. The same as it would under colder than standard conditions.
    d. Lower than it would under standard conditions.

39. Assume an altimeter indicates an altitude of 2,500 feet MSL with an altimeter setting of 29.52" Hg. What is the approximate pressure altitude? (Do without reference to a chart)
    a. 2,900 feet.
    b. 2,540 feet.
    c. 2,400 feet.
    d. 2,100 feet.

40. The location of the static vent which would provide the most accurate measurement of static pressure under variable flight conditions is one installed
    a. In the pitot head which encounters relatively undisturbed air.
    b. In the cockpit where it is not influenced by variable angle of attack.
    c. On one side of the airplane and covered by a fine screen.
    d. On each side of the airplane where the system will compensate for variation of airplane attitude.

41. Pitot-static system errors are generally the greatest in which range of airspeed?
    a. Maneuvering speed.
    b. High airspeed.
    c. Low airspeed.
    d. Cruising airspeed.

42. One of the possible results of using the emergency alternate source of static pressure in an unpressurized airplane is that the: (use "high to low" rule – due to aerodynamic effects pressure is lower inside)
    a. Altimeter may indicate an altitude lower than the actual altitude being flown.
    b. Vertical velocity indicator may indicate a continuous descent.
    c. Altimeter may indicate an altitude higher than the actual altitude being flown.
    d. Airspeed indicator may indicate less than normal.

ATP
43. While maintaining FL 310, you observe the OAT is 15° colder than standard. What is the relationship between true altitude and pressure altitude?
    a. True altitude would be lower than 31,000 feet.
    b. Pressure altitude is lower than true altitude.
    c. It would be impossible to determine the relationship.
    d. They are both the same, 31,000 feet.

44. You are maintaining a constant pressure altitude and the outside air temperature is warmer than standard for that altitude. What is the density altitude with respect to pressure altitude?
    a. Lower.
    b. Higher.
    c. Impossible to determine.
    d. Same.

45. Enroute to FL 250, the altimeter is set correctly. On descent, a pilot fails to reset it to a local altimeter setting of 30.57. If the field elevation is 650 feet, and the altimeter is functioning properly, what will it indicate after landing?
    a. Sea level.
    b. 585 feet.
    c. 715 feet.
    d. 1,300 feet.

46. If the outside air temperature at a given altitude is warmer than standard, the density altitude is:
    a. Lower than true altitude.
    b. Higher than pressure altitude.
    c. Higher than true altitude but lower than pressure altitude.
    d. Lower than pressure altitude, but approximately equal to the true altitude.

47. What causes variations in altimeter settings between weather reporting points?
    a. Unequal heating of the earth's surface.
    b. Variation of terrain elevation creating barriers to the movement of an air mass.
    c. Coriolis force reacting with friction.
    d. Friction of the air with the earth's surface.

CHAPTER VI
MOISTURE

The reader should demonstrate an understanding of moisture to the extent that he can answer the following questions (as stated below, or as multiple-choice, etc.):

1. Name the three states of water.
2. Name the three primary influences of moisture.
3. Define "evaporation"; "latent heat of vaporization".
4. Define "sublimation"; "latent heat of fusion".
5. Distinguish between latent and sensible heat.
6. Define "condensation"; "dewpoint".
7. The warmer air is, the (less, more) moisture it can hold.
8. Define "relative humidity". "Saturated" means ________ % relative humidity.
9. What are the two basic methods for changing relative humidity?
10. When the temperature and dew point are close together, 5°F or less, what kind of weather is likely?
11. Define "hygroscopic nuclei". Give examples.
12. What basic instrument is used to determine relative humidity?
13. What is the effect of high relative humidity on the density of the air? What does this imply about aircraft performance?
14. What three factors will result in low density altitude?
15. Define "precipitation".
16. State the average value for the saturation-adiabatic process.
17. The primary process for cloud formation is cooling by (radiation, adiabatic expansion, conduction, convection).
18. What are the usual forms of precipitation?
19. Define "virga".
20. Define "supercooled water".
21. Describe the "ice-crystal process". Name another way in which droplets grow in sufficient size to produce rain.
22. Clouds usually need to be at least _________ feet thick in order to produce rain.
23. What factor primarily determines the vertical extent of clouds?
24. Explain the formation of dew; frost.
25. Describe the katabatic wind phenomenon.
26. Name the three types of hydroplaning; state the cause of each.

CHAPTER VI
MOISTURE

Moisture exists in our atmosphere in three states — gaseous water vapor, liquid water, and ice. There are three primary influences with regard to moisture — its life sustaining quality for our bodies and other life forms, its effect on visibility, and its capacity to hold heat. The need for moisture to sustain life is obvious. Perhaps just as obvious to any pilot is the fact that he is not in accordance with visual flight rules when he is flying in a cloud, due to visibility restrictions. The heat carrying capacity of water vapor is not at all obvious to most of us, but it is very important as we shall now see.

A. Latent Heat and Changes of State

Moisture is received into the atmosphere through a process called evaporation. The sun's energy received on a water surface is partially used to what we might call "loosen the molecular bonds" between water molecules. The energy absorbed is carried with free molecules and called latent heat of vaporization. Thus, evaporation, the changing of liquid water to water vapor, is a cooling process in the sense that heat is taken from the surroundings in order for this change of phase to occur. Latent heat of fusion works much the same way, as heat is absorbed to change ice to liquid water. Sublimation, the change of ice directly to water vapor, or vice versa, requires the sum of latent heat of fusion and latent heat of vaporization to occur. Our bodies are evaporative coolers. Suppose you have just been out jogging and you are now just "walking it off". If there is little moisture in the air, then the perspiration on your skin will evaporate into the adjacent air. The heat necessary to change the perspiration to vapor is taken from your skin, thus you are being cooled. If there is sufficient wind, the moisture escaping your skin as vapor will be carried away, allowing for a greater rate of evaporation. Had the air been quite still and contained substantial moisture already, then little moisture would have left your skin and you would not have cooled off very much. This can be a dangerous situation, leading to possible exhaustion, collapse, and even death.

As we have seen, when liquid water changes to water vapor there is a carrying of heat into the atmosphere (Figure 6-1). This latent heat was referred to as "the other kind of heat" in the chapter on temperature. It differs from sensible heat quite naturally by the fact that it cannot be sensed, that is, measured with a thermometer. However, when water vapor changes back to liquid, the process of condensation, latent heat is returned to the atmosphere as sensible heat and, therefore, can be measured by a thermometer — recall that temperature is a measure of the degree of molecular activity and, hence, the presence of sensible heat rather than the amount of sensible heat.

B. Relative Humidity and Dew Point

Now how can we initiate the condensation process? Well, one way is to cool air down to its dew point, the dew point being the temperature to which air must be cooled to become saturated, that is, to reach 100% relative humidity. The amount of moisture in the air is dependent on the temperature of the air. As the temperature of the air increases, the capacity of the air to accept water is greatly increased. In fact, air at 20°C can hold about twice as much mass of water as air at 9°C, 17.3 grams compared to about 8.7 grams per cubic meter.

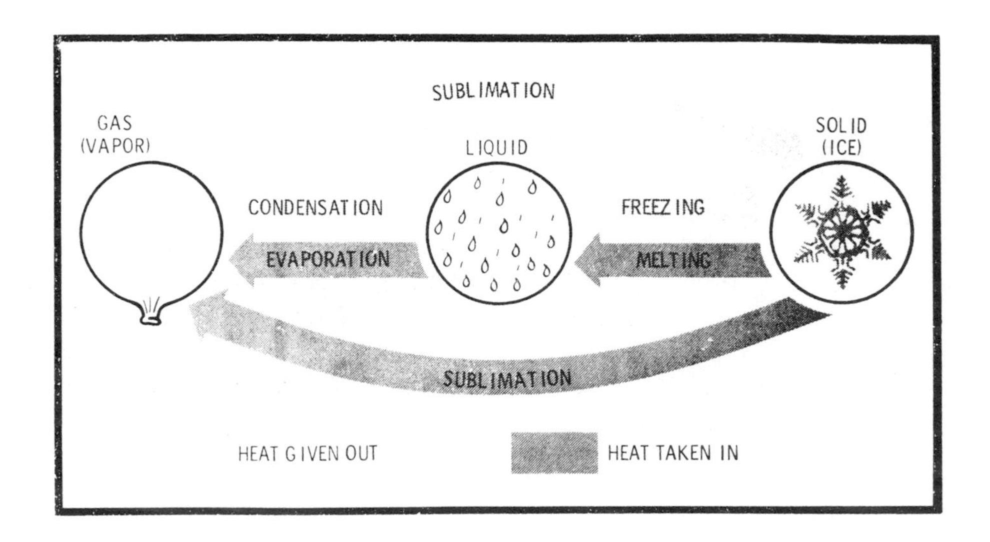

Figure 6-1. Change of state. Water vapor may change to liquid and then to ice or vice versa. It also may change directly to ice without going through the liquid state, or ice may change directly to water vapor. Heat is absorbed or surrendered during each change even when temperature does not change. Of the six changes of state shown, three absorb heat indicated by the dark arrows and three release heat shown by the lighter arrows.

Now relative humidity is the ratio of the amount of water vapor in the air to the maximum amount it could hold at that temperature and pressure. So if the relative humidity is 50% when the temperature is 20°C, the air holds half the mass of water vapor it could normally hold at that temperature, namely 8.7 grams as compared to 17.3. Air which holds all the moisture it can normally hold at a given temperature is said to be saturated, or has 100% relative humidity. Air at 9°C holding 8.7 grams per cubic meter is saturated. Let's prove this using the definition of relative humidity. Now the relative humidity at 9°C will be the ratio of the amount of water vapor being held by the air at 9°C to the maximum amount it can hold at that temperature and pressure, which is also 8.7 grams per cubic meter. The ratio of 8.7 to 8.7 is 1 which is 100% (recall that 100% means 1, 50% means .5, that is, 1/2, etc.).

Now back to dew point, which we defined to be the temperature to which air must be cooled to become saturated, that is, to reach 100% relative humidity. If air at 20°C contains 8.7 grams per cubic meter, then what is its dew point? Well, the dew point is 9°C, because when the air is cooled down to 9°C, then 8.7 grams is all it can hold and, hence, the air is saturated at that temperature. Thus, we can change the relative humidity of air by cooling the air (Figure 6-2). We can also change the relative humidity of air by increasing the amount of moisture in the air by, say, passing a volume of air over a lake. Remember that relative humidity is the ratio of the mass of water vapor in the air to the maximum it can hold at that temperature and pressure.

C. Significance of Temperature-Dew Point Spread

Now when the temperature and dew point are close together, 5°F (2.7°C) or less, relative humidity is quite high. For example, if the temperature is 75°F (24°C) and the dew point is 70°F (21.3°C), we need only cool the air 5°F in order for the air to be saturated. Suppose we do cool the air down to 70°F. Does this mean that rain is likely? Well, not usually, because when one says that relative humidity is high, one is usually referring to near the surface. We will soon see that for rain to occur there must be an abundance of moisture aloft. High relative humidity could well result in low clouds or fog, which we will discuss later. But even with a relative humidity of 100% near the surface, the air may be perfectly clear.

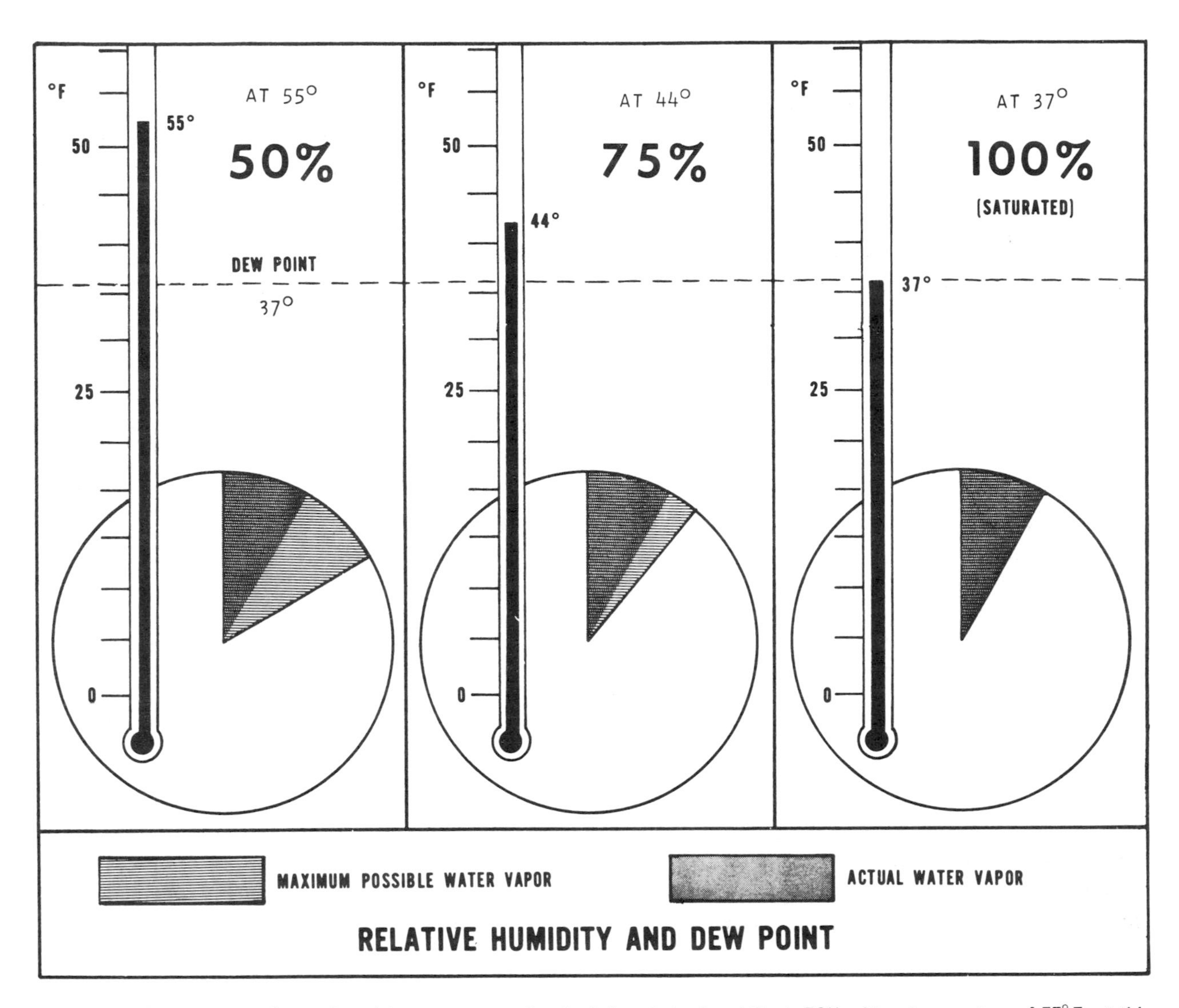

Figure 6-2. Relative humidity varies with temperature. On the left, relative humidity is 50% with a temperature of 55° F; at this temperature, the air could hold twice as much water vapor as it actually contains. Note that as temperature cools on the center and right, actual water vapor content does not change, but relative humidity increases reaching 100% on the right. Dewpoint changes with change in pressure, but here we have held pressure constant. Note that dewpoint remained constant at 37° F. When relative humidity on the right is 100%, temperature and dewpoint are the same.

D. Hygroscopic Nuclei

The slightest wind may initiate condensation on so-called hygroscopic nuclei, resulting in a dense fog or low clouds when the relative humidity is high. These nuclei are solid particles that are suspended in the air and are too small to be seen. Examples of such matter are sea salt, combustion byproducts, and dust. They are always in the air to a certain degree. Some of this matter has a great affinity, or attraction, for water vapor. If the air has little of this type of matter in a certain location, relative humidity could be 100% and the sky remain clear as we have already mentioned. The reverse is also true, for in the presence of an abundance of this highly water-attractive matter we may have condensation, or sublimation, when the relative humidity is only, say, 97%.

When the temperature decreases, the average motion of molecules decreases. Thus, water vapor molecules can more easily collect on nuclei. When the temperature increases, the air can hold more water vapor since the faster average molecular speed will make it more difficult for water vapor molecules to collect on nuclei.

E. Hygrometers

Relative humidity may be determined by several instruments, all generally classified as hygrometers. They use the principle of anything from chemical absorption to the shrinking or stretching of the human hair. A carbon humidity element has been used in the U.S. rawinsonde mentioned in the chapter on pressure. We do not determine relative humidity by taking the ratio of the temperature and dew point. For example, while we agree subjectively that a temperature of 75°F and a dew point of 70°F is indicative of high relative humidity, its value is not the ratio of 70 to 75. A chart is available to convert temperature and dew point to relative humidity.

F. Humidity, Density, and Performance

One further fact about relative humidity (which surprises many pilots), namely, that usually the higher the relative humidity, the lighter, less dense, the air. In the chapter on pressure we stated that water vapor varies from near 0 to 5% by volume. As water vapor takes up more of a given volume of air, some of the other elements must be reduced. Water vapor replaces nitrogen primarily and since water vapor has a smaller molecular weight than nitrogen, the volume of air is less dense with a greater amount of moisture. Before we added water vapor, and thus replaced nitrogen, maybe the relative humidity was, say, 50%, but then maybe we added enough water vapor to increase the relative humidity to, say, 80%. Air at 80% relative humidity is therefore lighter, less dense, than at 50% relative humidity. We would feel more uncomfortable at 80% than at 50% relative humidity, temperature remaining the same, because air holds most of the moisture it can hold at 80% and little moisture can escape our skins. This is a respiratory problem, however, not a "weight of the air" problem.

What is the effect of high relative humidity on an airplane? Well, less performance since the air is less dense. Any factor which reduces density, the mass of air in a given volume, will reduce the ability of the propeller to provide thrust and the wing to provide lift. We have already seen that low pressure and high temperature cause air to expand, thus thinning the air. Combining all three effects, when we try to take off on a hot, moist day when a low pressure system has moved in, we will find ourselves requiring much more runway.

G. The Condensation Process

We return to our primary concern, the condensation process. It is important to note the word condensation, as opposed to precipitation, the basic distinction being that precipitation is essentially a condensation or sublimation product that falls to the earth's surface.

Let's assume that we have evaporated a certain volume of water from an ocean. Taking a parcel of this air we let it rise in the atmosphere. We will show how this can occur a little later. Now this parcel cools adiabatically as it ascends, at the dry adiabatic rate of 5.5°F per 1,000 feet (10°C per km) until the air is saturated. But what happens when condensation occurs? Well, latent heat of vaporization is released as sensible heat. So if ascent continues, the cooling rate will be modified by the rate at which latent heat is being realized as sensible heat. An average value of 3.0°F per 1,000 feet (about 6°C per km) might be used for this so-called saturation-adiabatic process. This change is variable, however, for as air rises and begins to condense there is less and less of the moisture originally available for condensation, thus not as much latent heat returning as sensible heat to retard the cooling rate; as air continues to rise, there is also a larger volume, due to expansion, to keep saturated. So, the saturation-adiabatic rate approaches the dry adiabatic rate with increasing altitude. We have, however, shown one example of how the condensation process is initiated (Figure 6-3).

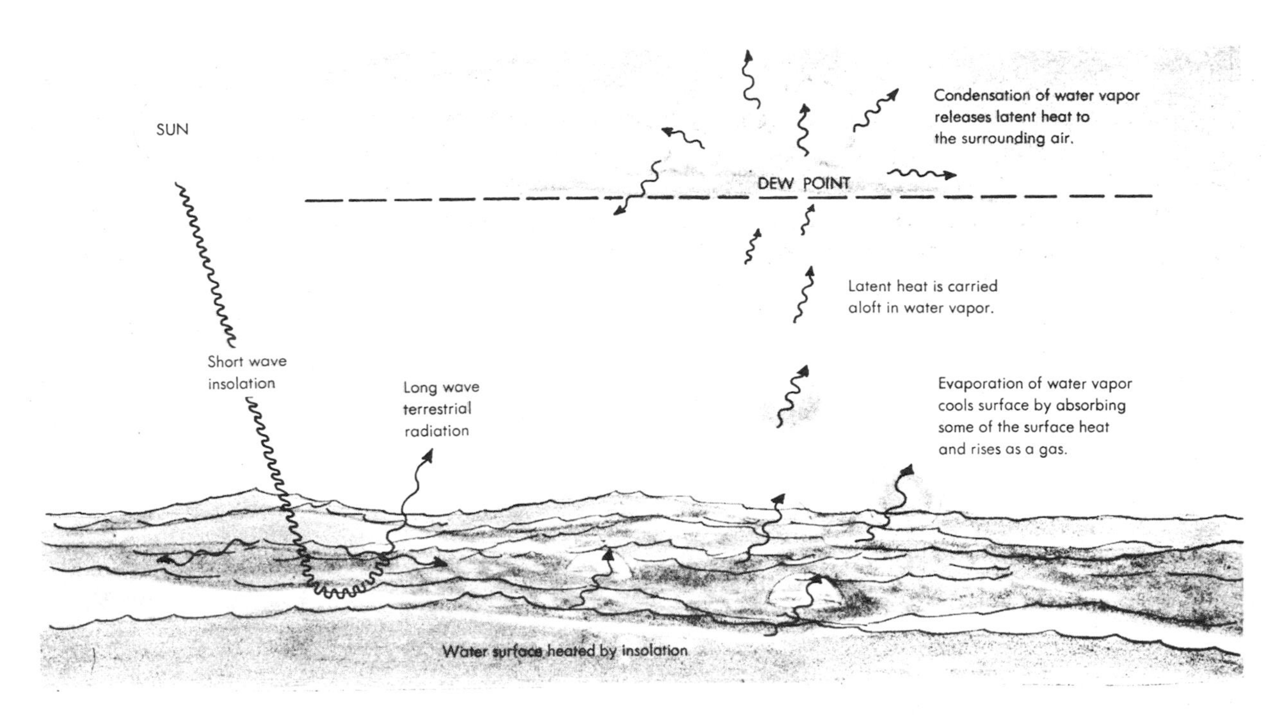
SUN
Condensation of water vapor releases latent heat to the surrounding air.
DEW POINT
Latent heat is carried aloft in water vapor.
Short wave insolation
Long wave terrestrial radiation
Evaporation of water vapor cools surface by absorbing some of the surface heat and rises as a gas.
Water surface heated by insolation

Figure 6-3. Evaporation and Condensation.

H. Cloud Formation

Our initial condensation product as air rises is a cloud, suspended minute water and/or ice particles. We mentioned that relative humidity increases as we add moisture to air or as we cool the air or, more often, by both processes acting simultaneously. Cooling is the more common process, primarily from adiabatic expansion. Radiational cooling, and air cooling by conduction (such as air lying stationary over a cold surface or moving over a cold surface), occur to a lesser degree.

I. Precipitation Forms and Processes

Our next higher-order type of condensation product is precipitation, which we defined as a condensation or sublimation product that falls to the earth's surface. In aviation, the forms usually considered are drizzle, rain, snow, ice pellets (formerly called sleet), hail, and ice crystals. These condensation products reach the earth's surface when their weight provides sufficient downward force to overcome updrafts. Virga has been seen by all of us as dark streamers of falling water drops beneath clouds but which level off or evaporate before reaching the surface. A pilot might fly through this "high level precipitation" when at a lower altitude there is no precipitation (Figure 6-4).

In order for rain to occur there must be some process which causes cloud particles to join and form raindrops. It takes many cloud droplets to form one large raindrop. One process by which rain is generated is the so-called ice-crystal process. In order to describe this process we need to mention the concept of supercooled water, liquid water droplets at temperatures below freezing. In fact, supercooled water is necessary for the various kinds of structural ice to occur, clear, rime, and mixed; all of this will be studied in greater detail in the chapter on icing. Why should liquid water exist at temperatures below freezing? Well, this phenomenon is not completely understood. Cloud droplets do appear to be absent of many of the impurities that water on the surface possesses, and this is believed to be a prime factor. Supercooled water has been observed at altitudes where the temperature is $-40°F$ ($-40°C$), and even higher in thunderstorms. Our ice-crystal process requires the presence of both supercooled water and ice crystals. Water evaporates from liquid droplets and sublimates on ice particles until they grow sufficiently in weight to fall.

There are clouds in which little or no ice is present. In such cases droplet collisions occur, resulting in droplets of sufficient weight to fall to the earth's surface. In mid and upper latitudes the freezing level is so low that the ice-crystal process is probably the initiator of precipitation, with collisions occurring as droplets fall (Figure 6-5). Clouds also need to be quite thick in order for precipitation to occur. Clouds less than 4,000 feet (1.2 km) thick are not too likely to produce rain, but clouds more than 4,000 feet thick are quite likely to produce rain. We do not mean to imply that precipitation will never occur when a cloud is less than 4,000 feet thick and beneath the freezing level, for drizzle is fairly frequent in such situations, but rain is of course more significant and usually requires these conditions.

At one point in the discussion we mentioned that we would let air rise. Well, it does not just automatically do so. One way to provide lift is to enter a low pressure area, for air spirals inward and upward in a counterclockwise fashion. Let's move a low over a source of water, say, the Great Lakes. As moisture is lifted, it cools adiabatically and condenses with sufficient lift. When air condenses to form a cloud, latent heat returns as sensible heat and if this air is warmer than the surrounding air (which may or may not be the case), then the warmer air being less dense will continue rising, boosting air above its position upward. If the air above is sufficiently moist, it too will condense as cloud and release sensible heat. One can see how this chain reaction can produce a cloud of great vertical extent, certainly capable of rain and perhaps even more violent forms of weather, such as the thunderstorm. The height to which the cloud extends will be determined largely by the stability of the air, a subject to be covered in another chapter.

We will mention at this point an exception to the high-good weather rule. Recall that a cold core high is a shallow system changing to lower pressure aloft. Suppose that you are flying through a cold core high at temperatures below freezing. If rain is being generated from moisture and lower pressure aloft and you encounter this precipitation, you will have a serious icing problem. Icing will be discussed in greater detail in a later chapter.

J. Dew and Frost

We now consider dew and frost. Dew forms when the temperature of a collecting surface, such as a car, is above freezing but below the dew point of the adjacent air. Recall that all matter radiates energy and, hence, at night the car will cool due to a net loss of radiation. The cooling of the adjacent air is by conduction. Thus, moisture collects on the car in a liquid state called dew. Note that dew is not a cloud since it is not suspended in air, other than indirectly on the car, of course, and it is not a form of precipitation since it did not fall to the earth's surface in droplet form. Dew forms on many different objects, such as leaves and grass. If, after dew forms, the temperature of the collecting surface drops below freezing, this dew will become merely frozen dew, but not frost. Frost is a sublimation product. Thus, for frost, water vapor must change directly to ice, and this will occur if the temperature of the collecting surface and the dew point of the surrounding air are both below freezing, again with the collecting surface temperature below that of the dew point of the adjacent air. If we left out the requirement that the dew point of the surrounding air be below freezing, we would have acquired dew first, subsequent freezing causing frozen dew.

Frost which forms on car windshields at night sublimates as we drive down the highway, the friction of air passing over the frost generating the heat necessary for sublimation. Frost

Figure 6-4. Virga. Precipitation from the cloud evaporates in drier air below and does not reach the ground.

effects in the upper atmosphere are of no major consequence, but frost can be very dangerous on takeoffs and landings. We will consider this form of ice further in the chapter on icing.

K. Another Look at the Katabatic Wind

As moist air rises up the western slopes of the Appalachians, it cools dry adiabatically at the rate of 5.5°F per 1,000 feet. If the air is moist enough and the terrain high enough, the air will be cooled down to its dew point, clouds will form, and rain will fall. However, as soon as condensation begins, latent heat returns as sensible heat, retarding the cooling rate to about 3.0°F per 1,000 feet as we saw earlier. When the air reaches the top of the mountains and starts its descent on the other side, it warms adiabatically at the rate of 5.5°F per 1,000 feet all the way down – we are assuming that most of the moisture condensed out on its way up, so we use the dry adiabatic lapse rate on the downward trek. Thus, air that started out at 75°F (23°C) on the western side of the Appalachians might well reach a temperature of 95°F (35°C) on the eastern side (Figure 6-6).

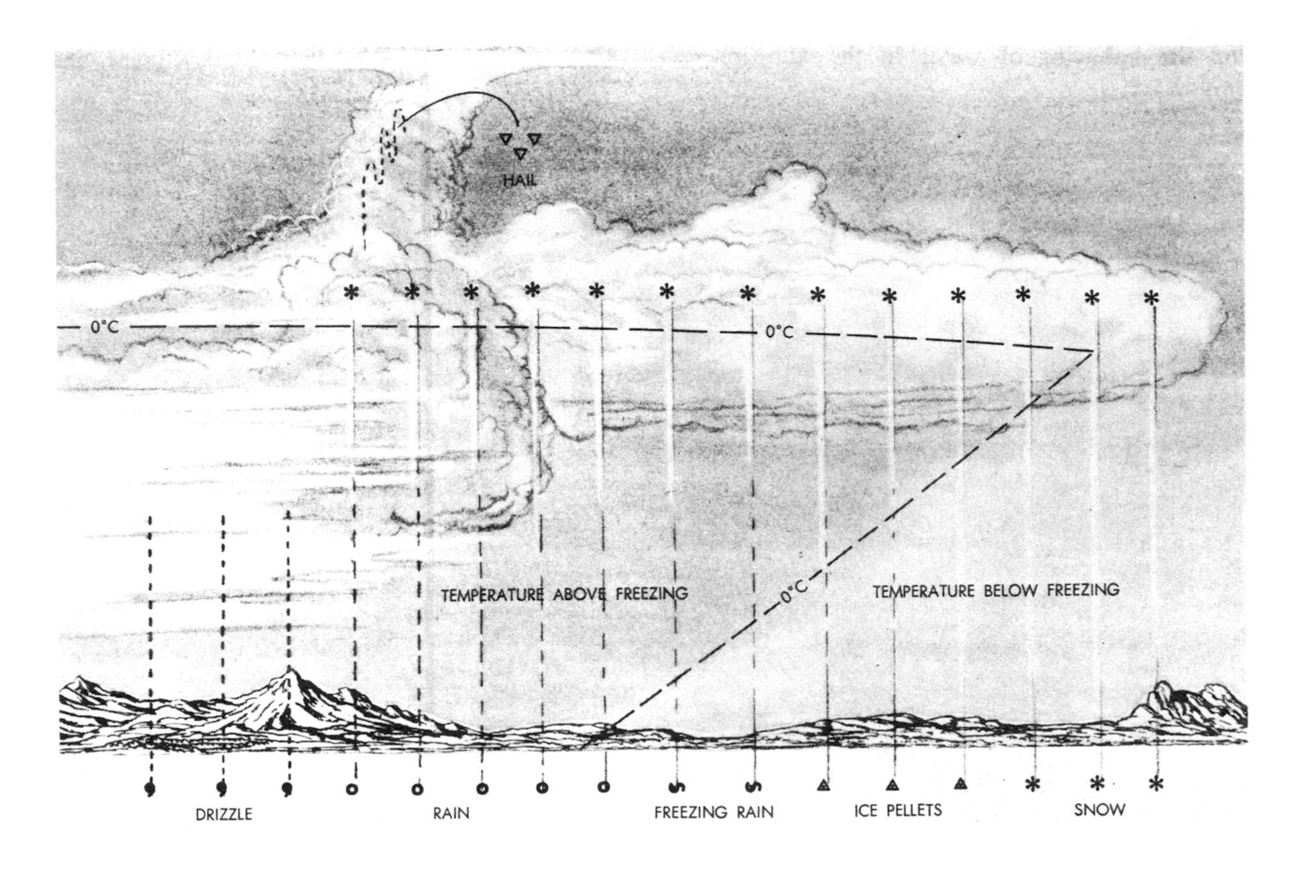

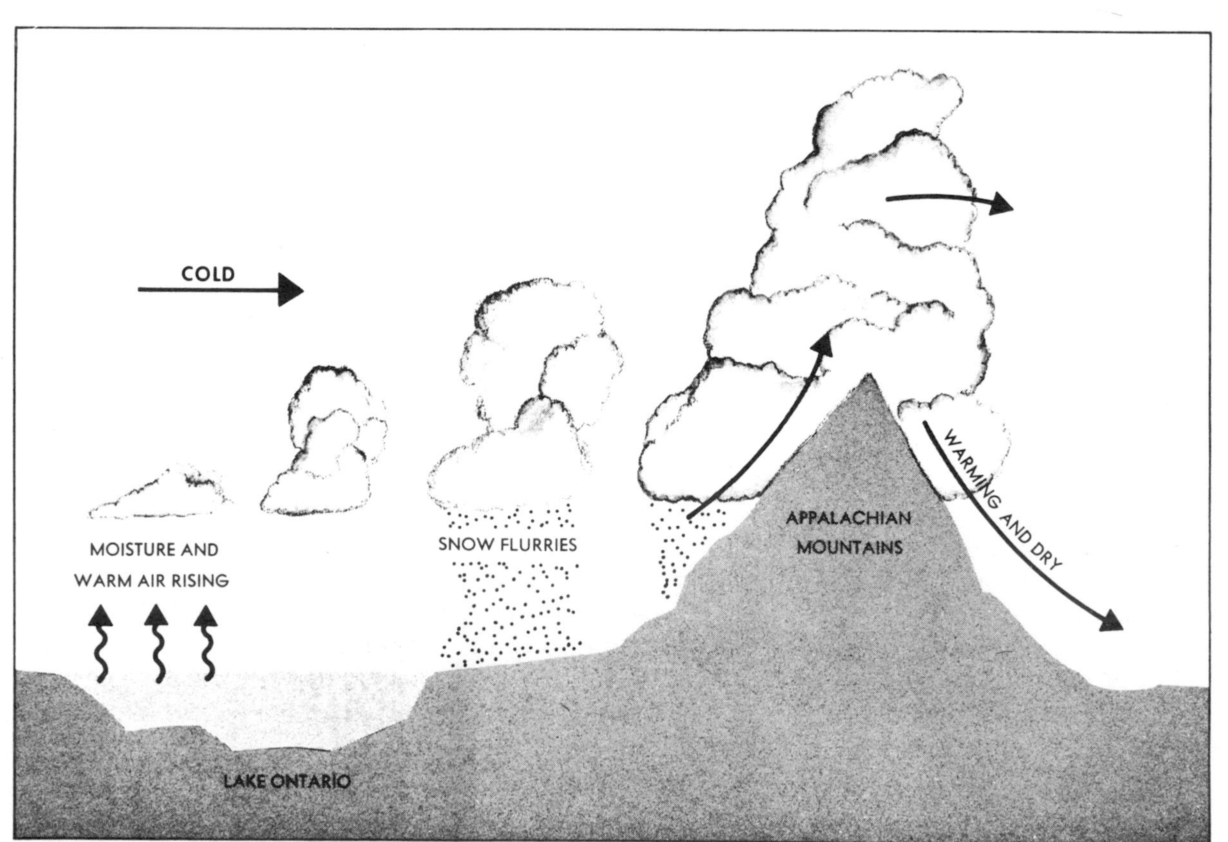

Figure 6-6. Strong cold winds across the Great Lakes absorb water vapor and may carry showers as far eastward as the Appalachians.

L. Hydroplaning

Hydroplaning is caused by the tires of an airplane being effectively lifted off a runway surface by the presence of a film of water. When this happens braking action and directional control will all but disappear. The greater the depth of water and the speed of the airplane the more the lifting effect of the water; in extreme conditions, tires may be forced completely off the runway surface.

Dynamic hydroplaning occurs at high speeds when there is standing water on the runway, whether the runway surface is itself smooth or not. As little as one-tenth of an inch of water can result in this type of hydroplaning. You would expect dynamic hydroplaning when you first land, due to the higher speed at initial contact; once started, the airplane will continue to hydroplane at even slower speeds.

Viscous hydroplaning requires a smooth runway surface. Due to the viscous properties of water, this type of hydroplaning will occur with a more shallow layer of water and at a slower speed. Reverted rubber hydroplaning is caused by a pilot locking the brakes for a prolonged period of time on a wet surface. An airplane's tires are effectively held off a runway surface in such a case by the steam created by friction.

MOISTURE
Review Questions

PRI

1. Listed below are factors which change density altitude.
   a. Decreasing barometric pressure.
   b. Increasing barometric pressure.
   c. Decreasing temperature.
   d. Increasing temperature.
   e. Decreasing relative humidity.
   f. Increasing relative humidity.

   Select the factors which increase the density altitude at a given airport.
   a. A, D, E.
   b. B, C, E.
   c. A, D, F.
   d. B, C, F.

2. Wisps or streaks of precipitation that evaporate before reaching the ground are referred to as
   a. A foehn gap.
   b. Cirrocumulus clouds.
   c. Virga.
   d. Roll clouds.

3. The temperature to which moist air must be cooled to become saturated is defined as
   a. Sublimation.
   b. Condensation nuclei.
   c. Relative humidity.
   d. Dewpoint.

INS

4. Which conditions result in the formation of frost?
   a. The freezing of dew.
   b. The collecting surface's temperature is at or below freezing and small droplets of moisture fall on the collecting surface.
   c. The temperature of the collecting surface is at or below the dewpoint of the adjacent air and the dewpoint is below freezing.
   d. Small drops of moisture falling on the collecting surface when the surrounding air temperature is at or below freezing.

5. What is meant by the term dewpoint?
   a. The temperature at which condensation and evaporation are equal.
   b. The temperature at which dew will always form.
   c. The temperature to which air must be cooled to become saturated.
   d. The spread between actual temperature and the temperature during evaporation.

6. The amount of water vapor which air can hold largely depends on
   a. The dewpoint.
   b. Air temperature.
   c. Stability of air.
   d. Relative humidity.

7. Clouds, fog, or dew will always form when
   a. Water vapor condenses.
   b. Relative humidity exceeds 100%.
   c. Water vapor is present.
   d. The dewpoint is higher than the temperature.

COM

8. Reverted rubber hydroplaning (airplane skimming on wet runway) occurs when the pilot
   a. Locks the wheel brakes for a prolonged period.
   b. Overcontrols the rudder.
   c. Intermittently applies wheel brakes for short periods.
   d. Lands in an excessive crosswind.

9. Dynamic hydroplaning (airplane skimming on wet runway) occurs at
   a. Slow speeds with only a thin film of water on the runway.
   b. High speeds with standing water on the runway.
   c. Slow speeds with standing water on the runway.
   d. High speeds with only a film of water on the runway.

10. Viscous hydroplaning (airplane skimming on wet runway) occurs at
    a. Slow speeds with only a thin film of water on a runway with a smooth acting surface.
    b. High speeds with standing water on the runway.
    c. Slow speeds with standing water on the runway.
    d. Only high speeds with a thin film of water on the runway.

CFI

11. Which statement is true regarding actual air temperature and dewpoint temperature spread?
    a. The temperature and dewpoint spread decreases as the relative humidity decreases.
    b. The temperature and dewpoint spread decreases as the relative humidity increases.
    c. Temperature and dewpoint spread are not related to relative humidity.
    d. The temperature and dewpoint spread increases as the relative humidity increases.

ATP

12. What is the process by which ice can form on a surface directly from water vapor on a cold, clear night?
    a. Sublimation
    b. Evaporation
    c. Supersaturation
    d. Condensation

13. What are the processes by which moisture is added to unsaturated air?
    a. Heating and sublimation.
    b. Evaporation and sublimation.
    c. Heating and condensation.
    d. Supersaturation and evaporation.

CHAPTER VII

CLOUD CLASSIFICATION

The reader should demonstrate an understanding of clouds to the extent that he can answer the following questions (as stated below, or as multiple-choice, etc.):

1. Name the four families of clouds, the types of each, and altitude range of their bases.
2. Describe each cloud type and the hazards associated with each.
3. Explain the meaning of "castellanus", "fractus", "mamma", and "standing lenticular".

CHAPTER VII

CLOUD CLASSIFICATION

Clouds provide visual clues of turbulence, precipitation, and visibility. Much of the nature of cloud formation was considered in the chapter on moisture; you may wish to review that unit at this time. Our purpose in this chapter is to classify the most common clouds and to give a description of many of them (Figures 7-1 through 7-13). Four families of clouds are usually considered – high, middle, low, and vertical development. The bases of high clouds, cirrus–Ci, cirrostratus–CS, and cirrocumulus–CC, usually range between 16,500 feet (5 km) to 45,000 feet (14 km). Bases of middle clouds, altostratus–AS, and altocumulus–AC, usually range between 6,500 (2 km) and 23,000 feet. Bases of low clouds, stratus–St, stratocumulus–Sc, and nimbostratus (nimbus means rain cloud)–Ns, usually range from near the surface to 6,500 feet (the FAA classifies Ns as middle cloud). Clouds of vertical development, cumulus–Cu, and cumulonimbus–Cb, have bases which usually range from near the surface to 12,000 feet (4 km).

Other words are used with the above classifications to allow for distinct variations in cloud structure. Castellanus is often used with altocumulus, called, appropriately enough, altocumulus-castellanus, castellanus referring to castle-tower protrusions from the altocumulus cloud. Fractus is often used with stratus and cumulus, such as fracto-stratus to indicate torn or ragged fragments of the stratus cloud. Mamma is often used with cumulonimbus clouds, such as cumulonimbus mammatus to refer to udder-like protuberances hanging from beneath the main cloud, usually indicative of very turbulent air.

Figure 7-1. CIRRUS. Cirrus are thin, feather-like ice crystal clouds in patches or narrow bands. Larger ice crystals often trail downward in well-defined wisps called "mares' tails." Wispy, cirrus-like, these contain no significant icing or turbulence.

Figure 7-2. CIRROCUMULUS. Cirrocumulus are thin clouds, the individual elements appearing as small white flakes or patches of cotton. May contain highly supercooled water droplets. Slight turbulence and icing.

A very important variation of the altocumulus cloud is the standing lenticular altocumulus cloud, which will be discussed in greater detail in the chapter on turbulence. The presence of these clouds in mountainous areas indicates mountain-wave turbulence which, in the author's opinion, is second in severity only to thunderstorm turbulence.

CLOUDS
Review Questions

PRI

1. Cumulonimbus clouds can best be described as
   a. Thin, white, featherlike clouds in patches or narrow bands.
   b. White or gray layers or patches of solid clouds, usually appearing in waves.
   c. Dense clouds, dark at lower levels, extending many thousands of feet upward.
   d. Fluffy, white clouds appearing in layers and sometimes producing steady precipitation.

INS

2. Clouds characterized by their lumpy, billowy appearance belong to which family of clouds?
   a. High clouds.
   b. Middle clouds.
   c. Low clouds.
   d. Clouds with extensive vertical development.

3. The suffix nimbus, used in naming clouds, means
   a. A cloud with extensive vertical development.
   b. A rain cloud.
   c. A middle cloud containing ice pellets.
   d. An accumulation of clouds.

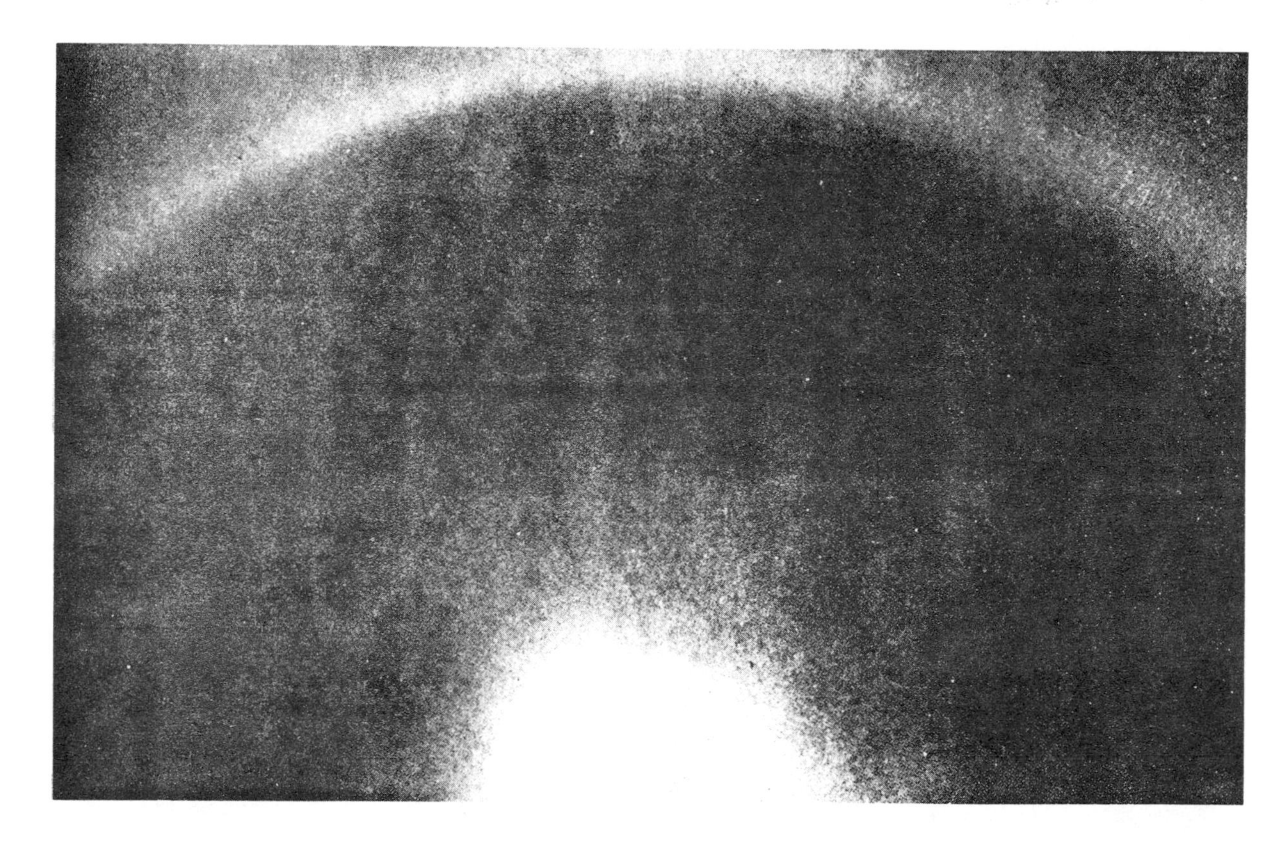

Figure 7-3. CIRROSTRATUS. Cirrostratus is a thin whitish cloud layer appearing like a sheet or veil. Cloud elements are diffuse, sometimes partially striated or fibrous. Due to their ice crystal makeup, these clouds are associated with halos—large luminous circles surrounding the sun or moon. No turbulence and little if any icing. The greatest problem flying in cirriform clouds is restriction to visibility. They can make the strict use of instruments mandatory.

4. Nimbostratus, a gray or dark massive cloud layer, belongs to which family of clouds?
   a. High clouds.
   b. Middle clouds.
   c. Low clouds.
   d. Clouds with extensive vertical development.

ATP

5. Clouds are divided into four families according to their
   a. Outward shape.
   b. Height range.
   c. Composition.
   d. Origin.

Figure 7-4. ALTOCUMULUS. Altocumulus are composed of white or gray colored layers or patches of solid cloud. The cloud elements may have a waved or roll-like appearance. Some turbulence and small amounts of icing.

Figure 7-5. ALTOSTRATUS. Altostratus is a bluish veil or layer of clouds. It is often associated with altocumulus and sometimes gradually merges into cirrostratus. The sun may be dimly visible through it. Little or no turbulence with moderate amounts of ice.

Figure 7-6. NIMBOSTRATUS. Nimbostratus is a gray or dark massive cloud layer, diffused by more or less continuous rain, snow, or ice pellets. This type is classified as a low cloud. Very little turbulence, but can pose a serious icing problem if temperatures are near or below freezing.

Figure 7-7. STRATUS. Stratus is a gray, uniform, sheet-like cloud with relatively low bases. When associated with fog or precipitation, the combination can become troublesome for visual flying. Little or no turbulence, but temperatures near or below freezing can create hazardous icing conditions.

Figure 7-8. STRATOCUMULUS. Stratocumulus bases are globular masses or rolls unlike the flat, sometimes indefinite, bases of stratus. They usually form at the top of a layer mixed by moderate surface winds. Sometimes, they form from the breaking up of stratus or the spreading out of cumulus. Some turbulence, and possible icing at subfreezing temperatures. Ceiling and visibility usually better than with low stratus.

Figure 7-9. CUMULUS. Fair weather cumulus clouds form in convective currents and are characterized by relatively flat bases and dome-shaped tops. Fair weather cumulus do not show extensive vertical development and do not produce precipitation. More often, fair weather cumulus indicates a shallow layer of instability. Some turbulence and no significant icing.

Figure 7-10. TOWERING CUMULUS. Towering cumulus signifies a relatively deep layer of unstable air. It shows considerable vertical development and has billowing cauliflower tops. Showers can result from these clouds. Very strong turbulence; some clear icing above the freezing level.

Figure 7-11. CUMULONIMBUS. Cumulonimbus are the ultimate manifestation of instability. They are vertically developed clouds of large dimensions with dense boiling tops often crowned with thick veils of dense cirrus (the anvil). Nearly the entire spectrum of flying hazards are contained in these clouds including violent turbulence. They should be avoided at all times! This cloud is the thunderstorm cloud and is discussed in detail in chapter 13, "Thunderstorms."

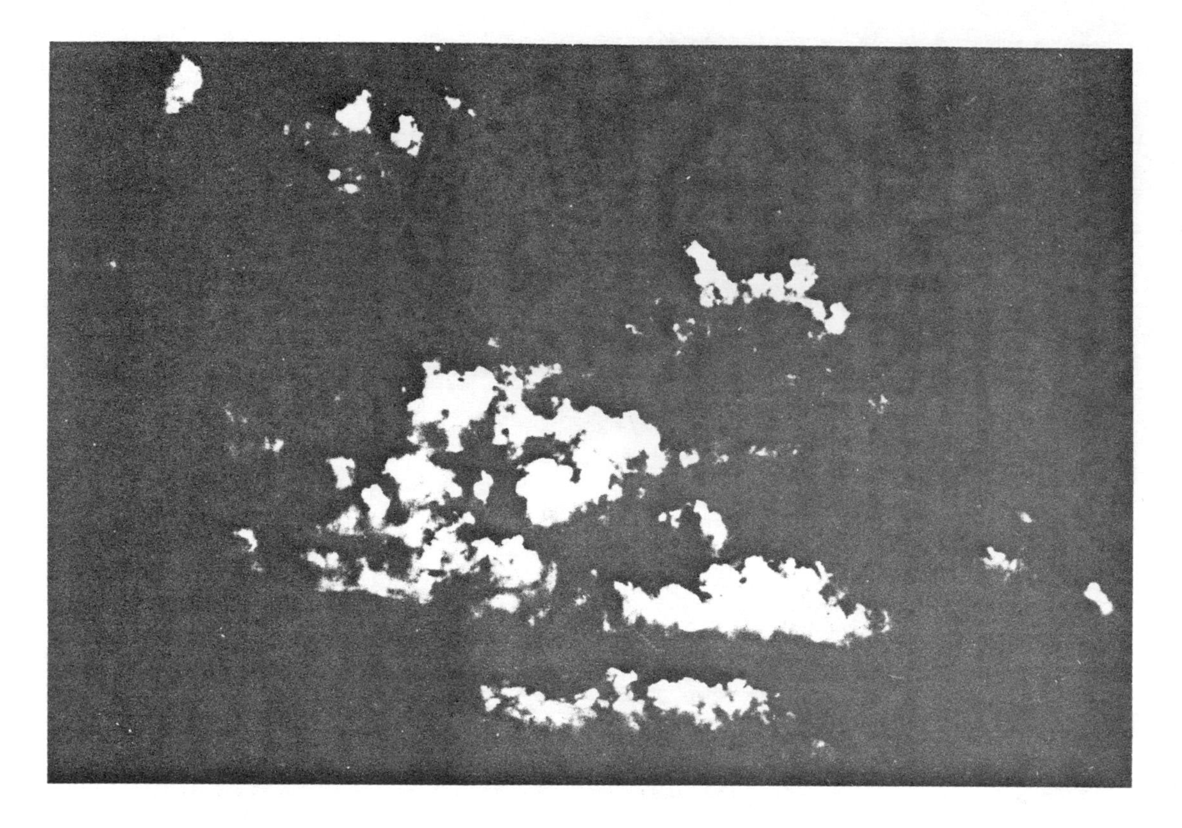

Figure 7-12. ALTOCUMULUS CASTELLANUS. Altocumulus castellanus are middle level convective clouds. They are characterized by their billowing tops and comparatively high bases. They are a good indication of mid-level instability. Rough turbulence with some icing.

Figure 7-13. STANDING LENTICULAR ALTOCUMULUS CLOUDS. Standing lenticular altocumulus clouds are formed on the crests of waves created by barriers in the wind flow. The clouds show little movement, hence the name standing. Wind, however, can be quite strong blowing through such clouds. They are characterized by their smooth, polished edges. The presence of these clouds is a good indication of very strong turbulence and should be avoided.

CHAPTER VIII

STABILITY AND SOARING

The reader should demonstrate an understanding of stability and soaring to the extent that he can answer the following questions (as stated below, or as multiple-choice, etc.):

1. Define "unstable atmosphere". What are the two primary components of the definition?
2. Name five ways in which air may be forced upward.
3. Mountain-wave turbulence occurs primarily in (stable, unstable) air.
4. A temperature inversion represents a(n) (absolutely unstable, conditionally unstable, absolutely stable) lapse rate.
5. What is the effect of an increase in moisture on the stability of air?
6. List the effects of stable and unstable air on clouds, turbulence, precipitation, and visibility.
7. Suppose the temperature and dew point at the surface are 91°F and 64°F, respectively. Estimate the bases of convective clouds.
8. Under what conditions does an absolutely unstable lapse rate occur?
9. What type of stability is represented by the standard lapse rate?
10. State some stabilizing and destabilizing influences.
11. What is the effect of mixing on stability?
12. How does a subsidence inversion occur?
13. Compare thermal updrafts and downdrafts.
14. How can dust and smoke help identify a thermal?
15. What is the primary hazard associated with dust devil soaring?
16. Compare the effects of concave and convex bases on cumulus clouds.
17. What is the effect of wind/wind shear on thermals?
18. What sky conditions and wind are conducive to thermal soaring?
19. Which is better for thermal soaring, a TI of –5 or –10?
20. Compute the thermal index for a surface temperature of 25°C at the 900-millibar level in Figure 8-16.
21. What minimum surface temperature is required for good thermals to develop from the surface to 10,000 feet in Figure 8-16?
22. What is the effect of the passage of a cold front on soaring?
23. When do sea breezes occur (day or night)? What is a visual clue to the location of the leading edge of a sea breeze?
24. Which cloud best identifies favorable sea breeze soaring?
25. Soaring is best ____________ (upwind, downwind) of a ridge or hill. What is the optimal wind speed over a ridge or hill for soaring?
26. List three conditions for a strong mountain wave.
27. What is the greatest potential hazard in mountain wave soaring?

CHAPTER VIII

STABILITY AND SOARING

A. Stability

Stability is a very misunderstood concept. Most pilots believe an unstable atmosphere is one in which air is ascending or descending. This is just not necessarily true. An unstable atmosphere is one in which if air is displaced vertically, it will continue to move vertically. Notice the requirement that there must be some outside action to start the vertical movement. Then, if the air when started moving vertically is warmer than the surrounding air, it will rise (or descend, if colder than the surrounding air). So when you hear the word unstable, think of two things – outside force first, then continued movement, depending on the temperature of the surrounding air.

1. Lifting Mechanisms

One outside force providing initial vertical displacement is higher terrain (Figure 8-1). As air flows up terrain it will be accelerated upward even faster if that air becomes warmer than the surrounding air. We will see how air can become warmer than surrounding air shortly. Other common ways air is displaced vertically are: a low pressure system moving into an area, cold air moving over warm air and sinking (forcing the warm air upward), intense heating of the earth's surface (called convection), convergence, that is, dryer air meeting moist air (the more dense dry air forcing moist air upward), and fronts, as we shall see later. The examples cited initiate upward movement; a high, for example, initiates downward movement, in which case air will continue to descend if in its descent it is colder than the surrounding air, an unstable condition.

2. Stable Air and Mountain Wave Turbulence

A good example of stable air is in mountain-wave turbulence. Air is lifted to the top of a mountain. If that air is colder than the surrounding air, it will descend and continue to oscillate up and down for some distance, similar to holding on to a rope and inducing a wave with a sharp jerk (Figure 8-2). The mountain generates the wave just by being an obstacle in the wind flow. Thus, mountain-wave turbulence initially occurs in stable air, for the air when displaced vertically begins to return to its normal position in a slowly dissipating or dampening wave. However, should these waves continue downstream into unstable air, large updrafts and precipitation may occur as you will soon see.

3. Air Movement After Initial Displacement

We now investigate the second part of the stability concept, namely, how air may or may not continue its vertical displacement after some outside force initiates vertical movement. The key to the discussion is the actual vertical distribution of temperature, that is, the degree of stability is determined by observing the actual lapse rate. To illustrate we take a balloon and fill it with air. Our outside force for initial vertical displacement will be for us to physically carry this balloon up to a height of, say, 4,000 feet. Now the air in our balloon will cool by expansion at the rate of 5.5°F per 1,000 feet, so that if the temperature at the surface had been 90°F, then the temperature inside the balloon would be $90° - 4 \times 5.5° = 68°$F at the altitude of 4,000 feet, assuming the air in the balloon does not become saturated as it rises. Now if the air at 4,000 feet is standard, that is, if the temperature is $90° - 4 \times 3.5° = 76°$ (since the standard lapse rate is 3.5°F per 1,000 feet), then the air inside the balloon would be colder than the

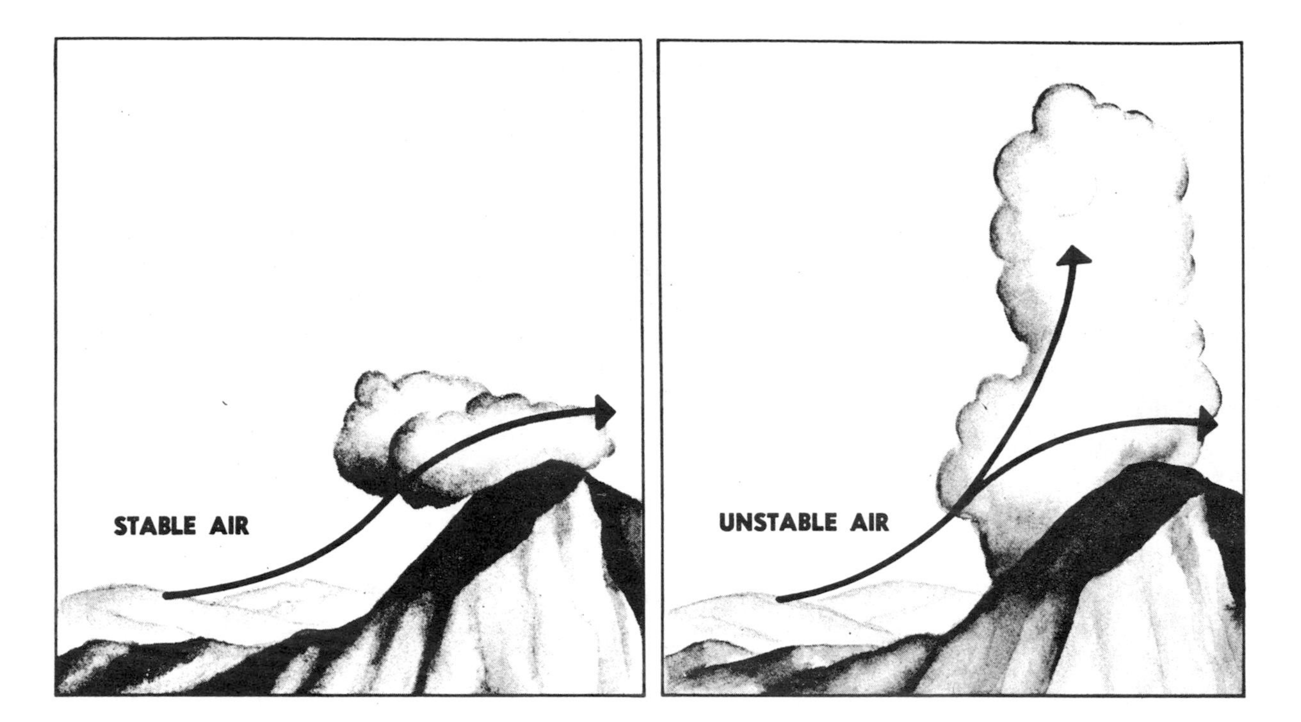

Figure 8-1. When stable air (left) is forced upward, air tends to retain horizontal flow and any cloudiness is flat and stratified. When unstable air is forced upward, the disturbance grows, and any resulting cloudiness shows extensive vertical development.

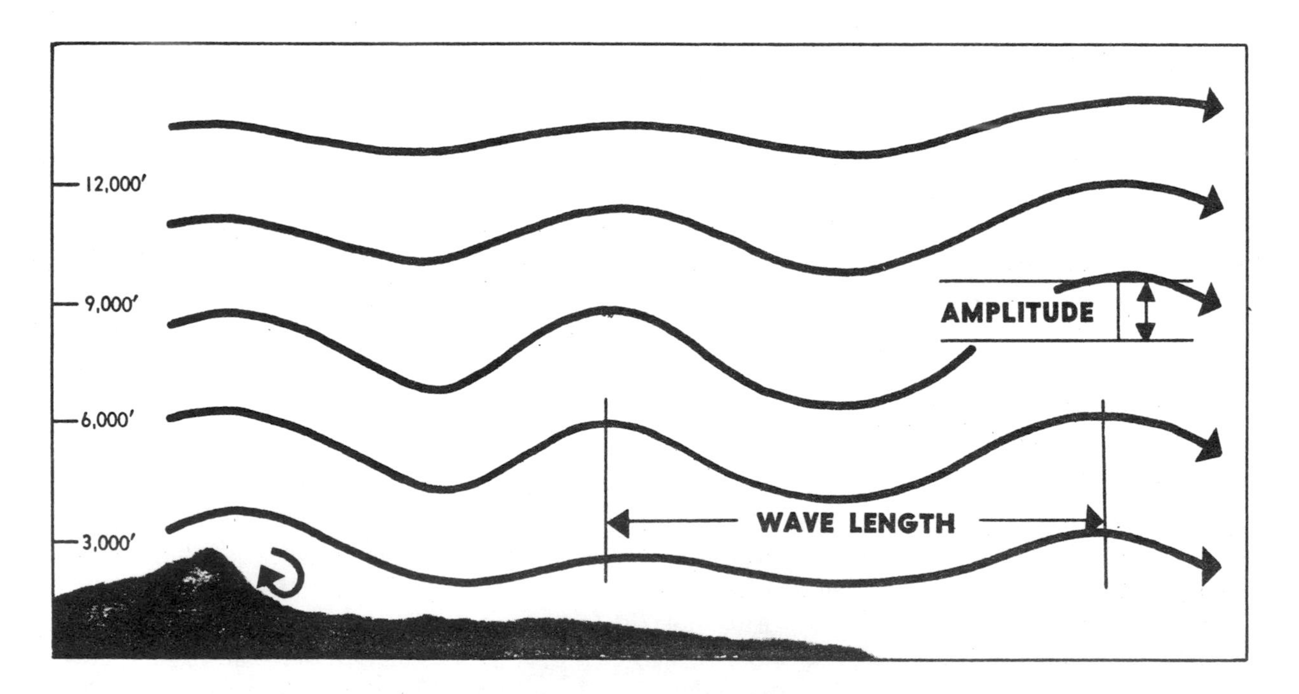

Figure 8-2. Wave length and amplitude.

surrounding air and, hence, would tend to return to its original position. Suppose that the air had been saturated before we lifted it. Then the rate of cooling would have been by the saturation–adiabatic rate of near 3°F per 1,000 feet, so that the temperature inside the balloon would have been 90° – 4 × 3°F = 78°, which would be warmer than the temperature of the surrounding air and, hence, the balloon would continue to rise.

Now suppose that the actual temperature lapse rate had not been standard. Suppose that the temperature of air at 4,000 feet had been 80°F instead of the standard 76°F. Then the saturated-air balloon's temperature of 78°F is colder than 80°F, thus the balloon would have tended to descend to the surface. Thus, the presence of moisture, usually indicated by relative humidity, does not automatically mean air is unstable. However, it is important to note the effect of greater moisture content in the balloon. In our first balloon, with relatively dry air, the temperature at 4,000 feet was 68°F. This is 12°F below the 80°F air temperature we are now assuming exists at 4,000 feet. In our saturated-air balloon the temperature at 4,000 feet was 78°F, 10°F closer to the 80°F temperature of the surrounding air. Thus, while greater moisture content does increase the tendency toward instability, due to the warmer temperature just seen, moist air is not automatically unstable air; the key is the temperature of the surrounding air. In fact, the temperature of the surrounding air does not have to decrease at all – recall the definition of inversion from the chapter on temperature as an increase in temperature with height.

We now see how air can be neutrally stable, that is, neither stable nor unstable. Suppose the temperature at 4,000 feet had been 68°F. Then our relatively dry air balloon would have had the same temperature when lifted to 4,000 feet and, hence, would not have tended to move further. Hence, when displaced, this balloon tends to stay where it is and the air is said to be neutrally stable relative to the dry adiabatic lapse rate.

A 4,000 foot temperature inversion represents an absolutely stable condition, for regardless of how much the air inside the balloon cools as it ascends, its temperature will be colder than the surrounding air and it will tend to descend. At the other extreme, a temperature lapse rate greater than 5.5°F per 1,000 feet would result in an absolutely unstable situation since both the dry and the saturation-adiabatic lapse rates would result in a warmer temperature of the air in the balloon at 4,000 feet, due to lapse rates of 5.5°F and 3.0°F per 1,000 feet, respectively.

Conditionally unstable means that the actual lapse rate of temperature is between the dry adiabatic and saturation-adiabatic rates. The standard lapse rate of 3.5°F per 1,000 feet was an example of this, for in the dry adiabatic case we obtained a temperature of 68°F on lifting and for the saturation-adiabatic case we obtained a temperature of 78°F, the standard lapse rate yielding a temperature of 76°F at 4,000 feet. This case illustrated conditional instability, the possibility of air becoming unstable if there was sufficient moisture and lift.

We briefly repeat a couple of key concepts at this point. There must be an outside force present to start vertical displacement, further displacement depending on the temperature of the surrounding air. And the greater the moisture content of air, the greater the tendency of air toward instability.

4. Factors Affecting Layer Stability

Normally, an absolutely unstable lapse rate only occurs in a shallow layer of air near the earth's surface on very hot days when winds are light. We have already mentioned that a temperature inversion is one way that an absolutely stable layer of air is created. The most frequent type of temperature inversion occurs at night (when the sky is clear and winds are light), due to the earth cooling most rapidly at the surface. While other types of temperature inversions do exist (as we shall soon see), the atmosphere is primarily in a conditionally unstable state.

We can generalize the above discussion by stating that a layer's stability is affected by warming or cooling (aloft and/or at the surface). It should be clear that warm air above cool air or a cold surface is a stabilizing influence while cold air above warm air or a warm surface is a destabilizing influence. Mixing of the air will also influence the lapse rate. For example, with warm air above cold air at the surface, after mixing the warm air will be cooler and the cold air warmer, thus steepening the lapse rate (a destabilizing influence). We will address these ideas more when we discuss air mass modification in the next chapter.

Up to now, our layer of air has remained stationary and our parcel, the balloon, has moved through it. What happens when a layer of air ascends or descends in the atmosphere? It turns out that a lapse rate of, say, 4°F per 1,000 feet can be reduced to 2°F, then 0°F, and then even have an increase in temperature with height, if the layer descends far enough. As the layer descends it is compressed, which can occur only if the distance travelled between the original top and the new top is greater than the distance travelled between the original bottom and the new bottom. Since the top travels a greater distance, then due to adiabatic heating the top will have a greater temperature increase than the bottom. An inversion created in this manner is called a subsidence inversion. If we reverse the process, that is, lift the layer of air, then the layer will acquire its original lapse rate.

We mention one last "type" of stability, convective instability. A layer of air is said to be convectively unstable if, after sufficient lift of the layer to cause saturation throughout the layer, its lapse rate exceeds the moist adiabatic rate. Instability results because, after a parcel of air begins to rise after the layer has become saturated, then the parcel will cool at the moist adiabatic rate and, since we are assuming that the lapse rate of the layer exceeds the moist adiabatic rate, the temperature of the parcel is greater than the surrounding air and, hence, the parcel will continue to rise freely. However, a layer does not always become unstable after becoming saturated throughout; in many cases the lapse rate of the layer does not exceed the moist adiabatic rate after the necessary lift of the layer to become saturated. But, high moisture content in low levels and dry air aloft favor convective instability. Conversely, dry air in low levels and high moisture content aloft favor stability. We will accept these results without proof.

When we discuss layer stability we are primarily concerned with vertically displaced layers of air and the idea of stability is with regard to the possible change in the lapse rate of the layer. Parcel stability refers to the movement of a parcel (think of a balloon) within a layer due to differences in temperature between the parcel and the layer. We combined both concepts in the above example.

When we consider parcel stability in two adjoining layers of air, we may well find that the upper layer chokes, or caps off, parcel instability in the lower layer. For example, on a hot summer afternoon local heating will cause bubbles of air to rise, and, if moisture condenses as the bubbles start to rise, then a cloud starts to form. If there is a subsidence inversion in the layer above, then the bubble will stop its ascent. But, in general, just because a cloud is not rising, this does not mean that air is not rising, for if the parcel's temperature had been warmer than the temperature of the layer above, it would have continued to rise. The fact that there is no cloud extending into a second layer may be due only to the fact that all the available moisture in the parcel has condensed out to form cloud. Thus, the air above the cloud top might continue to rise, resulting in very turbulent air. Many pilots are surprised when this occurs.

5. *A Stability Table*

Table 8-1 summarizes the relationship between stability and clouds, turbulence, precipitation, and visibility.

TABLE 8-1. A STABILITY TABLE

	e. Stable	f. Unstable
a. Clouds	Stratiform	Cumuliform
b. Turbulence	Smooth	Rough
c. Precipitation	Steady or Continuous	Showery
d. Visibility	Fair to Poor	Good, except for blowing obstructions

a. With regard to clouds, "strata" refers to layers, so stratiform clouds tend to form in layers. The greater the stability of air, the greater the tendency of air to maintain its original position when displaced and, hence, in general the less the vertical extent of the cloud. "Cumulus" means "heap", and once air starts to move in unstable air, it continues the movement, resulting in the heaped cloud of great vertical extent classified as cumuliform. A technique for forecasting the base of cumuliform clouds in thousands of feet is to take the temperature minus the dew point at the surface and divide by 4.5 (the dry adiabatic lapse rate is 5.5°F per 1,000 feet and the dewpoint lapse rate is 1°F per 1,000 feet and, hence, the two lapse rates approach each other at 4.5°F per 1,000 feet). For example, if the temperature is 88 and the dewpoint is 70, then (88-70)÷4.5 is 4 and, hence, 4,000 feet would be the approximate base for afternoon convective clouds, which would be cumuliform. Recall that 5.5°F per 1,000 feet is also 3°C per 1,000 feet and 10°C per km. Thus, the two rates – temperature and dew point – also approach each other at the rates of 2.5°C per 1,000 ft. and 8°C per km. For a more technical discussion of the relationship between temperature, dew point, dew point lapse rate, relative humidity, pressure and density see Appendix D.

b. With regard to turbulence the results are obvious with exceptions, such as the mountain-wave turbulence that occurs primarily when air is stable.

c. Precipitation falling from thick stratiform clouds is more or less continuous or steady. Cumulus clouds of sufficient vertical extent often result in thunderstorms, the ultimate manifestation of a cumulus cloud, with lightning, thunder, hail, and possible tornadoes, but with regard to the nature of the precipitation it is best classified as showery. We have all observed the difference in the two classifications of precipitation, the steady precipitation more characteristic of the winter season, with showery precipitation move frequent in summer.

d. Air which is stable tends to retard vertical displacement. If the relative humidity is low and there is little extraneous material in the air, such as dust, smoke, etc., then visibility will be fair. But with high relative humidity and a great concentration of foreign matter, visibility can be quite poor; in fact, visibility can be reduced to zero, both horizontally and vertically. In unstable air, as air continues to move after being initially displaced, whatever moisture and/or foreign matter concentration exists will be reduced. The main exception would be the blowing obstruction, such as rain beneath a thunderstorm, or blowing sand or snow.

e. A final reminder that stable air does not always mean good flying weather. Surface subsidence inversions from stationary highs, called stagnant, have caused many air pollution problems over the years, resulting in more deaths than aircraft accidents.

f. And a reminder that unstable air is not a problem unless there is an initial lifting force.

B. Soaring

The following figures will further illustrate the ideas of lift and stability, as applied in the mostly recreational sport of soaring (Figures 8-3 through 8-24). A combination of favorable weather and pilot skill may permit a non-powered aircraft to reach altitudes in excess of 40,000 feet. The figures will be grouped relative to four classes: Thermal Soaring, Sea Breeze Soaring, Ridge or Hill Soaring, and Mountain Wave Soaring.

Continued on page 8-11

1. Thermal Soaring

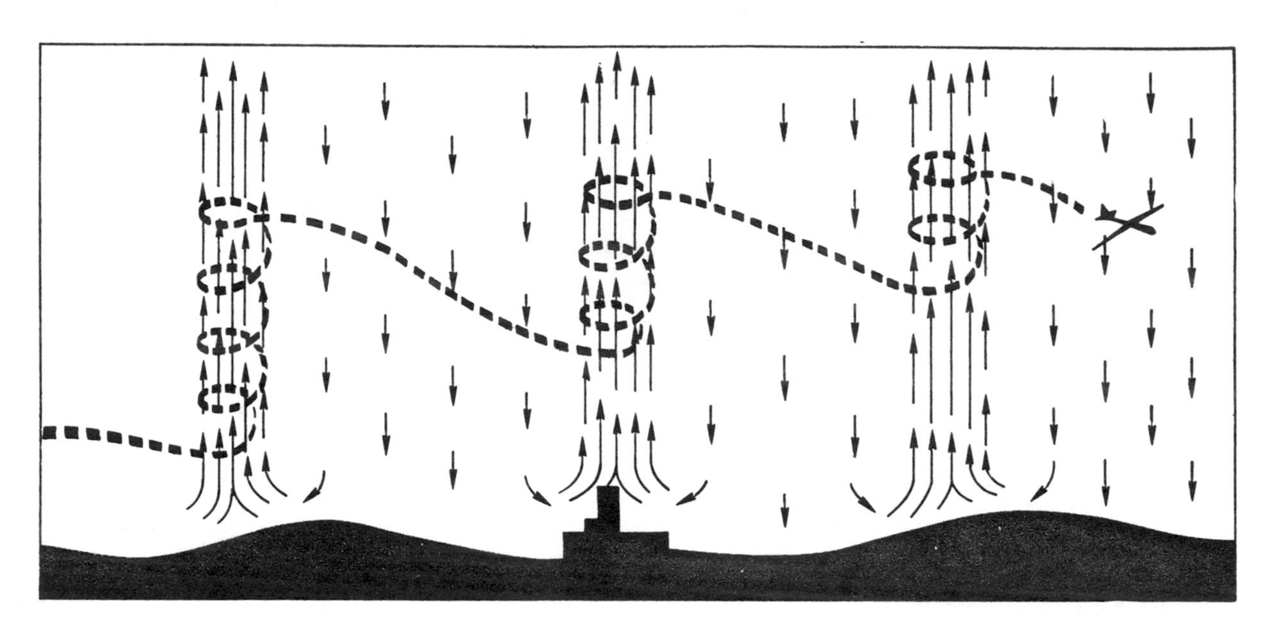

Figure 8-3. Thermals generally occur over a small portion of an area while downdrafts predominate. Updrafts in the thermal usually are considerably stronger than the downdrafts. Sailplane pilots gain altitude in thermals and hold altitude loss in downdrafts to a minimum.

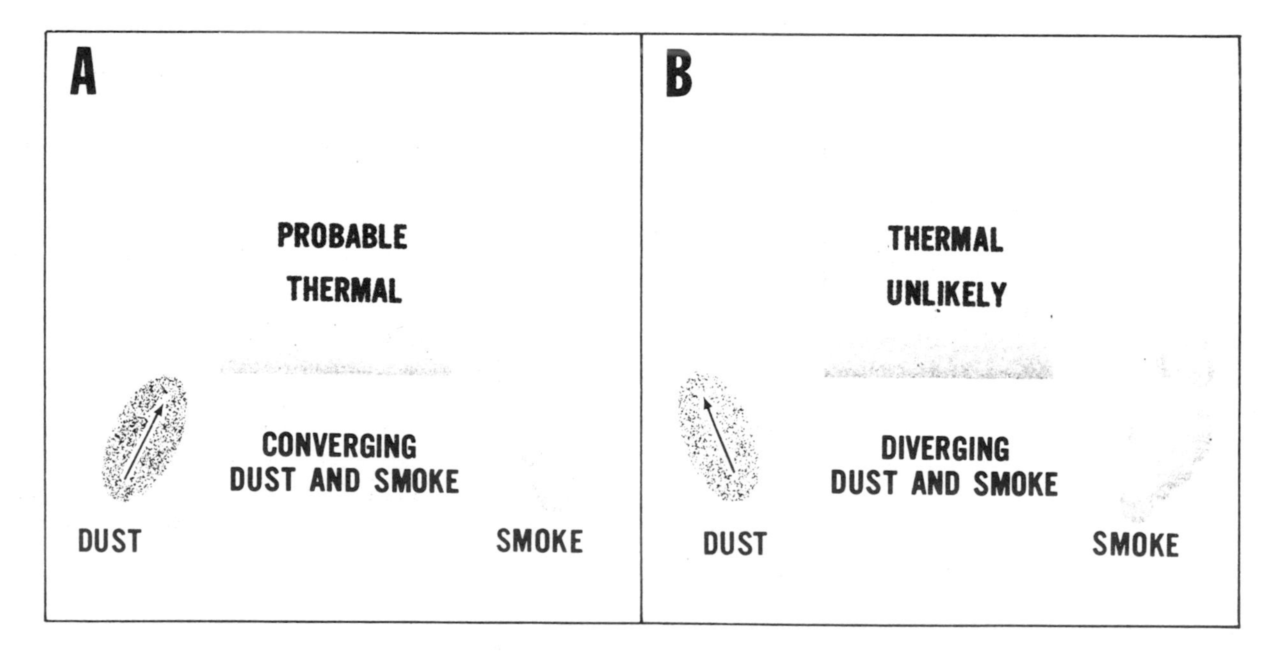

Figure 8-4. Using surface dust and smoke movement as indications of a thermal. When you have sighted an area which you think will heat rapidly, look for dust or smoke movement at the surface as an indicator of surface wind. Converging dust or smoke streamers (left) enhance the probability of a thermal. Diverging streamers reduce the likelihood of a thermal.

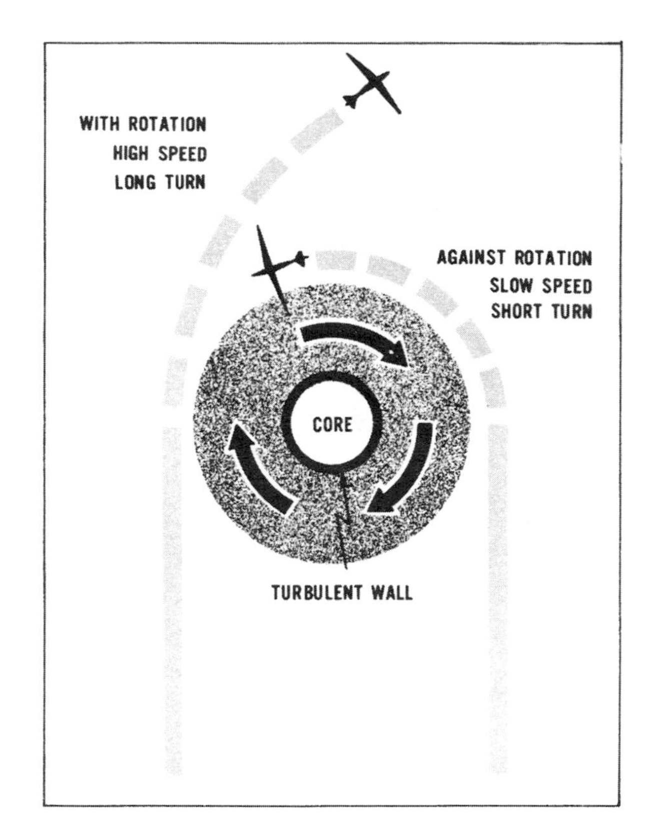

Figure 8-5. Horizontal cross section of a dust devil rotating clockwise. If the aircraft approaches the dust devil with the direction of rotation as on the left, increasing tailwind reduces airspeed and may result in loss of altitude or even a stall. When the pilot regains equilibrium, his circling speed is the sum of his airspeed and the tangential speed of the vortex; his radius of turn may be too great to remain in the thermal. When approaching against the rotation, the aircraft gains airspeed; circling speed is slowed as the tangential speed of the vortex is subtracted from airspeed. The pilot has much more freedom and latitude for maneuvering. At the center is a core providing little or no lift. Immediately surrounding the core is a turbulent wall.

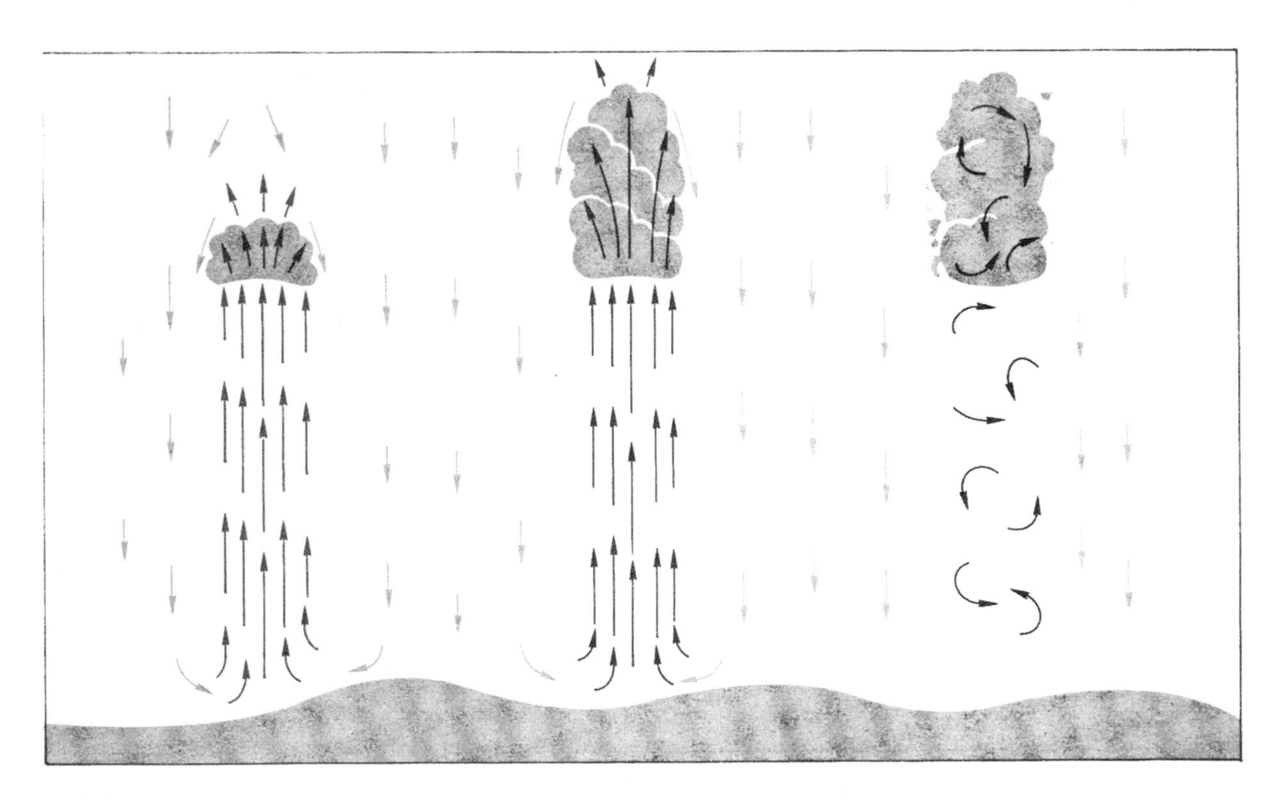

Figure 8-6. Cumulus clouds grow only with active thermals, as shown left and center. On the right, the thermal has subsided and the cloud is decaying. Look for a thermal only under a cumulus with a concave base and sharp upper outlines. A cumulus with a convex base or fragmentary outline is dissipating; the thermal under it has subsided. Most often, a cloud just beginning to grow as on the left is the better choice because of its longer life expectancy.

Figure 8-7. Photograph of a dying cumulus. Note the indistinct edges and cloud fragments. The base appears to be convex. One would expect little or no lift beneath this cloud. In contrast, note the top of the cumulus in the lower left corner. Edges are more defined, and a thermal is more likely under this cloud.

Figure 8-8. Altocumulus castellanus clouds. Most often, they develop in an unstable layer aloft, and thermals do not extend from the ground upward to these clouds. Convection with these clouds may be used for lift if the pilot is able to attain altitude to the base of the unstable layer. Smoke lying near the ground indicates stability in the lower levels.

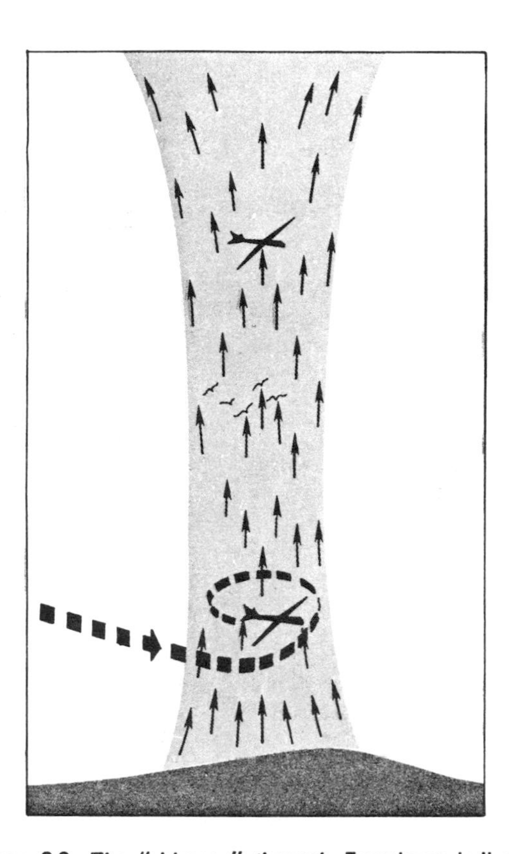

Figure 8-9. The "chimney" thermal. Experience indicates that the chimney thermal, which is continuous from the ground upward, is the most prevalent type. A sailplane can find lift in such a thermal beneath soaring birds or other soaring aircraft.

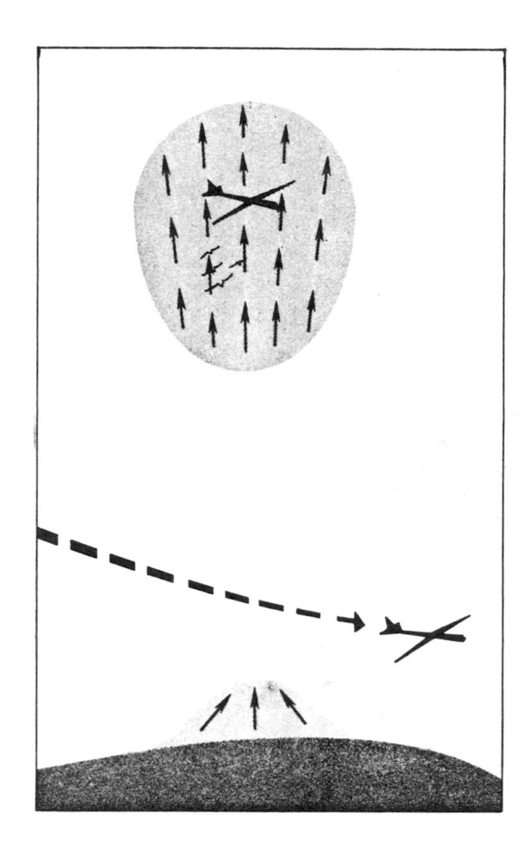

Figure 8-10. Thermals may be intermittent "bubbles". Frequency of the bubbles ranges from a few minutes to an hour or more. A soaring pilot will be disappointed when he seeks lift beneath birds or sailplanes soaring in this type thermal.

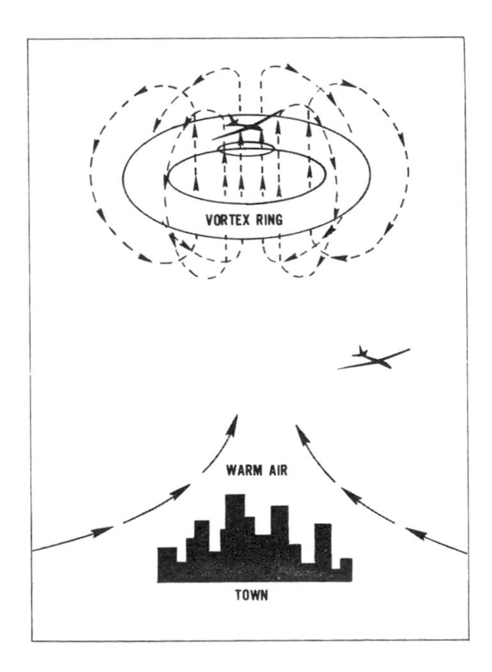

Figure 8-11. A bubble thermal sometimes develops a vortex ring, resembling a smoke ring blown straight upward. The center of the ring provides excellent lift. A pilot finds only weak lift or possibly sink in the fringes of the ring.

A sailplane must have auxiliary power to become airborne such as a winch, a ground tow, or a tow by a powered aircraft. Once the sailcraft is airborne and the tow cable released, performance of the craft depends on the weather and the skill of the pilot. Forward thrust comes from gliding downward relative to the air the same as thrust is developed in a power-off glide by a conventional aircraft. Therefore, to gain or maintain altitude, the soaring pilot must rely on upward motion of the air.

To a sailplane pilot, "lift" means the rate of climb he can achieve in an up-current, while "sink" denotes his rate of descent in a downdraft or in neutral air. "Zero sink" means that upward currents are just strong enough

Continued on page 8-13

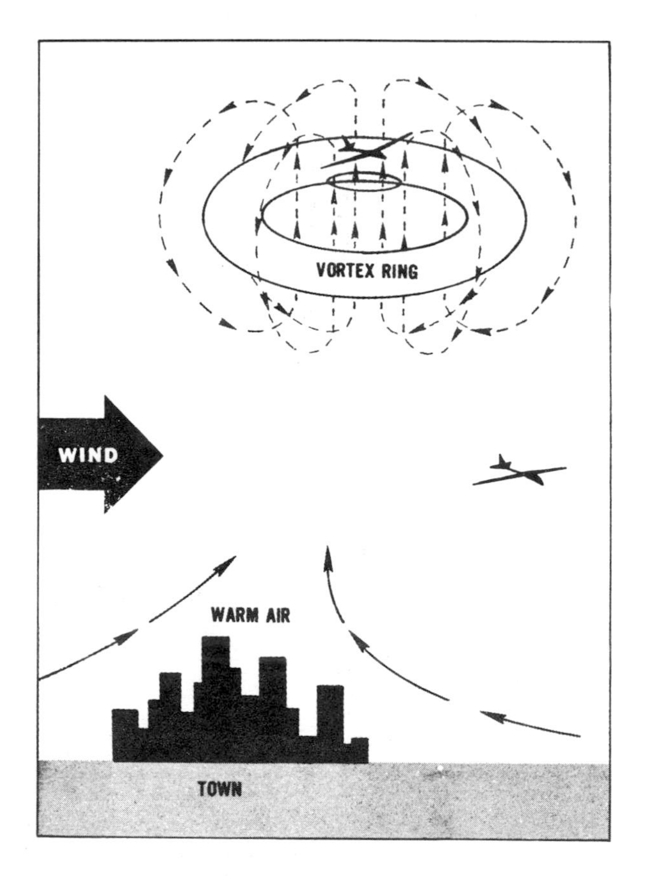

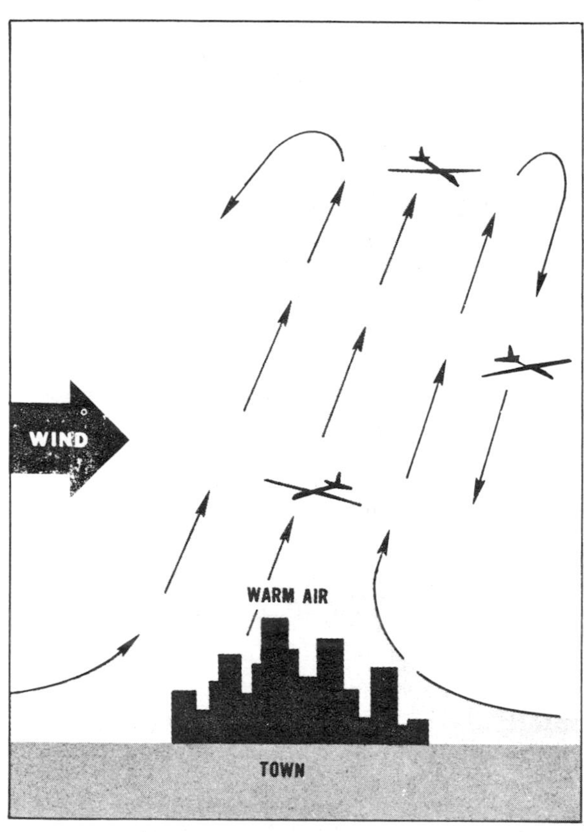

Figure 8-12. Wind causes thermals to lean. A pilot seeking lift beneath soaring birds, other aircraft, or cumulus clouds should enter the thermal upwind from the higher level visual cue.

Figure 8-13. Photograph of cumulus clouds severed by wind shear. Locating thermals and remaining in them under these clouds would be difficult.

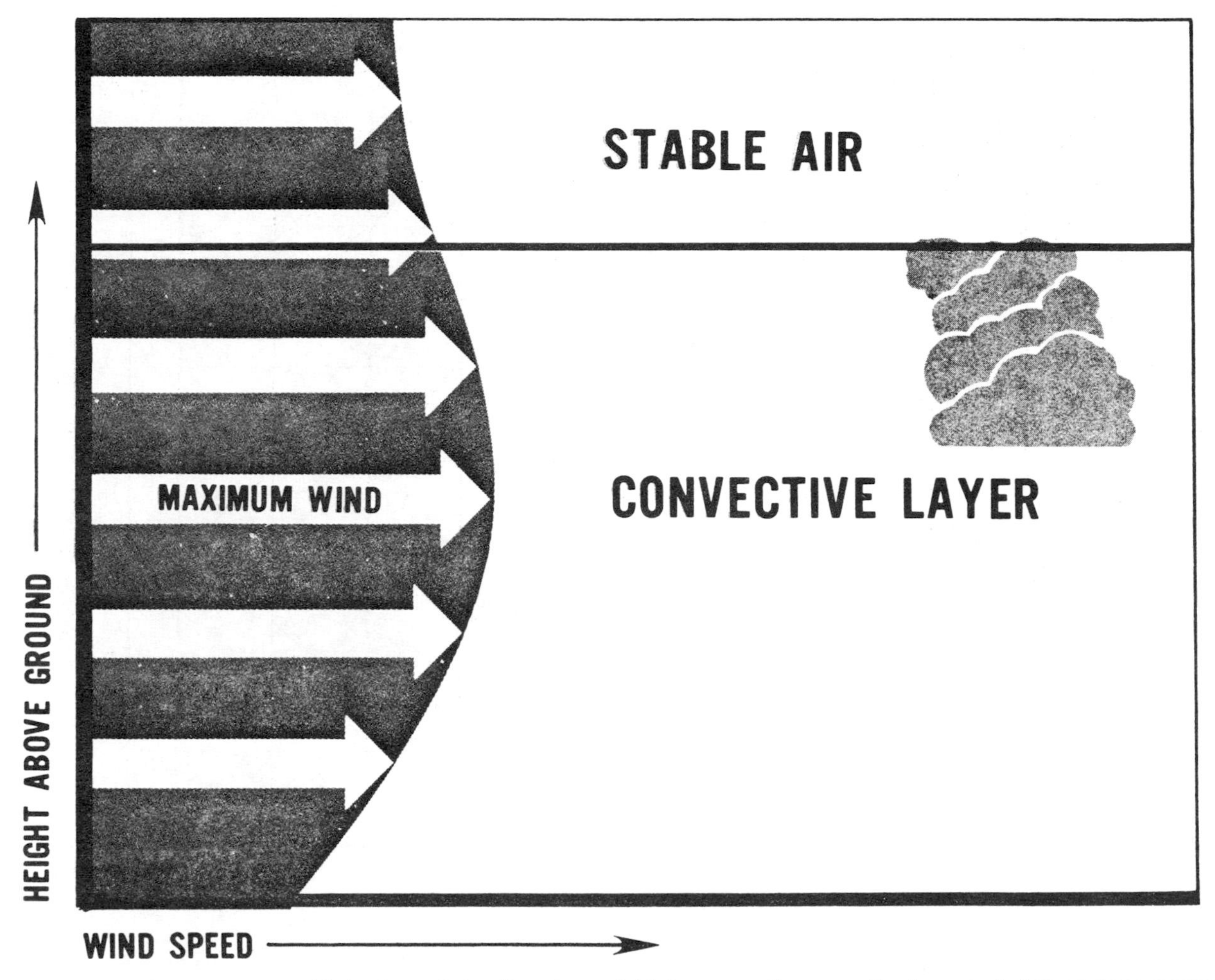

Figure 8-14. Conditions favorable for thermal streeting. A very stable layer caps the convective layer, and wind reaches a maximum within the convective layer. If cumulus clouds mark thermal streets, the top of the convective layer is about the height of the cloud tops. Thermal streets provide the most favorable type of thermals for cross-country soaring.

to enable him to hold altitude but not to climb. Sailplanes are highly efficient machines; a sink rate of a mere 2 feet per second provides an airspeed of about 40 knots, and a sink rate of 6 feet per second gives an airspeed of about 70 knots. Some two-place training craft have somewhat higher sink rates.

In lift, a sailplane pilot usually flies 35 to 40 knots with a sink rate of about 2 feet per second. Therefore, if he is to remain airborne, he must have an upward air current of at least 2 feet per second. There is no point in trying to soar until weather conditions favor vertical speeds greater than the minimum sink rate of the aircraft. These vertical currents develop from several sources, and these sources categorize soaring into five classes: (1) Thermal Soaring, (2) Frontal Soaring, (3) Sea Breeze Soaring, (4) Ridge or Hill Soaring, and (5) Mountain Wave Soaring.

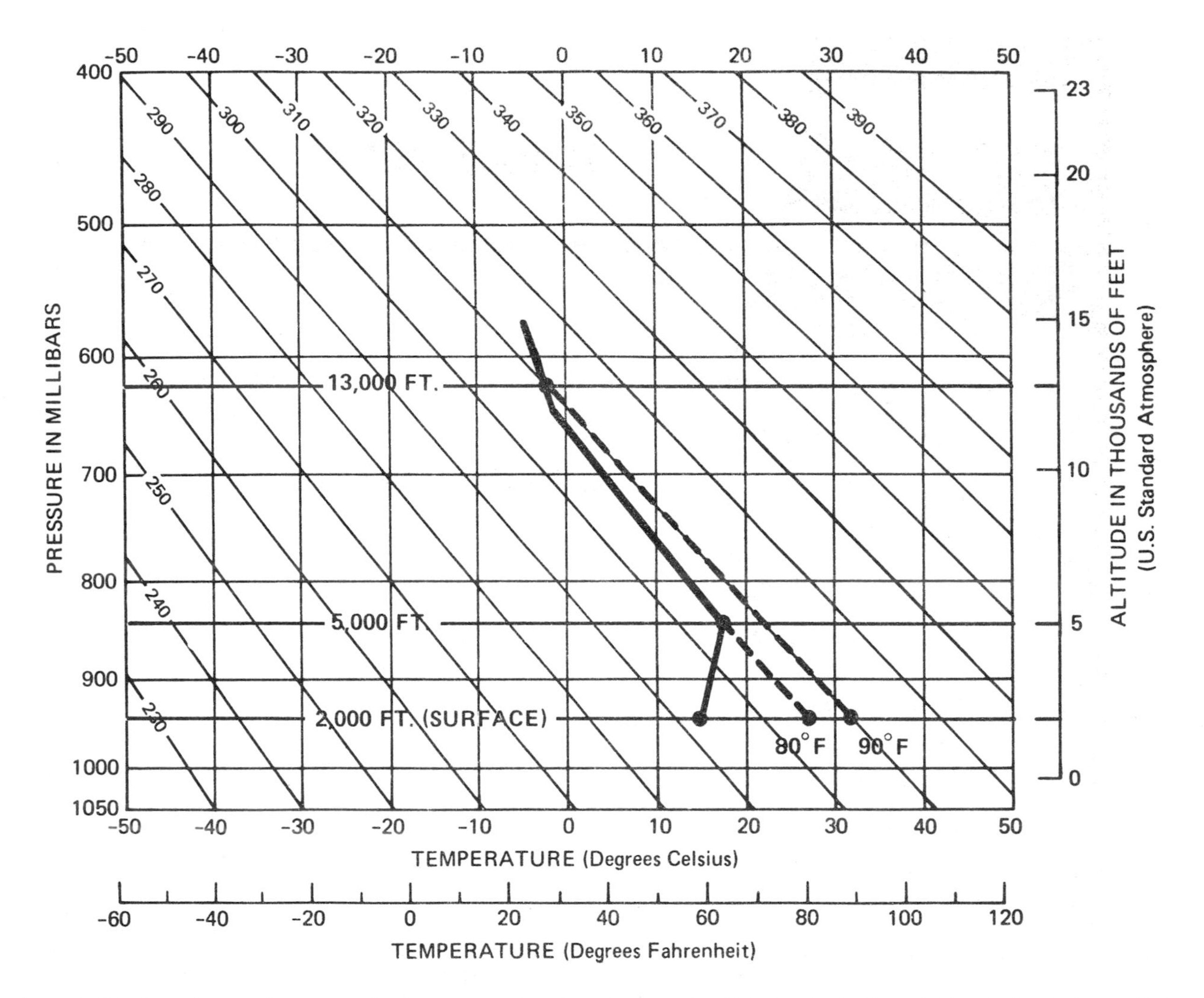

Figure 8-15. An early morning upper air observation plotted on the pseudo-adiabatic chart. The solid black line is the vertical temperature profile or existing lapse rate from the surface to about 15,000 feet ASL. Some altitude lines are projected across the chart from the altitude scale on the right to aid in interpretation. If thermals are to develop, the lapse rate must become equal to or greater than the dry adiabatic rate of cooling — that is, the line representing the lapse rate must slope parallel to or slope more than the dry adiabats. Since it does not, the air in the early morning was stable. By the time the surface temperature reached 80° F, convection occurred to 5,000 feet; the existing lapse rate then was parallel to the dry adiabat following the dashed line from the surface to 5,000 feet; the air was unstable in the lower levels. By the time the temperature reached the afternoon maximum of 90° F, the air was unstable to 13,000 feet; the existing lapse rate in the heat of the day was dry adiabatic and the air unstable to 13,000 feet ASL. This is the maximum height you could expect thermals on this particular day.

1. *Thermal Soaring*

What is a thermal? A thermal is simply the updraft in a small-scale convective current. Clear skies and surface winds of 5 knots or less would be most conducive to the formation of thermals.

All pilots scan the weather pattern for convective activity. Remember that turbulence is proportional to the speed at which the aircraft penetrates adjacent updrafts and downdrafts. The fast moving powered aircraft experiences "pounding" and tries to avoid convective turbulence. The slower moving soaring pilot enjoys a gradual change from thermals to areas of sink. He chases after local convective cells using the thermals for lift.

A soaring aircraft is always sinking relative to the air. To maintain or gain altitude, therefore, the soaring pilot must spend sufficient time in thermals to overcome the normal sink of the aircraft as well as to regain altitude lost in downdrafts. He usually circles at a slow airspeed in a thermal and then darts on a beeline to the next thermal as shown in figure 8-3.

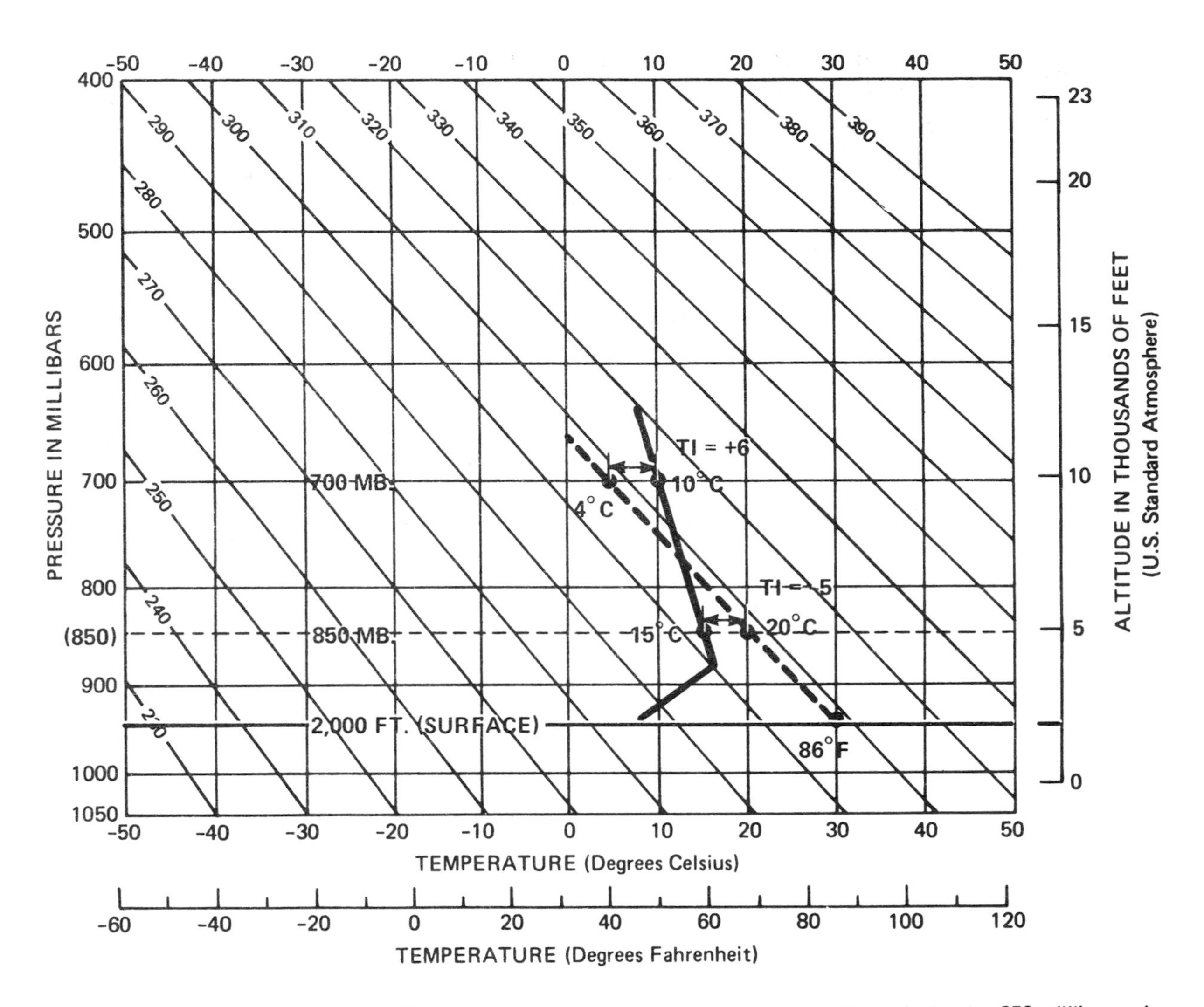

Figure 8-16. Computing the thermal index (TI). From an early morning upper air observation, obtain the 850-millibar and 700-millibar temperatures — 15°C and 10°C respectively, in this example. Obtain a forecast maximum temperature, 86°F, and plot it at the surface elevation. Draw a dry adiabat, the dashed line, upward along the 700-millibar level. This dry adiabat is the temperature profile of a rising column of air. To find the TI at any level, subtract the temperature of the rising column at that level from the temperature of the original sounding at the same level. The TI at 850 millibars is –5 (15 – 20 = –5). At 700 millibars, the TI is +6 (10 – 4 = +6).

Low-level heating is prerequisite to thermals; and this heating is mostly from the sun, although it may be augmented by man-made heat sources such as chimneys, factories, and cities. Cool air must sink to force the warm air upward in thermals. Therefore, in small-scale convection thermals and downdrafts coexist side by side. The net upward displacement of air must equal the net downward displacement. Fast rising thermals generally cover a small percentage of a convective area while slower downdrafts predominate over the remaining greater portion as diagrammed in Figure 8-3.

Since thermals depend on solar heating, thermal soaring is restricted virtually to daylight hours with considerable sunshine. Air tends to become stable at night due to low-level cooling by terrestrial radiation, often resulting in an inversion at or near the surface. Stable air suppresses convection, and thermals do not form until the inversion "burns off" or lifts sufficiently to allow soaring beneath the inversion. The earliest that soaring may begin varies from early forenoon to early afternoon, the time depending on the strength of the inversion and the amount of solar heating. Paramount to a pilot's soaring achievement is his skill in diagnosing and locating thermals.

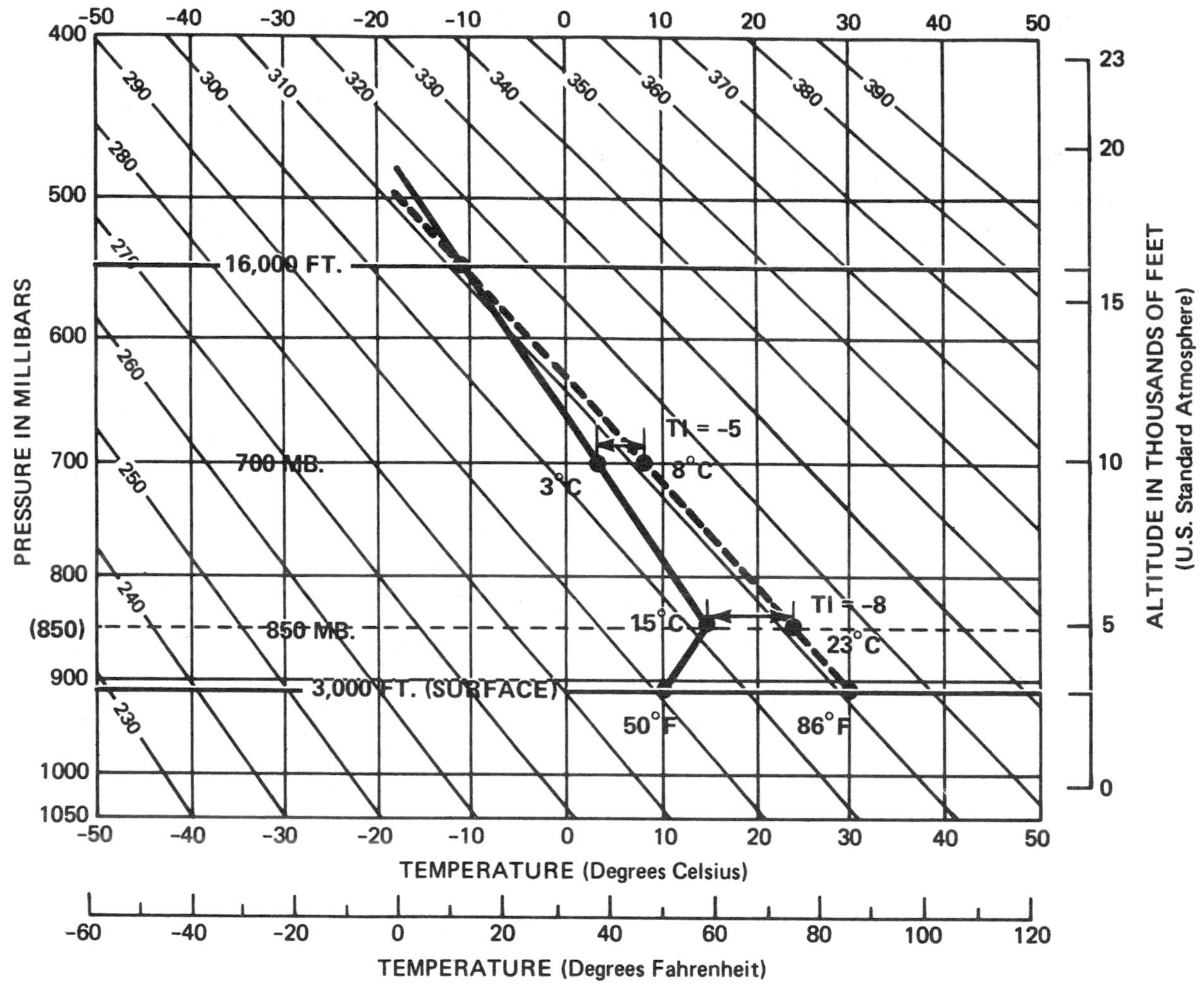

Figure 8-17. Another example of computing TIs and maximum height of thermals. See discussion in caption of figure 8-16. By the time of maximum heating, excellent lift should be available in lower levels and moderate lift above 10,000 feet. Although thermals should continue to 16,000 feet, you could expect weak lift above 12,000 or 13,000 feet because of the small difference between temperatures in the thermal and in the surrounding air.

2. *Frontal Soaring*

Warm air forced upward over cold air above a frontal surface can provide lift for soaring. However, good frontal lift is transitory, and it accounts for a very small portion of powerless flight. Seldom will you find a front parallel to your desired cross-country route, and seldom will it stay in position long enough to complete a flight. A slowly moving front provides only weak lift. A fast moving front often plagues the soaring pilot with cloudiness and turbulence.

A front can on occasion provide excellent lift for a short period. You may on a cross-country be riding wave or ridge lift and need to move over a flat area to take advantage of thermals. A front may offer lift during your transition. Frequently, the air behind a cold front provides excellent soaring for several days, for a rising barometer would indicate both decreasing clouds and winds.

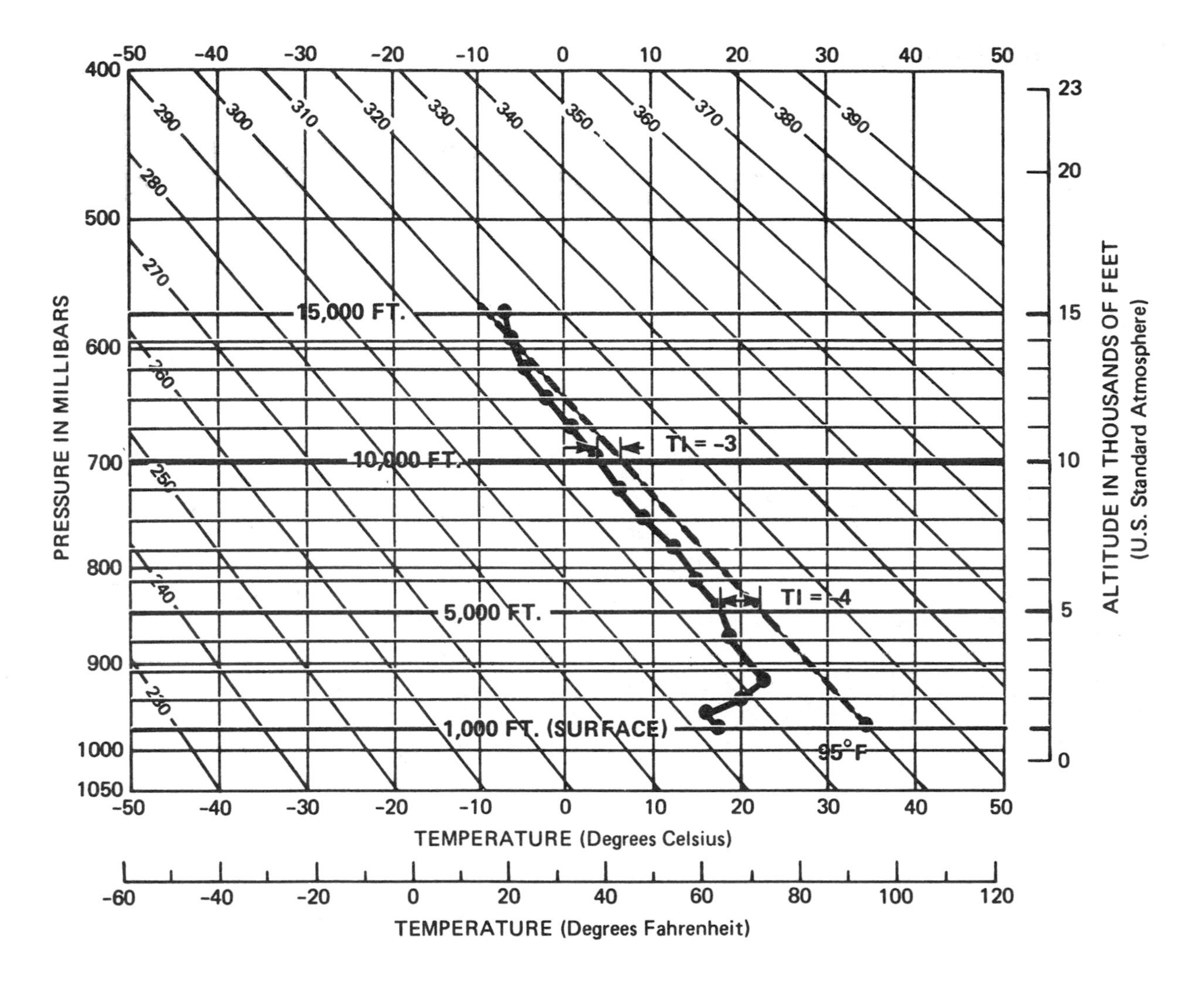

Figure 8-18. An upper air observation made from an aircraft called an airplane observation or APOB. Maximum height of thermals and TIs are computed the same as in preceding examples except that TIs are for indicated altitudes instead of pressure levels. The APOB may be used in lieu of or as a supplement to the forecast.

3. *Sea Breeze Soaring*

In many coastal areas during the warm seasons, a pleasant breeze from the sea occurs almost daily. Caused by the heating of land on warm, sunny days, the sea breeze usually begins during early forenoon, reaches a maximum during the afternoon, and subsides around dusk after the land has cooled. The leading edge of the cool sea breeze forces warmer air inland to rise as shown in Figure 8-19. Rising air from over land returns seaward at higher altitude to complete the convective cell.

A sailplane pilot operating in or near coastal areas often can find lift generated by this convective circulation. The transition zone between the cool, moist air from the sea and the warm, drier air inland is often narrow and is a shallow, ephemeral kind of pseudo-cold front. A difference in visibility between land and sea is a visual clue to the leading edge of the sea breeze.

2. *Sea Breeze Soaring*

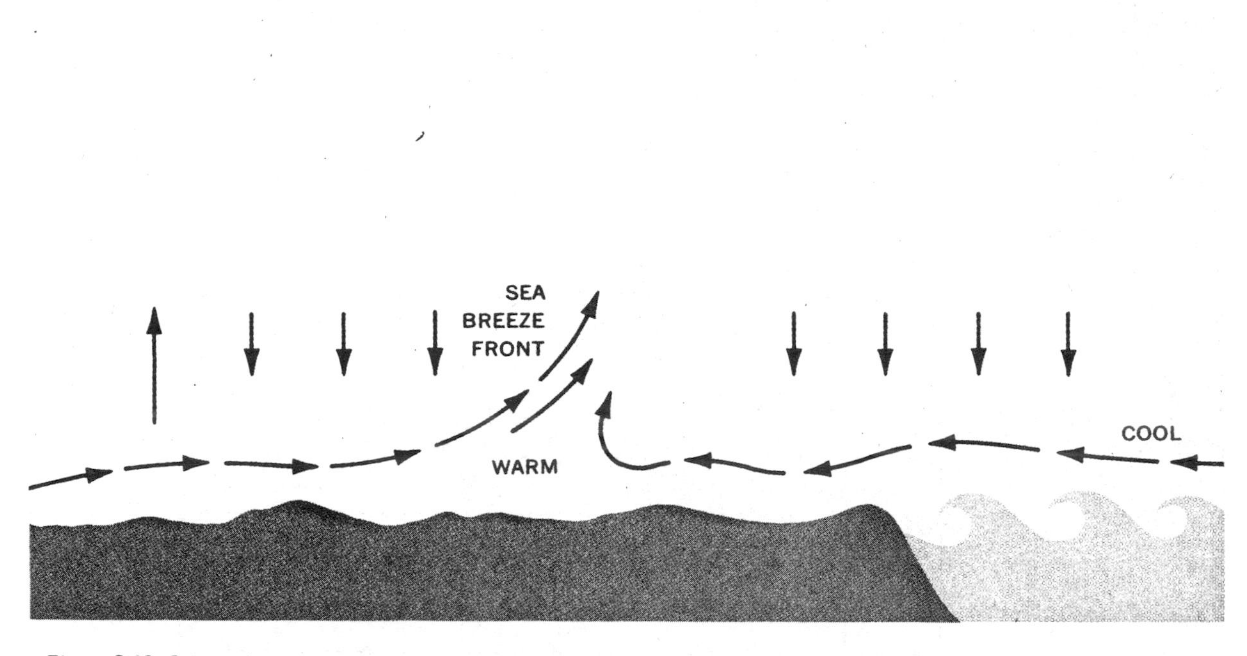

Figure 8-19. Schematic cross section through a sea breeze front. If the air inland is moist, cumulus often marks the front.

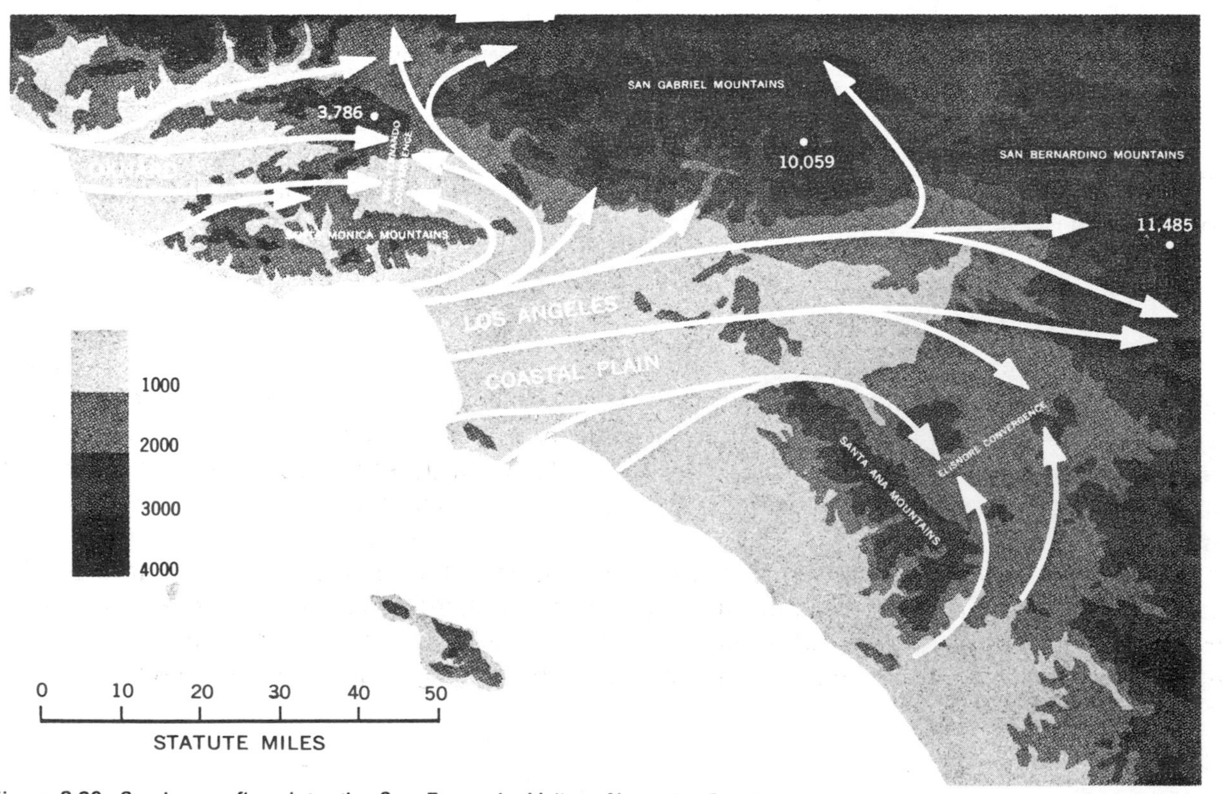

Figure 8-20. Sea breeze flow into the San Fernando Valley. Note the San Fernando convergence zone, upper left, and the Elsinore convergence zone, lower right.

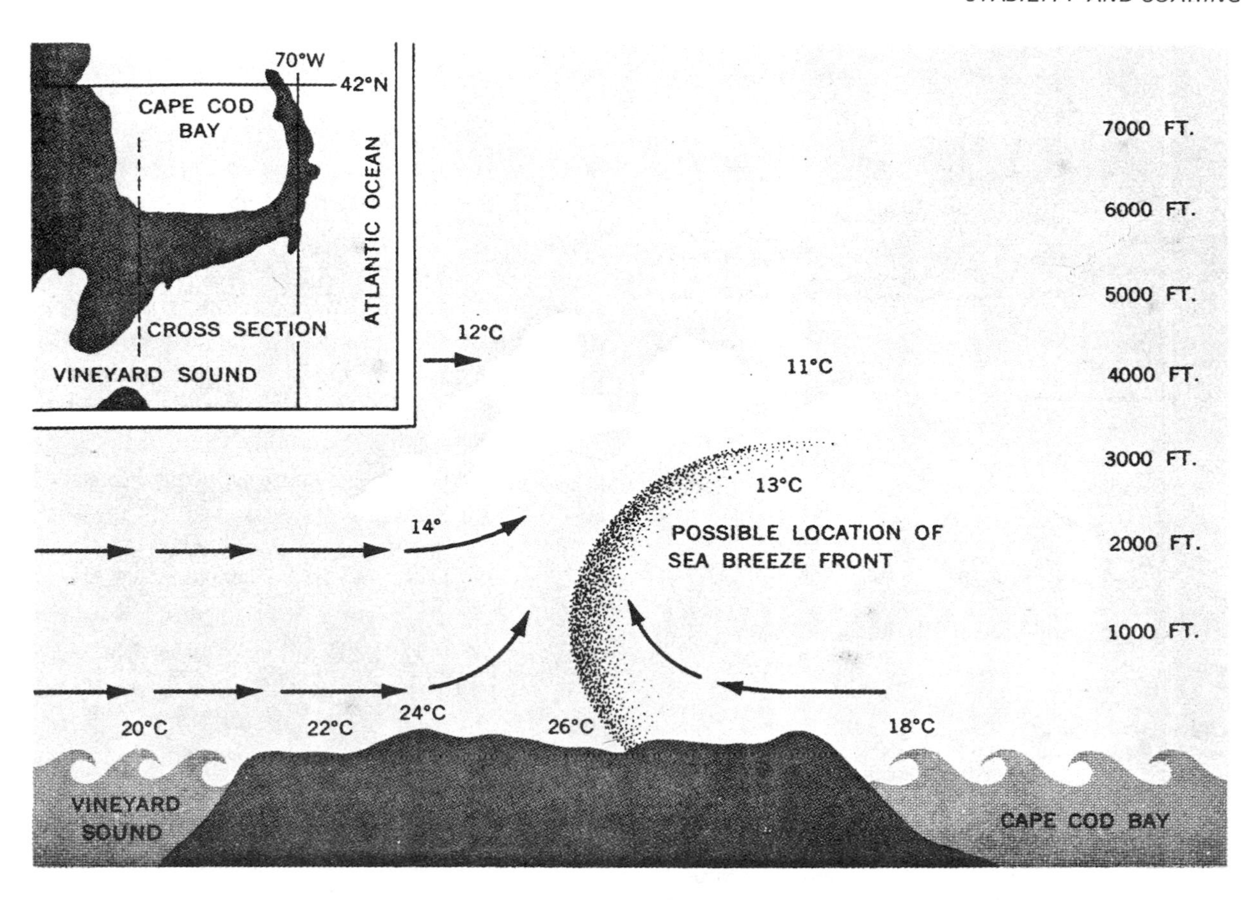

Figure 8-21. Sea breeze convergence zone, Cape Cod, Massachusetts. Sea breezes from opposite coasts converge over the cape.

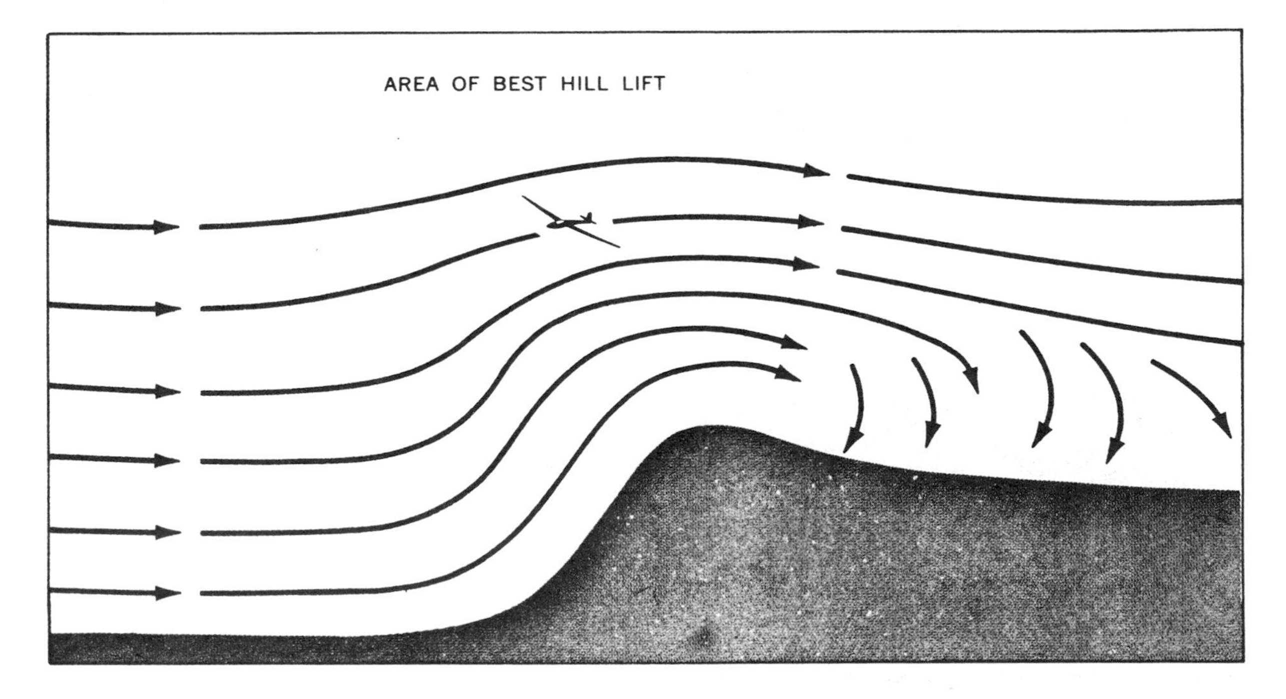

Figure 8-22. Schematic cross section of airflow over a ridge. Note the area of best lift. Downdrafts predominate leeward in the "wind shadow."

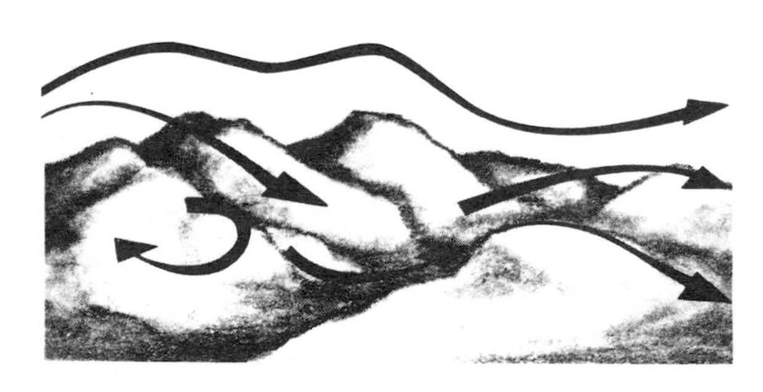

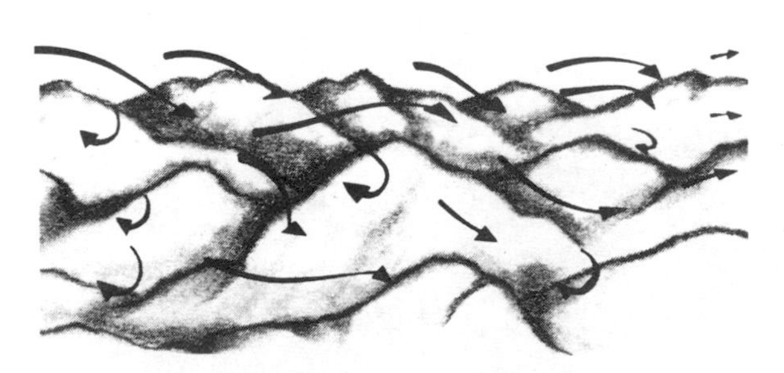

Figure 8-23. Windflow over various types of terrain. The many deviations from these patterns depend on wind speed, slope profile, and terrain roughness.

4. Ridge or Hill Soaring

Wind blowing toward hills or ridges flows upward, over, and around the abrupt rises in terrain. The upward moving air creates lift which is sometimes excellent for soaring. Figure 8-22 is a schematic showing area of best lift. Ridge or hill soaring offers great sport to the sailplane pilot who accepts the challenge and can wait for proper wind and stability combinations.

To create lift over hills or ridges, wind direction should be within about 30 to 40 degrees normal to the ridge line. A sustained speed of 15 knots or more usually generates enough lift to support a sailplane. Height of the lift usually is two or three times the height of the rise from the valley floor to the ridge crest. Strong winds tend to increase turbulence and low-level eddies without an appreciable increase in the height of the lift.

Stability affects the continuity and extent of lift over hills or ridges. Stable air allows relatively streamlined upslope flow. A pilot experiences little or no turbulence in the steady, uniform area of best lift shown in Figure 8-22. Since stable air tends to return to its original level, air spilling over the crest and downslope is churned into a snarl of leeside eddies, also shown in Figure 8-22. Thus, stable air favors smooth lift but troublesome leeside low-altitude turbulence.

When the airstream is moist and unstable, upslope lift may release the instability generating strong convective currents and cumulus clouds over windward slopes and hill crests. The initially laminar flow is broken up into convective cells. While the updrafts produce good lift, strong downdrafts may compromise low altitude flight over rough terrain. As with thermals, the lift will be transitory rather than smooth and uniform.

4. Mountain Wave Soaring

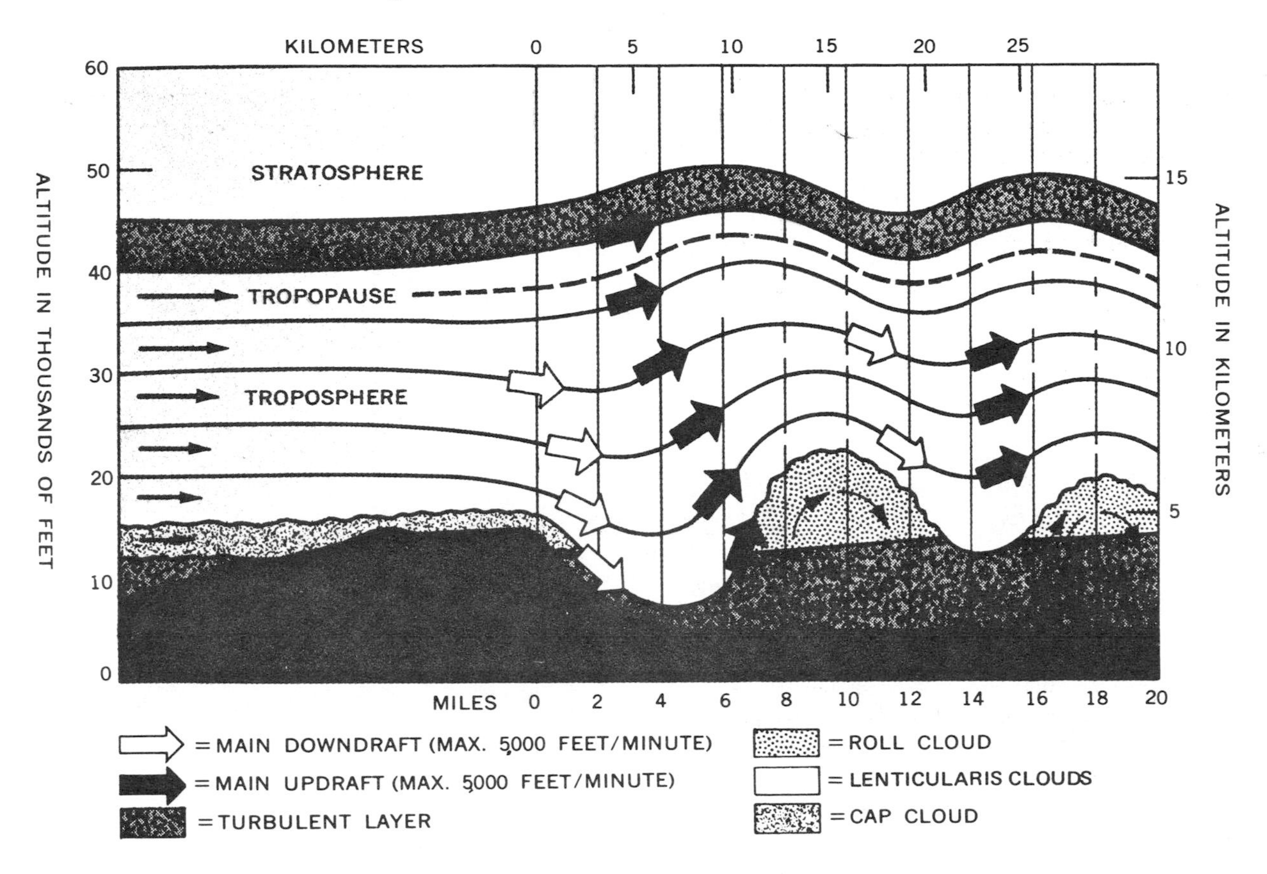

Figure 8-24. Schematic cross section of a mountain wave. Best lift is upwind from each wave crest for about one-third the distance to the preceding wave crest.

5. Mountain Wave Soaring

Soaring flights to above 35,000 feet have frequently been made in mountain waves. Once a soaring pilot has reached the rising air of a mountain wave, he has every prospect of maintaining flight for several hours. While mountain wave soaring is related to ridge or hill soaring, the lift in a mountain wave is on a larger scale and is less transitory than lift over smaller rises in terrain. Figure 8-24 is a cross section of a typical mountain wave. The greatest potential danger from vertical and rotor-type currents will usually be encountered on the leeward side when flying into the wind.

When strong winds blow across a mountain range, large "standing" waves occur downwind from the mountains and upward to the tropopause. The waves may develop singly; but more often, they occur as a series of waves downstream from the mountains. While the waves remain about stationary, strong winds are blowing through them. Air dips sharply immediately to the lee of a ridge, then rises and falls in a wave motion downstream.

A strong mountain wave requires:

a. Marked stability in the airstream disturbed by the mountains. Rapidly building cumulus over the mountains visually marks the air unstable; convection, evidenced by the cumulus, tends to deter wave formation.

b. Wind speed at the level of the summit should exceed a minimum which varies from 15 to 25 knots depending on the height of the range. Upper winds should increase or at least remain constant with height up to the tropopause.

c. Wind direction should be within 30 degrees normal to the range. Lift diminishes as winds more nearly parallel the range.

STABILITY
Review Questions

PRI

1. Suppose conditionally unstable air with high moisture content and very warm surface temperatures are forecast. From these conditions you should expect
   a. Continuous heavy precipitation.
   b. Fog and drizzle.
   c. Strong updrafts and cumuliform clouds.
   d. Smooth air and excellent weather for flying.

2. The weather condition normally associated with unstable air is
   a. Fair to poor visibility.
   b. Good visibility, except in blowing sand or snow.
   c. Stratiform clouds.
   d. Continuous precipitation.

INS

3. What are the characteristics of unstable air?

Visibility	Type of Precipitation	Type of Clouds
a. Poor	Intermittent	Cumulus
b. Poor	Steady	Stratus
c. Good	Showers	Cumulus
d. Good	Steady	Stratus

4. What are characteristics of unstable air?
   a. Turbulence and good surface visibility.
   b. Turbulence and poor surface visibility.
   c. Nimbostratus clouds and good surface visibility.
   d. Nimbostratus clouds and poor surface visibility.

5. What type of clouds will be formed if very stable moist air is forced upslope?
   a. Vertical clouds with increasing height.
   b. Layer-like clouds with little vertical development.
   c. First layer clouds and then vertical clouds.
   d. First vertical clouds and then layer clouds.

6. What are some characteristics of unstable air?
   a. Poor visibility, intermittent rain, and clear icing.
   b. Good visibility, intermittent rain, and rime icing.
   c. Poor visibility, showers, and clear icing.
   d. Good visibility, showers, and cumuliform clouds.

7. What is a characteristic of stable air?
   a. Stratiform clouds.
   b. Unlimited visibility.
   c. Fair weather cumulus clouds.
   d. Temperature decreases rapidly with altitude.

8. Moist, stable air flowing upslope can be expected to
   a. Dissipate cloudiness.
   b. Produce low stratus and fog.
   c. Develop convective turbulence.
   d. Cause showers and thunderstorms.

9. A temperature inversion forms
   a. Only in summer.
   b. Only in winter.
   c. An unstable layer of air.
   d. A stable layer of air.

CFI

10. Whether clouds will be predominantly stratiform or cumuliform is determined by the
    a. Degree of stability of the air being lifted.
    b. Percent of moisture content of the air.
    c. Source of lift.
    d. Temperature of the air being lifted.

11. The conditions necessary for the formation of stratus clouds are a cooling action and
    a. Stable, moist air.
    b. Unstable, moist air.
    c. Unstable air containing excess condensation nuclei.
    d. Stable or unstable air.

12. Suppose an airport has an elevation of 2,000 feet. Assuming standard lapse rates, if the temperature at this airport is 70°F and the surface dewpoint temperature is 52°F, the base of the clouds formed by a lifting process would be located at approximately
    a. 8,000 feet MSL.
    b. 6,000 feet MSL.
    c. 4,000 feet MSL.
    d. 3,000 feet MSL.

13. If clouds form as a result of very stable moist air being forced to ascend a mountain slope, the clouds will be
    a. Cumulus type with considerable vertical development and turbulence.
    b. Stratus type with little vertical development and little or no turbulence.
    c. Cumulonimbus with considerable vertical development and heavy rains.
    d. Cirrus type with no vertical development or turbulence.

ATP

14. Stability of the atmosphere can be determined by the measurement of the
    a. Actual temperature lapse rate.
    b. Atmospheric pressure at various levels.
    c. Wind velocity and atmospheric pressures.
    d. Surface temperature.

15. Which atmospheric process tends to increase the stability of an air mass?
    a. Sublimation from ice or snow to the lower layers of an air mass.
    b. Orographic lifting of an air mass.
    c. Subsidence of a relatively thick layer of air.
    d. Advection of a cold air mass over a warmer surface.

16. Steady precipitation, in contrast to showery, preceding a front is an indication of
    a. Cumuliform clouds with moderate turbulence.
    b. Stratiform clouds with moderate turbulence.
    c. Cumuliform clouds with little or no turbulence.
    d. Stratiform clouds with little or no turublence.

17. Moisture and vertical movement have what effect on the stability of an air mass?
    a. Sinking of an air mass and addition of water vapor to the lower layers tend to decrease its stability.
    b. Lifting of an air mass and removal of water vapor from the lower layers tend to decrease its stability.
    c. Sinking of an air mass and removal of water vapor from the lower layers tend to increase its stability.
    d. Lifting of an air mass and addition of water vapor to the lower layers tend to increase its stability.

CHAPTER IX
AIR MASSES

The reader should demonstrate an understanding of air masses to the extent that he can answer the following questions (as stated below, or as multiple-choice, etc.):

1. Define "air mass".
2. An air mass is a (high, low, low or high). Explain.
3. Define "source region".
4. How are air masses modified?

CHAPTER IX
AIR MASSES

A. Air Mass Defined

In the chapter on wind we discussed masses of air that inhabit various latitudes between the equator and the pole. We define an air mass as a horizontally large body of air, roughly 1,000 miles (1.6 km) in diameter at the surface, possessing fairly uniform characteristics horizontally (Figure 9-1); the characteristics with which we are concerned are temperature and humidity.

B. Air Masses as Highs

We can reason that an air mass is a high but not a low. (In meterology some lows are considered air masses; however, in aviation meterology we will use the definition as stated above for air masses, thereby eliminating the low and simplifying later discussions, particularly with regard to fronts.) A low is not considered to be an air mass because it fails at least one of our definition requirements. First, lows are usually less extensive in size, the hurricane being a notable exception. But even a large low fails the requirement of a uniform horizontal distribution of temperature and moisture. If one were to take a horizontal section out of a low, one would usually find quite a contrast in moisture content due to the drawing in of air from both land and water sources. Recall that air spirals inward and upward in a low, thus bringing in air with varying amounts of moisture, depending primarily on whether the flow of air is from over a water source or a land area. Temperatures will also vary considerably for the same reasons, with the added complication of latent heat being released as sensible heat when the air is lifted and cools to condensation adiabatically. For these reasons a low is not considered an air mass in this book.

Recall that we said that an air mass is a high; we did not say that every high is an air mass. A good example of a high which is not an air mass is the high which develops to give the sea breeze in the daytime. The flow of air, as you recall from the chapter on wind, is from a water source to land in the daytime due to differential heating effects over land and water. The flow of air from the relatively cool water to the warmer land reflects a high pressure circulation, the air spiralling downward and outward in a clockwise manner from the cool air. The differential temperature effects probably occur no higher than a couple of thousand feet. Thus, we have sort of a "bubble high" over the water, one which does not fit the size requirements for an air mass. Again, we repeat that an air mass is a high, not a low, but that not every high is an air mass. Most highs which invade mid-latitudes from the north or south are of sufficient size to qualify as an air mass. Since in large highs air spirals downward from higher altitudes where there are of course no land masses and oceans, temperature and moisture characteristics are quite uniform horizontally at any given altitude.

C. Source Region

We now consider the concept of source region, defining this to be the location or area over which an air mass takes on its identifying characteristics of temperature and moisture. In the chapter on temperature we made the statement that conduction was not a primary mechanism for the overall atmospheric distribution of heat. While that is still true, conduction is of considerable importance in the life history of an air mass relative to two surfaces (the source region and the surface over which the air mass travels), and also in the time it takes to

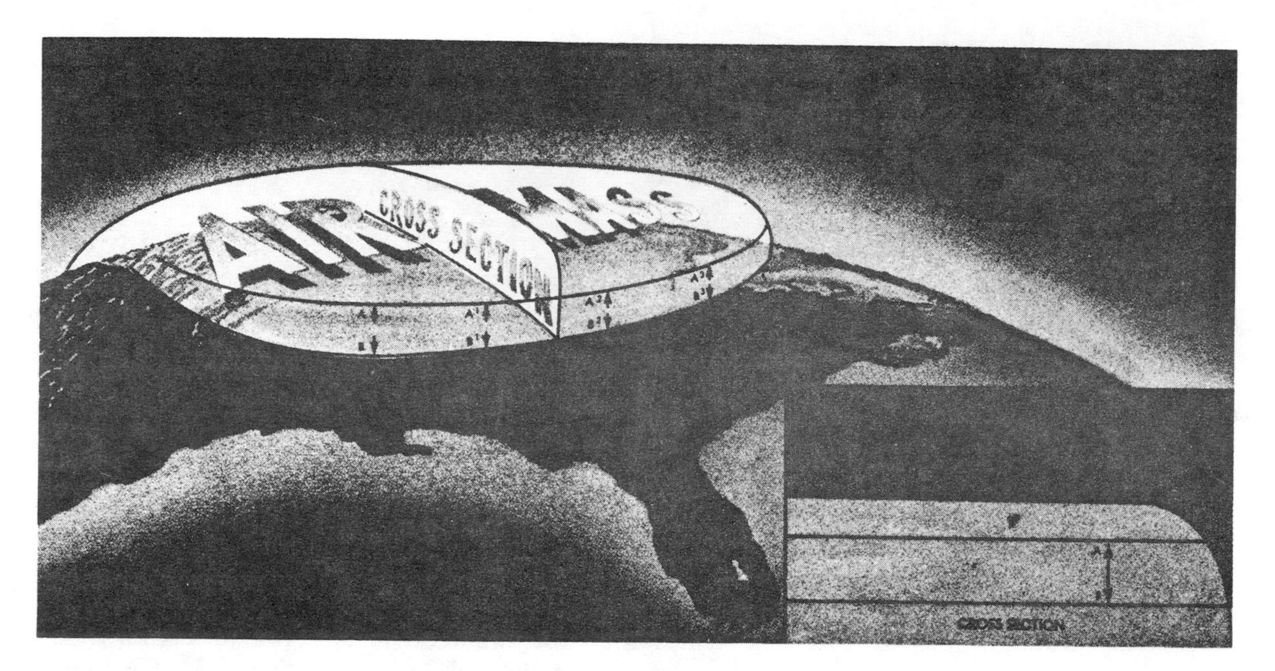

Figure 9-1. Horizontal uniformity of an air mass. (Properties of air at A[1], A[2], etc., are about the same as those at A; properties at B[1], B[2], etc., are about the same as those at B.)

accomplish that travel. In its source region, an air mass will take on the temperature and moisture characteristics of the surface over which it lies over a period of several days, even weeks.

An air mass that originates over a large continental Polar (cP) area, such as Alaska or Canada, will be very cold and dry. Over a large maritime Polar (mP) region, such as the ocean region well northeast of the state of Washington, the air will be less cold but very moist. We reason similarly for the more southerly latitudes. Air masses that move out of the Tropics are warm, and the moisture content is primarily dependent on whether the air mass moves from some land mass such as Mexico (continental Tropical – cT) or from some water source (maritime Tropical – mT).

D. Air Mass Modification

The surfaces over which air masses move are of significant modifying influence on them. For example, a maritime Polar air mass will have its moisture content changed considerably due to the vertical lift and subsequent precipitation over the Rocky Mountains. The temperature of the lower layers of the air mass will also be changed due to the katabatic effect. The degree of modification will also depend on the time that the air mass is in its source region and over new terrain, and on lapse rate changes due to subsidence aloft.

It should be obvious that air masses are modified by cooling or warming from below and by the addition or removal of moisture. Let's now see how these modifying influences affect the stability of air. Once we identify these effects we will know basically the kind of weather we can expect by reference to our chart on clouds, precipitation, turbulence, and visibility from Chapter VIII on stability. As a continental Polar air mass moves into mid-latitudes, it will typically be traveling over a warmer surface. From Chapter VIII we know that warming from below is a destabilizing influence near the ground where the warming is taking place. However, this just means that the lapse rate is steeper in the first two or three-thousand feet and therefore that the air is less stable, but not necessarily unstable. Whether this lower layer is stable or unstable depends on the actual lapse rate created by conduction, advection and convection.

Due to some surface areas being warmer than others, vertical currents are likely whether the lower layer is stable or unstable; of course, the vertical currents will be stronger if the lower layer is unstable. And how high these vertical currents rise depends on the degree of surface heating and the lapse rate in the upper levels. It may be that these vertical currents are capped off at 3,000 feet, or thermals may rise to 15,000 feet or more during the afternoon hours. And at night, due to surface cooling, no currents may be present at all. And just as there are daily differences, seasonal differences will also occur, primarily between summer and winter. Thus, there are many possibilities with regard to the smoothness of the air, but since we are assuming that the air is dry, skies will be clear to at most partly cloudy. Of course, if a continental Polar air mass passes over a water source, such as the Great Lakes, then snow is quite likely on the lee side of the lakes in winter. And a continental Polar air mass that moves into an area and stagnates (remains essentially stationary for days) can cause serious pollution and VFR flight problems, due to the subsidence of the air mass in conjunction with night-time cooling.

A maritime Polar air mass is more likely to be conditionally unstable as it moves out of its source region over a new underlying surface. The air is likely to be less cold than the continental type, but at the same time possess greater water content. The maritime Polar air mass is therefore likely to be conditionally unstable, that is, to have a lapse rate between the dry adiabatic and saturation-adiabatic rates. As you recall, this meant that as long as air is not saturated, less than 100% relative humidity, then the air would be stable; if air is lifted sufficiently to become satuated, then the air would be unstable. As this air mass moves in off the Pacific and is warmed from below, it is likely to be quite unstable and as a result yield all of the characteristics of unstable air.

Let's now consider Tropical air masses. A continental Tropical air mass will start out as warm and dry. As such an air mass moves northward, it will typically be cooled from below in winter, more so by an over-the-land trajectory than over water. Water content will increase if the air mass moves over water. The cooler air in the lower layers being more dense than the warmer air aloft will tend to retain its lower level position. Even if this air is lifted by higher terrain, that air will tend to return to its lower altitude due to the relatively warm air above. If the air is dry and terrain effects are minimal, the air will be essentially clear and smooth. If the air is quite moist as a result of passing over a water source, then stratiform clouds and fog are likely, and if the clouds thicken appreciably due to lifting over sufficiently higher terrain, continuous or steady precipitation will be likely.

The maritime Tropical air mass will be similar in its weather to the continental Tropical air mass that passes over a water source, as just described above. And if the maritime Tropical air mass moves over higher terrain, such as from the Gulf of Mexico up over the western slopes of the Appalachians, then as it loses much of its moisture content, it will have characteristics after descent on the eastern slopes much as the continental Tropical air mass that moves over land, that is, the air will be essentially clear and smooth. If the maritime Tropical air mass is warmed (in summer) rather than cooled from below, then the weather associated with unstable air is most probable.

We could have spent more time in a formal air mass classification system. However, the discussion should indicate that the concepts of air mass, stability, and resulting weather are highly related. Due to the large extent of air masses, as a pilot you should be aware of the nature of the types of air masses mentioned as they move into your area. And just as importantly, you need to be aware of the modifying influences of terrain in your area on these air masses. Even though you may be flying in the same air mass for several days, you are likely to find changes. We now turn our attention to the related and important chapter on fronts.

AIR MASSES
Review Questions

PRI

1. A moist, cold air mass that is being warmed from below is characterized, in part, by
   a. Smooth air.
   b. Fog and drizzle.
   c. Continuous heavy precipitation.
   d. Showers and thunderstorms.

2. Which of the following would decrease the stability of an air mass?
   a. Decrease in water vapor.
   b. Cooling from below.
   c. Warming from below.
   d. Sinking of the air mass.

CFI

3. Consider the following air mass characteristics:
   A. Smooth air (above the friction level) and fair to poor visibility.
   B. Turbulence up to about 10,000 feet and good visibility, except in areas of precipitation.
   C. Cumuliform clouds.
   D. Stratiform clouds.
   E. Stable lapse rate.
   F. Unstable lapse rate.

   A moist air mass which is colder than the surface over which it passes frequently has which of the above characteristics?
   a. A, D, and E.
   b. A, D, and F.
   c. B, C, and F.
   d. B, D, and E.

CHAPTER X
FRONTS

The reader should demonstrate an understanding of fronts to the extent that he can answer the following questions (as stated below, or as multiple-choice, etc.):

1. Define "polar front".
2. Define "stationary front"; "cold front"; "warm front"; "occluded front".
3. State the average slope of a cold front; a warm front.
4. What are the primary factors upon which the type of weather along a front is dependent? Compare warm and cold fronts with regard to these factors.
5. Define "embedded thunderstorm".
6. What is one of the particularly disturbing features of the stationary front?
7. List the sequence of events with the approach and passage of a cold front; a warm front.
8. In flying through a frontal surface, wind corrections should always be made to the ______.
9. Why should a new altimeter setting be obtained, when available, after crossing a front?
10. Distinguish between the cold and warm front occlusion.
11. A front never extends from or through a (high, low).
12. How does the wind flow in the upper levels affect frontal wave development?

CHAPTER X

FRONTS

This chapter will concentrate on the so-called battleground between air masses, called a front. Whenever two air masses meet, there is a zone of transition between them where there is usually a distinct difference in temperature and moisture characteristics. Such a contrast may result in extremely violent weather. In any given location this zone of transition is an influence on local weather for varying amounts of time as we shall soon see.

A. The Polar Front

As we mentioned in the chapter on wind, mid-latitudes offer the prime location for the meeting of air masses moving from polar and tropical source regions. The polar front was defined as the boundary along which these air masses meet. Thus, the polar front is a more or less continuous line extending around the globe, its exact position determined by which air masses, polar or tropical, penetrate the furthest in mid-latitudes (Figure 10-1).

B. Life History of Fronts

On a local scale, the polar front can be best sub-classed as a stationary, cold, warm, or occluded front. In fact, we may describe the life history of a front in terms of this sub-classification. Assume that a polar continental air mass moves from central Canada into Tennessee, where it meets a tropical maritime air mass from the Gulf of Mexico. Assume for the moment that these two giants muscle each other to a standoff. Such a standoff could occur due to winds in the upper levels, for if these winds were from the southwest, then the northward component of that wind flow would help the tropical maritime air mass hold its position against the polar continental air mass. Now as these two air masses maintain their relative north-south positions, we take a closer look at the narrow boundary between them. As long as the boundary stays as is, we have what is called a stationary front, a zone of transition between two air masses in which neither air mass is dominating. We now give an example of cold, warm, and occluded frontogenesis, that is, the genesis, or the beginning of such fronts.

Due to many meteorological factors, including differential heating effects and terrain features, individual tongues of cold air will move southward and tongues of warm air will move northward. This gives a series of minor waves on the original boundary. These "frontal waves" form primarily on slow-moving cold fronts and stationary fronts. We look at one of these waves. As cold air spills southward and warm air moves northward to the right, a counterclockwise circulation pattern is induced, that is, a shallow low develops at the top of the wave; many such lows may become very large systems with sufficient time. Our main concern right now, however, is the frontal concept, not the low. The leading edge of the tongue of cold air spilling southward on the back side of the low is called a cold front. Note that the cold front is a boundary between two air masses and is called cold because in its southward movement cold air is replacing warm air. The warm front is the boundary between air masses and termed warm because on the front side of the low in our example, warm air is replacing cold air.

Note that we are considering only the surface position of fronts at this point; we shall consider the vertical structure a little later. Now as the cold front continues to move south-eastward, it catches up with the warm front near the low. Wherever this occurs we have what is called an occluded front, of which there are two types as we shall see later. When occlusion occurs, cold air has pushed all the way underneath the warm air along the north to south

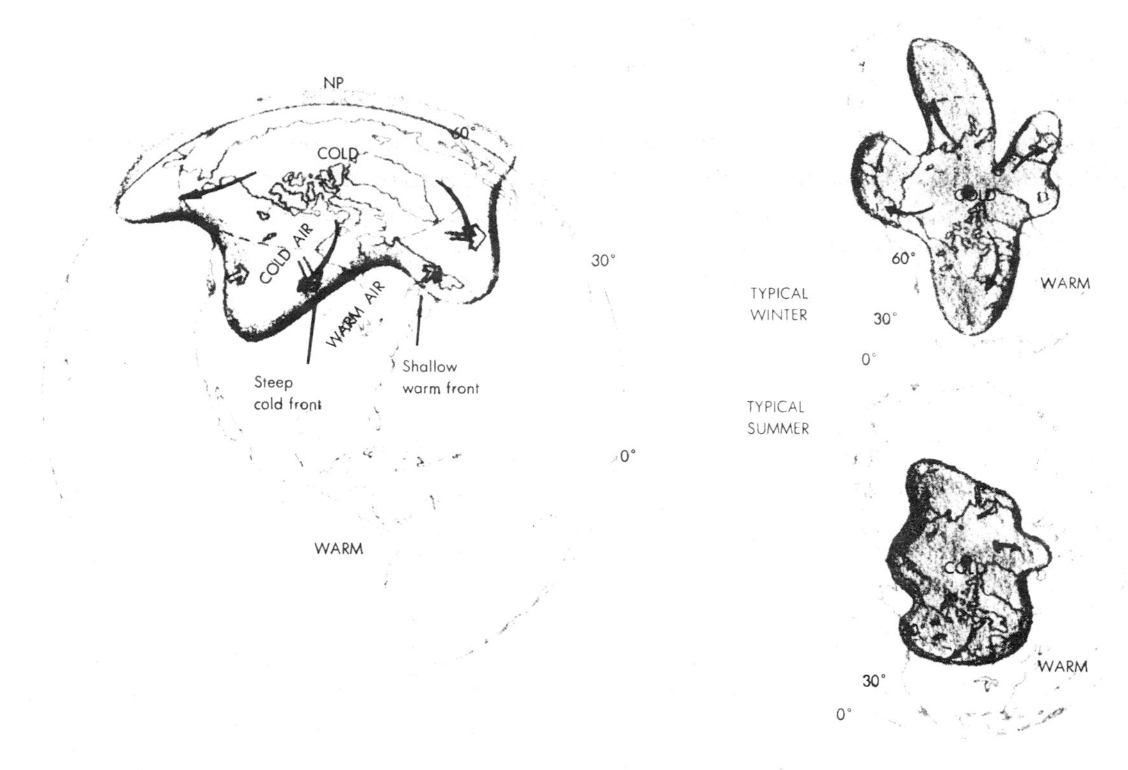

Figure 10-1. Wave-Like Shape of Polar Front.

occlusion line and has caught up to the cold air ahead of the warm front. The low is now cut off from the surface by the rejoining of low-level cold air, cutting off the circulation of the air and causing the low to more or less die. The north-south occlusion line disappears and the cycle has been completed, with a similar east-west boundary line as before between the air masses. We have just described frontolysis, the degeneration or dissipation of fronts.

Meteorologically, we have perhaps over-simplified, but we hope you have been able to visualize this wave-theory on frontal formation (Figures 10-2 and 10-3). And note particularly that while our fronts, cold, warm, and occluded, were always the boundary between air masses, they extended from a low. Even the stationary front lay in a zone of low pressure, since by definition the area between two highs must be a zone of lower pressure – recall that a high is a center of pressure surrounded on all sides by lower pressure. So even though a front lies in a zone of low pressure, the front itself is considered to be the boundary between air masses.

We described the polar front earlier as a more or less continuous line through mid-latitudes. We certainly do not mean to imply that there is one line across the U.S. running from, say, Los Angeles to North Carolina, marking the boundary between cold air masses to the north and warm air masses to the south. We have just seen that even if such a line existed, it would have a series of frontal waves developing along the boundary. When an exceptionally cold air mass moves southward, instead of being matched in force by an air mass to the south it may well push rapidly from northwest to southeast across the U.S. This will especially be the case when, in addition, winds in the upper levels push the air mass in that direction. In such a case, assuming this large air mass is in the midwest, there will be a huge break in the continuous line we termed the polar front. The northern extremes of the air mass may reach into central Canada, the southern extremes extending well into the Gulf of Mexico. Even the eastern and

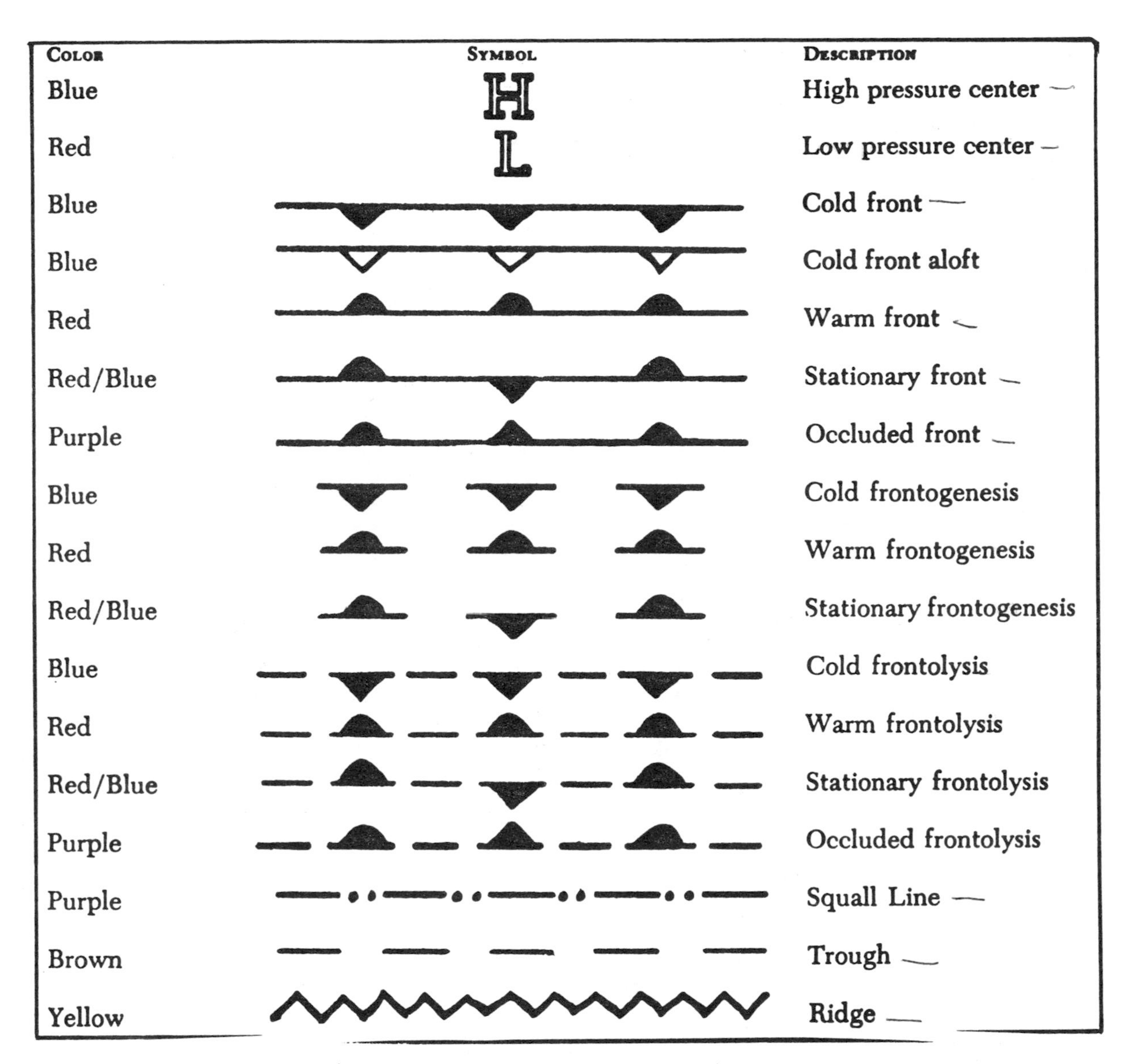

Color	Symbol	Description
Blue	H	High pressure center
Red	L	Low pressure center
Blue		Cold front
Blue		Cold front aloft
Red		Warm front
Red/Blue		Stationary front
Purple		Occluded front
Blue		Cold frontogenesis
Red		Warm frontogenesis
Red/Blue		Stationary frontogenesis
Blue		Cold frontolysis
Red		Warm frontolysis
Red/Blue		Stationary frontolysis
Purple		Occluded frontolysis
Purple		Squall Line
Brown		Trough
Yellow		Ridge

Figure 10-2. List of symbols on surface analyses.

western extremes may be so large as to provide a break across the central U.S. Such an air mass has "plugged the gap" between the polar and tropical source regions. Air masses of this size have occurred in the U.S. Thus, we should be able to now view the polar front more realistically, not just as an idealized solid continuous line extending from coast to coast (Figure 10-4).

C. Vertical Structure of Fronts

We now discuss the vertical structure of fronts. To do so, we first need to recall that a front is the boundary between air masses. Since two air masses not only meet at the ground but aloft as well, we actually have a frontal surface that extends vertically from the zone of contact at the surface. This surface will be curved toward the cold air. Picture a huge drop of molasses on a table. As the drop spreads outward, it exhibits a curved surface between itself and the

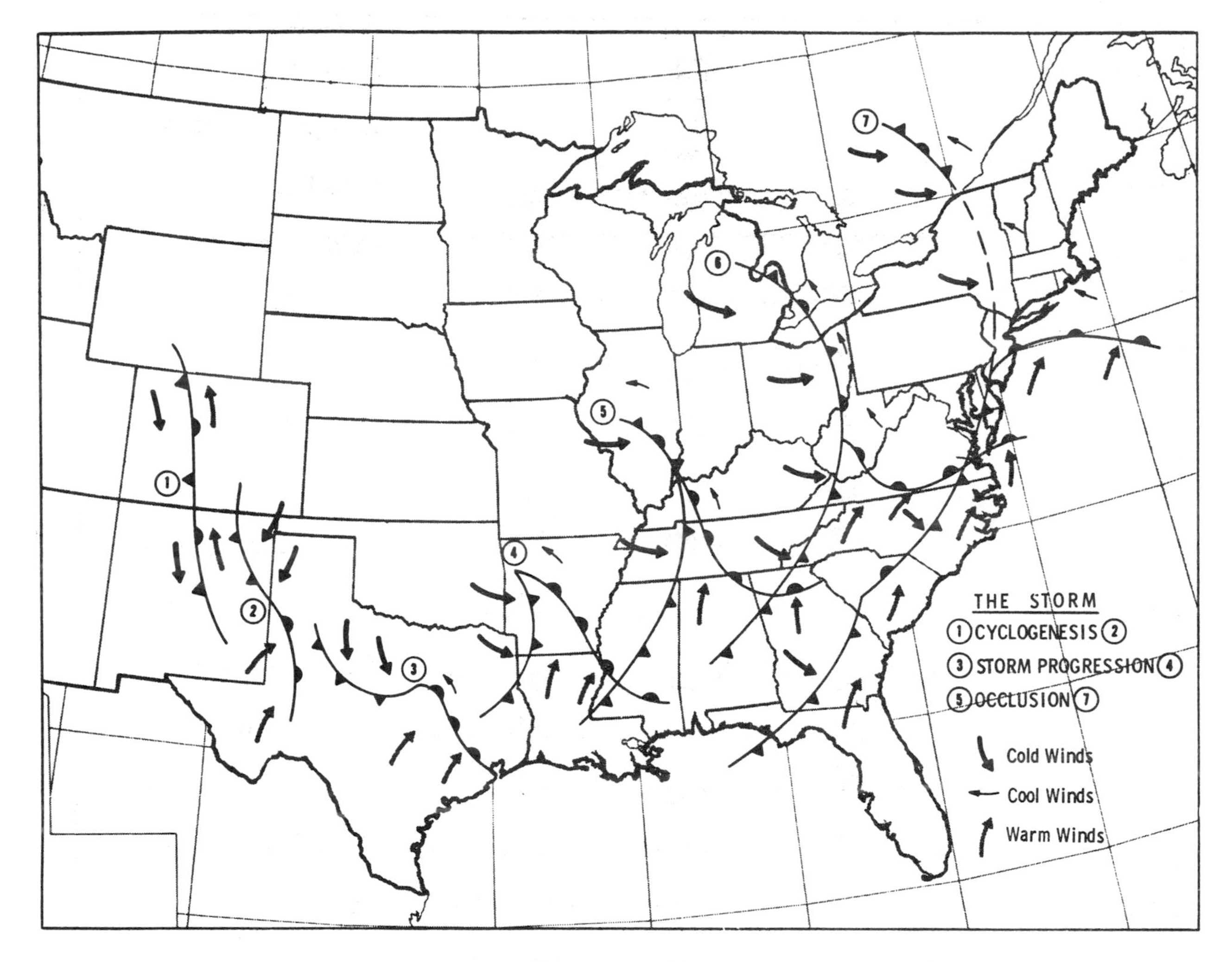

Figure 10-3. Stages in the occluding process.

surrounding air. The curvature is slight along the table top due to friction with the surface, but is more pronounced with height toward the molasses. This should help you visualize a cold air mass as it moves southward. The height of a front depends on the heights of the adjacent air masses.

1. *The Cold Front*

The average slope of a cold front is about 1/80. This means that for each 80 miles back into the cold air the frontal surface is found one mile higher. The word slope implies a straight line (like a block wedge) whereas our frontal surface is curved, but the approximation is useful.

What kind of weather can we expect to have with the cold front? Well, the weather will be dependent on such factors as the speed of movement of the air mass, the slope of the front, the amount of moisture and the stability of the warm air being replaced, and the upper wind flow. Note that the frontal boundary provides a lifting mechanism for warm, moist air, the cold air moving much as a wedge, lifting the warmer air to the south (Figure 10-5). The faster the frontal movement and the steeper the slope of the front, the more rapidly warmer air will be lifted. If

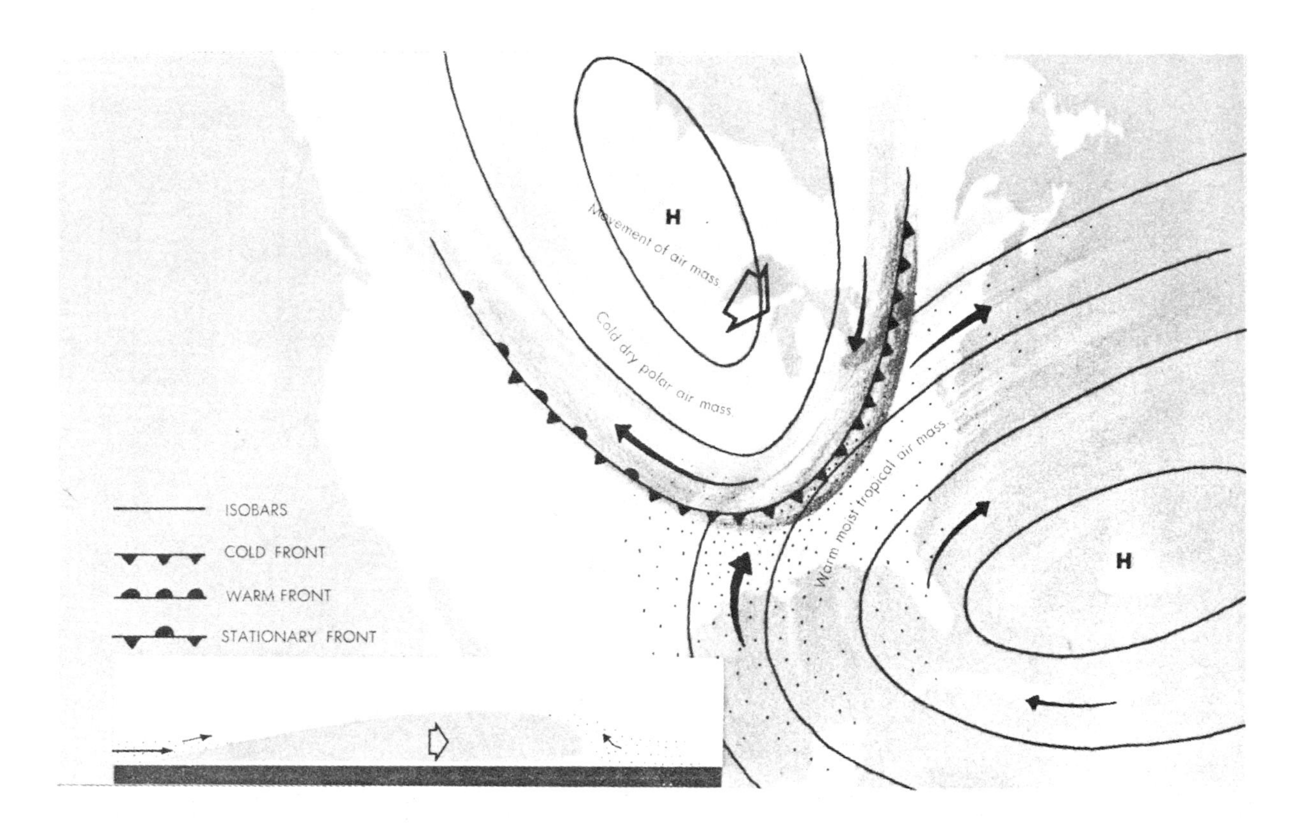

Figure 10-4. The Polar Front in North America.

the warm air is sufficiently moist and unstable over a substantial height, then all the weather characteristics of unstable air will occur — cumuliform clouds, showery precipitation, turbulence, and good visibility except in blowing obstructions (Figure 10-6).

As will be seen in a later chapter, violent thunderstorms will develop along, or in advance, or for slow-moving cold fronts, behind the front (Figures 10-7 through 10-9). Even without sufficient moisture, the air is likely to be rough as it is forced up the frontal surface, even though there may be no clouds to so indicate. If, in addition, there is a low or trough in the upper levels above the frontal surface, such pressure systems will compound the lifting effect of the frontal surface, accelerating the air to higher altitudes, resulting in even more violent weather.

2. *The Warm Front*

To visualize the warm front, we first freeze the motion of a cold front, our wedge of cold air. Assume our wedge is at a 45° angle for simplicity; realistically, it would be more like one-half a degree, sloping back over the cold air. We now turn the wedge over and place it next to its old position. Now assuming the cold air retreats northward or moves in an easterly direction, the warm air just moves in and rather gently glides up the retreating cold air. Note that we did not say that the warm air was forcing cold air to retreat, for cold air is more dense than warm air, and all other things being equal should force out the warm air. But if the winds in the upper levels were from the southwest, this would force the cold air mass to move northeasterly, allowing the warm air to replace, not displace the cold air.

The average slope of a warm front is about 1/100; note that the slope of the front is away from the warm air and toward the cold air. The same factors which affect cold frontal weather

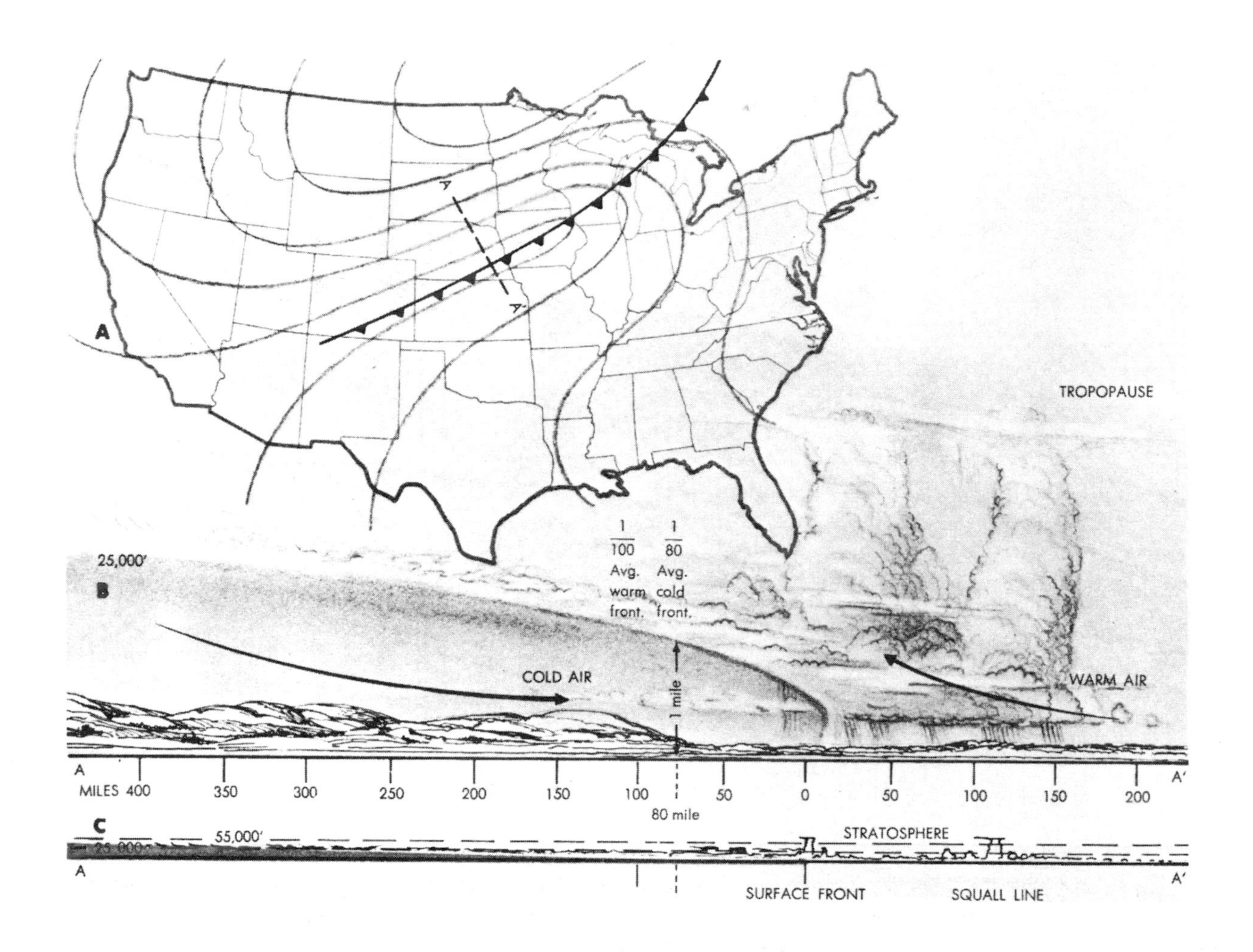

Figure 10-5. Cold Front as shown on Surface Weather Chart.

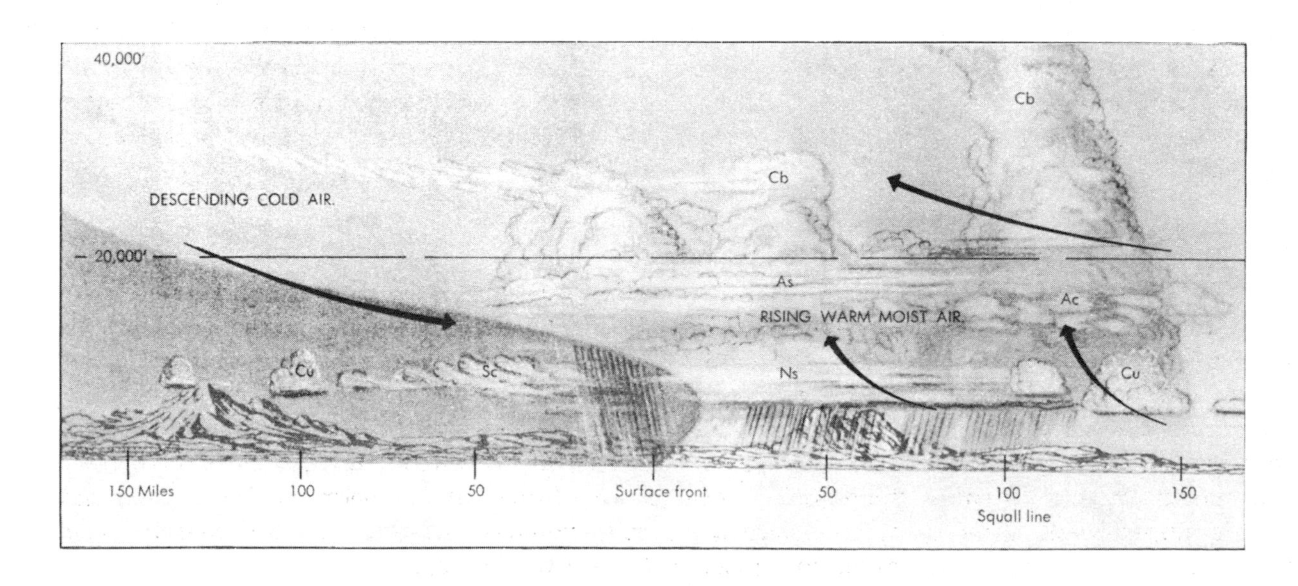

Figure 10-6. Fast-Moving Cold Front and Squall Line.

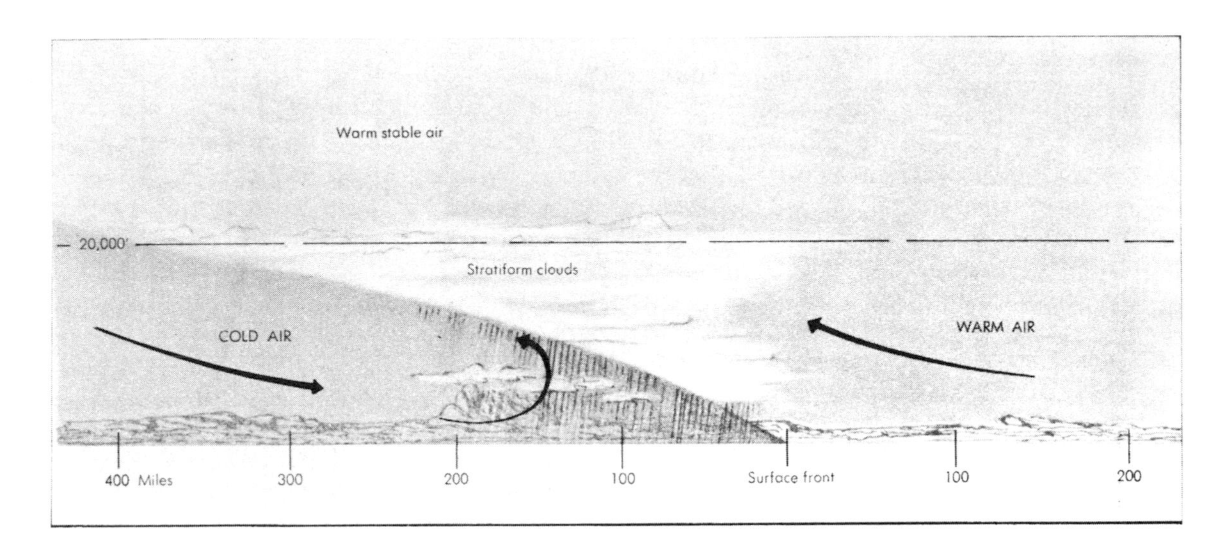

Figure 10-7. Slow-Moving Cold Front and Stable Air.

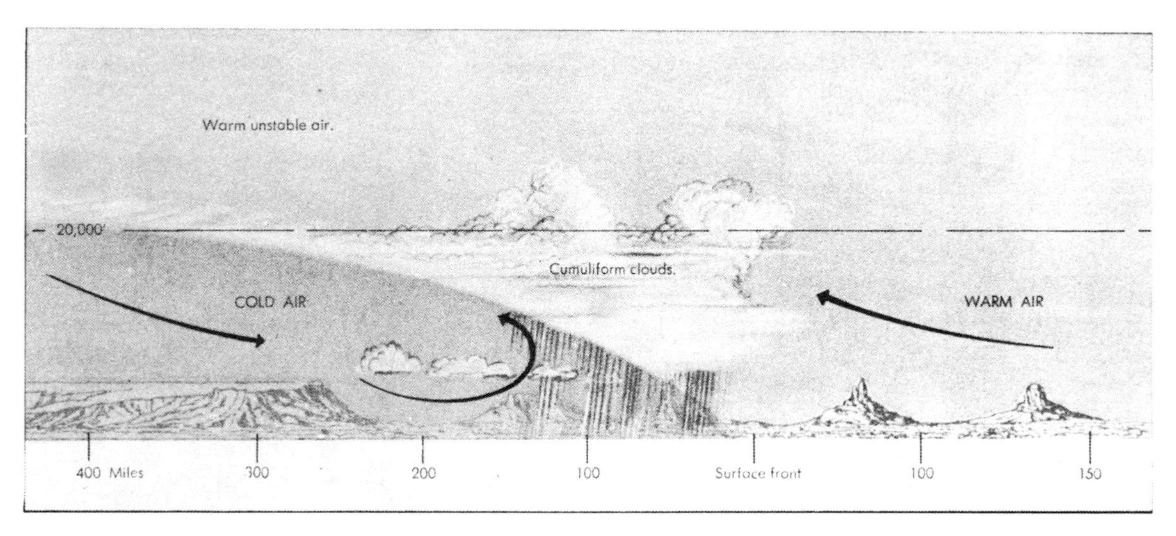

Figure 10-8. Slow-Moving Cold Front and Conditionally Unstable Air.

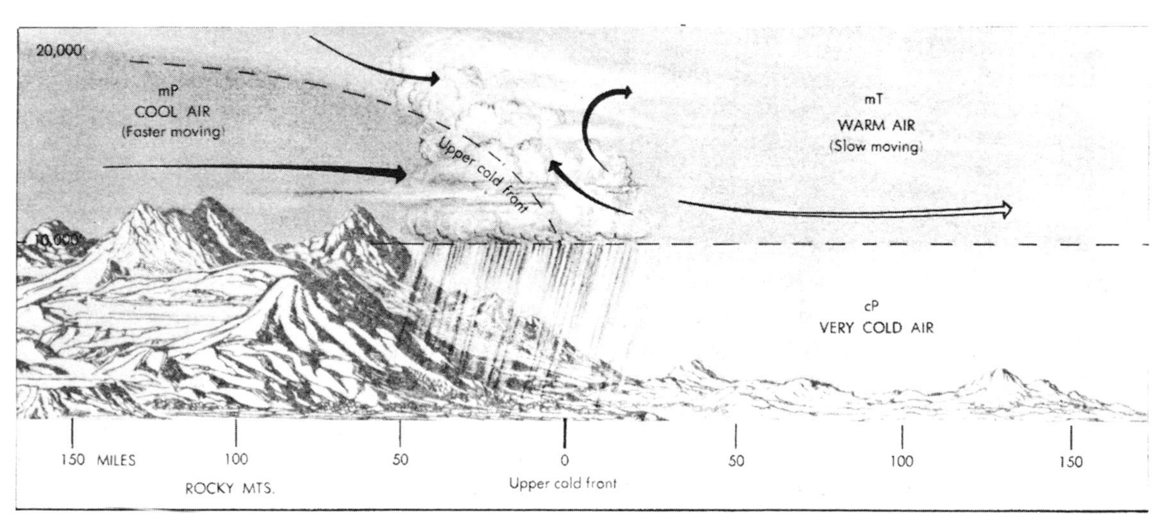

Figure 10-9. Upper Cold Front.

affect warm frontal weather. The slope and speed of movement of the warm front are not as great as for the cold front. These factors, together with the fact that the warm air falls in behind the retreating cold air, lead us to conclude that the weather associated with the warm front is usually, but not always, less violent than that experienced with a cold front.

Because air tends to "stack up" behind the warm front as the cold air retreats, there is usually a broad area of cloudiness with the warm front, provided that there is sufficient moisture. If, in addition to moisture content, the air is unstable, thunderstorms may develop, for even the slightest lift of air as it falls behind and up the retreating cold air will result in cumuliform clouds, if the air is indeed unstable. While thunderstorms associated with a warm front are usually less severe than those with a cold front, they are usually hidden from view in the broad mass of clouds; such thunderstorms are called embedded (Figures 10-10 and 10-11).

The weather associated with a stationary front is primarily dependent on the amount of moisture available and the stability of the warm air. As we have already noted, any frontal zone lies in an area of relatively low pressure and, hence, the lift that occurs with a low-circulation pattern will usually result in rough air. One of the particularly disturbing features of the stationary front is that it may persist in a given area for several days, even weeks in mid-latitudes, resulting in variable weather on a day-to-day basis. No front is perfectly stationary, so quasi-stationary is perhaps a better description. Just a 50-mile oscillation each day can alternate cold air mass and warm air mass weather at a given location.

3. Occluded Fronts

The occluded front, because of its definition as a cold front catching up with a warm front near the associated low, can and usually does yield the combined weather effects of both cold and warm fronts. At this point we describe the two kinds of occluded fronts, the cold and the warm. We begin by picturing a low in Arkansas centered between three highs located in Kansas, Louisiana, and Kentucky. Picture a cold front extending from the low in Arkansas along the Texas-Oklahoma border and a warm front extending from the low into Mississippi. Assume that in the upper levels, at around 30,000 feet, winds are northwesterly shifting to southwesterly across the low in Arkansas.

We now set this typical frontal pattern in motion. The northwesterly winds will force the cold front southeastward into southeast Texas. As the front approaches some location in southeast Texas, pressure will be falling, the temperature will be warm, the dewpoint will probably be high due to the proximity of the Gulf of Mexico, and the winds will be southerly. Assuming the warm air along the frontal zone is unstable, a huge line of thunderstorms may lie along the front. As the front passes, pressure rises sharply, the temperature becomes colder, the dewpoint lowers, reflecting dryer air, and the winds shift to the north. Since air must spiral in a counterclockwise manner toward the low from north winds north of the front to south winds south of the front, the wind shift at our location will be south to west to north, that is, clockwise.

While the cold front moves southeastward, the warm front moves northeastward, driven by the southwesterly winds in the upper levels. Actually, the primary push is on the air mass in Kentucky, its apparent retreat due to the upper level winds, and not being forced out by the warmer air in Louisiana. At a point, say, Memphis, Tennessee in advance of the front, pressure will again be dropping, the temperature will be cool, the relative humidity low, and the winds easterly as air continues to spiral counterclockwise in toward the low around the fronts. High cirriform clouds will be overhead with lower clouds to the south, again, assuming high moisture content to the south. As the warm front arrives, there will likely be a broad zone of low clouds

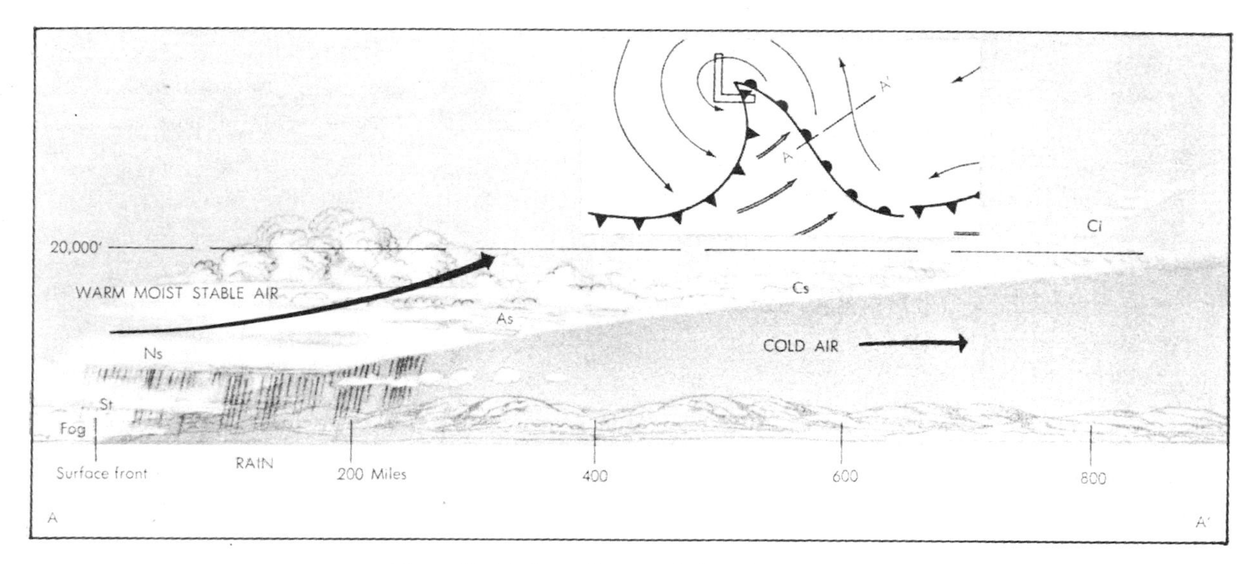

Figure 10-10. Warm Front and Moist Stable Air.

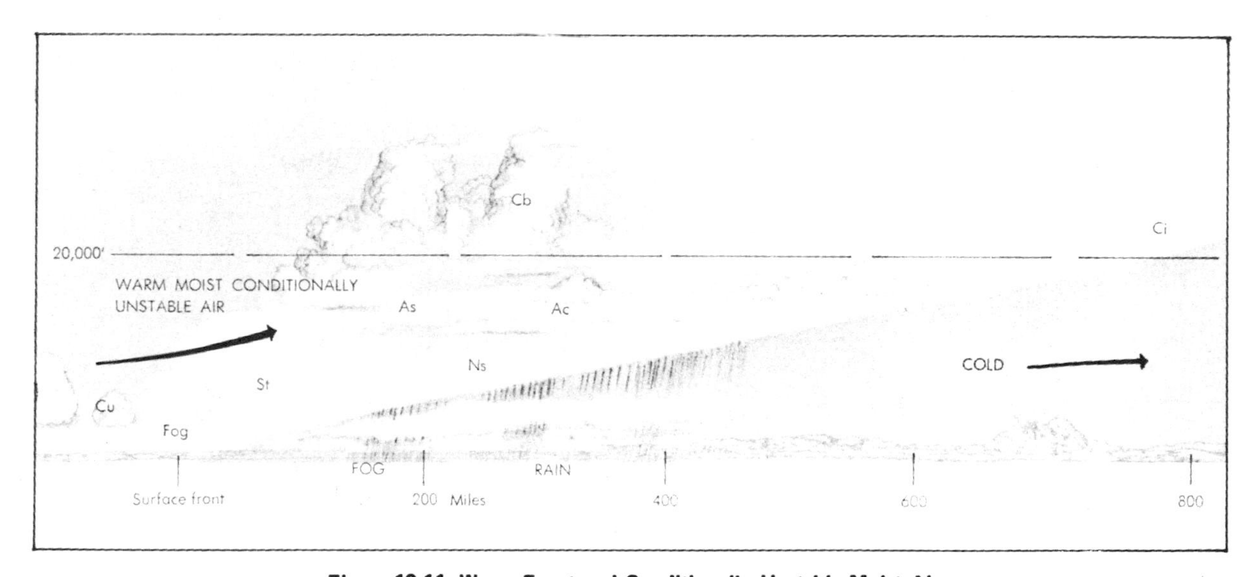

Figure 10-11. Warm Front and Conditionally Unstable Moist Air.

and fog, with steady drizzle or rain and possible thunderstorms if the warm air is unstable. After the front passes, pressure will rise, the temperature will increase, the relative humidity will increase, and the winds will be southerly, having again shifted clockwise, east to south.

Note that since the wind shifts in a clockwise manner in both cases, then anytime you cross a front, warm or cold, going north or south, you should always make wind corrections to the right in order to maintain a desired course. Draw a picture to convince yourself of this fact. Also, in crossing a front, a new altimeter setting should be obtained in order to compensate for changes in temperature and pressure.

We are getting close to what we initially set out to do, namely, describe two types of occluded fronts. We know that the air behind the cold front is colder than the air ahead of the cold front. We also know that the air ahead of the warm front is colder than the air behind the warm front. Picture these statements in your mind. Since the air in both Kansas and Kentucky is cold, which is the coldest? Well, assume that the air mass in Kentucky had moved from central Canada southeastward a couple of days ago and that the air mass in Kansas has just moved in from central Canada due southward. Then the air in Kansas is colder than the air in Kentucky, and the air in Louisiana is warmer than both.

We will call the air in Kentucky cool. Now as our cold front moves southeastward, it will begin to catch up with the warm front near the low. When the air behind the cold front catches the warm front, the leading edge of the warm air, the cold air will have pushed all the warm air in its path upward and now be impacting the cool air ahead of the warm front. The cold air will force the cool air upward also, and we have the cold-front occlusion (Figure 10-12).

We now illustrate the warm-front occlusion. Assume the air mass in Kansas had come not from central Canada due southward but rather from western Canada over the Rockies into Kansas. Because of the katabatic effect this air could be cool compared to the air in Kentucky, which we now call cold. Again, the air to the south in Louisiana is warm compared to these two air masses. Now as the cold front again catches up with the warm front, the cool air impacts the cold air in advance of the warm front. The cool air rides up and over the cold air (Figure 10-13).

D. Fronts and Lows

Not all lows in mid-latitudes have fronts extending from them, but most of them do. One exception would be a small bubble low lying inside an air mass due to small-scale local temperature differences, a nighttime land-breeze over one of the Great Lakes, for example. Since a front is the boundary between air masses, there would be no front extending from this low,

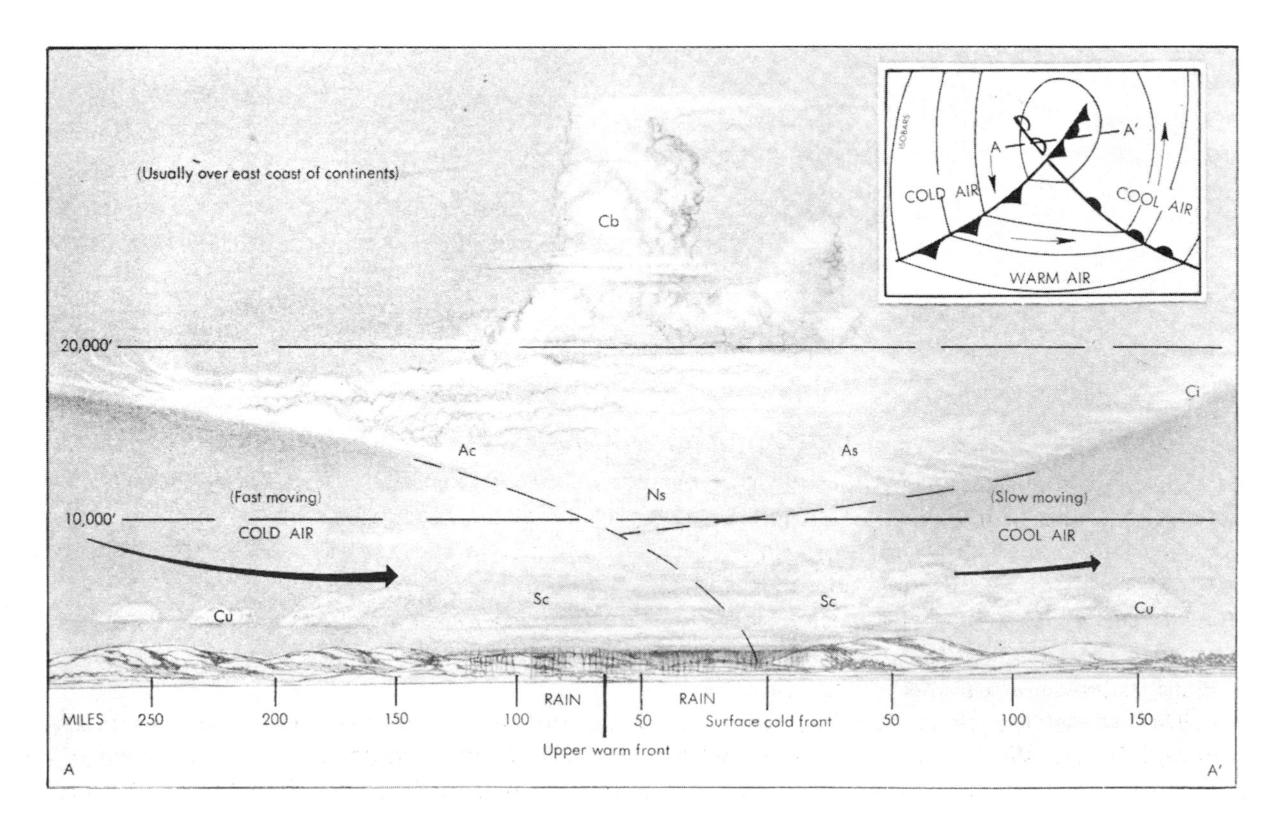

Figure 10-12. Vertical Cross-Section of a Cold Front Occlusion.

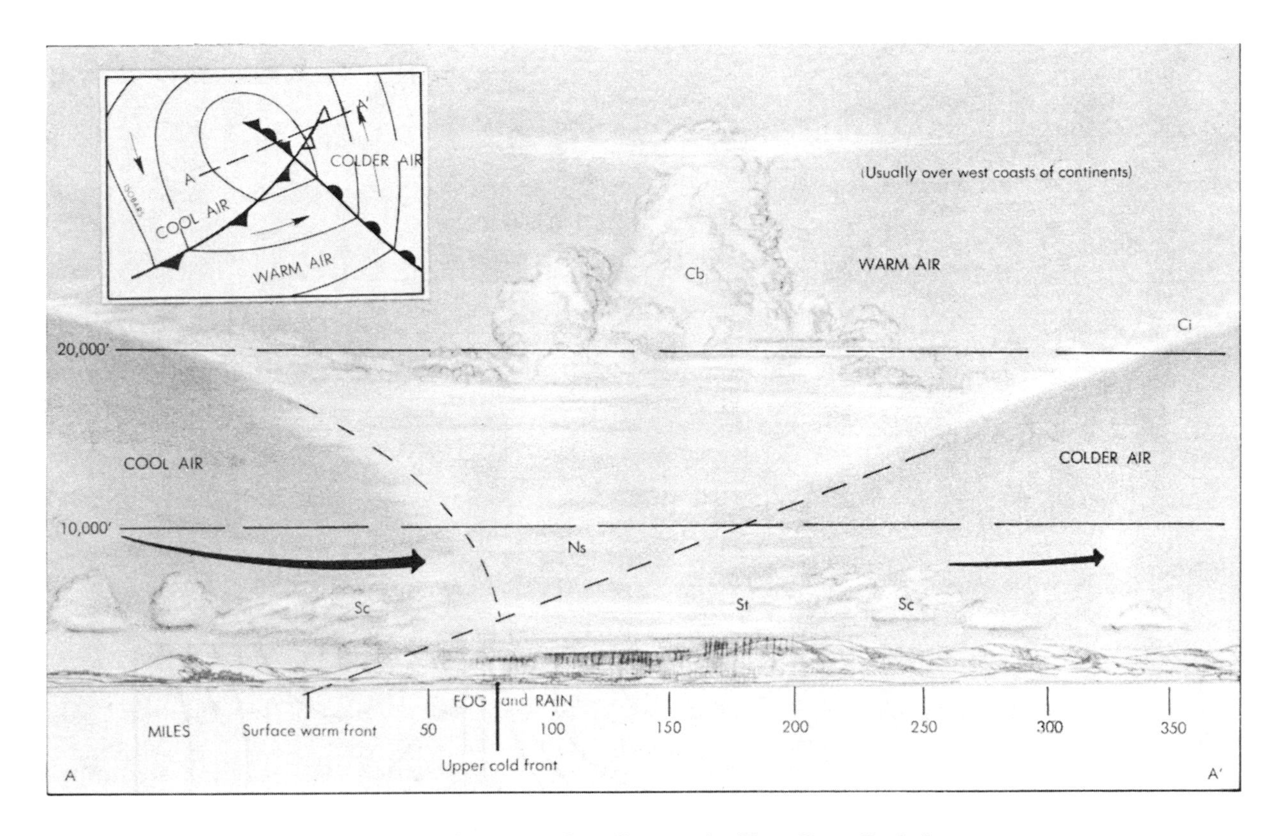

Figure 10-13. Vertical Cross-Section of a Warm Front Occlusion.

being embedded within the air mass; that is, a front does not extend from or through an air mass, only between air masses. Since an air mass is a high, fronts do not extend into or from such highs, nor do they extend into any high, air mass or not, because a front lies in a zone of low pressure. The next time you see a weather map, check out these facts.

The wind flow in the upper levels is the primary reason frontal waves develop. Divergence in the upper levels (directly above the surface low) that is stronger than convergence in the low will intensify the low (strengthens—pressure drops).

FRONTS
Review Questions

PRI

1. The zone of transition between two air masses of different density is referred to as a
   a. Front.
   b. Foehn.
   c. Frontolysis.
   d. Frontogenesis.

2. When a cold front overtakes a warm front, the two of them join together to form
   a. A squall line.
   b. A stationary front.
   c. An occluded front.
   d. A dewpoint front.

3. Regarding the characteristics and weather associated with a warm front, which of the following is a true statement?
   a. The presence of thunderstorms along a warm front is usually easy to detect, since they are not embedded in cloud masses.
   b. The frontal zone may have zero ceilings and zero visibilities over a wide area.
   c. Colder air is overtaking and replacing warmer air and this usually produces wide bands of precipitation ahead of the warm front surface position.
   d. Squall lines sometimes develop 300 miles ahead of warm fronts.

4. An advancing warm front that has moist and unstable air is characterized, in part, by
   a. A wall of turbulent clouds known as a "squall line".
   b. Stratiform clouds and smooth air.
   c. Thunderstorms embedded in the cloud masses.
   d. Tornadic activity and extensive electrical discharges.

INS

5. Which weather phenomenon is always associated with the passage of a frontal system?
   a. A wind shift.
   b. An abrupt decrease in pressure.
   c. Clouds, either ahead or behind the front.
   d. A abrupt decrease in temperature.

CFI

6. Which statement is true regarding a cold front occlusion?
   a. The air ahead of the warm front is colder than the air behind the overtaking cold front.
   b. The air ahead of the warm front is warmer than the air behind the overtaking cold front.
   c. The air between the warm front and cold front is colder than either the air ahead of the warm front or the air behind the overtaking cold front.
   d. The air ahead of the warm front has the same temperature as the air behind the overtaking cold front.

7. The factor which determines whether the type of cloudiness associated with a front will be predominantly stratiform or cumuliform is the
   a. Relative humidity of the air behind the front.
   b. Degree of stability of the air being lifted.
   c. Dewpoint of the air being lifted.
   d. Pressure of the air behind the front.

ATP

8. Frontal waves normally form on
   a. Stationary or occluded fronts.
   b. Rapidly moving cold fronts.
   c. Slow moving warm fronts or occluded fronts.
   d. Slow moving cold fronts or stationary fronts.

9. Frontal activity can produce fogs which are a result of
   a. Nocturnal cooling.
   b. Evaporation of surface moisture.
   c. Saturation due to evaporation of precipitation.
   d. Adiabatic cooling.

10. In which direction should a pilot correct the aircraft heading to maintain a desired course when flying through a frontal system on a flight from St. Louis to New York?
    a. To the left when flying from a cold to a warm front.
    b. To the right when flying from a warm to a cold front; to the left when flying from a cold to a warm front.
    c. To the left when flying from a warm to a cold front.
    d. To the right regardless of the type of frontal system.

CHAPTER XI
TURBULENCE

The reader should demonstrate an understanding of turbulence to the extent that he can answer the following questions (as stated below, or as multiple-choice, etc.):

1. What are the best actions to keep from exceeding design load limits and to avoid stalls?
2. When are convective currents most active?
3. Describe air motion in pressure systems.
4. What is a cause/factor for one out of every three weather-related accidents?
5. Why should you carry an extra margin of airspeed at low altitudes in hilly terrain?
6. In mountainous areas, winds in excess of _____ knots may cause severe turbulence.
7. Describe clouds and turbulence in mountains when the air is moist and unstable.
8. When air is stable in mountainous areas, how strong may downdrafts be on the lee side of a mountain?
9. What type of cloud presence indicates the existence of mountain-wave turbulence?
10. Explain why clouds in mountains appear to be stationary when winds may be quite strong.
11. Where do "rotors" occur?
12. In order to be better prepared for flight in the mountains, what meteorological information would be especially helpful?
13. Name some procedures to keep in mind as you approach an area of known mountain-wave turbulence.
14. Why should you always expect turbulence when crossing a front?
15. Describe the location of turbulence when a temperature inversion is present, and the effect of a 25 knot wind or greater along the top of the inversion.
16. What causes the disturbance known as wake turbulence?
17. The strength of vortices is governed by what wing factors?
18. What is the usual hazard resulting from wake turbulence?
19. How wide and deep are wake turbulence vortices?
20. At what distance below the path of a generating aircraft do vortices level off?
21. What is the effect of a crosswind on the lateral movement of vortices?
22. Describe the vortex avoidance procedures for landing behind a large aircraft on the same runway; on a parallel runway closer than 2,500 feet; on a crossing runway.
23. Describe the vortex avoidance procedures for landing behind a departing large aircraft on the same runway; on a crossing runway.
24. What procedures should be used when departing behind a large aircraft; making intersection takeoffs on the same runway?
25. What procedures should be used when departing or landing after a large aircraft executing a low approach, missed approach, or touch-and-go landing; when flying enroute?
26. Describe the jet stream in terms of speed, thickness, length, and width.
27. State the variation in height of the tropopause.
28. How may one often locate the tropopause?
29. State a rule that depends on temperature to avoid turbulence near the tropopause.
30. Define wind shear.
31. What degree of vertical shear may result in moderate or greater turbulence?
32. What degree of horizontal shear may result in severe turbulence?
33. Under what conditions does a headwind or tailwind adversely affect aircraft performance? What is the effect on the aircraft for low-altitude operations?
34. Define maneuvering speed.
35. Define CAT.
36. With what is most CAT associated?
37. State the most common altitudes for CAT in winter; in summer.
38. State the approximate dimensions of "patches of CAT".
39. What are the best means for identifying CAT?
40. Why are forecasts of CAT difficult to make?
41. What are your best means of coping with CAT?

CHAPTER XI

TURBULENCE

(See Appendix C, pages 15-34, for AC 00-30A and AC 00-50A)

Turbulence may be thought of as simply the roughness of the air as we fly our airplane. The degree of turbulence will depend on such factors as the size of the aircraft and its speed. A study of the general circulation in the chapter on wind involved various air motions, such as updrafts, downdrafts, convergence, etc., all of which contribute to possible turbulence. In fact, one wonders if a flight could be completed without turbulence. As a matter of fact, it would be quite unusual to complete an entire flight without any turbulence. We can only hope that turbulence will be neither too frequent nor too severe.

Turbulence can result from horizontal as well as vertical changes in the air flow. In general, initially adjusting power for a given airspeed (maneuvering speed), keeping the wings level, and accepting changes in altitude are your best actions to keep from exceeding design load limits and to avoid stalls. To more fully understand turbulence we now look at some of its causes: convection, pressure systems, obstructions to wind flow, thunderstorms, convergence, fronts, cold air advection, temperature inversions, wake turbulence, the jet stream, the tropopause, and wind shear.

A. Convection

We observed in the chapter on stability how convective currents may develop in the lower extremes of a stable air mass, causing updrafts and downdrafts in the lower layers. This vertical pattern resulted primarily from the nature of the earth's surface, that is, differential heating effects were created due to the contrast of land versus water, paved roads versus green vegetation, etc. (Figures 11-1 and 11-2). Such currents are most active on summer afternoons when winds are light.

B. Pressure Systems

Pressure systems — highs, lows, troughs, and ridges — will cause turbulence due to the pattern of air circulation. Air in highs and ridges will be descending, as well as changing direction horizontally. For lows and troughs, air motion will be vertically upward and at a faster rate.

C. Obstructions

1. Buildings

Buildings will alter the vertical and horizontal flow of wind. Do not underestimate such effects at an airport on an airplane that may be taking off or landing. Air circulating around, over, or through nearby hangars may render an aircraft uncontrollable, particularly when the wind speed is 25 knots or more (Figure 11-3).

2. Low-Level Terrain

Flying at low levels over rough terrain is usually more of a discomfort than a potential danger; to avoid rough air you should usually climb. There is one particular case in which operating at low levels over rough terrain is a danger, however, and that is in takeoffs and landings. In fact, about one out of every three weather-related accidents has an unfavorable wind as a cause or factor. Air that rolls over terrain represents just one type of unfavorable wind, but one that can be very dangerous in certain situations (Figure 11-4).

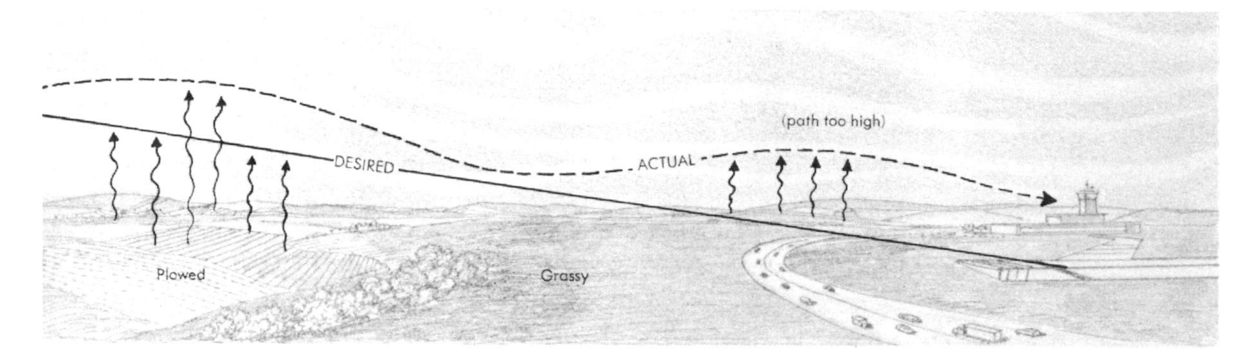

Figure 11-1. Updrafts may cause Pilots to Overshoot.

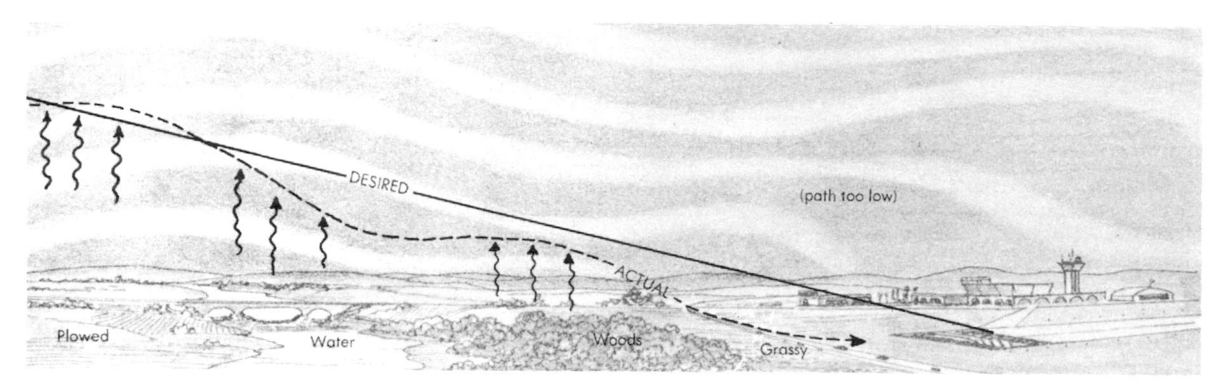

Figure 11-2. Downdrafts may cause Pilots to Undershoot.

Figure 11-3. Obstructions at an airport may induce turbulence in the takeoff, landing, and taxi areas. Each air traffic specialist should become familiar with areas where obstructions cause turbulence at his airport of concern.

As a student pilot, the author approached a cross-country airport located in hilly terrain. I let the airspeed get .too low, and as the air rolled over the rough terrain I was thrown into a steep angle of bank and the aircraft stalled, all of this on final approach at a very low altitude. Without a reduction in angle of attack and an application of power, the airplane, and myself, would have become accident statistics (an extra margin of airspeed increases momentum — a stall is less likely — and makes your controls more effective).

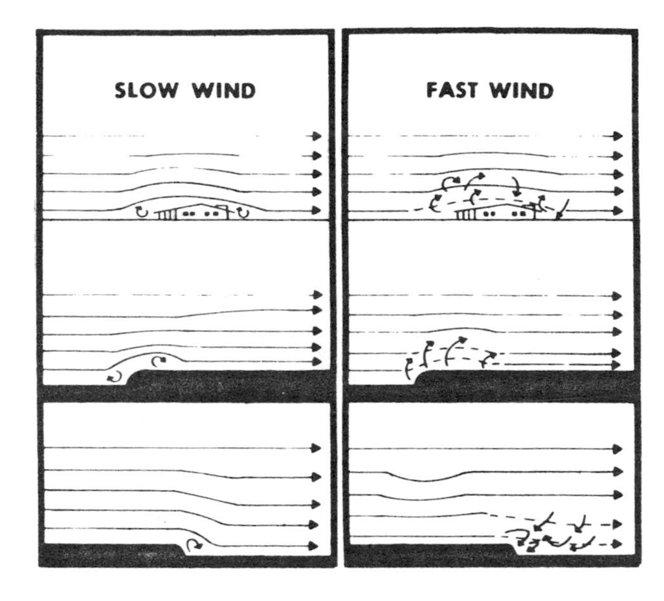

Figure 11-4. Obstruction to wind flow cause irregular eddies along and downwind from obstructions. The degree of turbulence depends on the size and shape of an obstruction, the speed of the wind, and the stability of the air.

3. Mountains

Flying in mountainous areas presents a special problem; winds in excess of 50 knots can cause severe turbulence. If the air is unstable and moist, large cumuliform clouds may develop on the windward side of the mountain, the side facing the wind, and vertical swells of air may rise to heights of 50,000 feet (15 km) or more, particularly if thunderstorms develop. But even if you do not see any clouds, still assume that these vertical currents are there, and that will be the case if the air is relatively dry and unstable.

If the air is stable, the mountain will induce strong waves in the wind flow, resulting in strong downdrafts on the lee side, say, 5,000 feet (1.5 km) per minute. Many light aircraft have tried to climb from low altitudes to the tops of mountains in such situations. It would be easy to say that the pilot was unlucky and that the wind blew him into the mountain. But this is too easy an out. In reality, he flew his aircraft into the mountain. He was not COVERED. He probably failed to COmprehend the weather factors involved. Had he done so, he would have Visualized the potential danger, air spilling downward on the lee side of the mountain. He could have then Evaluated the potential for reduced rate of climb on his aircraft, well in advance, and Responded by climbing well downwind of the mountain, for example. After climbing, he would have then needed to reEvaluate and possibly Determine a new course of action, for his climb may have been of insufficient altitude, placing his aircraft in the middle of the wave formation over the top of the mountain, for example (Figure 11-5).

The presence of mountain-wave turbulence is often marked by the altocumulus standing-lenticular cloud that forms at the tops or crests of the wave formation. The word "standing" is misleading for the word implies stationary, whereas the air may be moving at speeds of 50 knots or more.

The reason for the stationary appearance is the adiabatic effect. If the air is sufficiently moist, then the slight lift of the moisture as it approaches wave crests causes the moisture to condense, forming the flat, lens-shaped clouds along the crests. Then as the air starts downward on the lee side of the wave crests, that air is warmed adiabatically and the visible moisture evaporates. This remarkable phenomenon is easily seen by time-lapse photography, but not by the naked eye. Thus, while the cloud appears to be stationary, it is actually continually forming and dissipating adiabatically. With winds of 50 knots or more this is obviously a situation to be avoided. Many pilots are unaware of these facts, to which a lot of airplane parts on mountain

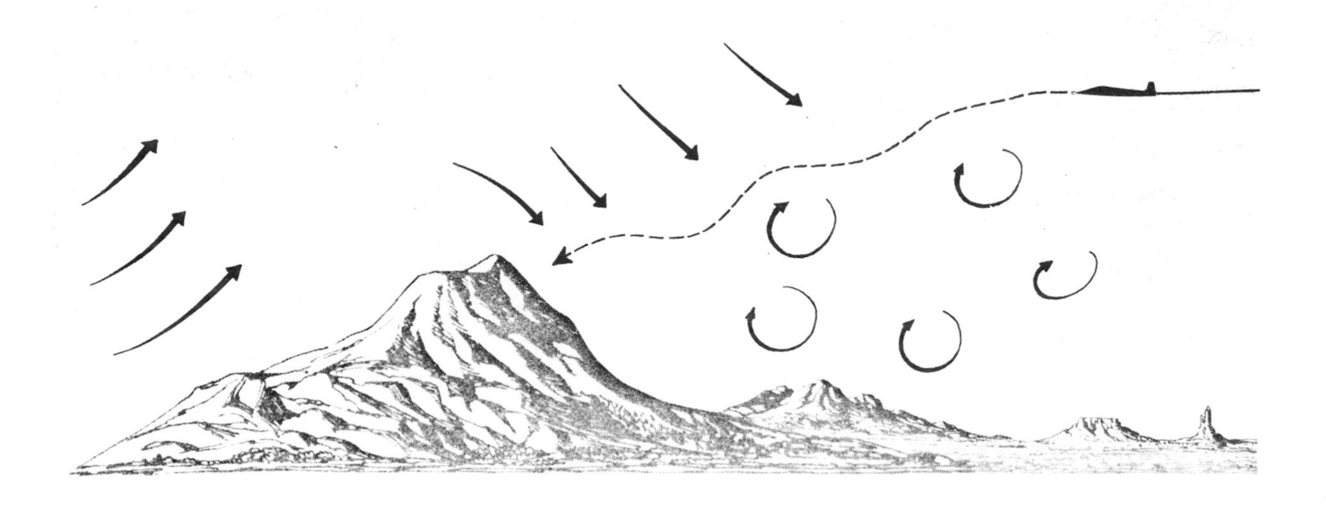

Figure 11-5. Wind Flow over Mountain Ranges produces Turbulence.

slopes will attest. Once again, if sufficient moisture is not available, you will not have these visual clues; assume that mountain wave exists anyway (Figures 11-6 and 11-7).

Another area of turbulence is the rotary circulation about halfway to two-thirds of the way down the mountain, underneath the wave crests. The circulation is caused by the interaction of the clockwise circulation of the lower wave crests with the ground. In the presence of moisture this rotation will cause a roll cloud to develop. But again, in the absence of sufficient moisture, the rotary cloud may not be there, but we assume that the circulation is (Figure 11-8).

Knowledge of cloud reports, wind direction and speed, and the stability of the air will assist you in having a better idea of what to expect when you approach mountains, but the situation can always change and you must keep yourself COVERED. Some procedures to keep in mind are to begin your climb well in advance, perhaps 100 miles (160 km) downwind if mountain wave is known to exist, 50 miles (80 km) otherwise. Climb to an altitude at least one and one-half times the height of any mountain. For example, if a mountain is 6,000 feet (1.8 km) high, then one-half of 6,000 is 3,000 so fly at least as high as 9,000 feet (2.7 km). You should consider approaching mountain ranges at a 45° angle in order to allow for a quick change in direction should the air become very rough. In extreme cases you may have to choose between going around the mountain range and going back.

D. Thunderstorms

Turbulence with thunderstorms may be extreme, due largely to adjacent updrafts and downdrafts. You will gain a greater appreciation for the potential for turbulence near thunderstorms when we study that chapter. Small-scale gusts embedded in large-scale drafts will be particularly important.

E. Convergence

We have mentioned convergence before, the merging of wind from different directions, resulting in considerable roughness of the air. When converging air is of different densities, the lighter air will be propelled upward, adding to the degree of turbulence.

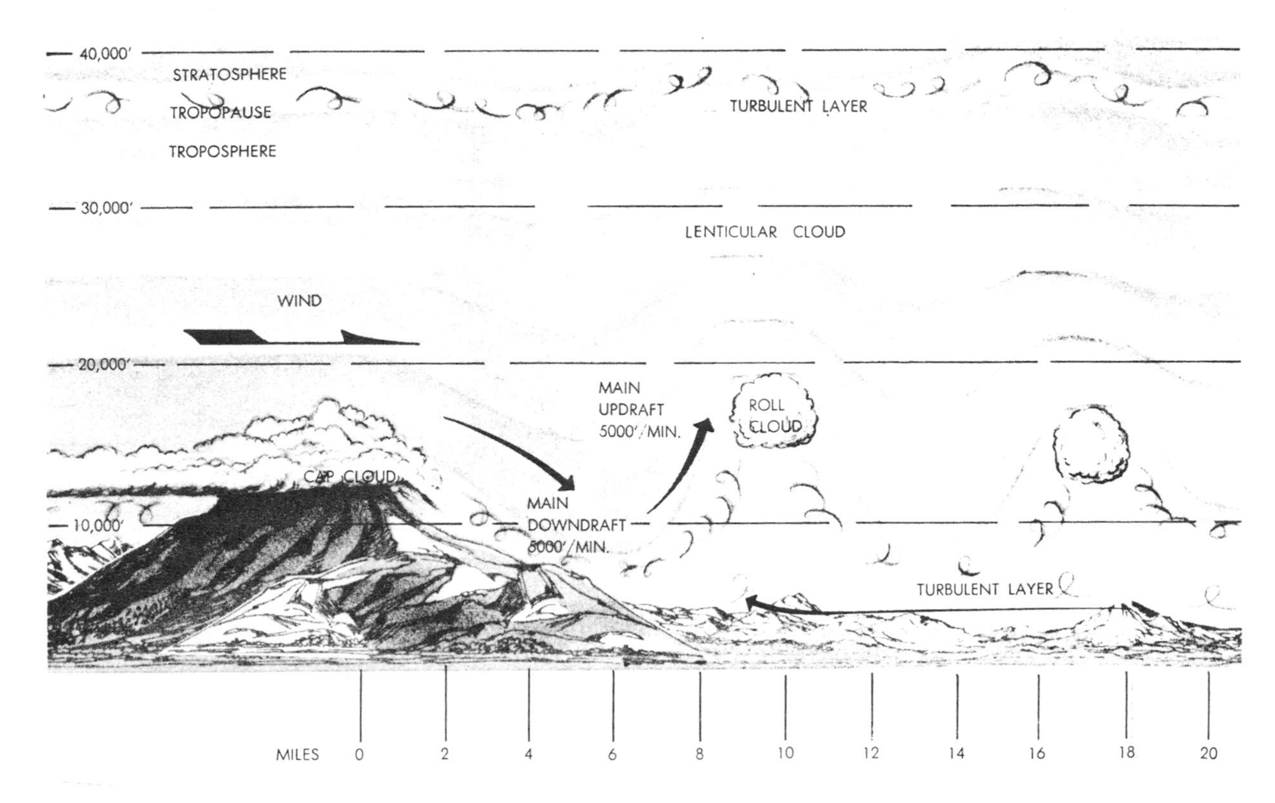

Figure 11-6. Typical Cloud Formation, Main Downdraft and Updraft in Mountain Waves.

Figure 11-7. Standing Lenticular Clouds Associated with a Mountain Wave.

Figure 11-8. Photographs of standing wave rotor clouds. They are named standing because they have very little movement, but the wind flowing through the clouds can be extremely strong.

F. Fronts

We mentioned in the chapter on fronts how air is forced up a frontal boundary, particularly for a cold front. Add to this the fact that a front is always accompanied by a wind shift, since it is the boundary between air masses, and you can readily see that turbulence may be significant. Always expect turbulence when crossing a front, and above the surface as well as near the surface.

G. Cold Air Advection

Cold air advection is the horizontal movement of cold air over a warm surface or warm air. The sinking of denser, colder air will result in vertical motion sufficient to give moderate turbulence.

H. Temperature Inversions

Earlier, we defined a temperature inversion as an increase in temperature with height. If the wind speed is at least 25 knots along the top of the inversion, this may well induce a series of rotary circulations that are hazardous for lower-level operations, particularly for takeoffs and landings (Figure 11-9).

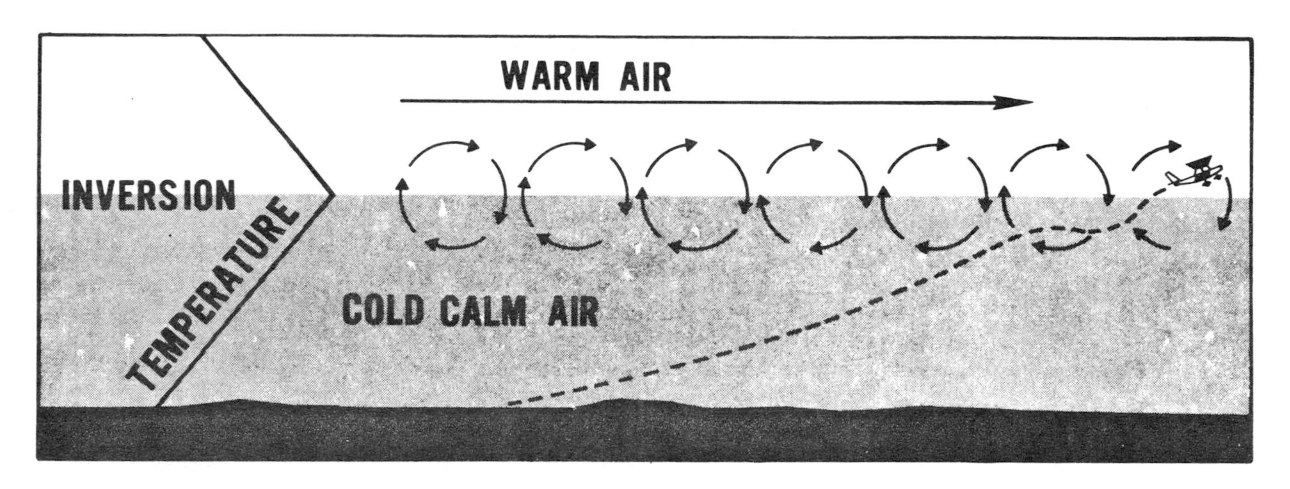

Figure 11-9. Wind shear turbulence in a zone between relatively calm wind below an inversion and strong wind above the inversion. As explained in Figure 11-22, the eddies roll along in the shear zone much as rollers do in a roller bearing. This condition is most common at night or in early morning.

I. Wake Turbulence

The following five pages is a direct reprint on wake turbulence from the Airman's Information Manual. You may be unfamiliar with a few terms, depending on your flight experience, but you will be able to comprehend all the important ideas. Since some new information on wake turbulence is discovered fairly frequently, and since portions of the AIM are published several times a year, the most recent AIM should be consulted for the latest information. Reprints/future editions of this text will contain the most current information at the time of publication.

Section 3. WAKE TURBULENCE

7-41. GENERAL

a. Every aircraft generates a wake while in flight. Initially, when pilots encountered this wake in flight, the disturbance was attributed to "prop wash." It is known, however, that this disturbance is caused by a pair of counter rotating vortices trailing from the wing tips. The vortices from larger aircraft pose problems to encountering aircraft. For instance, the wake of these aircraft can impose rolling moments exceeding the roll control authority of the encountering aircraft. Further, turbulence generated within the vortices can damage aircraft components and equipment if encountered at close range. The pilot must learn to envision the location of the vortex wake generated by larger (transport category) aircraft and adjust the flight path accordingly.

b. During ground operations and during takeoff, jet engine blast (thrust stream turbulence) can cause damage and upsets if encountered at close range. Exhaust velocity versus distance studies at various thrust levels have shown a need for light aircraft to maintain an adequate separation behind large turbojet aircraft. Pilots of larger aircraft should be particularly careful to consider the effects of their "jet blast" on other aircraft, vehicles, and maintenance equipment during ground operations.

7-42. VORTEX GENERATION

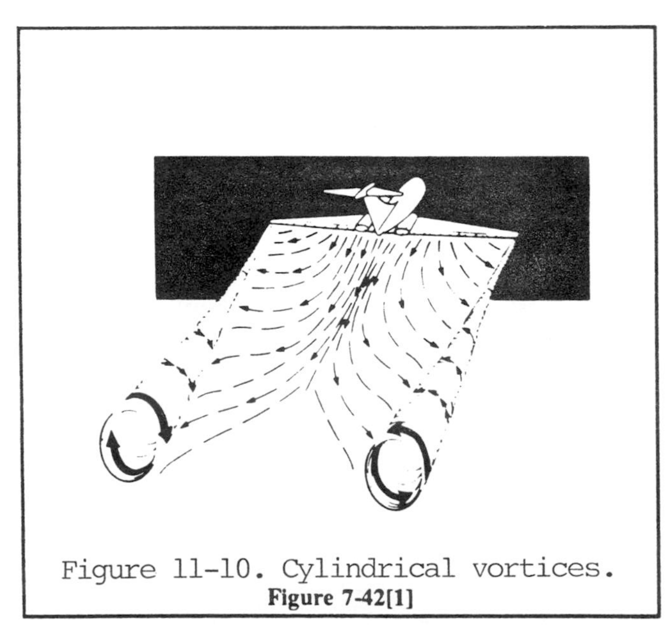

Figure 11-10. Cylindrical vortices.
Figure 7-42[1]

Lift is generated by the creation of a pressure differential over the wing surface. The lowest pressure occurs over the upper wing surface and the highest pressure under the wing. This pressure differential triggers the roll up of the airflow aft of the wing resulting in swirling air masses trailing downstream of the wing tips. After the roll up is completed, the wake consists of two counter rotating cylindrical vortices. (**See Figure 7-42[1].**) Most of the energy is within a few feet of the center of each vortex, but pilots should avoid a region within about 100 feet of the vortex core.

7-43. VORTEX STRENGTH

a. The strength of the vortex is governed by the weight, speed, and shape of the wing of the generating aircraft. The vortex characteristics of any given aircraft can also be changed by extension of flaps or other wing configuring devices as well as by change in speed. However, as the basic factor is weight, the vortex strength increases proportionately. Peak vortex tangential speeds up to almost 300 feet per second have been recorded. The greatest vortex strength occurs when the generating aircraft is HEAVY, CLEAN, and SLOW.

b. INDUCED ROLL

1. In rare instances a wake encounter could cause in-flight structural damage of catastrophic proportions. However, the usual hazard is associated with induced rolling moments which can exceed the roll control authority of the encountering aircraft. In flight experiments, aircraft have been intentionally flown directly up trailing vortex cores of larger aircraft. It was shown that the capability of an aircraft to counteract the roll imposed by the wake vortex primarily depends on the wing span and counter control responsiveness of the encountering aircraft.

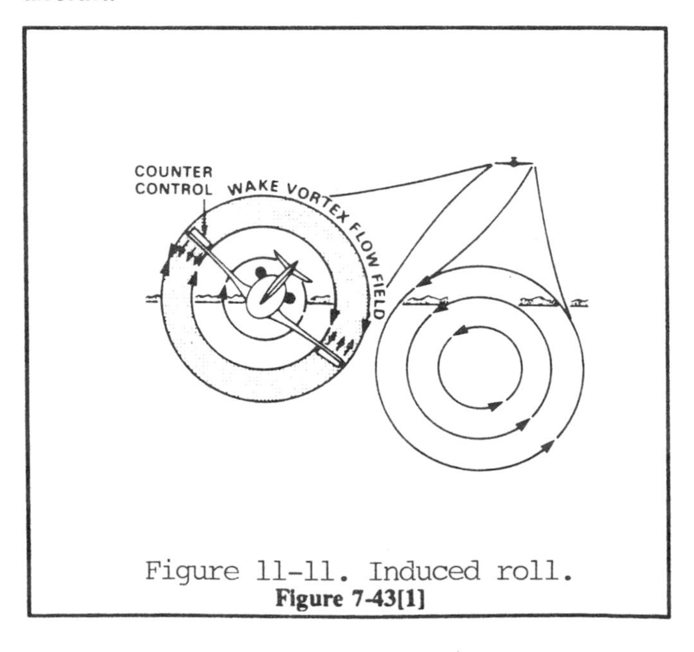

Figure 11-11. Induced roll.
Figure 7-43[1]

2. Counter control is usually effective and induced roll minimal in cases where the wing span and ailerons of the encountering aircraft extend beyond the rotational flow field of the vortex. It is more difficult for aircraft with short wing span (relative to the generating aircraft) to counter the imposed roll induced by vortex flow. Pilots of short span aircraft, even of the high performance type, must be especially alert to vortex encounters. **(See Figure 7-43[1].)**

3. The wake of larger aircraft requires the respect of all pilots.

7-44. VORTEX BEHAVIOR

a. Trailing vortices have certain behavioral characteristics which can help a pilot visualize the wake location and thereby take avoidance precautions.

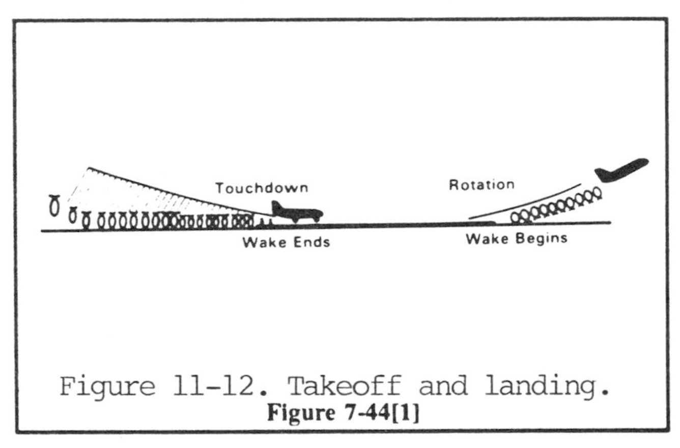

Figure 11-12. Takeoff and landing.
Figure 7-44[1]

1. Vortices are generated from the moment aircraft leave the ground, since trailing vortices are a by-product of wing lift. Prior to takeoff or touchdown pilots should note the rotation or touchdown point of the preceding aircraft. **(See Figure 7-44[1] - Wake Begins/Ends.)**

2. The vortex circulation is outward, upward and around the wing tips when viewed from either ahead or behind the aircraft. Tests with large aircraft have shown that the vortices remain spaced a bit less than a wing span apart, drifting with the wind, at altitudes greater than a wing span from the ground. In view of this, if persistent vortex turbulence is encountered, a slight change of altitude and lateral position (preferably upwind) will provide a flight path clear of the turbulence.

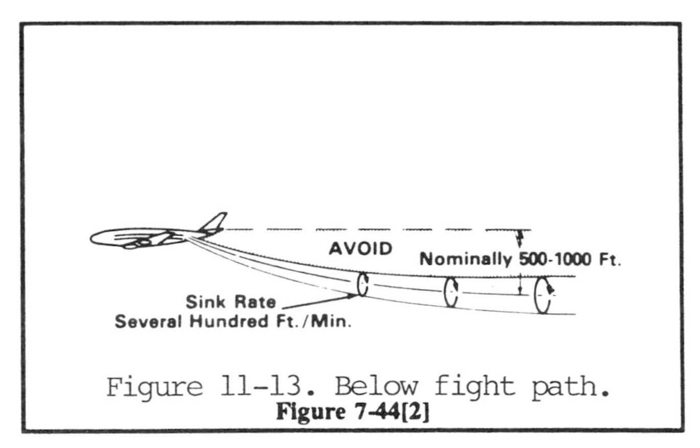

Figure 11-13. Below fight path.
Figure 7-44[2]

3. Flight tests have shown that the vortices from larger (transport category) aircraft sink at a rate of several hundered feet per minute, slowing their descent and diminishing in strength with time and distance behind the generating aircraft. Atmospheric turbulence hastens breakup. Pilots should fly at or above the preceding aircraft's flight path, altering course as necessary to avoid the area behind and below the generating aircraft. **(See Figure 7-44[2] - Vortex** Sink Rate.) However, vertical separation of 1,000 feet may be considered safe.

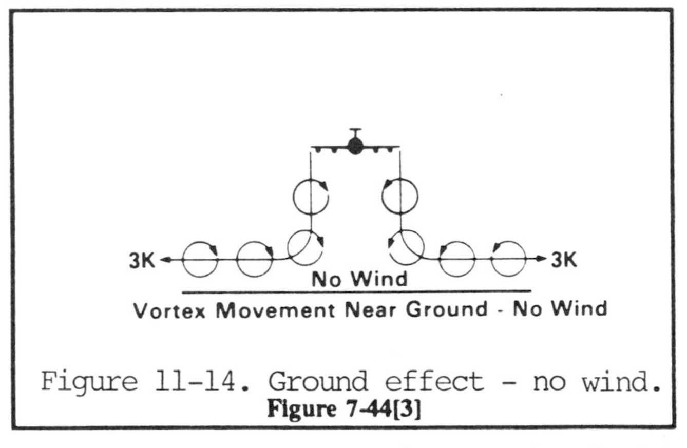

Figure 11-14. Ground effect - no wind.
Figure 7-44[3]

4. When the vortices of larger aircraft sink close to the ground (within 100 to 200 feet), they tend to move laterally over the ground at a speed of 2 or 3 knots. **(See Figure 7-44[3]**)

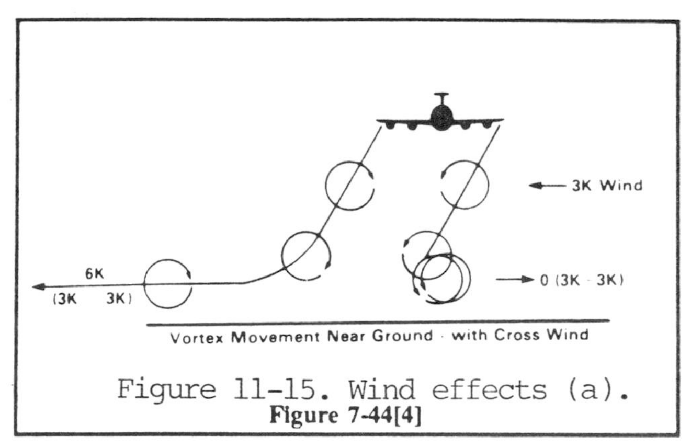

Figure 11-15. Wind effects (a).
Figure 7-44[4]

b. A crosswind will decrease the lateral movement of the upwind vortex and increase the movement of the downwind vortex. Thus a light wind with a cross runway component of 1 to 5 knots could result in the upwind vortex remaining in the touchdown zone for a period of time and hasten the drift of the downwind vortex toward another runway. (See **Figure 7-44[4] - Vortex Movement in Ground Effect** with Wind.) **Similarly, a tailwind condition can move the vortices of the preceding aircraft forward into the touchdown zone. THE LIGHT QUARTERING TAILWIND REQUIRES MAXIMUM CAUTION. Pilots should be alert to large aircraft upwind from their approach and takeoff flight paths.** (See Figure 7-44[5])

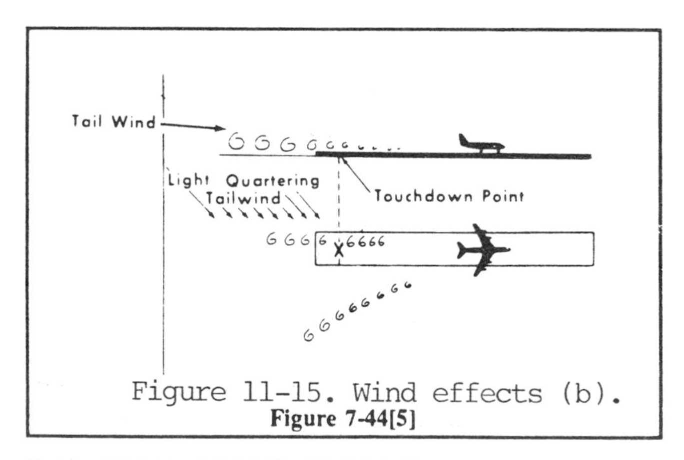

Figure 11-15. Wind effects (b).
Figure 7-44[5]

7-45. OPERATIONS PROBLEM AREAS

a. A wake encounter can be catastrophic. In 1972 at Fort Worth a DC-9 got too close to a DC-10 (two miles back), rolled, caught a wingtip, and cartwheeled coming to rest in an inverted position on the runway. All aboard were killed. Serious and even fatal GA accidents induced by wake vortices are not uncommon. However, a wake encounter is not necessarily hazardous. It can be one or more jolts with varying severity depending upon the direction of the encounter, weight of the generating aircraft, size of the encountering aircraft, distance from the generating aircraft, and point of vortex encounter. The probability of induced roll increases when the encountering aircraft's heading is generally aligned with the flight path of the generating aircraft.

b. AVOID THE AREA BELOW AND BEHIND THE GENERATING AIRCRAFT, ESPECIALLY AT LOW ALTITUDE WHERE EVEN A MOMENTARY WAKE ENCOUNTER COULD BE HAZARDOUS.

c. Pilots should be particularly alert in calm wind conditions and situations where the vortices could:

1. Remain in the touchdown area.
2. Drift from aircraft operating on a nearby runway.
3. Sink into the takeoff or landing path from a crossing runway.
4. Sink into the traffic pattern from other airport operations.
5. Sink into the flight path of VFR aircraft operating on the hemispheric altitude 500 feet below.

d. Pilots of all aircraft should visualize the location of the vortex trail behind larger aircraft and use proper vortex avoidance procedures to achieve safe operation. It is equally important that pilots of larger aircraft plan or adjust their flight paths to minimize vortex exposure to other aircraft.

7-46. VORTEX AVOIDANCE PROCEDURES

a. Under certain conditions, airport traffic controllers apply procedures for separating IFR aircraft. The controllers will also provide to VFR aircraft, with whom they are in communication and which in the tower's opinion may be adversely affected by wake turbulence from a larger aircraft, the position, altitude and direction of flight of larger aircraft followed by the phrase "CAUTION - WAKE TURBULENCE." WHETHER OR NOT A WARNING HAS BEEN GIVEN, HOWEVER, THE PILOT IS EXPECTED TO ADJUST HIS OR HER OPERATIONS AND FLIGHT PATH AS NECESSARY TO PRECLUDE SERIOUS WAKE ENCOUNTERS.

b. The following vortex avoidance procedures are recommended for the various situations:

1. **Landing behind a larger aircraft - same runway:** Stay at or above the larger aircraft's final approach flight path - note its touchdown point - land beyond it.
2. **Landing behind a larger aircraft - when parallel runway is closer than 2,500 feet:** Consider possible drift to your runway. Stay at or above the larger aircraft's final approach flight path - note its touchdown point.

3. **Landing behind a larger aircraft - crossing runway:** Cross above the larger aircraft's flight path.

4. **Landing behind a departing larger aircraft - same runway:** Note the larger aircraft's rotation point - land well prior to rotation point.

5. **Landing behind a departing larger aircraft - crossing runway:** Note the larger aircraft's rotation point - if past the intersection - continue the approach - land prior to the intersection. If larger aircraft rotates prior to the intersection, avoid flight below the larger aircraft's flight path. Abandon the approach unless a landing is ensured well before reaching the intersection.

6. **Departing behind a larger aircraft:** Note the larger aircraft's rotation point - rotate prior to larger aircraft's rotation point - continue climb above the larger aircraft's climb path until turning clear of his wake. Avoid subsequent headings which will cross below and behind a larger aircraft. Be alert for any critical takeoff situation which could lead to a vortex encounter.

7. **Intersection takeoffs - same runway:** Be alert to adjacent larger aircraft operations, particularly upwind of your runway. If intersection takeoff clearance is received, avoid subsequent heading which will cross below a larger aircraft's path.

8. **Departing or landing after a larger aircraft executing a low approach, missed approach or touch-and-go landing:** Because vortices settle and move laterally near the ground, the vortex hazard may exist along the runway and in your flight path after a larger aircraft has executed a low approach, missed approach or a touch-and-go landing, particular in light quartering wind conditions. You should ensure that an interval of at least 2 minutes has elapsed before your takeoff or landing.

9. **En route VFR (thousand-foot altitude plus 500 feet):** Avoid flight below and behind a large aircraft's path. If a larger aircraft is observed above on the same track (meeting or overtaking) adjust your position laterally, preferably upwind.

7-47. HELICOPTERS

In a slow hover-taxi or stationary hover near the surface, helicopter main rotor(s) generate downwash producing high velocity outwash vortices to a distance approximately three times the diameter of the rotor. When rotor downwash hits the surface, the resulting outwash vortices have behavioral characteristics similar to wing tip vortices produced by fixed wing aircraft. However, the vortex circulation is outward, upward, around, and away from the main rotor(s) in all directions. Pilots of small aircraft should avoid operating within three rotor diameters of any helicopter in a slow hover taxi or stationary hover. In forward flight, departing or landing helicopters produce a pair of strong, high-speed trailing vortices similar to wing tip vortices of larger fixed wing aircraft. Pilots of small aircraft should use caution when operating behind or crossing behind landing and departing helicopters.

7-48. PILOT RESPONSIBILITY

a. Government and industry groups are making concerted efforts to minimize or eliminate the hazards of trailing vortices. However, the flight disciplines necessary to ensure vortex avoidance during VFR operations must be exercised by the pilot. Vortex visualization and avoidance procedures should be exercised by the pilot using the same degree of concern as in collision avoidance.

b. Wake turbulence may be encountered by aircraft in flight as well as when operating on the airport movement area. (See WAKE TURBULENCE in the Pilot/Controller Glossary.)

c. Pilots are reminded that in operations conducted behind all aircraft, acceptance of instructions from ATC in the following situations is an acknowledgment that the pilot will ensure safe takeoff and landing intervals and accepts the responsibility of providing his own wake turbulence separation.

1. Traffic information,

2. Instructions to follow an aircraft, and

3. The acceptance of a visual approach clearance.

d. For operations conducted behind heavy aircraft, ATC will specify the word "heavy" when this information is known. Pilots of heavy aircraft should always use the word "heavy" in radio communications.

7-49. AIR TRAFFIC WAKE TURBULENCE SEPARATIONS

a. Because of the possible effects of wake turbulence, controllers are required to apply no less than specified minimum separation for aircraft operating behind a heavy jet and, in certain instances, behind large nonheavy aircraft.

1. Separation is applied to aircraft operating directly behind a heavy jet at the same altitude or less than 1,000 feet below:

(a) Heavy jet behind heavy jet - 4 miles.

(b) Small/large aircraft behind heavy jet - 5 miles.

2. Also, separation, measured at the time the preceding aircraft is over the landing threshold, is provided to small aircraft:

(a) Small aircraft landing behind heavy jet - 6 miles.

(b) Small aircraft landing behind large aircraft - 4 miles.

7-49a2b NOTE– See Aircraft Classes in Pilot/Controller Glossary.

3. Additionally, appropriate time or distance intervals are provided to departing aircraft:

(a) Two minutes or the appropriate 4 or 5 mile radar separation when takeoff behind a heavy jet will be:

- from the same threshold

- on a crossing runway and projected flight paths will cross

- from the threshold of a parallel runway when staggered ahead of that of the adjacent runway by less than 500 feet and when the runways are separated by less than 2,500 feet.

7-49a3a NOTE– Pilots, after considering possible wake turbulence effects, may specifically request waiver of the 2-minute interval by stating, "request waiver of 2-minute interval" or a similar statement. Controllers may acknowledge this statement as pilot acceptance of responsibility for wake turbulence separation and, if traffic permits, issue takeoff clearance.

b. A 3-minute interval will be provided when a small aircraft will takeoff:

1. From an intersection on the same runway (same or opposite direction) behind a departing large aircraft,

2. In the opposite direction on the same runway behind a large aircraft takeoff or low/missed approach.

7-49b2 NOTE– This 3-minute interval may be waived upon specific pilot request.

c. A 3-minute interval will be provided for all aircraft taking off when the operations are as described in b(1) and (2) above, the preceding aircraft is a heavy jet, and the operations are on either the same runway or parallel runways separated by less than 2,500 feet. Controllers may not reduce or waive this interval.

d. Pilots may request additional separation i.e., 2 minutes instead of 4 or 5 miles for wake turbulence avoidance. This request should be made as soon as practical on ground control and at least before taxiing onto the runway.

7-49d NOTE– FAR 91.3(a) states: "The pilot in command of an aircraft is directly responsible for and is the final authority as to the operation of that aircraft."

7-50 thru 7-60. RESERVED

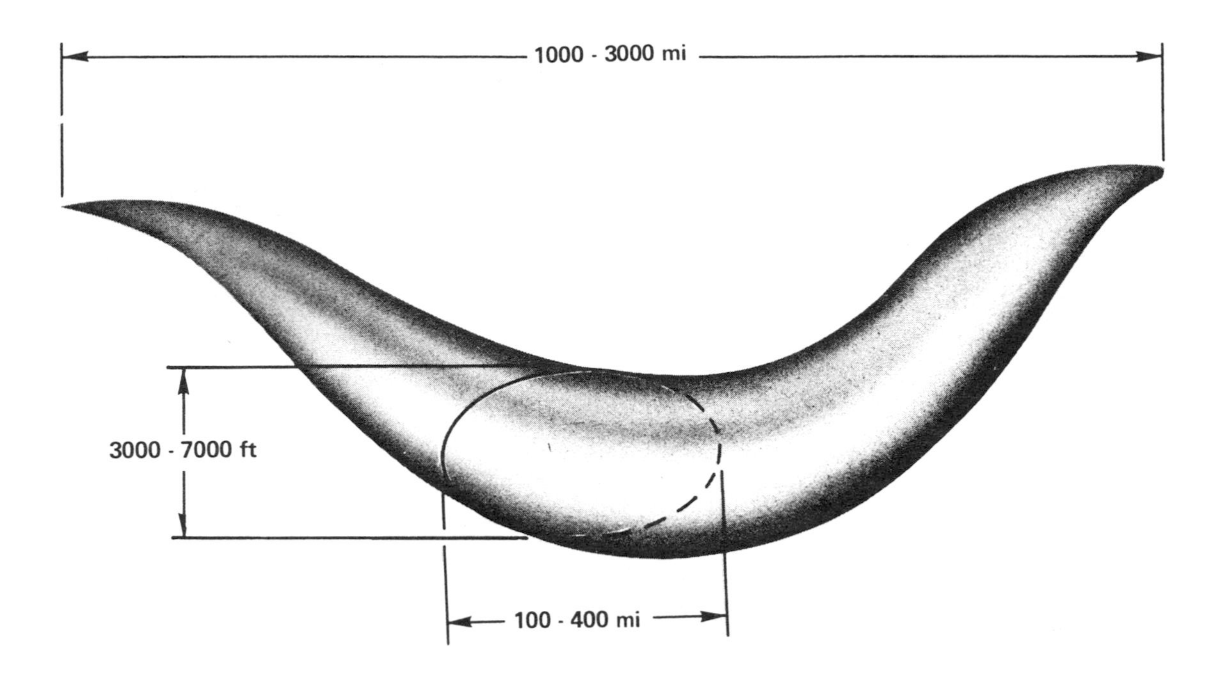

Figure 11-16. A Jet Stream Segment.

J. Jet Stream

Depending on an aircraft's orientation to the jet stream, turbulence may range from light to extreme. The worst turbulence associated with the jet is the wind shear that occurs along the edges, particularly on the polar side below the core (Figures 11-16 and 11-17).

The location of the polar jet and the polar front are highly correlated (Figure 11-18). Both move southward in the winter. Due to the higher vertical extent of the tropopause toward the equator, the core of the jet rises and the speed of the core winds increases.

K. Tropopause

The tropopause represents the boundary between the troposphere and the stratosphere, varying in height from 20,000 feet (6 km) over the poles to about 65,000 feet (20 km) over the equator (Figure 11-19). Often there is an inversion above the tropopause. Thus, one may locate the position of the tropopause by an abrupt change in temperature. Coupled with the presence of a jet stream, the air near the tropopause may be very rough. To reduce turbulence you should climb when the temperature is rising (should place you in the stratosphere, above the jet) and descend when temperatures are decreasing (should place you below the jet).

L. Wind Shear (See Chapter 15 for more information.)

The above discussions all involved the notion of vertical and/or horizontal changes in the air flow. We may classify all of these under a general heading and simply say that turbulence is caused by wind shear. Any sudden change in wind speed and/or direction in a short vertical and/or horizontal direction will be our definition of wind shear (Figures 11-20 and 11-21).

A vertical shear of more than 6 knots per 1,000 feet (300 m) can generate moderate or greater turbulence. A horizontal shear of 40 knots in a distance of about 150 miles (240 km) can produce severe turbulence. For example, on a weather chart if you note winds of 30 knots from the west over a certain location at some altitude and winds of 70 knots from the west over another location 150 miles to the south at the same altitude, you may expect horizontal wind

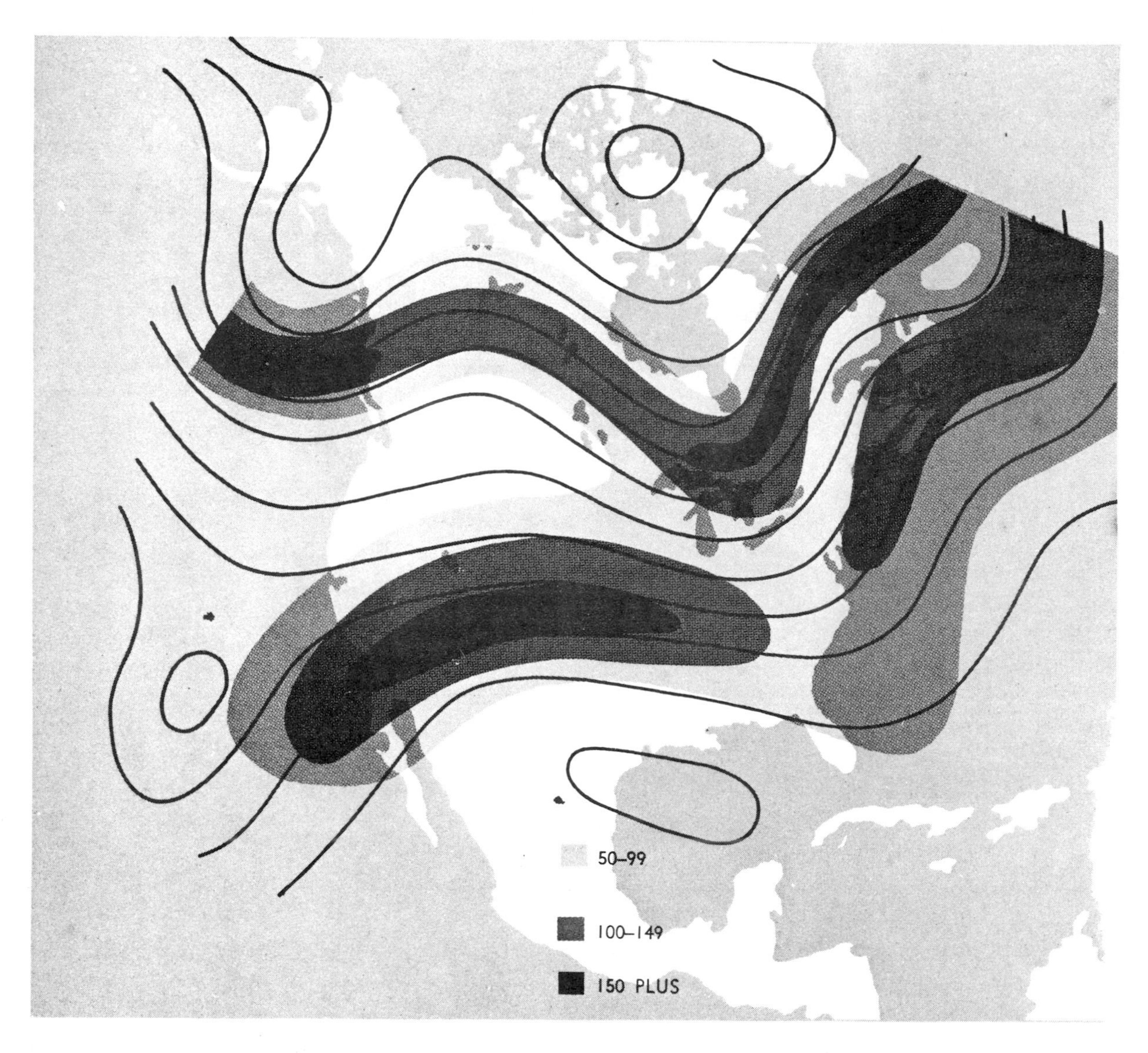

Figure 11-17. Multiple jet streams. Note the "segments" of maximum winds embedded in the general pattern. Turbulence usually is greatest on the polar sides of these maxima. Note both a polar jet and sub-tropical jet, which form in tropopause overlaps — see Figure 11-19.

shear somewhere, but not necessarily continuously, between the two points. Flying directly across such a wind field will produce a horizontal tearing stress on your aircraft.

Similarly, and perhaps more obviously, if the winds were as stated over the first location, that is, 30 knots from the west, but 50 knots from the east at the second location, then the difference in speed is only 20 knots, but we add 30 and 50 to obtain 80 knots of wind shear due to the fact that the winds are exactly opposite in direction. This is a common occurrence between adjacent air masses (Figure 11-22).

1. Headwinds and Tailwinds

A wind shear that results from a decreasing headwind, or an increasing tailwind, or a change from a headwind to a tailwind, will adversely affect the performance of the aircraft. We will now try to explain why this happens.

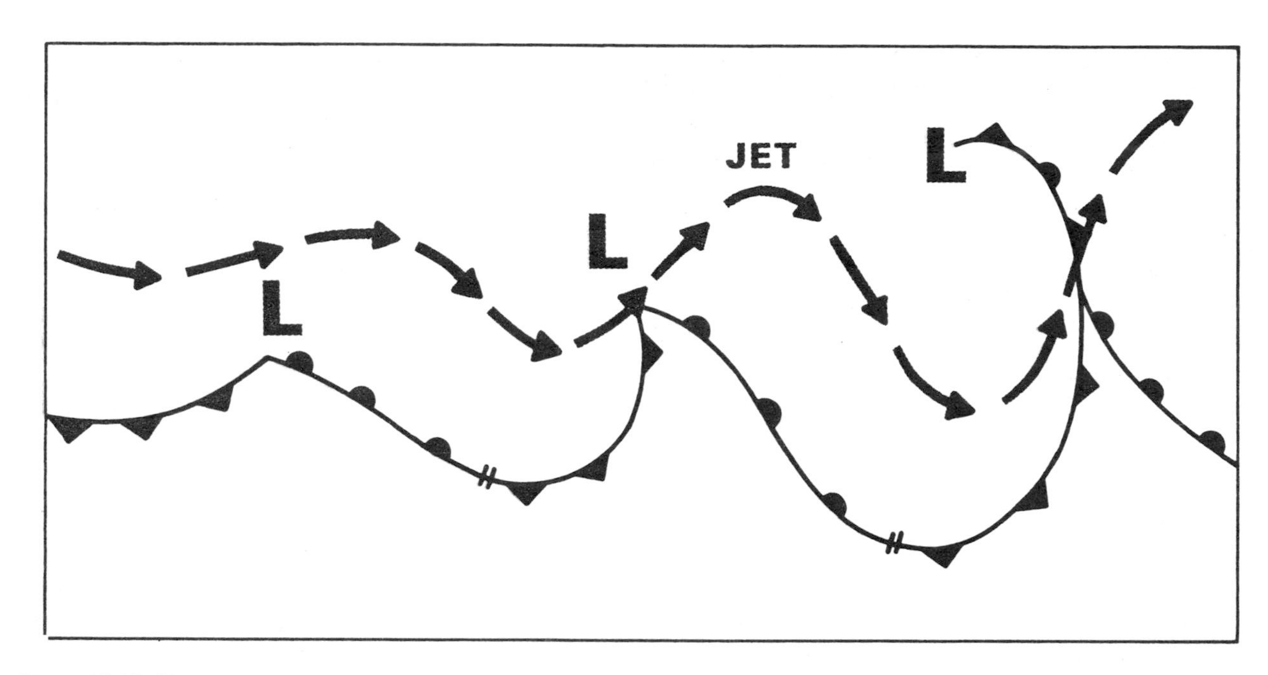

Figure 11-18. Mean jet positions relative to surface systems. Cyclogenesis (development) of a surface low usually is south of the jet as shown on the left. The deepening low moves nearer the jet, center. As it occludes, the low moves north of the jet, right; the jet crosses the frontal system near the point of occlusion.

First, assume that our headwind is constant and that we add power to reach a cruise airspeed. The aircraft picks up speed as long as thrust exceeds drag. When drag equals thrust our speed remains constant. As you recall from an earlier chapter, this result is in accordance with Newton's second law, for thrust and drag are forces acting opposite to each other along our flight path, and when they are equal the net force along the flight path is zero. With $F = 0$, $a = 0$, since $a = F/m$. But acceleration equal zero means the aircraft is flying at a constant speed, and that fact is reflected by the airspeed indicator as a result of the impact of air on the pitot tube.

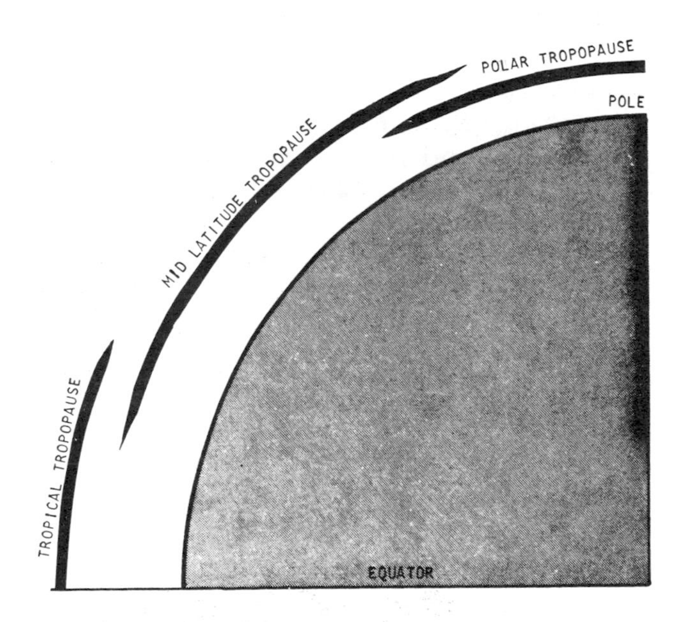

Figure 11-19. The tropopause is discontinuous and is in several discrete overlapping layers. Generally it is lower in cold air than in warm air. Thus the layers lower stepwise from the equator to the poles and are lower in winter than in summer.

Now assume that instead of having a constant headwind, the headwind component decelerates. But since $a = F/m$, there must be less force on the pitot tube and the result is a reduction in indicated airspeed. Now we did not change power, so when the headwind component stops decelerating and maintains a constant speed, we will once again have the same thrust, which is again equal to drag, and our indicated airspeed will be the same as before.

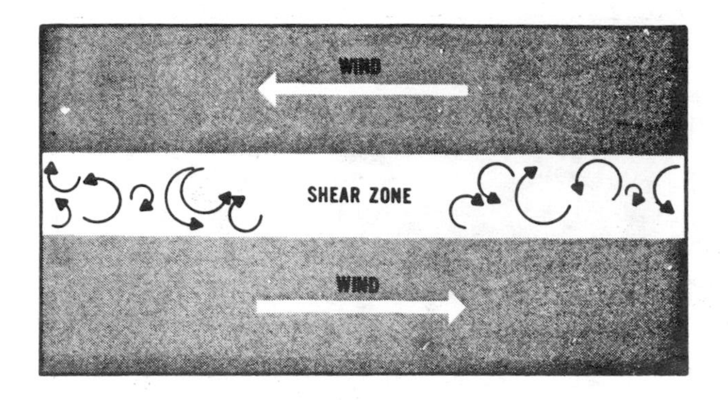

Figure 11-20. Two wind currents in differing directions produce friction between the two. The friction zone is called a shear zone and is a zone of small scale eddies. Shear may be vertical, horizontal, or any direction between.

Whether the headwind is, say, 20 knots or 10 knots, then as we saw above, indicated airspeed will be the same if the power setting and the density of the air remain the same. The danger is not in having a constant 20 knot headwind or a constant 10 knot headwind, for since indicated airspeed is the same, the mass of air encountered by the airplane is the same and, hence, aircraft performance is the same. You may wish to review our discussion of the boat in water that illustrated this fact in the chapter on airspeed.

Again, the danger is not in the difference between constant headwind components; the danger is in the deceleration process of changing from, say, a 20 knot headwind to a 10 knot headwind. As we observed, in that process the air was in a sense accelerating away from the airplane so there was a loss of indicated airspeed, which meant the aircraft was impacting less air and therefore aircraft performance was less. The aircraft will pitch downward in such a situation.

Keep yourself COVERED. Try to COmprehend the factors involved in the headwind deceleration process. Visualize less air impacting the airplane. Evaluate this as a loss of aircraft performance. Respond by applying more power. This response is particularly important for low altitude operations. Suppose you are on final approach and all of a sudden the headwind component begins to decelerate. You will lose indicated airspeed. Your aircraft controls will be less effective and you may stall, being at a low airspeed already, and you have very little altitude in which to recover. After responding by adding power, you will need to Evaluate your situation again and possibly Determine a new course of action, such as a go-around.

Remember, when a headwind decelerates, or a tailwind accelerates, or there is change from a headwind to a tailwind, you are in danger for low altitude operations while such changes are occurring. We conclude that an extra margin of airspeed is wise for such low airspeed operations as takeoffs and landings, particularly when the air is gusty. Many pilots use a rule-of-thumb of

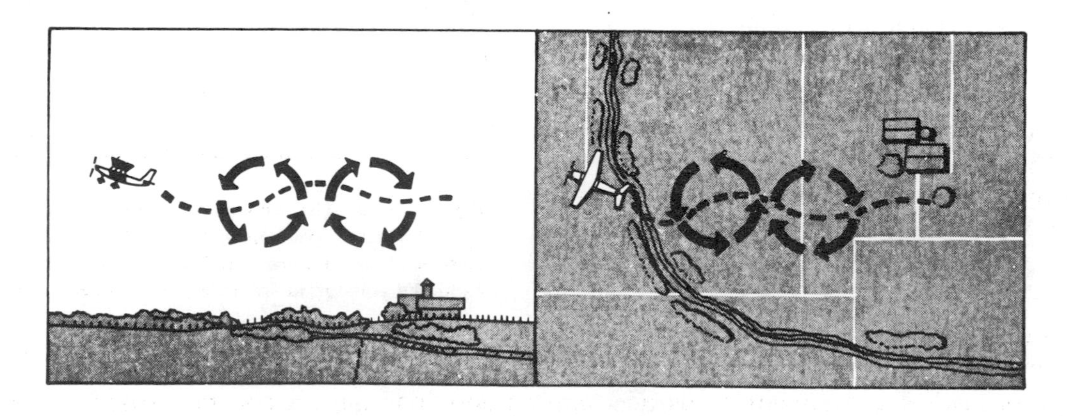

Figure 11-21. Turbulent eddies and how they affect flight path. On the left, eddies displace the aircraft vertically. On the right, eddies displace the aircraft horizontally.

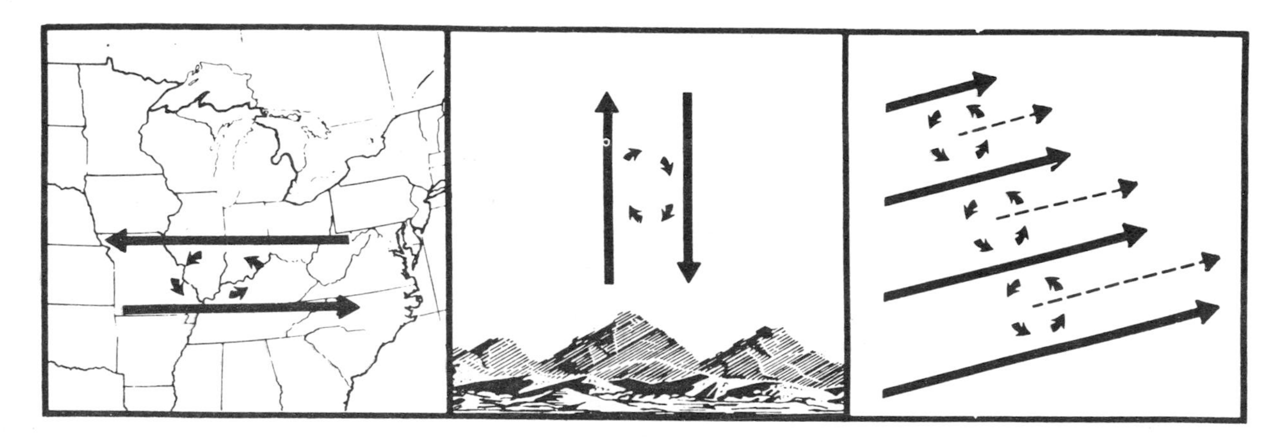

Figure 11-22. Wind shear with resultant turbulent eddies. On the left are winds of opposite directions. In the center, eddies develop between an updraft and a downdraft. The right shows a speed gradient (wind is from the same direction but increases with distance either vertically or horizontally). Eddies form anywhere within the current if shear is sufficiently strong. They roll along at about the average wind speed as indicated by the dashed arrows. Visualize these as tubular eddies similar to rollers in a roller bearing rolling between the bearing surfaces.

one-half the gust speed added to their normal speed in such operations. Do not forget the effects of friction of the air with the earth's surface below 2,000 feet (610 m). This is a natural wind shear and it is always present in varying degrees. And recognize that while an increasing headwind component or a decreasing tailwind component or a change from a tailwind to a headwind component will increase performance while the change is occurring (and the airplane will pitch upward), your aircraft will perform after the change as it did before the change, again assuming that the power setting and the density of the air have not changed.

At cruise altitudes, any losses in indicated airspeed, reflecting decreased aircraft performance, are not usually a problem since you have an extra margin of airspeed and plenty of altitude to recover; this statement is made for normal operations only. One example at cruise altitudes and speeds where you may have a problem with wind shear is in steep turns, for your stall speed will be high in such turns; if a stall does occur, it is usually not critical unless you do not act promptly to recover. Stall recoveries are a very important part of flight training, which you will soon discover if you have not already.

2. *CAT*

High-level wind shear, above approximately 15,000 feet (4.6 km), is sometimes called CAT (Clear Air Turbulence). However, despite its name, it refers to turbulence in clouds as well, although most CAT does occur in clear air (about 80% of the time). Most CAT is associated with the jet stream, most encounters near 30,000 feet (9.1 km) in winter and 34,000 feet (10.4 km) in summer. "Patches of CAT" are quite variable, but 2,000 feet (610 m) thick, 20 miles (32 km) wide, and 50 miles (15 km) or more long would be descriptive (Figure 11-23). Pilot reports are the best means yet available for identifying areas of CAT (Figure 11-24). Forecasts of CAT are difficult, due largely to the lack of sufficient upper-level data. Changes in altitude and flying recommended airspeeds are your best means of coping with CAT.

3. *Turbulence Criteria and Location*

We conclude this chapter with a set of turbulence criteria (Table 11-1) and a set of locations where these criteria are likely to be met. You will help your fellow pilots by filing reports of turbulence, and also by reporting negative turbulence in areas of forecast turbulence.

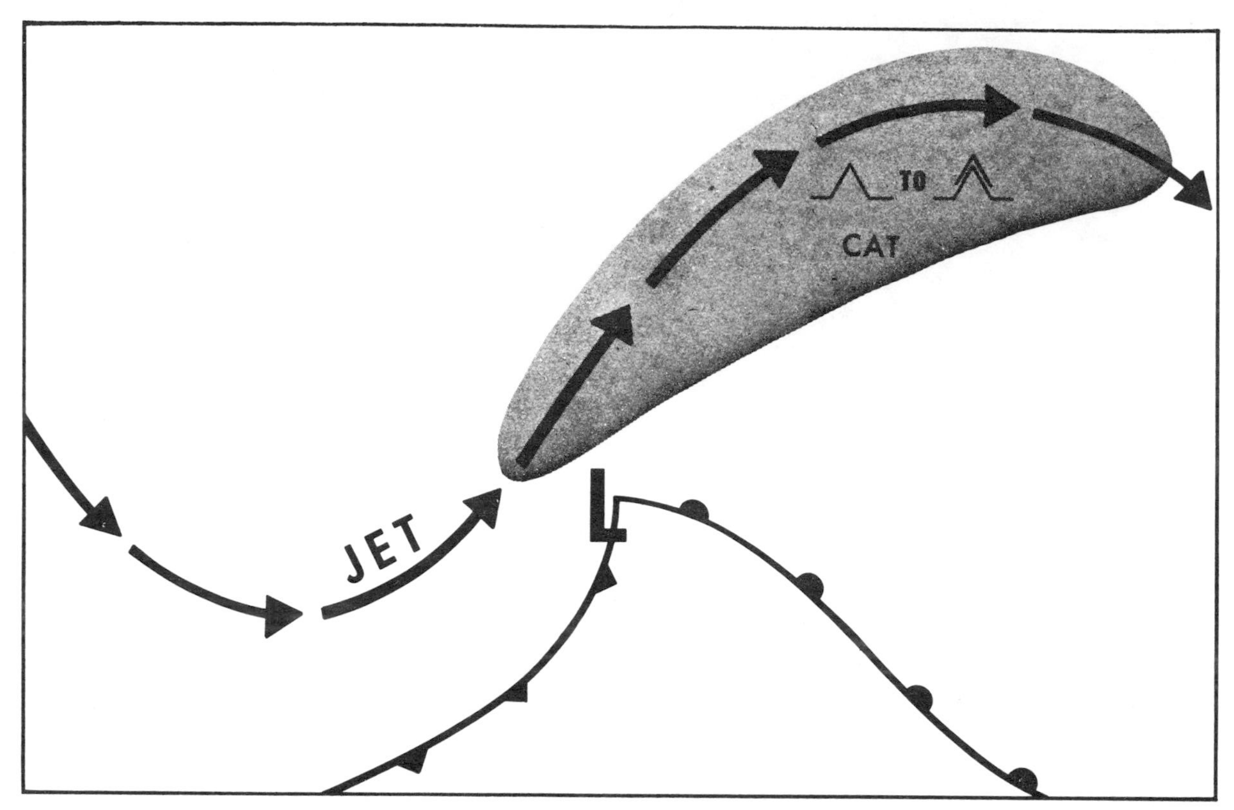

Figure 11-23. A frequent CAT location is along the jet stream north and northeast of a rapidly deepening surface low.

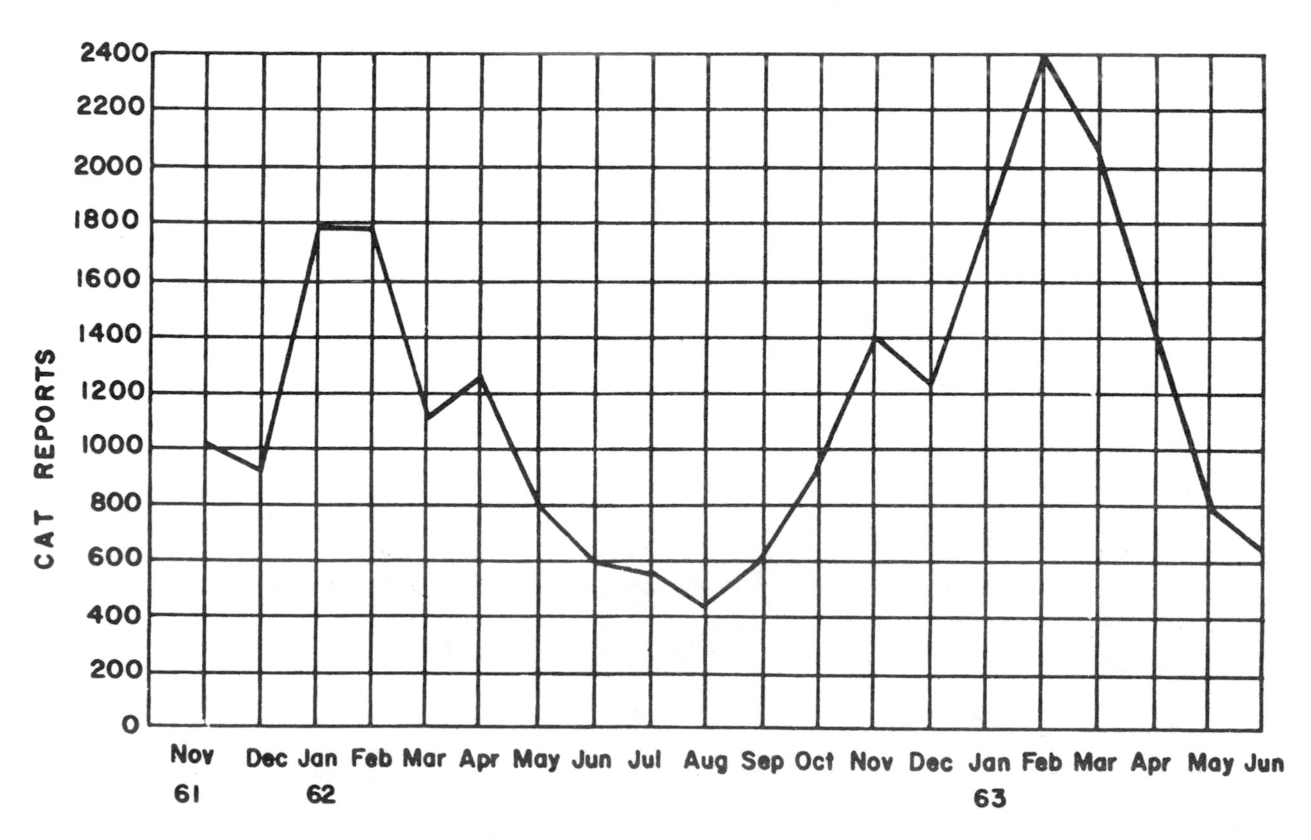

Figure 11-24. Moderate or greater CAT by months over the U.S. The greatest number of encounters are during January and February. Jet-stream winds are lower altitude, more frequent, and stronger during the winter than in summer.

TABLE 11-1. TURBULENCE REPORTING CRITERIA TABLE.

Intensity	Aircraft Reaction	Reaction Inside Aircraft	Reporting Term Definition
LIGHT	Turbulence that momentarily causes slight, erratic changes in altitude and/or attitude (pitch, roll, yaw). Report as Light Turbulence. or Turbulence that causes slight, rapid and somewhat rhythmic bumpiness without appreciable changes in altitude or attitude. Report as Light Chop.	Occupants may feel a slight strain against seat belts or shoulder straps. Unsecured objects may be displaced slightly. Food service may be conducted and little or no difficulty is encountered in walking.	Occasional—Less than 1/3 of the time. Intermittent—1/3 or 2/3. Continuous—More than 2/3. Note—Pilots should report location(s), time (GMT), intensity, whether in or near clouds, altitude, type of aircraft and, when applicable, duration of turbulence. Duration may be based on time between two locations or over a single location. All locations should be readily identifiable. Example: a. Over Omaha, 1232Z, Moderate Turbulence, in cloud, Flight Level 310, B707. b. From 50 miles south of Albuquerque to 30 miles north of Phoenix, 1210Z to 1250Z, occasional Moderate Chop, Flight Level 330, DC8.
MODERATE	Turbulence that is similar to Light Turbulence but of greater intensity. Changes in altitude and/or attitude occur but the aircraft remains in positive control at all times. It usually causes variations in indicated airspeed. Report as Moderate Turbulence; or Turbulence that is similar to Light Chop but of greater intensity. It causes rapid bumps or jolts without appreciable changes in aircraft altitude or attitude. Report as Moderate Chop.	Occupants feel definite strains against seat belts or shoulder straps. Unsecured objects are dislodged. Food service and walking are difficult.	
SEVERE	Turbulence that causes large, abrupt changes in altitude and/or attitude. It usually causes large variations in indicated airspeed. Aircraft may be momentarily out of control. Report as Severe Turbulence.	Occupants are forced violently against seat belts or shoulder straps. Unsecured objects are tossed about. Food service and walking are impossible.	
EXTREME	Turbulence in which the aircraft is violently tossed about and is practically impossible to control. It may cause structural damage. Report as Extreme Turbulence.		

a. Light Turbulence

(1) In mountainous areas even with light winds.

(2) In and near small cumulus clouds.

(3) In clear air convective currents over heated surfaces.

(4) With weak shears in the vicinity of:

(a) Troughs aloft.

(b) Low aloft.

(c) Jet streams.

(d) The tropopause.

(5) In the lower 5,000 feet of the troposphere:

(a) When winds are near 15 knots.

(b) Where the air is colder than the underlying surface.

b. Moderate Turbulence

(1) In mountainous areas with a wind component of 25 to 50 knots perpendicular to and near the level of the ridge:

(a) At all levels from the surface to 5,000 feet above the tropopause with preference for altitudes:

Within 5,000 feet of the ridge level.

At the base of the relatively stable layers below the base of the tropopause.

Within the tropopause layer.

(b) Extending outward on the lee of the ridge for 150 to 300 miles.

(2) In and near other towering cumuliform clouds.

(3) In the lower 5,000 feet of the troposphere:

(a) When surface winds exceed 25 knots.

(b) Where heating of the underlying surface is unusually strong.

(c) Where there is an invasion of very cold air.

(4) In fronts aloft.

(5) Where vertical wind shears exceed 6 knots per 1,000 feet, and/or horizontal wind shears exceed 18 knots per 150 miles.

c. Severe Turbulence

(1) In mountainous areas with a wind component exceeding 50 knots perpendicular to and near the level of the ridge:

(a) In 5,000 foot layers:

At and below the ridge level in rotor clouds or rotor action.

At the tropopause.

Sometimes at the base of other stable layers below the tropopause.

(b) Extending outward on the lee of the ridge for 50 to 150 miles.

(2) In and near growing and mature thunderstorms.

(3) Occasionally in other towering cumuliform clouds.

(4) Fifty to 100 miles on the cold side of the center of the jet stream, in troughs aloft, and in lows aloft where vertical shears exceed 6 knots per 1,000 feet and horizontal wind shears exceed 40 knots per 150 miles.

d. Extreme Turbulence

(1) In mountain wave situations, in and below the level of well developed rotor clouds. (Sometimes it extends to the ground.)

(2) In growing severe thunderstorms (most frequently in organized squall lines) indicated by:

(a) Large hailstones (3/4 inch or more in diameter).

(b) Strong gradients in radar echo, or

(c) Almost continuous lightning.

TURBULENCE
Review Questions

PRI

1. Which statement describes the normal characteristics of standing lenticular clouds?
   a. The clouds have dense boiling tops. They contain violent turbulence and are considered the most hazardous of the cloud types.
   b. The clouds have billowing tops and comparatively high bases, producing continuous rain.
   c. The clouds are gray or dark, containing very little turbulence and are not a hazard to flight.
   d. The clouds are almond or lens-shaped and show little or no movement, but may contain strong winds and turbulence.
2. An almond or lens-shaped cloud which appears stationary, but which may contain winds of 50 knots or more, is referred to as:
   a. An inactive frontal cloud.
   b. A funnel cloud.
   c. A lenticular cloud.
   d. A stratus cloud.
3. While flying on the leeward side of a mountain range, you observe almond or lens-shaped clouds. These clouds are referred to as
   a. Cirrocumulus clouds.
   b. Roll clouds.
   c. Cirrus clouds.
   d. Lenticular clouds.
4. If severe turbulence is encountered, the airplane's airspeed should be reduced to
   a. Maneuvering speed.
   b. The minimum steady flight speed in the landing configuration.
   c. Normal operation speed.
   d. Maximum structural cruising speed.
5. Low-level wind shear occurs
   a. When surface winds are 15 knots and there is no change in wind direction and wind speed with height.
   b. After a warm front has passed.
   c. When there is a low-level temperature inversion with strong winds above the inversion.
   d. When surface winds are light and variable.
6. Low-level wind shear is best described as
   a. Deflection of wind currents as the result of Coriolis force.
   b. A downward motion of the air associated with continuous winds blowing with an easterly component due to the rotation of the earth.
   c. A change in wind direction and/or speed in a very short distance in the atmosphere.
   d. A violently rotating column of air extending from a cumulonimbus cloud.
7. In strong wind conditions, flight over a mountainous area within close proximity to the peaks may be hazardous because of
   a. Violent downdrafts on the windward side.
   b. Violent downdrafts on the leeward side.
   c. Strong turbulence associated with stratus clouds.
   d. Wind shear on the windward side.

INS

8. Standing lenticular clouds (ACSL), in mountainous areas, indicate
   a. Turbulence.
   b. Unstable air.
   c. An inversion.
   d. Light variable winds.

9. The presence of altocumulus standing-lenticular clouds is a good indication of
   a. Heavy rain.
   b. Very strong turbulence.
   c. Heavy icing conditions.
   d. An approaching storm.

10. Where does wind shear occur?
    a. Only at higher altitudes, usually in the vicinity of jet streams.
    b. At any level, and it can exist in both a horizontal and vertical direction.
    c. Primarily at lower altitudes in the vicinity of mountain waves.
    d. Only in the vicinity of thunderstorms.

11. Where does wind shear occur?
    a. Wind shear of any significance occurs only in connection with the jet stream.
    b. Wind shear may be associated with either a wind shift or a windspeed gradient at any level in the atmosphere.
    c. It occurs primarily at lower altitudes in the vicinity of mountain waves.
    d. It occurs only when there is a strong temperature inversion, or when the jet stream is associated with a strong low.

12. Hazardous wind shear is commonly encountered near the ground in the vicinity of thunderstorms and with warm afternoon temperatures:
    a. During periods when the wind velocity is stronger than 35 knots.
    b. Near mountain valleys when the lapse rate is greater than normal.
    c. During periods of strong low level temperature inversion.
    d. On windward side of a hill or mountain.

13. Moderate turbulence exists when
    a. Rapid and somewhat rhythmic bumpiness is experienced without appreciable changes in altitude or attitude.
    b. Large, abrupt changes in altitude or attitude occur but the airplane may only be out of control momentarily.
    c. Changes in altitude or attitude occur but the aircraft remains in positive control at all times.
    d. Continued flight in this environment may result in structural damage.

COM

14. Which type of approach and landing is recommended during gusty wind conditions?
    a. A power-off approach and power-off landing.
    b. A power-on approach and power-on landing.
    c. A power-off approach and power-on landing.
    d. A power-on approach and power-off landing.

15. Why can turbulent air cause an increase in stalling speed?
    a. The true airspeed is abruptly increased.
    b. The load factor is suddenly decreased.
    c. The angle of attack is decreased.
    d. The angle of attack is increased.

16. The indicated airspeed on the final approach to a landing should be faster than normal when
    a. Atmospheric conditions are below standard.
    b. Landing at airports above 5,000 feet MSL.
    c. Making a power approach.
    d. Turbulent conditions exist.

17. Turbulent air can cause an increase in stalling speed when there is
    a. A decrease in angle of attack.
    b. A sudden decrease in load factor.
    c. An abrupt increase in true airspeed.
    d. An abrupt change in relative wind.

18. The principal cause of hazardous conditions associated with the wake turbulence of large airplanes is the
    a. High speeds at which large airplanes operate.
    b. Tornado-like vortices generated by the wingtips.
    c. Propeller or jet "wash."
    d. Laminar flow airfoil used on airplane designs.

19. During a takeoff made behind a departing large jet airplane, the pilot can minimize the hazard of wingtip vortices by
    a. Extending the takeoff roll and not rotating until well beyond the jet's rotation point.
    b. Maintaining extra speed on takeoff and climbout.
    c. Remaining below the jet's flight path until able to turn clear of its wake.
    d. Being airborne prior to reaching the jet's rotation point and climbing above its flight path.

20. What effect would a light crosswind have on the wingtip vortices generated by a large airplane that had just taken off?
    a. Both vortices would move downwind at a greater rate than if the surface wind was directly down the landing runway.
    b. The downwind vortex would tend to remain on the runway longer than the upwind vortex.
    c. A light crosswind would rapidly dissipate the strength of both vortices.
    d. The upwind vortex would tend to remain on the runway longer than the downwind vortex.

21. Choose the correct statement regarding wake turbulence.
    a. The greatest vortex strength is produced when the generating airplane is heavy, clean, and fast.
    b. The primary hazard is loss of control because of induced roll.
    c. Vortex generation begins with the initiation of the takeoff roll.
    d. Vortices tend to remain level for a period of time.

22. Which pilot action would be most appropriate for minimizing the hazards of wingtip vortices if cleared for takeoff behind a large jet?
    a. Be airborne prior to reaching the point where the jet rotated, and climb above its flight path.
    b. Maintain the ground run until past the point where the jet took off, and climb below the jet's flight path.
    c. Taxi into position on the runway and hold until the vortices subside.
    d. Take off and climb at maximum speed to attain positive aircraft control in the event turbulence is encountered.

23. Hazardous vortex turbulence that might be encountered behind large aircraft is created only when that aircraft is
    a. Using high power settings.
    b. Operating at high airspeeds.
    c. Heavily loaded.
    d. Developing lift.

24. The loss of aircraft control, which may occur if a light airplane is flown into the wake of a large airplane, is caused principally by
    a. The tornado-like vortices produced by the wingtips of the large airplane.
    b. High-speed sound waves similar to those produced by sonic "booms."
    c. Turbulence created by the propellers or jet exhaust of the large airplane.
    d. Meteorological factors which create wind shear.

25. If wake turbulence is encountered, the probability of induced roll increases when the encountering aircraft's
    a. Airspeed is slower than that of the generating aircraft.
    b. Altitude is higher than that of the generating aircraft.
    c. Heading is aligned with the flight path of the generating aircraft.
    d. Heading is perpendicular to the flight path of the generating aircraft.

26. Which statement is true relating to the effect of low-level wind shear on airplane performance?
    a. A headwind which shears to a tailwind causes the airplane to pitch up.
    b. A headwind which shears to a tailwind causes an initial increase in airspeed.
    c. A tailwind which shears to a headwind causes the airplane to pitch up.
    d. A tailwind which shears to a headwind causes an initial decrease in airspeed.

27. Which statement is true relating to the effect of low-level wind shear on airplane performance?
    a. A tailwind which shears to a headwind causes an initial decrease in airspeed.
    b. A tailwind which shears to a headwind causes the airplane to pitch down.
    c. A headwind which shears to a tailwind causes an initial increase in airspeed.
    d. A headwind which shears to a tailwind causes the airplane to pitch down.

28. Which statement is true relating to the effect of low-level wind shear on airplane performance?
    a. A headwind which shears to a tailwind causes the airplane to pitch up.
    b. A headwind which shears to a tailwind causes an initial decrease in airspeed.
    c. A tailwind which shears to a headwind causes the airplane to pitch down.
    d. A tailwind which shears to a headwind causes an initial decrease in airspeed.

29. During departure, under conditions of suspected low-level wind shear, a sudden decrease in headwind will cause
    a. A loss in airspeed equal to the decrease in wind velocity.
    b. A gain in airspeed equal to twice the amount of decrease in wind velocity.
    c. A loss in airspeed equal to twice the amount of decrease in wind velocity.
    d. A gain in airspeed equal to the decrease in wind velocity.

30. During departure, when low-level wind shears to a tailwind or rapidly decreasing headwind, aircraft performance will
    a. Increase.
    b. Decrease.
    c. Remain unchanged.
    d. Initially increase, then decrease.

CFI

31. If a strong temperature inversion is encountered immediately after takeoff or during an approach to a landing, a potential hazard exists because of turbulence created by
    a. Strong convective currents.
    b. Cold air overriding calm, warm air.
    c. Wind shear.
    d. Strong surface winds.

32. Which one of the following types of clouds would indicate areas of convective turbulence?
    a. Altocumulus standing-lenticular clouds.
    b. Nimbostratus clouds.
    c. Cirrus clouds.
    d. Towering cumulus clouds.

33. Suppose a strong temperature inversion exists near the surface. Should this phenomenon be considered hazardous to aircraft?
    a. No; a temperature inversion near the surface creates smooth air and good visibilities.
    b. No; a sudden change in outside air temperature would be the only evidence of a temperature inversion.
    c. Yes; a potential hazard exists due to strong, steady downdrafts.
    d. Yes; a potential hazard exists due to turbulence created by wind shear.

34. Which statement is true regarding "standing mountain waves"?
    a. They are always indicated by the presence of a lenticular cloud formation.
    b. They are sometimes marked by stationary, lens-shaped clouds.
    c. They are found on the windward side of the mountain.
    d. They are generally stationary over the mountain.

35. When turbulence causes changes in altitude and/or attitude but aircraft control remains positive, that turbulence should be reported as
    a. Moderate.
    b. Severe.
    c. Light.
    d. Very light.

36. Turbulence that causes unsecured objects in the cockpit or cabin to become dislodged should be reported as
    a. Severe.
    b. Moderate.
    c. Light.
    d. Very light.

37. Turbulence that causes a pilot to feel a slight strain against seatbelts should be reported as
    a. Very light.
    b. Light.
    c. Moderate chop.
    d. Moderate.

38. Turbulence that causes a pilot to feel definite strains against the seatbelt or shoulder straps should be reported as
    a. Severe.
    b. Moderate.
    c. Light.
    d. Very light.

39. When unsecured objects are tossed about the cockpit or cabin due to turbulence, that turbulence should be reported as
    a. Very light.
    b. Moderate.
    c. Light.
    d. Severe.

ATP

40. Hazardous wind shear is commonly encountered near the ground
    a. During periods when the wind velocity is stronger than 35 knots and near mountain valleys.
    b. During periods of strong temperature inversions and near thunderstorms.
    c. Near mountain valleys and on the windward side of a hill or mountain.
    d. Near thunderstorms and during periods when the wind velocity is stronger than 35 knots.

41. What is the recommended action a pilot should take with respect to temperature indications to cross a jet stream core to minimize the effects of CAT?
    a. If temperature rises—climb; if temperature decreases—descend.
    b. Climb to a higher altitude when the temperature rises or decreases.
    c. If temperature rises—descend; if temperature decreases—climb.
    d. Descend to a lower altitude when the temperature rises or decreases.

42. Which feature is associated with the tropopause?
    a. Absolute upper limit of cloud formation.
    b. Abrupt change in temperature lapse rate.
    c. Constant height above the earth.
    d. Absence of wind and turbulent conditions.

43. In general terms, what is the migration pattern, level, and strength of the jet stream during the winter months in the middle and high latitudes?
    a. Shift toward the south, core rises to a higher altitude, and speed increases.
    b. Shift toward the north, core rises to a higher altitude, and speed decreases.
    c. Shift toward the south, core descends to a lower altitude, and speed increases.
    d. Shift toward the north, core descends to a lower altitude, and speed decreases.

44. Where will the area of strongest turbulence be encountered when departing a jet stream?
    a. Above the core on the polar side.
    b. Above the core on the equatorial side.
    c. Below the core on the polar side.
    d. Below the core on the equatorial side.

CHAPTER XII
ICING

The reader should demonstrate an understanding of icing to the extent that he can answer the following questions (as stated below, or as multiple-choice, etc.):

1. What are the two main types of aircraft icing?
2. What is necessary for structural icing to form? List some consequences of such icing. State the effects on flying ability of an aircraft, fuel consumption, and stall speed.
3. State the usual meteorological conditions necessary for the occurrence of induction icing and state the effects on aircraft.
4. State the range in the accretion rate of ice.
5. What effect on power is an accumulation of half an inch (1.25 cm) of structural ice, and what does this imply about operating an aircraft?
6. State factors on which the accretion rate depends.
7. State the three kinds of structural ice.
8. How does clear ice form?
9. Describe the appearance of clear ice and state reasons for that appearance.
10. Where are the droplets necessary for clear ice formation usually found? What is the usual temperature range for clear ice formation?
11. Clear ice accounts for approximately what percentage of icing occurrences?
12. Describe the appearance and formation process of rime ice.
13. Compare clear and rime ice with regard to ease of removal and weight.
14. Where and at what temperatures are droplets for rime ice usually found?
15. Rime ice accounts for approximately what percentage of icing occurrences?
16. Why is icing at temperatures below -20°C usually less a problem?
17. What is mixed ice and how often does it occur?
18. Describe the meteorological conditions under which mixed icing occurs.
19. State the effects of ice on the propeller. What is one means of reducing the amount of ice on the propeller and how would you know such a means is effective?
20. State the effect of an iced-over static port on the altimeter, rate-of-climb indicator, and indicated airspeed.
21. State the effect on indicated airspeed when both the pitot tube and static port are iced over.
22. State the effects of icing on radio antennas.
23. State the relationship between icing and aircraft control surfaces, landing gear, and braking.
24. Define "anti-icer"; "deicer". Give examples of each and state where they are used.
25. State three types of induction icing. Describe the formation of each.
26. What is one good means of preventing throttle icing. Where and why is such icing especially critical?
27. Why is evaporative ice less a problem in fuel-injected engines?
28. State the meteorological conditions conducive to maximum evaporative ice formation.
29. Why is evaporative icing less a problem for cold temperatures (below 32°F, 0°C)?
30. What is the main means for preventing carburetor ice?
31. State the sequence of events on RPM, from the beginning of evaporative ice formation to its removal by use of carburetor heat.
32. Carburetor heat is primarily a(n) (anti-icer, deicer).
33. State an emergency procedure for eliminating carburetor ice.
34. State two adverse effects of the use of carburetor heat during operations requiring full power.
35. Whose recommendations should be followed for the use of carburetor heat?
36. State the effects of carburetor ice on aircraft with a constant speed propeller.
37. How should carburetor heat be regulated on higher output engines, especially those with superchargers?
38. Under what conditions is frost most likely to occur at cruise altitudes?
39. Why is frost a special concern on takeoffs and landings?
40. Relative to icing, at what angle should you cross a front?
41. What causes most of the icing in a frontal area?
42. What is your chief means of avoiding ice in a frontal area?
43. How should you conduct changes in altitude relative to possible icing?
44. Compare the zone of icing in warm and cold fronts.
45. Why might it be preferable to fly in a cloud as opposed to above or beneath it?
46. Compare dry and wet snow relative to icing.
47. What do wet snow and ice pellets suggest about temperatures aloft?
48. Why are fronts, lows, thunderstorms, and orographic lifting especially dangerous relative to icing?

CHAPTER XII

ICING

Aircraft icing is one of the most serious of weather hazards. We classify aircraft icing into two main types — structural and induction. Structural icing requires visible liquid moisture and below freezing temperatures and can cause: incorrect readings on aircraft instruments, restricted vision, loss of flying ability of the airplane, loss of radio communication, and loss of proper usage of brakes, landing gear, and control surfaces.

Loss of flying ability occurs due to a loss of lift and thrust and increases in weight and drag; fuel consumption is also greater. Due to a change in the shape of the wing, the critical angle of attack will be lower, resulting in an even higher stall speed (Figure 12-1). Induction icing can occur when outside temperatures are quite warm and moisture is in the vapor state; this kind of icing causes a loss of power, partial or total.

Actually, structural ice can form in air that is a few degrees above freezing. Since an aircraft flies largely by inducing a pressure change in the air in which it flies, a lower pressure will allow the air to expand and cool. We indicated earlier that impurities in the air appear to be less when supercooled water is in abundance. The aircraft is itself a large impurity, and when it encounters supercooled water the accretion rate of ice can be great. In fact, the accretion rate usually averages as little as one-half inch (1.25 cm) per hour to as much as 2 to 3 inches (5 to 7.5 cm) in less than five minutes. As little as one-half inch of ice can reduce the lifting power of some aircraft by as much as 50%. Thus, one should avoid abrupt maneuvers. The accretion rate will depend primarily on the size of drops, assuming there is sufficient liquid water content. Airspeed and the structure of the airfoil also influence the accretion rate. The upper limit of airspeed is about 400 knots for significant icing; at speeds above 400 knots, heat created by friction and compression will all but stop the ice formation. Avoid too low an airspeed, for the extra weight and change in shape of the wing will cause an increase in stall speed. There are three kinds of structural ice — clear, rime, and mixed.

A. Clear Ice

Clear ice forms when large drops strike the aircraft surface and slowly freeze. The relatively slow freezing allows the drops to spread out before freezing, giving this ice its clear, glassy appearance, similar to the glaze on trees during freezing rain. The large droplets necessary for this kind of ice are most often found in cumuliform clouds and freezing rain. In fairly smooth air, the relatively slow freezing will most usually occur at temperatures between 0 °C and −15 °C, but can occur at temperatures much colder in cumulonimbus clouds. Clear ice accounts for about 10% of icing occurrences.

B. Rime Ice

Rime ice, a milky, opaque, rough kind of ice, is formed when small drops strike the aircraft surface and freeze rapidly. The rapid freezing causes drops to freeze before they can spread out, trapping substantial amounts of air; this accounts for its appearance and makes it brittle and fairly easy to remove. Rime ice is not as heavy as clear ice. The small droplets necessary for rime ice formation are most often found in stratiform clouds or drizzle at temperatures of −10°C to −20°C. While rime ice is lighter than clear ice, it can significantly alter the shape of the wing, causing a substantial loss in lifting power, mentioned before. Rime ice occurs about 70% of the time that icing is present.

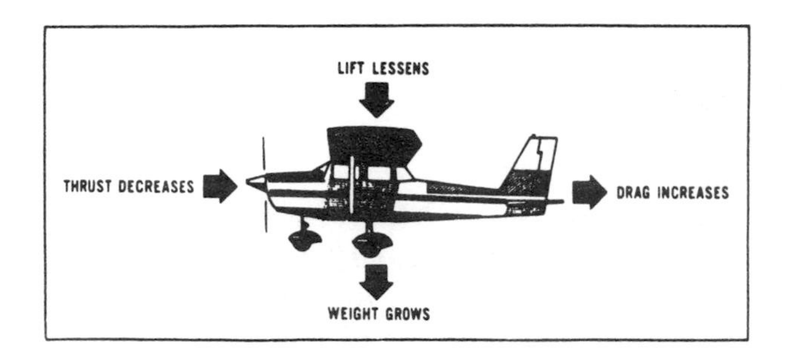

Figure 12-1. Aircraft icing and the four basic forces. Increased drag and decreased thrust work together to decrease forward speed. Less lift and added weight tend to force the aircraft downward. Thus each of the four effects decreases efficiency.

At altitudes where temperatures are below −20°C, icing is usually less a problem due to less moisture as a result of colder temperatures, and also because most of the moisture is in the form of ice crystals that do not adhere to the aircraft.

C. Mixed Ice

Mixed ice is just as the name implies, a combination of rime and clear ice, and occurs about 20% of the time that icing is present. This type of ice usually forms when supercooled water drops vary in size or when supercooled water is mixed with snow or ice particles.

D. Icing of Propellers

We have mentioned the reduction in lifting power resulting from ice accumulations. This would be primarily a result of ice on the wings. Certainly, ice on the propeller will reduce its efficiency (Figure 12-2). And, of course, if the propeller is rendered useless, a completely ice-free wing is of no value to us. Varying the speed of the propeller may help sling off some of the ice. You will likely hear considerable noise as ice particles are slung from propellers against the fuselage, a disturbing but not in itself hazardous situation.

E. Pitot-Static System

We know from a previous discussion that the airspeed indicator operates off the pitot-static system; the altimeter and vertical velocity indicators operate off the static port only. Assume that only the static port ices over. Then as the aircraft descends, air trapped in the static system contracts (expands when climbing), thus causing a greater differential in pressure between the static port and ram air pressure on the pitot tube; this greater pressure differential translates as a false increase in airspeed.

If the ram air input and the drain hole of the pitot system become blocked, trapping the pressure in the system, indicated airspeed will generally not change during level flight, even when the actual airspeed is varied by large power changes. This is because the airspeed indicator will be operating much as an altimeter, that is, only off the static port (Figure 12-3). If the drain hole remained open with the ram air input blocked, the indicated airspeed would generally drop to zero (air is not trapped in the system, but there is no impact air).

F. Radio Antenna Icing

Icing of the radio antenna causes reduction in communication, sometimes a total loss. The antenna may vibrate and howl in the wind until it breaks off.

G. Control Surfaces, Brakes, and Landing Gear

Very little ice is required to freeze the various control surfaces of small aircraft. Taxiing in water, slush, or mud can freeze the landing gear. Just as in a car, for maximum effectiveness of the brakes use them on and off for brief periods of time. Do not lock the brakes. After taking off and raising the gear, you would be wise to recycle the gear to break or shake off some of the slush or mud.

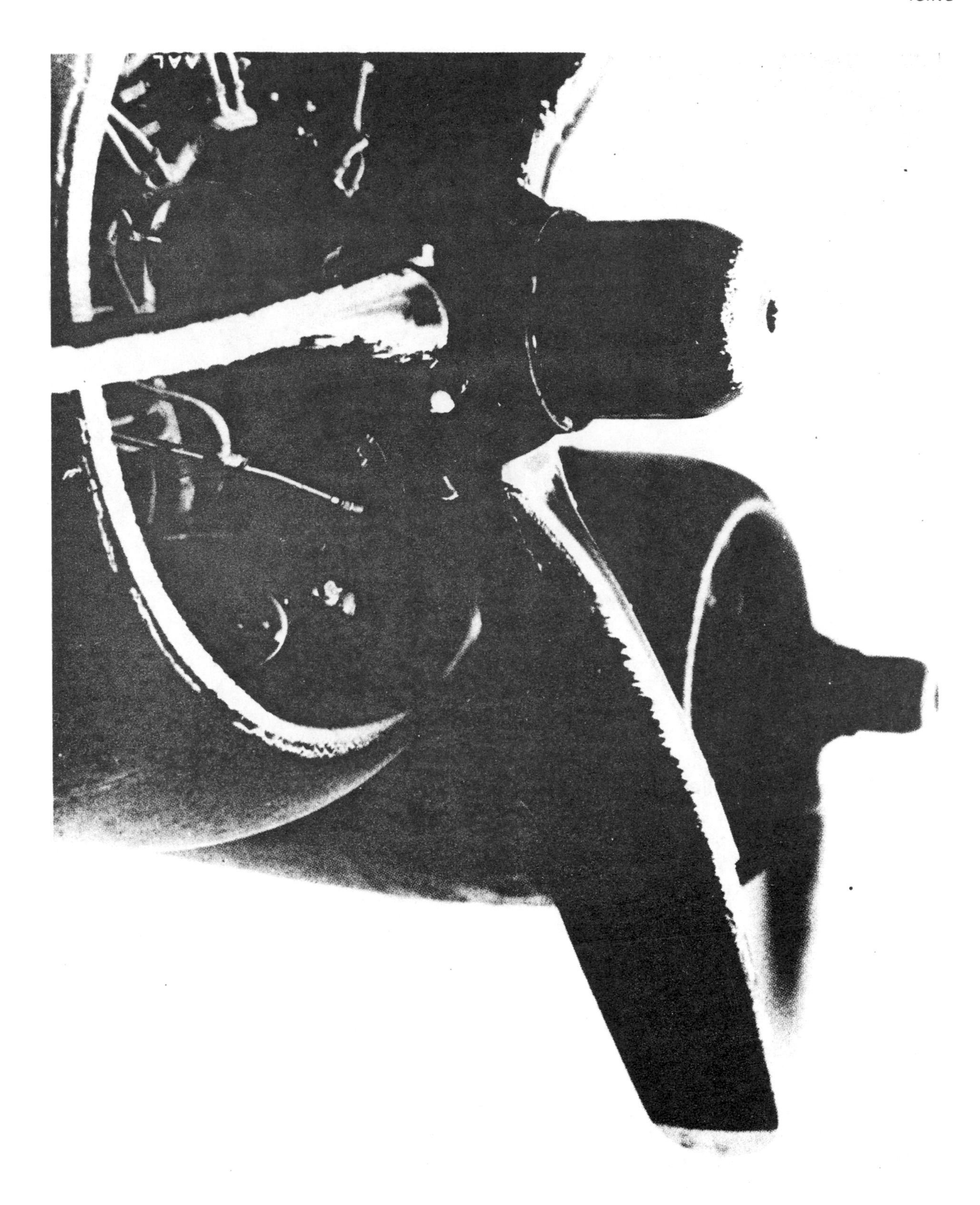

Figure 12-2. Propeller icing. Ice may form on propellers just as on any airfoil. It reduces propeller efficiency and may induce severe vibrations.

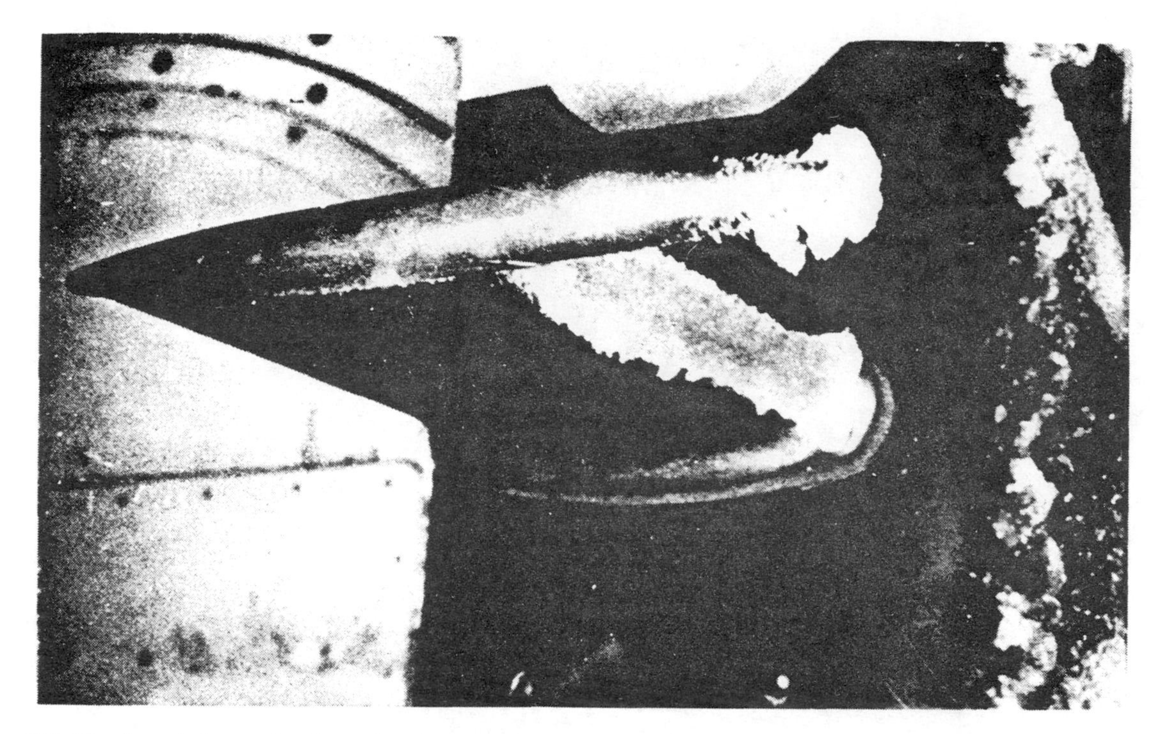

Figure 12-3. Mixed rough ice on the pitot tube of an aircraft. This icing is also a form of instrument icing. Mixed icing on airfoils is most effective of all types in spoiling their aerodynamic efficiency.

H. Anti-icing and Deicing

With regard to ice removal or prevention, several methods are available, which we will only briefly mention. Equipment used to prevent ice formation are called anti-icers and equipment used to remove ice are called deicers. Fluids on propellers are good anti-icers. Inflatable boots on the leading edges of wings are deicers, as ice is allowed to accumulate to about a quarter inch (.6 cm) and then cracked off by inflation of the boot. Heat applied through electrical means or hot air along the leading edges of the wing, propellers or windshield, are both anti-icing and deicing methods.

I. Induction Icing

As we have already indicated, induction icing may cause partial or total loss of the engine or engines. Induction icing is really of three types – impact ice, throttle ice, and carburetor ice. We will try to briefly describe these without getting too involved with the structure of the powerplant. Impact ice forms at the air intake, as well as at other parts of the induction system, and requires freezing temperatures and some form of water particles, liquid or solid, to pose a problem. It affects both fuel injected engines and engines equipped with carburetors. Throttle ice most often forms when the throttle is in the closed position. This type of icing is especially likely when taxiing or when making an approach for landing. One should open the throttle occasionally with the cockpit throttle control in order to reduce the likelihood of such icing. This is especially important in landing approaches if more power is needed, for you will have none available if the carburetor throttle is stuck.

1. Carburetor Ice

When one speaks of induction icing, one is usually referring to carburetor ice. There is less danger of such icing in the fuel-injected engine because vaporization takes place in or near cylinders. However, this can be a major problem with carburetors, particularly the common float type. The process is as follows: the vaporization of fuel, as the evaporation of water, will remove heat from the surroundings, that is, from the air in the carburetor. Add to this a carburetor venturi effect, and this process can result in a temperature drop of as much as 60°F (33°C). So if the outside air temperature is, say, 70°F (21°C), there would be sufficient cooling for the water vapor in the carburetor to sublimate on various parts of the carburetor. This is particularly dangerous with high humidity, say, 68°F (20°C) for our dewpoint. The air would be almost saturated before cooling and the maximum amount of moisture is available for ice formation. The most likely locations for carburetor ice are on the throttle valve and venturi (Figure 12-4).

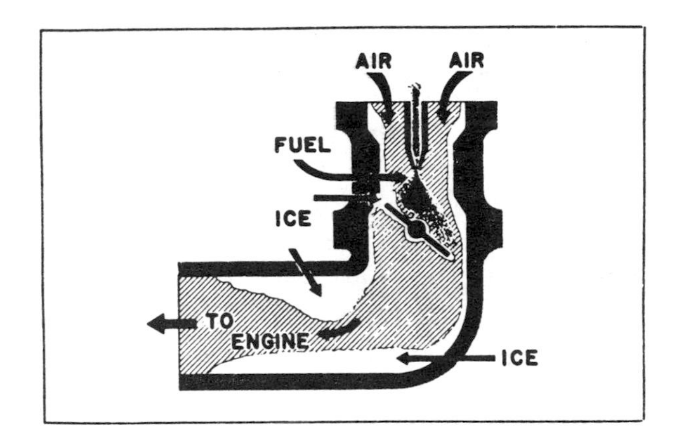

Figure 12-4. Carburetor icing, a form of induction system icing. This type icing occurs under a wide range of atmospheric conditions. More accidents result from carburetor icing than from any other kind.

2. Carburetor Heat

Be alert for carburetor icing when the outside air temperature is between 32°F (0°C) and 70°F (21°C) and the relative humidity is 50% or greater. At temperatures below 32°F, there is less moisture available for formation of carburetor ice; in fact, for approximately each 20°F (11°C) decrease in temperature the capacity of air for water is cut in half. Carburetor heat is our main means for preventing this kind of icing. When the possibility of carburetor icing is suspected, apply full carburetor heat (there are several types of these) to bring in hot exhaust air to be mixed with fuel.

Carburetor ice is first indicated on the usual variable-speed propeller trainer aircraft by a decrease in RPM. As carburetor heat is applied, there will be another decrease in RPM due to the less dense hot air, and the engine will generate less power as we saw in the chapter on temperature. The mixture can be leaned slightly to compensate, but this procedure is best left to the more experienced pilot, the lesser experienced pilot having his hands full with other considerations. As the ice begins to melt, we have our first increase in RPM, followed by a second increase after carburetor heat is turned off.

Carburetor heat is primarily an anti-icer, not a deicer. Some ice that has already formed may be removed, but in most cases this will be too late. For example, if in partially closed throttle operations a loss in RPM is noted, then application of carburetor heat will be to no avail if the throttle is stuck since there will not be a generation of sufficient hot exhaust air. In an emergency situation, the mixture could be leaned until the engine runs so rough that the engine backfires. Such action could remove the ice and, upon increasing the mixture, you may find that engine power comes back quickly. Again, this is usually an emergency consideration only. Thus, whenever carburetor ice is likely, apply full carburetor heat before closing the throttle; periodically open the throttle to make sure it is not sticking and to generate more hot exhaust air for the carburetor heat system. Since a loss of power occurs with the use of carburetor heat, avoid using it for prolonged periods of time during full power operations. Not only will power be lost, but higher cylinder pressures will occur, resulting in possible engine damage. Always follow the manufacturer's recommendations for the use of carburetor heat.

For aircraft with a constant speed propeller, the same sequence of events of evaporative ice on RPM will apply, except that it will be manifold pressure that is involved rather than RPM. When operating higher output engines, especially those with superchargers, the use of carburetor heat (or alternate air) should be regulated by reference to the carburetor air or mixture temperature gage. Carburetor air temperature at the inlet should be kept to about 85°F (30°C) to 90°F (32°C) to compensate for the drop that will come from evaporation. The mixture temperature, after the drop, should be maintained at 35°F (2°C) to 40°F (4°C). The above discussion will have more meaning when you study the engine and systems in more depth.

J. Frost

We now discuss frost, the formation of which was described in the chapter on moisture. At cruise airspeeds the formation of frost is of no major consequence and not that frequent. When it does occur at higher altitudes it is often as a result of a cold airplane, below freezing, descending into a layer of warm, moist air. Continued flight in the warm air will result in the melting or sublimation of the frost. Frost is more of a problem on takeoffs and landings when the aircraft is at low speeds. While frost does not substantially change the shape of the wing, the roughness of its surface causes a disruption in the smooth flow of the air over the wing, resulting in enough loss of lift (and increase in drag) that when combined with a heavily loaded aircraft and high density altitude, can prevent the aircraft from becoming airborne. At the very least, a heavy coat of frost can cause a ten percent increase in stall speed (Figure 12-5). Other forms of ice on the aircraft on the ground should be removed, for the change in shape as well as the extra weight on a wing will greatly affect its lifting capacity. The effect on control surfaces, brakes, and landing gear has already been mentioned.

K. Weather Systems

Certainly, icing will be more prevalent during winter months, due to colder temperatures and widespread cloudiness in mid-latitudes. Freezing rain with numerous winter fronts is a particularly dangerous situation. Keep yourself COVERED. Comprehend or understand the factors at work in frontal zones. Visualize or picture the vertical structure of fronts. Evaluate your sitation — do you turn around and go back, do you circumnavigate, that is, detour the system, do you fly above the front at high altitudes, or do you cross the front? Much of this evaluation should be conducted before leaving the ground. If your response or reaction as a result of your evaluation is to cross the front, then do so at a 90° angle to lessen your time in the frontal area. From your mental picture of warm and cold fronts you realize that there is warmer air above, but note that this is only relatively warm air compared to colder air below and, hence, may or may not be at temperatures above freezing. You may also find warmer air below as a result of warm surface temperatures.

If you are in freezing rain at, say, 3,000 feet (910 m), you will find warmer temperatures aloft — snow melts in above freezing air, falling as raindrops, and then freezes when encountering freezing temperatures. And freezing rain is your chief hazard concerning icing in the frontal area. Knowledge of where the warm air is or can most likely be found should be gained from your preflight briefing. That way you will minimize your time in freezing rain in finding warm air. You should ascend or descend at as large an angle as practical, but not at an excessive airspeed since accretion rates are greater with increasing speeds. Keep enough airspeed to avoid a

Figure 12-5. Frost on an aircraft. Always remove ice or frost before attempting takeoff.

stall, however. Also, realize that the zone of icing will usually be much wider in a warm frontal area due to the larger mass of clouds than with the cold front (Figures 12-6 and 12-7).

There are times when flying in the base of a cloud is preferable to flying below the cloud or at a higher level. Suppose the temperature is −2°C in freezing rain just below a cloud, becoming even colder with lower altitudes. In the base of the cloud the temperature could well be 0°C to 4°C due to the release of latent heat as sensible heat, with temperatures decreasing above this shallow layer. Such a layer of warm air can be a real life saver.

Dry snow is of no concern to us since it will not stick to the aircraft, nor are ice pellets. Wet snow will collect on the aircraft and it is evidence of above freezing temperatures at your altitude. Ice pellets are evidence of warmer temperatures aloft, as snow melts in a warm layer to form rain which, when passing through a layer of below freezing temperatures, freezes as ice pellets.

A very important point that needs to be recalled here is that so strong are the forces involved, that air can be highly super-saturated, which means that the mass of moisture for potential icing is enormous. Thus, fronts, lows, thunderstorms, and ascending air in mountains, called orographic lifting, are areas where icing is particularly dangerous. The combination of a front crossing a mountain range is particularly hazardous. Even in the absence of a front, icing in mountains up to about 5,000 feet (1.5 km) above the tops, and higher in cumiliform clouds, can result in hazardous icing.

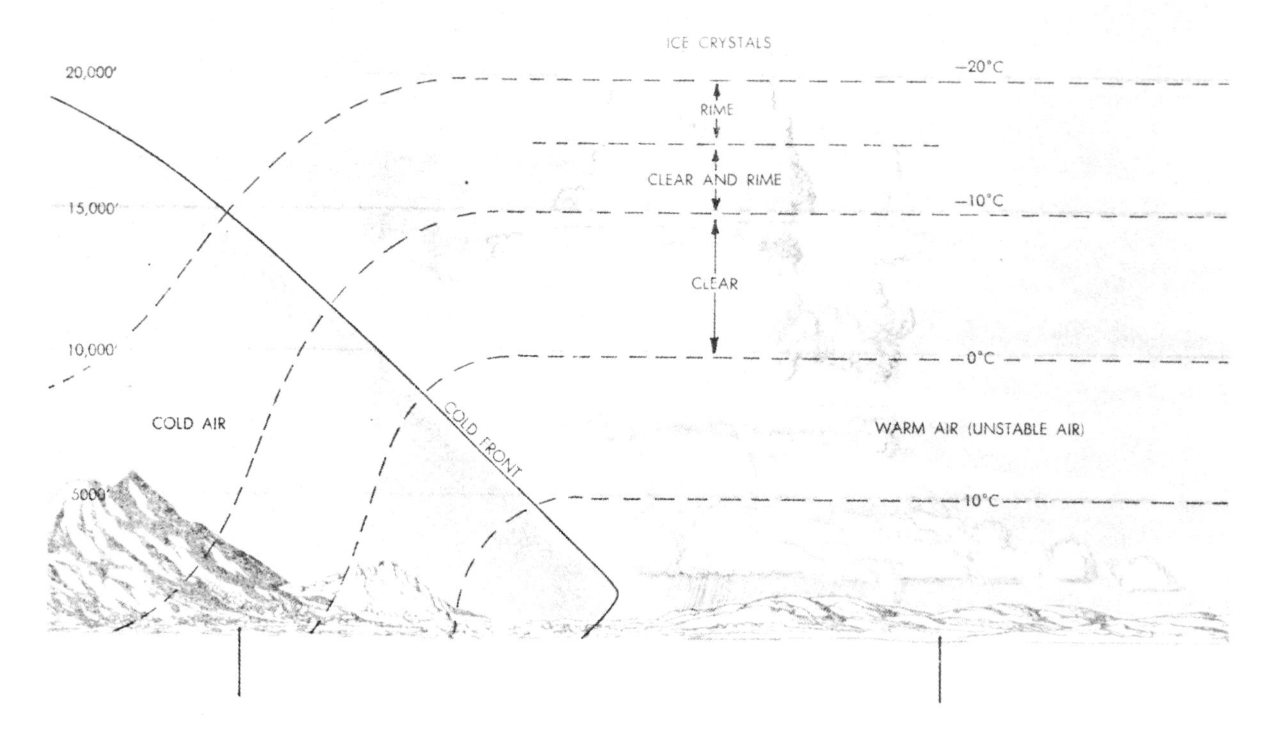

Figure 12-6. Cold Front Icing Zone.

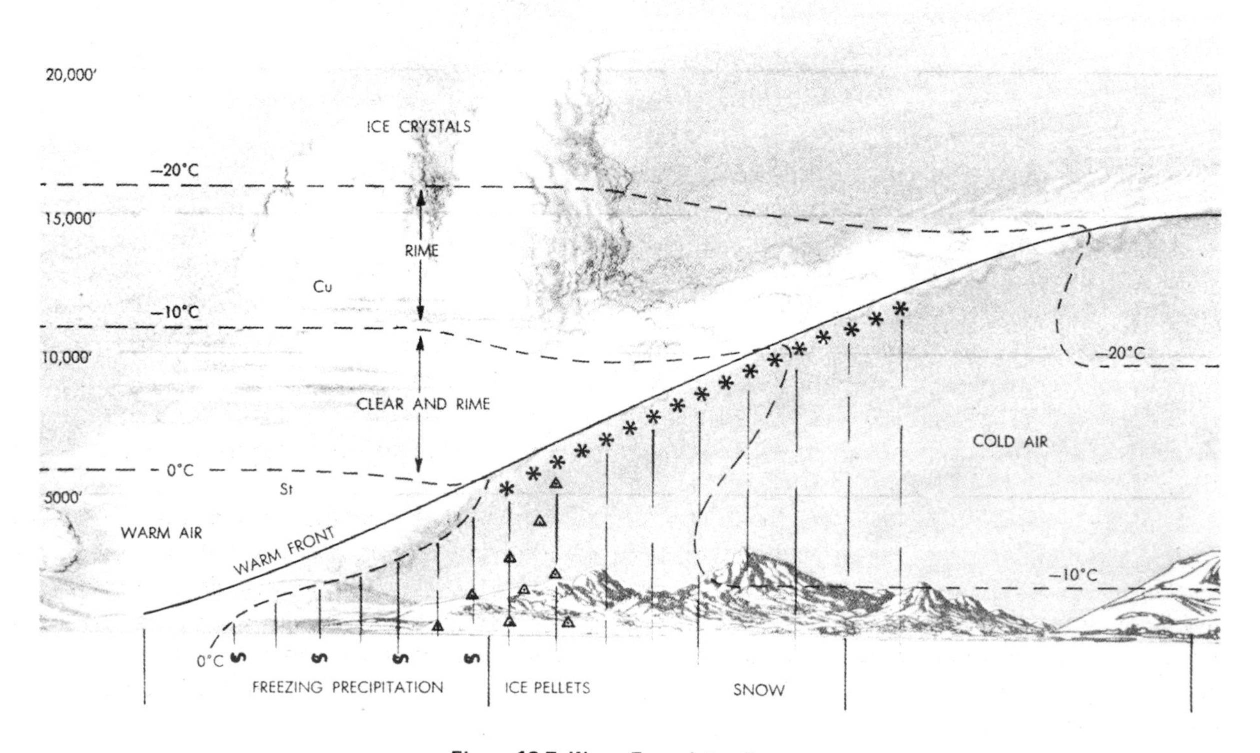

Figure 12-7. Warm Front Icing Zone.

TABLE 12-1. ICING CRITERIA TABLE.

INTENSITY	ICE ACCUMULATION
TRACE	Ice becomes perceptible. Rate of accumulation slightly greater than rate of sublimation. It is not hazardous even though deicing/anti-icing equipment is not utilized, unless encountered for an extended period of time (over 1 hour).
LIGHT	The rate of accumulation may create a problem if flight is prolonged in this environment (over 1 hour). Occasional use of deicing/anti-icing equipment removes/prevents accumulation. It does not present a problem if the deicing/anti-icing equipment is used.
MODERATE	The rate of accumulation is such that even short encounters become potentially hazardous and use of deicing/anti-icing equipment or diversion is necessary.
SEVERE	The rate of accumulation is such that deicing/anti-icing equipment fails to reduce or control the hazard. Immediate diversion is necessary.

We try to avoid icing if at all possible, particularly when we do not have deicing or anti-icing equipment. But even the presence of this equipment does not guarantee safety, as Table 12-1 clearly shows. In such cases you will need to change course and/or altitude. As in flying turbulence, file reports when you receive icing, and when you do not receive icing in areas of forecast icing.

ICING
Review Questions

PRI

1. Select the true statement concerning aircraft structural icing.
   a. It is impossible for weather forecasters to identify regions where icing is possible.
   b. Rime ice is the most common type of ice encountered in cumuliform clouds.
   c. The most rapid accumulations of clear ice are usually at temperatures from 0°C to –15°C.
   d. The most common type of icing encountered in lower level stratus clouds is clear ice.

2. Select the true statement regarding aircraft structural icing.
   a. It is unnecessary for an aircraft to fly through rain or cloud droplets for structural ice to form.
   b. Clear ice is most likely to form on an airplane when flying through stratified clouds or light drizzle.
   c. In order for structural ice to form, the temperature at the point where moisture strikes the aircraft must be 0°C (32°F) or colder.
   d. Rime ice gradually freezes on an airplane's surface becoming a smooth sheet of solid ice.

3. Which statement is true regarding frost which has not been removed from the lifting surfaces of an airplane before flight?
   a. It would present no problems since frost will blow off when the airplane starts moving during takeoff.
   b. It may cause the airplane to become airborne with a lower angle of attack and at a lower indicated airspeed.
   c. It may prevent the airplane from becoming airborne.
   d. It will change the curvature of the wing (camber) thereby increasing lift during the takeoff.

4. The type of ice which forms on an aircraft surface depends on
   a. An inversion aloft.
   b. The increase in flight altitude.
   c. The temperature/dewpoint spread.
   d. The size of the water drops or droplets that strike the aircraft surface.

5. The most rapid accumulation of clear ice on an airplane in flight may occur with temperatures between 0°C to –15°C in
   a. Ice fog.
   b. Any clouds or dry snow.
   c. Cumuliform clouds.
   d. Stratiform clouds.

6. Hazardous in-flight structural icing, with which a pilot should be familiar includes frost, rime, and clear ice. Which statement is true concerning this hazard? (Frost is not structural ice).
   a. Frost may form in flight when a cold aircraft descends from a zone of subzero temperatures to a zone of above freezing temperatures and high relative humidity.
   b. Clear ice is a milky, opaque, and granular deposit of ice with a rough surface.
   c. Rime ice is a transparent ice with a glossy surface.
   d. Cumuliform type clouds are less apt to produce serious ice formation than other type clouds.

INS

7. You encounter wet snow during a cross-country flight. What does this indicate regarding temperatures in the area?
   a. You are flying from a warm air mass into a cold air mass.
   b. You are in an "inversion" with colder air below.
   c. The temperature is above freezing at your altitude.
   d. The temperature is above freezing at higher altitudes.

8. The presence of ice pellets at the surface is evidence that
   a. A cold front has passed.
   b. There are thunderstorms in the area.
   c. Temperatures are above freezing at some higher altitude.
   d. You can climb to a higher altitude without encountering more than light icing.

9. What is indicated if you encounter ice pellets at 8,000 feet?
   a. Freezing rain at a higher altitude.
   b. You are approaching an area of thunderstorms.
   c. You will encounter hail if you continue your flight.
   d. The formation of low clouds or fog.

10. What are the characteristics of rime ice, and what conditions are most favorable for its formation?
    a. Opaque, rough appearance, tending to spread back over an aircraft surface. Most frequently encountered in cumuliform clouds at temperatures slightly below freezing.
    b. Smooth appearance and builds forward from leading surfaces into a sharp edge. Most common in cumuliform clouds at temperatures of –20°C to –25°C.
    c. Milky, granular appearance, forming on leading edges, and accumulating forward into the airstream. Stratiform clouds and temperatures of –10°C to –20°C are most conducive to its formation.
    d. Transparent appearance and tendency to take the shape of the surface on which it freezes. Stratiform clouds and temperatures only slightly below freezing promote its formation.

11. What is an operational consideration if you fly into rain which freezes on impact?
    a. You have flown into an area of thunderstorms.
    b. Temperatures are above freezing at some higher altitude.
    c. You have flown through a cold front.
    d. If you descend, you will fly out of this icing condition.

12. In which conditions would you most likely encounter clear structural icing, and how would it normally appear?
    a. Cumuliform clouds; large water droplets; temperatures between 0 and –15°C. Appears smooth and tends to spread back over an aircraft surface.
    b. Stratiform clouds; small water droplets; temperatures between –10°C and –20°C. Appears granular and tends to accumulate forward into the airstream.
    c. Cumuliform clouds; small water droplets; temperatures –20°C to –25°C. Appears transparent and tends to take the shape of the surface on which it freezes.
    d. Stratiform clouds; large water droplets; temperatures well below freezing. Appears opaque and builds forward from leading surfaces into a sharp edge.

13. If high humidity or visible moisture is present, in what temperature range would you expect to encounter the most severe icing? (Structural ice requires visible moisture).
    a. 0°C to –10°C.
    b. –10°C to –15°C.
    c. –15°C to –25°C.
    d. –20°C to –30°C.

14. In which environment is aircraft structural ice most likely to have the highest accumulation rate?
    a. Cumulus clouds.
    b. Cirrus clouds.
    c. Stratus clouds.
    d. Freezing rain.

15. If the pitot cover is not removed on an aircraft with the static source on the side of the fuselage, which instrument(s) would be affected?
    a. Vertical speed and airspeed only.
    b. Altimeter and vertical speed only.
    c. Airspeed, altimeter, and vertical speed.
    d. Airspeed only.

COM

16. If the static pressure ports iced over while descending from altitude, the airspeed indicator would read
    a. Zero.
    b. High.
    c. Low.
    d. Correctly.

17. If the ram air input to the pitot head of the pitot system becomes blocked (drain hole open), the indicated airspeed will generally
    a. Decrease as altitude is increased.
    b. Remain unchanged.
    c. Increase as altitude is increased.
    d. Drop to zero.

18. If the ram air input and the drain hole of the pitot system become blocked, trapping the pressure in the system, the indicated airspeed will generally
    a. Vary excessively during level flight when the actual airspeed is varied.
    b. Decrease during climbs.
    c. Not change during level flight, even when the actual airspeed is varied by large power changes.
    d. Increase during descents.

19. How is an airplane's performance affected by frost on the wings?
    a. Lift is decreased; drag is decreased.
    b. Lift is increased; drag is decreased.
    c. Lift is decreased; drag is increased.
    d. Lift is increased; drag is increased.

20. If necessary to take off from a slushy runway, the freezing of landing gear mechanisms can be minimized by
    a. Retracting the gear immediately to prevent freezing.
    b. Delaying gear retraction.
    c. Increasing the airspeed to V_{le} before retraction.
    d. Recycling the gear.

21. In which sections of the carburetor would icing most likely occur?
    a. Main air bleed and main discharge nozzle.
    b. Venturi and on the throttle valve.
    c. Float chamber and fuel inlet screen.
    d. Accelerator pump and main metering jet.

22. The first indication of carburetor icing in airplanes equipped with constant-speed propellers would most likely be a
    a. Rough running engine followed by loss of RPM.
    b. Decrease in revolutions per minute.
    c. Decrease in manifold pressure.
    d. Rough running engine followed by an increase in manifold pressure.

23. Which statement is true regarding throttle ice in engine induction systems?
    a. Throttle ice is formed at cruise power settings.
    b. Throttle ice is usually formed in induction systems when the throttle is closed.
    c. Throttle ice affects both fuel injection engines and engines equipped with carburetors.
    d. Throttle ice occurs only in combination with impact ice.

24. Which statement is true regarding induction system icing?
    a. Throttle ice is usually formed in the induction system of fuel injection engines.
    b. Impact ice affects both fuel injection engines and engines equipped with carburetors.
    c. Fuel ice is usually formed in the induction system of fuel injection engines.
    d. Induction system icing affects only engines equipped with carburetors.

25. The low temperature that causes carburetor ice in an engine equipped with a float-type carburetor is normally the result of the
    a. Freezing temperature of the air entering the carburetor.
    b. Vaporization of fuel and expansion of air in the carburetor.
    c. Low volatility of the fuel.
    d. Compression of air at the carburetor venturi.

26. In an aircraft equipped with a float-type carburetor and a constant-speed propeller, carburetor icing would probably first be detected by
    a. A drop in manifold pressure.
    b. A drop in manifold pressure and engine RPM.
    c. Detonation.
    d. A drop in engine RPM.

27. In an airplane equipped with a constant-speed propeller and a manifold pressure (MP) gauge, the presence of carburetor ice can be verified by applying carburetor heat and noting an immediate
    a. Decrease in MP with no further change in MP.
    b. Increase in MP and then a gradual increase in MP.
    c. Increase in MP and then a gradual decrease in MP.
    d. Decrease in MP and then a gradual increase in MP.

28. Which conditions should alert a pilot to the possibility of induction icing?
    a. Any temperature below freezing with a relative humidity less than 50%.
    b. A temperature between 32°F and 70°F with a relative humidity greater than 50%.
    c. A temperature between 0°F and 32°F with a relative humidity between 30% and 50%.
    d. A temperature between 32°F and 50°F with a relative humidity less than 50%.

29. In an airplane equipped with a manifold pressure gauge, a tachometer, a cylinder head temperature gauge, and an exhaust gas temperature indicator, the first indication of induction icing will be noted by a decrease in
    a. Cylinder head temperature.
    b. RPM.
    c. Manifold pressure.
    d. Exhaust gas temperature.

30. When operating higher output engines, especially those with superchargers, the use of carburetor heat should be regulated by reference to the
    a. Degree of roughness at which the engine is operating.
    b. Manifold pressure or RPM indicator.
    c. Cylinder head temperature gauge.
    d. Carburetor air or mixture temperature gauge.

31. Which statement is true regarding the use of carburetor heat or alternate air during flight?
    a. It is preferable to use carburetor heat or alternate air as a prevention, rather than as a deicer.
    b. Full carburetor heat should be continuously used when the temperature is below 32°F.
    c. Partial heat should be used in airplanes that are not equipped with some instrumentation to determine the effect of the heat.
    d. Partial carburetor heat should be used when the temperature is below 32°F.

32. An increase in carburetor air temperature while operating at the same altitude with the same RPM and MP, will produce
    a. More horsepower.
    b. Less horsepower.
    c. Fluctuating horsepower.
    d. The same horsepower.

33. If carburetor heat is used in such a manner as to provide too much heat at the carburetor intake, it will cause
    a. The engine to idle too fast.
    b. A decrease in fuel consumption.
    c. A loss of RPM and a reduction of maximum power.
    d. Excessive cylinder head temperatures.

ATP

34. What determines the type of structural icing that can form on the surface of an aircraft?
    a. Rate at which water freezes upon contact with aircraft.
    b. Temperature of the air and the aircraft surface.
    c. Size of the water droplets and outside air temperature.
    d. Percent of relative humidity and outside air temperature.

35. Freezing rain encountered during climb is normally evidence that
    a. There exists a layer of warmer air above.
    b. You can climb to a higher altitude without encountering more than light icing.
    c. A cold front has passed.
    d. There are thunderstorms in the area.

CHAPTER XIII
THUNDERSTORMS

The reader should demonstrate an understanding of thunderstorms to the extent that he can answer the following questions (as stated below, or as multiple-choice, etc.):

1. In mid-latitudes, in which seasons are thunderstorms most frequent and why? In which geographical region(s) of the U.S. are thunderstorms most frequent?
2. What factors must be present in order for a thunderstorm to form?
3. Describe the steps in the formation of a convective thunderstorm.
4. Concerning thunderstorms, what information should you request before taking off?
5. State the primary stages of thunderstorm development.
6. Define "thunderstorm".
7. What is the sign that a thunderstorm has reached the mature stage?
8. How does the anvil top relate to the movement of a thunderstorm?
9. State the two classifications of thunderstorms.
10. How long do air mass thunderstorms usually last? How far across are they?
11. State the types of air mass thunderstorms.
12. Under what conditions do convective thunderstorms form?
13. What provides the lift for orographic thunderstorms?
14. What is the chief characteristic of the steady-state thunderstorm? How long may these storms last and why?
15. What usually accounts for the development and movement of steady-state thunderstorms?
16. Define "squall line".
17. Relative to fronts, where do squall lines develop?
18. Can squall lines develop in air masses? Explain.
19. What is usually considered to be the number one hazard associated with thunderstorms?
20. What is the main danger associated with thunderstorm drafts?
21. Which is larger and stronger, updrafts or downdrafts?
22. Where (in thunderstorms) are updrafts the strongest? What would be an average updraft rate?
23. Where are downdrafts the strongest?
24. Why are thunderstorm downdrafts particularly hazardous in mountainous terrain?
25. When and at what altitude should you attempt to fly underneath a thunderstorm? What is the best rule with regard to flying beneath thunderstorms?
26. Define "plow wind". What are its typical horizontal and vertical limits?
27. When landing and encountering the plow wind (or "first gust"), in which direction should you turn? Why?
28. Describe the plow wind, in terms of horizontal and vertical speeds and changes in direction.
29. Define "gust".
30. Define "entrainment".
31. How far away from the center of a storm is turbulence likely to be encountered in clear air?
32. With regard to turbulence, what are some rules for flying over thunderstorms?
33. Does the general appearance of a thunderstorm always indicate the degree of turbulence to be expected?
34. What percent of thunderstorms should you assume contain hail?
35. At what altitudes is hail most likely?
36. What kind of icing is the most hazardous with thunderstorms? Where (at what temperature) does the most severe icing occur?
37. Pressure changes near a thunderstorm may cause how much error in the altimeter? Describe the sequence of pressure changes as a thunderstorm approaches, arrives, and moves out.
38. Is lightning usually a serious hazard? Describe how lightning occurs.
39. Explain the contraction LTGICCCCG.
40. Describe "thunder".
41. At what thunderstorm altitudes is a lightning strike most likely?
42. What is the major effect of precipitation static?
43. Define "funnel cloud"; "tornado"; "waterspout".
44. What is the chief use of radar?
45. On what factors is the image on a weather radar dependent?
46. Thunderstorm information will be provided by ATC on their radar only under what condition?
47. Compare warm and cold front thunderstorms with regard to bases and severity.
48. What is one of the chief dangers of warm front thunderstorms?

CHAPTER XIII

THUNDERSTORMS

The thunderstorm does not occur frequently at any one given spot, but when it does occur it is capable of the most violent kinds of weather. In mid-latitudes, thunderstorms form most frequently in the spring, summer and fall, and during these periods, most often in the southeastern portions of the U.S. Warmer temperatures are available to initiate convection in the summer. Convection and fronts combine to lift air in the transition seasons of spring and fall. While fronts are more numerous in the winter, particularly cold fronts, the colder air reduces the total number of thunderstorms. The availability of moisture, combined with warmer temperatures, leads to the greatest frequency of thunderstorms in the southeastern U.S. (Figure 13-1).

A. Factors Necessary for Development

In order for a thunderstorm to form, three factors must be present: 1) a source of lift, 2) unstable air, and 3) high moisture content. The convective thunderstorm forms as follows, assuming the presence of all three factors. The intense heating of the air near the surface causes parcels, or bubbles of air to start to rise. If the air is unstable, it will continue to rise; after sufficient lift, water vapor begins to condense as it cools adiabatically. Thus, clouds begin to form, and at the same time we know that latent heat will be released as sensible heat, a pocket or layer of warm air which if warmer than the surrounding air will start to rise, forcing moist air above to higher levels, resulting in even more condensation and, hence, more clouds and sensible heat. Therefore, the thunderstorm develops as a chain reaction, convection being responsible for the first link and release of sensible heat being responsible for subsequent links. In this way a thunderstorm is a giant heat machine (Figure 13-2).

We repeat that all three factors mentioned above are necessary. Suppose the air is unstable with high moisture content but there is no source of lift. Then there may be nothing other than clear skies and smooth air – remember that unstable air is air that, when lifted, will continue to rise because it is warmer than the surrounding air; the rise will continue until that air is capped, or choked off by a stable layer above. There have been many days, particularly in the summertime, when one extra degree of temperature would have changed a clear day into one engulfed with thunderstorms.

Suppose we have a day on which a source of lift is available and the air is unstable, but there is little moisture content. Then, again taking the convection example, bubbles of air start to rise, and continue to do so because the air is unstable; but with low moisture content there may be no clouds, or a few fair-weather cumulus, or maybe just a few towering cumulus. Just a bit more moisture content and thunderstorms could have been widespread. Just a 5° change in wind direction relative to a water source could be responsible for the lack of sufficient moisture. Note, however, that the air will still be quite rough.

Our last case will be a source of lift combined with sufficient moisture content, but with stable rather than unstable air. Air begins to rise in such a situation, but since the air is stable, the rising air becomes cooler than the surrounding air and tends to return to its original position. Thin clouds may form in the process, but because the air is stable, that moisture is trapped and, hence, kept from developing the necessary vertical extent for a thunderstorm. Had the actual lapse rate of the air been just a bit greater, thunderstorms could again have been numerous. Thus, the difference between no thunderstorms, just a few thunderstorms, and numerous thunderstorms is often quite small.

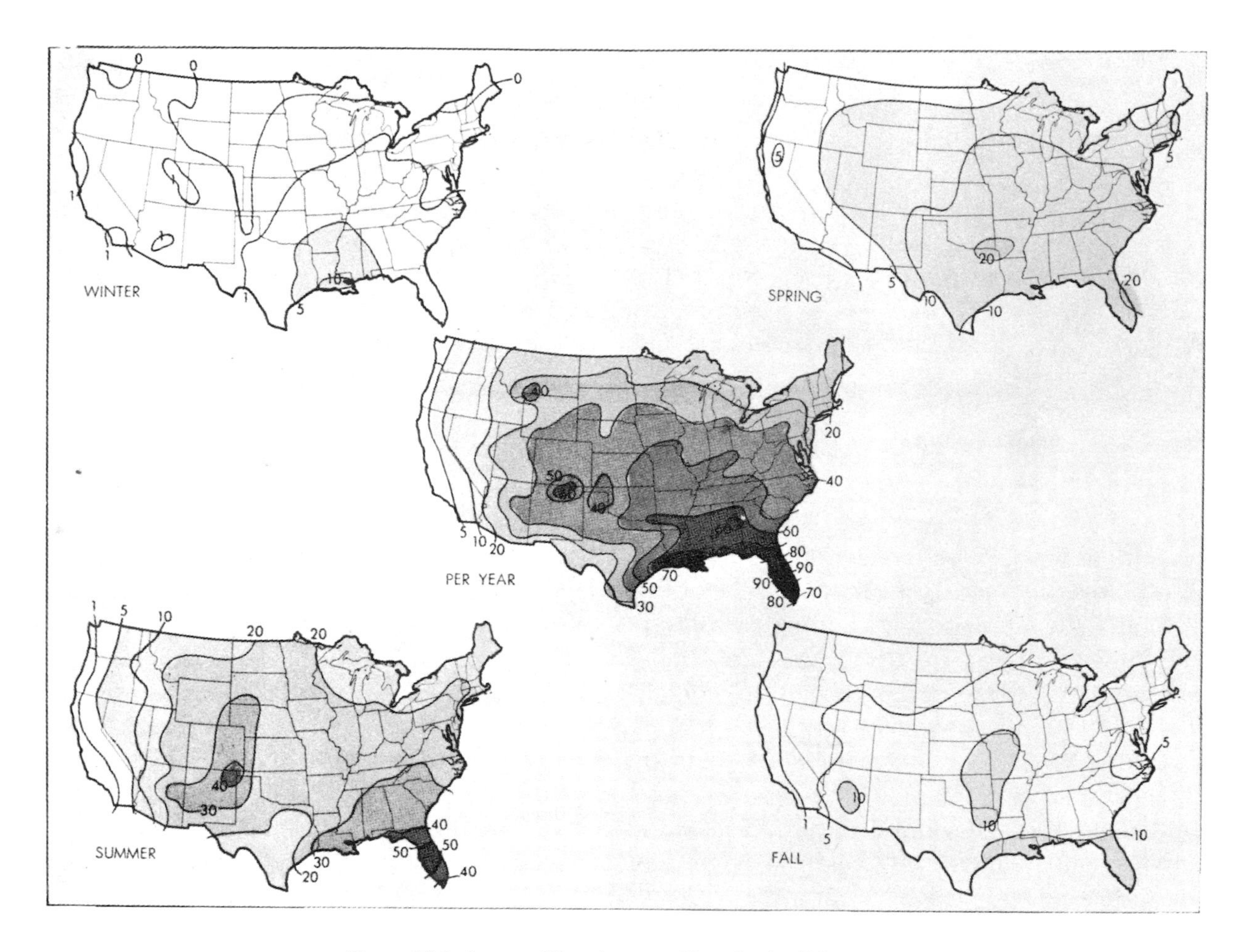

Figure 13-1. Average Thunderstorm Days in the U.S.

Forecasters cannot consistently forecast temperatures accurate to the degree, or wind direction within 5°. Thus, when you receive a forecast of thunderstorms and the air is smooth and clear, remember that the forecaster may have only missed his forecast by one degree of surface temperature. And more important than defending the weather forecaster is your keeping yourself COVERED. For suppose you did receive a forecast of numerous thunderstorms and the air is clear and smooth as you fly at, say, 2 p.m. Then by COmprehending, or understanding weather factors – lift, stability, and moisture – you Visualize or picture moist, unstable air, although you cannot really see the moisture in the vapor state. You then Evaluate your situation by coming to the tentative conclusion that there is yet to be a source of lift. Since it is only 2 p.m., the maximum daily temperature is yet to be reached, so you anticipate that by 4 p.m. there may be numerous thunderstorms, as forecast. You then Respond or react by watching for cloud development, first, small cumulus, then towering cumulus, then the cumulonimbus. You then Evaluate or re-Evaluate your situation and Determine a course of action – circumnavigate the developing storms, land at a nearby airport, etc.

Note that you could have anticipated all of the above events on the ground. For example, you could have asked the weather briefer what the source of lift for the thunderstorms was to be, and if he told you convection, you could have planned your flight for earlier or later in the day in order to avoid these afternoon and early evening thunderstorms. Never leave the ground into an area of forecast thunderstorms without identifying the necessary source of lift. It is also

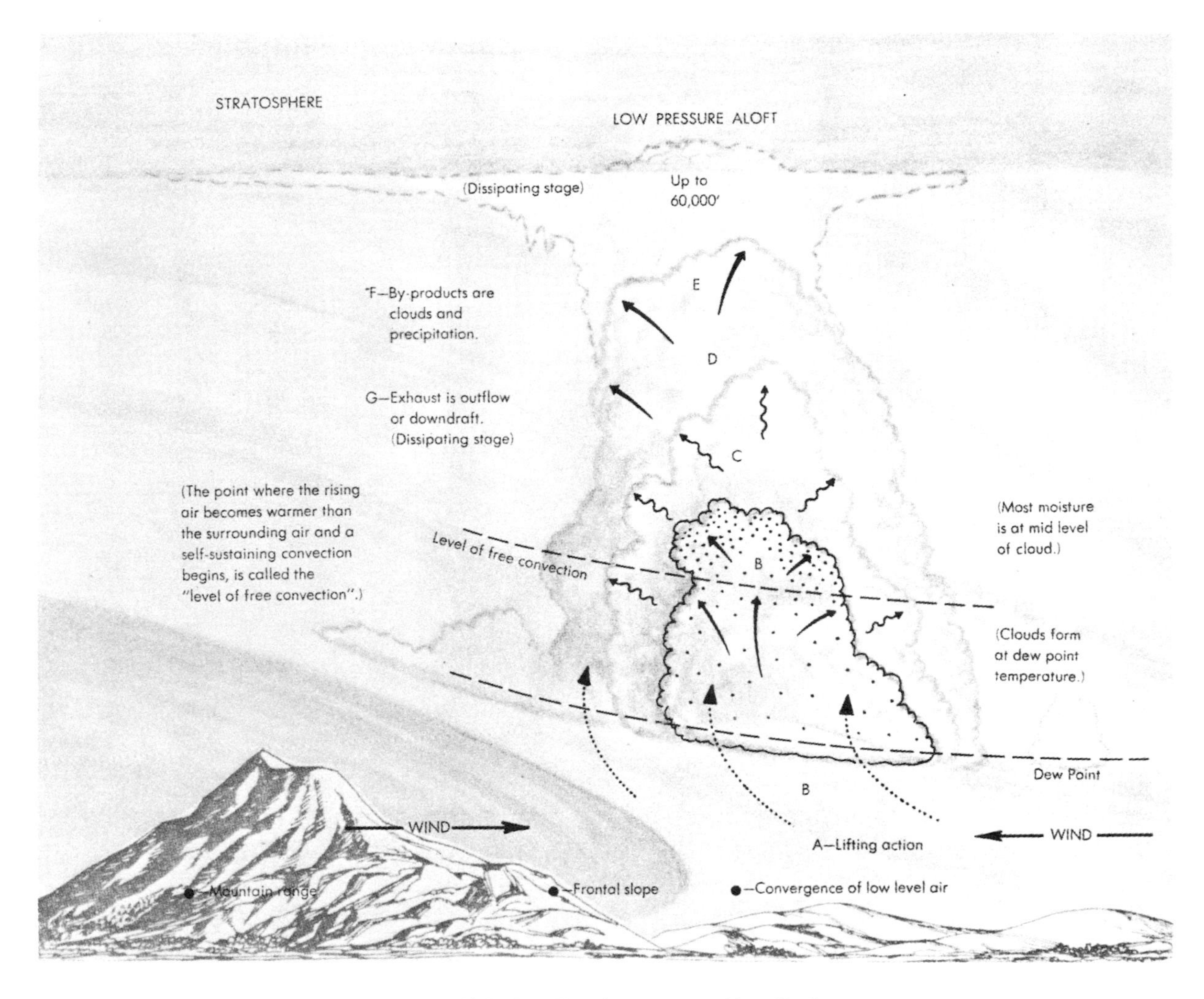

Figure 13-2. The Thunderstorm as a Heat Engine.

highly recommended that you ask the weather briefer for some indication of the degree of stability and moisture content, for in reality the occurrence of thunderstorms does depend on the degree to which all of these factors are present.

B. Stages of Development

We now discuss the stages of thunderstorm development with a very ideal model. Each thunderstorm is different in its structure and severity. We point out as best we can the similarities between these storms. Thunderstorm cell development is traditionally considered to occur in three stages: 1) the cumulus or growth stage, 2) the mature or fully developed stage, and 3) the dissipating or degenerating stage (Figure 13-3). The cell is the heart of the thunderstorm, and while the mass of cloudiness associated with the thunderstorm may extend for many miles, individual cells may be as small as only a few hundred feet across. We will consider a thunderstorm to be a local storm, composed of one or usually more than one cell, that develops from a cumulonimbus cloud. A thunderstorm is a cumulonimbus cloud with lightning; lightning causes thunder even though you may not hear it.

The cumulus stage is marked by an updraft that may involve vertical speeds of 2,000 to 3,000 feet (600 to 900 m) or more per minute. As air ascends, moisture condenses. Rain at the

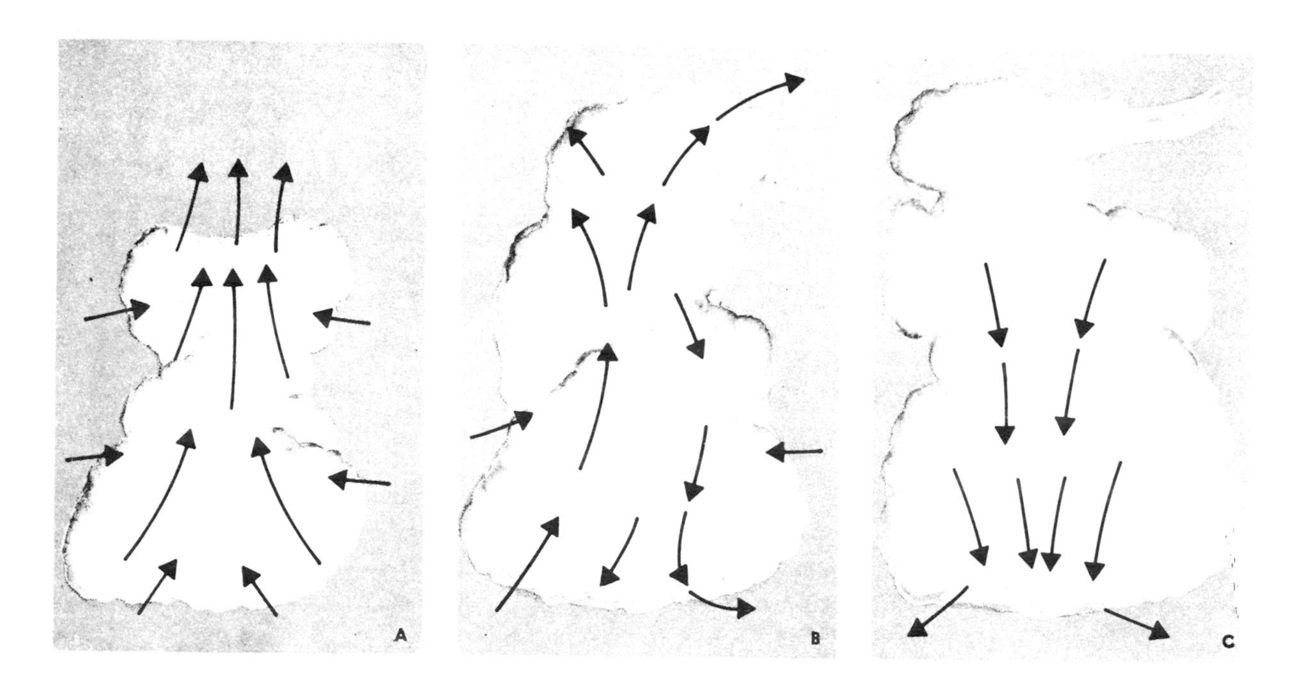

Figure 13-3. The stages of a thunderstorm. (A) is the cumulus stage; (B), the mature stage; and (C), the dissipating stage. Arrows depict air flow.

earth's surface is usually considered to be the sign that a thunderstorm has reached the fully mature stage. The updrafts which marked the cumulus stage may grow to 5,000 to 8,000 feet (1.5 to 2.4 km) per minute in the early part of the mature stage.

When clouds reach heights of anywhere between 25,000 and 80,000 feet (7.6 and 24.4 km), the updrafts can no longer support the huge volume of water, whether it be in liquid or frozen form. At higher levels, the relative humidity might be, say, 600%, the force of the updraft being so strong that much more moisture is forced into a given volume than could normally be held in still air. As precipitation begins to fall, the air can be said to be "dragged down" with it, creating downdrafts possibly as high as 12,000 to 15,000 feet (3.7 to 4.6 km) per minute. Entrainment, the mixing of surrounding dry air with the falling rain, accelerates the downdraft; this is due to the evaporation of some of the raindrops into the dryer air, causing the air to become cooler. A huge anvil top is also indicative of the mature stage, the anvil being blown by winds in the upper levels in the direction of movement of the storm. The dissipating stage is marked by the termination of rain and downdrafts at the earth's surface.

C. Air Mass Thunderstorms

We will classify thunderstorms as being of two types, air mass and steady-state. The air mass thunderstorm forms in the interior of an air mass – recall that the boundary between air masses is called a front. In these huge highs, air is sinking down and spiralling outward in a clockwise manner. The sink rates are not great, as noted in an earlier chapter, but there is descent of the air. As a result, these thunderstorms will tend to be slower moving and, indeed, may become so slow moving that huge volumes of moisture reach the earth's surface. These thunderstorms tend to be isolated or scattered about over a large geographical area. But this does not mean that air mass thunderstorms are always less severe than steady-state thunderstorms. They usually are, but extreme turbulence, hail, and tornadoes do occur with air mass thunderstorms. Because of their tendency to remain stationary, they usually last anywhere from 20 minutes to an hour and a half, as the supply of moisture is exhausted and the downdraft chokes off the main updraft. An individual air mass thunderstorm may extend from 5 to 30 miles (8 to 48 km) across. The three types of air mass thunderstorms usually considered are convective, orographic, and nocturnal, the only difference being in the source of lift.

1. Convective

We have already discussed the convective thunderstorm to a certain degree. They tend to form in the late afternoon on hot, moist days when winds are light. When winds are strong, local pockets of temperature and moisture are dispersed and, hence, the storms are less likely to develop.

2. Orographic

The orographic thunderstorm acquires its name from lifting over mountainous terrain. When combined with afternoon heating, these storms can become quite severe.

3. Nocturnal

The nocturnal thunderstorm receives its source of lift primarily from radiational cooling of cloud tops, to a lesser degree, but much the same as the cooling that takes place at the surface on a clear night. The tops of the clouds, becoming much colder, will tend to descend, forcing the moist air below upward. If the air is conditionally unstable, thunderstorms will develop, ones which are potentially severe even though developing at night.

D. Steady-State Thunderstorms

The steady-state thunderstorm is characterized by movement. A strong updraft is maintained in the direction of movement of the storm, for the bulk of precipitation falls toward the rear (Figure 13-4). As long as the air is unstable and moist in the direction of movement, this kind of storm may last for several hours. Weather systems usually account for the development and movement of these storms, such as fronts, lows or troughs, and large-scale convergence. And since the movement of these thunderstorms is in a more or less continuous line, we just call them squall line thunderstorms.

Another common name for the line along which thunderstorms develop is the instability line. So if there is an area of unstable air with sufficient lift and moisture, then thunderstorms develop and the instability line becomes a squall line. One common instability line lies in Western Texas, Oklahoma, and Kansas. The southeasterly flow of air from the Gulf of Mexico meets a southwesterly flow of air from southwest Texas. The air from the Gulf is, as we have seen, lighter than the dryer air from the land mass area of southwest Texas. Thus, this convergence will cause the more moist air to rise, and if the air is conditionally unstable, thunderstorms will develop. One can determine the location of this line before thunderstorms develop by noting the difference in dew point, a 20°F difference not being uncommon; thus, this line is sometimes called a dewpoint front. The word front is not really correct since the circulation on the eastern side of the line is clockwise, but counterclockwise on the western side. This cannot be a front, for we have an air mass to the east but a low to the west, and we know that a front is always the boundary between air masses. The word front is used to signify that any significant weather is caused by the difference in dew point across the line of convergence.

Squall lines often develop along, ahead, or even behind fronts (most frequently with fast-moving cold fronts). Sometimes a squall line will be one hundred to three hundred miles (160 to 480 km) ahead of a cold front, for once these intense thunderstorms develop, the power

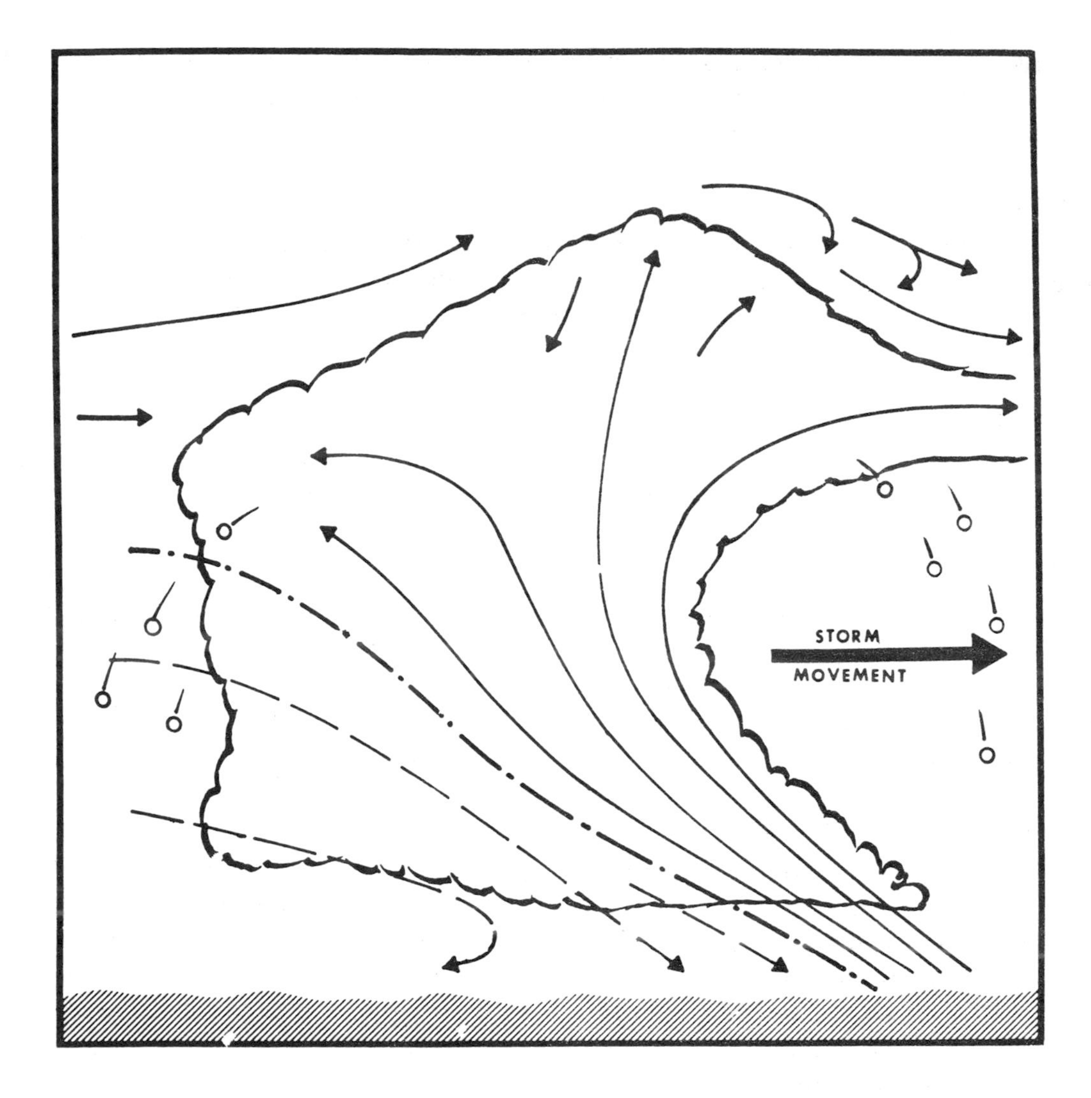

Figure 13-4. Schematic of the mature stage of a steady-state thunderstorm cell, showing a sloping updraft with the downdraft and precipitation outside the updraft not impeding it. The steady-state mature cell may continue for many hours and deliver the most violent thunderstorm turbulence.

is so enormous that the line itself becomes a separate entity, moving out in advance of the front, usually parallel to the front due to upper level winds.

Squall lines will develop well into the interior of an air mass under certain conditions. Picture a tropical maritime air mass over the southeastern U.S. in the summer. The vertical extent of this air mass might be 10,000 feet (3 km), with a trough above running, say, along the Mississippi-Alabama border. During the hot afternoon hours, convective activity reaching 10,000 feet will be extended higher by the trough. A fully developed squall line often develops in advance of the trough and moves with the trough. Had the trough not been there, then typical air mass thunderstorms would have been present, scattered randomly about, had of course the air been sufficiently unstable. So whenever a squall line develops well into the interior of an air mass, you can be reasonably sure that there is a trough aloft and that the thunderstorms are the long-lasting steady-state thunderstorms, rather than the more short-lived air mass type. Therefore, the squall line thunderstorm is the most severe. All the hazards possible with thunderstorms, such as turbulence, hail and tornadoes, will be more severe with these storms.

E. Thunderstorm Hazards

We now discuss the hazards of thunderstorms and how you might best cope with them (also see MICROBURSTS in Ch. 15).

1. Turbulence

Turbulence is the number-one hazard with thunderstorms. It is always present, in varying degrees. We will consider two factors concerning turbulence, drafts and gusts. Drafts, whether up or down, are not that hard on the aircraft structure in the interior of the draft. Of course, large altitude variations will occur. The real danger with drafts is crossing adjacent updrafts and downdrafts. Such extreme wind shear has literally snapped more than one aircraft in two. The most common visible sign of extreme turbulence in thunderstorms is the cumulonimbus cloud with very frequent lightning and roll clouds. It is difficult to comprehend the amount of power contained by an average thunderstorm, but we have all felt its wrath at one time or another (Figure 13-5).

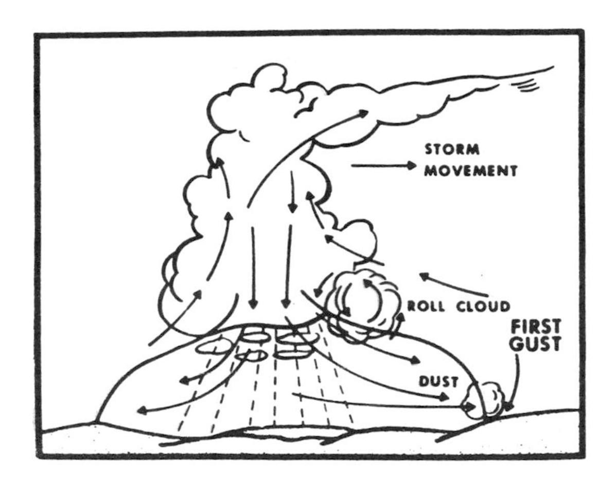

Figure 13-5. Cross section of a thunderstorm, showing location of surface wind gusts, roll cloud, and other turbulent areas relative to the movement of the storm. Surface gusts and rotor motion can be extremely hazardous to aircraft landing, taking off, and flying at low altitude.

Research has indicated that updrafts are larger and stronger, both horizontally and vertically, than downdrafts. There is some evidence that updrafts are strongest in the middle and higher portions of the thunderstorm. Updrafts of more than 6,000 feet (1.8 km) per minute have been recorded. The strongest downdrafts usually occur in the middle layers of the storm; while the updraft is typically stronger than the average downdraft of about 2,000 feet (600 m) per minute, downdrafts of more than 12,000 feet (3.7 km) per minute have been encountered. These downdrafts are very hazardous for aircraft attempting to fly beneath thunderstorms, and for aircraft at low levels many miles away. Never fly underneath a thunderstorm in mountainous terrain, for peaks may be hidden by clouds and downdrafts may push your aircraft into the mountains. Attempt to fly underneath a thunderstorm only if you are flying over flat terrain or open water and at an altitude which will keep you half way between the surface and the base of the thunderstorm. The best rule is to not attempt to fly underneath a thunderstorm.

The initial downdraft we are discussing is sometimes called the plow wind. This plow wind may affect low level operations as high as 6,500 feet (2 km) above the ground for an average distance of about 10 miles (16 km) (Figure 13-6). However, the author observed such an effect over a distance of several hundred miles, as follows. As a weather officer at Carswell AFB, Texas, I observed a cluster of steady-state thunderstorms which moved from extreme southern Oklahoma across the Red River into northeastern Texas. Before this movement, wind reports were southerly at 10 to 15 knots from northeast Texas to Brownsville, Texas, the extreme southern tip of Texas several hundred miles away. The next report of surface winds showed that they had shifted to the north all the way to Brownsville. The winds at Brownsville were only about ten knots, but if one had been practicing takeoffs and landings, this windshift would have been totally unexpected, and a 20 knot wind shear from the south at 10 to the north at 10 would have caused a significant loss in performance. This would have been noticed by a sharp drop in indicated airspeed.

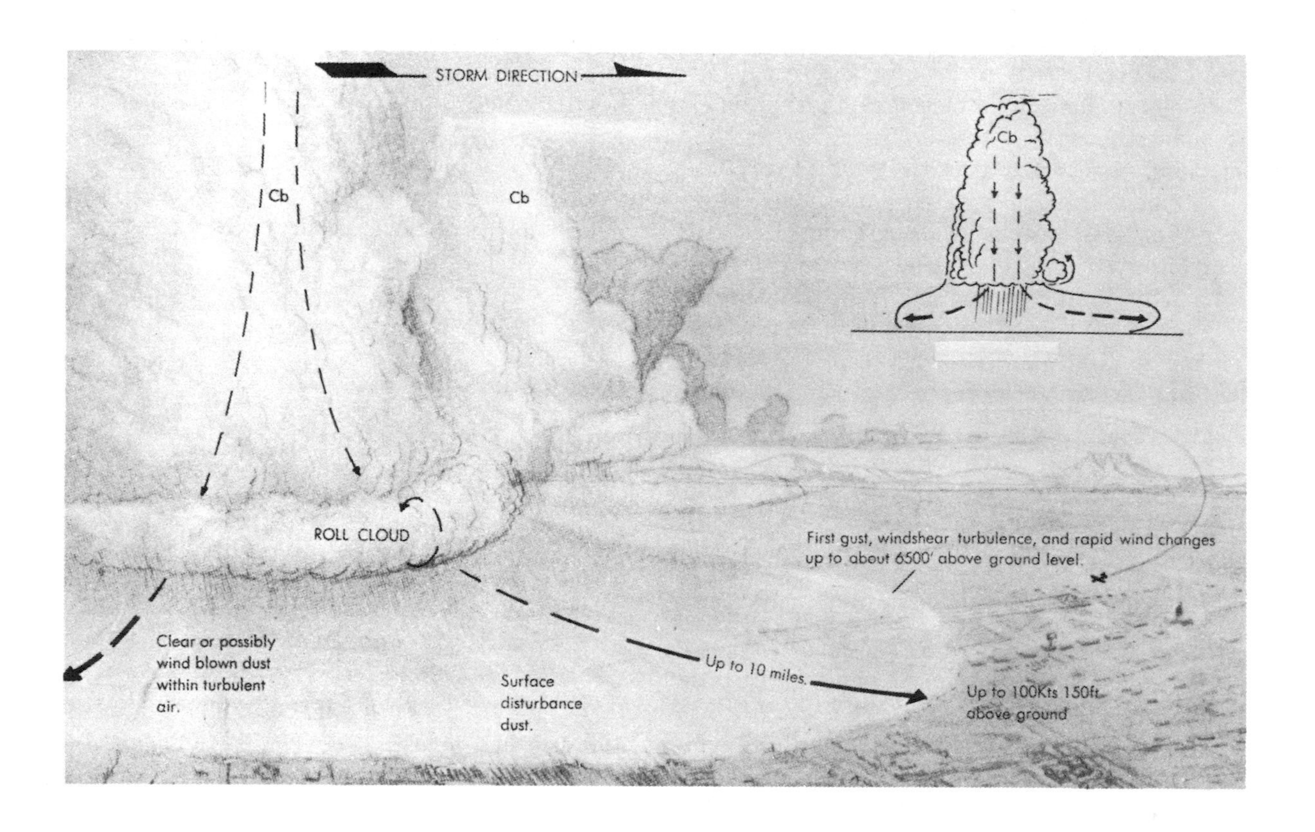

Figure 13-6. First Gust Hazards.

An adverse shear on final approach can be an extremely dangerous situation. Studies of accidents involving large jet airliners in the 1970s have revealed downdrafts of extreme intensity near the surface. Researchers Fujita and Byers proposed the name "downburst" for these intense downdrafts. Thunderstorms producing downbursts have appeared as "spearhead echoes" on radar.

When you are practicing or flying at low levels anywhere near thunderstorms, be alert to the plow wind and fly with an extra margin of indicated airspeed. Not only do you have the problem with the shear effect, but possibly a strong crosswind, particularly when landing or taking off. If you do find it necessary to abandon approach, for example, turn toward the "plow". Why? Well, this plow wind acts much like a front and the wind will shift clockwise from, say, south to west to north. So in a planned approach to the south, turn right to face this gusty, likely variable west to north wind. You will note a sudden increase in performance, reflected by an increase in airspeed and rate of climb. Again, we emphasize that when wind speed is constant from a given direction, your indicated airspeed and, hence, aircraft performance will be the same, whether proceeding upwind or downwind. For constant conditions, in turning from downwind to upwind or vice versa, a change in momentum is involved, but in shallow turns the effect on indicated airspeed and, hence, performance is minimal, and normally not noticeable. But where wind shear is involved, there is a dramatic effect for the reasons given in the chapter on turbulence – you may wish to review wind shear in that chapter at this time.

The plow wind does not only spread out in advance of the storm. A thunderstorm may be moving away from an airport, but the downdraft, spreading out in all directions, can still affect the airport, even though to a lesser degree than forward. The plow wind is a high circulation, that is, air spirals downward and outward in a clockwise fashion, thunderstorm updrafts being a low circulation.

The plow wind may reach speeds of 100 knots or more, and there may be an increase of as much as 50 percent in wind speed between the surface and 1,500 feet (460 m); most of the increase will occur in the first 150 feet (46 m). About two out of every three plow winds result in wind shifts of more than 90° and about one out of every 10 exceeds 180°. An average speed and direction change from prevailing conditions for the plow wind would be about 15 knots and 45°.

Because of the gustiness of the plow wind, it is often called the first gust or gust front. We have spoken rather loosely, mostly intuitively, about the gust to this point. We will consider gusts to be sudden, irregular, random and brief fluctuations in the air flow. These gusts are often embedded in drafts. One may be in an updraft and all of a sudden experience a downward gust, giving a very sharp bump or jolt. The sizes of these gusts range from a few inches to several hundred feet in diameter and, hence, can cause great difficulties in controlling the aircraft.

We have already mentioned that turbulence occurs with adjacent updrafts and downdrafts, for even though the downdraft will usually be occurring in the central portion of the storm, air is being drawn inward and upward around the perimeter of the storm, a process known as entrainment. And the storm may be marked by several areas of updrafts and downdrafts, for remember we are illustrating only one model of these storms; as we have already indicated, each storm has its own identity. Turbulence may well extend outward from the center to 20 miles (32 km) in clear air. Turbulence is also extensive near the tops of the storms. It is recommended that at least 5,000 feet (1.5 km) of vertical clearance be given tops, and 10,000 feet (3 km) when wind speeds exceed 100 knots at the top. Storms higher than 40,000 feet (12 km) should be circumnavigated. While the vertical and horizontal extents of a thunderstorm are indicative of dangerous turbulence, the appearance of the storm does not always indicate the degree of turbulence that can be expected. It is best to assume that there is no correlation between the appearance and degree of turbulence. Just assume extreme turbulence for any thunderstorm and stay away from them if at all possible.

2. *Hail*

The last statement above also applies to hail, produced primarily during the mature stage. Since most thunderstorms contain hail at some stage of their development, just assume they all do. Do not try to equate the appearance of the storm with the hail potential it may possess (Figure 13-7). A hailstone may range from pea-size to the size of a grapefruit. Hailstones as large as two pounds have been found, and stones near one-half of a foot (15 cm) in diameter have been observed near 30,000 feet (9 km). Most large hailstones are created by numerous trips of ice particles in updrafts and downdrafts, growing larger and larger on successive trips through levels containing large amounts of liquid water. Most hail occurs between the altitudes of 10,000 and 30,000 feet.

Hail has been observed in clear air as high as 45,000 feet (13.7 km). Stay 5 miles (8 km) away from approaching severe thunderstorms to avoid possible hail. Hailstones with a diameter of one-half inch can cause great damage to an aircraft. The ingestion of huge amounts of hail or liquid water can stop a jet engine, as well as cause structural damage due to the extreme force present. This is another reason for following the manufacturer's recommendations for proper airspeed in such circumstances. At least one of the larger jet airliners is thought to have been forced down due to hail ingestion in the engines.

3. *Icing*

Icing was discussed in the previous chapter. Clear ice is the greatest hazard, but rime and mixed ice also occur. The most severe icing occurs near the freezing level to temperature levels

Figure 13-7. Photograph of hail damage to aircraft.

of −25°C. However, icing at temperature levels colder than −40°C are not uncommon. Of course, icing will be more severe the longer the time spent in storms, thunderstorm clusters or squall lines being more dangerous than individual storms.

4. Altimetry

Effects on altimeters may result in 1,000 feet (300 m) of error. Since an approaching thunderstorm is an intense small-scale low, surface pressure lowers initially, then rises with the onset of the plow wind, and then gradually returns to normal as the storm moves out.

5. Lightning

We may be oversimplifying, but we will say that lightning results from a separation of electrical charges, positive and negative, and in nature's own unique way lightning offers the ladder by which a proper balance can be established between these charges (Figure 13-8). In aviation you may see the acronym LTGICCCCG, which means lightning in the cloud, cloud-to-cloud and cloud-to-ground. Severe thunderstorms are often accompanied by cloud-to-ground strikes; however, the majority of lightning discharges never reach the ground. Thunder is the noise one hears as a result of the air being heated and expanding more rapidly than the speed of sound, nature's own answer to the man-made sonic boom.

Anytime an airplane flies near a thunderstorm, an encounter with lightning is possible. Most thunderstorms will reach an altitude where the temperature is approximately −20°C before lightning begins, the greatest frequency occurring just prior to the period of heaviest rainfall. Lightning strikes have occurred at all levels within a thunderstorm, although most frequently within 2,500 feet (760 m) of the freezing level, from −5°C to 5°C in terms of temperature. However, the majority of commercial airline incidents have occurred at lower altitudes in non-stormy clouds and in areas outside of active thunderstorm cells. Strikes of this type are probably due to charge buildups on the aircraft. Structural damage can occur, but the damage is usually small perforations, or pinholes. Fuel ignition is very unlikely, but can be catastrophic when it does occur; JP-4 forms a highly explosive mixture and is significantly more vulnerable to explosion than gasoline or

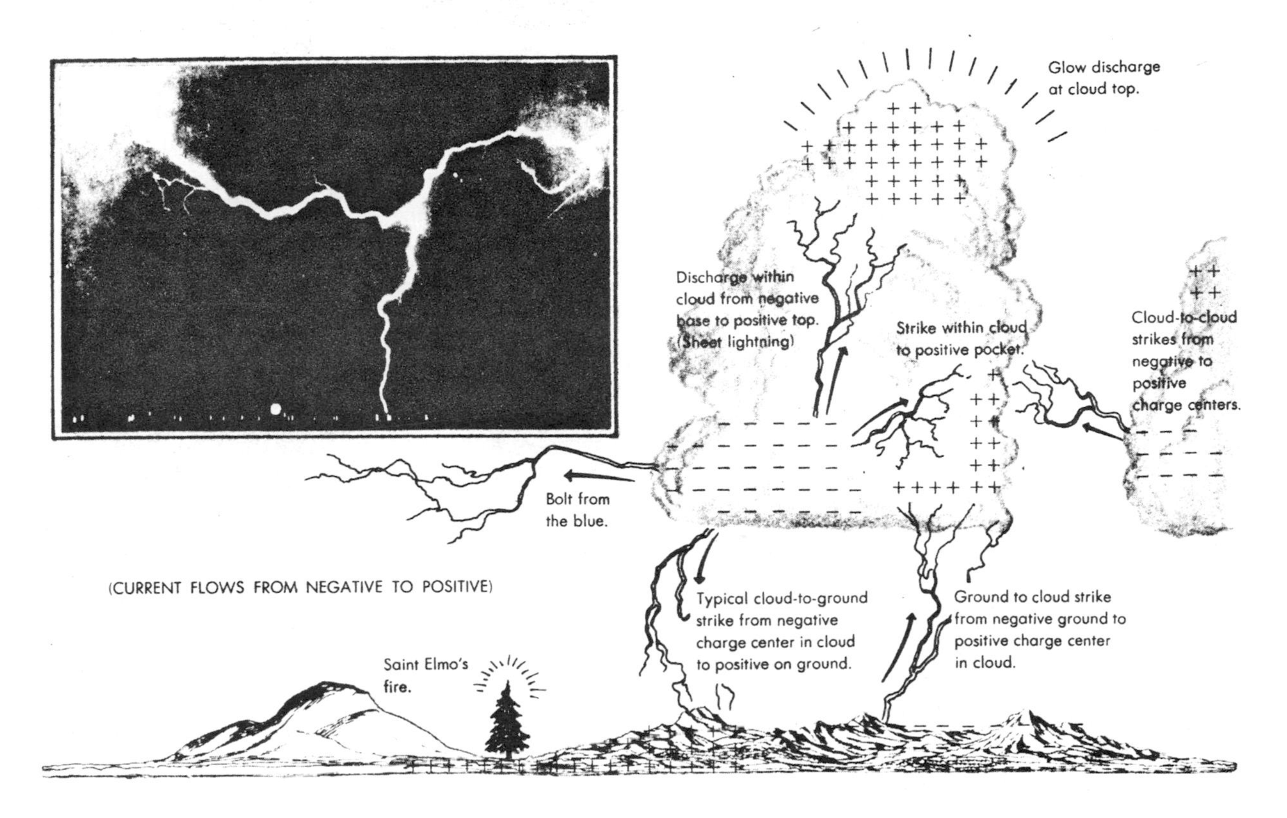

Figure 13-8. Lightning Variations.

kerosene jet fuel. Some damage has occurred to radios, flight instruments, radar, and electrical systems. You would be best advised to assume that all instruments are incorrect near frequent lightning.

There have been instances of pilots receiving burns on hands and face from lightning. Temporary blindness can result due to frequent flashes. Turning up the cockpit lights, wearing sunglasses, and avoiding staring outside for prolonged periods of time can reduce this bothersome lightning effect. Up to this point we have been discussing what might be called electrical-field lightning. No amount of anti-lightning hardware can eliminate it. You avoid lightning by avoiding thunderstorms. But, again, lightning is less a hazard than hail. We now discuss a form of lightning which results basically from the effects of friction, commonly known as precipitation static.

6. Precipitation Static

Whenever an aircraft flies through an area of liquid or solid particles, whether it be clouds, rain, ice particles, sand or dust, the friction of impact with these particles generates static electricity, usually called precipitation static. This static electricity is discharged primarily from sharp metallic points on the aircraft. The visual appearance is in the form of a corona, circular about a propeller and like an elongated rubber band along the leading edge of a wing. The corona can best be seen at night and can occur at any level within the thunderstorm, but like so many other thunderstorm hazards, appears to be most frequent and intense near the freezing level. The major effect of precipitation static is radio interference, which can be reduced with antistatic hardware.

Figure 13-9. Tornado.

Figure 13-10. Waterspout.

7. *Tornadoes*

Thunderstorms are also responsible for reduced ceilings and visibilities, a discussion of which is contained in the next chapter. We conclude this section on thunderstorm hazards with a brief study of tornadoes. We will classify the tornado as the ultimate form of turbulence, relative to its effects on aircraft. Thus, we still say that, in general, turbulence is the chief thunderstorm hazard and that the tornado is the worst form of turbulence. First, we mention the funnel cloud, a violent rotating column of air that extends or hangs from a cumulonimbus cloud but does not reach the ground. A tornado is just a funnel cloud which reaches the ground, and a tornado over water is called a waterspout (Figures 13-9 and 13-10).

Tornadoes are known to develop frequently in the southwestern portion of a low along a cold front extending from the low, with a jet stream cutting across the low in the upper levels. By no means are tornadoes limited to this set of circumstances. They can occur with an isolated air mass thunderstorm, as indicated earlier. However, they do occur most frequently with the steady-state thunderstorm. Every state of the U.S. has recorded tornadoes, although they are most frequent in the midwest, less frequent in the eastern portions of the U.S., and infrequent in the west (Figure 13-11).

The average diameter of a tornado is a few hundred yards. They usually occur in families and may extend upward to 35,000 feet (11 km) or more, although an average height would be about 10,000 feet (3 km). This is another reason for not flying underneath thunderstorms, even if you can see all the way across. Tornadoes may appear to move fairly slowly, but some of these storms move 70 mph (61 kts) or more. Tornadoes usually spiral in a counterclockwise manner, but anticyclonic tornadoes do occur, particularly with families of tornadoes. The pressure drop across a distance of 100 yards (90 m) in a tornado can be equivalent to 15 inches

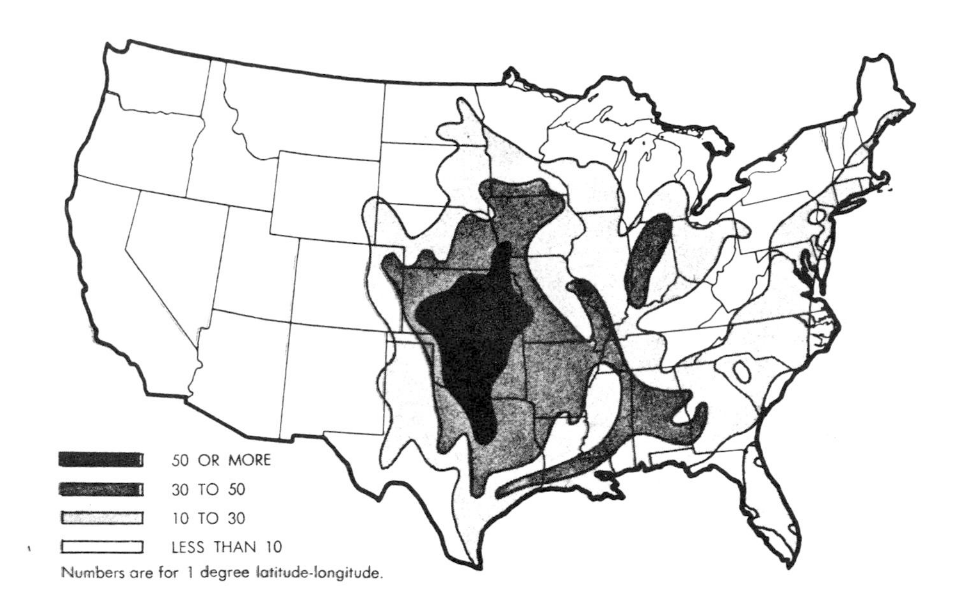

Figure 13-11. Tornado Frequency (1916 through 1963).

(38 cm) of mercury. It is a small wonder that wind speeds of 230 mph (200 kts) or more can occur inside these storms. Avoid thunderstorms with low-level roll clouds and tubular appendages – remember also that a tornadic-like circulation may be present as far as 10 to 20 miles (16 to 32 km) away from the main cloud, and that it may not even be visible (Figure 13-12). Again, it is best to stay at least 20 miles away from all thunderstorms.

8. Use of Radar

Radar is of limited value to the pilot. Its chief value is in depicting general areas of thunderstorms. The image on the weather radar screen is a function of the size and number of precipitation particles, size being of more importance than number. Thus, weather radar is of no use in determining areas of low ceilings and visibilities, again, since it is calibrated for precipitation. The weather radar is of limited value in depicting individual thunderstorm hazards, such as hail, turbulence, lightning, and tornadoes. If you have an aircraft with airborne radar you are very fortunate, but you should make absolutely sure that you are aware of the manufacturer's interpretation guidelines for that radar. What we will call a "weak" radar may show the opposite of what a "strong" weather radar would show for depicting hail, for example. More than one pilot has made interpretive mistakes. And don't forget that behind one radar image there may be other storms that are not visible – the storms in front reflect most of the radar energy (Figure 13-13).

Use weather radar for weather avoidance (Figure 13-14). Use radar for thunderstorm penetration only when you are an experienced pilot, you know your radar and its limitations, and you have no other choice. Radar used by air traffic controllers is designed primarily to show aircraft, although it can depict some of the heavier thunderstorm activity; such information will be on a time-permitting basis only. Because only the heavier thunderstorms will appear on the ATC radar, you may be asked to fly a heading which would take you into a thunderstorm. Advise ATC, for the presence of that storm may not be known to them. Operation Raincheck is one opportunity you have to observe ATC as it works aircraft on a large-scale basis. It is highly recommended that you take advantage by attending. Check with your local Flight Standards District Office for availability and times.

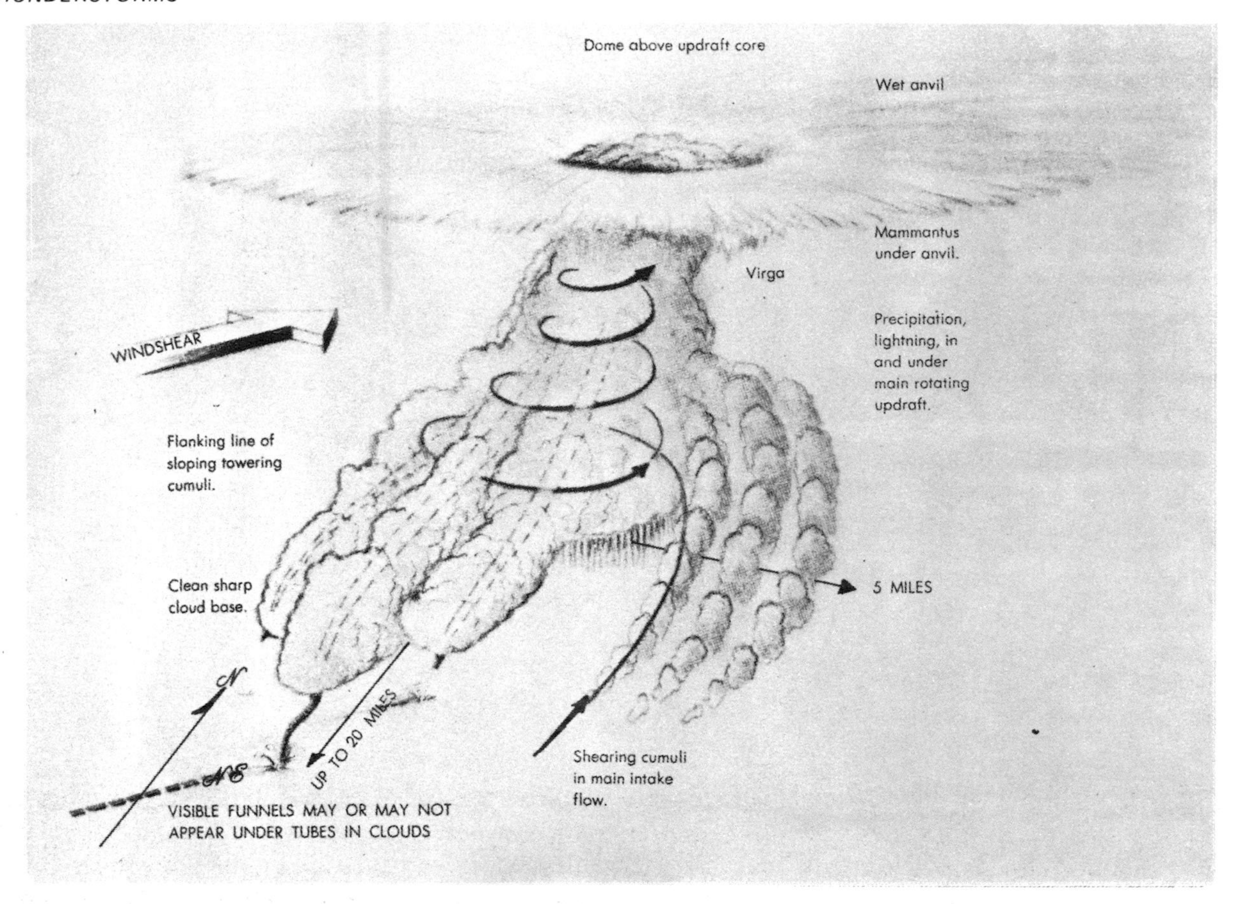

Figure 13-12. Tornado Funnel Cloud.

9. *Additional Tips When Flying Thunderstorms*

We now add some pointers on flying thunderstorms that have not already been mentioned. The bases of cold front thunderstorms are generally lower and more severe than those with a warm front, but remember that thunderstorms with a warm front are often embedded within a large cloud mass. Listen for crash static in headphones for an indication of warm front thunderstorms. In flying over thunderstorms, be aware that due to huge updrafts the thunderstorm may be at your planned flight level or higher by the time you get there. Thunderstorms have been known to develop to the mature stage in less than 20 minutes. The stratosphere usually acts as a cap on the tops of thunderstorms due to a temperature inversion above the tropopause. The presence of the inversion is usually indicated by a haze layer of ice crystals. Flying through this haze layer is relatively safe, compared to your other alternatives. With regard to squall lines be aware that, especially along fronts, one squall line may develop and move out and then later another squall line may also develop.

Be leery of statements made by pilots that they encountered no problems when they passed through an area of thunderstorms. They may have been lucky, but you might not be so fortunate. While in the thunderstorm area fly below the recommended penetration airspeed, but not so slow as to stall or lose control. Avoid heavy rain areas, for these are probably where the most intense cells are located. And lastly, try to keep yourself composed. To keep yourself COVERED you must keep your wits about you. See Do's and Dont's of Thunderstorm Flying in the section on Safety of Flight in the chapter on Flight Safety for some tips from the Airman's Information Manual concerning flying thunderstorms.

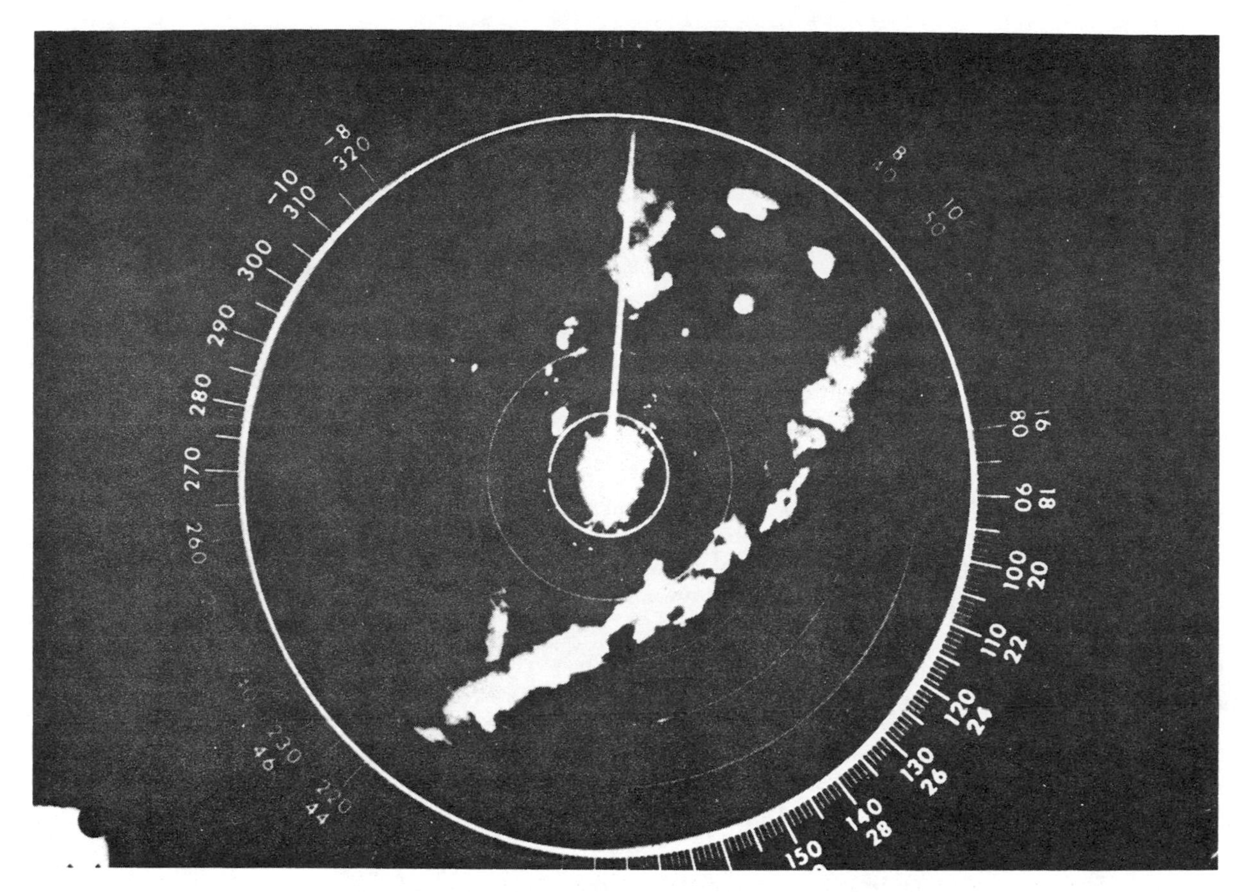

Figure 13-13. Radar Photograph of a Line of Thunderstorms.

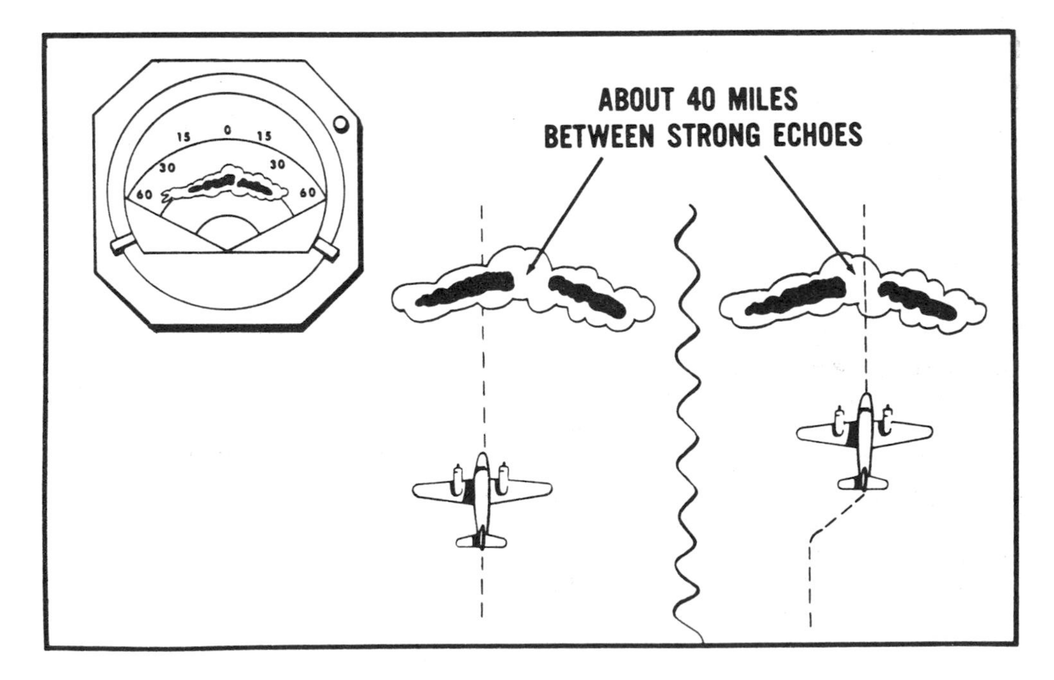

Figure 13-14. Airborne radar aids the pilot in avoiding turbulence and hail caused by thunderstorms. A pilot should avoid a strong echo by at least 20 miles; thus, he would need a 40-mile corridor between adjacent strong echoes.

THUNDERSTORMS
Review Questions

PRI

1. Hail, an in-flight hazard, is most likely to be associated with
   a. Cirrocumulus clouds.
   b. Stratocumulus clouds.
   c. Cumulonimbus clouds.
   d. Cumulus clouds.

2. Of the following cloud types, which is most likely to produce hail?
   a. Cumulus.
   b. Stratocumulus.
   c. Cirrocumulus.
   d. Cumulonimbus.

3. HAIL is considered to be an in-flight hazard. Which statement is true concerning this hazard?
   a. Hail is usually produced by cirrostratus clouds.
   b. Subtropical and tropical thunderstorms contain more hail than thunderstorms in northern latitudes.
   c. Large hailstones are entirely composed of clear ice.
   d. Hailstones may be thrown outward from a storm cloud for as much as 5 miles.

4. Which statement is true regarding the in-flight hazard called HAIL?
   a. Hail is usually produced by cirrocumulus clouds.
   b. Large hailstones usually do not have alternating layers of clear and cloudy ice.
   c. Subtropical and tropical thunderstorms contain more hail than thunderstorms in northern latitudes.
   d. Large hail is most commonly found in thunderstorms which have strong updrafts and large liquid water content.

5. Consider the following statements with relation to HAIL as an in-flight hazard to aircraft, and select those which are correct.
   A. There is a useful correlation between the external visual appearance of thunderstorms and the amount of hail within them.
   B. Large hail is most commonly found in thunderstorms which have strong updrafts and large liquid water content.
   C. Hail may be found at any level within a thunderstorm, but not in the clear air outside of the storm cloud.
   D. Hail is usually produced during the mature stage of the thunderstorm's life span.
   E. Hailstones may be thrown upward and outward from a storm cloud for as much as 5 miles.

   The true statements are
   a. B, D, E.
   b. A, B, C.
   c. A, B, C, D, E.
   d. A, B, D, E.

6. Select the statement which is correct in regard to the life cycle of thunderstorms.
   a. Throughout the dissipating stage of a thunderstorm the updrafts continue to develop.
   b. The beginning of rain at the earth's surface indicates the dissipating stage of the thunderstorm.
   c. The beginning of rain at the earth's surface indicates the mature stage of the thunderstorm.
   d. The initial stage of a thunderstorm is always a nimbus cloud which means "rain cloud."

7. A squall line, which may precede a cold front, will often be characterized by
   a. Widespread fog and extremely cold surface temperature.
   b. Thunderstorms and turbulence.
   c. Milder weather conditions than the cold front itself.
   d. Fog, low stratus clouds, and steady drizzle.

8. In regard to flying in the vicinity of thunderstorms, you should be aware that
   a. Avoidance of lightning and hail is assured by flying in the clear air outside the confines of a thunderstorm cell.
   b. The overhanging anvil of a thunderstorm points in the direction from which the storm has moved.
   c. The most severe conditions, such as heavy hail, destructive winds, and tornadoes are generally associated with squall line thunderstorms.
   d. Avoidance of severe turbulence is assured by circumnavigating thunderstorms and clearing edges of the storms by 5 miles.

9. A squall line is usually associated with
   a. A fast-moving cold front.
   b. A fast-moving warm front.
   c. A stationary front.
   d. An occluded front.

10. The most severe weather conditions, such as destructive winds, heavy hail, and tornadoes, are generally associated with
    a. Fast-moving fronts.
    b. Squall line thunderstorms.
    c. Slow-moving warm fronts.
    d. Slow-moving cold fronts.

11. Thunderstorms are produced by which type clouds?
    a. Stratocumulus.
    b. Altostratus.
    c. Cumulonimbus.
    d. Nimbostratus.

12. Tornadoes are more likely to occur with which type thunderstorms?
    a. Air mass thunderstorms.
    b. Steady-state thunderstorms associated with cold fronts or squall lines.
    c. Squall line thunderstorms that form ahead of warm fronts.
    d. Tropical thunderstorms during the mature stage.

INS

13. What visible signs indicate extreme turbulence in thunderstorms?
    a. Cumulonimbus clouds, very frequent lightning, and roll clouds.
    b. Base of the clouds close to the surface, heavy rain, and hail.
    c. Low ceiling and visibility, hail, and precipitation static.
    d. Lightning, roll clouds, low ceilings and visibility, and precipitation static.

14. What conditions are necessary for the formation of a thunderstorm?
    a. Frontal activity, cumulus clouds, and sufficient moisture.
    b. Cumulus clouds, unbalance of static electricity, and turbulence.
    c. Sufficient heat, moisture, and electricity.
    d. Lifting action, unstable air, and sufficient moisture.

15. Which weather phenomenon is always associated with a thunderstorm?
    a. Lightning.
    b. Heavy rain showers.
    c. Supercooled raindrops.
    d. Hail.

COM

16. Suppose hazardous low-level wind shear is encountered during the initial climb after takeoff. Select the true statement.
    a. The wind direction will always change from a headwind to a tailwind when flying through wind shear.
    b. When passing through wind shear the groundspeed will usually remain constant.
    c. Low-level wind shear may be associated with a thunderstorm's gust front that precedes the actual storm by 15 nautical miles.
    d. The pilot should decrease power to compensate for the increase in lift.

CFI

17. The conditions necessary for the formation of cumulonimbus clouds are a lifting action and
    a. Unstable air containing excess condensation nuclei.
    b. Unstable, moist air.
    c. Unstable air containing excess condensation nuclei.
    d. Stable or unstable air.

18. Which of the following is considered to be the most hazardous condition associated with thunderstorms?
    a. Lightning.
    b. Static electricity.
    c. St. Elmo's Fire.
    d. Wind shear and turbulence.

ATP

19. Which of the following features do you normally associate with the "cumulus stage" of thunderstorm formation?
    a. Heavy rain at the surface.
    b. Continuous updraft.
    c. Frequent lightning.
    d. Roll cloud.

20. Which weather phenomenon signals the beginning of the mature stage of a thunderstorm?
    a. The appearance of an anvil top.
    b. The start of rain at the surface.
    c. A sharp drop in temperature.
    d. Strong and gusty surface winds.

CHAPTER XIV
CEILING AND VISIBILITY

The reader should demonstrate an understanding of ceiling and visibility to the extent that he can answer the following questions (as stated below, or as multiple-choice, etc.):

1. What is the cause of one out of every four general aviation fatal accidents?
2. What are the two basic kinds of airspace?
3. What conditions must be satisfied before you can enter controlled airspace when the weather is less than VFR?
4. How is VFR determined?
5. Write down the VFR visibility and cloud minimums from memory.
6. Name and describe the types of controlled airspace.
7. A special VFR has meaning only relative to what kind of controlled airspace?
8. What are the cloud and visibility requirements for special VFR?
9. Will you always be granted a special VFR? Why or why not?
10. Who can usually receive a special VFR at night?
11. Why might special VFRs never be given?
12. What does "basic VFR" mean?
13. What does "clear of clouds" mean?
14. Why may flying VFR on top of an overcast be inadvisable?
15. What is the "minimum safe altitude for all operations?"
16. What is one especially important reason for becoming instrument rated?
17. Even though a private pilot is not required to file a VFR flight plan, what is a weather requirement for flight not in the vicinity of an airport?
18. Define "ceiling."
19. Give an example of an obscuration aloft.
20. Define "clear;" "scattered;" "broken;" and "overcast."
21. Define "thin;" "transparent."
22. List and define four kinds of visibility.
23. Distinguish between a cloud ceiling and an obscuration ceiling.
24. Upon what does the fog classification system primarily depend?
25. State the two basic ways fogs form.
26. How does cooling and/or the addition of moisture affect the temperature-dewpoint spread?
27. In what season would you expect the greatest likelihood of fog formation?
28. Under what conditions does radiation fog form? Describe the formation process.
29. State an adverse effect of a strong low-level temperature inversion.
30. Define "advection fog;" "sea fog."
31. What is the primary location for the occurrence of advection fog?
32. Which two fogs are considered to be the most significant to the pilot? Compare them.
33. Define "upslope fog."
34. Define "steam fog." Describe the possible kinds of weather resulting from such a fog.
35. Describe the "precipitation-induced fog."
36. Define "haze." Where is it most commonly found?
37. Define "smoke."
38. State the effect of stable air on the problems encountered with haze or smoke.
39. Compare fog, haze, and smoke in terms of how they are dispersed.
40. Describe the effects of blowing dust and sand.
41. Which precipitation forms most often restrict visibility? How does the absence of windshield wipers affect visibility?

CHAPTER XIV

CEILING AND VISIBILITY

There are many certificated pilots who are unaware of just what VFR really means. Proof of this fact is that about one out of every four general aviation fatal accidents is due to pilots trying to continue VFR into adverse weather. In order to better understand what VFR really means, we first briefly discuss the concept of airspace. Concerning VFR requirements, we consider two basic kinds of airspace, controlled and uncontrolled. The FAA designates various blocks of airspace as controlled for such reasons as safety and volume of traffic. Airspace not designated as controlled is called, appropriately enough, uncontrolled airspace. To you as a pilot you do not enter controlled airspace when the weather is less than VFR unless you are instrument rated, the aircraft is instrument equipped, and you are on an instrument flight plan. You will see below that VFR weather is determined by cloud and visibility requirements for various combinations of altitude and airspace.

A. VFR Visibility and Cloud Minimums

1. Types of Controlled Airspace

We will describe operations related to Table 14-1 after we very briefly discuss the types of controlled airspace listed below:

Think of control area as usually beginning at 1,200 ft AGL (outlined by blue tint on aeronautical charts) and that it is primarily associated with Federal Airways or routes.

For the entire continental U.S., continental control area begins at 14,500 ft MSL, excluding the first 1,500 ft AGL (which would only occur in mountainous areas).

Control zones are normally circular areas with a radius of five miles of an airport (actual area is as designated on an appropriate navigation chart), based at the surface and extending upward to the base of the continental control area (extends indefinitely upward if continental control area not present).

Positive control area extends from 18,000 ft to 60,000 ft MSL.

Terminal control area extends vertically to altitudes designated on navigation charts, in layers around certain primary airports, with lateral limits designated on charts.

Think of transition area as usually beginning at 700 ft AGL (outlined by magenta tint on aeronautical charts) and that it is primarily associated with airports having approved instrument approaches.

2. Control Zone

To fly in terminal control area or positive control area, one must meet requirements other than weather. Since our emphasis is on weather requirements we will not delve into these other restrictions. The type of controlled airspace we wish to pay special attention to here is called control zone. Suppose one wished to fly VFR through a control zone at 7,500 feet, an altitude which we will assume to be at least 1,200 feet above the ground. Then even if there is a ceiling below you (a term we will define shortly; think of it now as a cloud layer hiding the ground, an "overcast"), you are VFR at your altitude as long as you stay 1,000 feet above and 500 feet

TABLE 14-1. VFR VISIBILITY AND CLOUD MINIMUMS

Altitude	Controlled Airspace	Uncontrolled Airspace
More than 1,200 ft AGL and at or above 10,000 ft MSL	Visibility 5 miles, 1,000 ft above and below clouds,1 mile horizontal.	Same as controlled.
More than 1,200 ft AGL but less than 10,000 ft MSL	Visibility 3 miles, 1,000 ft above and 500 ft below clouds, 2,000 ft horizontal.	Visibility 1 mile (5 at night), 1,000 ft above and 500 ft below clouds, 2,000 ft horizontal.
At or below 1,200 ft AGL	Same as above with the additional requirement that when operating beneath a ceiling in a control zone you must have at least a 1,000 foot ceiling.*	Same as above at night. Visibility 1 mile, clear of clouds during daytime.

*If you do not have at least a 1,000 foot ceiling and visibility of 3 miles when operating beneath a ceiling in a control zone, you may request a special VFR from Air Traffic Control before entering the control zone. If granted, special VFR weather minimums are: visibility 1 mile and clear of clouds (same as for uncontrolled airspace below 1,200 ft AGL).

below clouds, 2,000 feet horizontally from clouds at your flight level, and your visibility is at least three miles. But if a pilot wishes to fly VFR beneath a ceiling in a control zone, then in addition to the above requirements, that ceiling must be at least 1,000 feet above the ground (Figures 14-1 and 14-2).

3. *Special VFR*

Normally, you would be operating at low levels beneath a ceiling in a control zone only for the purposes of taking off and landing. In such situations, if you cannot maintain the required three miles visibility and/or the ceiling is below 1,000 feet, you may request a special VFR which, if granted, changes your weather minimums to "clear of clouds" and one mile visibility. You must request the special VFR from Air Traffic Control before entering the control zone. There is no certainty of being granted a special VFR; ATC may be working a large number of instrument approaches and the whole purpose of having a control zone in the first place is to keep non-instrument pilots out when weather is bad.

Non-instrument rated pilots can never receive a special VFR at night; an instrument rating is a minimum requirement. There are airports at which you will never receive a special VFR, day or night, instrument rated or not (control zone outlined by "T" on navigation charts). These are usually very high volume airports; they want non-instrument pilots away from the airport when weather is poor and instrument pilots flying prescribed approaches all the way to the runway.

We mentioned "when the weather is poor", above, for taking off and landing in a control zone. How poor? Well, remember you are within limits as long as the ceiling is at least 1,000 feet and visibility at least three miles. Remember this 1,000 and 3 requirement. It is sometimes called "basic VFR" and refers to operating beneath a ceiling in a control zone (operation of an airport rotating beacon during daylight indicates weather below basic VFR); special VFR has meaning only for control zones and changes weather requirements to clear of clouds and one mile visibility. Clear of clouds means just simply not in clouds.

4. *Legal VFR and Common Sense*

We need to make a very important point at this time. Even though you may be VFR at 7,500 feet while passing through a control zone with a ceiling beneath you, is this necessarily safe? What happens if your engine quits in a single-engine airplane? Then in your descent to land you will have to pass through the ceiling below, and if the ceiling is caused by an overcast, you will have to enter clouds. At this point you are no longer VFR. You say who cares about being VFR at a time like this? Well, it is true that you have an emergency and that regulations give

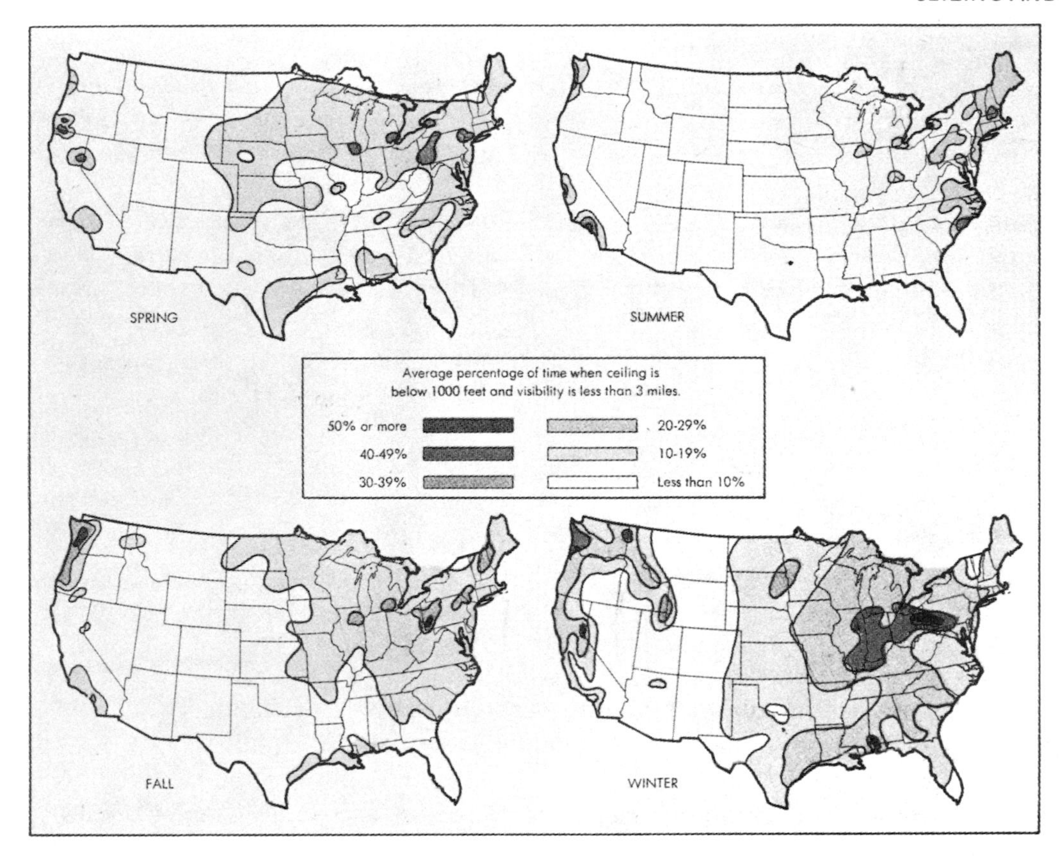

Figure 14-1. Seasonal Restrictions to Visibility.

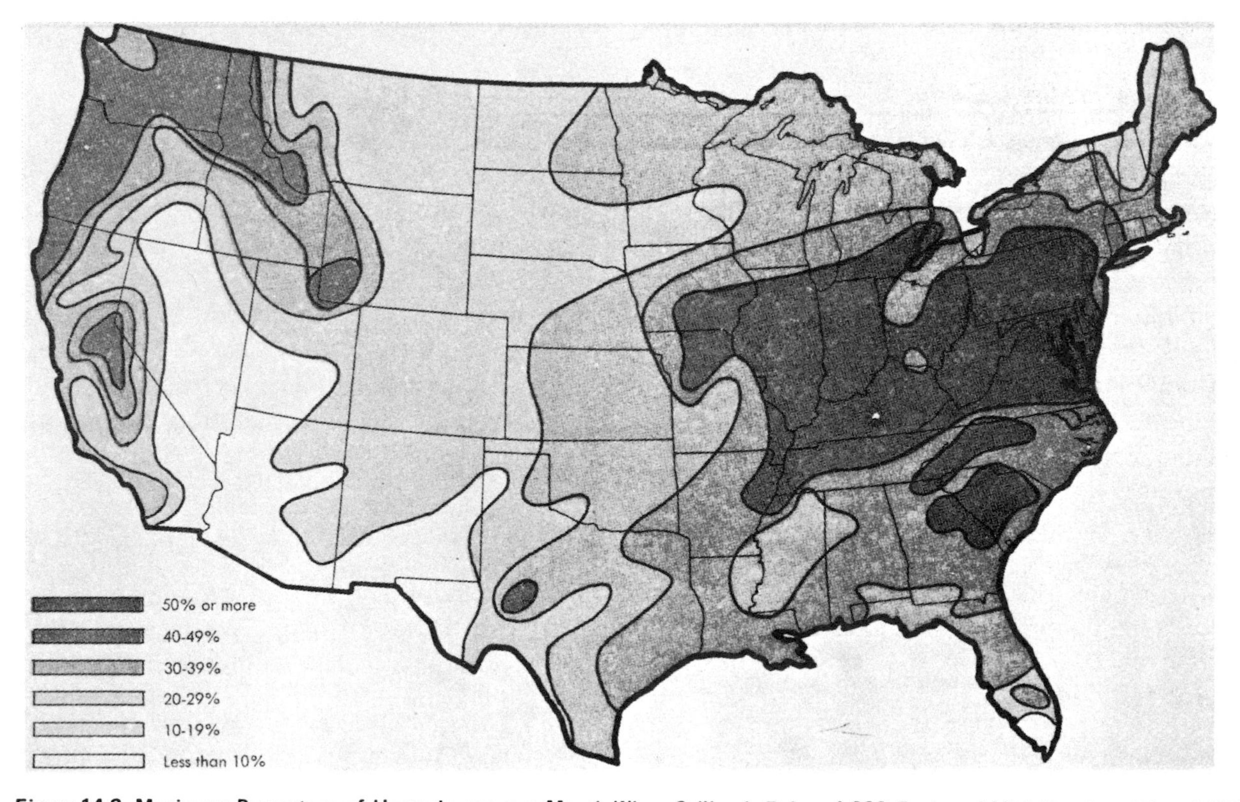

Figure 14-2. Maximum Percentage of Hours in any one Month When Ceiling is Below 1,000 Feet, and Visibility Less Than 3 Miles.

you the right to deviate from the rules to the extent necessary to meet that emergency. Landing safely is the all important matter right now. You should declare an emergency to ATC or to a Flight Service Station — so that they can help you land, before you fly into other aircraft or natural or man-made obstacles in or beneath clouds.

Thus, while you may be legally VFR in flying on top of a solid cloud deck, such action may not be wise, for the minimum safe altitude for all operations is an altitude such that should a power unit fail, you can make a landing without causing damage to life or property below. That is a regulation. Therefore, while you may be legally VFR, are you legal with regard to the minimum safe altitude regulation? The FAA has the right to judge you on how you exercise the privilege of authority over the aircraft, to the extent of revoking your pilot certificate.

Common sense dictates that you not routinely fly on top of a solid deck of clouds. As long as clouds do not form an overcast below, you have a chance to fly between them as you descend to a safe altitude. Keep yourself COVERED. Comprehend the factors which lead to ceiling and visibility problems. Visualize the weather pattern in which you may be flying. Evaluate your aircraft's position in relation to weather patterns. Respond by seeking a safe altitude. Then Evaluate again and Determine a new course of action, if necessary, even to the extent of declaring an emergency.

Flying on top of an overcast is not advisable even if your aircraft does not fail. You must still land eventually, and if the overcast is still there, you have a problem. This is one of the chief reasons for becoming instrument rated. While flying through clouds is not an everyday experience for a pilot on an instrument flight plan, should flight in clouds occur, ATC will provide the service necessary so that your flight can be conducted safely. In fact, when a pilot files an instrument flight plan, ATC works on the basic assumption that you are not VFR from start to end of flight.

While a private pilot is not required to file a VFR flight plan, it is still a regulation that he consider all available weather information for flight not in the vicinity of an airport. Do not fear weather, but keep yourself COVERED.

5. Ceiling Precisely Defined

Before proceeding any further we define more precisely what we mean by ceiling and visibility. Ceiling is: 1) the height above the earth's surface of the lowest layer of cloud or obscuring phenomena aloft that is reported as broken or overcast but not thin, or 2) vertical visibility when the sky and/or clouds are totally hidden by a surface-based obscuration. We will now discuss this definition part by part.

First, note that a ceiling is always relative to the earth's surface. We already know what a cloud is. Smoke would be an example of an obscuring phenomenon aloft. A clear sky means less than one-tenth sky coverage, scattered means one-tenth to five-tenths, broken means six-tenths to nine-tenths, and overcast means more than nine-tenths sky coverage. A summation principle is used on cloud layers. For example, maybe actual sky coverage is three-tenths at 2,000 feet and three-tenths at 4,000 feet. But as a weather observer looks upward from the ground, if these layers do not overlap at all, he will record six-tenths at 4,000 feet by summing the layers, and six-tenths is broken and a ceiling. For safety you should assume the worst at 4,000 feet, anywhere from six to nine-tenths actual sky coverage at that level, and not count on part of the cloud layer coming from a lower cloud deck. A cloud or obscuration layer is denoted as thin anytime at least half of the layer is transparent (means that sky or higher cloud can be seen through that layer).

6. *Visibility Defined*

We concern ourselves with four kinds of visibility: prevailing, flight, slant, and vertical. Prevailing visibility is a surface visibility and is the greatest distance objects can be seen and identified through at least half the horizon circle, the half not being necessarily a continuous half. Flight visibility is sometimes called air-to-air visibility, how far one can see horizontally from a flight altitude; this is the visibility normally used in the table on VFR visibility and cloud minimums. Slant range visibility is the visibility a pilot can expect to have on final approach to a runway, for his vision is slanted from his altitude to the surface.

Vertical visibility is important since it is the ceiling value when sky and/or clouds are totally hidden by a surface-based obscuration, such as fog. If the ceiling is reported as 500 obscured, then that means the weather observer could see approximately 500 feet vertically up into a fog, for example. Note the difference between a 500 foot cloud ceiling and a 500 foot obscuration ceiling. Beneath 500 feet below clouds you may have unrestricted visibility. While you may have 500 feet of vertical visibility in, say, a fog, and can see the ground directly beneath you, it is likely that you cannot identify the end of the runway until at a lower altitude. Thus, the obscuration ceiling reduces your slant range visibility (Figure 14-3). When the word obscuration is used by itself, it means surface based. "Obscuration aloft" denotes the non-surface based type. We have already devoted a chapter to a discussion of clouds. Depending on the depth and density of clouds, your visibility could be zero in all directions. Again, you are never VFR when in a cloud; you should be on an instrument flight plan in such a case.

B. Obscuring Phenomena

The remainder of this chapter will be concerned with a discussion of types of obscuring phenomena. We begin with a discussion of fog.

1. *Fog Formation*

Fog is a cloud on the ground and may be composed of water and/or ice particles. We will see that there are several types of fog, the classification being primarily dependent on the way fog forms. The two basic ways fog forms are by cooling air to near its dewpoint and by adding moisture. There are several ways air can be cooled and moisture content increased.

Cooling and/or addition of moisture will decrease the temperature-dewpoint spread. The presence of abundant condensation nuclei will often cause fog to form when the spread is as much as 5°F. Thus, in winter near coastal industrial areas, we would expect the greatest likelihood of fog formation. But fog can occur anywhere as we shall soon see.

Fog can be a real problem, for as a surface-based obscuration it may be responsible for a ceiling below 1,000 feet (as a result of the vertical visibility through it). You may have heard the statement "zero-zero." Well, this means zero vertical visibility and zero prevailing visibility, in the context of surface weather reports. Visualize yourself trying to land in such a situation. This zero-zero condition may occur instantly from a totally clear condition. It is primarily a hazard during takeoffs and landings, but for enroute flying it can also be a problem in maintaining visual reference to the ground, particularly if a forced landing results.

2. *Radiation Fog*

Radiation fog, often called ground fog, forms on clear nights with light winds when there is sufficient moisture, primarily in flat, inland regions. On a clear night the earth's surface cools, for it radiates heat outward and receives only indirect solar radiation. Without a cloud layer to absorb and reflect this heat downward, much of the heat is lost to space. Thus, the earth's

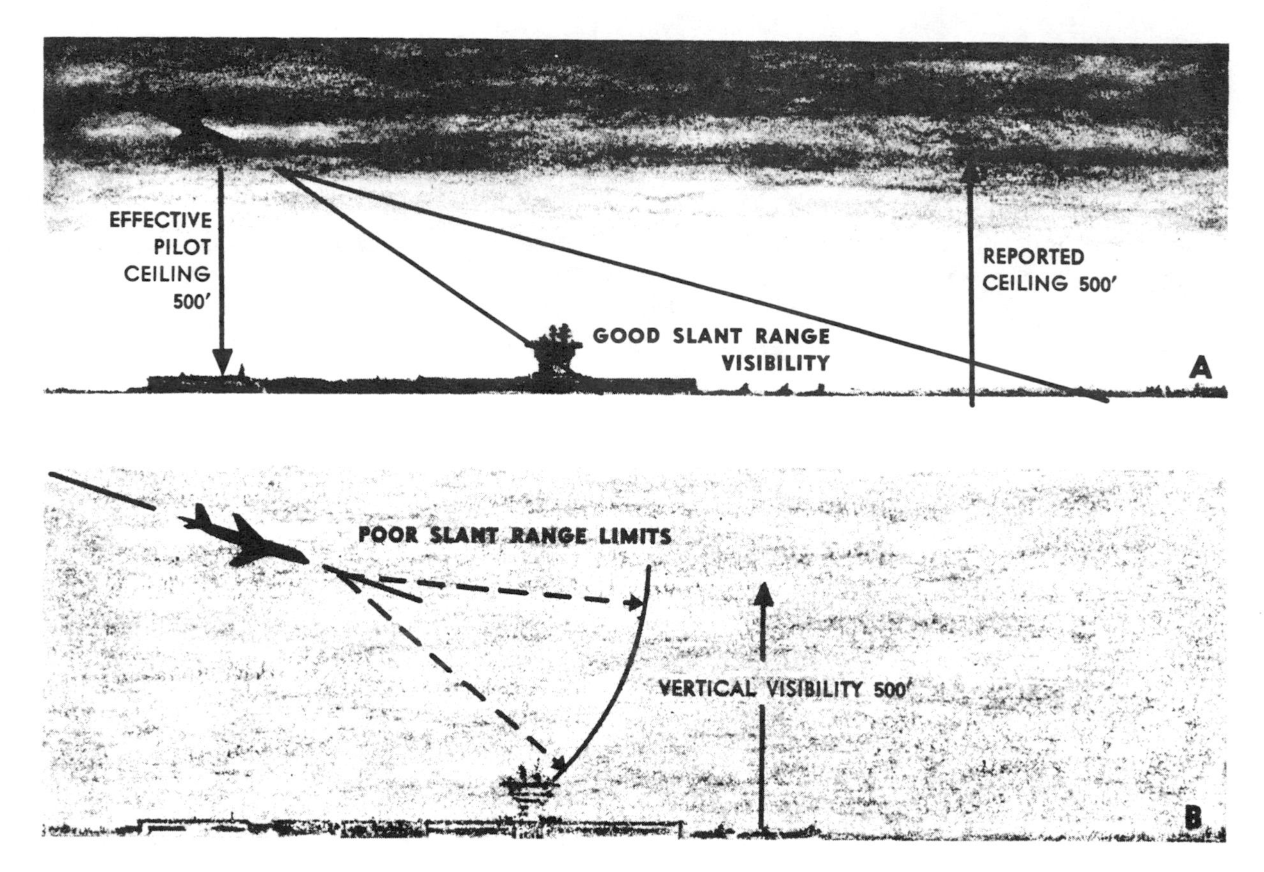

Figure 14-3. Difference between the ceiling caused by a surface-based obscuration (B) and the ceiling caused by a layer aloft (A). When visibility is not restricted, slant range vision is good upon breaking out of the base of a layer aloft.

surface becomes quite cool and cools the air above by conduction. This cooling will become less and less with height. With very cool temperatures at the surface, we may well form an inversion during the first, say, 2,000 feet. If enough moisture is present, the cooling of air beneath the inversion can cause fog to form. If winds are practically calm, radiation fog will be very shallow. But if winds are, say, five knots or less, then the slight mixing of air can cause the fog to become deeper. A very strong wind will tend to mix the air too much, lessening the cooling effect and dispersing the available moisture.

When a strong low-level inversion does form, fog is "trapped beneath the inversion." Recall that an inversion is an increase in temperature with height and, if portions of our fog were to be lifted, that part would cool adiabatically, and being cooler than the surrounding air would return to near the surface. During early morning hours, the earth's surface will be heated by solar radiation which will in turn heat the air adjacent to the ground, causing that portion of the fog to lift or evaporate. This is sometimes stated as fog "burning off." We would be left with a stratus cloud that is still trapped by the upper portions of the inversion. As more and more heat at the earth's surface is transferred upward, the inversion is broken and the stratus layer partially evaporates and partially disperses due to vertical and/or horizontal movement.

Other than warm fronts, strong low-level inversions generally cause the worst ceiling and visibility problems. You may well be in the clear above the inversion. But, again, you will have to land eventually. Many pilots get up early on a day when there is a strong inversion and just write off the whole day for flight purposes. But 90% of the time by noon, there will not even be a ceiling. There are times, infrequently, in which a very strong inversion will persist throughout

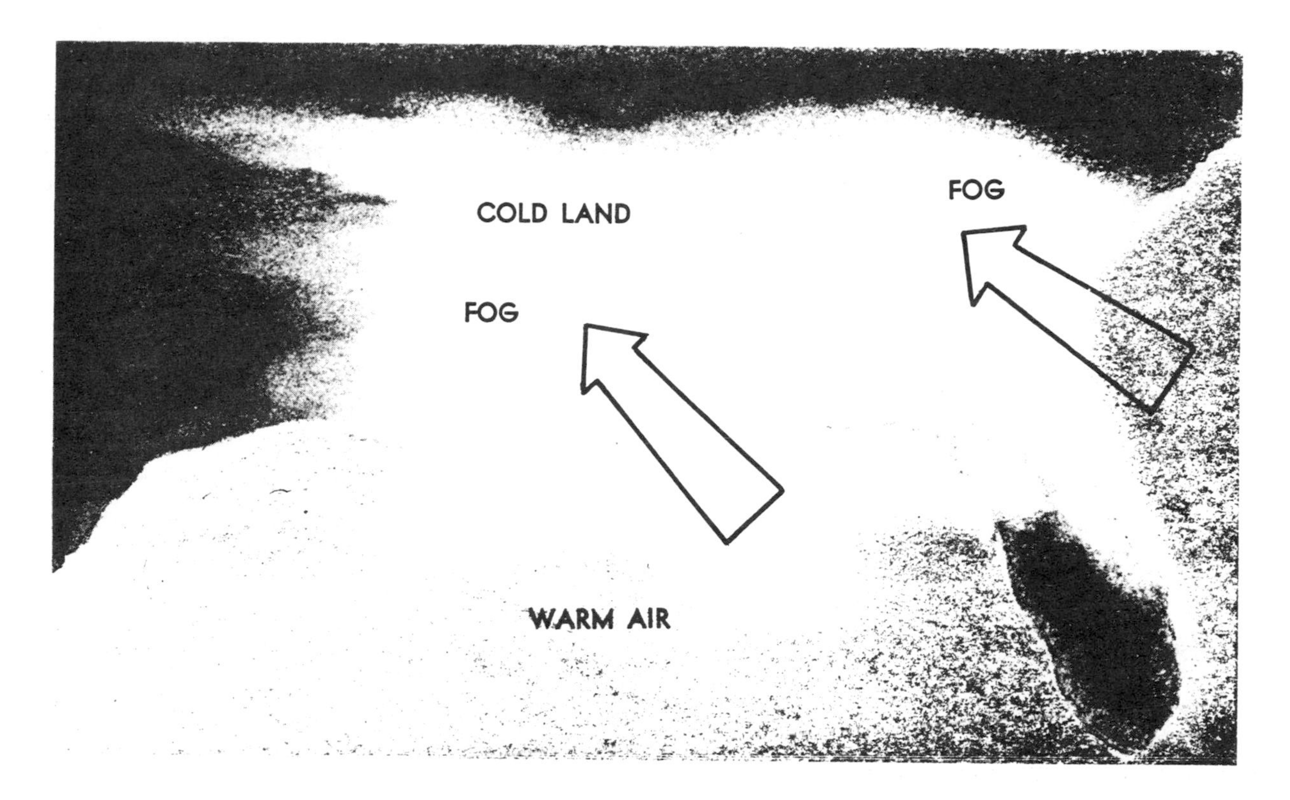

Figure 14-4. Advection fog over the southeastern United States and Gulf Coast. The fog often may spread to the Great Lakes and northern Appalachians.

the whole day. On an instrument flight plan, a pilot flies through this shallow condition quite rapidly and may find beautiful flying weather above. This would not be the case, however, if the fog and/or low clouds are associated with weather systems.

3. *Advection Fog*

An advection fog forms whenever moist air moves over colder ground or water. Note that advection means movement. The moisture that was responsible for radiation fog was already in the area (recall that winds were necessarily light). But for advection fog to form, moisture must move in from some source. The cold ground or water may have resulted from nighttime radiation, or from the cooling of wintertime air masses, or in the case of water, from colder water upwelling from below. An advection fog formed at sea is called a sea fog.

From the description given, one would expect advection fog primarily along coastal areas during winter. However, such fog can occur deep within the interior of a continental area. A good example is in winter when moisture is brought throughout the eastern portions of the U.S. by strong southerly winds off the Gulf of Mexico. Advection fogs have formed from Alabama all the way to the Great Lakes (Figure 14-4).

In contrast with radiation fog, advection fog occurs with stronger winds, may occur when skies are cloudy, usually covers a greater geographical area, usually lasts longer, and may occur anytime, day or night. Radiation and advection fogs are considered to be the most significant fogs to the pilot. There are several other types, however, which can on occasion be of significance.

4. Upslope Fog

Upslope fog forms when moist, stable air is cooled adiabatically when forced above higher terrain. Note the requirement for air to be stable. Had the air been unstable, then this air when lifted above the slopes would have been warmer than the surrounding air and would have continued to rise, dispersing the moisture and the cooling effect. Upslope fog is common along the western slopes of the Rockies.

5. Steam Fog

Steam fog occurs as a result of the movement of cold air over warm water. Picture cold air over a heated swimming pool – this is an example of a steam fog. On a larger scale, a cold air mass moving over the Great Lakes can result in an extensive steam fog. When the air above such a surface is very cold, that air will be unstable. Heating is provided by warm, moist water and, hence, the air will continue to rise. If the air is unstable through a large vertical extent, turbulence and thunderstorms may well develop. If only the lower layers of air are unstable with stable air above, a dense, low cloud mass is likely with moderate icing potential.

6. Ice Fog

We will consider ice fog to be similar to steam fog, except that the air is so cold that water vapor sublimates as ice crystals. In such a case icing potential is eliminated, unless there is a mixture of ice crystals and supercooled water, in which case mixed icing is probable.

7. Precipitation-Induced Fog

Precipitation-induced fog is caused by the evaporation of rain or drizzle, evaporation occurring while precipitation is falling and/or after it has reached the ground. This type of fog is especially widespread when it occurs with warm fronts, although it may occur with or without the presence of fronts. We now discuss other restrictions to visibility: haze, smoke, dust, sand, and precipitation.

8. Haze

Haze is a suspension of dry particles, other than smoke or dust, too small to be seen individually but which collectively restrict visibility. Salt particles are one type of dry particles which constitute haze. Haze can extend to very high levels in stable air, but is most often found below 15,000 feet. Above haze, horizontal visibility is usually quite good, but vertical visibility is reduced. Visibility is variable within the haze layer, the greatest restriction occurring when flying facing the sun.

9. Smoke

Smoke is a suspension of particles produced by combustion. Like haze, it is most prevalent when air is stable. When smoke is trapped beneath an inversion, visibility is especially reduced (Figure 14-5).

Recall that fog primarily evaporates, but haze or smoke will have to be dispersed by wind. Thus, haze or smoke may persist for days in areas where a high has stagnated. In such a high, air sinks slowly downward, acting as a cap or lid on the haze or smoke.

Figure 14-5. Smoke Trapped in Stagnant Air under an Inversion.

10. Blowing Dust and Sand

Blowing dust and sand can restrict visibility to a great degree. When air is unstable and winds are strong, dust may be lifted to heights of 15,000 feet. Sand is usually not lifted more than about 50 feet. However, visibility can approach zero in all directions in blowing sand.

The precipitation forms of rain, drizzle and snow most often restrict visibility, rain less than drizzle or snow. Remember, when rain falls on windshields with no wipers, visibility drops, and can be near zero in heavy or blowing snow. Drizzle usually occurs when air is stable and, when combined with fog, can present serious visibility problems. Rain does not usually lower prevailing visibility to less than one mile except in heavy showers. But without windshield wipers, visibility from the cockpit can be less than 50% of the prevailing visibility. When obtaining a briefing, always determine the factors responsible for rain, snow, fog, clouds, etc.

CEILING AND VISIBILITY
Review Questions

PRI

1. When the Aviation Weather Reports for your destination show an air temperature of 55°F to be within 4° of the dewpoint temperature and the spread between the two is decreasing, it is likely that upon arrival you would encounter
   a. Fog or low clouds.
   b. Thunderstorms and cold frontal-type weather.
   c. An increase in pressure altitude.
   d. Freezing precipitation or icing conditions.

2. Select the true statement concerning a temperature inversion.
   a. A temperature inversion normally develops with a decrease in the temperature as height is increased.
   b. A temperature inversion occurs when unstable air rapidly transfers heat from the surface upward.
   c. A temperature inversion often develops near the ground on clear, cool nights when the wind is light.
   d. A temperature inversion is usually indicated by the base of a line of cumulus clouds.

3. Advection fog is formed as a result of
   a. Moist air condensing as it moves over a cooler surface.
   b. The ground cooling adjacent air to the dewpoint temperature on clear, calm nights.
   c. The addition of moisture to a mass of cold air as it moves over a body of water.
   d. Moist, unstable air being cooled as it is forced up a sloping land surface.

4. Radiation fog is most likely to occur under which of the following conditions?
   a. Warm, moist air flowing from a body of water over a cold surface with an 8 to 10 knot wind causing mixing and condensation.
   b. Warm, moist air being forced upslope by light winds resulting in the air being cooled and condensed.
   c. Low temperature/dewpoint spread, calm wind conditions, the presence of hygroscopic nuclei, low overcast, and favorable topography.
   d. A clear sky, little or no wind, and small temperature/dewpoint spread.

5. A temperature inversion would most likely result in which of the following weather conditions?
   a. Clouds with extensive vertical development above an inversion aloft.
   b. Good visibility in the lower levels of the atmosphere and poor visibility above an inversion aloft.
   c. An increase in temperature as altitude is increased.
   d. A decrease in temperature as altitude is increased.

INS
6. What types of fog depend upon a wind in order to exist?
   a. Radiation fog and ice fog.
   b. Precipitation fog and steam fog.
   c. Uplsope fog and downslope fog.
   d. Advection fog and upslope fog.

7. What situation is most conducive to the formation of radiation fog?
   a. Warm, moist air over low, flatland areas on clear, calm nights.
   b. Moist, tropical air moving over cold, offshore water.
   c. The movement of cold air over much warmer water.
   d. A warm, moist air mass on the windward side of mountains.

8. At times, fog is prevalent in industrial areas because of
   a. Atmospheric stabilization around cities.
   b. An abundance of condensation nuclei from combustion products.
   c. The high rate of evaporation from water used by factories.
   d. A high concentration of steam from industrial plants.

9. In which situation is advection fog most likely to form?
   a. A warm, moist air mass on the windward side of mountains.
   b. An air mass moving inland from the coast in winter.
   c. A light breeze blowing colder air out to sea.
   d. Warm, moist air settling over a warmer surface under no-wind conditions.

10. In what localities is advection fog most likely to occur?
    a. Coastal areas.
    b. Mountain slopes.
    c. Level inland areas.
    d. Mountain valleys.

11. How is ceiling defined? The height of the
    a. Highest layer of clouds or obscuring phenomena aloft that covers over six-tenths of the sky.
    b. Lowest layer of clouds that contributed to the overall overcast.
    c. Lowest layer of clouds which is at least thin broken.
    d. Lowest layer of clouds or obscuring phenomena aloft that is reported as broken or overcast.

12. The most frequent type of ground or surface-based temperature inversion is that produced by
    a. Terrestrial radiation on a clear, relatively still night.
    b. Warm air being lifted rapidly aloft in the vicinity of mountainous terrain.
    c. The movement of colder air under warm air, or the movement of warm air over cold air.
    d. Widespread sinking of air within a thick layer aloft resulting in heating by compression.

COM

13. A special VFR clearance to enter a control zone requires that while in the control zone the pilot remain
    a. Clear of all clouds.
    b. At least 2,000 feet from all clouds.
    c. At least 1,000 feet from all clouds.
    d. At least 500 feet from all clouds.

14. No person may operate an airplane in a control zone under a special VFR clearance at night unless that person
    a. Enters the Airport Traffic Area at or above 1,500 feet AGL and maintains that altitude until descending for a landing.
    b. Holds at least a commercial pilot certificate and an instrument rating.
    c. Uses the runway which is served by an operating Visual Approach Slope Indicator.
    d. Meets the applicable requirements for instrument flight and the airplane is equipped as required for instrument flight.

15. Special VFR minimums apply to operations within what type airspace?
    a. Control Zones.
    b. Control Areas.
    c. Airport Traffic Areas.
    d. Restricted Areas.

16. What is the minimum flight visibility and proximity to cloud requirements for VFR flight, at 6,500 feet MSL, in a Control Area?
    a. 1,000 feet under or 500 feet over; 1 mile visibility.
    b. 1,000 feet over or 500 feet under; 1 mile visibility.
    c. 1,000 feet under or 500 feet over; 3 miles visibility.
    d. 1,000 feet over or 500 feet under; 3 miles visibility.

17. To operate an airplane VFR outside controlled airspace at an altitude of more then 1,200 feet AGL but less than 10,000 feet MSL, the minimum flight visibility during daytime operations is:
    a. 5 miles.
    b. 3 miles.
    c. 2 miles.
    d. 1 mile.

18. What is the minimum basic VFR flight visibility for all flights at or above 10,000 feet MSL except when less than 1,200 feet AGL?
    a. 5 miles.
    b. 3 miles.
    c. 3 miles during daylight hours, and 5 miles during hours of darkness.
    d. 1 mile in uncontrolled airspace, and 3 miles in controlled airspace.

19. Operation of an airport rotating beacon during the hours of daylight would mean
    a. That weather in the control zone is below basic VFR weather minimums.
    b. That takeoffs and landings only are authorized at the present time.
    c. Nothing to the pilot because these beacons operate continuously.
    d. Right-hand traffic is in effect.

CFI

20. When an air mass is stable, which of the following conditions is most likely to exist?
    a. Dust, smoke, and haze at the higher levels with good visibility at the surface.
    b. Moderate to severe turbulence in lower levels.
    c. Towering cumulus and cumulonimbus clouds.
    d. Smoke, dust, haze, etc., concentrated in lower levels with resulting poor visibility.

21. Radiation fog is most likely to occur with which of the following conditions?
    a. Warm, moist air flowing over a cold surface with an 8 to 10 knot wind causing mixing and condensation.
    b. Warm, moist air being forced upslope by light winds resulting in the air being cooled and condensed.
    c. Low temperature/dewpoint spread, calm wind conditions, the presence of hygroscopic nuclei, low overcast, and favorable topography.
    d. High humidity during the early evening, cool cloudless night with light winds, and favorable topography.

22. What causes radiation fog to form?
    a. Moist, unstable air being cooled as it is forced up a sloping land surface.
    b. The addition of moisture to a mass of cold air as it moves over a body of water.
    c. The ground cooling the adjacent air to the dewpoint temperature during conditions of calm air and clear sky.
    d. Moist air being warmed by the ground over which it passes.

23. Which of the following is the cause of advection fog?
    a. Moist, stable air being cooled adiabatically as it moves up sloping terrain.
    b. Warm, moist air moving over colder ground or water.
    c. Terrestrial radiation cooling the ground which in turn cools the air in contact with it.
    d. Saturation of cool air as the precipitation falling through it is evaporated.

ATP
24. Which situation is most conducive to the formation of advection fog?
    a. A light breeze blowing colder air out to sea.
    b. A warm, moist air mass on the windward side of the mountains.
    c. An air mass moving inland from the coast in wintertime.
    d. Warm, moist air settling over a cool surface under no-wind conditions.

25. Clouds, fog, or dew will always form when
    a. The temperature and dewpoint are equal.
    b. Water vapor condenses.
    c. The dewpoint is higher than the temperature.
    d. Relative humidity exceeds 100%.

26. The localities in which radiation fog would most likely occur are
    a. Level inland areas.
    b. Mountain slopes.
    c. Coastal areas.
    d. Mountain valleys.

27. Fogs produced by frontal activity are generally a result of saturation due to
    a. Evaporation of surface moisture.
    b. Nocturnal cooling.
    c. Evaporation of precipitation.
    d. Adiabatic cooling.

CHAPTER XV
FLIGHT SAFETY

The reader should demonstrate an understanding of flight safety to the extent that he can answer the following questions (as stated below, or as multiple-choice, etc.):

1. What is required by regulation should you be unable to complete a flight as planned?
2. In order to learn more about weather, what habit should you develop?
3. Describe pilots who never use a FSS.
4. What is said to exist at an airport with a FSS but no operating control tower? What information may pilots obtain? Is this data considered control information? Is pilot participation mandatory?
5. What is said to exist at at an airport with an operating control tower? State the dimensions of such an area.
6. What does ATIS mean and what kind of information is available? What is the meaning of the absence of visibility and/or ceiling data from ATIS?
7. Where is UNICOM assigned? What kind of information is received? How do you determine its availability?
8. What is a good practice to follow at airports having no communication facilities?
9. What information should you give to the briefer when telephoning for weather?
10. What is EFAS and for what is it designed? What frequency is used and what two words are always used when making a radio call for EFAS?
11. Describe TWEB and PATWAS.
12. Name and describe the types of preflight briefings.
13. Name the two basic types of in-flight weather advisories. What is the purpose of this service?
14. For what general kinds of weather will SIGMETS be issued? AIRMETS?
15. Up to 12 hours (and even beyond) which is more likely to be correct, a forecast ceiling 3,000 feet or more and visibility 3 miles or greater, or conditions below 1,000 feet and/or 1 mile?
16. If poor weather is forecast to occur within 3 to 4 hours, what is the probability of occurrence?
17. When are forecasts of poor flying conditions during the first few hours of the forecast period the most reliable?
18. What kind of weather is the most difficult to forecast accurately?
19. Which is more difficult to forecast, surface visibility or ceiling height?
20. List weather that forecasters CAN predict at least 75 percent of the time.
21. List weather that forecasters CANNOT predict with an accuracy which satisfies present aviation operational requirements.
22. State four reasons why pilots fly into unfavorable weather.
23. List ten weather signposts and their warnings.
24. What pertinent information should a weather briefing include?
25. Where can you obtain a weather briefing?
26. State the most frequently cited cause of accidents in which both pilots and weather were involved.
27. Approximately what percent of weather-involved accidents occur during daylight hours and in VFR conditions? What percent of accidents involved pilots who had less than 100 hours total flight time in the type of aircraft flown?
28. Describe a pilot most likely to have been involved in a nonfatal, weather-involved general aviation accident.
29. Read and interpret weather teletype data and charts.

CHAPTER XV

FLIGHT SAFETY

A. General

In this chapter we will include a partial description of FAA weather services that are available to you. You will learn how to use these services and some of the limitations of aviation weather forecasts. Additional information on severe weather is included, on wind shear, microbursts, and thunderstorms. Also included is a special study on nonfatal, weather-involved general aviation accidents.

Before we present the above topics, a few general comments should be made. First and foremost, you should always have an alternate plan of action in mind should you be unable to complete your flight as planned; this is a regulation. So you should always be skeptical to a degree. Be logical where weather is concerned. Some people will not walk outside when there is lightning around but they will fly into areas where lightning is occurring.

In order to learn more about weather, establish the habit of asking weather briefers why certain weather conditions exist or are forecast to exist. Only then will you be able to begin to keep yourself COVERED. You must first COmprehend the weather factors involved before you can Visualize various patterns. Then you can Evaluate your weather situation as improving or deteriorating and Respond by perhaps flying to an alternate airport. Even then you must continue to Evaluate, and possibly Determine new courses of action. Do not wait until you are in the air to go through this process. Do as much of it as you can on the ground. A wise pilot will often Determine to stay on the ground.

Keep in mind that weather is good much more often than bad. But bad weather is often invisible and, hence, catches many pilots unaware. If weather does become a problem, try to stay relaxed and think logically rather than reacting emotionally. Use your radio to stay continually updated on weather and to let the aviation system know when you are in trouble.

Above all, make use of FAA Flight Service Stations. They are your primary means for weather support. In fact, the primary thrust of this book has been to give you the basic knowledge of weather concepts necessary to be able to use the Flight Service Station (FSS). Pilots who never use the FSS are either careless or do not possess a sufficient understanding of meteorological principles to know what to ask for or how to use what data they do receive.

At airports having a FSS but no operating control tower, an airport advisory area is said to exist (extends ten statute miles from the airport). Pilots may obtain such advisory information as winds, runway most favorable, other known traffic, etc. This data is not control information and pilot participation is not mandatory, but strongly recommended. At airports with an operating control tower, and with or without a FSS, only an airport traffic area is said to exist (radius of five statute miles extending from the surface up to but not including 3,000 feet (900 m) above the surface — again, exists only where there is an operating control tower). A tower telling a pilot to "follow the Bonanza to final approach for landing" is an example of control information.

Automatic Terminal Information Service (ATIS) is a recording of non-control information that is considered essential for safety. Pilots should always listen to this recorded data when available. Absence of visibility and ceiling means that they are at least five miles and 5,000 feet (1.5 km), respectively.

UNICOM (think of as a CB radio to be used within approximately ten miles of the airport) is assigned to Aeronautical Advisory Stations at certain airports not served by a control tower.

UNICOM Is not intended to be used for Air Traffic Control purposes, except for:
(1) Revision to proposed departure time.
(2) Takeoff, arrival, or flight plan cancellation time.
(3) ATC clearance, provided arrangements are made between the ATC facility and the UNICOM licensee to handle such messages.
The information received using UNICOM will be of the same nature as from a FSS when there is no operating control tower.

There are, of course, many small airports having no communication facilities whatsoever. In such a case you should either observe other traffic or fly over the airport (at least 500 feet — 150 m — above traffic pattern altitude) in order to observe wind devices for direction and to note any traffic pattern indicators. Also, pilots should ''self announce'', a procedure by which pilots use the MULTICOM frequency 122.9 to announce their intentions (and to listen to other pilots) in the vicinity of such an airport (within approximately ten miles).

B. Safety of Flight

1. Airman's Information Manual and AFM 51-12

a. National Weather Service Aviation Products

(1) Weather service to aviation is a joint effort of the National Weather Service (NWS), the Federal Aviation Administration (FAA), the military weather services, and other aviation oriented groups and individuals. The NWS maintains an extensive surface, upper air and radar weather observing program; a nationwide aviation weather forecasting service; and also provides pilot briefing service. The majority of pilot weather briefings are provided by FAA personnel at Flight Service Stations (FSSs). Surface weather observations are taken by the NWS and NWS certified FAA, contract and supplemental observers and by automated observing systems.

(2) Aviation forecasts are prepared by Weather Service Forecast Offices (WSFOs). These offices prepare and distribute terminal forecasts for specific airports in 50 states and the Caribbean. These forecasts, which are amended as required, are valid for 24 hours. The last 6 hours are given in categorical outlook terms. WSFOs also prepare route forecasts and synopses for Pilots Automatic Telephone Weather Answering Service (PATWAS), Transcribed Weather Broadcast (TWEB), and briefing purposes. A centralized aviation forecast program originating from the National Aviation Weather Advisory Unit (NAWAU) in Kansas City was implemented in November 1982. In the conterminous U.S., all In-flight Advisories and all Area Forecasts are issued by NAWAU. Area Forecasts are prepared and amended as required, while In-flight Advisories are issued only when conditions warrant. Winds aloft forecasts are provided for flight planning purposes.

b. FAA Weather Services

(1) The FAA maintains a nationwide network of Flight Service Stations (FSSs) to serve the weather needs of pilots. In addition, NWS meteorologists are assigned to most Air Route Traffic Control Centers (ARTCCs) as part of the Center Weather Service Unit (CWSU). They provide advisory service and short-term forecasts (nowcasts) to support the needs of the FAA and other users of the system.

(2) The primary source of preflight weather briefings is an individual briefing obtained from a briefer at the FSS or NWS. These briefings, which are tailored to your specific flight, are available 24 hours a day through the local FSS or through the use of toll free lines (INWATS). Numbers for these services can be found in the Airport/Facility Directory under the FAA and

NWS Weather and Aeronautical Information Telephone Numbers section and the DOT, Federal Aviation Administration or Department of Commerce, National Weather Service listing under U.S. Government in your local telephone directory. NWS pilot briefers do not provide aeronautical information (NOTAMs, flow control advisories, etc.) nor do they accept flight plans.

(3) Other sources of weather information are as follows:

(a) The A.M. Weather telecast on the PBS television network.

(b) The Transcribed Weather Broadcast (TWEB), telephone access to the TWEB (TEL-TWEB) and Pilots Automatic Telephone Weather Answering Service (PATWAS) provide continuously updated recorded weather information for short or local flights.

(c) The Interim Voice Response System (IVRS) is now available in many metropolitan areas. Using this system, a limited weather data base can be accessed by telephone. Detailed instructions on how to use IVRS are contained in *"A Pilot's Guide to IVRS"*. A copy can be obtained from any FAA Flight Standards District Office or Flight Service Station.

(d) Weather and aeronautical information is also available from numerous private industry sources on an individual or contract pay basis. Information on how to obtain this service should be available from local pilot organizations.

(e) The Direct User Access Terminal System (DUATS) can be accessed by pilots with a current medical certificate toll free in the 48 contiguous states via personal computer. Pilots can receive alpha-numeric preflight weather data and file domestic VFR and IFR flight plans. DUATS is described more fully in Chapter XVI.

(4) In-flight weather information is available from any FSS within radio range. En route Flight Advisory Service (EFAS) is provided to serve the nonroutine weather needs of pilots in flight.

c. Preflight Briefing

(1) Flight Service Stations (FSSs) are the primary source for obtaining preflight briefings and in-flight weather information. In some locations, the Weather Service Office (WSO) provides preflight briefings on a limited basis. Flight Service Specialists are qualified and certificated by the NWS as Pilot Weather Briefers. They are not authorized to make original forecasts, but are authorized to translate and interpret available forecasts and reports directly into terms describing the weather conditions which you can expect along your flight route and at your destination. Available aviation weather reports and forecasts are displayed at each FSS and WSO. Some of the larger FSSs provide a separate display for pilot use. Pilots should feel free to use these self briefing displays where available, or to ask for a briefing or assistance from the specialist on duty. Three basic types of preflight briefings are available to serve your specific needs. These are: Standard Briefing, Abbreviated Briefing, and Outlook Briefing. You should specify to the briefer the type of briefing you want, along with appropriate background information. This will enable the briefer to tailor the information to your intended flight. The following paragraphs describe the types of briefings available and the information provided in each.

(3) STANDARD BRIEFING — You should request a Standard Briefing any time you are planning a flight and you have not received a previous briefing or have not received preliminary information through mass dissemination media; e.g., TWEB, PATWAS, IVRS, etc. The briefer will automatically provide the following information in the sequence listed, except as noted, when it is applicable to your proposed flight.

(a) Adverse Conditions.

(b) VFR Flight Not Recommended — When VFR flight is proposed and sky conditions or visibilities are present or forecast, surface or aloft, that in the briefer's judgment would make flight under visual flight rules doubtful, the briefer will describe the conditions, affected locations, and use the phrase *"VFR flight is not recommended."* This recommendation is advisory in nature. The final decision as to whether the flight can be conducted safely rests solely with the pilot.

(c) Synopsis — A brief statement describing the type, location and movement of weather systems and/or air masses which might affect the proposed flight.

(d) Current Conditions. This element will be omitted if the proposed time of departure is beyond two hours, unless the information is specifically requested by the pilot.

(e) En Route Forecast.

(f) Destination Forecast. Any significant changes within 1 hour before and after the planned arrival are included.

(g) Winds Aloft.

(h) Notices to Airmen (NOTAMs).

(i) ATC Delays — Any known ATC delays and flow control advisories which might affect the proposed flight.

(j) Pilots may obtain other information from FSS briefers upon request.

(4) Abbreviated Briefing — Request an Abbreviated Briefing when you need information to supplement mass disseminated data, update a previous briefing, or when you need only one or two specific items. Provide the briefer with appropriate background information, the time you received the previous information, and/or the specific items needed. You should indicate the source of the information already received so that the briefer can limit the briefing to the information that you have not received, and/or appreciable changes in meteorological conditions since your previous briefing. To the extent possible, the briefer will provide the information in the sequence shown for a Standard Briefing. If you request only one or two specific items, the briefer will advise you if adverse conditions are present or forecast. Details on these conditions will be provided at your request.

(5) Outlook Briefing — You should request an Outlook Briefing whenever your proposed time of departure is six or more hours from the time of the briefing. The briefer will provide available forecast data applicable to the proposed flight. This type of briefing is provided for planning purposes only. You should obtain a Standard or Abbreviated Briefing prior to departure in order to obtain such items as current conditions, updated forecasts, winds aloft and NOTAMs.

(6) Inflight Briefing — You are encouraged to obtain your preflight briefing by telephone or in person before departure. In those cases where you need to obtain a preflight briefing or an update to a previous briefing by radio, you should contact the nearest FSS to obtain this information. After communications have been established, advise the specialist of the type

briefing you require and provide appropriate background information. You will be provided information as specified in the above paragraphs, depending on the type briefing requested. In addition, the specialist will recommend shifting to the flight watch frequency when conditions along the intended route indicate that it would be advantageous to do so.

(7) Following any briefing, feel free to ask for any information that you or the briefer may have missed. It helps to save your questions until the briefing has been completed. This way, the briefer is able to present the information in a logical sequence, and lessens the chance of important items being overlooked.

d. En Route Flight Advisory Service (EFAS)

(1) EFAS is a service specifically designed to provide en route aircraft with timely and meaningful weather advisories pertinent to the type of flight intended, route of flight and altitude. In conjunction with this service, EFAS is also a central collection and distribution point for pilot reported weather information. It is normally available throughout the conterminous U.S. and Puerto Rico from 6 a.m. to 10 p.m. EFAS provides communications capabilities for aircraft flying at 5,000 feet above ground level on a common frequency of 122.0 MHz. High altitude EFAS, now in the process of being implemented in some areas of the country, provides a separate discrete frequency within each Air Route Traffic Control Center's area for use only by aircraft flying at an altitude of 18,000 feet MSL and above. EFAS is provided by specially trained specialists in selected FSSs controlling one or more remote communications outlets covering a large geographical area.

(2) FSSs which provide En Route Flight Advisory Service are listed regionally in the Airport/Facilities Directories.

(3) EFAS is not intended to be used for filing or closing flight plans, position reporting, to get a complete preflight briefing, or to obtain random weather reports and forecasts. In such instances, the flight watch specialist will provide the name and radio frequency of the FSS to contact for such services. Pilot participation is essential to the success of EFAS through a continuous exchange of information on winds, turbulence, visibility, icing, etc., between pilots in-flight and flight watch specialists on the ground. Pilots are encouraged to report good as well as bad and expected as well as unexpected flight conditions to flight watch facilities.

e. In-Flight Weather Advisories

(1) Currently listed as Severe Weather Forecasts Alerts (AWW), convective SIGMETs (WST), SIGMETs (WS), Center Weather Advisory (CWA), or AIRMETs (WA). These advisories are issued individually; however, the information contained in them is also included in relevant portions of the Area Forecast (FA). When these advisories are issued subsequent to the FA, they automatically amend appropriate portions of the FA until the FA itself has been amended. In-flight advisories serve to notify en route pilots of the possibility of encountering hazardous flying conditions which may not have been forecast at the time of the preflight briefing. Whether or not the condition described is potentially hazardous to a particular flight is for the pilot to evaluate on the basis of experience and the operational limits of the aircraft.

f. Categorical Outlooks

(1) Categorical outlook terms, describing general ceiling and visibility conditions for advanced planning purposes, are defined as follows:

(a) LIFR (Low IFR) — Ceiling less than 500 feet and/or visibility less than 1 mile.

(b) IFR — Ceiling 500 to less than 1,000 feet and/or visibility 1 to less than 3 miles.

(c) MVFR (Marginal VFR) — Ceiling 1,000 to 3,000 feet and/or visibility 3 to 5 miles inclusive.

(d) VFR — Ceiling greater than 3,000 feet and visibility greater than 5 miles; includes sky clear.

g. Weather Radar Services

(1) The National Weather Service operates a network of radar sites for detecting coverage, intensity, and movement of precipitation. The network is supplemented by FAA and DOD radar sites. Local warning radars augment the network by operating on an as needed basis to support warning and forecast programs.

(2) A clear radar display (no echoes) does not mean that there is no significant weather within the coverage of the radar site. Clouds and fog are not detected by the radar; however, when echoes are present, turbulence can be implied by the intensity of the precipitation and icing is implied by the presence of the precipitation at temperatures at or below zero degrees Celsius. Used in conjunction with other weather products, the radar provides invaluable information for weather avoidance and flight planning.

h. ATC In-Flight Weather Avoidance Assistance

(1) ATC Radar Weather Display

(a) Areas of radar weather clutter result from rain or moisture. Radars cannot detect turbulence. The determination of the intensity of the weather displayed is based on its precipitation density. Generally, the turbulence associated with a very heavy rate of rainfall will normally be more severe than any associated with a very light rainfall rate.

(b) ARTCCs are phasing in computer generated digitized radar displays to replace broadband radar display. This new system, known as Narrowband Radar, provides the controller with two distinct levels of weather intensity by assigning radar display symbols for specific precipitation densities measured by the narrowband system.

(2) Weather Avoidance Assistance

(a) To the extent possible, controllers will issue pertinent information on weather or chaff areas and assist pilots in avoiding such areas when requested. Pilots should respond to a weather advisory by either acknowledging the advisory or by acknowledging the advisory and requesting an alternative course of action as follows:

- Request to deviate off course by stating the number of miles and the direction of the requested deviation. In this case, when the requested deviation is approved, the pilot is expected to provide his own navigation, maintain the altitude assigned by ATC and to remain within the specified mileage of his original course.
- Request a new route to avoid the affected area.
- Request a change of altitude.
- Request radar vectors around the affected areas.

(b) For obvious reasons of safety, an IFR pilot must not deviate from the course or altitude or flight level without a proper ATC clearance. When weather conditions encountered are so severe that an immediate deviation is determined to be necessary and time will not permit approval by ATC, the pilot's emergency authority may be exercised.

(c) When the pilot requests clearance for a route deviation or for an ATC radar vector, the controller must evaluate the air traffic picture in the affected area, and coordinate with other controllers (if ATC jurisdictional boundaries may be crossed) before replying to the request.

(d) It should be remembered that the controller's primary function is to provide safe separation between aircraft. Any additional service, such as weather avoidance assistance, can only be provided to the extent that it does not derogate the primary function. It's also worth noting that the separation workload is generally greater than normal when weather disrupts the usual flow of traffic. ATC radar limitations and frequency congestion may also be a factor in limiting the controller's capability to provide additional service.

(e) It is very important, therefore, that the request for deviation or radar vector be forwarded to ATC as far in advance as possible. Delay in submitting it may delay or even preclude ATC approval or require that additional restrictions be placed on the clearance. Insofar as possible the following information should be furnished to ATC when requesting clearance to detour around weather activity:

- Proposed point where detour will commence.
- Proposed route and extent of detour (direction and distance).
- Point where original route will be resumed.
- Flight conditions (IFR or VFR).
- Any further deviation that may become necessary as the flight progresses.
- Advise if the aircraft is equipped with functioning airborne radar.

(f) To a large degree, the assistance that might be rendered by ATC will depend upon the weather information available to controllers. Due to the extremely transitory nature of severe weather situations, the controller's weather information may be of only limited value if based on weather observed on radar only. Frequent updates by pilots giving specific information as to the area affected, altitudes, intensity and nature of the severe weather can be of considerable value. Such reports are relayed by radio or phone to other pilots and controllers and also receive widespread teletypewriter dissemination.

(g) Obtaining IFR clearance or an ATC radar vector to circumnavigate severe weather can often be accommodated more readily in the enroute areas away from terminals because there is usually less congestion and, therefore, offer greater freedom of action. In terminal areas, the problem is more acute because of traffic density, ATC coordination requirements, complex departure and arrival routes, adjacent airports, etc. As a consequence, controllers are less likely to be able to accommodate all requests for weather detours in a terminal area or be in a position to volunteer such routing to the pilot. Nevertheless, pilots should not hesitate to advise controllers of any observed severe weather and should specifically advise controllers if they desire circumnavigation of observed weather.

i. Pilot Weather Reports (PIREP)

(1) FAA air traffic facilities are required to solicit PIREPs when the following conditions are reported or forecast: Ceilings at or below 5,000 feet; Visibility at or below 5 miles; Thunderstorms and related phenomena; Icing of light degree or greater; Turbulence of moderate degree or greater; and Windshear.

(2) Pilots are urged to cooperate and promptly volunteer reports of these conditions and other atmospheric data such as: Cloud bases, tops and layers; Flight visibility; Precipitation; Visibility restrictions such as haze, smoke and dust; Wind at altitude; and Temperature aloft.

(3) PIREPs should be given to the ground facility with which communications are established, i.e., EFAS, FSS, ARTCC or terminal ATC. One of the primary duties of EFAS facilities, radio call "FLIGHT WATCH", is to serve as a collection point for the exchange of PIREPs with en-route aircraft.

(4) If pilots are not able to make PIREPs by radio, reporting upon landing of the in-flight conditions encountered to the nearest FSS or Weather Service Office will be helpful. Some of the uses made of the reports are:

(a) The ATCT uses the reports to expedite the flow of air traffic in the vicinity of the field and for hazardous weather avoidance procedures.

(b) The FSS uses the reports to brief other pilots, to provide in-flight advisories, and weather avoidance information to en route aircraft.

(c) The ARTCC uses the reports to expedite the flow of en route traffic, to determine most favorable altitudes, and to issue hazardous weather information within the center's area.

(d) The NWS uses the reports to verify or amend conditions contained in aviation forecasts and advisories. In some cases, pilot reports of hazardous conditions are the triggering mechanism for the issuance of advisories. They also use the reports for pilot weather briefings.

(e) The NWS, other government organizations, the military, and private industry groups use PIREPs for research activities in the study of meteorological phenomena.

(f) All air traffic facilities and the NWS forward the reports received from pilots into the weather distribution system to assure the information is made available to all pilots and other interested parties.

j. Turbulence

(1) When the wind blows across a valley or canyon, a downdraft will occur on the lee side, while an updraft will result on the windward side (figure 15-1). If flight through the canyon is required, the safest path is to fly near to the side of the pass or canyon which affords an upslope wind, since additional lift is provided.

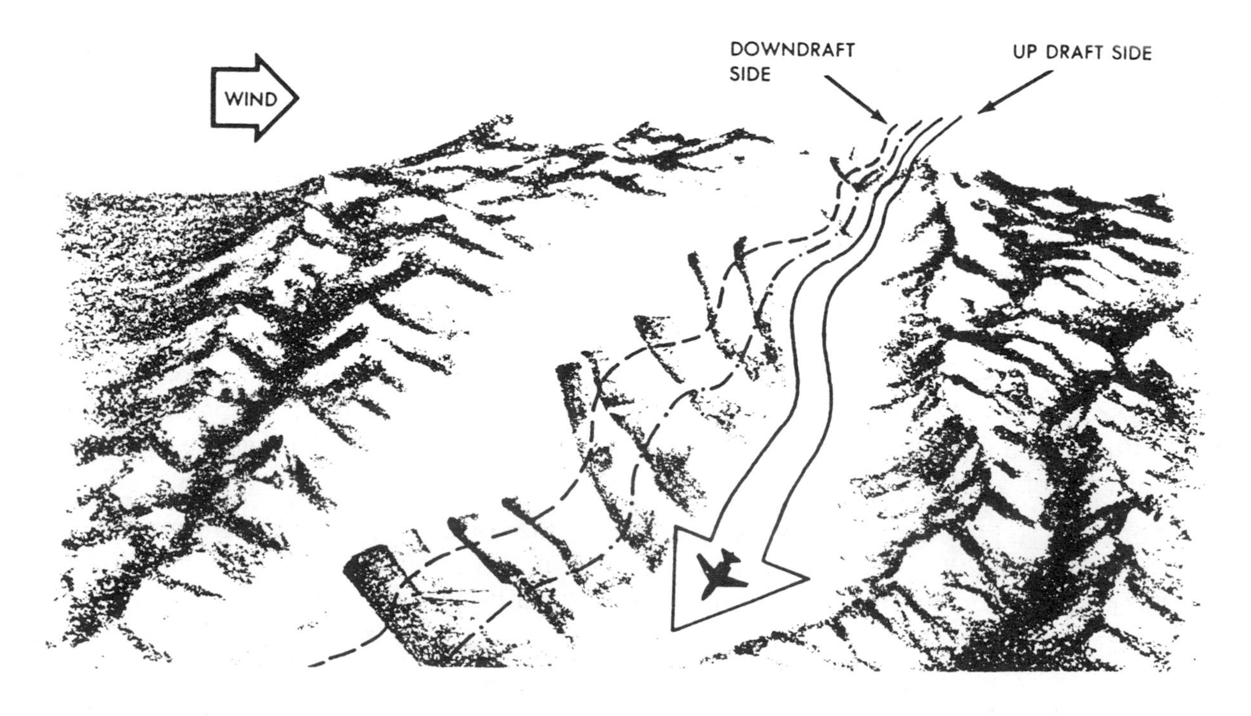

Figure 15-1. In A Valley Or Canyon, Safest Path Is on Upslope Wind Side.

(2) If the wind blows across a narrow canyon or gorge, it will veer down into the canyon (figure 15-2). Turbulence will be found near the middle and downwind side of the canyon or gorge. Aircrews must avoid the downwind side of *narrow* canyons, because a rate-of-descent of such magnitude may be established that the aircraft may continue descending and crash.

WIND
AVOID THIS SIDE OF NARROW GORGE OR CANYON BECAUSE OF TURBULENCE OR DOWN-DRAFTS.

Figure 15-2. Avoid Downwind Side of Narrow Gorge.

(3) The occurrence of CAT can extend to very high levels and can be associated with other windflow patterns which produce shears. A sharp trough aloft, especially one moving at a speed greater than 20 knots, can have clear air turbulence in or near the trough, even though the wind speeds can be rather low as compared with the speeds near the jet stream. However, the winds on opposite sides of the trough can have a difference of 90 degrees or more in direction (figure 15-3A). CAT can occur in the circulation around a closed low aloft, particularly if the flow is merging or splitting (figure 15-3B) and is to the northeast of a cutoff low aloft (figure 15-3C). When anticipating or encountering CAT, fly the turbulence-penetration air speed recommended in the flight manual for your aircraft. Ordinarily, this will reduce the effect of turbulence. However, if the intensity of the turbulence requires further action, climb, descend, and/or change course to exit the turbulent zone, using the information provided by the weather forecaster during preflight or pilot-to-metro facilities. Make very gradual climbs, descents, and turns to minimize additional stress on the aircraft. Finally, always make inflight reports of CAT or other turbulence encounters.

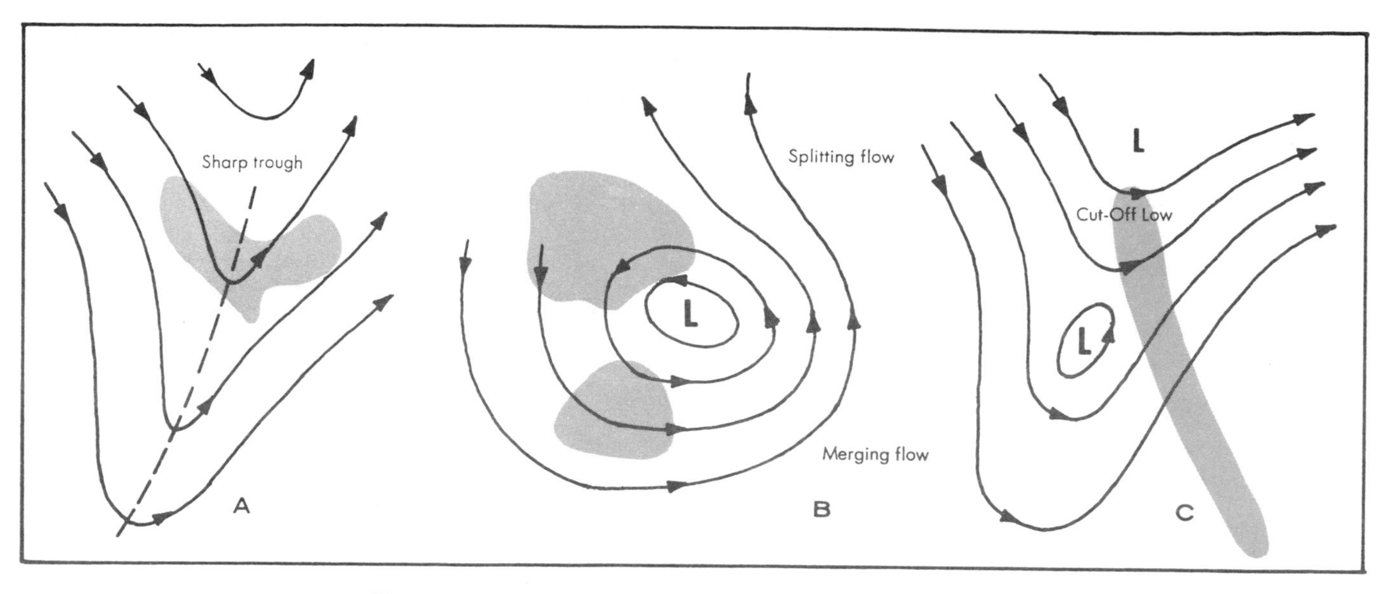

Figure 15-3. Wind Patterns Associated With High-Level CAT.

(4) The thunderstorm downdraft may produce the most dangerous of shear conditions associated with the outflow of a thunderstorm. For example, in figure 15-4A an aircraft passing through the gust front and downdraft along path AB would encounter not only a rapid change in the horizontal wind field but also a downward vertical motion. The downward vertical motion can add or subtract 100 feet per minute or more to the descent or ascent rate of the aircraft and cause it to crash. In figure 15-4B the departing aircraft would experience both a downdraft and tailwind while still near the ground. The resulting loss of lift could prove disastrous to the aircrew.

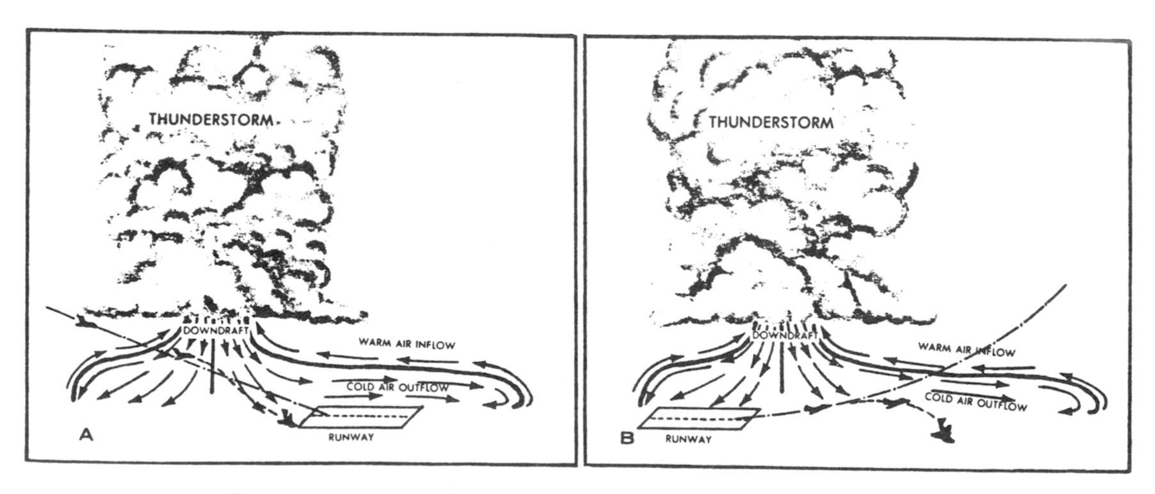

Figure 15-4. Wind Shear Associated With Thunderstorm Downdraft.

(5) Low level wind shear occurs with a cold front after the front passes the aerodrome. Because cold fronts have a greater slope and normally move faster than warm fronts, the duration of low level wind shear at a station is usually less than two hours.

Wind shear associated with a warm front is more dangerous to aerodrome operations. Strong winds aloft, associated with the warm front, may cause a rapid change in wind direction and speed where the warm air overrides the cold, dense air near the surface (figure 15-5). Warm frontal wind shear may persist six hours or more over an airfield ahead of the front because of the front's shallow slope and slow movement. Further, low ceilings and visibilities frequently associated with warm fronts may compound aircrew problems.

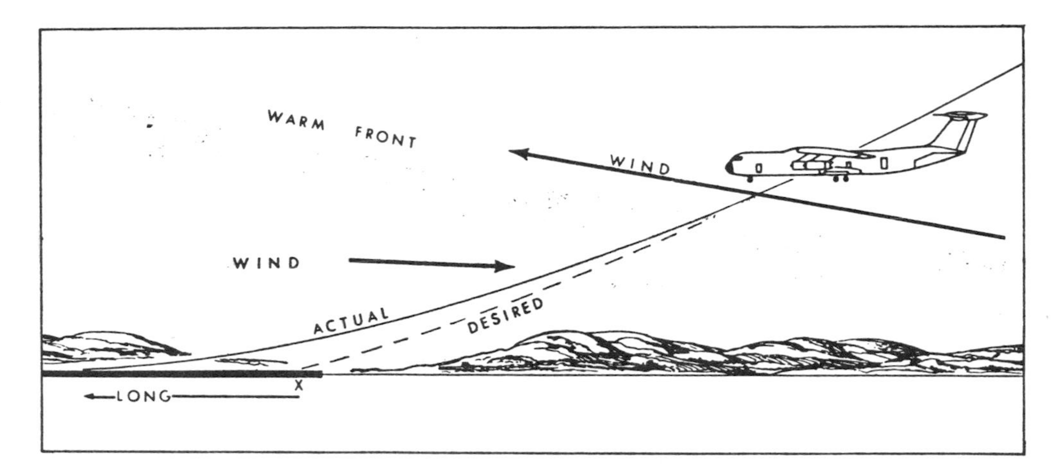

Figure 15-5. Wind Shear In A Warm Front.

(6) The low level jet often forms just above a radiation inversion. It starts to form at sundown, reaches maximum intensity just before sunrise, and is destroyed by daytime heating (usually by 10 A.M. local time). The low level jet is observed in all parts of the world at all times of the year. In the United States it is common in the Great Plains and central states. The jet promotes convergence, and in conjunction with sufficient moisture will often cause night-time thunderstorms in these areas.

The cooling of the earth creates a calm, stable dome of cold air 300-1,000 feet thick, termed an inversion layer. The low level jet occurs just above the top of the inversion layer, and while speeds of 30 knots are common, windspeeds in excess of 65 knots have been reported. Anytime a radiational inversion is present, the possibility of low level wind shear exists (figure 15-6).

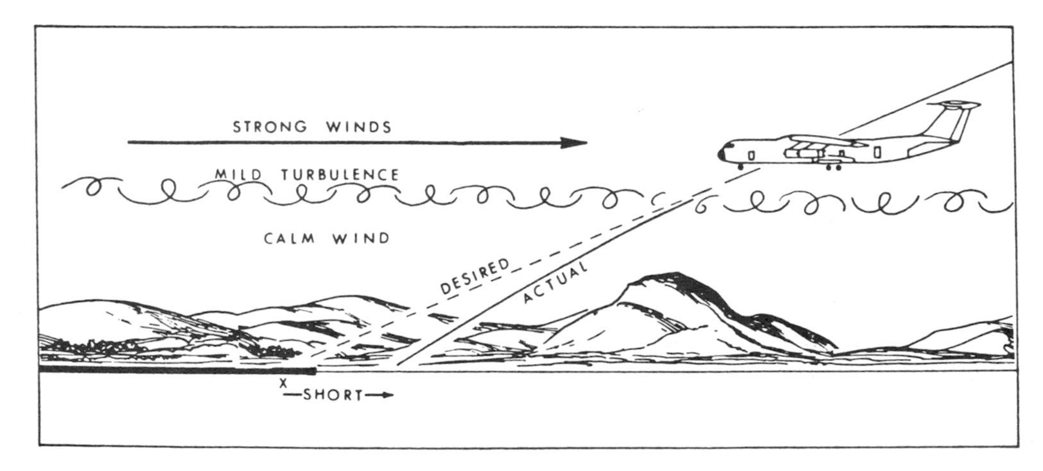

Figure 15-6. Wind Shear During Radiation Inversion.

(7) Certain aerodromes are infamous for the treacherous winds that frequently exist. These winds are caused by funneling, i.e., the terrain is such that the prevailing winds force a large mass of air to be channeled through a narrow space (such as a canyon) where it is accelerated

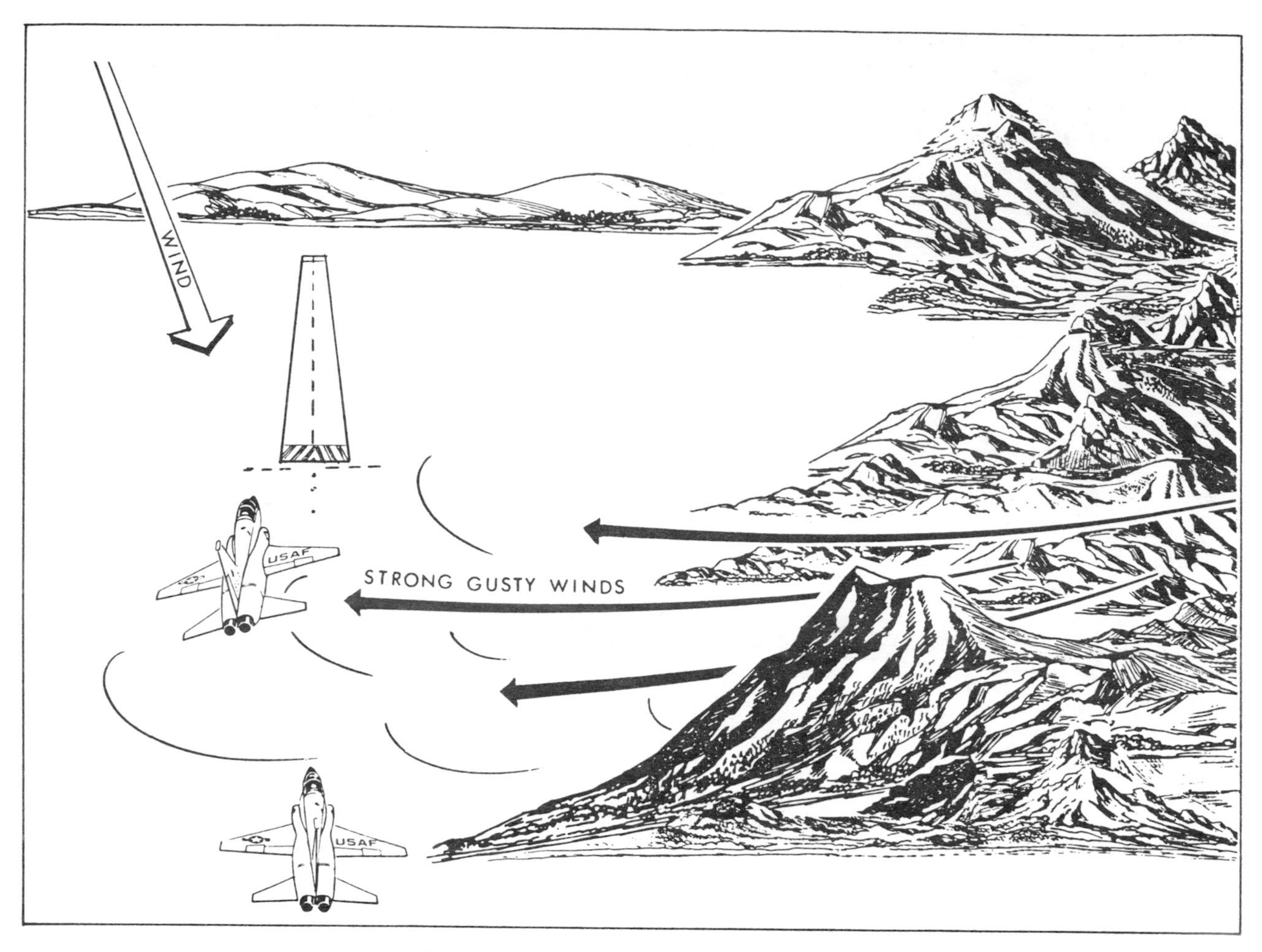

Figure 15-7. A Wind Funneling Condition.

and then spills out into the flightpath of aircraft. These winds sometimes reach velocities of 80 knots. Caution is required when conducting operations near mountains or along straits and channels (figure 15-7).

(8) Two less prominent sources of low level wind shear deserve brief discussion: gusty or strong surface winds and land/sea breezes. Fluctuations of 10 knots or more from the mean sustained wind speed or strong winds blowing past buildings and structures near a runway can produce localized areas of shear. This type of shear can be particularly hazardous to light aircraft. Observing the local terrain and requesting pilot reports of conditions near the runway are the best means for anticipating wind shear from this source.

Land and sea breezes commonly occur near large lakes, bays, or oceans. The flow to or from the water is caused by differential heating and cooling of land and sea surfaces. The sea breeze is a small scale frontal boundary and can reach speeds of 15 to 20 knots. It can move inland 10 to 20 miles, reaching its maximum penetration in mid or late afternoon. The depth of the breeze is approximately 2,000 feet. Land breezes occur at night because the land becomes cooler than the water. The land breeze has less intensity than the sea breeze, and unless aircraft penetrate it on a long, low approach over water, there is little threat to flying safety.

(9) A recent cold frontal passage or an impending warm frontal passage is an indicator of possible windshear.

Another indicator to the aircrews is an abnormal power setting and rate of descent required for the approach. The rate of descent required to maintain a given glide slope is based on groundspeed. Therefore, a tailwind during approach requires a reduced power setting and an increased rate of descent in order to maintain the glidepath. Conversely, a headwind during approach requires an increased power setting and a decreased rate of descent in order to maintain the glidepath.

The best indication to the aircrew that the aircraft is encountering wind shear occurs on the instrument panel. Fluctuations in the indicated airspeed and the vertical velocity indicator always accompany wind shear. Another indication is a large difference between indicated airspeed and groundspeed. Any rapid changes in the relationship between the two represent a wind shear.

Crews in aircraft equipped with an inertial navigation system (INS) can compare the wind at the initial approach altitude with the reported runway surface wind to see if there is a wind shear situation between the aircraft and the runway. Remember: INS winds are in degrees true; tower winds are in degrees magnetic. This will make little difference at airfields where variation is only a few degrees, but it makes a considerable difference when variation is 20 degrees or greater.

(10) The most important elements for the aircrew in coping with a wind shear environment are the crew's awareness of an impending wind shear encounter and the crew's decision to avoid the encounter or to *immediately* respond if an encounter occurs.

During takeoff, if a tailwind shear or gust front from a thunderstorm is anticipated, it is important to penetrate the shear at a relatively high airspeed. A departure route that will provide the most favorable wind condition (generally a headwind) should be used. If necessary, delay the takeoff until the shear situation subsides.

Aircrews must be alert to the possibility of wind shear on all approaches. During an approach, the comparison of actual groundspeed and a *reference groundspeed* is a proven procedure for detecting and coping with low level wind shear. Reference groundspeed is computed by subtracting the surface headwind component or adding the tailwind component to approach true airspeed.

If actual groundspeed exceeds reference groundspeed, a decreasing tailwind condition will occur. If actual groundspeed is less than reference groundspeed, a decreasing headwind condition will occur.

If a decreasing tailwind condition is experienced, maintain computed approach speed. If a decreasing headwind condition is experienced, maintain the reference groundspeed to provide the necessary energy to penetrate the shear safely or go around.

If wind shear is encountered on final approach, do not hesitate to go around if the approach profile and airspeed cannot be reestablished. A go-around is often the aircrew's best course of action. If necessary, delay the approach until the shear situation has subsided or divert to a suitable alternate. Finally, when wind shear is encountered, pass this information, including magnitude of airspeed change and altitude of occurrence, to air traffic control and the weather station so that other aircrews may be informed.

k. Microbursts

(1) Relatively recent meteorological studies have confirmed the existence of microburst phenomena. Microbursts are small-scale intense downdrafts which, on reaching the surface, spread outward in all directions from the downdraft center. This causes the presence of both

vertical and horizontal wind shears that can be extremely hazardous to all types and categories of aircraft, especially at low altitudes. Due to their small size, short life-span, and the fact that they can occur over areas without surface precipitation, microbursts are not easily detectable using conventional weather radar or wind shear alert systems.

(2) Parent clouds producing microburst activity can be any of the low or middle layer convective cloud types. Note, however, that microbursts commonly occur within the heavy rain portion of thunderstorms, and in much weaker, benign-appearing convective cells that have little or no precipitation reaching the ground.

(3) An important consideration for pilots is the fact that the microburst intensifies for about 5 minutes after it strikes the ground.

(4) Characteristics of microbursts include:

(a) Size — The microburst downdraft is typically less than 1 mile in diameter as it descends from the cloud base to about 1,000-3,000 feet above the ground. In the transition zone near the ground, the downdraft changes to a horizontal outflow that can extend to approximately 2 1/2 miles in diameter.

(b) Intensity — The downdrafts can be as strong as 6,000 feet per minute. Horizontal winds near the surface can be as strong as 45 knots resulting in a 90 knot shear (headwind to tailwind change for a traversing aircraft) across the microburst. These strong horizontal winds occur within a few hundred feet of the ground.

(c) Visual Signs — Microbursts can be found almost anywhere that there is convective activity. They may be embedded in heavy rain associated with a thunderstorm or in light rain in benign-appearing virga. When there is little or no precipitation at the surface accompanying the microburst, a ring of blowing dust may be the only visual clue of its existence.

(d) Duration — An individual microburst will seldom last longer than 15 minutes from the time it strikes the ground until dissipation. The horizontal winds continue to increase during the first 5 minutes with the maximum intensity winds lasting approximately 2 to 4 minutes. Sometimes microbursts are concentrated into a line structure and, under these conditions, activity may continue for as long as an hour. Once microburst activity starts, multiple microbursts in the same general area are not uncommon and should be expected.

(5) Microburst wind shear may create a severe hazard for aircraft within 1,000 feet of the ground, particularly during the approach to landing and landing and take-off phases. The aircraft may encounter a headwind (performance increasing), followed by a downdraft and tailwind (both performance decreasing), possibly resulting in terrain impact.

(6) Pilots should heed wind shear PIREPS, as a previous pilot's encounter with a microburst may be the only indication received. However, since the wind shear intensifies rapidly in its early stages a PIREP may not indicate the current severity of a microburst. Flight in the vicinity of suspected or reported microburst activity should always be avoided. Should a pilot encounter one, a wind shear PIREP should be made at once (figures 15-8 and 15-9).

l. Thunderstorms

(1) Turbulence, hail, rain, snow, lightning, sustained updrafts and downdrafts, icing conditions — all are present in thunderstorms. While there is some evidence that maximum turbulence exists at the middle level of a thunderstorm, recent studies show little variation of turbulence intensity with altitude.

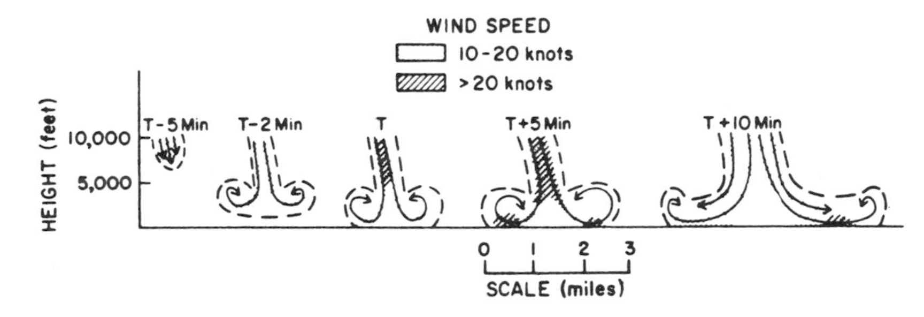

Figure 15.8 Vertical Cross Section Of The Evolution Of A Microburst Wind Field. T is the time of initial divergence at the surface. The shading refers to the vector wind speeds. Figure adapted from Wilson et al., 1984. Microburst Wind Structure and Evaluation of Doppler Radar for Wind Shear Detection, DOT/FAA Report No. DOT/FAA/PM-84/29, National Technical Information Service, Springfield, VA 37 pp.

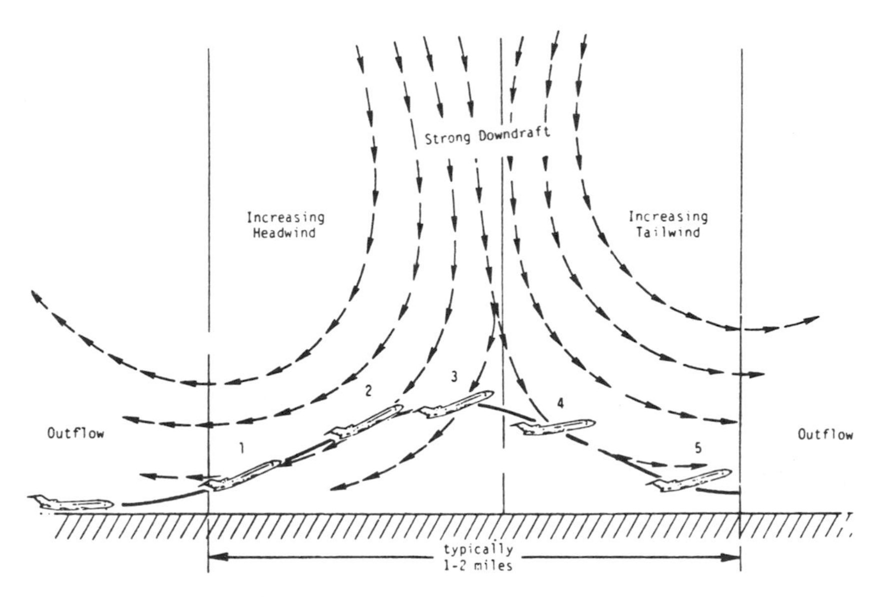

Figure 15.9 A Microburst Encounter During Takeoff. The airplane first encounters a headwind and experiences increasing performance (1), this is followed in short succession by a decreasing headwind component (2), a downdraft (3), and finally a strong tailwind (4), where 2 through 5 all result in decreasing performance of the airplane. Position (5) represents an extreme situation just prior to impact. Figure courtesy of Walter Frost. FWG Associates, Inc., Tullahoma, Tennessee.

(2) There is no useful correlation between the external visual appearance of thunderstorms and the severity or amount of turbulence or hail within them. The visible thunderstorm cloud is only a portion of a turbulent system whose updrafts and downdrafts often extend far beyond the visible storm cloud. Severe turbulence can be expected up to 20 miles from severe thunderstorms. This distance decreases to about 10 miles in less severe storms.

(3) Weather radar, airborne or ground based, will normally reflect the areas of moderate to heavy precipitation (radar does not detect turbulence). The frequency and severity of turbulence generally increases with the radar reflectivity which is closely associated with the areas of highest liquid water content of the storm. NO FLIGHT PATH THROUGH AN AREA OF STRONG OR VERY STRONG RADAR ECHOS SEPARATED BY 20-30 MILES OR LESS MAY BE CONSIDERED FREE OF SEVERE TURBULENCE.

(4) Turbulence beneath a thunderstorm should not be minimized. This is especially true when the relative humidity is low in any layer between the surface and 15,000 feet. Then the lower altitudes may be characterized by strong out-flowing winds and severe turbulence.

(5) The probability of lightning strikes occurring to aircraft is greatest when operating at altitudes where temperatures are between minus 5 degrees centigrade and plus 5 degrees centigrade. Lightning can strike aircraft flying in the clear in the vicinity of a thunderstorm.

(6) The National Weather Service recognizes only 2 classes of intensities of Thunderstorms as applied to aviation surface weather observations:

(a) Thunderstorm T;

(b) Severe T+.

(7) National Weather Service radar systems are able to objectively determine radar weather echo intensity levels by use of Video Integrator Processor (VIP) equipment. These thunderstorm intensity levels are on a scale of one to six. (See RADAR WEATHER ECHO INTENSITY LEVELS in the Glossary.)

Example:

Alert provided by an ATC facility to an aircraft:

(aircraft identification) LEVEL FIVE INTENSE WEATHER ECHO BETWEEN TEN O'CLOCK AND TWO O'CLOCK, ONE ZERO MILES MOVING EAST AT TWO ZERO KNOTS, TOP FLIGHT LEVEL THREE NINE ZERO.

Example:

Alert provided by a Flight Service Station:

(aircraft identification) LEVEL FIVE INTENSE WEATHER ECHO TWO ZERO MILES WEST OF ATLANTA VOR TWO FIVE MILES WIDE, MOVING EAST AT TWO ZERO KNOTS, TOPS FLIGHT LEVEL THREE NINE ZERO.

aa. Do's and Don'ts of Thunderstorm Flying

(1) Above all, remember this: never regard any thunderstorm as "light" even when radar observers report the echoes are of light intensity. Avoiding thunderstorms is the best policy. Following are some Do's and Don'ts of thunderstorm avoidance:

(a) Don't land or take off in the face of an approaching thunderstorm. A sudden wind shift or low level turbulence could cause loss of control.

(b) Don't attempt to fly under a thunderstorm even if you can see through to the other side. Turbulence under the storm could be disastrous.

(c) Don't try to circumnavigate thunderstorms covering 6/10 of an area or more either visually or by airborne radar.

(d) Don't fly without airborne radar into a cloud mass containing scattered embedded thunderstorms. Scattered thunderstorms not embedded usually can be visually circumnavigated.

(e) Do avoid by at least 20 miles any thunderstorm identified as severe or giving an intense radar echo. This is especially true under the anvil of a large cumulonimbus.

(f) Do clear the top of a known or suspected severe thunderstorm by at least 1,000 feet altitude for each 10 knots of wind speed at the cloud top. This should exceed the altitude capability of most aircraft.

(g) Do remember that vivid and frequent lightning indicates a severe thunderstorm.

(h) Do regard as severe any thunderstorm with tops 35,000 feet or higher whether the top is visually sighted or determined by radar.

(2) If you cannot avoid penetrating a thunderstorm, following are some Do's before entering the storm:

(a) Tighten your safety belt, put on your shoulder harness if you have one and secure all loose objects.

(b) Plan your course to take you through the storm in a minimum time and hold it.

(c) To avoid the most critical icing, establish a penetration altitude below the freezing level or above the level of minus 15 degrees centigrade.

(d) Turn on pitot heat and carburetor or jet inlet heat. Icing can be rapid at any altitude and cause almost instantaneous power failure or loss of airspeed indication.

(e) Establish power settings for reduced turbulence penetration airspeed recommended in your aircraft manual. Reduced airspeed lessens the structural stresses on the aircraft.

(f) Turn up cockpit lights to highest intensity to lessen danger of temporary blindness from lightning.

(g) If using automatic pilot, disengage altitude hold mode and speed hold mode. The automatic altitude and speed controls will increase maneuvers of the aircraft, thus increasing structural stresses.

(h) If using airborne radar, tilt your antenna up and down occasionally. Tilting it up may detect a hail shaft that will reach a point on your course by the time you do. Tilting it down may detect a growing thunderstorm cell that may reach your altitude.

(3) Following are some Do's and Don'ts during thunderstorm penetration:

(a) Do keep your eyes on your instruments. Looking outside the cockpit can increase danger of temporary blindness from lightning.

(b) Don't change power settings; maintain settings for reduced airspeed.

(c) Do maintain a constant attitude; let the aircraft "ride the waves." Maneuvers in trying to maintain constant altitude increases stresses on the aircraft.

> NOTE: When the pilot anticipates that he will be required to 'ride the waves' as described above, it is important that he notify ATC as soon as possible to insure that appropriate action is taken to insure separation from other aircraft. When this situation occurs suddenly, and the pilot executes his emergency authority to maintain a constant attitude, regardless of altitude, he should advise ATC as soon as possible.

(d) Don't turn back once you are in the thunderstorm. A straight course through the storm most likely will get you out of the hazards most quickly. In addition, turning maneuvers increase stresses on the aircraft.

bb. Key to Aviation Weather Observations and Forecasts (Tables 15-1 and 15-2).

TABLE 15-1. AVIATION WEATHER OBSERVATIONS

LOCATION IDENTIFIER TYPE & TIME OF REPORT*	SKY AND CEILING	VISIBILITY WEATHER AND OBSTRUCTION TO VISION	SEA-LEVEL PRESSURE	TEMPERATURE AND DEW POINT	WIND	ALTIMETER SETTING	REMARKS AND CODED DATA
MCI SA 0758	15 SCT M25 OVC	IR-F	132	/58/56	/1807	/993/	R01VR20V40

SKY AND CEILING

Sky cover contractions are in ascending order. Figures preceding contractions are heights in hundreds of feet above station. Sky cover contractions are:

CLR	Clear: Less than 0.1 sky cover.
SCT	Scattered: 0.1 to 0.5 sky cover.
BKN	Broken: 0.6 to 0.9 sky cover.
OVC	Overcast: More than 0.9 sky cover.
–	Thin (When prefixed to the above symbols).
–X	Partial obscuration: 0.1 to less than 1.0 sky hidden by precipitation or obstruction to vision (bases at surface).
X	Obscuration: 1.0 sky hidden by precipitation or obstruction to vision (bases at surface).

Letter preceding height of layer identifies ceiling layer and indicates how ceiling height was obtained. Thus:

E	Estimated height	V	Immediately following numerical value, indicates a variable ceiling.
M	Measured		
W	Indefinite		

VISIBILITY

Reported in statute miles and fractions.
(V – Variable)

WEATHER AND OBSTRUCTION TO VISION SYMBOLS:

A	Hail	K	Smoke
BD	Blowing dust	L	Drizzle
BS	Blowing snow	R	Rain
D	Dust	RW	Rain showers
F	Fog	S	Snow
GF	Ground fog	SG	Snow grains
H	Haze	SW	Snow showers
IC	Ice crystals	T	Thunderstorms
IF	Ice fog	T+	Severe thunderstorm
IP	Ice pellets	ZL	Freezing drizzle
IPW	Ice pellet showers	ZR	Freezing rain

Precipitation intensities are indicated thus:
– Light, (no sign) Moderate, + Heavy.

WIND

Direction in tens of degrees from true north, speed in knots. 0000 indicates calm, G indicates gusty. Peak speed of gusts follows G or Q when gusts or squall are reported. The contraction WSHFT followed by GMT time group in remarks indicates windshift and its time of occurrence. (Knots × 1.15 statute mi/hr.)

EXAMPLES: 3627, 360 Degrees, 27 knots: 3627G40, 360 Degrees, 27 knots, peak speed in gusts 40 knots.

ALTIMETER SETTING

The first figure of the actual altimeter setting is always omitted from the report.

RUNWAY VISUAL RANGE (RVR)

RVR is reported from some stations. For planning purposes, the value range during 10 minutes prior to observations and based on runway light settings are reported in hundreds of feet. Runway identification precedes RVR report.

PILOT REPORTS (PIREPs)

When available, PIREPs in fixed-format may be appended to weather observations. PIREPs are designated by UA or UUA for urgent PIREPs.

DECODED REPORT

Kansas City international record observation completed at 0758 GMT, 1500 feet scattered clouds, measured ceiling 2500 feet overcast, visibility 1 mile, light rain, fog, sea-level pressure 1013.2 millibars, temperature 58°F, dewpoint 56°F, wind 180°, 7 knots, altimeter setting 29.93 inches. Runway 01, visual range 2000 feet lowest, 4000 feet highest in the past 10 minutes.

***TYPE OF REPORT**

SA – a scheduled record observation.
SP – an unscheduled special observation indicating a significant change in one or more elements.
RS – a scheduled record observation that also qualifies as a special observation. The designator for all three types of observations is followed by a 24 hour clock-time-group in Coordinated Universal Time (UTC or Z).

TABLE 15-2. AVIATION WEATHER FORECASTS

TERMINAL FORECASTS contain information for specific airports on expected ceiling, cloud heights, cloud amounts, visibility, weather and obstructions to vision and surface wind. They are issued 3 times/day and are valid for 24 hours. The last six hours of each forecast are covered by a categorical statement indicating whether VFR, MVFR, IFR or LIFR conditions are expected. Terminal forecasts will be written in the following form:

CEILING: Identified by the letter "C"
CLOUD HEIGHTS: In hundreds of feet above the station (ground)
SKY COVER AMOUNT: Includes any obscuration
CLOUD LAYERS: Stated in ascending order of height
VISIBILITY: In statute miles but omitted if over 6 miles
WEATHER AND OBSTRUCTION TO VISION: Standard weather and obstruction to vision symbols are used
SURFACE WIND: In tens of degrees and knots; omitted when less than 6; WND — in categorical outlook means winds 25 knots or stronger.

EXAMPLE OF TERMINAL FORECAST

DCA 221010: DCA Forecast 22nd day of month — valid time 10Z-10Z. **10 SCT C18 BKN 5SW–3415G25 OCNL C8 X ½SW**: Scattered clouds at 1000 feet, ceiling 1800 feet broken, visibility 5 miles, light snow showers, surface wind 340 degrees, 15 knots. Gusts to 25 knots, occasional ceiling 8 hundred feet sky obscured, visibility ½ mile in moderate snow showers. **12Z C50 BKN 3312G22**: At 12Z becoming ceiling 5000 feet broken, surface wind 330 degrees, 12 knots. Gusts to 22. **04Z MVFR CIG**: Last 6 hours of FT; after 04Z marginal VFR due to ceiling.

AREA FORECASTS are 12-hour aviation forecasts, plus a 6-hour categorical outlook prepared 3 times/day giving general descriptions of cloud cover, weather and frontal conditions for an area the size of several states. Heights of cloud tops, and icing are referenced ABOVE SEA LEVEL (ASL); ceiling heights, ABOVE GROUND LEVEL (AGL); bases of cloud layers are ASL unless indicated. Each SIGMET or AIRMET affecting an FA area will also serve to amend the Area Forecast.

SIGMET, AIRMET and CWA messages warn airmen in flight of potentially hazardous weather such as squall lines, thunderstorms, fog, icing, and turbulence. SIGMET concerns severe and extreme conditions of importance to all aircraft. AIRMET concerns less severe conditions which may be hazardous to some aircraft or to relatively inexperienced pilots. CWA messages concerns both SIGMET and AIRMET type conditions described in greater detail and relating to a specific ARTCC area. All are broadcast by FAA on NAVAID voice channels.

WINDS AND TEMPERATURES ALOFT (FD) FORECASTS are 6, 12 and 24-hour forecasts of wind direction (nearest 10° true N) and speed (knots) for selected flight levels. Temperatures aloft (°C) are included for all but the 3000 foot level.

EXAMPLES OF WINDS AND TEMPERATURES ALOFT (FD) FORECASTS:
FD WBC 121745
BASED ON 121200Z DATA
VALID 130000Z FOR USE 1800-0300Z. TEMPS NEG ABV 24000

	3000	6000	9000	12000	18000	24000	30000	34000	39000
BOS	3127	3425-07	3420-11	3421-16	3516-27	3512-38	311649	292451	283451
JFK	3026	3327-08	3324-12	3322-16	3120-27	2923-38	284248	285150	285749

At 6000 feet ASL over JFK wind from 330° at 27 knots and temperature minus 8°C.

TWEB (CONTINUOUS TRANSCRIBED WEATHER BROADCAST) — Individual route forecasts covering a 25 nautical mile zone either side of the route. By requesting a specific route number, detailed en route weather for a 12 or 18-hour period (depending on forecast issuance) plus a synopsis can be obtained.

PILOTS . . . report in-flight weather to nearest FSS. The latest surface weather reports are available by phone at the nearest pilot weather briefing office by calling at H + 10.

C. FAA Form For Standard Briefing.

DEPARTMENT OF TRANSPORTATION

FEDERAL AVIATION ADMINISTRATION

GENERAL AVIATION ACCIDENT PREVENTION PROGRAM

Pilot's Weather Briefing Guide

PILOTS: Help us to help you get a better weather briefing.
You will get faster service, and you can assist the weather briefer by telling him:

1. That you are a pilot. If you are a student, private, or commercial pilot—say so. The weather briefer needs to know that you are a pilot, not someone who calls just to find out the general weather picture.
2. The type of airplane you are planning to fly. (light single engine, high performance multi-engine, and jets all present different briefing problems.)
3. Your destination. If you plan to stop somewhere enroute or deviate from the normal course, you should tell the briefer your intentions.
4. Your estimated departure time.
5. Whether or not you can go IFR. (Are you instrument rated, current, and is your airplane properly equipped?)

A pre-flight weather briefing should include:

1. Weather synopsis (position of lows, fronts, ridges, etc.).
2. Current weather conditions.
3. Forecast weather conditions.
4. Alternate routes (if necessary).
5. Hazardous weather.
6. Forecast winds aloft.
7. NOTAM information.

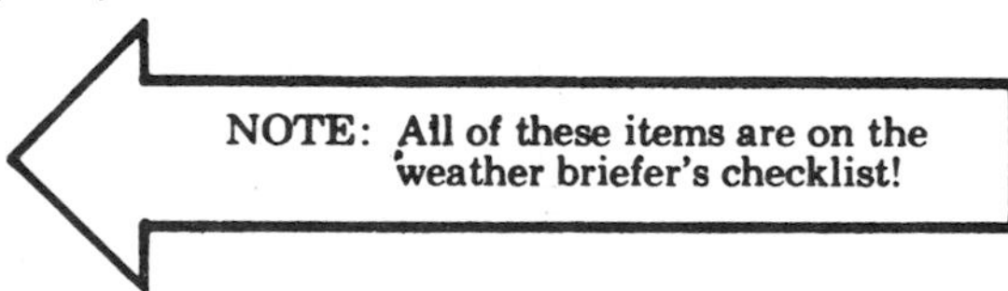

NOTE: All of these items are on the weather briefer's checklist!

USE THE OTHER SIDE OF THIS SHEET TO WRITE DOWN THE WEATHER INFORMATION AS THE BRIEFER GIVES IT TO YOU.

★★★★★

We hope the weather for your flight is good all the way to your destination. But, if it looks a little doubtful, it may be a good idea to give some thought to the following:

The flight visibility through a wet windshield is not nearly as good as the ground visibility reported on the station teletype sequence.

Veteran pilots always have an "out," (or alternate course of action) if the weather is worse than expected. If you don't have a good "out," don't go!

The weather between reporting points may not be as good as the weather shown at reporting stations, especially during marginal weather conditions.

Any thunderstorms that may be in the vicinity of your route will add greatly to the weather hazard.

The possibility of icing in the clouds may prevent you from going IFR safely, even as a last resort.

If the cloud ceilings will keep you at low-level, have you considered the tall T. V. and radio towers? Some of these may not show on your charts!

The risk of flying in marginal weather at night is a great deal higher than during the day.

If there is anything about your weather briefing that is not clearly understood, or if you would like additional information, it will be cheerfully discussed with you. The primary job of the people in the Flight Service Stations is to help you plan a safe flight.

FAA Form 8000–20 (3–72) REMEMBER! A SAFE FLIGHT IS NO ACCIDENT

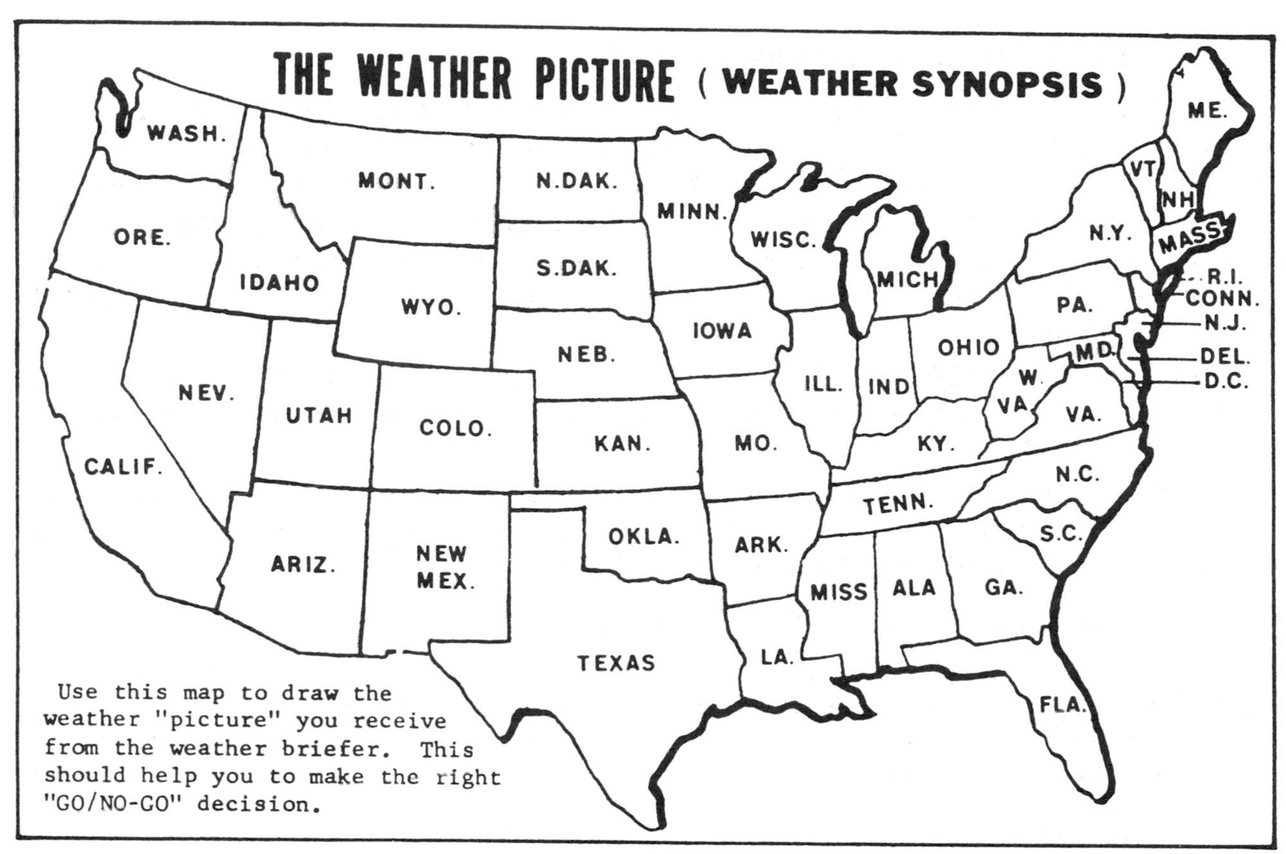

Use the space below to jot down the weather data as you receive it.

ENROUTE WEATHER

LOCATION	REPORTED
	FORECAST

DESTINATION WEATHER

LOCATION	REPORTED
	FORECAST

NOTAMS/SPECIAL INFO

WINDS ALOFT

LOCATION	FORECAST

SURFACE WINDS

HAZARDOUS WEATHER

LOCATION	TYPE

D. Why Do Pilots Fly Into Unfavorable Weather?

1. GO-ITUS — "I gotta get there. . ."; "I don't have time to wait." This is a condition that usually converts HOT PILOTS into COLD BODIES, and is a most difficult "disease" to cure. This attitude can be controlled only through sound reasoning and judgment by the individual.

2. MISINTERPRETATION OF FORECASTS AND REPORTS — "It looks like VFR. . ."; "aw, it's good enough." Applicant performance on FAA written examinations indicate that the problem lies not in reading the data, but in knowing just what it means in terms of expected weather conditions.

3. FAILURE TO KEEP ABREAST OF WEATHER CHANGES — Weather conditions do change, and the best way to keep informed enroute is to listen to in-flight advisories and scheduled broadcasts.

4. IGNORING IN-FLIGHT WEATHER SIGNS — "It's just a little shower. . ."; "just a few puffs of clouds." Rarely does weather suddenly go bad with no warning. Signs of deteriorating weather should be learned and observed by the VFR pilot.

ARE YOU "WEATHER WISE" OR OTHERWISE?

DOES A STATION REPORT OF VFR CEILING MEAN ENROUTE VFR? NO, the ceiling reported is the height above the reporting point only. It must also be related to the surrounding and enroute terrain to determine if adequate VFR separation can be maintained between stations. Additionally, unreported conditions between stations may be lower than those reported at the stations.

IS REPORTED VISIBILITY THE SAME AS VISIBILITY ALOFT? NO, the reported visibility is the visibility at the surface only. Conditions aloft may restrict flight visibility more or less than that reported. Cockpit visibility in precipitation is further reduced by rain, drizzle, or snow spreading over the windshield. Forward visibility in a light snowfall may be zero due to the relative horizontal movement of the snow. Sunlight reflecting off haze or dust aloft reduces the visibility considerably.

WHAT CAN BE LEARNED FROM TEMPERATURE REPORTS? High temperatures reduce takeoff and landing performance. Low temperatures reflect the approximate freezing level and the areas of possible icing in precipitation. Sudden temperature changes reveal the relative position of a front and its associated weather.

WHAT IS THE SIGNIFICANCE OF DEWPOINT? Specifically, a dewpoint value relatively close (2° — 5°) to the air temperature is indicative of the probability of fog, low clouds, or precipitation.

WHY SHOULD THE REPORTED WIND DATA BE NOTED? The velocity and direction of the surface wind should be related to the runway at the point of intended landing to determine the degree of crosswind. Wind data also reflects the degree of turbulence to expect. A sudden shift in direction often reveals the position of frontal weather relative to a station.

OF WHAT VALUE IS THE ALTIMETER SETTING? Correct cruising altitudes and adequate vertical clearance are dependent on the application of altimeter settings. A rapid and continual drop in pressure (altimeter setting) forewarns of approaching inclement weather.

WHAT IS A PIREP AND WHERE IS IT FOUND? A PIREP is a report of weather conditions at flight altitude, particularly between stations, seen by the pilot instead of the ground observer. Reports are often broadcast, and a pilot report summary is disseminated hourly to stations by teletype. Cloud base and top reports are found in the Remarks section of sequence reports.

ARE YOU GETTING THE REAL PICTURE FROM FORECASTS AND REPORTS? Only when the above are considered in analyzing forecasts and reports will you have the full story.

WHAT ARE SOME OF THE WEATHER SIGNPOSTS AND THEIR WARNINGS?

Blowing Dust — turbulence, poor visibility at low levels, particularly into the sun.

Low Layer of Haze — possible fog or stratus cloud in early morning or late evening; poor visibility, particularly into the sun.

Light Puffs of Clouds at Low Levels — probable fog or stratus cloud, particularly in early morning or late evening.

Ragged Cloud Base — turbulence, erratic visibility, possible precipitation.

Bulbous Cloud Base — turbulence, possible precipitation, conducive to TORNADOS.

Roll-Type Clouds — DANGEROUS turbulence, dust and poor visibility, hazardous landing conditions, subsequent precipitation.

Line of Heavy Dark Clouds — SEVERE turbulence, dust and poor visibility, hazardous landing conditions, precipitation, hail.

Opening in Wall of Dark Clouds (SUCKER HOLE) — DANGEROUS turbulence, possible precipitation and poor visibility as the hole is entered.

Gradual Lowering and Thickening of the Ceiling — inadequate terrain clearance, possible widespread precipitation, and fog.

Near Freezing Temperature — poor visibility in precipitation with ice forming on the windshield as well as the aircraft structure.

THE 180° TURN IS AVIATION'S BEST SAFETY DEVICE — IF USED PRIOR TO BEING ENVELOPED BY ADVERSE WEATHER. DON'T BE A "PUSHER" IN THE HOPE THAT THE WEATHER WILL GET BETTER!

E. How Reliable Are Aviation Weather Forecasts?

How good are our forecasts and services? A pilot should understand the limitations as well as the capabilities of present day meteorology. The meteorologist understands many atmospheric behaviors and has watched them long enough to know that his knowledge of the atmosphere certainly is not complete.

Pilots who understand the limitations of observations and forecasts usually are the ones who make the most effective use of the weather forecast service. The safe pilot continually views aviation forecasts with an open mind. He knows that weather always is changing and consequently that the older the forecast, the greater chance that some part of it will be wrong. The weather-wise pilot looks upon a forecast as professional advice rather than an absolute surety. To have complete faith in weather forecasts is almost as bad as having no faith at all.

Recent studies of aviation forecasts indicate the following:

1. Up to 12 hours — and even beyond — a forecast of good weather (ceiling 3,000 feet or more and visibility 3 miles or greater) is much more likely to be correct than a forecast of conditions below 1,000 feet or below 1 mile.

2. If poor weather is forecast to occur within 3 to 4 hours, the probability of occurrence is better than 80 percent.

3. Forecasts of poor flying conditions during the first few hours of the forecast period are most reliable when there is a distinct weather system, such as a front, a trough, precipitation, etc. There is a general tendency to forecast too little bad weather in such circumstances.

4. The weather associated with fast-moving cold fronts and squall lines is the most difficult to forecast accurately.

5. Errors occur when attempts are made to forecast a specific time that bad weather will occur. Errors are made less frequently, of course, when forecasting that bad weather will occur during some period of time.

6. Surface visibility is more difficult to forecast than ceiling height. Visibility in snow is the most difficult of all visibility forecasts. Skill in these forecasts leaves much to be desired.

Available evidence shows that forecasters CAN predict the following at least 75 percent of the time:

The passage of fast-moving cold fronts or squall lines within plus or minus 2 hours, as much as 10 hours in advance.

The passage of warm fronts or slow-moving cold fronts within plus or minus 5 hours, up to 12 hours in advance.

The rapid lowering of ceiling below 1,000 feet in pre-warm front conditions within plus or minus 200 feet and within plus or minus 4 hours.

The onset of a thunderstorm 1 to 2 hours in advance, if radar is available.

The time rain or snow will begin, within plus or minus 5 hours.

Forecasters CANNOT predict the following with an accuracy which satisfies present aviation operational requirements:

The time freezing rain will begin.

The location and occurrence of severe or extreme turbulence.

The location and occurrence of heavy icing.

The location of the initial occurrence of a tornado.

Ceilings of 100 feet or zero before they exist.

The onset of a thunderstorm which has not yet formed.

The position of a hurricane center to nearer than 80 miles for more than 24 hours in advance.

"WEATHER TO GO, OR NO GO – THAT IS THE QUESTION"
IF IN DOUBT, DON'T!

F. Special Study, Nonfatal, Weather-Involved General Aviation Accidents – NTSB Report AAS-76-3

This study is based on 7,856 nonfatal, weather-involved general aviation accidents which have occurred from 1964 through 1974.

During the 11 year study period, "inadequate preflight planning preparation and/or planning" was the most frequently cited cause in which both pilots and weather were involved. Statistics reveal that most of the nonfatal, weather-involved general aviation accidents occurred during the landing regime, either during the landing roll or during leveloff and touchdown, when unfavorable wind conditions existed, and the weather was VFR. Unfavorable winds were cited five times more frequently as a cause or factor than were low ceilings, and 16 times more frequently than was thunderstorm activity. Statistics also reveal that a pilot was 12 times more likely to encounter weather as predicted than to encounter weather worse than predicted.

As a result of its findings, the National Transportation Safety Board urges general aviation pilots to attend the various safety seminars, clinics and courses of instruction sponsored by both government and industry. For familiarization purposes, there is no substitute for visiting National Weather Service and Federal Aviation Administration facilities to determine what data are available and the means by which they can be obtained. The Board urges all pilots to postpone any flight until a timely and thorough preflight weather briefing can be obtained and reiterates that if there is any doubt – DON'T GO.

From 1964 through 1974, weather was the most frequently cited causal factor for nonfatal general aviation accidents. From 1964 through 1974, 7,856 nonfatal, weather-involved accidents resulted in injuries to 3,637 persons. Serious injuries resulted in 16.2 percent of these accidents. Although slightly more than half of the accidents resulted in no injuries, there was substantial damage to the aircraft. Although complete data on economic losses are not available, data are available for 1970 through 1972. Based on these data, hull damage alone cost $8,000,000 for that three year period.

Statistics show that from 1964 through 1974, there were 54,039 general aviation accidents. 47,093 were nonfatal accidents, 16.7% of which were weather-involved and 14.5% of which were nonfatal, weather-involved. Nonfatal accidents comprised 75% of the total weather-involved accidents.

From 1965 to 1968, the percentage of nonfatal accidents which were weather-involved increased dramatically and that percentage has remained at a relatively high level. On the other hand, the accident rate per 100,000 hours flown (all nonfatal accidents) has been downward over most of the period from 1964 through 1974.

About 50% of the nonfatal, weather-involved accidents have occurred during pleasure flying and more than 12% during non-commercial business flying; the remainder occurred during other kinds of flying.

Almost 89% of the accidents examined in this study occurred during daylight hours and in visual flight rules (VFR) conditions; only about 7% occurred at night. By contrast, for fatal, weather-involved accidents from 1964 through 1972, 60% occurred during daylight hours, 36% occurred at night, and only 40% occurred in VFR conditions.

The accidents examined in this study occurred during 56 phases of flight. The phase most frequently coded was the landing roll (fixed wing) – 22.8%. The next six phases in descending order of frequency were: leveloff/touchdown – 17.9%; initial climb – 11.6%; normal cruise – 7.6%; final approach (VFR) – 6.6%; takeoff run – 5.8%; and taxi from landing – 4.5%.

1. *Pilot Data*

General Aviation pilots with low total flight times were frequently involved in weather accidents. In about 84% of the accidents for which data were available, the pilots involved had less than 100 hours total flight time.

It is significant to note that in more than 95% of the cases examined, the pilot involved had less than 100 hours in the type of aircraft flown. Nearly 67% of these pilots had 10 hours or less in type.

Information concerning pilot time during the 90 days before the accident was available in a majority of the cases examined. In 75% of the cases examined, the pilots involved had less than 50 hours of flight time during the 90 days before the accident. More than 21% of those pilots had less than 10 hours of flight time during that 90 day period.

2. Pilot Certification

Since 1964, private pilots have been involved in more weather accidents than any other type of pilot and, as would be anticipated, most pilots involved in nonfatal, weather-involved accidents held private pilot certificates. Of the nonfatal, weather-involved accidents, 46.8% of the pilots held private certificates while 41.5% of the total pilot population had private certificates. Most pilots were in the single engine, land category. The Board's statistics also show that there were more than 40 pilots who either had no certificates or were flying with certificates which had expired.

3. Geographical Distribution of Accidents and Pilots Involved Compared with Total Active Pilots

Accident exposure to various geographical areas was determined by separating the nonfatal, weather-involved accidents into FAA regions and comparing the number of accidents in each region with the number of pilots in that particular region. The New England Region had the best record with 252 accidents, and Alaska was next with 355 accidents. However, when those figures are compared with the numbers of active pilots in the respective regions, it can be seen that the record in Alaska is the worst for any FAA region and that the Alaskan record is about six times worse than that of the Southern region, which has the best record. The Great Lakes and Western regions rank No. 1 and No. 2 both in numbers of accidents and in numbers of active pilots, but their records are equivalent and rank high when compared to the other regions.

4. Pilot Age

The ages of pilots in nonfatal, weather-involved accidents were categorized into groups. The peak group was 31 to 35 years; however, both the 26 to 30 year group and the 36 to 40 group were near the peaks. By contrast, pilots involved most often in fatal, weather-involved accidents were between 41 and 46 years old. Updated statistics on pilot age were solicited from the FAA in order to determine the age group to which most pilots belong. These data indicate that at least since 1968 there have been more pilots in the younger age group than in the 36 to 40 year group.

5. Pilot-Involvement As A Cause Or Factor

During the 11 years studied, "Inadequate Preflight Preparation and/or Planning" was the most frequently cited cause in which both pilot and weather were involved. The next most frequently cited causes in order of frequency were: "Failed to obtain/maintain flying speed," "improper leveloff," "failed to maintain directional control," and "selected unsuitable terrain."

6. Flight Plans

More than 80% of the pilots in the 7,856 nonfatal, weather-involved accidents did not file flight plans of any type, less than 14% filed VFR flight plans, and less than 5% filed instrument flight rules (IFR) flight plans. In more than 88% of the cases, the weather at the accident site was VFR. By comparison, from 1964 to 1974, for fatal, weather-involved accidents the weather at the accident site was VFR only in 40% of the accidents and IFR in 54% of the accidents.

7. Weather Phenomena As A Cause/Factor

Unfavorable wind was the most frequently cited weather factor in nonfatal, weather-involved accidents. Low ceiling and updrafts or downdrafts were the next most frequently cited factors, followed by conditions conducive to carburetor icing, fog, rain, sudden windshifts, and thunderstorm activity.

Information concerning weather forecasts was not available in many of the accidents examined because most of them occurred during VFR conditions and in the immediate vicinity of the airport. However, based on available data, a pilot was 12 times more likely to encounter weather as predicted than to encounter weather worse than predicted. As the Board indicated in its previous special study, experienced pilots generally are aware that forecasts cannot be considered completely accurate, but they also know that they cannot be ignored. Forecasts should be treated as the best professional advice available.

8. Weather Briefings

Information concerning weather briefings was also not available in many accident cases examined, probably because most accidents occurred during VFR conditions and in the vicinity of airports. Nevertheless, for those cases in which the pilot was briefed, most briefings were received from Flight Service Station personnel by telephone. In those cases for which information was available, more than 45% of the pilots were not briefed. The Board's statistics also show that of the 2,704 pilots who were not briefed, 2,615 also did not file flight plans. Of the 2,698 pilots who were briefed on weather, 1,093 also filed plans of some type.

9. Time of Year

During April and May, more nonfatal, weather-involved accidents occurred than during any other time of the year. Beginning in November, the accident trend began to rise and peaked in May. From May, the trend was downward until the low point was reached in November. The largest rise in accidents was from February to March and the largest drop in accidents was from August to September.

10. Statistical Summary

Based only on the statistics presented, a pilot most likely to have been involved in a nonfatal, weather-involved general aviation accident:

1. Received a preflight briefing from a Flight Service Station which utilized Weather Service Forecasts which were reasonably accurate.
2. Was proposing a pleasure flight.
3. Had less than 100 total flight-hours.
4. Had less than 100 hours in type of aircraft (and most probably had less than 10 hours in type.)
5. Had less than 50 hours in the 90 days before the accident (and probably had less than 10 hours.)
6. Had a private pilot certificate with a single-engine, land rating.
7. Had not filed a flight plan.
8. Was between the ages of 31 and 35.
9. Crashed in VFR conditions with unfavorable winds involved.
10. Was accompanied by at least one passenger.

FLIGHT SAFETY
Review Questions

1. Where are Airport Traffic Areas in effect?
   a. At all airports.
   b. Only at airports that have an operating control tower.
   c. Only at airports within a control zone.
   d. At all airports that have a Flight Service Station on the field.

2. Airspace within a horizontal radius of 5 statute miles from the geographical center of any airport at which a control tower is operated and which extends from the surface up to, but not including, 3,000 feet above the surface, is defined as
   a. An Airport Advisory Area.
   b. A Control Area.
   c. An Airport Traffic Area.
   d. A Control Zone.

3. An Airport Traffic Area is that airspace within a horizontal radius of 5 statute miles from the geographical center of an airport, at which a control tower is in operation, and extends
   a. From the surface upward to 5,000 feet.
   b. Upward from 700 feet above the surface.
   c. Upward from 1,200 feet above the surface.
   d. From the surface upward to, but not including, 3,000 feet.

4. To determine if UNICOM is available at an airport without a control tower, you should refer to
   a. The Automatic Terminal Information Service (ATIS).
   b. The appropriate Airport/Facility Directory.
   c. Graphic Notices and Supplemental Data.
   d. The Notices to Airmen (NOTAMS).

5. AIRMETs are issued as a warning of weather conditions hazardous
   a. to all airplanes.
   b. particularly to heavy airplanes.
   c. to VFR operations only.
   d. particularly to light airplanes.

6. In regard to UNICOM, select the true statement from the following:
   a. UNICOM may not be used for any communication, other than providing known traffic, runway in use, and wind conditions.
   b. UNICOM radio frequencies are assigned to Aeronautical Advisory Stations at certain airports not served by a control tower.
   c. To obtain the correct UNICOM frequency for a particular airport, a pilot should contact the nearest FSS.
   d. UNICOM radio stations are assigned by the FAA to control traffic at nontower airports.

7. SIGMETs are issued as a warning of weather conditons hazardous
   a. to all airplanes.
   b. particularly to light airplanes.
   c. to VFR operations only.
   d. particularly to heavy airplanes.

8. Automatic Terminal Information Service (ATIS) is the continuous broadcast of recorded information
   a. Alerting pilots of radar identified aircraft when their aircraft is in unsafe proximity to terrain or obstruction.
   b. Concerning nonessential information to relieve frequency congestion.
   c. Concerning noncontrol information in selected high activity terminal areas.
   d. Concerning sky conditions limited to ceilings below 1,000 feet and visibilities less than 3 miles.

9. Pilots of aircraft arriving or departing certain high activity terminal areas can receive continuous broadcasts concerning essential but routine information by using
   a. Aeronautical Advisory Stations (UNICOM).
   b. Automatic Terminal Information Service (ATIS).
   c. Aeronautical Multicom Service.
   d. Radar Traffic Information Service.

10. The continuous broadcast of recorded noncontrol information in selected high activity terminal areas is referred to as
    a. Terminal Radar Service Area (TRSA).
    b. Terminal Control Area (TCA).
    c. Automatic Terminal Information Service (ATIS).
    d. Transcribed Weather Broadcasts (TWEBs).

11. When telephoning a weather briefing facility for preflight weather information, pilots should
    a. Identify themselves as pilots.
    b. Tell the number of hours they have flown within the preceding 90 days.
    c. State the number of occupants on board and the color of the aircraft.
    d. State that they possess a current medical certificate.

12. When telephoning a weather briefing facility for preflight weather information, you should state
    a. That you possess a current medical certificate.
    b. Your intended route, destination, and type of aircraft.
    c. The color of the aircraft and number of occupants on board.
    d. Your total flight time.

13. When telephoning a weather briefing facility for preflight weather information, you should state
    a. The number of hours you have flown within the preceding 90 days.
    b. That you possess a current medical certificate.
    c. Whether you intend to fly VFR only.
    d. The color of the aircraft and number of occupants on board.

14. How are forecast winds aloft stated with respect to direction and speed? Direction relative to
    a. magnetic north and speed in knots.
    b. magnetic north and speed in miles per hour.
    c. true north and speed in knots.
    d. true north and speed in miles per hour.

15. When you telephone a weather briefing facility for preflight weather information, you should
    a. Identify yourself as a pilot (student, private, or commercial).
    b. State your intended route and destination.
    c. Identify the radio communications equipment aboard the aircraft.
    d. State the number of persons aboard and the color of the aircraft.

    Which of the above statements are true?
    1. a, b, and c.
    2. a and d.
    3. a, b, c, and d.
    4. a and b.

16. When telephoning a weather briefing facility for preflight weather information, you should give at least the
    a. Total pilot flight time.
    b. Proposed departure time and estimated time enroute.
    c. Aircraft color and number of occupants aboard.
    d. Expiration date of your medical certificate.

Use the following reports to answer questions 17 through 24.

```
INK SA 1854 CLR 15 106/77/63/1112G18/000
BOI SA 1854 150 SCT 30 181/62/42/1304/015
LAX SA 1852 7 SCT 250 SCT 6HK 129/60/59/2504/991 LAX 6/38
MDW RS 1856 -X M7 OVC 11/2R+F 990/63/61/3205/980/RF2 RB12
JFK RS 1853 W5 X 1/2F 180/68/64/1804/006/R04RVR22V30 TWR VSBY1/4
```

17. What is the surface wind condition at INK?
    a. 120° at 11 knots gusting to 18 knots.
    b. 11° at 12 knots gusting to 18 knots.
    c. 110° at 12 knots gusting to 18 knots.
    d. 180° at 11 or 12 knots; gusty.

18. Determine the sky condition and ceiling, if any, for MDW.
    a. Measured ceiling 700 overcast.
    b. Sky obscured; ceiling not measurable.
    c. Overcast 1,100 feet.
    d. Measured ceiling 700 and overcast at 1,100 feet.

19. What is the visibility, weather, and obstruction to vision at MDW?
    a. Visibility 11 miles, occasionally 2 miles, with rain + fog.
    b. Visibility 1½ miles, heavy rain, and fog.
    c. 11 miles visibility, except when rain + fog reduce it to 2 miles.
    d. Visibility 1½ miles, rain, and heavy fog.

20. What is the sky condition and ceiling at LAX?
    a. Scattered; 700 to 2,500 feet; no ceiling.
    b. Ceiling 700 feet; 2,500 feet scattered.
    c. Ceiling 600 feet; variable scattered 700 to 25,000 feet.
    d. 700 feet scattered; 25,000 feet scattered; no ceiling.

21. Which of the reporting stations is reporting the lowest sea level pressure?
    a. MDW
    b. INK
    c. LAX
    d. JFK

22. What is the reported visibility at JFK for the purpose of determining if a landing may be made on RWY 4R after a successful instrument approach?
    a. 5 miles
    b. ¼ mile
    c. ½ mile
    d. 2,200 - 3,000 feet RVR

23. What is the reported ceiling at JFK?
    a. 5,000 feet
    b. Indefinite — 500 feet
    c. 500 feet overcast
    d. 0 — sky obscured

24. What is the visibility, weather, and obstructions to vision reported at LAX?
    a. Visibility 6 to 38 miles; snow crystals in thunderstorms.
    b. Unlimited visibility; no obstructions to vision.
    c. Visibility 6 miles, haze, and smoke.
    d. Visibility 6 to 38 miles; scattered patches of haze and smoke.

25. The absence of a visibility entry in a Terminal Forecast specifically implies that the surface visibility
    a. exceeds basic VFR minimums.
    b. exceeds 6 miles.
    c. is at least 15 miles in all directions from the center of the runway complex.
    d. is at least 1 mile above the minimum visibility requirement for an approach to the primary instrument runway.

26. Omission of a wind entry in a Terminal Forecast specifically implies that the wind is expected to be less than
    a. 5 knots.
    b. 10 knots.
    c. 6 knots.
    d. 12 knots.

27. From which primary source should you obtain information regarding the weather expected to exist at your destination at your estimated time of arrival?
    a. Low Level Prog Chart.
    b. Weather Depiction Chart.
    c. Terminal Forecast (FT).
    d. Radar Summary and Weather Depiction Chart.

Use the following terminal forecasts to answer questions 28 through 34.

MEM 251010 C5 X 1/2S-BS 3325G35 OCNL C0 X 0S+BS.
16Z C30 BKN 3BS BRF SW-. 22Z 30 SCT 3315. 00Z CLR. 04Z VFR WIND..
BUF FT RTD 251615 1620Z 100 SCT 250 SCT 1810. 18Z 50 SCT 100 SCT
1913 CHC C30 BKN 3TRW AFT 20Z. 03Z 100 SCT C250 BKN. 09Z VFR..
FTW FT AMD 1 251410 1425Z C8 OVC 4F OVC 0CNL BKN.
15Z 20 SCT 250-BKN. 19Z 40 SCT 120 SCT CHC C30 BKN 3TRW. 04Z MVFR CIG F..

28. What is the latest time the FTW FT AMD 1 is valid?
    a. 1410Z, 26th day of the month.
    b. 1425Z, 26th day of the month.
    c. 1000Z, 26th day of the month.
    d. 0400Z, 26th day of the month.

29. What weather conditions are forecast to accompany the thundershowers at BUF?
    a. Ceiling 3,000 broken; 3 miles visibility.
    b. 5,000 feet scattered; 10,000 feet scattered; wind 190° at 13 knots.
    c. 10,000 scattered; ceiling 25,000 broken.
    d. 500 to 1,000 scattered with a chance of 300 feet ceiling broken.

30. What conditions are forecast for FTW in the 6-hour categorical outlook portion of the amended terminal forecast?
    a. 4,000 scattered; 12,000 scattered.
    b. Marginal VFR conditions with low ceilings and fog.
    c. Chance of ceiling 3,000 broken with 3 miles visibility in thundershowers.
    d. 2,000 scattered, 25,000 thin broken and, after 1900Z, 4,000 scattered 12,000 scattered with a chance of ceiling 3,000 broken, 3 miles visibility in thundershowers.

31. What are the lowest ceilings forecast for MEM during the period from 1000Z to 1600Z on the 25th?
    a. 100 feet
    b. 1,000 feet
    c. 0
    d. 500 feet

32. What condition is expected to cause the low visibility at MEM?
    a. Lowering of ceiling to 0.
    b. Gusty winds and blowing sand.
    c. Smoke plus blowing sand.
    d. Blowing snow.

33. When are thundershowers expected in the vicinity of BUF?
    a. After 2000Z.
    b. After 1913Z.
    c. Between 1800Z and 2000Z.
    d. Between 1800Z and 0300Z.

34. What conditions are forecast for MEM in the 6-hour categorical outlook portion of the FT?
    a. Clouds 3,000 scattered; wind 330° at 15 knots.
    b. Clear.
    c. Ceiling 3,000; 3 miles visibility; wind 330° at 15 knots.
    d. VFR conditions; wind 25 knots or stronger.

35. Pilot participation in the airport advisory service program is
    a. Not mandatory, but strongly recommended.
    b. Mandatory for all aircraft landing at the primary airport.
    c. Not mandatory, except for aircraft on a VFR flight plan.
    d. Mandatory for all aircraft entering this area.

36. What type of facility is located within an Airport Advisory Area?
    a. An operating control tower.
    b. A Flight Service Station.
    c. An Automatic Terminal Information Service.
    d. An Approach Control.

37. How many miles from an airport does an Airport Advisory Area extend?
    a. 5 statute miles.
    b. 10 statute miles.
    c. 10 nautical miles.
    d. 15 nautical miles.

38. If the visibility is included in an ATIS broadcast it indicates a visibility less than
    a. 1 mile.
    b. 2 miles.
    c. 3 miles.
    d. 5 miles.

39. If the sky condition or ceiling is omitted in an ATIS broadcast it indicates that the ceiling is
    a. 2,000 feet or more.
    b. 3,000 feet or more.
    c. 4,000 feet or more.
    d. 5,000 feet or more.

40. Which statement is true regarding the use of airborne weather-avoidance radar for the recognition of certain weather conditions?
    a. The clear area between intense echoes indicates that visual sighting of storms can be maintained when flying between the echoes.
    b. The radar scope provides no assurance of avoiding instrument weather conditions.
    c. The avoidance of hail is assured when flying between and just clear of the most intense echoes.
    d. Areas of light rain, snow, and minute cloud droplets return significant echoes on the radar scope.

41. Assume that a pilot operating VFR is assigned a vector and an altitude by ATC. The pilot should
    a. Not enter clouds, but should deviate so as to maintain VFR conditions; advising ATC is not necessary.
    b. Enter clouds if the sky condition is observed as scattered.
    c. Enter clouds if instrument rated.
    d. Not enter clouds, and should advise ATC that VFR conditions cannot be maintained.

Use the following Area Forecast to answer questions 42 through 45.

```
DFWH FA Ø41Ø4Ø
HAZARDS VALID UNTIL Ø423ØØ
OK TX AR LA TN MS AL AND CSTL WTRS
FLT PRCTNS...TURBC...TN AL AND CSTL WTRS
          ...ICG...TN
          ...IFR...TX
TSTMS IMPLY PSBL SVR OR GTR TURBC SVR ICG AND LLWS
NON MSL HGTS NOTED BY AGL OR CIG
THIS FA ISSUANCE INCORPORATES THE FOLLOWING AIRMETS STILL IN
EFFECT...NONE.

DFWS FA Ø41Ø4Ø
SYNOPSIS VALID UNTIL Ø5Ø5ØØ
AT 11Z RDG OF HI PRES ERN TX NWWD TO CNTRL CO WITH HI CNTR
OVR ERN TX. BY Ø5Z HI CNTR MOVS TO CNTRL LA.

DFWI FA Ø41Ø4Ø
ICING AND FRZLVL VALID UNTIL Ø423ØØ
TN
FROM SLK TO HAT TO MEM TO ORD TO SLK
OCNL MDT RIME ICGIC ABV FRZLVL TO 1ØØ. CONDS ENDING BY 17Z.
FRZLVL 8Ø CHA SGF LINE SLPG TO 12Ø S OF A IAH MAF LINE.

DFWT FA Ø41Ø4Ø
TURBC VALID UNTIL Ø423ØØ
TN AL AND CSTL WTRS
FROM SLK TO FLO TO 9ØS MOB TO MEI TO BUF TO SLK
OCNL MDT TURBC 25Ø-38Ø DUE TO JTSTR. CONDS MOVG SLOLY EWD
AND CONTG BYD 23Z.

DFWC FA Ø41Ø4Ø
SGFNT CLOUD AND WX VALID UNTIL Ø423ØØ...OTLK Ø423ØØ-Ø5Ø5ØØ
IFR...TX
FROM SAT TO FSX TO BRO TO MOV TO SAT
VSBY BLO 3F TIL 15Z.
OK AR TX LA MS AL AND CSTL WTRS
8Ø SCT TO CLR EXCP VSBY BLO 3F TIL 15Z OVR PTNS S CNTRL TX.
OTLK...VFR.
TN
CIGS 3Ø-5Ø BKN 1ØØ VSBYS OCNLY 3-5F BCMG AGL 4Ø-5Ø SCT TO
CLR BY 19Z. OTLK...VFR.
```

42. What hazards are forecast in the Area Forecast for Tennessee (TN), Alabama (AL), and the coastal waters?
    a. Ceilings 3,000 to 5,000 ft. and visibilities 3 to 5 mi. in fog.
    b. Thunderstorms with severe or greater turbulence, severe icing, and low level wind shear.
    c. Moderate rime icing above the freezing level to 10,000 ft.
    d. Moderate turbulence from 25,000 to 38,000 ft. due to the jetstream.

43. What type obstructions to vision, if any, are forecast for the entire area from 2300Z until 0500Z the next day?
    a. None of any significance, VFR is forecast.
    b. Visibility 3 to 5 mi. in fog.
    c. Visibility below 3 mi. in fog over south-central Texas.
    d. Visibilities sometimes below 3 mi. in fog.

44. What sky condition and type obstructions to vision, if any, are forecast for all the area except TN from 1040Z until 2300Z?
    a. None of any significance, VFR is forecast.
    b. Ceilings 3,000 to 5,000 ft. broken, visibility 3 to 5 mi. in fog.
    c. 8,000 ft. scattered to clear except visibility below 3 mi. in fog until 15Z over south-central Texas.
    d. Generally ceilings 3,000 to 8,000 ft. to clear with visibilities sometimes below 3 mi. in fog.

45. The forecast ceiling and visibility for the entire area from 2300Z through 0500Z is ceiling
    a. Less than 500 ft. and/or visibility less than 1 SM.
    b. 500 to less than 1,000 ft. and/or visibility 1 to less than 3 SM.
    c. 1,000 to 3,000 ft. and/or visibility 3 to 5 SM.
    d. Greater than 3,000 ft. and visibility greater than 5 SM.

Use the following Winds and Temperatures Aloft Forecast to answer questions 46 through 49.

FD WBC 151745
BASED ON 151200Z DATA
VALID 1600Z FOR USE 1800-0300Z. TEMPS NEG ABV 24000

FT	3000	6000	9000	12000	18000	24000	30000	34000	39000
ALS			2420	2635−08	2535−18	2444−30	245945	246755	246862
AMA		2714	2725+00	2625−04	2531−15	2542−27	265842	256352	256762
DEN			2321−04	2532−08	2434−19	2441−31	235347	236056	236262
HLC		1707−01	2113−03	2219−07	2330−17	2435−30	244145	244854	245561
MKC	0507	2006+03	2215−01	2322−06	2338−17	2348−29	236143	237252	238160
STL	2113	2325+07	2332+02	2339−04	2356−16	2373−27	239440	730649	731960

46. What wind is forecast for St. Louis (STL) at 18,000 ft.?
    a. 030° at 56 kt.
    b. 235° at 06 kts.
    c. 230° at 56 kts.
    d. 235° at 06 gusting to 16 kts.

47. Determine the wind and temperature aloft forecast for Denver (DEN) at 30,000 ft.
    a. 023° at 53 kts., temperature 47° C.
    b. 230° at 53 kts., temperature −47° C.
    c. 235° at 34 kts., temperature −7° C.
    d. 235° at 34 kts., temperature −47° F.

48. Interpret the wind and temperature aloft forecast for 3,000 ft. at Kansas City (MKC).
    a. 050° at 7 kts., temperature missing.
    b. 360° at 5 kts., temperature −7° C.
    c. 360° at 50 kts., temperature +7° C.
    d. 005° at 7 kts., temperature missing.

49. What wind is forecast for St. Louis (STL) at 34,000 ft.?
    a. 073° at 6 kts.
    b. 730° at 61 kts.
    c. 306° at 19 kts.
    d. 230° at 106 kts.

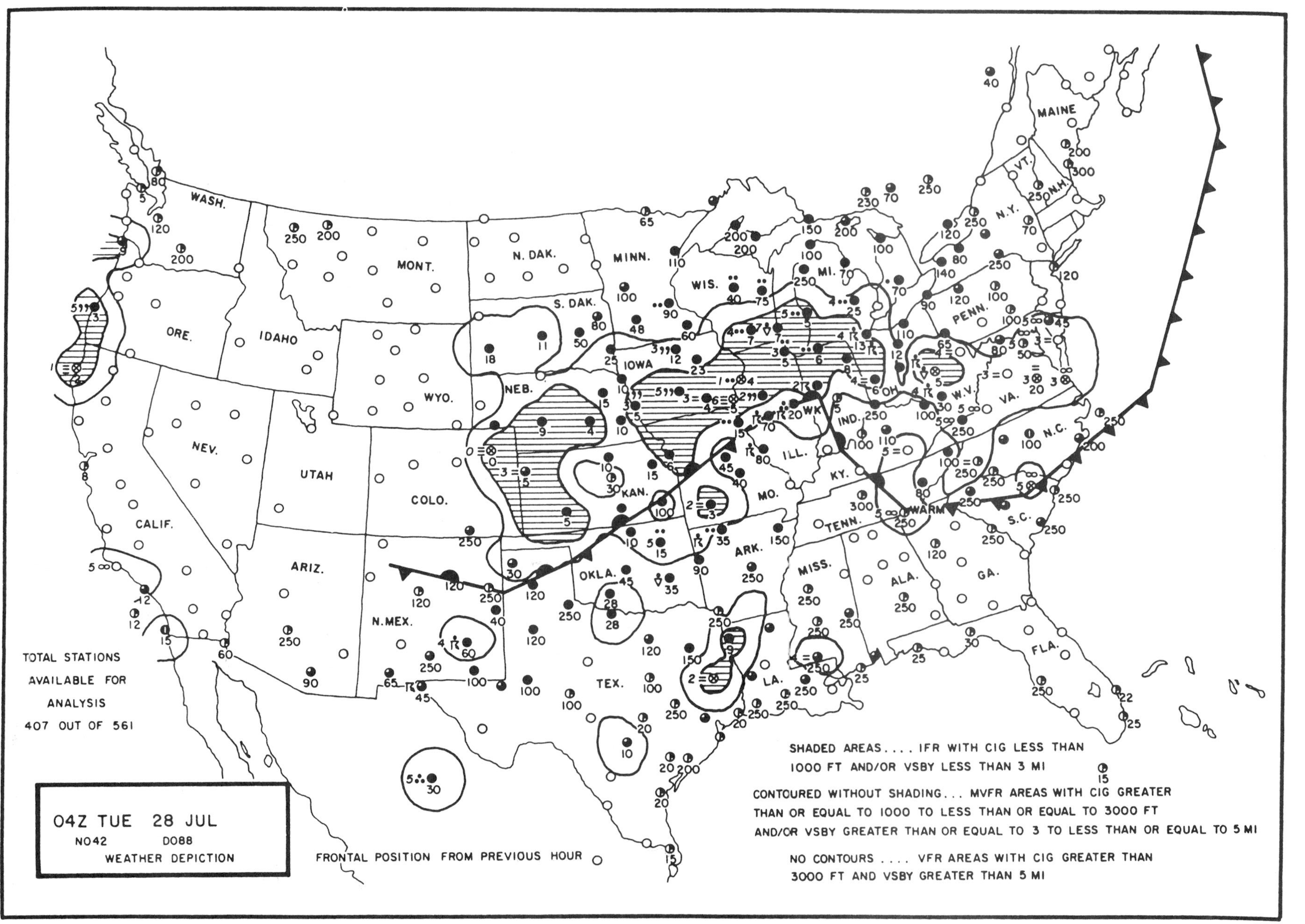
TOTAL STATIONS AVAILABLE FOR ANALYSIS 407 OUT OF 561
04Z TUE 28 JUL
NO42
D088
WEATHER DEPICTION
FRONTAL POSITION FROM PREVIOUS HOUR
SHADED AREAS . . . . IFR WITH CIG LESS THAN 1000 FT AND/OR VSBY LESS THAN 3 MI
CONTOURED WITHOUT SHADING . . . MVFR AREAS WITH CIG GREATER THAN OR EQUAL TO 1000 TO LESS THAN OR EQUAL TO 3000 FT AND/OR VSBY GREATER THAN OR EQUAL TO 3 TO LESS THAN OR EQUAL TO 5 MI
NO CONTOURS . . . . VFR AREAS WITH CIG GREATER THAN 3000 FT AND VSBY GREATER THAN 5 MI
WARM
WASH.
ORE.
CALIF.
IDAHO
NEV.
MONT.
WYO.
UTAH
ARIZ.
COLO.
N. MEX.
N. DAK.
S. DAK.
NEB.
KAN.
OKLA.
TEX.
MINN.
IOWA
MO.
ARK.
LA.
WIS.
MI.
ILL.
IND.
OH
KY.
TENN.
MISS.
ALA.
GA.
FLA.
S.C.
N.C.
VA.
W.V.
PENN.
N.Y.
VT.
N.H.
MAINE

Weather Depiction Chart.

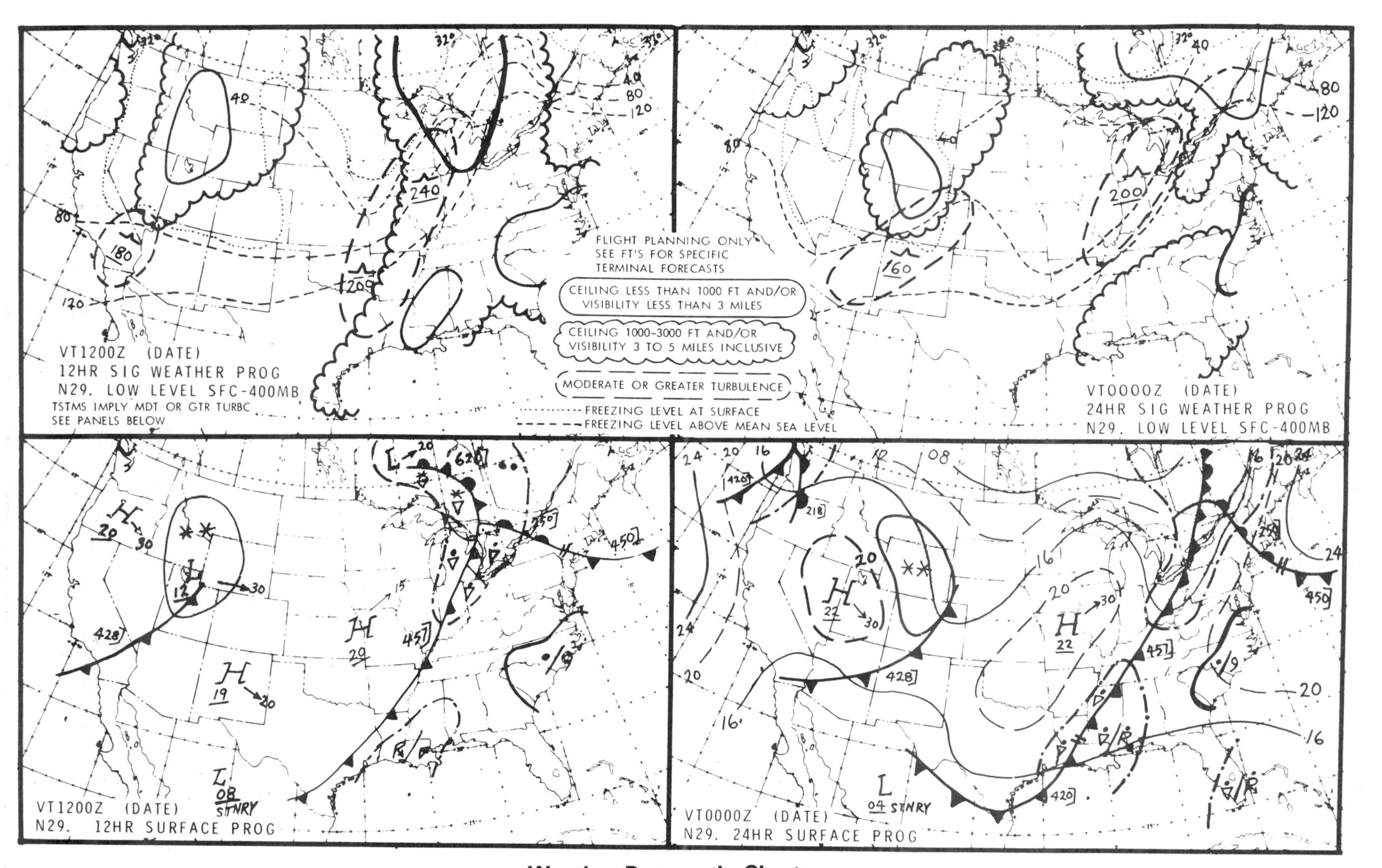

Weather Prognostic Chart.

50. Of what value is the weather depiction chart to the pilot? (See p. 15-37)
    a. To determine general weather conditions on which to base flight planning.
    b. For a forecast of cloud coverage, visibilities, and frontal activity.
    c. To determine the frontal trends and air mass characteristics.
    d. For an overall view of thunderstorm activity and forecast cloud heights.

51. The IFR weather in eastern Texas is due to (See p. 15-37)
    a. Intermittent rain.
    b. Fog.
    c. Dust devils.
    d. Sandstorm.

52. What is the status of the front that extends from New Mexico to Indiana? (See p. 15-37)
    a. Stationary.
    b. Occluded.
    c. Retreating.
    d. Dissipating.

53. According to the depiction chart, the current weather for a flight from central Arkansas to southest Alabama is (See p. 15-37)
    a. Broken clouds at 2,500 ft.
    b. Visibility from 3 to 5 mi.
    c. Broken to scattered clouds at 25,000 ft.
    d. Fog and low ceilings to move in from the gulf.

54. What weather is forecast for the Gulf Coast area just ahead of the cold front during the first 12 hr.? (See p. 15-38)
    a. Marginal VFR to IFR with intermittent thundershowers and rain showers.
    b. IFR with moderate or greater turbulence over the coastal areas.
    c. Thunderstorm cells moving northeastward ahead of the front.
    d. Rain and drizzle dissipating, clearing along the front.

55. Interpret the weather symbol depicted in lower California on the 12-hour Significant Weather Prog. (See p. 15-38)
    a. Moderate turbulence, surface to 18,000 ft.
    b. Thunderstorm tops at 18,000 ft.
    c. Base of clear air turbulence, 18,000 ft.
    d. Moderate turbulence, 180 mb level.

56. The band of IFR weather associated with the cold front in the western states is forecast to move (See p. 15-38)
    a. Southeast at 30 kts. with moderate snow showers.
    b. Northeast at 12 kts. with the front and producing snow showers.
    c. Eastward at 30 kts. with the low and producing snow showers.
    d. Eastward at 30 kts. with continuous snow.

57. At what altitude is the freezing level found over southeastern Oklahoma on the 12-hour Significant Weather Prog? (See p. 15-38)
    a. Surface.
    b. 4,000 ft.
    c. 8,000 ft.
    d. 10,000 ft.

Use the following pilot weather report and PIREP FORM to answer questions 58 through 62.

```
UA /OV OKC-TUL /TM 18ØØ /FL 12Ø /TP BE9Ø /SK Ø12 BKN Ø55 /
/Ø72 OVC Ø89 /CLR ABV /TA -9/WV Ø921/TB MDT Ø55-Ø72 /ICG LGT-MDT
CLR Ø72-Ø89.
```

BACK

Encoding Pilot Weather Reports (PIREP)

1. UA - **Routine PIREP, UUA - Urgent PIREP**
2. /OV - **Location:** Use 3-letter NAVAID idents only.
   a. Fix: /OV ABC, /OV ABC 090025
   b. Fix to fix: /OV ABC-DEF, /OV ABC-DEF 120020, /OV ABC 045020-DEF 120005, /OV ABC-DEF-GHI
3. /TM - **Time:** 4 digits in GMT: /TM 0915.
4. /FL - **Altitude/Flight Level:** 3 digits for hundreds of feet. If not known, use UNKN: /FL095, /FL310, /FLUNKN.
5. /TP - **Type aircraft:** 4 digits maximum, If not known use UNKN /TP L329,/TP B727 /TP UNKN.
6. /SK - **Cloud layers:** Describe as follows:
   a. Height of cloud base in hundreds of feet. If unknown, use UNKN.
   b. Cloud cover symbol.
   c. Height of cloud tops in hundreds of feet.
   d. Use solidus (/) to separate layers.
   e. Use a space to separate each sub element.
   f. Examples: /SK 038 BKN, /SK 038 OVC 045, /SK 055 SCT 073/085 BKN 105, /SK UNKN OVC
7. /WX - **Weather:** Flight visibility reported first. Use standard weather symbols, intensity is not reported: /WX FV02 R H, /WX FV01 TRW.
8. /TA - **Air temperature in Celsius:** If below zero, prefix with a hyphen: /TA 15, /TA -06.
9. /WV - **Wind:** Direction and speed in six digits. /WV 270045, /WV 280110.
10. /TB - **Turbulence:** Use standard contractions for intensity and type (use CAT or CHOP when appropriate). Include altitude only if different from /FL. /TB EXTRM, /TB LGT-MDT BLO-090.
11. /IC - **Icing:** Describe using standard intensity and type contractions. Include altitude only if different than /FL: /IC LGT-MDT RIME, /IC SVR CLR 028-045.
12. /RM - **Remarks:** Use free form to clarify the report. Most hazardous element first: /RM LLWS -15KT SFC-003 DURGC RNWY 22 JFK.

Refer to FAAH 7110.10 for expanded explanation of TEI coding.

Examples of Completed PIREPS

UA /OV RFD 170030/TM 1315/FL160/TP PA60 /SK 025 OVC 095/180 OVC /TA -21/WV 270048

UA /OV DHT 360015-AMA-CDS/TM 2116/FL050/TP PA32 /SK UNKN OVC/WX FV03 R /TB LGT/TA 04/RM HVY RAIN

FRONT

PIREP FORM

Pilot Weather Report → = *Space Symbc*

3-Letter SA Identifier

— — — → 1. **UA** → ____ *Routine Report* **UUA** → ____ *Urgent Report*

2. **/OV** → Location:
3. **/TM** → Time:
4. **/FL** Altitude/Flight Level:
5. **/TP** → Aircraft Type:

Items 1 through 5 are mandatory for all PIREPs

6. **/SK** → Sky Cover:
7. **/WX** → Flight Visibility and Weather:
8. **/TA** → Temperature *(Celsius)*:
9. **/WV** → Wind:
10. **/TB** → Turbulence:
11. **/IC** → Icing:
12. **/RM** → Remarks:

FAA FORM 7110-2 (1-85) Supersedes Previous Edition

58. The base and tops of the overcast layer reported are
    a. 1,200 ft. AGL and 5,500 ft. AGL.
    b. 1,200 ft. MSL and 5,500 ft. MSL.
    c. 7,200 ft. AGL and 8,900 ft. AGL.
    d. 7,200 ft. MSL and 8,900 ft. MSL.

59. The winds and temperature at 12,000 ft. MSL are
    a. 090° at 21 kts. and −9° C.
    b. 010° at 20 kts. and 1° F.
    c. 009° at 121 MPH and 90° F.
    d. 100° at 2 MPH and 12° F.

60. If the field elevation is 614 ft. MSL, what is the height above ground level of the base of the ceiling?
    a. 586 ft. AGL.
    b. 1,200 ft. AGL.
    c. 6,586 ft. AGL.
    d. 7,200 ft. AGL.

61. The intensity of turbulence reported by the pilot is
    a. Light.
    b. Moderate.
    c. Severe.
    d. Extreme.

62. The intensity and type of icing reported by the pilot is
    a. Light rime.
    b. Light to moderate clear.
    c. Moderate rime.
    d. Moderate to severe clear.

63. To get a complete weather briefing for the planned flight, the pilot should request
    a. An outlook briefing.
    b. A general briefing.
    c. An abbreviated briefing.
    d. A standard briefing.

64. Which type of briefing should a pilot request to supplement mass disseminated data?
    a. An outlook briefing.
    b. A supplemental briefing.
    c. An abbreviated briefing.
    d. A standard briefing.

65. A weather briefing that is provided when the information requested is 6 or more hours in advance of the proposed departure time is
    a. An outlook briefing.
    b. A forecast briefing.
    c. A prognostic briefing.
    d. An abbreviated briefing.

CHAPTER XVI

THE DUAT SYSTEM AND WEATHER FORECASTING

The reader should demonstrate an understanding of the DUAT system and weather forecasting to the extent that he/she can answer the following questions (as stated below, or as multiple-choice, etc.):

1. Describe the DUAT system.
2. List and describe the types of forecasts.
3. State why a knowledge of one's own local area is important.
4. Define forecasting skill.
5. State how much skill is typical in temperature and precipitation forecasts.
6. Calculate and interpret probability forecasts.
7. State what forecasting method best locates the position of large-scale highs, lows, and fronts.
8. Describe the benefits of observing clouds in making weather forecasts.
9. Where (with what) should a forecast begin?
10. What causes highs to build and lows to deepen?
11. What causes highs to weaken and lows to fill?
12. Where must upper level divergence or convergence occur in order to affect surface highs or lows.
13. State the location of strongest and weakest divergence and convergence associated with a jet core/max.
14. Why is the 500-mb chart important?
15. What are long waves and how do they move?
16. What are short waves and how do they move?
17. What is cold (and warm) advection?
18. Describe how vorticity is related to convergence and divergence?
19. What can be expected when an upper level trough is to the west of your location?
20. How does one identify significant moisture on upper level charts?
21. What are the three keys to any weather forecast?
22. What is the benefit of combining the DUAT system with forecasting principles?
23. In regard to weather forecasting, how should one stay COVERED?

CHAPTER XVI

THE DUAT SYSTEM AND WEATHER FORECASTING

The Direct User Access Terminal (DUAT) system will be described in section A. The DUAT system will allow pilots to become more independent where weather flight planning is concerned, but this independence must be accompanied by extra care/knowledge in using the system. Therefore, some additional information, on weather forecasting, is included in Section B.

A. The FAA and Three Vendors Bring Flight Planning, via DUAT, to the Home Computer — Free*.

BY MARC E. COOK

*Reprinted from *AOPA Pilot* 1989 all rights reserved

The face of flight planning is about to change again. For the last few years, the Federal Aviation Administration has been phasing out smaller flight service stations and incorporating the weather reporting and flight plan filing services into centralized automated FSSs. The AFSS combines taped weather information, accessed through Touch-Tone telephones, with automated flight planning. Instrument flight plans can be filed over the telephone onto a tape without having to wait for a live briefer; VFR plans can be closed the same way. But many pilots have found that the additional services offered by the AFSS were offset somewhat by the extra time it took to actually talk to a briefer.

While the FAA has been systematically closing low volume FSSs and begun relying more and more upon AFSSs, another method of weather briefing and flight planning has started to gain popularity. Computerized flight planning, offered by a number of vendors, has become many pilots' favorite source for the same sort of information previously offered solely by FSSs. What's more, many of these vendors supply weather information that is simply not available from the FAA over the telephone, including dedicated radar sites and real-time weather pictures.

With a home computer and a modem — a device that allows your computer to talk to other computers via a telephone line — pilots can have weather briefings, tailored flight plans, and, more recently, access to the FAA's flight plan computer. Although these computer-planning vendors provide more services than can be obtained from an FSS over the telephone, use of them costs money — sometimes not an insignificant amount, depending on how frequently you use the system.

Now, with the Direct User Access Terminal (DUAT) system, scheduled to go on line this September, the FAA has married the best parts of the new computerized systems with one of the most endearing aspects of the old FSS arrangement: price. Using a toll-free telephone number, the basic DUAT services will be absolutely free. Anyone with a pilot's certificate — access codes are given out only to licensed pilots, an FAA restriction — and an IBM or compatible computer can get on the system. Moreover, computers capable of receiving basic text over the modem (such as the Apple Macintosh, which otherwise doesn't generally talk to IBMs) also can gain access to DUAT.

Rather than attempt to construct the DUAT services in-house, the FAA has turned to three contractors to provide the weather briefing and flight plan filing functions. Contel Federal Systems, Lockheed Data Plan, and Data Transformation Corporation will be the three vendors for the DUAT program. Thanks to FAA specifications, the basic weather briefing and flight plan filing programs will be nearly identical among the three vendors. ''We're locked into what the FAA wants to do with the basics, to maintain consistency from one vendor to another and one region to another,'' says Lockheed's Dan Woods.

So how does a pilot, computer in hand, step up to DUAT? First, the hardware. DUAT has been designed to work with a broad range of computers and modems. A simple 300-baud modem and a basic communications program, together with most any kind of computer — the DUAT system prompts the user for screen width and depth, so it can be tailored to even the simplest of computers — is all that's needed. The sign-on will be simple, with the 800 numbers available through the FAA and the vendors' various promotional campaigns. Then the system will ask for an access code and password; for first-time users, the program will walk you through sign-up, issue an access code, and ask you to make up a password. For anyone now using modem services like CompuServe, adding DUAT to the list will be straightforward.

We tried two of the three vendors' DUAT systems several months before they were scheduled to go on line. Once in the system, the pilot can retrieve a wealth of weather information. In addition to the standard record observations, terminal forecasts, area forecasts, and winds aloft forecasts, DUAT will provide pireps and airmets, significant clouds and weather advisories, and convective outlooks. Virtually every bit of textual information the local FSS briefers have available can be retrieved through DUAT. Moreover, the FAA has specified a simple menu format that list your options at each step; even those shy of computers should have no trouble working through the system.

Not only is all the information there, but there are several different ways to ask for it. For example, you can enter the departure, destination, and enroute points of your intended flight, and DUAT will supply a weather briefing for that route. DUAT will provide in that briefing information within 25 nautical miles either side of your chosen course. Moreover, you can ask for only parts of that route, or even smaller parts of the local weather picture; the last includes reports within a 25-nm radius. Finally, should you be interested in only the winds aloft forecast, for example, there is an option to do only that. Users have 20 minutes per session to check the weather and file flight plans. Of course, if the computer cuts you off before you're finished, you can log on again immediately; the 20-minute rule was mandated by the FAA, and all vendors' DUAT will include it.

The only criticism that could be leveled at DUAT would have to do with the format. Because the FAA required DUAT to provide the same information, in the same form, as the FSSs receive, it remains encoded. If you remember the sequences and codes, that's really not a problem, but users accustomed to the FSS briefer explaining the weather in plain English might find themselves afloat in a sea of nearly incomprehensible letters and numbers. All three vendors intend to address this issue by offering extra-cost expanded briefings, with the FAA-standard contractions converted into real English, and documentation provided with the DUAT service will include a list of the services and their codes. Finally, a help menu will be available by entering a question mark, so should a pilot find himself too bogged down to continue, the system will help bail him out.

It pays to know your three-letter identifiers well, too; forecasts depict area weather by defining a line from point to point. For instance, occasional ceilings of less than 1,000 feet from Washington, D.C., to Baltimore to Pittsburgh might read: FRM DCA TO 20NE BWI TO 10SW HAR TO AGC TO CKB TO IAD TO DCA . . . OCSL CIG BLO 10. This style of depiction might suffice for those with bank-vault-tight memories, but the weekend pilot will probably find himself spending more time looking up identifiers than he'd like to. Not surprisingly, DUAT will include a menu item to decode identifiers, which will make consulting a map or airport guide unnecessary. Also, at least two of the vendors will offer at additional cost an expanded forecast that will automatically decode the identifiers.

One of the more exciting prospects of DUAT is the on-line flight plan filing. A few aviation computer-service vendors have recently begun offering direct filing, but it will be part and parcel of the DUAT system. When you ask for the weather briefing, DUAT stores that route of flight and will automatically enter it into a standard flight plan form. Then you simply enter the number of passengers, usable fuel, requested altitude, and any alternates. Your airplane's particulars are stored into a permanent file with your access code and password, so filing flight plans requires entering only the essentials listed above.

While the weather briefing and flight plan filing services will be the basics of DUAT, the individual vendors will be allowed to offer, at additional cost, a range of additional services. Among these services: color radar maps (with the appropriate home computer hardware, of course), customized flight planning, prognostic charts, and costs analysis. At press time, all three vendors were tight-lipped about precisely what will be offered as an add-on package to DUAT, and only one, Data Transformation, would take a guess at the surcharge for additional information packages: between 50 and 75 cents per minute.

Although no one was willing to put all the cards on the table at press time, all gladly talked about the potential for growth built into the DUAT format. ''We have designed our DUAT system to be highly modular, and the potential to link some of our other weather information systems to it is great,'' says George J. Jakabcin, manager of Contel Federal's weather systems sales. ''We intend to leave DUAT free to grow. The market will tell us which services it wants and which services it doesn't, and we will continue to expand DUAT's capabilities as necessary.''

One of the areas open for expansion has to do with graphics. Basic DUAT screens will offer text only, but a value-added package is free to use whichever format its maker decides. ''We're working hard to get graphics into the system,'' says Bill Barto, project manager at Data Transformation. ''We think that there are quite a few computers out there with graphics capabilities, and we want to take advantage of that.'' Similarly, Dan Woods, describing Lockheed's interest in graphics capabilities, states, ''It's an important point. By providing maps and pictorial displays of weather, we can greatly enhance the value of the information. We are working on software that we can provide to the user to make our graphics compatible with a large number of computers. Software for the Macintosh is next.''

Incorporating graphics into the system is also viewed as a method to make DUAT easier to use for all levels of users. ''We're very concerned that the system be easy to use,'' says Jakabcin. ''Pilots are used to talking to someone on the telephone, and we want DUAT to be just as simple.'' The question inevitably will come up about whether inexperienced pilots will be able to adequately decode and interpret the textual information provided by DUAT. Jakabcin demurred, ''I think it might make all pilots pay a bit more attention to the specifics of the weather, and, perhaps more importantly, DUAT information is consistent across the country and not open to subjective interpretation.'' Lockheed's Woods agrees, ''Standardization is very important, not only in terms of the weather information presented, but also in the way it is retrieved.''

So with the basic DUAT poised to go on line this fall, what will be the next step? Interactive flight plan filing could be in the cards, with certain preferred IFR routes built into the database. With this feature, a pilot could enter his departure and destination points, and the computer would generate a flight path using FAA-preferred routing. Victor airways, and automatic inclusion of SIDs and STARs. Ultimately, the DUAT system could be expanded to include realtime radar returns and other high-level, high-tech features that currently are the province of the airline set.

So while the basic DUAT services by themselves aren't earth-shattering, the concept and format of the system points to something of a revolution in the way we check weather and file flight plans. And the FAA's insistence that the primary services remain free will mean that even the most optioned-out briefing package could cost less with DUAT than is now the case.

It's too early to tell which of the three DUAT vendors has an edge. Because the basic DUAT functions are identical among the three, the decision must come down to the quality of each vendor's value-added packages. Brand loyalty will play something of a role, according to Lockheed. ''We have Data-Plan users now, and we expect most of them to stay on when we go to DUAT,'' says Woods. Likewise, Contel's Jakabcin says, ''We expect overlap of pilots now using the WXBrief package Contel now produces.'' Of course, there's no reason a pilot can't sign up with all three, a prospect that none of the vendors seem to mind. ''We enjoy the competition,'' says one spokesman.

Open competitiveness is something the FAA has fostered within the DUAT program, with hopes that the resulting weather dissemination system will be as good as it can be. In conjunction with the National Airspace System Plan and the FAA's push to replace aging computers in approach control facilities and enroute centers, DUAT represents an important step forward in consolidating and revising our airspace system. The future, come September, is here.

For more information:

Contel Federal Systems
12015 Lee Jackson Highway
Fairfax, Virginia 22033
800/345-DUAT

Data Transformation Corporation
559 Greentree Road
Turnersville, New Jersey 08012
609/228-3232

Lockheed DataPlan, Incorporated
121-A Albright Way
Los Gatos, California 95030
408/866-7611; 800/767-DUAT

B. Weather Forecasting

1. *General*

This section will not make you a weather expert where forecasting is concerned. Weather forecasting is both an art and a science. The science part is knowing the variables (factors), such as temperature, moisture, pressure, etc., that result in weather patterns. The art is in according the proper weight to each factor, that is, recognizing which factor(s) is(are) the most dominant in a given situation. Ultimately, it either rains or it does not, for example. Many times the difference can be one degree of temperature, five percent relative humidity, or a ten degree shift in wind direction. Again, one degree of temperature can make the difference in whether there are only towering cumulus clouds or cumulonimbus clouds (thunderstorms).

2. *Using Existing Forecasts*

a. Forecasting Methods

Most forecasts are classified as climatological, persistence, trend, analogue, or numerical. A climatological forecast is a forecast of average conditions (usually based on at least thirty years of data). A persistence forecast is one that calls for no change; for example, if the high temperature is 96 °F today, then persistence would call for a high temperature of 96 °F the next day, whereas climatology would call for a temperature of perhaps 98 °F (assumed average for that day over many years).

The trend method is frequently used in forecasting frontal movement. If a front moved 10 and 12 miles in successive hours, the trend method would call for a 14 mile movement in the next hour. Persistence and trend methods are best used for time scales of but a day or less.

The analogue method is essentially one of recognizing an existing weather pattern as being similar to one that has occurred in the past. One often hears weather forecasters state that with certain types of weather patterns certain results occur. An example would be a particular positioning of highs and lows — when high pressure exists in the upper levels over the southeastern United States in summer, it is quite hot below at the surface, but when high pressure in the upper levels is over the southwestern part of the United States, surface temperatures in the southeast are cooler at the surface due to a northwesterly flow of cooler air in the upper levels over the southeast.

Numerical forecasts are "computer-driven" forecasts, that is, they are generated by computers that represent the atmosphere by mathematical models. The models are sets of equations in which current data (temperature, pressure, etc.) is fed in and future data is predicted at fixed time intervals, say every five minutes. This process can be repeated until 12, 24, 36, and 48 hour forecasts have been generated. This method is only as good as the equations and the density of data. Since reporting stations are many miles apart, we often find that many smaller-scale phenomena are missed, that is, large highs and lows are identified while local rain, fog, or winds between reporting stations are not depicted.

It should be obvious from the above that a knowledge of one's own local area is indispensable. Most television meteorologists find that it takes several years in any given region to develop a real "feel" for that area. Most forecasts are for large areas and must be localized by meteorologists *and* pilots.

b. The Reliability of Weather Forecasts

This issue was first addressed in section E of Chapter XV — you may wish to review that section before continuing. Reliability is related to skill, that is, one can be correct on the next day's high temperature, for example, by just calling for the same temperature as the current day (persistence) or for the average temperature (climatology). Has one shown forecasting skill by using persistence or climatology? The accepted definition of skill is that accuracy must be based on methods other than persistence or climatology. If the high temperature is 96 °F today and is forecast to be 96 °F the next two days (persistence) then no skill has been shown if the forecast is accurate. Skill levels for both temperature and precipitation are quite high up to twelve hours, and fairly high out to two days. Forecasts of temperature are fairly good from two to five days and precipitation out to three days. Beyond these limits, skill levels drop off until they become minimal for periods beyond one month.

At this point we might make a comment about "normal" and "average". One might say that it is normal not to be average. For example, the average high temperature for a given day might be 96 °F, but how often would one expect to get exactly 96 °F on that day? One year, the high temperature might be 94 °F, another year 100 °F, etc., that is, it is normal not to be average. Admittedly, this use of these words in this manner is only to illustrate that it is not unusual to have weather variations. Some locations could have the same average temperature but quite different variations — maybe two locations have an average high of 96 °F, but one has ranged historically between 90 °F and 100 °F while another location has varied from 85 °F to 110 °F.

c. Probability Forecasts

One often hears a forecast of a 20% chance of rain, or 50%, etc., especially on television and radio. We might see how this is done, using the following example. Suppose that during a particular time of the year that three-fourths of the time fronts pass a given location and one-fourth of the time they do not. Stated another way, there is a 75% chance that a front will pass and a 25% chance that it will not. Suppose further that when the front passes, rain occurs 80% of the time, but when it does not pass, rain still occurs 40% of the time. What is the probability of rain? Note that the 80% and 40% depend on whether or not the front passes. What we do is to reduce the 80% by multiplying by 75%, reduce the 40% by multiplying by 25%, and then add the results. Well, three-fourths of 80% is 60% and one-fourth of 40% is 10%; the sum of 60% and 10% is 70%, and that is our probability of rain in this example. The numbers may change, but the method does not. The computer can keep a huge volume of records concerning weather situations and update the percentages year after year (either a front passed or it did not and either it rained or it did not in each case). And, incidentally, the 70% chance of rain is the probability that *any given location* in the forecast area will receive at least a trace of rain, that is, one would expect rain on 7 days out of each 10 days under a 70% probability of rain forecast.

3. Making Forecasts

This section will be fairly short, for one of the primary means of forecasting, numerical, is beyond the scope of this text. Several courses in advanced math and physics are necessary to understand numerical weather prediction. It would be difficult to improve upon the large-scale positions of highs, lows, and fronts by this method. However, the following techniques and other information should be beneficial.

a. Local Knowledge

It is possible to improve upon local forecasts, for recall that numerical forecasts are based on observations many miles apart. Some locations may be perfectly clear, yet some areas may be "fogged in". In one case at Carswell AFB, Texas, fog moved in from over Lake Worth and a pilot "landed on top of the fog", thereby overshooting the runway — skies were absolutely clear everywhere else. Again, local knowledge is very important.

b. Clouds

Do not forget to look at clouds for clues, such as the cloud patterns associated with advancing cold and warm fronts. Even stability can be indicated by clouds. Towering cumulus clouds by 2:00 p.m. indicate that the air is likely to be unstable and that thunderstorms are possible later in the day.

c. Weather Charts

This author uses the following primary charts in making forecasts: surface and weather depiction charts, the radar summary, current and forecast 850,700 and 500-mb constant pressure charts, current and forecast jet stream positions, significant weather prog charts, severe weather charts, satellite charts, vorticity, and stability charts. Many other charts are also available. Your DUAT menu should include all charts and data found in Aviation Weather Services and other charts and data prepared by the National Weather Service. It is important to always start making one's own forecast by beginning with the "official" forecast, from the FSS or the National Weather Service.

Again, it takes a vast amount of knowledge of meteorology (several courses) and a study of many weather situations to become a competent weather forecaster. It is the opinion of this author that the best approach for a pilot is to practice forecasting so that he/she may be better prepared to understand the forecasts received in order that he/she may choose an appropriate action in the event a forecast does not turn out as expected. However, the assumption that the average pilot is a better forecaster than meteorologists who routinely make forecasts can be quite dangerous, so request weather assistance anytime you are unsure.

The DUAT system provides the pilot with a natural laboratory in which to study weather patterns. We now list but a very few features to look for in putting weather information together. First, note from Figure 5-3 that for deep systems (see p. 4-18, 20) high pressure (hill) is associated with warm air aloft and that low pressure (valley) is associated with colder air aloft. Thus,

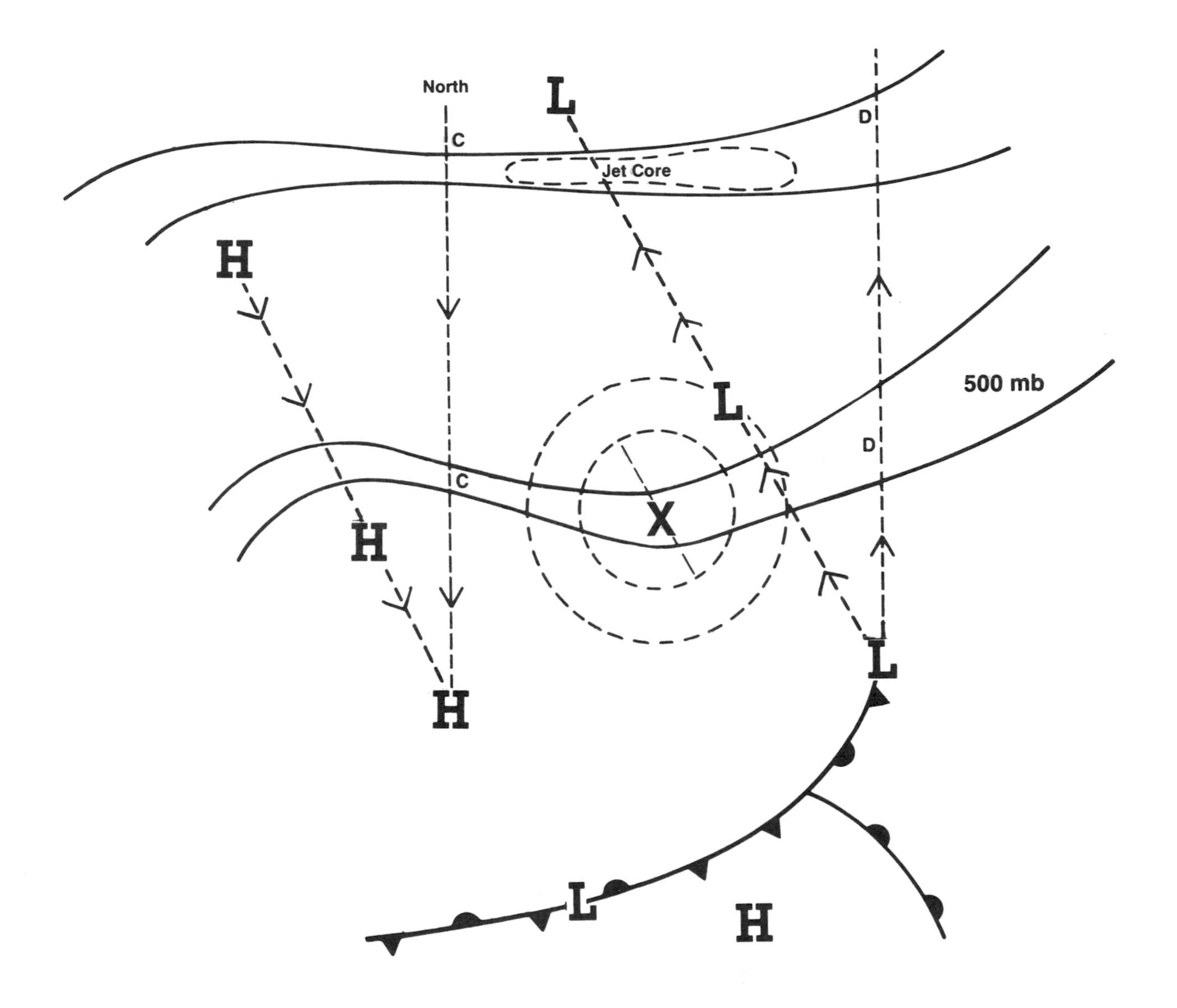

Figure 16-1. Relationship between surface and upper level pressure systems, vorticity, and the jet stream. Highs and lows tilted in the vertical. Strongest divergence (D) and convergence (C) north of jet core, above surface low and high, respectively – secondary surface high and low south of jet core. Short wave in 500-mb trough (X marks maximum vorticity – decreases radially), with divergence downwind (above surface low) and convergence upwind (above surface high). Maximum rate of descent is along the axis of the high — maximum ascent rate along axis of low. As the high and low move easterly (from left to right), they are strengthened by the vertical components of descent and ascent, respectively, in their paths.

high pressure systems are usually tilted to the southwest in the upper levels and lows to the northwest; note this when you look at highs and lows on the surface and upper level charts; the tilt will be more apparent the higher you look (Figure 16-1).

High pressure systems will tend to build (become stronger — pressure rises) when upper level convergence brings in mass at a faster rate in the upper levels than is leaving at the surface. This convergence must take place directly above the high at the surface (the high in the upper levels is still tilted toward the warm air). If more mass is leaving at the surface than is arriving aloft, the high will weaken (pressure drops). Similar statements can be made about lows. When upper level divergence takes mass away in the upper levels at a greater rate than air is converging at the surface, then the low is deepening (intensifying – pressure drops). If more mass is coming in at the surface than is leaving in the upper levels, then the low fills (weakens — pressure rises). The upper level divergence must take place directly above the low at the surface (the low in the upper levels is still tilted toward the colder air).

Jet stream effects are, for the most part, beyond the scope of this text. These effects are important, however, leading to the development of severe storms and the intensification of surface highs and lows. In fact, the jet stream often causes highs and lows to ''build down'' from higher levels to the surface. The strongest area of divergence is north and downwind of the core; the weakest divergence is south and upwind of the core. The strongest area of convergence is north and upwind of the core; the weakest convergence is south and downwind of the core. (Figure 16-1.)

The 500-mb chart is very important. Recall from earlier in the text that about half the weight of the atmosphere is above and below its average height of approximately 18,000 feet (5.5 km.) Thus, winds at this level tend to push, or steer pressure systems. We have yet to discuss the notion of long and short-waves, at the 500-mb level or higher. Figure 11-17 illustrates long waves reaching from the Pacific Northwest (trough), over Montana (ridge), through the Great Lakes (trough), and out into the Atlantic (ridge). These long-waves can wrap around the entire globe. They usually move slowly eastward, but at times they are stationary, and on occasion move westward.

Embedded within these long-waves are shorter waves that move more quickly through troughs than ridges. These short waves are often called ''impulses'', wind-shift lines, or even kinks in the overall windflow pattern. Figure 16-1 shows one of these short-waves in a trough at the 500-mb level. When these short-waves add a sufficient amount of cold air (cold air advection) to a trough, the trough will often deepen (intensify). If these short-waves add a sufficient amount of warm air (warm air advection) to a ridge, the ridge will often build (intensify). This can be seen on upper level charts when the wind flow cuts across isotherms (recall that these are lines of constant temperature).

Even if convergence and divergence are identified (Figure 4-14), they are not easily quantified. However, they are mathematically related to vorticity, which is at least easier to quantify. It turns out that upper level divergence and lower level convergence are typical downwind of a vorticity maximum, and upwind of a vorticity maximum there is upper level convergence and lower level divergence. Note in Figure 16-1 that the air is diverging (spreading out directionally) downwind of the vorticity maximum and converging upwind of the vorticity maximum. Whenever a trough in a long wave is to the west of your location, short-waves will usually move through the trough, generating one line of thunderstorms after another along the short-wave line. The timing of these impulses is very difficult; again vorticity is your best clue (on progs).

Finally, look for moisture on upper level charts — the station symbol is usually shaded when the temperature and dew point are within 5 °C of each other. Chapter VIII provided some stability considerations (which you may wish to review at this time), many of which are reflected in a stability chart found in Aviation Weather Services. Recall that moisture, stability, and a source of lift are the three keys to any forecast.

In summary, with DUAT and the *brief* discussion on weather forecasting (again, meteorologists take several courses related to forecasting) you now have some basic tools with which to *learn.* Once again, keep yourself COVERED — COmprehend the principles of weather forecasting, Visualize present and future (forecast) weather patterns, Evaluate your proximity to such patterns, Respond accordingly, and, most importantly, when forecasts go awry re-Evaluate (hopefully, you anticipated what might go wrong before you took off) and Determine a new course of action (using your judgment and that of the weather support system).

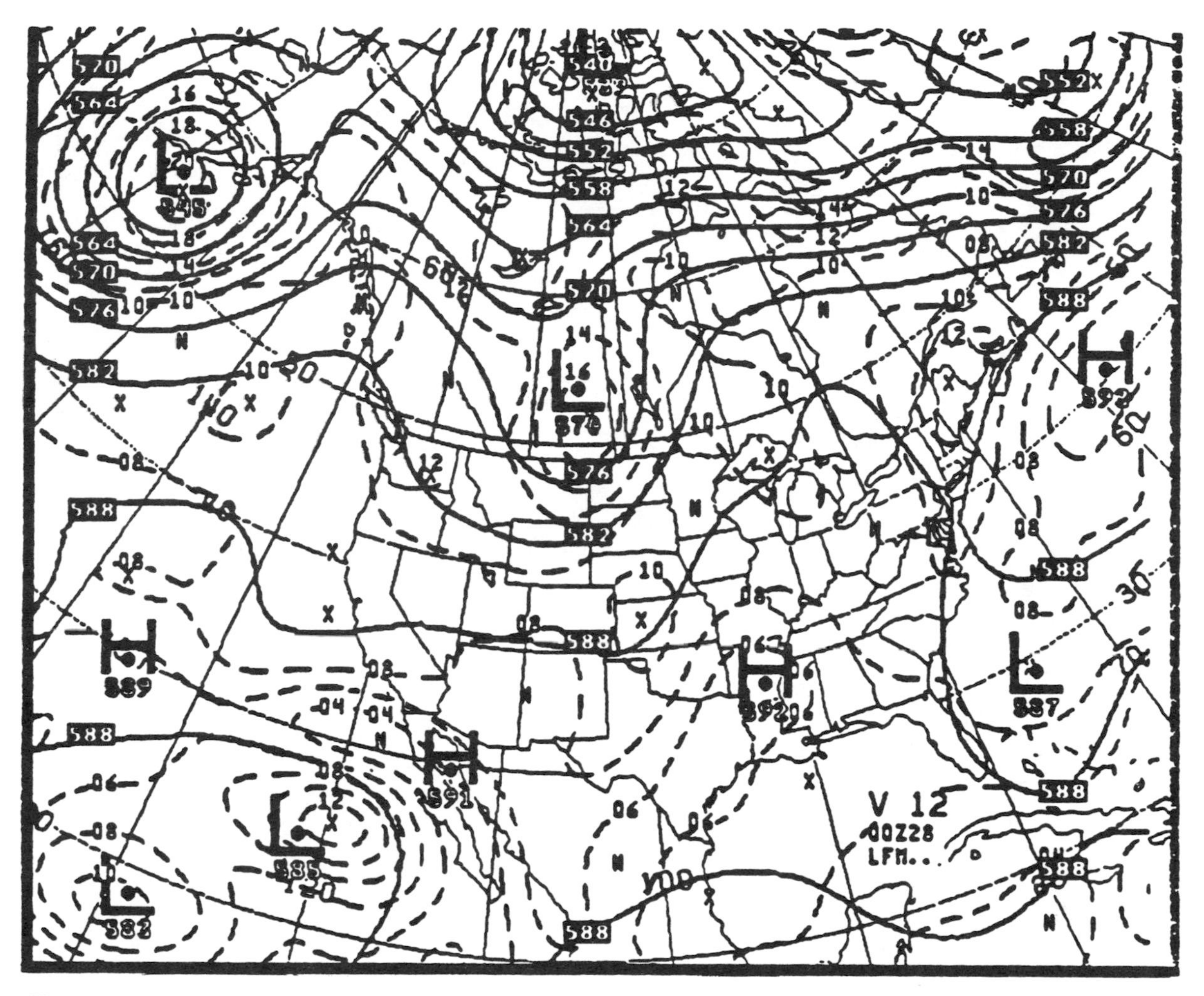

Figure 16.2 – 12HR FCST 500MB Heights/Vorticity Valid OOZ Sat 28 Jul 90

d. Further Applications of Vorticity

We conclude this chapter with another application of vorticity. Consider Figure 16-2, a National Weather Service forecast of 500MB heights and vorticity (not described in Aviation Weather Services). Centers of positive absolute vorticity are indicated by "X" and negative absolute vorticity by "N". Recall from the text that vorticity is taken to be positive in counterclockwise, or cyclonic motion, and negative in clockwise, or anticyclonic motion. Also recall that upper level divergence and lower level convergence are typical downwind of a vorticity maximum. Note the vorticity maximum along the Washington-Oregon border on the chart above. If the atmosphere is sufficiently moist and unstable downwind of the X, then cloudiness and precipitation could be expected in that area, as well as intensification of any fronts (even frontogenesis is possible, if a front is not already present).

Remarkably, vorticity can be used in summarizing weather. First, we acknowledge that the sun is the primary driving force for all of the earth's weather, for temperature differences cause pressure differences which initiate wind flow. Of course, the Coriolis force and the law of conservation of angular momentum influence wind direction and speed, respectively, due to the rotation of the earth on its axis and its spherical shape. And land-sea differences also influence the wind as well. Changes in wind direction and speed (wind shear) contribute to the life-history of highs and lows, changes which can be quantified using vorticity.

THE DUAT SYSTEM AND WEATHER FORECASTING
Review Questions
(Not in FAA documents)

1. Weather forecasting is a(n)
   a. Art.
   b. Science.
   c. Both a. and b.
   d. None of the above.

2. Which of the following is *not* a forecasting method?
   a. Analogue.
   b. Television.
   c. Persistence.
   d. Climatology.

3. What forecast method is frequently used in forecasting frontal movement?
   a. Trend.
   b. Numerical.
   c. Climatology.
   d. Resistance.

4. What forecast method is ''computer driven''?
   a. Qualitative.
   b. Trend.
   c. Analogue.
   d. Numerical.

5. Skill cannot occur with which forecasting method?
   a. Persistence.
   b. Climatology.
   c. Both a. and b.
   d. Numerical.

6. Skill levels in forecasting temperature and precipitation are quite high up to
   a. 6 hours.
   b. 12 hours.
   c. One day.
   d. Two days.

7. A 70% chance of rain means
   a. There is a 70% chance that rain will occur somewhere in the forecast area.
   b. Rain is expected to occur in 70% of the area.
   c. Rain is very unlikely.
   d. Any given location in the forecast area has a 70% chance of receiving rain.

8. In the upper levels of the atmosphere, highs are usually tilted to
   a. The Southwest.
   b. The Northeast.
   c. There is no tilt.
   d. The Northwest.

9. Stronger upper level divergence than lower level convergence will cause a
   a. High to build.
   b. High to weaken.
   c. Low to deepen.
   d. Low to fill.

10. The strongest area of convergence associated with a jet core/max is located
    a. North and upwind of the core.
    b. North and downwind of the core.
    c. South and upwind of the core.
    d. South and downwind of the core.

11. Which of the following is *not* another name for a short wave?
    a. Kink.
    b. Impulse.
    c. Wind shift.
    d. Amplitude.

12. Downwind of a vorticity maximum expect
    a. Upper level divergence and lower level divergence.
    b. Upper level convergence and lower level convergence.
    c. Upper level divergence and lower level convergence.
    d. Upper level convergence and lower level divergence.

13. The station symbol on upper level charts is often shaded when the temperature and dew point are within how many degrees of each other?
    a. 5 °C.
    b. 5 °F.
    c. 0 °C.
    d. 0 °F.

14. What is a key to any weather forecast?
    a. Source of lift.
    b. Moisture.
    c. Stability.
    d. All of the above.

CONTENTS

Section 1—THE AVIATION WEATHER SERVICE PROGRAM

Section 2—SURFACE AVIATION WEATHER REPORTS

Section 3—PILOT AND RADAR REPORTS AND SATELLITE PICTURES

Section 4—AVIATION WEATHER FORECASTS

Section 5—SURFACE ANALYSIS CHART

Section 6—WEATHER DEPICTION CHART

Section 7—RADAR SUMMARY CHART

Section 8—SIGNIFICANT WEATHER PROGNOSTICS

Section 9—WINDS AND TEMPERATURES ALOFT

Section 10—COMPOSITE MOISTURE STABILITY CHART

Section 11—SEVERE WEATHER OUTLOOK CHART

Section 12—CONSTANT PRESSURE CHARTS

Section 13—TROPOPAUSE DATA CHART

Section 14—TABLES AND CONVERSION GRAPHS

ILLUSTRATIONS

Figure | Page

Section 3—PILOT AND RADAR REPORTS AND SATELLITE PICTURES

Section 4—AVIATION WEATHER FORECASTS

Section 5—SURFACE ANALYSIS

Section 6—WEATHER DEPICTION CHART

Section 7—RADAR SUMMARY CHART

Section 8—SIGNIFICANT WEATHER PROGNOSTICS

Section 9—WINDS AND TEMPERATURES ALOFT

Figure | Page

Section 10—COMPOSITE MOISTURE STABILITY CHART

Section 11—SEVERE WEATHER OUTLOOK CHART

Section 12—CONSTANT PRESSURE CHARTS

Section 13—TROPOPAUSE DATA CHART

TABLES

Table | Page

Section 2—SURFACE AVIATION WEATHER REPORTS

Section 3—PILOT AND RADAR REPORTS AND SATELLITE PICTURES

Section 4—AVIATION WEATHER FORECASTS

Section 5—SURFACE ANALYSIS

Section 6—WEATHER DEPICTION CHART

Section 7—RADAR SUMMARY CHART

Table Page

Section 8—SIGNIFICANT WEATHER PROGNOSTICS

Section 10—COMPOSITE MOISTURE STABILITY CHART

Section 11—SEVERE WEATHER OUTLOOK CHART

Section 12—CONSTANT PRESSURE CHARTS

Section 14—TABLES AND CONVERSION GRAPHS

INTRODUCTION

The rapid expansion of air transportation makes necessary a move toward mass briefings to meet aviation demands. As a results, you, the pilot, must become increasingly self-reliant in getting your weather information. On occasion, you may need to rely entirely on self-briefing.

This advisory circular, AC 00-45C, explains weather service in general and the details of interpreting and using coded weather reports, forecasts, and weather charts, both observed and prognostic. Many charts and tables apply directly to flight planning and in-flight decisions.

This advisory circular is an excellent source of study for pilot certification examinations. Its 14 sections contain information needed by all pilots, from the student pilot to the airline transport pilot.

AC 00-45 is updated periodically to reflect changes brought about by the latest service demands, techniques, and capabilities. The purchase of an updated copy is a wise investment for any active pilot.

Comments and suggestions for improving this publication are encouraged and should be directed to:

National Weather Service Coordinator, AAC-909
Federal Aviation Administration
Mike Monroney Aeronautical Center
P.O. Box 25082
Oklahoma City, OK 73125

Advisory Circular, AC 00-45C, supersedes AC 00-45B, Aviation Weather Service, revised 1979.

Section 1

THE AVIATION WEATHER SERVICE PROGRAM

Weather service to aviation is a joint effort of the National Weather Service (NWS), the Federal Aviation Administration (FAA), the Department of Defense (DOD) weather service, and other aviation oriented groups and individuals. Because of international flights and a need of world-wide weather, foreign weather services also have a vital input into our service.

This section follows the development and flow of observations, reports, and forecasts through the service to the users, as depicted in figure 1-1.

OBSERVATIONS

Weather observations are measurements and estimates of existing weather both at the surface and aloft. When recorded and transmitted, an observation becomes a report and these reports are the basis of all weather analyses and forecasts. Note in figure 1-1 that high speed communications and automated data processing have improved the flow of weather reports to the aviation user.

Surface Observations

Surface aviation observations include weather elements pertinent to flying. A network of airport stations provides routine up-to-date aviation weather reports. Most of the stations in the network are either NWS or FAA; however, the military services and contracted civilians are also included. A major change in the surface weather observation network is underway with the installation of

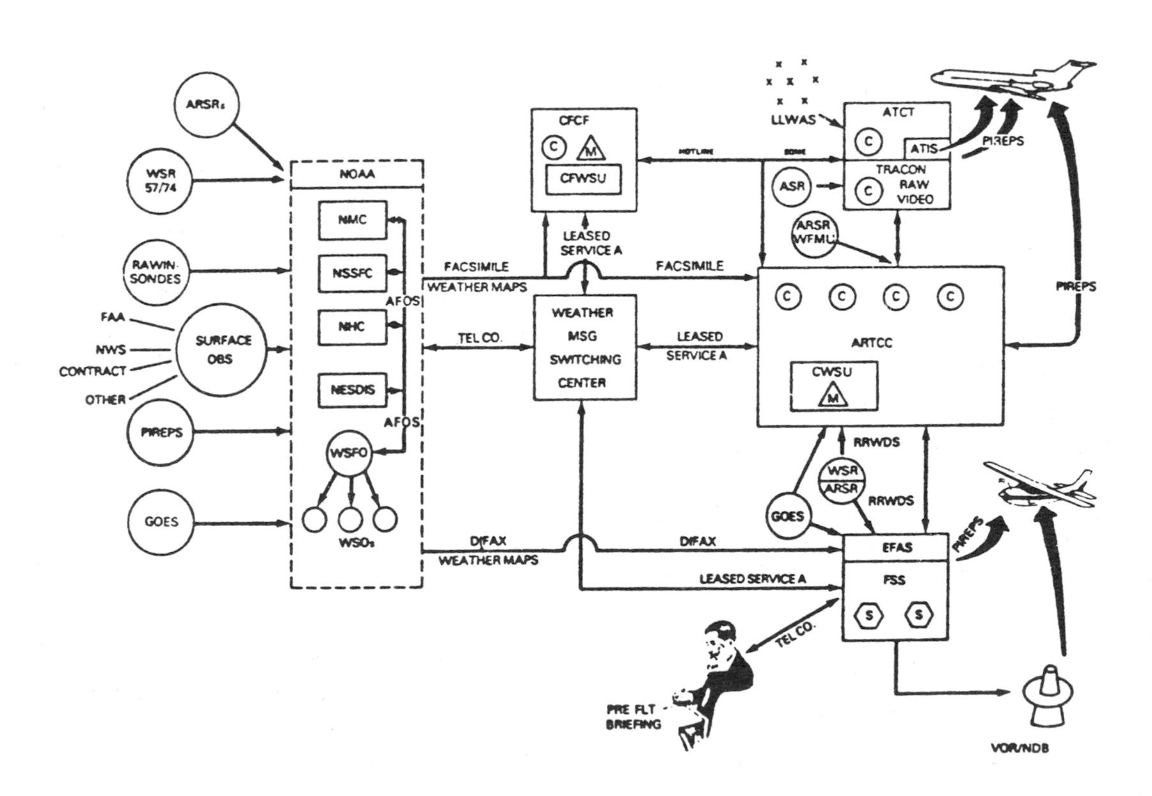

FIGURE 1-1. Data flow in the aviation weather network. Note the important feedback of pilot reports.

automated weather observing stations across the country. These automated stations are expected to become a major part of the network. Existing types of automated observations are discussed in Section 2.

Radar Observations

Precipitation reflects radar signals and the reflected signals are displayed as echoes on the radar scope. NWS radar covers nearly all the U.S. east of the Rocky Mountains. Radar coverage over the remainder of the U.S. is largely by Air Route Traffic Control radars. Thus, except for some western mountainous terrain, radar coverage is nearly complete over the contiguous 48 states. Figure 1-2 maps the radar observing network.

The new Radar Remote Weather Display System (RRWDS) is a significant improvement over previous radar remote systems. This system is specifically designed to provide *real-time* radar weather information from many different radars, which makes it very useful to the FSS specialist and the ARTCC meteorologist. The RRWDS display is similar to the color video display systems of private enterprise. It is connected to FAA and Air Force Air Route Surveillance radars as well as NWS weather radars (figure 1-2A). This gives briefers access to real-time radar weather information from areas of the country where it was previously not available.

Satellite Observations

Visible and infrared images of clouds are available from weather satellites in orbit. Satellite pictures are an important additional source of weather observations. GOES and NOAA satellite products are available through the facsimile network and directly from NWS Satellite Field Service Stations (SFSS). For more information on satellite products, see Section 3 "Satellite Pictures".

Upper Air Observations

Another important source of observed weather data is from radiosonde balloons and PIREPs. Upper air observations from radiosonde taken twice daily at specified stations furnish temperature, humidity, pressure, and wind, often to heights above 100,000 feet. Pilots themselves are a vital source of upper air weather observations. In fact, aircraft in flight are the only means of directly observing turbulence, icing, and the height of cloud tops.

Low Level Wind Shear Alert System (LLWAS)

This system provides pilots and controllers with information on hazardous surface wind conditions (on or near the airport) that creates unsafe landing or departure conditions. The system is a real-time, computer controlled, surface wind sensor system which evaluates wind speed and direction from sensors on the airport periphery with center field wind data. During the time that an alert is posted, air traffic controllers provide wind shear advisories to all arriving and departing aircraft.

NATIONAL OCEANIC AND ATMOSPHERIC ADMINISTRATION (NOAA)

NOAA collects and analyzes data and prepares forecasts on a national, hemispheric, and global basis. Following is a description of those facilities tasked with this duty.

National Environmental Satellite Data and Information Service (NESDIS)

The National Environmental Satellite Data and Information Service (NESDIS) located in Washington D.C. directs the weather satellite program and works in close cooperation with NWS meteorlogists at National Meteorological Center (NMC) and the Satellite Field Service Stations (SFSS). Satellite cloud photographs are available at field stations from NMC via facsimile or directly from a SFSS. Figures 3-4 and 3-5 are examples of GOES satellite pictures received from a SFSS.

National Meteorological Center (NMC)

The National Meteorological Center (NMC) of the NWS, located in Washington D.C. is the hub of weather processing. From worldwide weather reports it prepares guidance forecasts and charts of observed and forecast weather for use by various forecast facilities as described below. Many of the charts are computer prepared. Others are computer outputs adjusted and annotated by meteorologists. A few are manually prepared by forecasters.

Some NMC products are specifically for aviation. For example, NMC prepares the wind and temperatures aloft forecast. Figure 1-3 is the network of forecast winds and temperatures for the contiguous 48 states. Figure 1.3A shows the Alaskan and Hawaiian network of forecast winds and temperatures.

National Hurricane Center (NHC)

The NWS National Hurricane Center (NHC) located in Miami FL develops hurricane forecasting techniques and issues hurricane forecasts for the Atlantic, the Caribbean, the Gulf of Mexico and adjacent land areas. Hurricane warning centers at San Francisco and Honolulu issue warnings for the eastern and central Pacific.

National Severe Storms Forecast Center (NSSFC)

The NWS National Severe Storms Forecast Center (NSSFC) issues forecasts of severe convective

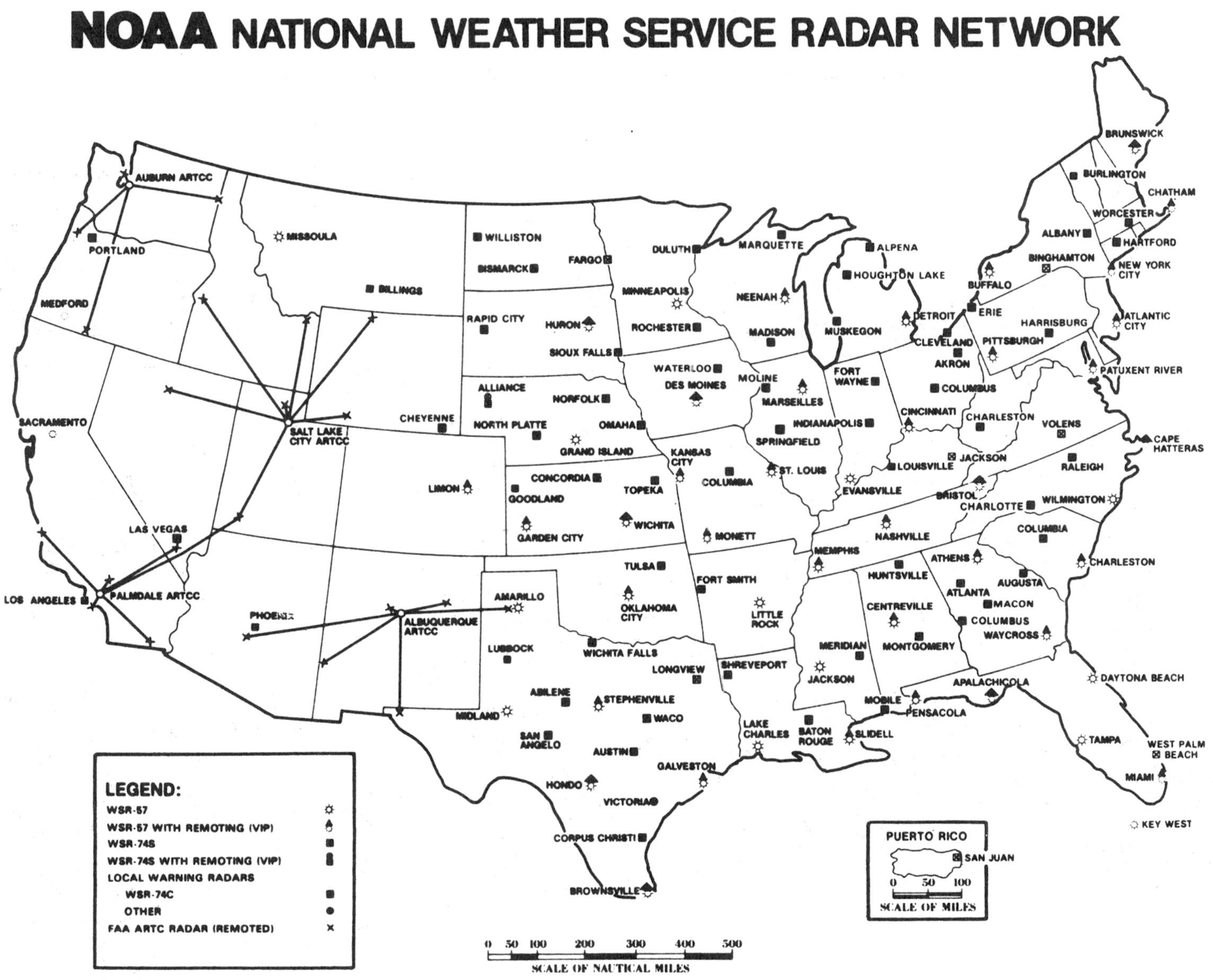

NOAA NATIONAL WEATHER SERVICE RADAR NETWORK

LEGEND:

- WSR-57
- WSR-57 WITH REMOTING (VIP)
- WSR-74S
- WSR-74S WITH REMOTING (VIP)
- LOCAL WARNING RADARS
  - WSR-74C
  - OTHER
- FAA ARTC RADAR (REMOTED)

FIGURE 1-2. The radar observing network.

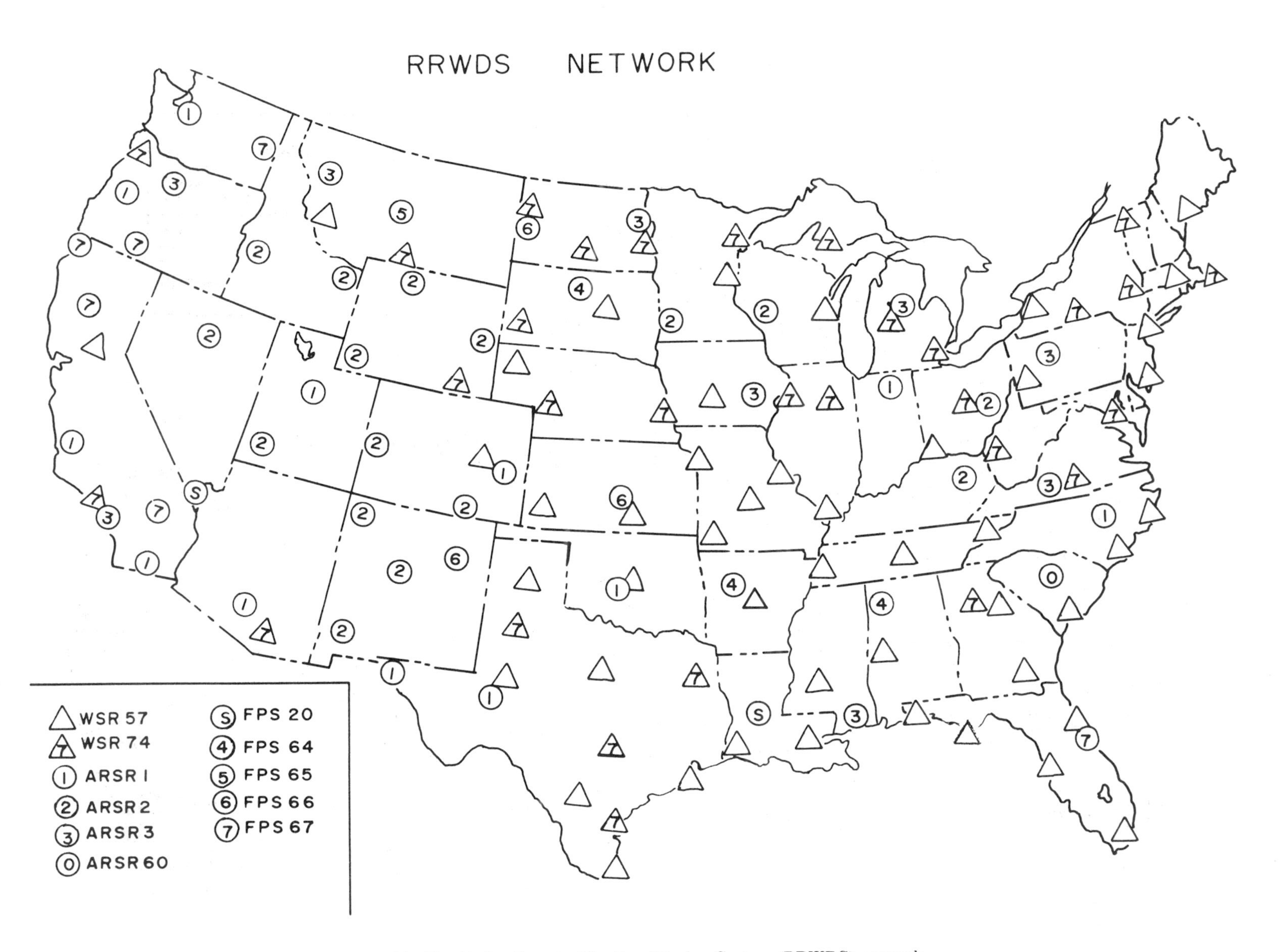

FIGURE 1-2A. The Radar Remote Weather Display System (RRWDS) network.

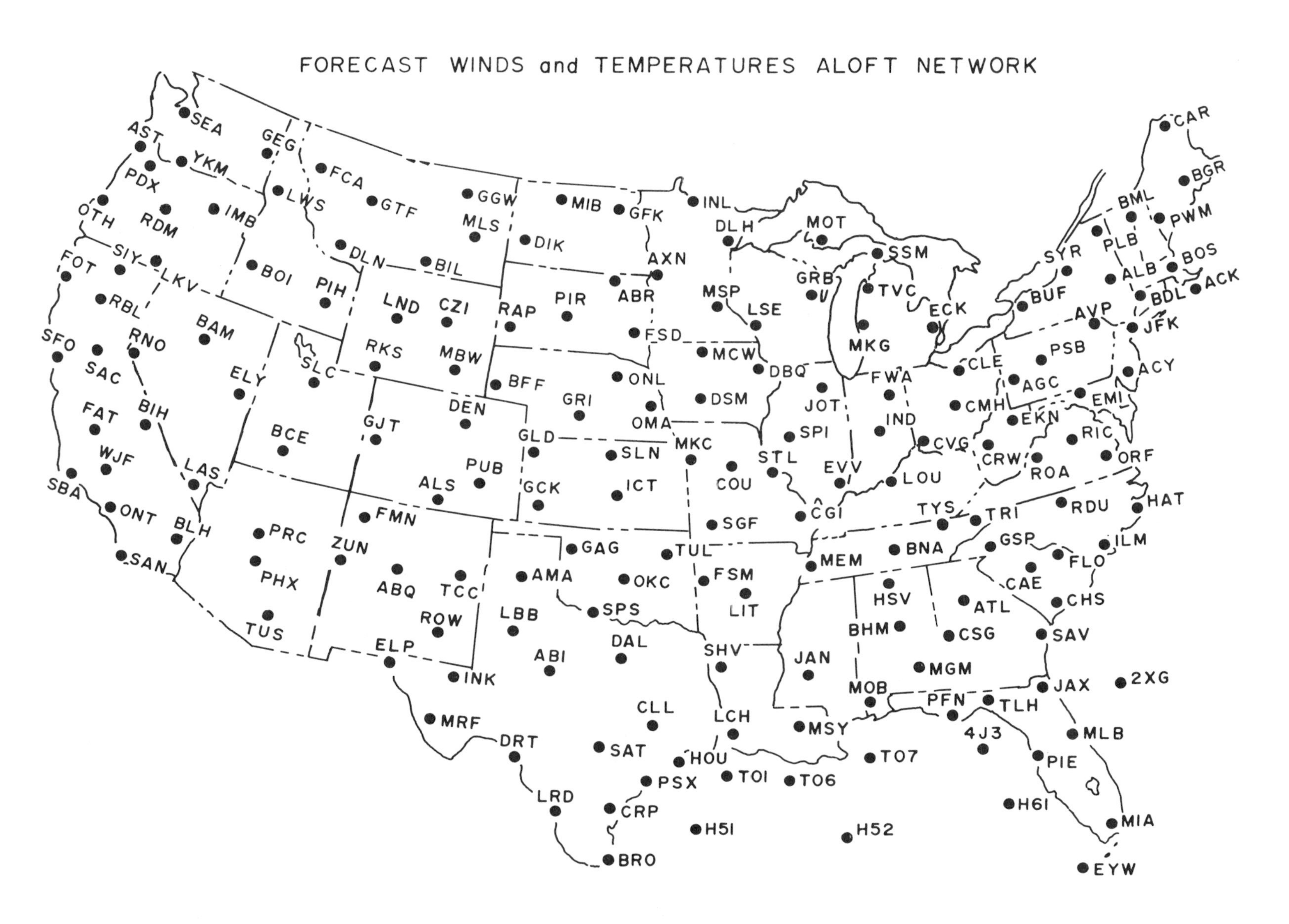

FIGURE 1-3. The forecast winds and temperatures aloft network.

FD LOCATIONS - ALASKA and HAWAII

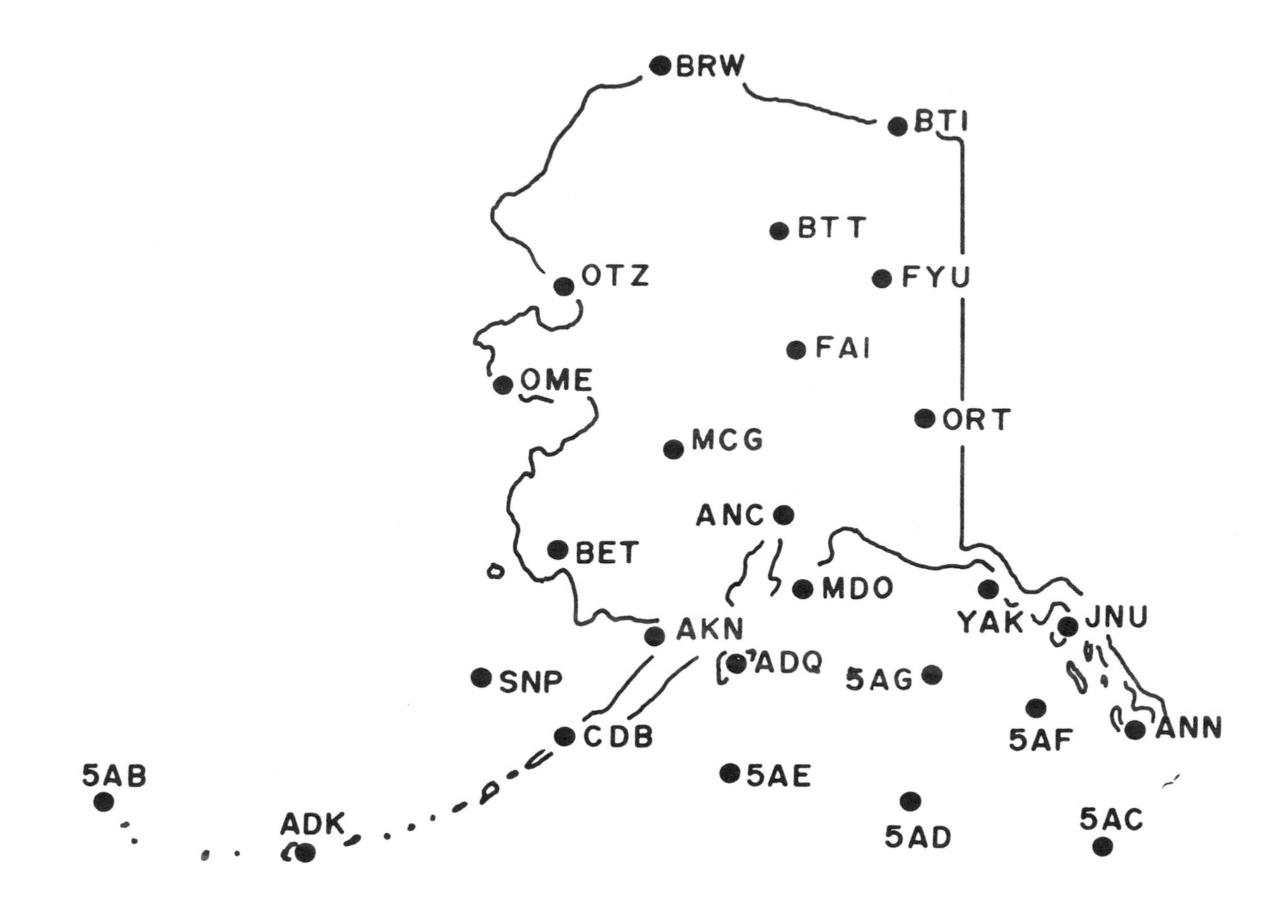

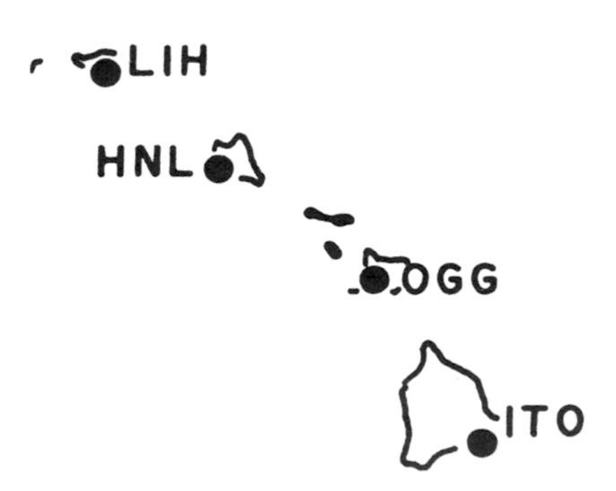

FIGURE 1-3A. The forecast winds and temperatures aloft network for Alaska and Hawaii.

storms, such as severe weather watches and convective outlooks, for the contiguous 48 states. It is located at Kansas City MO near the heart of the area most frequently affected by severe thunderstorms.

National Aviation Weather Advisory Unit (NAWAU)

This is a National Weather Service aviation dedicated unit located in Kansas City MO. Meteorologists in this unit prepare and issue area forecasts and inflight advisories (convective and non-convective SIGMETS and AIRMETS) for the contiguous 48 states (figure 1-5). These two products were formerly prepared and issued by designated WSFOs.

Weather Service Forecast Office (WSFO)

A Weather Service Forecast Office (WSFO) issues various public and aviation oriented forecasts and weather warnings for their area of responsibility. In support of aviation, products include terminal forecasts as well as Transcribed Weather Broadcast (TWEB) synopses and route forecasts. Figure 1-4 and 1-4A show locations of WSFOs and the airports for which each office prepares terminal forecasts. Figures 1-6 and 1-7 show TWEB routes.

Weather Service Office (WSO)

A Weather Service Office (WSO) prepares and issues public forecasts and warnings and provides general weather service for their local areas. A WSO

FIGURE 1-4. Locations of WSFOs, their areas of responsibility, and airports for which each prepares terminal forecasts.

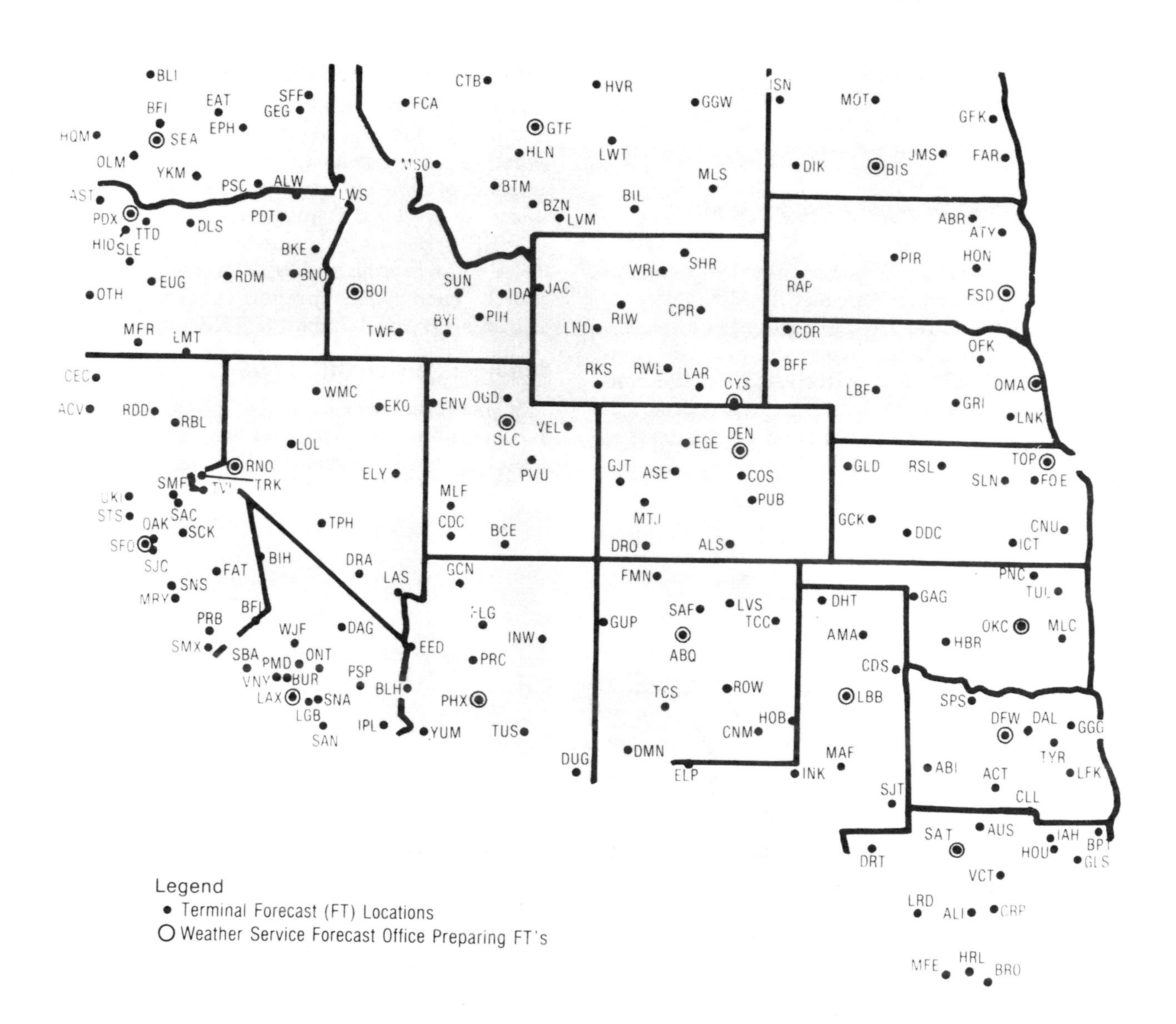

FIGURE 1-4. Continued.

may also amend terminal forecasts for a period of two hours or less when unexpected significant changes in the weather occur.

SERVICE OUTLETS

A weather service outlet is defined as any facility, either government or non-government, that provides aviation weather service. This section discusses only FAA and NWS outlets.

Flight Service Stations (FSS)

The FAA Flight Service Station (FSS) provides more aviation weather briefing service than any other government service outlet. It provides preflight and inflight briefings, makes scheduled and unscheduled weather broadcasts, and furnishes weather advisories to known flights in the FSS area. Because of the tremendous number of flight operations, selected FSSs also provide transcribed weather briefings. By listening to the recordings, you can assess further need for more detailed person to person briefing. There are two types of recordings - (1) Transcribed Weather Broadcast (TWEB) and (2) Pilot's Automatic Telephone Weather Answering Service (PATWAS).

The TWEB is a continous broadcast on low/medium frequencies (200 to 415 kHz) and selected VORs (108.0 to 117.95 MHz). The TWEB is based on a route of flight concept with the order and content of the TWEB transcription as follows:

1. Synopsis
2. Flight Precautions
3. Route Forecasts
4. Outlook (Optional)

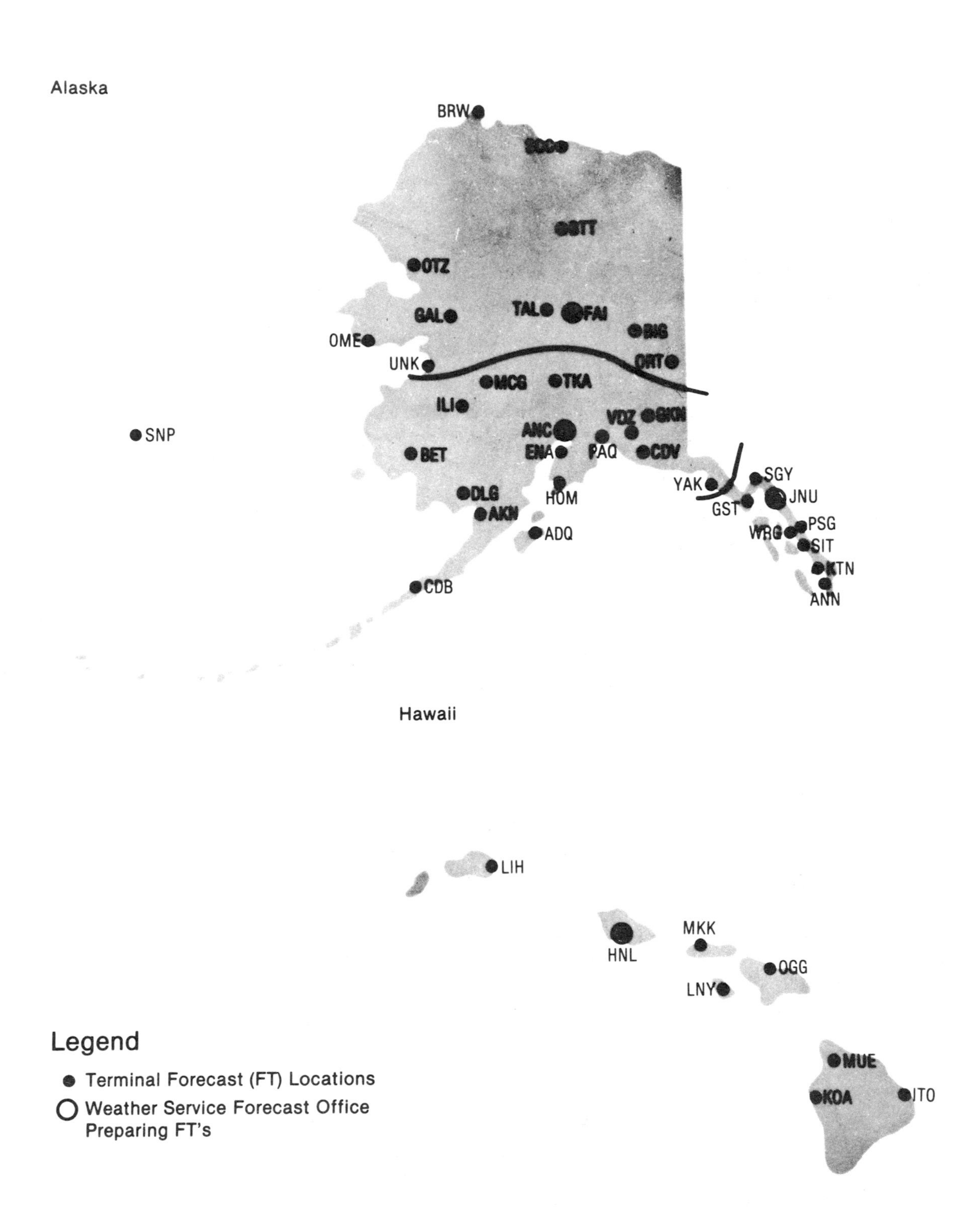

FIGURE 1-4A. Locations of WSFOs in Alaska and Hawaii, their areas of responsibility, and airpots for which each prepares terminal forecasts.

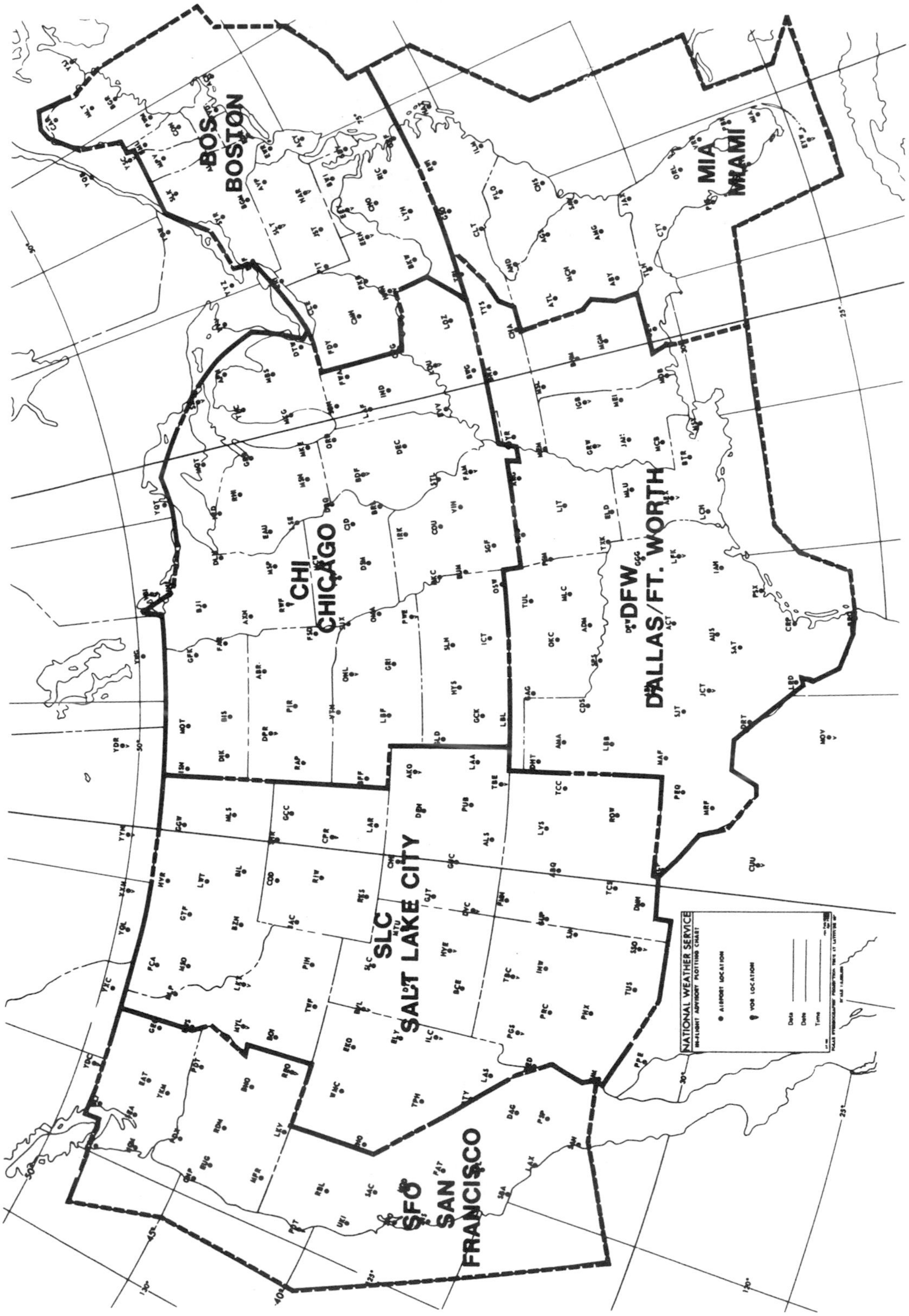

FIGURE 1-5. Locations of the area forecasts.

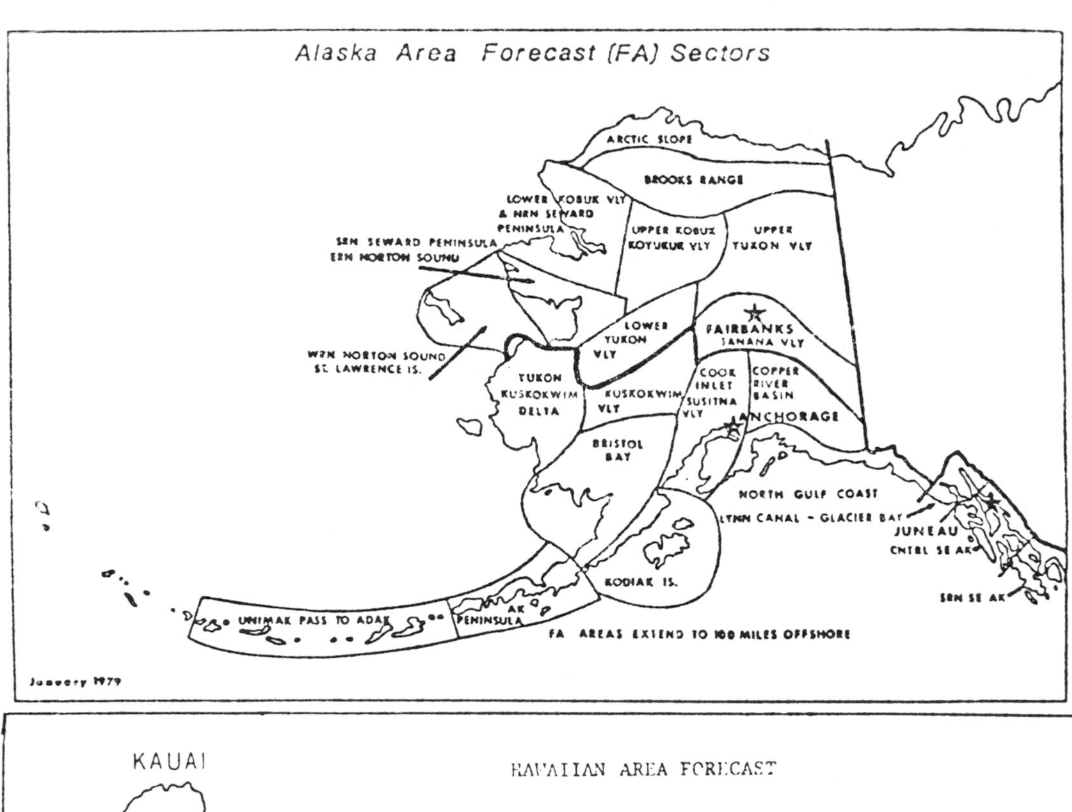

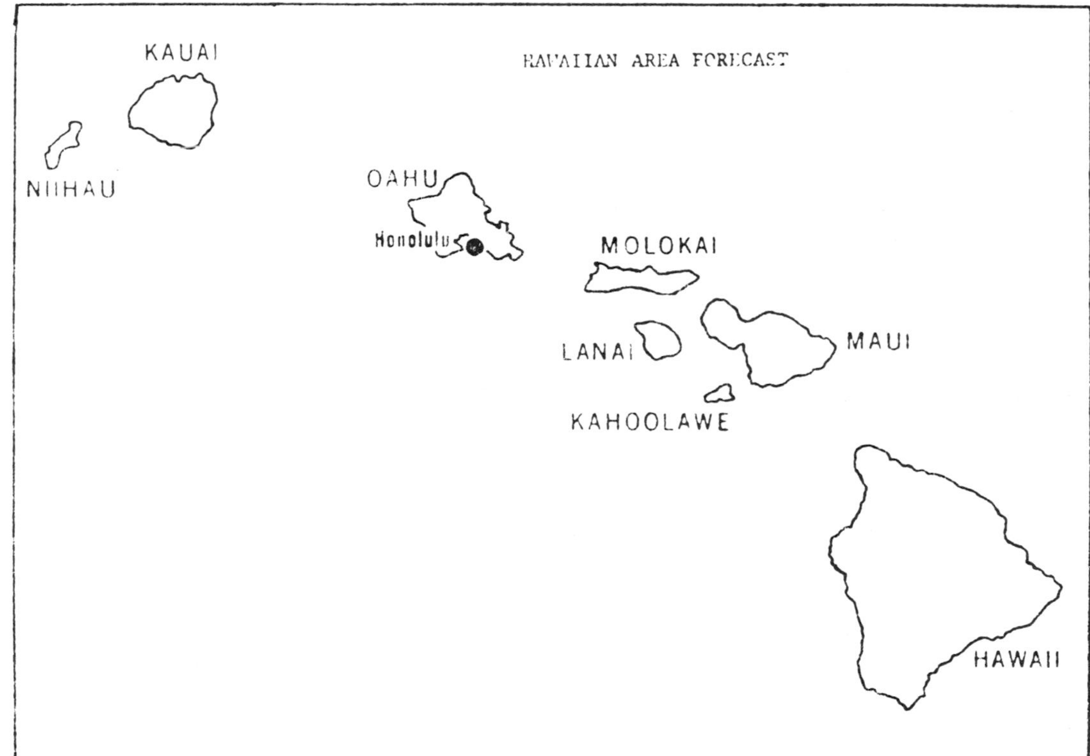

FIGURE 1-5A. Locations of the area forecasts in Alaska and Hawaii.

FIGURE 1-6. Numbered routes for which TWEB route forecasts are prepared.

5. Winds Aloft Forecast
6. Radar Reports
7. Surface Weather Reports
8. Pilot Reports
9. Notice to Airmen (NOTAMs)

The first five items are forecasts prepared by the NWS and are discussed in detail in Section 4. The synopsis and route forecast are prepared specifically for the TWEB by the WSFOs. Flight precautions, outlook and winds aloft are adapted respectively from inflight advisories, area forecasts and the NMC winds aloft forecast. Radar reports and pilot reports are discussed in Section 3. Surface reports are the subject of Section 2.

PATWAS is a recorded telephone briefing service with the forecast for the local area - usually within a 50 nautical mile radius of the station. A few selected stations also include route forecasts similar to the TWEB.

The order and content of the PATWAS recording are as follows:

1. Introduction (describing PATWAS area)
2. Adverse Conditions
3. Recommendation (VFR flight not recommended, if appropriate)
4. Synopsis
5. Current Conditions
6. Surface Winds

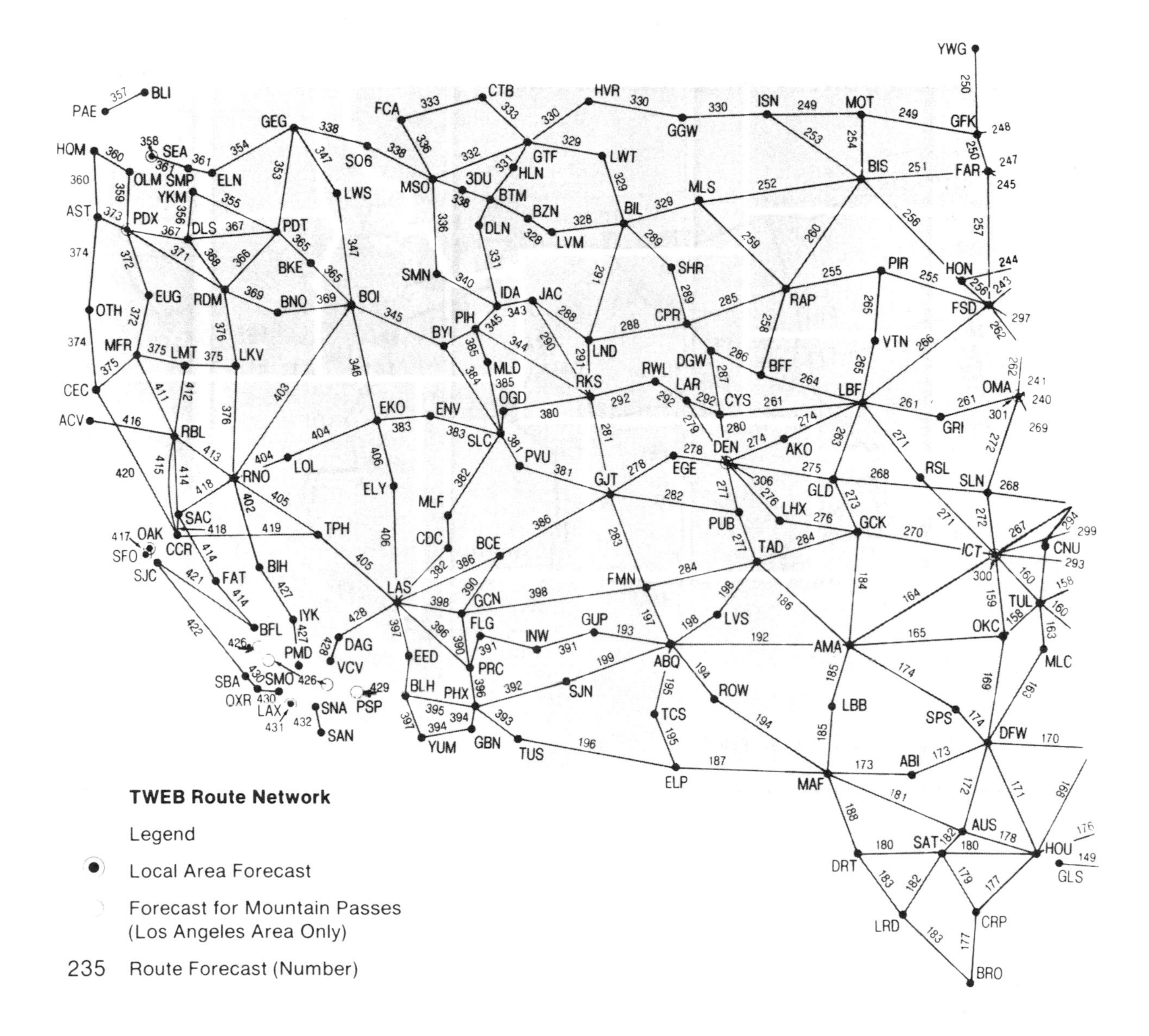

FIGURE 1-6. Continued.

7. Forecast
8. Winds Aloft
9. NOTAMS
10. Military Training Activity
11. Closing Announcements

FAA facilities providing PATWAS have operational procedures that place a high operational priority on PATWAS. This insures the information is current and accurate. Detailed PATWAS information is usually prepared at selected time intervals between 0500 and 2200 local time with updates issued as needed. A general outlook for the PATWAS area is available between 2200 and 0500 local time.

Figure 1-6 shows TWEB routes for which forecasts are prepared. Figure 1-7 shows cross-country TWEB routes. The Airman's Information Manual gives basic flight information and Air Traffic Control (ATC) procedures. The Airport Facility Directory lists PATWAS telephone numbers of FSS and NWS briefing offices.

The Hazardous In-flight Weather Advisory Service (HIWAS) is a continuous broadcast service of in-flight weather urgent PIREPs, CWAs, and AWWs over selected VORs. Also, hazardous weather not yet covered by an advisory will be included. In areas where HIWAS is already being utilized, controllers and specialists have discontinued their routine broadcast of in-flight advisories but continue broadcasting a short alerting message.

The Enroute Flight Advisory Service (Flight Watch) is a weather service on a common frequency

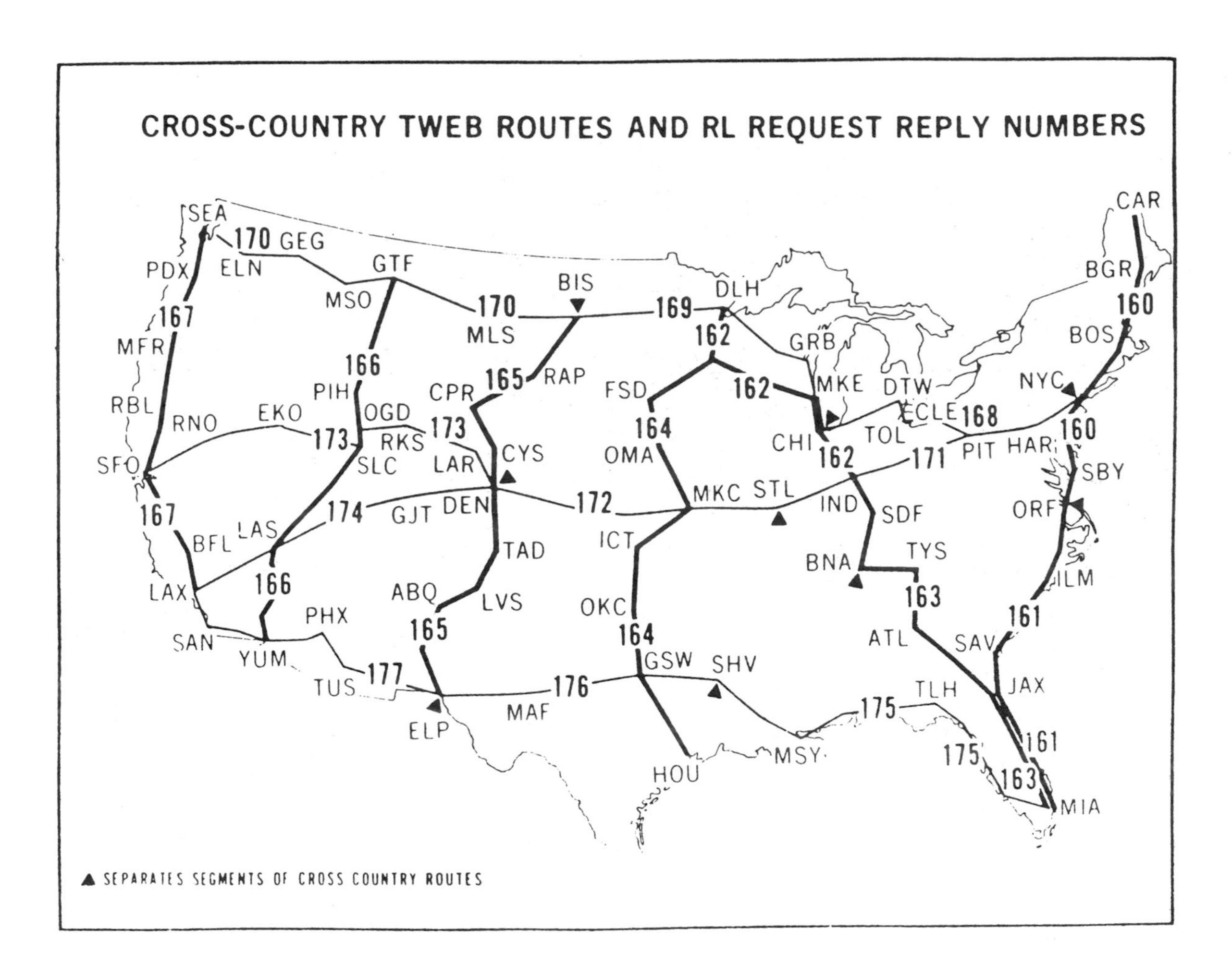

FIGURE 1-7. Cross country numbered routes for which route forecasts are available.

of 122.0 MHz from selected FSSs. The Flight Watch specialist maintains a continuous weather watch, provides time-critical assistance to enroute pilots facing hazardous or unknown weather, and may recommend alternate or diversionary routes. Additionally, Flight Watch is a focal point for rapid receipt and dissemination of pilot reports. Figure 1-8 indicates the sites where EFAS and associated outlets are located. To avail yourself of this service, call "FLIGHT WATCH" on 122.0 MHz (example "JACKSONVILLE FLIGHT WATCH, THIS IS. . .".

Air Traffic Control Command Center (ATCCC)

This operational facility is located at FAA headquarters in Washington D.C. Its objective is to manage the flow of air traffic on a system-wide basis, to minimize delays by watching capacity and demand, and to achieve maximum utilization of the airspace. Because weather is the overwhelming reason for air traffic delays and reroutings, this facility is supported by full-time NWS meteorologists whose function is to advise ATCCC flow controllers by continuously monitoring the weather throughout the system and anticipating weather developments that might affect system operations.

Air Route Traffic Control Center (ARTCC)

An ARTCC is a radar facility established to provide air traffic control service to aircraft operating on IFR flight plans within controlled airspace and principally during the enroute phase of flight.

Center Weather Service Unit (CWSU)

All FAA facilities within an ARTCC boundary are supported by a CWSU. This unit is a joint agency aviation weather support team located at each ARTCC. The unit is composed of National Weather Service (NWS) meteorologists and FAA controllers, the latter being assigned as Weather Coordinators. The primary task of the CWSU meteorologist is to provide FAA facilities within the ARTCC area of responsibility with accurate and timely weather information. This information is based on a continuous analysis and interpretation of - (1) real-time weather data at the ARTCC through the use of radar, satellite and PIREPs and (2) various NWS products such

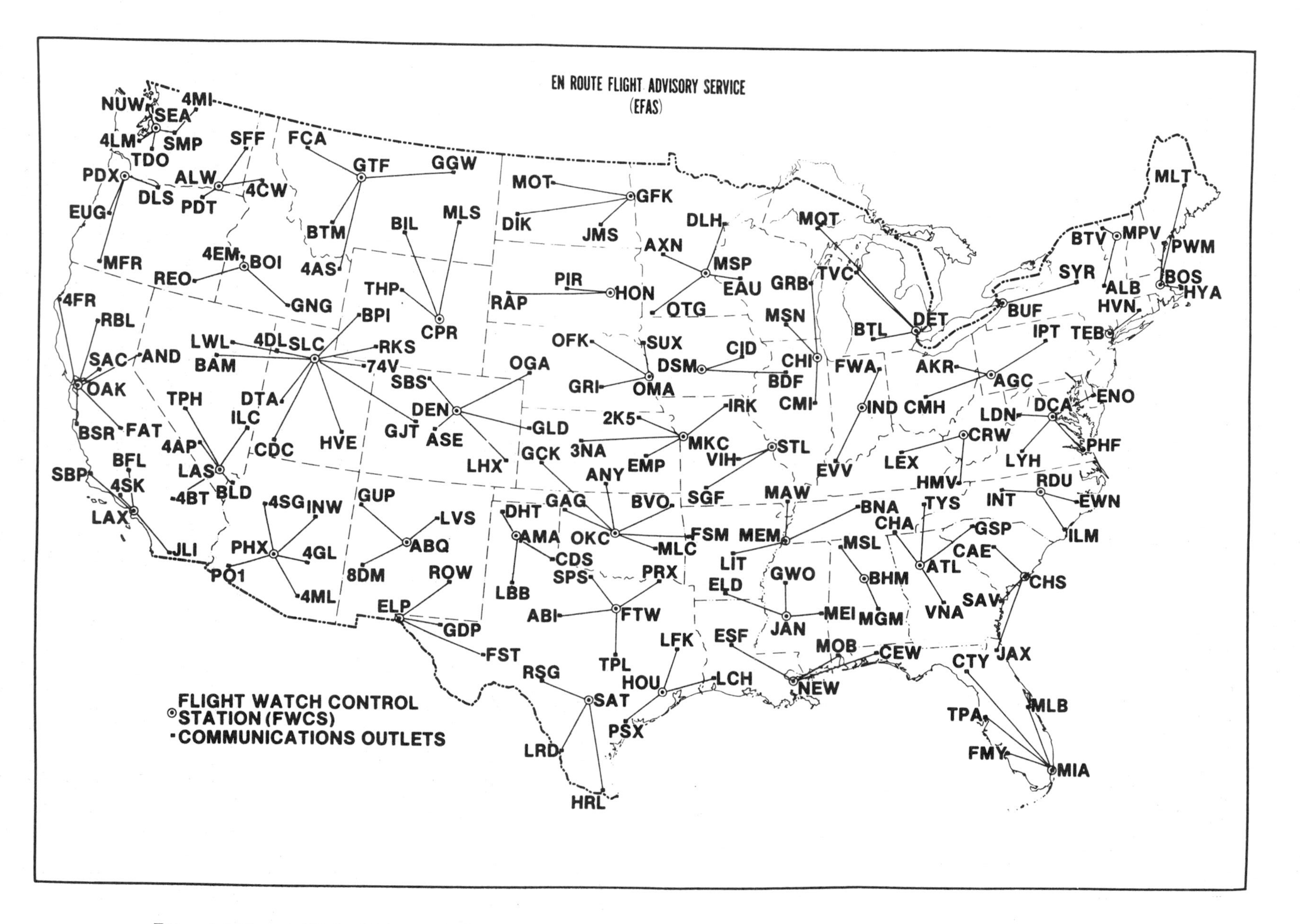

FIGURE 1-8. Enroute Flight Advisory (Flight Watch Facilities). An aircraft at 5,000 feet can receive a transmission to a distance of about 80 miles from any central or remote site.

as terminal and area forecasts, in-flight advisories, etc. The flow or exchange of weather information between the CWSU meteorologists and the FAA facilities is the primary task of the Weather Coordinator.

Similar to CWSU in the ARTCCs, there is a Central Flow Weather Service Unit (CFWSU) located in the Central Flow Control Facility (CFCF) in the ATCCC. The on-duty meteorologist in the CFWSU has the responsibility of the weather coordination on the national level.

Terminal Control Facility

The FAA terminal controller informs arriving and departing aircraft of pertinent local weather conditions. The controller becomes familiar with and remains aware of current weather information needed to perform air traffic control duties in the vicinity of the terminal. The responsibility for reporting visibility observations is shared with the NWS at many ATCT facilities. At other tower facilities, the controller has the full responsibility for observing, reporting, and classifying aviation weather elements.

Automatic Terminal Information Service (ATIS) is provided at most major airports to inform pilots, as they approach the terminal area, of the current weather and other pertinent local airport information.

Weather Service Office

NWS Offices provide weather briefings in areas not served by Flight Service Stations and provide local warnings to aviation. They furnish backup assistance to FAA service outlets.

Weather Service Forecast Office

NWS Forecast Offices provide some selective pilot briefings and supply backup service to FAA outlets. When getting a briefing from a FSS, you may, if necessary, request a telephone "patch in" to the WSFO forecaster. A few WSFOs also make and record PATWAS.

AM WEATHER

A fifteen minute weather program is broadcast Monday through Friday mornings nationally on approximately 250 Public Broadcast Television Stations.

Professional meteorologists from the National Weather Service and the National Environment Satellite Data and Information Service provide weather information primarily for pilots to enable them to make a better go or no-go flight decision.

National and Regional Weather Maps are provided, along with satellite sequences, radar reports, winds aloft, radar reports, and weatherwatches. Extended forecasts are provided daily and on Fridays to cover the weekend. AM WEATHER broadcasts also serve many other interest areas that depend upon forecasts.

The program draws upon the U.S. weather observation network, on geostationary and polar orbiting satellite data, and on computer analysis to produce daily forecasts with 85 to 90% accuracy.

COMMUNICATION SYSTEMS

As noted earlier, high speed communications and automated data processing have improved the flow of weather data and products through the aviation weather network. Examining figure 1-1, the flow of weather information between the NOAA weather facilities is accomplished through the Automation of Field Operations and Services (AFOS) communications system. Alphanumeric and graphic products are displayed on a cathode ray tube (CRT), similar to a television screen, which eliminates the need for the slower teletypewriters and facsimile machines.

The flow of alphanumeric weather information to the FAA Service Outlets is accomplished through the Leased Terminal Equipment (LTE) which also displays data on a CRT and so eliminates the need for teletypewriters. The Leased Terminal Equipment is replacing the Leased Service A System (LSAS) in Figure 1.1.

Exchange of weather information between the NWS and FAA Service Outlets is generally accomplished in two ways. Graphic products (weather maps) are received by FAA Service Outlets from NMC in Washington, D.C. over a low speed facsimile system. Alphanumeric information is exchanged through the Weather Message Switching Center (WMSC) in Kansas City, MO. This switching facility serves as the gateway for the flow of alphanumeric information from one communication system to another (i.e. between the various FAA facilities, NWS, and other users).

USERS

The ultimate users of the aviation weather service are pilots and dispatchers. Maintenance personnel also may use the service in protecting idle aircraft against storm damage. As a user of the service, you also contribute to it. Send pilot weather reports (PIREPs) to help your fellow pilots, briefers and forecasters. The service can be no better or more complete than the information that goes into it.

In the interest of safety, you should get a complete briefing before each flight. If you have L/MF radio, you can get a preliminary briefing by listening to the TWEB at your home or place of business. If you don't have a radio and PATWAS is available, dial PATWAS for a briefing. Many times the weather situation may be complex and you may not completely comprehend the recorded message. If you need additional information after listening to the TWEB or PATWAS, you should contact the FSS or WSO for a more complete briefing tailored for your specific flight.

How to Get a Good Weather Briefing

When requesting a briefing, make known you are a pilot. Give clear and concise facts about your flight:

1. Type of flight VFR or IFR
2. ACFT Ident or pilot's name
3. ACFT type
4. Departure point
5. Route-of-flight
6. Destination
7. Altitude
8. Estimated time of departure
9. Estimated time en route or EST time of arrival

With this background, the briefer can proceed directly with the briefing and concentrate on weather relevant to your flight.

The weather briefing you receive depends on the type requested. A STANDARD briefing should include:

1. Hazardous weather if any (you may elect to cancel at this point)
2. Weather synopsis (positions and movements of lows, highs, fronts, and other significant causes of weather)
3. Current weather
4. Forecast weather (enroute and destination)
5. Forecast winds aloft
6. Alternate routes (if any)
7. Aeronautical information (NOTAMs)

An ABBREVIATED briefing will be provided when the user requests information 1) to supplement mass disseminated data, 2) to update a previous briefing or 3) be limited to specific information.

An OUTLOOK briefing will be provided when the briefing is six or more hours in advance of proposed departure. Briefing will be limited to applicable forecast data for the proposed flight.

The FSSs and WSOs are to serve you. You should not hesitate to discuss factors that need elaboration or to ask questions. You have a complete briefing only when *you* have a clear picture of the weather to expect. It is to your advantage to make a final weather check immediately before departure if at all possible.

Request/Reply Service

The request/reply service is available at all FSSs, WSOs and WSFOs. You may request through the service any reports or forecasts not routinely available at your service outlet. These include route forecasts used in TWEB and PATWAS, recorder briefings and RADAR plots (see figure 3-3). You can request a forecast for any numbered route shown in figure 1-6 or any of the longer cross-country routes shown in figure 1.7.

Have an Alternate Plan of Action

When weather is questionable, get a picture of expected weather over a broader area. Preplan a route to take you rapidly away from the weather if it goes sour. When you fly into weather through which you cannot safely continue, you must act quickly. Without preplanning, you may not know the best direction to turn; a wrong turn could lead to disaster. A preplanned diversion beats panic. Better be safe than sorry.

Section 2

SURFACE AVIATION WEATHER REPORTS

When an observation is reported and transmitted, it is a weather *report.* A surface aviation weather report contains some or all of the following elements:

1. Station designator
2. Type and time of report
3. Sky condition and ceiling
4. Visibility
5. Weather and obstructions to vision
6. Sea level pressure
7. Temperature and dew point
8. Wind direction, speed, and character
9. Altimeter setting
10. Remarks and coded data

Those elements not occurring at observation time or not pertinent to the observation are omitted from the report. When an element should be included but is unavailable, the letter "M" is transmitted in lieu of the missing element. Those elements that are included are transmitted in the above sequence.

Following are five (5) reports as transmitted. These reports are used in discussing the above 10 elements. If you have this reference in a loose leaf binder, you will find it helpful to remove this page and keep it before you as you proceed through the discussion.

INK SA 1854 CLR 15 106/77/63/1112G18/000
BOI SA 1854 150 SCT 30 181/62/42/1304/015
LAX SA 1852 7 SCT 250 SCT 6HK 129/60/59/
2504/991
MDW RS 1856 -X M7 OVC 1 1/2R+F
990/63/61/3205/
980/RF2 RB12
JFK RS 1853 W5 X 1/4F 180/68/64/1804/006/
R04RVR22V30 SFC VSBY 1/2

STATION DESIGNATOR

The station designator is the three-letter location identifier for the reporting station. These five reports are from Wink TX (INK), Boise ID (BOI), Los Angeles CA (LAX), Chicago Midway Airport IL (MDW), and John F. Kennedy Airport, New York City NY (JFK).

TYPE AND TIME OF REPORT

The two basic types of reports are:

1. Record observation (SA), reports taken on the hour and
2. Special reports (RS or SP), observations taken when needed to report significant changes in weather.

Record observations (SA) are transmitted in sequenced collectives and are indentified by sequence headings. The first three reports are of this type (INK, BOI and LAX). A record special is a record observation that reports a significant change in weather. It is identified by the letters "RS" as shown in the reports from MDW and JFK. A special "SP" is an observation taken other than on the hour to report a significant change in weather. All reports transmitted must convey the time in Greenwich Mean Time and the type of observation.

SKY CONDITION AND CEILING

A clear sky or a layer of clouds or obscuring phenomena *aloft* is reported by one of the first seven *sky cover designators* in table 2-1. A layer is defined as clouds or obscuring phenomena with the base at approximately the same level. Height of the base of a layer precedes the sky cover designator. Height is in hundreds of feet *above ground level.*

Note that INK is reporting sky clear. No height precedes the designator since no sky cover is reported. BOI reports a scattered layer at 15,000 feet above the station. Figures 2-1 and 2-2 illustrate single layers of scattered clouds.

When more than one layer is reported, layers are in ascending order of height. For each layer above a lower layer or layers, the sky cover designator for that layer represents the *total sky* covered by that layer and all lower layers. In other words, the summation concept of cloud layers is used. LAX reports two layers, a scattered layer at 700 feet and a higher layer at 25,000 feet. Total coverage of the two layers does not exceed 5/10 coverage, so the upper layer also is reported as scattered. Figure 2-3 and 2-4 illustrate cloud over of multiple layers.

286-959 - 91 - 2 : QL 3

TABLE 2-1. Summary of sky cover designators

Designator	Meaning		Spoken
CLR	Clear. (Less than 0.1 sky cover.)		CLEAR
SCT	Scattered layer Aloft. (0.1 through 0.5 sky cover.)		SCATTERED
BKN*	Broken Layer Aloft. (0.6 through 0.9 sky cover.)		BROKEN
OVC*	Overcast Layer Aloft. (More than 0.9, or 1.0 sky cover.)		OVERCAST
–SCT	Thin scattered.	At least ½ of the sky cover aloft is transparent at and beclow the level of the layer aloft.	THIN SCATTERED
–BKN	Thin Broken.		THIN BROKEN
–OVC	Thin Overcast.		THIN OVERCAST
X*	Surface Based Obstruction. (All of sky is hidden by surface based phenomena.)		SKY OBSCURED
–X	Surface Based Partial Obscuration. (0.1 or more, but not all, of sky is hidden by surface based phenomena.		SKY PARTIALLY OBSCURED

*Sky condition represented by this designator will constitute a ceiling layer.

"Transparent" sky cover is clouds or obscuring phenomena aloft through which blue sky or higher sky cover is visible. As explained in table 2-1, a scattered, broken or overcast layer may be reported as "thin". To be classified as thin, a layer must be half or more transparent, and remember that sky cover of a layer includes all sky cover below the layer. For example, if at LAX the sky had been visible through half or more of the total sky cover reported by the higher layer, the report would have been

LAX SA 1854 7 SCT 250-SCT etc.

Any phenomenon *based at the surface* and hiding all or part of the sky is reported as SKY OBSCURED* or SKY PARTIALLY OBSCURED* as explained in table 2-1. An obsuration or partial obscuration may be caused by precipitation, fog, dust, blowing snow, etc. No height value precedes the designator for partial obscurations since vertical visibility is not restricted overhead. A height value precedes the designator for a total obscuration and denotes vertical visibility into the phenomenon.

Ceiling is defined as:

1. Height of the lowest layer of clouds or obscuring phenomena aloft that is reported as broken or overcast and not classified as thin, or

*Descriptions in capital letters are the usual phraseology in which these reports are broadcast.

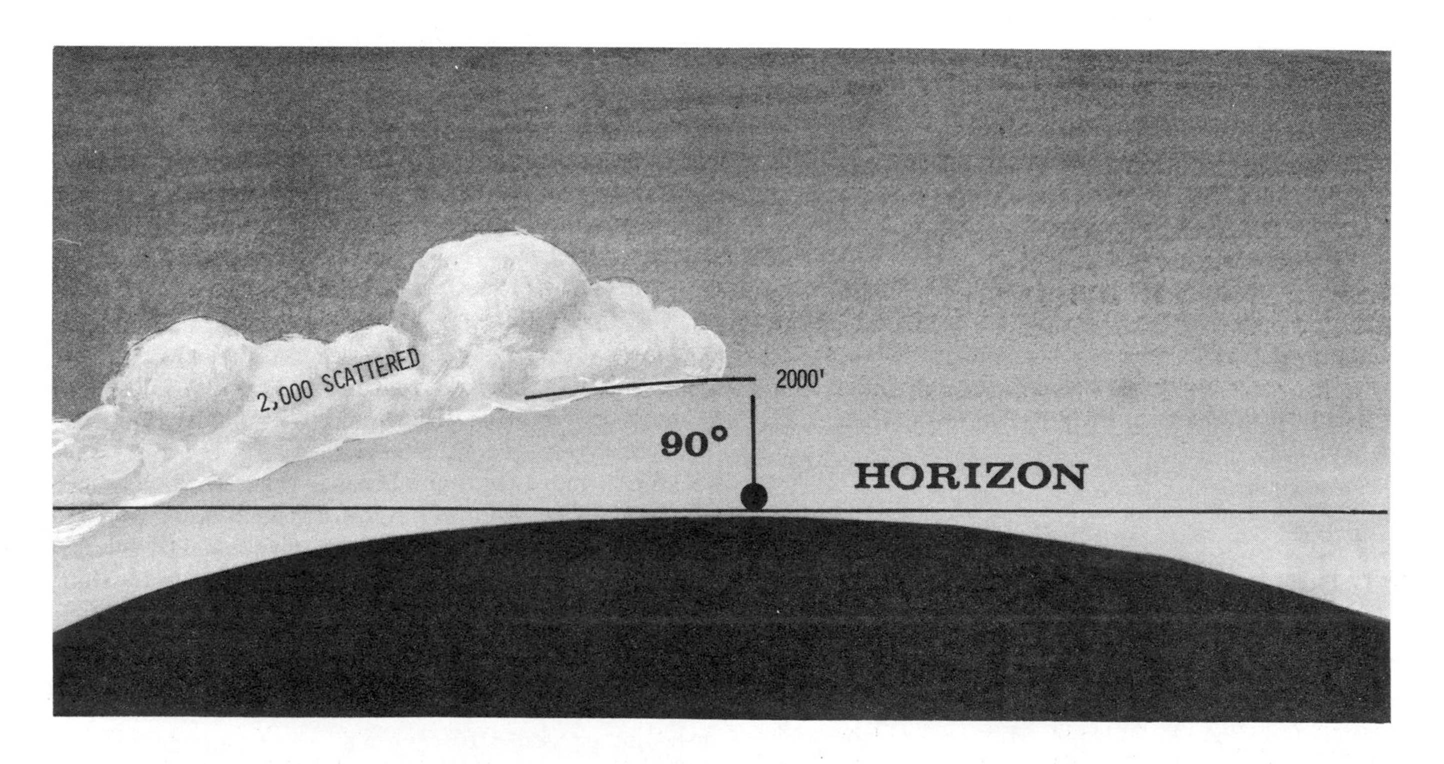

FIGURE 2-1. Scattered sky cover by a single advancing layer. Scattered is 5/10 or less sky cover (5/10 in this example).

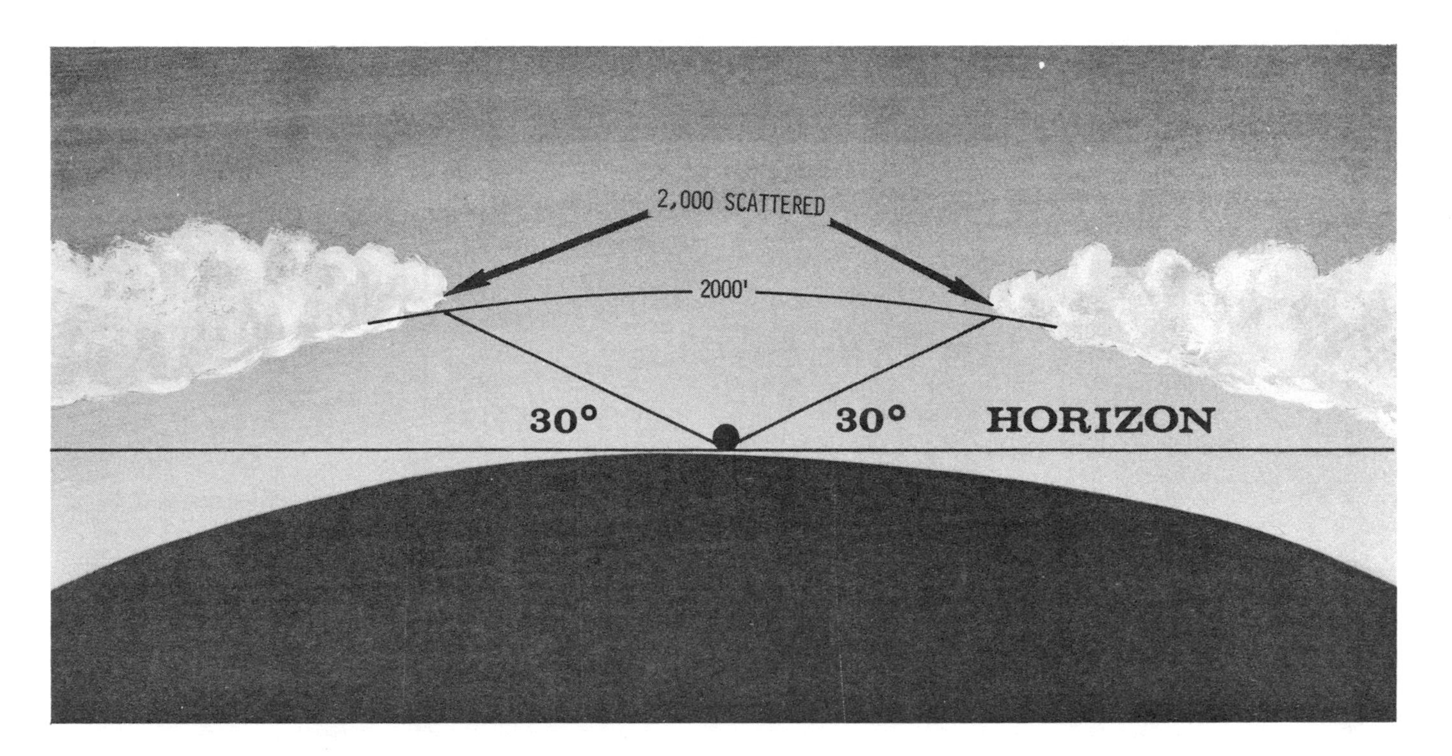

FIGURE 2-2. Scattered sky cover by a single layer surrounding the station (5/10 covered in this example).

2. Vertical visibility into surface-based obscuring phenomena that hides all the sky.

Now look at the reports from MDW and JFK. MDW reports a partial obscuration and an overcast at 700 feet. The overcast constitutes a ceiling at 700 feet. Note also that the height of this ceiling layer is preceded by the letter "M'. JFK reports a total obscuration and the height value preceding the sky cover designator represents 500 feet vertical visibility into the obscuring phenomenon. Height of the ceiling value is preceded by the letter "W". The "M" and "W" are "ceiling designators".

A *ceiling designator* always precedes the height of the ceiling layer. Table 2-2 lists and explains ceiling designators. At MDW the ceiling height was measured. JFK had an indefinite ceiling which was vertical visibility into a surface based total obscuration.

The sky cover and ceiling as determined from the ground represent as nearly as possible what the pilot should experience in flight. In other words, a pilot flying at or above the reported ceiling layer aloft should see less than half the surface below him. The pilot descending through a surface based total obscuration should first see the ground directly below him from the height reported as vertical visibility into the obscuration. However, because of the differing viewing points of the pilot and the observer, these surface reported values do not always exactly agree with what the pilot sees. Figure 2-6 illustrates the effect of an obscured sky on the vision from a descending aircraft.

TABLE 2-2. Ceiling designators

Coded	*Meaning*	*Spoken*
M	Measured. Heights determined by ceilometer, ceiling light, cloud detection radar, or by the unobscured portion of a landmark protruding into the ceiling layer. (Figure 2-5 illustrates the principle of the ceilometer.)	MEASURED CEILING
E	Estimated. Heights determined from pilot reports, balloons, or other measurements not meeting criteria for measured ceiling.	ESTIMATED CEILING
W	Indefinite. Vertical visibility into a surface based obstruction. Regardless of method of determination, vertical visibility is classified as an indefinite ceiling.	INDEFINITE CEILING

The letter "V" appended to the ceiling height indicates a variable ceiling. The range of variability is shown in remarks. Variable ceiling is reported only when it is critical to terminal operations. As an example,

M12V OVC and in remarks CIG10V13

means MEASURED CEILING ONE THOUSAND TWO HUNDRED VARIABLE OVERCAST, CEILING VARIABLE BETWEEN ONE

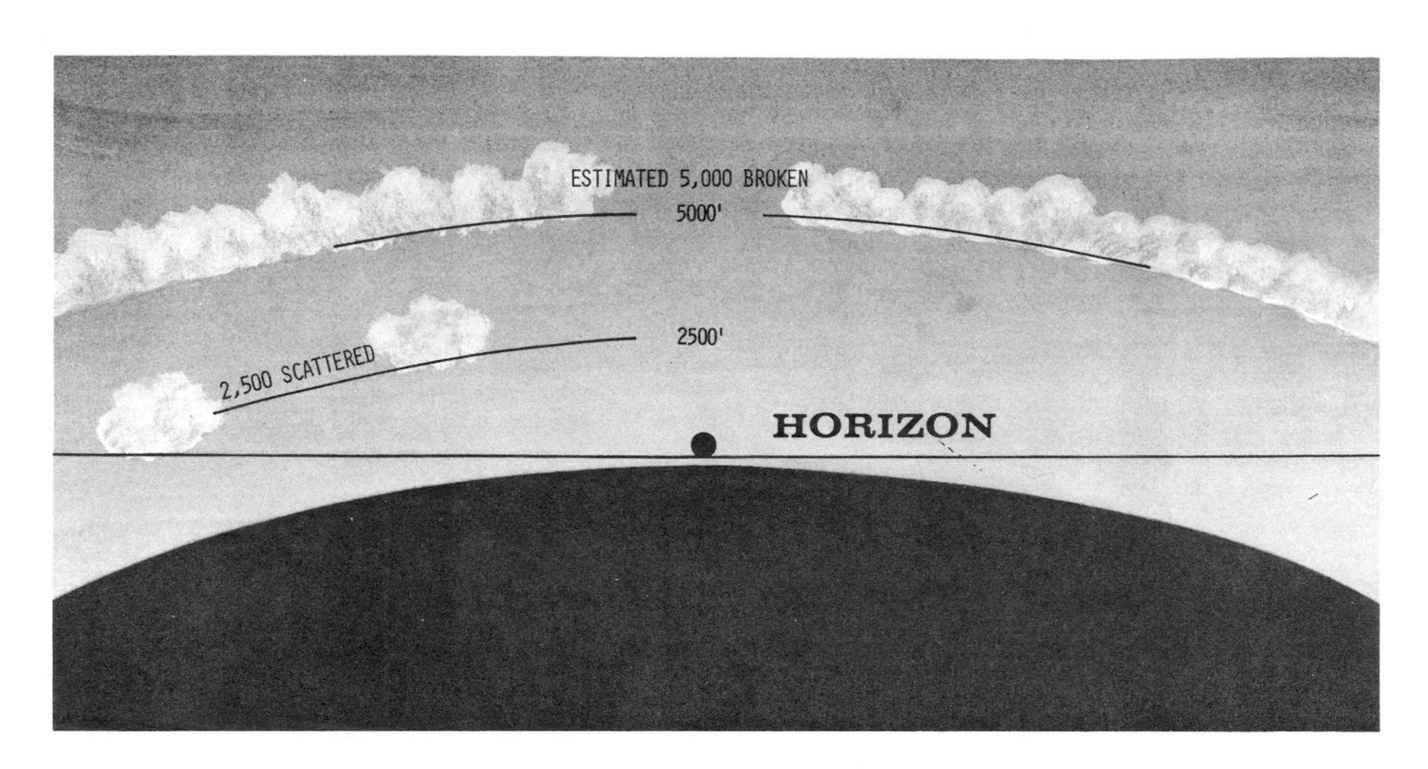

FIGURE 2-3. Summation of cloud cover in multiple layers.

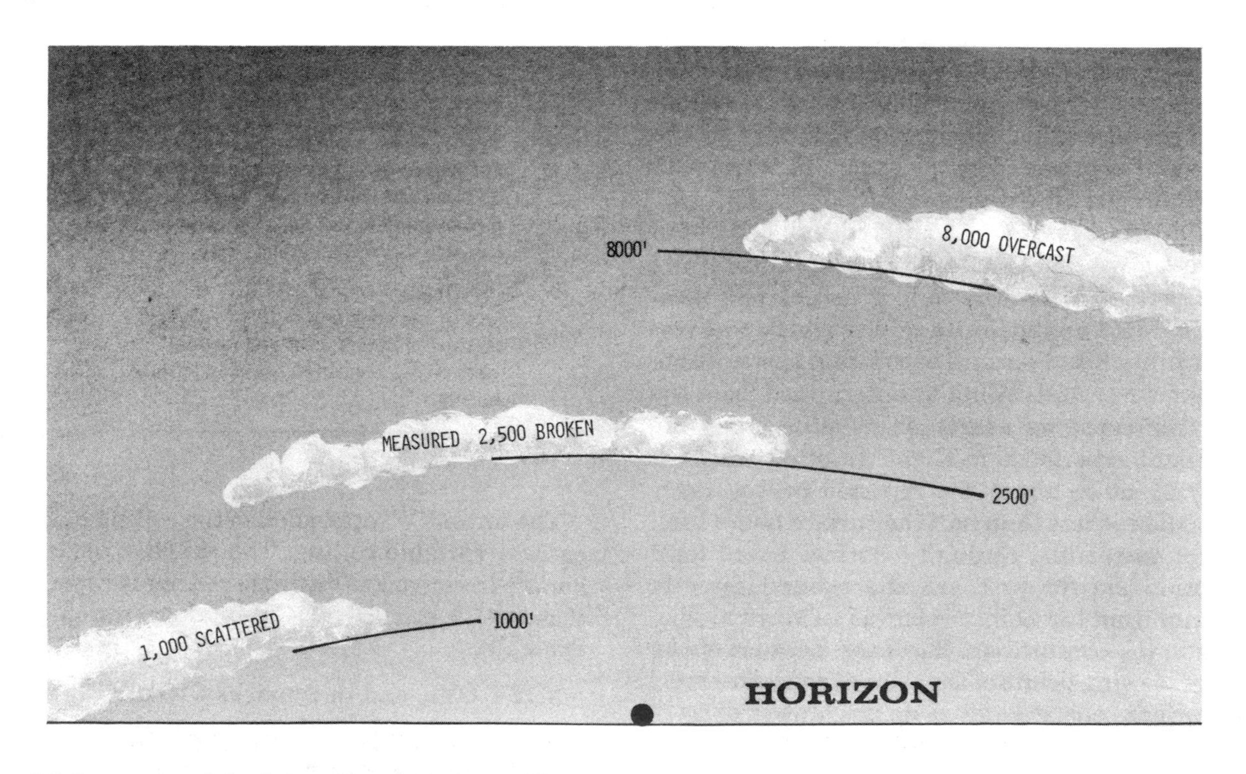

FIGURE 2-4. Summation of cloud cover in multiple layers. Note that at the height of the upper layer, sky cover is reported as overcast even though the upper layer itself covers less than 1/2 of the sky (10 SCT M25 BKN 80 OVC).

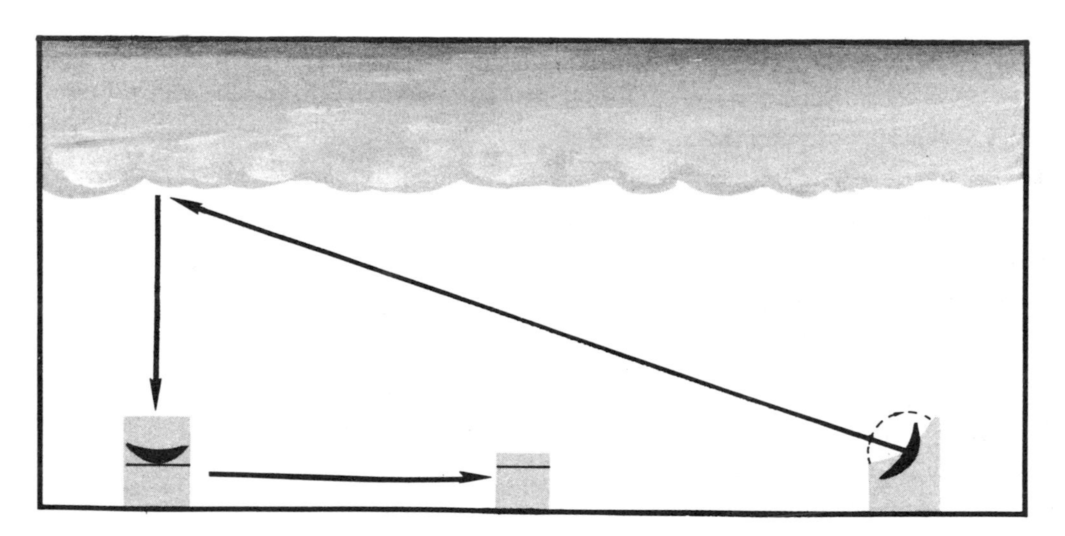

FIGURE 2-5. The rotating beam ceilometer. The projector beams a spot of modulated light on the cloud. The modulated light can be detected day or night. As the projector rotates, the spot moves along the cloud base. When the spot is directly over the detector, it excites a photoelectric cell measuring the angle of the light beam. Height of the cloud is then determined automatically by triangulation.

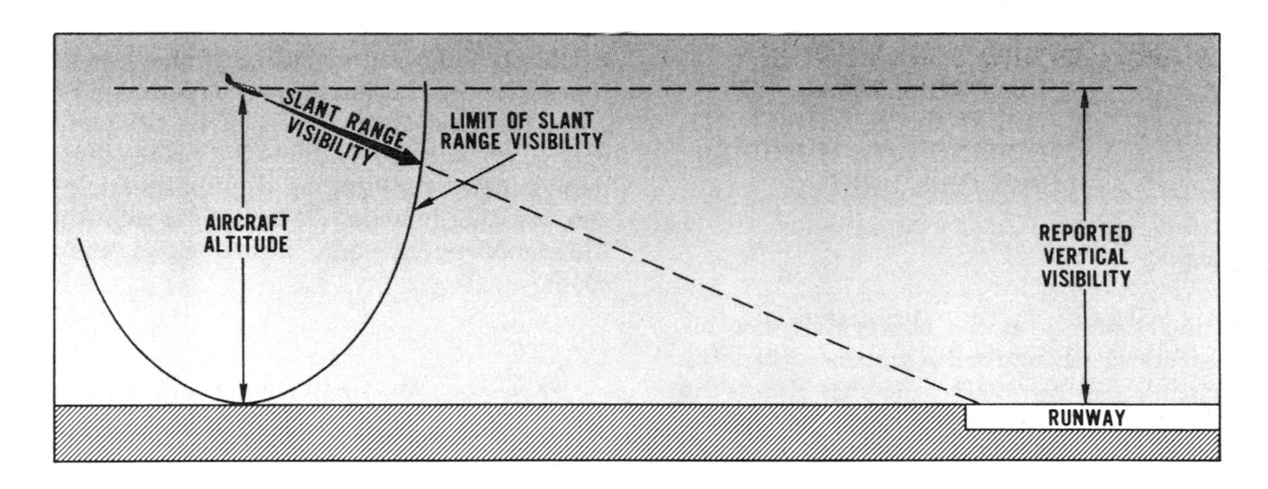

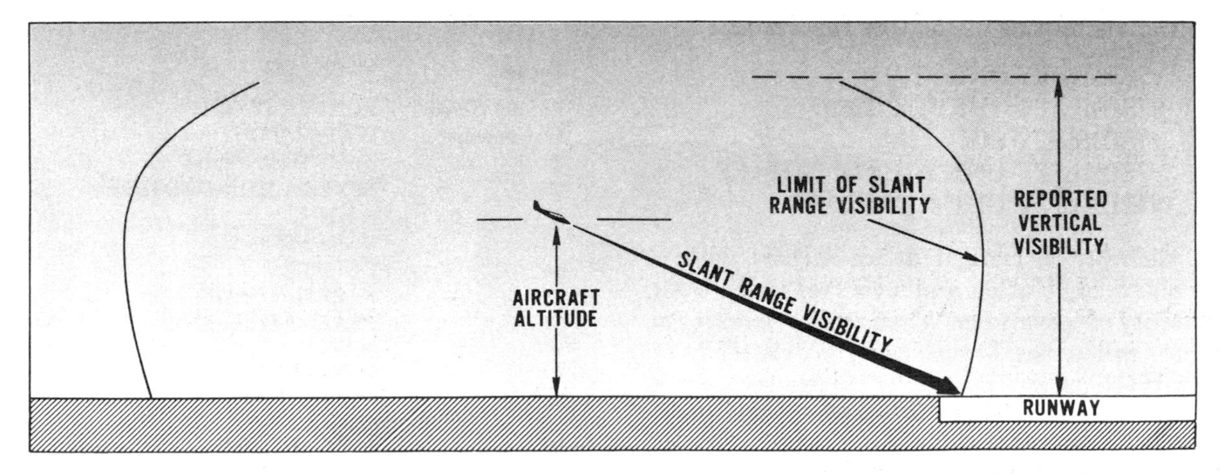

FIGURE 2-6. Vertical visibility is the altitude above the ground from which a pilot should first see the ground direclty below him (top). His real concern is slant range visibility which most often is less than vertical visibility. He usually must descend to a lower altitude (bottom) before he sees a representative surface and can fly by visual reference to the ground.

THOUSAND AND ONE THOUSAND THREE HUNDRED.

Now let's go back to our five reports and read them through sky and ceiling:

INK SA 1854 CLR	WINK, 1854 GREENWICH, CLEAR
BOI SA 1854 150 SCT	BOISE, 1854 GREENWICH, ONE FIVE THOUSAND SCATTERED
LAX SA 1852 7 SCT 250 SCT	LOS ANGELES, 1852 GREENWICH, SEVEN HUNDRED SCATTERED, TWO FIVE THOUSAND SCATTERED
MDW RS 1856 −X M7 OVC	CHICAGO MIDWAY, RECORD SPECIAL, 1856 GREENWICH, SKY PARTIALLY OBSCURED, MEASURED CEILING SEVEN HUNDRED OVERCAST
JFK RS 1853 W5 X	NEW YORK KENNEDY, RECORD SPECIAL, 1853 GREENWICH, INDEFINITE CEILING FIVE HUNDRED SKY OBSCURED

VISIBILITY

Prevailing visibility at the observation site immediately follows sky and ceiling in the report. Prevailing visibility is the greatest distance objects can be seen and identified through at least 180 degrees of the horizon. It is reported in statute miles and fractions.

Prevailing visibilities in the five reports are:

INK	VISIBILITY ONE FIVE
BOI	VISIBILITY THREE ZERO
LAX	VISIBILITY SIX
MDW	VISIBILITY ONE AND ONE-HALF
JFK	VISIBILITY ONE-QUARTER

When visibility is critical at an airport with a weather observing station and a control tower, both take visibility observations. When tower visibility is less than 4 miles, the lowest reported visibility of the 2 observations (surface, tower) is the prevailing visibility. The other is reported in remarks. Otherwise surface visibility is the reported prevailing visibility. Note that the report from JFK has a remark,

SFC VSBY1/2 meaning SURFACE VISIBILITY ONE-HALF.

The letter "V" suffixed to prevailing visibility denotes a variable visibility. The range of variability is shown in remarks. Variable visibility is reported only when critical to aircraft operations. As an example,

3/4V and in remarks VSBY1/2V1

means VISIBILITY THREE QUARTERS VARIABLE. . . VISIBILITY VARIABLE BETWEEN ONE-HALF AND ONE.

Visibility in some directions may differ significantly from prevailing visibility. These significant differences are reported in remarks. For example, prevailing visibility is reported as 11/2 miles with a remark,

VSBY NE21/2SW3/4

which means visibility to the northeast is 2-1/2 miles and to the southwest, it is 3/4 of a mile.

WEATHER AND OBSTRUCTIONS TO VISION

Weather and obstructions to vision when occurring at the station at observation time are reported immediately following visibility. If observed at a distance from the station, they are reported in remarks.

The term *weather* as used for this element refers only to those items in table 2-3 rather than to the more general meaning of all atmospheric phenomena. Weather includes all forms of precipitation plus thunderstorm, tornado, funnel cloud and waterspout.

TABLE 2-3. Weather symbols and meanings

Coded	*Spoken*
Tornado	TORNADO
Funnel Cloud	FUNNEL CLOUD
Waterspout	WATERSPOUT
T	THUNDERSTORM
T+	SEVERE THUNDERSTORM
R	RAIN
RW	RAIN SHOWER
L	DRIZZLE
ZR	FREEZING RAIN
ZL	FREEZING DRIZZLE
A	HAIL
IP	ICE PELLETS
IPW	ICE PELLET SHOWER
S	SNOW
SW	SNOW SHOWER
SP	SNOW PELLETS
SG	SNOW GRAINS
IC	ICE CRYSTALS

Precipitation is reported in one of three intensities. The intensity symbol follows the weather symbol with meanings as follows:

Light	−
Moderate	(no sign)
Heavy	+

No intensity is reported for hail (A) or ice crystals (IC).

A thunderstorm is reported as "T" and a severe thunderstorm as "T+". A *severe thunderstorm* is one in which surface wind is 50 knots or greater and/or surface hail is 3/4 inch or more in diameter.

Obstructions to vision include the phenomena listed in table 2-4. No intensities are reported for obstructions to vision.

TABLE 2-4. Obstructions to vision—symbols and meanings

Coded	*Spoken*
BD	BLOWING DUST
BN	BLOWING SAND
BS	BLOWING SNOW
BY	BLOWING SPRAY
D	DUST
F	FOG
GF	GROUND FOG
H	HAZE
IF	ICE FOG
K	SMOKE

Now referring back to our initial five reports, INK and BOI report no weather or obstructions to vision and no entries appear in the reports. Note at this point that by definition, obstructions to vision are only reported for visibilities of 6 miles or less whereas weather symbols will be used regardless of visibility. LAX reports two obstructions to vision, haze and smoke. MDW reports heavy rain as weather and fog as an obstruction to vision. JFK reports fog; is this weather or obstruction to vision?

There are two types of remarks concerning either a surface based obscuration or an obscuring phenomenon aloft. These remarks are discussed here.

When obscuring phenomenon is surface based and partially obscures the sky, a remark reports tenths of sky hidden. For example,

K6

means 6/10 of the sky is hidden by smoke. Now look at the report from MDW; how much of the sky is hidden and by what obscuring phenomenon?

Note the remark

RF2

which means 2/10 of the sky is hidden by rain and fog.

A layer of obscuring phenomenon aloft is reported in the sky and ceiling portion the same as a layer of cloud cover. A remark identifies the layer as an obscuring phenomenon. For example,

20 −BKN and a remark K20 −BKN

means a thin broken layer of smoke based at 2,000 feet above the surface.

SEA LEVEL PRESSURE

Sea level pressure is separated from the preceding elements by a space. It is transmitted in record hourly reports only. It is in three digits to the nearest tenth of a millibar with the decimal point omitted. Sea level pressure usually is greater than 960.0 millibars and less than 1050.0 millibars. The first 9 or 10 is omitted. To decode, prefix a 9 or 10 whichever brings it closer to 1000.0 millibars. Again going back to our five reports, sea level pressures are:

INK	1010.6 millibars
BOI	1018.1 millibars
LAX	1012.9 millibars
MDW	999.0 millibars
JFK	1018.0 millibars

TEMPERATURE AND DEW POINT

Temperature and dew point are in whole degrees Fahrenheit. They are separated from sea level pressure by a slash (/). If sea level pressure is not transmitted, temperature is separated from preceding elements by a space. Temperature and dew point are separated also by a slash. A minus sign precedes a temperature or dew point when below zero (0) degree F. From our five reports, we have:

INK...77/63	WINK...TEMPERATURE SEVEN SEVEN, DEW POINT SIX THREE
BOI...62/42	BOISE...TEMPERATURE SIX TWO, DEW POINT FOUR TWO
LAX...60/59	LOS ANGELES...TEMPERATURE SIX ZERO, DEW POINT FIVE NINER
MDW...63/61	CHICAGO MIDWAY...TEMPERATURE SIX THREE DEW POINT SIX ONE
JFK...68/64	NEW YORK KENNEDY...TEMPERATURE SIX EIGHT, DEW POINT SIX FOUR

Additional examples with minus values:

CAR. . . −4/−16 CARIBOU ME. . . TEMPERATURE MINUS FOUR, DEW POINT MINUS ONE SIX

FAR. . . 6/−8 FARGO ND. . . TEMPERATURE SIX, DEW POINT MINUS EIGHT

WIND

Wind follows dew point and is separated from it by a slash. Average one minute direction and speed are in four digits. The first two digits are direction *from* which the wind is blowing. It is in tens of degrees referenced to true north*, i.e., 01 is 10 degrees; 21 is 210 degrees; 36 is 360 degrees or north. The second two digits are speed in knots. A calm wind is reported as 0000.

If wind speed is 100 knots or greater, 50 is added to the direction code and the hundreds digit of speed is omitted. Example,

5908

means 090 degrees (09 + 50 = 59) at 108 knots.

A *gust* is a variation in wind speed of at least 10 knots between peaks and lulls. A *squall* is a sudden increase in speed of at least 15 knots to a sustained speed of 20 knots or more lasting for at least one minute. Gusts or squalls are reported by the letter "G" or "Q" respectively following the average one-minute speed and followed by the peak speed in knots. For example,

1522Q37

means wind 150 degrees at 22 knots with peak speed in squalls to 37 knots.

Winds decoded from our five reports are

INK WIND ONE ONE ZERO DEGREES AT ONE TWO PEAK GUSTS ONE EIGHT

BOI WIND ONE THREE ZERO DEGREES AT FOUR

LAX WIND TWO FIVE ZERO DEGREES AT FOUR

MDW WIND THREE TWO ZERO DEGREES AT FIVE

JFK WIND ONE EIGHT ZERO DEGREES AT FOUR

When any part of the wind report is *estimated* (direction, speed, peak speed in gusts or squalls), the letter "E" precedes the wind group. Example,

E1522G28

is decoded WIND ONE FIVE ZERO DEGREES ESTIMATED TWO TWO PEAK GUSTS ESTIMATED TWO EIGHT.

A few stations do not transmit sea level pressure, temperature and dew point; and these elements usually are not included in a special. When the elements are not transmitted, the wind group is separated from the preceding element by a space; i.e.,

CSM SP W5 X 2F 1705/990

is a special from Clinton-Sherman OK (CSM) *not* transmitting sea level pressure, temperature or dew point.

ALTIMETER SETTING

Altimeter setting follows the wind group and is separated from it by a slash. Normal range of altimeter settings is from 28.00 inches to 31.00 inches of mercury. The last three digits are transmitted with the decimal point omitted. To decode, prefix to the coded value either a 2 or a 3 whichever brings it closer to 30.00 inches. Examples.

996 means ALTIMETER TWO NINER NINER SIX, (29.96 inches)

013 means ALTIMETER THREE ZERO ONE THREE (30.13 inches)

REMARKS

Remarks, if any, follow altimeter setting separated from it by a slash. Certain remarks should be reported routinely and others the observer may include when considered significant to aviation. Often, some of the most important information in an observation may be the remarks portion.

Runway Visibility and Runway Visual Range

The first remark, when transmitted, should be runway visibility or runway visual range. Figure 2-7 illustrates the difference. The terms are defined as follows:

Runway visibility—the visibility from a particular location along an identified runway, usually determined by a transmissometer instrument. It is in miles and fractions of a mile. Figure 2-8 diagrams the principle of the transmissometer.

Runway visual range—the maximum horizontal distance down a specified instrument runway at which a pilot can see and identify standard high intensity runway lights. It is always determined using a transmissometer and is reported in hundreds of feet.

*Wind direction for the local station is *broadcast* in degrees magnetic.

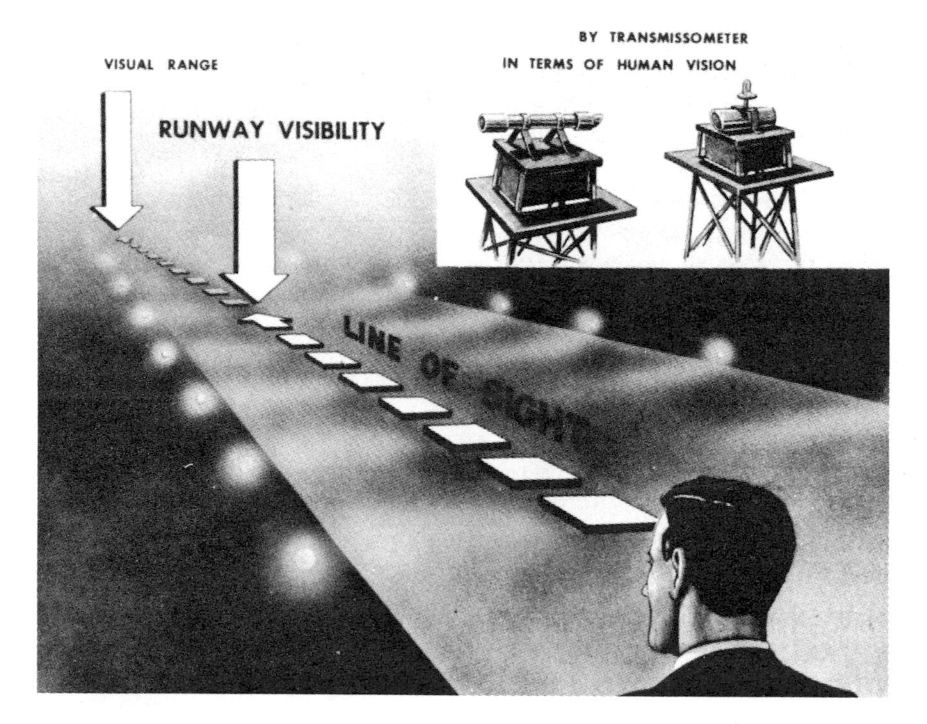

FIGURE 2-7. Difference between *runway visibility* and *runway visual range.* Runway visibility is the distance down the runway the pilot can see unlighted objects or unfocused lights of moderate intensity. Runway visual range is the distance he can see high intensity runway lights. Visual range usually is greater than visibility because the high intensity lights penetrate farther into the obscuring phenomena.

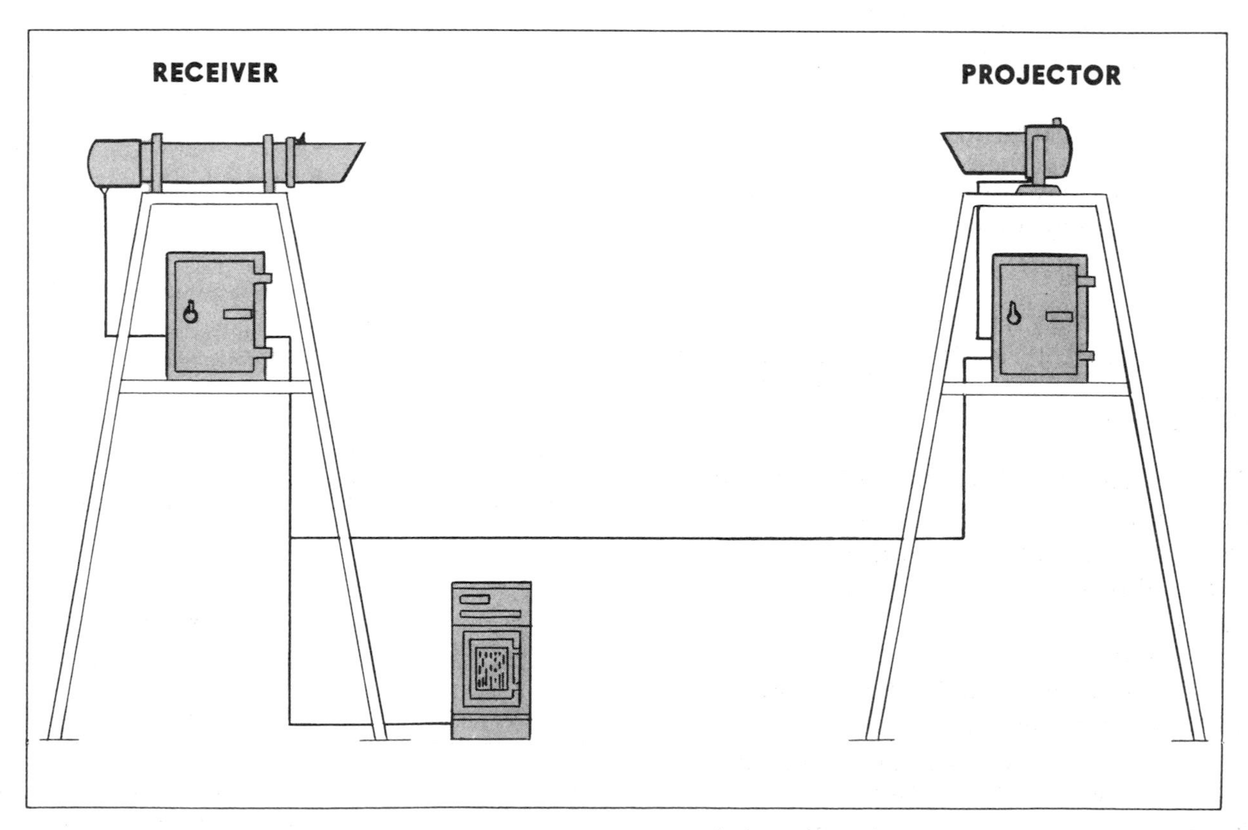

FIGURE 2-8. The transmissometer. The projector beams light toward the receiver. Obscuring phenomena in the path of the beam absorbs some of the light. A photoelectric cell in the receiver measures the amount of light penetrating through the obscuring phenomena. The amount received is converted into visibility.

The report consists of a runway designator and the contraction "VV" or "VR" followed by the appropriate visibility or visual range. Both the VV and the VR report are for a 10 minute period preceding observation time. The remark usually reports the 10 minute extremes separated by the letter "V". However, if the visual range or visibility has not changed significantly during the 10 minutes, a single value is sent indicating that the value has remained constant.

The following examples show several reports and their decoding:

R36VV11/2 RUNWAY THREE SIX, VISIBILITY VALUE ONE AND ONE-HALF. (Visibility remained constant during the 10 minutes period.)

R05LVV1V2 RUNWAY FIVE LEFT, VISIBILITY VALUE VARIABLE BETWEEN ONE AND TWO.

R18VR20V30 RUNWAY ONE EIGHT, VISUAL RANGE VARIABLE BETWEN TWO THOUSAND FEET AND THREE THOUSAND FEET.

R26RVR24 RUNWAY TWO SIX RIGHT, VISUAL RANGE TWO THOUSAND FOUR HUNDRED FEET. (Visual range remained constant during the 10 minute period.)

Runway visual range in excess of 6,000 feet is written 60+. VR less than the minimum value that can be observed by the instrument is encoded as the minimum suffixed by a minus sign. For example:

R36LVR10-V25

is decoded RUNWAY THREE SIX LEFT, VISUAL RANGE VARIABLE FROM LESS THAN ONE THOUSAND FEET TO TWO THOUSAND FIVE HUNDRED FEET.

Heights of Bases and Tops of Sky Cover Layers

Bases and tops of clouds or obscuring phenomena may be reported. These remarks originate from pilots (i.e., PIREPs). Therefore, heights are MSL. Examples,

/UA. . ./SK BKN 50
means top of broken layer 5,000 feet (MSL).

/UA. . ./SK OVC 30/60 OVC
means top of lower overcast 3,000 feet, base of higher overcast 6,000 feet (MSL).

PIREPs which are more than 15 minutes old are omitted unless it is considered to be operationally significant.

Clarification of Coded Data

Following, by category, are coded remarks clarifying or expanding on coded elements:

Sky and Ceiling

Coded Elements	*Coded Remarks*
FEW CU	Few cumulus clouds
HIR CLDS VSB	Higher clouds visible
BINOVC	Breaks in overcast
BRKS N	Breaks north
BKN V OVC	Broken layer variable to overcast
CIG 14V19	Ceiling variable between 1400 feet and 1900 feet
ACCAS ALQDS*	Altocumulus castellanus all quadrants
ACSL SW-NW*	Standing lenticular altocumulus southwest to northwest
ROTOR CLDS NW*	Rotor clouds northwest
VIRGA E-SE*	Virga (precipitation not reaching the ground) east through southeast
30 SCT V BKN	Scattered layer at 3000 feet variable to broken
SC BANK NW	Stratocumulus cloud bank northwest
TCU W*	Towering cumulus clouds west
CB N MOVG E*	Cumulonimbus north moving east
CBMAM OVHD-W*	Cumulonimbus mamma overhead to west
CONTRAILS N 420 MSL	Condensation trails north at 42,000 feet MSL
CLDS TPG MTNS SW	Clouds topping mountains southwest
RDGS OBSCD W-N	Ridges obscured west through north
CUFRA W APCHG STN	Cumulus fractus clouds west approaching station
LWR CLDS NE	Lower Clouds northeast

*These cloud types are highly significant and the observer should always report them. Figure 2-9 through 2-14 are photographs of these clouds and explains their significance. A pilot in flight should also report them when observed.

Obscuring Phenomena

Coded Elements	*Coded Remarks*
D5	Dust obscuring 5/10 of the sky
S7	Snow obscuring 7/10 of the sky
BS3	Blowing snow obscuring 3/10 of the sky

FK4	Fog and smoke obscuring 4/10 of the sky
K20 SCT	Scattered layer of smoke aloft based at 2,000 feet above the surface
THN F NW	Thin fog northwest from reporting station)

Visibility (Statute Miles)

VSBY S1W1/4	Visibility south 1, west 1/4
VSBY 1V3	Visibility variable between 1 and 3
TWR VSBY 3/4	Tower visibility 3/4
SFC VSBY 1/2	Surface visibility 1/2

Weather and Obstruction to Vision

Coded Elements	*Coded Remarks*
T W MOVG E FQT LTGCG	Thunderstorm west moving east, frequent lightning cloud to ground
RB30	Rain began 30 minutes after the hour
SB15E40	Snow began 15, ended 40 minutes after the hour
UNCONFIRMED TORNADO 15W OKC MOVG NE 2000	Unconfirmed tornado 15 (NM) west of Oklahoma City, moving northeast, sighted at 2000Z
T OVHD MOVG E	Thunderstorm overhead, moving east
OCNL DSNT LTG NW	Occassional distant lightning northwest
HLSTO 2	Hailstones 2 inches in diameter
INTMT R−	Intermittent light rain
DUST DEVILS NW	Dust devils northwest
OCNL RW	Occasional moderate rain shower
WET SNW	Wet Snow
SNOINCR 5	Snow increase 5 inches during past hour
R− OCNLY R+	Light rain occasionally heavy rain
RWU	Rain shower of unknown intensity
F DSIPTG	Fog dissipating
K DRFTG OVR FLD	Smoke drifting over field
KOCTY	Smoke over city
SHLW GFDEP 4	Shallow ground fog 4 feet deep
PATCHY GF S	patchy ground fog south

Wind

Coded Elements	*Coded Remarks*
WSHFT 30	Wind shifted at 30 minutes past the hour
WND 27V33	Wind variable between 270 degrees and 330 degrees
PK WND 3348/22	Peak wind within the past hour from 330 degrees at 48 knots occurred 22 minutes past the hour

Pressure

Coded Elements	*Coded Remarks*
PRESRR	Pressure rising rapidly
PRESFR	Pressure falling rapidly
LOWEST PRES 631 1745	Lowest pressure (sea level) 963.1 millibars at 1745 GMT
PRJMP 8/1012/18	Pressure jump (sudden increase) 0.08 inches began 1012 GMT, ended 1018 GMT

Freezing Level Data

Upper air (rawinsonde) observation stations append in remarks *freezing level data.* The coded remark is appended to the first record report transmitted after the information becomes available. Code for the remark is as follows:

RADAT UU (D) ($h_ph_ph_p$) ($h_ph_ph_p$) (/n)

(a) RADAT—a contraction identifying the remark as "freezing level data.".
(b) UU—relative humidity at the freezing level in percent. When more than one level is sent, "UU" is the highest relative humidity observed at any of the levels transmitted.
(c) (D)—a coded letter "L", "M", or "H" to indicate that relative humidity is for the "lowest", "middle", or "highest" level coded. This letter is omitted when only one level is sent.
(d) ($h_ph_ph_p$)—a height in hundreds of feet above MSL at which the upper air sounding crossed the zero (0) degree Celsius isotherm. No more than three levels are coded. If the sounding crosses the zero (0) degree Celsius isotherm more than three times, the levels coded are the lowest and the top two levels.

FIGURE 2-9. Towering Cumulus (TCU). The most direct significance of this cloud is that the atmosphere in the lower altitudes is unstable and conducive to turbulence.

FIGURE 2-10. Cumulonimbus (CB). The anvil portion of a CB is composed of ice crystals. The CB or thunderstorm cloud contains most types of aviation weather hazards; particularly turbulence, icing, hail, and low level wind shear (LLWS).

FIGURE 2-11. Cumulonimbus Mamma (CBMAM). This characteristic cloud can result from violent up and down currents and is often associated with severe weather. It indicates possible severe or greater turbulence.

FIGURE 2-12. Altocumulus Castellanus (ACCAS). ACCAS indicates unstable conditions aloft, but not necessarily below the base of the cloud. Note in this picture a surface based inversion shown by the trapped smoke, indicating stable conditions at the surface. Thus, rising air causing the ACCAS is originating somewhere above the surface based inversion. Compare with towering cumulus, a cloud representing unstable air and turbulence from the surface upward.

FIGURE 2-13. Virga. Virga is precipitation falling from a cloud but evaporating before reaching the ground. Virga results when air below the cloud is very dry and is common in the western part of the country. Virga associated with showers suggests strong downdrafts with possible moderate or greater turbulence.

FIGURE 2-14. Standing Lenticular Altocumulus (ACSL). These clouds are characteristic of the standing or mountain wave. A similar cloud is the Standing Lenticular Cirrocumulus (CCSL). CCSL are whiter and at higher altitude. Both are indicative of possible severe or greater turbulence.

(e) (/n)—indicator to show the number of crossings of the zero (0) degree Celsius isotherm, other than those coded. The indicator is omitted when all levels are coded.

Examples:

RADAT 87045	Relative humidity 87%, only crossing of zero (0) degrees C isotherm was 4,500 feet MSL.
RADAT 87L024105	Relative humidity 87% at the lowest (L) crossing. Two crossings occurred at 2,400 and 10,500 feet MSL.*
RADAT 84M019045051/1	Relative humidity 84% at the middle (M) crossing of the three coded crossings. Coded crossings were at 1,900, 4,500 and 5,100 feet. The 84% humidity was at 4,500 feet MSL. "/1" indicates one additional crossing and it was between 1,900 and 4,500 feet.
RADAT MISG	The sounding terminated below the first crossing of the zero (0) degree C isotherm—temperatures were all above freezing.
RADAT ZERO	The entire sounding was below zero (0) degree C.

*Temperature was below zero (0) degree C below 2,400 feet MSL; above zero (0) degrees C between 2,400 feet MSL and 10,500 feet MSL; and below zero (0) degrees C above 10,500 feet MSL.

Icing Data

When the rawinsonde observer determines definitely that icing is occurring on his instruments, he enters the data in the following code:

RAICG HHMSL (SNW)

(a) RAICG—indicates icing data follows.
(b) HH—height in hundreds of feet at which icing occurred. "MSL" is always appended to the height.
(c) (SNW)—used to indicate that snow is causing a reduced balloon ascension rate. (Omitted otherwise.)

Examples:

RAICG 12MSL—Icing at 1,200 feet MSL.

RAICG 24MSL SNW—Icing at 2,400 feet MSL in snow.

Other Information

A group or groups of numerically coded data may appear in remarks. These data are primarily of concern to the meteorologist and are not discussed here.

A printed arrow marks the end of weather information and signifies that the rest of the report is notice(s) to airmen (NOTAM). The NOTAM code is explained in the AIRMAN'S INFORMATION MANUAL.

REPORT IDENTIFIERS

A heading begins the record hourly collective on the local circuit identifying the type of message, the circuit number, data and time of observations making up the collective reports. For example,

SA21 271900

means surface aviation reports (SA), 21 is the circuit number, 27 is the day of the month, and the observations were made at 1900 GMT.

A slightly different heading begins each relay. It identifies the location of reporting stations by states and indicates the time the relay began. For example,

MO20192

means the relay is from Missouri, day of the month is the 20th (20), time of observations is 1900 GMT (19), and the relay began 12 minutes past the hour (12).

Relay designators other than states are INTERMEDIATE EAST, FAR EAST, NEAR NORTH, etc. The relay collectives are assembled by a centralized computer and are unique to each circuit.

Individual reports must each convey the time and type of report. Following are examples:

Example 1

INK SA 1100. . . . .

indicates a relayed report from Wink, Texas for 1100 GMT (all times transmitted in teletypewriter reports are GMT). The "SA" signifies a record hourly.

Example 2

INK SA COR 1100. . . . .

signifies a correction to the 1100 GMT record hourly report as originally transmitted. The correction may transmit the complete corrected report or it may contain only the corrected element or elements.

Example 3

INK SP 2315. . . . .

indicates a special report of an observation taken at 2315 GMT to report a significant change in weather.

Example 4

INK SP COR 2315. . . . .

indicates a correction to the above special report.

Example 5

BVO SW 1130. . . . .

indicates a Supplemental Aviation Weather Reporting Station (SAWRS) report by the contraction "SW". SAWRS reports are unscheduled and are made by non-Government observers at airports not served by a regularly reporting weather station. Observations are taken during commercial aircraft operations. Type and time are transmitted. This report was from Bartlesville OK at 1130 GMT.

READING THE SURFACE AVIATION WEATHER REPORT

Now that we have studied the individual elements and their decoding, let's read completely each of the five reports. Capitalized phrases are those elements which *normally* are broadcast by the station at or near the airport where the observation was made:

INK SA 1854 CLR 15 106/77/63/1112G18/000

WINK, WINK, 1854 GREENWICH, CLEAR, VISIBILITY ONE FIVE, pressure 1010.6 millibars, TEMPERATURE SEVEN SEVEN, DEW POINT SIX THREE, WIND ONE ONE ZERO DEGREES AT ONE TWO PEAK GUSTS ONE EIGHT, ALTIMETER THREE ZERO ZERO ZERO.

BOI SA 1854 150 SCT 30 181/62/42/1304/015

BOISE, BOISE, 1854 GREENWICH, ONE FIVE THOUSAND SCATTERED, VISIBILITY THREE ZERO, pressure 1018.1 millibars, TEMPERATURE SIX TWO, DEW POINT FOUR TWO, WIND ONE THREE ZERO DEGREES AT FOUR, ALTIMETER THREE ZERO ONE FIVE.

LAX SA 1852 7 SCT 250 SCT 6HK 129/60/59/2504/991

LOS ANGELES, LOS ANGELES, 1852 GREENWICH, SEVEN HUNDRED SCATTERED, TWO FIVE THOUSAND SCATTERED, VISIBILITY SIX, HAZE, SMOKE. pressure 1012.9 millibars, TEMPERATURE SIX ZERO, DEW POINT FIVE NINER, WIND TWO FIVE ZERO DEGREES AT FOUR, ALTIMETER TWO NINER NINER ONE.

MDW RS 1856 −X M7 OVC 1 1/2 R+F 990/63/61/3205/980/RF2 RB12

CHICAGO, CHICAGO MIDWAY, RECORD SPECIAL, 1856 GREENWICH, SKY PARTIALLY OBSCURED, MEASURED CEILING SEVEN HUNDRED OVERCAST, VISIBILITY ONE AND ONE-HALF, HEAVY RAIN, FOG, pressure 999.0 millibars, TEMPERATURE SIX THREE, DEW POINT SIX ONE, WIND THREE TWO ZERO DEGREES AT FIVE, ALTIMETER TWO NINER EIGHT ZERO, TWO TENTHS SKY OBSCURED BY RAIN AND FOG, rain began 12 minutes past the hour.

JFK RS 1853 W5 X 1/4F 180/68/64/1804/006/R04RVR22V30 SFC VSBY 1/2

NEW YORK, NEW YORK KENNEDY, RECORD SPECIAL, 1853 GREENWICH, INDEFINITE CEILING FIVE HUNDRED SKY OBSCURED, VISIBILITY ONE-QUARTER, FOG, pressure 1018.0 millibars, TEMPERATURE SIX EIGHT, DEW POINT SIX FOUR, WIND ONE EIGHT ZERO DEGREES AT FOUR, ALTIMETER THREE ZERO ZERO SIX, RUNWAY FOUR RIGHT VISUAL RANGE VARIABLE BETWEEN TWO THOUSAND TWO HUNDRED FEET AND THREE THOUSAND FEET, SURFACE VISIBILITY ONE HALF.

AUTOMATED SURFACE OBSERVATIONS

Automatic Meteorological Observing Station AMOS

The AMOS is a solid-state system capable of automatically observing temperature, dew point, wind direction and speed, pressure (altimeter setting), peak wind speed, and precipitation accumulation. The field sensors are tied in directly to the FAA observation network. It transmits a weather report whenever the station is polled by the circuits. At a staffed AMOS, the observer can manually enter additional information to give a more complete observation. Tables 2-1, 2-2, 2-3 and 2-4 are used to interpret sky condition and weather. Visibility is in statute miles. The intensity symbols for precipitation are light (−), moderate (no sign), and heavy (+).

Figure 2-15 is the breakdown of an unstaffed AMOS.

Example:

PGO AMOS 76/61/0308/007 PK WND 18 013
which decodes as follows:

Observation from Page OK, temperature 76 degrees F, dew point 61 degrees F, wind from 030 degrees at 8 knots, altimeter setting 30.07 inches, peak wind since the last hourly observation 18 knots, 13 hundredths of an inch of liquid precipitation since last synoptic observation.

Figure 2-16 is the breakdown of a staffed AMOS.
Example:

GLS AMOS SA 1356 E12 BKN 150 OVC 7TRW-F 79/77/1103/004 PK WND 15 008/TB32 NW MOVG W

MDO AMOS 33/29/3606/975 PK WND 08 001

MDO	STATION IDENTIFICATION: (Middleton Island AK) Identifies report using FAA identifiers.
AMOS	AUTOMATIC STATION IDENTIFIER
33	TEMPERATURE: (33 degrees F.) Minus sign indicates sub-zero temperatures.
29	DEW POINT: (29 degrees F.) Minus sign indicates sub-zero temperatures.
3606	WIND: (360 degrees true at 6 knots. Direction is first two digits and is reported in tens of degrees. To decode, add a zero to first two digits. The last digits are speed; e.g., 2524=250 degrees at 24 knots.
975	ALTIMETER SETTING: (29.75 inches) The tens digit and decimal are omitted from report. To decode prefix a 2 to code if it begins with an 8 or 9. Otherwise prefix a 3; e.g., 982=29.82, 017=30.17.
PK WND 08	PEAK WIND SPEED: (8 knots) Reported speed is highest detected since last hourly observation.
001	PRECIPITATION ACCUMULATION: (0.01 inches. Amount of precipitation since last synoptic time (00, 06, 12, 1800 GMT).

FIGURE 2-15. Decoding observations from unstaffed AMOS stations.

SMP SP 0056 AMOS -X M20 BKN 7/8L-FK 046/66/65/2723/967
PK WND 36 027/VSBY S 1/4

SMP	STATION IDENTIFICATION: (Stampede Pass WA) Identifies report using FAA identifiers.
SP	TYPE OF REPORT: (Special) SA=Record
0056	TIME OF REPORT: GMT
AMOS	AUTOMATIC STATION IDENTIFIER
-X M20 BKN	SKY & CEILING: (partly obscured sky, ceiling measured 2,000 feet broken) Figures are height in 100s of feet above ground. Number preceding an X is vertical visibility into a total obscuration in 100s of feet. Symbol after height is amount of sky cover. Letter preceding height indicates method used to determine height.
7/8	PREVAILING VISBILITY: (Seven-eights statute miles)
L-FK	WEATHER & OBSTRUCTIONS TO VISION: (Light drizzle, Fog, & Smoke) Algebraic signs indicate intensity.
046	SEA-LEVEL PRESSURE: (1004.6 millibars) Only the tens, units, and tenths digits are reported.
66	TEMPERATURE: (66 degrees F.) Minus sign indicates sub-zero temperatures.
65	DEW POINT: (65 degrees F.) Minus sign indicates sub-zero temperatures.
2723	WIND: (270 degrees true at 23 knots) Direction is first two digits and is reported in tens of degrees. To decode, add a zero to first two digits. The last digits are speed; e.g., 2524=250 degrees at 24 knots.
967	ALTIMETER SETTING: (29.67 inches) The tens digit and decimal are omitted from report. To decode prefix a 2 to code if it begins with an 8 or 9. Otherwise prefix a 3; e.g., 982=29.82, 017=30.17.
PK WND 36	PEAK WIND SPEED: (36 knots) Reported speed is highest detected since last hourly observation.
027	PRECIPITATION ACCUMULATION: (0.27 inches) Amount of precipitation since last synoptic time (00, 06, 12, 1800 GMT).
VSBY S 1/4	MISCELLANEOUS REMARKS & NOTAMS: (Visibility to south 1/4 mile) Remarks are given using contractions.

FIGURE 2-16. Decoding observations from staffed AMOS stations.

This Galveston TX observation contains similar information to the PGO example above but with the addition of type and time of observation, sky condition, visibility and weather, and remarks made by an observer.

Examples of a staffed AMOS observation:

GLS AMOS SA 1455 CLR 7 82/71/0311/010 PK WND 15 000 GLS AMOS SA 1755 11 SCT E100 OVC 7 80/73/0706/996 PK WND 13 028/TCU ALQDS RWU SW
GLS AMOS RS 1555 7 SCT 15 SCT E60 OVC 7R− 78/75/0506/006 PK WND 12 008

Automatic Observing Station (AUTOB)

The AUTOB is an AMOS with added capability to *automatically* report sky conditions, visibility and precipitation occurrence. AUTOB is polled at 20 minute intervals. The upper limit of cloud amount and height measurements is 6,000 feet AGL. Visibility in statute miles is determined by a backscatter sensor with reportable categories of 0 to 8 (see table 2-5). If a visibility report consisting of 3 values is encountered, it is decoded as shown in the following example:

"VB786", 7 = present visibility, 8 = maximum visibility during past 10 minutes, and 6 = minimum visibility during past 10 minutes.

TABLE 2-5. Reportable visibility categories.

Index of vis. (stat. mi.)	*When vis. is:*	*Index of vis. (stat. mi.)*	*When vis is:*
0	less than 15/16	5	4 1/2 - 5 1/2
1	1 - 1 7/8	6	5 1/2 - 6 1/2
2	2 - 2 7/8	7	6 1/2 - 7 1/2
3	3 - 3 1/2	8	above 7 1/2
4	3 1/2 - 4 1/2		

AUTOB may indicate no cloud layers in either a clear situation or during an inability to penetrate a surface-based obscuration. To distinguish the two, the following rules apply. If the visibility is less than 2 miles, either a partial obscuration "−X" or indefinite obscuration "WX" is reported. A "−X" implies some cloud returns and a "WX" implies no cloud returns. A vertical visibility value for "WX" is not measured. When visibility is 2 miles or greater and no cloud returns are detected a "CLR BLO 60" is used which indicates a clear sky below 6,000 feet. "E" is the ceiling designator. A maximum of 3 (lowest) cloud layers will be reported. Figure 2-17 is the breakdown of an AUTOB message.

Example:

DRT AUTOB 25 SCT E40 OVC BV5 P 58/52/ 1412/995 PK WND 16 004

This Del Rio TX observation indicates: sky condition of two thousand five hundred feet scattered, ceiling four thousand feet overcast, surface visibility between 4 1/2 and 5 1/2 statute miles, precipitation has occurred within 10 minutes of the observation, temperature 58 degrees F, dew point 52 degrees F, wind from 140 degrees at 12 knots, altimeter setting 29.95 inches, peak wind since the last hourly observation 16 knots, four hundredths (0.04 inches) of liquid precipitation since the last synoptic observation.

Examples of an AUTOB message:

DRT AUTOB CLR BLO 60 BV8 75/65/0905/ 991 PK WND 08 000
DRT AUTOB 25 SCT E31 OVC BV8 83/71/ 1408/989 PK WND 18 120
DRT AUTOB 5 SCT E10 OVC BV4 P 75/74/ 1306/978 PK WND 07 001
DRT AUTOB −X E5 BKN BV1 68/67/1805/ 992 PK WND 09 000
DRT AUTOB WX BV0 70/70/1401/995 PK WND 02 000
DRT AUTOB −X 8 SCT E13 OVC BV1 73/ 69/1307/000 PK WND 10 000 HIR CLDS DETECTED

A remark "HIR CLDS DETECTED" is included if clouds are detected above an overcast, with the higher clouds "HIR CLDS" being less than 6,000 feet. For example:

E30 OVC BV8 65/58/3606/988 PK WND 10 000 HIR CLDS DETECTED

means higher clouds are being detected above the overcast, but are less than 6,000 feet. Note that an AUTOB makes no distinction between a thin and opaque cloud layer. "E30 OVC" may be a thin overcast, but is reported as a ceiling.

Remote Automatic Meteorological Observing System (RAMOS)

The breakdown of a RAMOS message is shown in figure 2-18. Note the similarity to the unstaffed AMOS observation except that a 3-hour pressure change, maximum/minimum temperature, and 24-hour precipitation accumulation is also included at designated times.

ENV AUTOB E25 BKN BV7 P 33/29/3606/975 PK WND 08 001

ENV	STATION IDENTIFICATION: (Wendover UT) Identifies report using FAA identifiers.
AUTOB	AUTOMATIC STATION IDENTIFIER
E25 BKN	SKY & CEILING: (Estimated 2,500 feet broken) Figures are height in 100s of feet above ground. Contraction after height is amount of sky cover. Letter preceding height indicates ceiling. WX reported if visibility is less than 2 miles and no clouds are detected. NO CLOUDS WILL BE REPORTED ABOVE 6,000 FEET.
BV7	PRESENT VISIBILITY: Reported in whole statute miles from 0 to 8. Visibility is averaged over a 10 minute time period.
P	PRECIPITATION OCCURRENCE: (P=precipitation in past 10 minutes).
33	TEMPERATURE: (33 degrees F.) Minus sign indicates sub-zero temperatures.
29	DEW POINT: (29 degrees F.) Minus sign indicates sub-zero temperatures.
3606	WIND: (360 degrees true at 6 knots. Direction is first two digits and is reported in tens of degrees. To decode, add a zero to first two digits. The last digits are speed; e.g., 2524=250 degrees at 24 knots.
975	ALTIMETER SETTING: (29.75 inches) The tens digit and decimal are omitted from report. To decode prefix a 2 to code if it begins with an 8 or 9. Otherwise prefix a 3; e.g., 982=29.82, 017=30.17.
PK WND 08	PEAK WIND SPEED: (8 knots) Reported speed is highest detected since last hourly observation.
001	PRECIPITATION ACCUMULATION: (0.01 inches) Amount of precipitation since last synoptic time (00, 06, 12, 1800 GMT).

NOTE: If no clouds are detected below 6,000 feet and the visibility is greater than 2 miles, the reported sky condition will be CLR BLO 60.

FIGURE 2-17. Decoding observations from AUTOB stations.

P67 RAMOS SA 2356 Ø46/66/65/2723/967 PK WND 36
ØØØ2 Ø27 83 2ØØ43

P67	STATION IDENTIFICATION: (Lidgerwood ND) Identifies report using FAA identifiers.
RAMOS	AUTOMATIC STATION IDENTIFIER
SA	TYPE OF REPORT: (Record) SP=Special
2356	TIME OF REPORT: GMT
Ø46	SEA-LEVEL PRESSURE: (1ØØ4.6 millibars) Only the tens, units, and tenths digits are reported.
66	TEMPERATURE: (66 degrees F.) Minus sign indicates sub-zero temperatures.
65	DEW POINT: (65 degrees F.) Minus sign indicates sub-zero temperatures.
2723	WIND: (27Ø degrees true at 23 knots) Direction is first two digits and is reported in tens of degrees. To decode, add a zero to first two digits. The last digits are speed; e.g., 2524=25Ø degrees at 24 knots.
967	ALTIMETER SETTING: (29.67 inches) The tens digit and decimal are omitted from report. To decode prefix a 2 to code if it begins with an 8 or 9. Otherwise prefix a 3; e.g., 982=29.82, Ø17=3Ø.17.
PK WND 36	PEAK WIND SPEED: (36 knots) Reported speed is highest detected since last hourly observation.
ØØØ2	THREE-HOUR PRESSURE CHANGE: (Rising then falling, Ø.Ø2 millibars higher now than three hours ago.) *
Ø27	PRECIPITATION ACCUMULATION: (Ø.27 inches) Amount of precipitation since last synoptic time (ØØ, Ø6, 12, 18ØØ GMT).
83	TEMPERATURE (MAX OR MIN) MAX at ØØ & Ø6Z, MIN at 12 & 18Z
2ØØ43	PRECIPITATION ACCUMULATION IN PAST 24 HOURS: (ØØ.43 inches); first digit (2) is the group identifier

---

* First digit in code is barometer tendency (see figure 5-6).

Ø = ∧
1 = ⌐
2 = /
3 = ✓
4 = ∧, –, ∨
5 = ∨
6 = ∟
7 = \
8 = ∧

FIGURE 2-18. Decoding observations from RAMOS stations.

Section 3
PILOT AND RADAR REPORTS AND SATELLITE PICTURES

The preceding section explained the decoding of surface aviation weather reports. However, these spot reports only sample the total weather picture. Pilot and radar reports along with satellite pictures help fill the gaps between stations.

PILOT WEATHER REPORTS (PIREPS)

No observation is more timely than the one you make from your cockpit. In fact, aircraft in flight are the only means of directly observing cloud tops, icing and turbulence. Your fellow pilots welcome your PIREP as well as do the briefer and forecaster. Help yourself and the aviation weather service by sending pilot reports!

A PIREP is usually transmitted in a prescribed format (see figure 3-1). The letters "UA" identify the message as a pilot report. The letters "UUA" identify an urgent PIREP. Required elements for all PIREPs are message type, location, time, flight level, type of aircraft and at least one weather element encountered. When not required, elements without reported data are omitted. All altitude references are MSL unless otherwise noted, distances are in nautical miles and time is in GMT.

A PIREP is usually transmitted as part of a group of PIREPs collected by state or as a remark appended to a surface aviation weather report. The phenomenon is coded in contractions and symbols. For example (referring to Figure 3-1 as a guide):

> UA /OV MRB-PIT/TM 1600/FL 100 /TP BE55 /SK 024 BKN 032/042 BKN-OVC /TA -12/IC LGT-MDT RIME 055-080/RM WND COMP HEAD 020 MH310 TAS 180. *

The PIREP decodes as follows:

> Pilot report, Martinsburg to Pittsburgh at 1600Z at 10,000 feet. Type of aircraft is a Beechcraft Baron. First cloud layer has base at 2,400 feet broken top 3,200 feet. Second cloud layer base is 4,200 feet broken occasionally overcast with no tops reported. Outside air temperature is −12 degree Celsius. Light to moderate rime icing reported between 5,500-8,000 feet. Headwind component is 20 knots. Magnetic heading is 310 degrees and true air speed is 180 knots.

* NOTE that all heights are referenced to MSL.

BACK

Encoding Pilot Weather Reports (PIREP)

1. UA - **Routine PIREP, UUA - Urgent PIREP**
2. /OV - **Location:** Use 3-letter NAVAID idents only.
   a. Fix: /OV ABC, /OV ABC 090025.
   b. Fix to fix: /OV ABC-DEF, /OV ABC-DEF 120020, /OV ABC 045020-DEF 120005, /OV ABC-DEF-GHI.
3. /TM - **Time:** 4 digits in GMT: /TM 0915.
4. /FL - **Altitude/Flight Level:** 3 digits for hundreds of feet. If not known, use UNKN: /FL095, /FL310, /FLUNKN.
5. /TP - **Type aircraft:** 4 digits maximum, If not known use UNKN: /TP L329,/TP B727 /TP UNKN.
6. /SK - **Cloud layers:** Describe as follows:
   a. Height of cloud base in hundreds of feet. If unknown, use UNKN.
   b. Cloud cover symbol.
   c. Height of cloud tops in hundreds of feet.
   d. Use solidus (/) to separate layers.
   e. Use a space to separate each sub element.
   f. Examples: /SK 038 BKN, /SK 038 OVC 045, /SK 055 SCT 073/085 BKN 105, /SK UNKN OVC
7. /WX - **Weather:** Flight visibility reported first. Use standard weather symbols, intensity is not reported: /WX FV02 R H, /WX FV01 TRW.
8. /TA - **Air temperature in Celsius:** If below zero, prefix with a hyphen: /TA 15, /TA -06.
9. /WV - **Wind:** Direction and speed in six digits. /WV 270045, /WV 280110.
10. /TB - **Turbulence:** Use standard contractions for intensity and type (use CAT or CHOP when appropriate). Include altitude only if different from /FL. /TB EXTRM, /TB LGT-MDT BLO-090.
11. /IC - **Icing:** Describe using standard intensity and type contractions. Include altitude only if different than /FL: /IC LGT-MDT RIME, /IC SVR CLR 028-045.
12. /RM - **Remarks:** Use free form to clarify the report. Most hazardous element first: /RM LLWS -15KT SFC-003 DURGC RNWY 22 JFK.

Refer to FAAH 7110.10 for expanded explanation of TEI coding.

Examples of Completed PIREPS

UA /OV RFD 170030/TM 1315/FL160/TP PA60 /SK 025 OVC 095/180 OVC /TA -21/WV 270048

UA /OV DHT 360015-AMA-CDS/TM 2116/FL050/TP PA32 /SK UNKN OVC/WX FV03 R /TB LGT/TA 04/RM HVY RAIN

FRONT

PIREP FORM

Pilot Weather Report → = *Space Symbol*

3-Letter SA Identifier — — → 1. **UA** → ____ *Routine Report* **UUA** → ____ *Urgent Report*

2. **/OV** →	Location:
3. **/TM** →	Time:
4. **/FL**	Altitude/Flight Level:
5. **/TP** →	Aircraft Type:
Items 1 through 5 are mandatory for all PIREPs	
6. **/SK** →	Sky Cover:
7. **/WX** →	Flight Visibility and Weather:
8. **/TA** →	Temperature *(Celsius)*:
9. **/WV** →	Wind:
10. **/TB** →	Turbulence:
11. **/IC** →	Icing:
12. **/RM** →	Remarks:

FAA FORM 7110-2 (1-85) Supersedes Previous Edition

FIGURE 3-1. Pilot Reports Format.

The following example appended to an aviation weather report,

DSM SA 1755 M8 OVC 3R-F 132/45/44/3213/ 992/UA /OV DSM 320012/TM 1735/FL UKN /TP UKN /SK OVC 065/080 OVC 140. *

is decoded "... pilot report 12 nautical miles on 320 degrees radial from Des Moines VOR, at 1735 GMT, flight level and type of aircraft are unknown, top of the lower overcast 6,500 feet; base of a second overcast layer at 8,000 feet with top at 14,000 feet."

Note that PIREPs adhere to format shown in figure 3-1.

UA /OV OKC 063064/TM 1522/FL080 TP C172/TA -04 /WV 245040 /TB LGT /RM IN CLR *

PIREP decodes as follows:

Pilot report, 64 nautical miles on 63 degree radial from Oklahoma City VOR at 1522 GMT at flight altitude 8,000 feet. Type aircraft is a Cessna 172. Outside air temperature is minus 4 degrees Celsius, wind is 245 degrees at 40 knots, light turbulence and clear skies.

Most contractions in PIREP messages are self-explanatory. Icing and turbulence reports state intensities using standard terminology when possible. Intensity tables for turbulence and icing are in section 14. If a pilot's description of an icing or turbulence encounter cannot readily be translated into standard terminology, the pilot's description is transmitted verbatim.

To lessen the chance of misinterpretation by others, you are urged to report icing and turbulence in standard terminology (intensity tables for turbulence and icing, section 14). This PIREP stated,... PRETTY ROUGH AT 6,500, SMOOTH AT 8,500 PA24... Would a report of "light", "moderate", or "severe" turbulence at 6,500 have meant more to you?

Pilot reports of individual cloud layers, bases and tops, are usually in symbols and are often appended to surface aviation weather reports. Height of cloud base precedes the sky cover symbol and top follows the symbol. For example, 038 BKN 070 means base of broken layer at 3,800 feet and top 7,000 feet (all MSL).

Outside air temperature is given in 2 digits, degrees Celsius, with negative values preceded by a hyphen. Wind is given as six digits with the first 3 digits being direction and the last 3 digits being speed in knots.

The following excerpts may further assist you in reading transmitted pilot weather reports:

.../RM DURGD OAOI 150 01 080...*

means "... during descent on and off instruments from 15,000 feet; on instruments from 8,000 feet...

...FL100.../TA-02/WV 250015*

is decoded "... at 10,000 feet, temperature -2 degrees C, wind 250 degrees at 15 knots..."

...FL060/TP C-172/SK INTMTLY BL/TB MDT...*

states "... at 6000 feet, Cessna 172, intermittently between layers (contraction BL); moderate turbulence..."

UA /OV ABQ/TM 1845.. TIJERAS PASS CLOSED DUE TO FOG AND LOW CLDS. UNABLE VFR RTNG ABQ.*

is self-explanatory. Information of this type is helpful to others planning VFR flight in the area.

UA /OV/ TOL/TM 2200/FL 310 /TP B707 /TB MDT CAT 350-390*

means "... over Toledo at 2200 GMT and flight level 31,000, a Boeing 707 reported moderate clear air turbulence from 35,000 to 39,000.

Pilot reports of a non-meteorological nature sometimes help air traffic controllers. This "plain language" report stated:

"... 3N PNS LRG FLOCK OF GOOSEY LOOKING BIRDS HDG GNLY NORTH MAY BE SEAGULLS FORMATION LOUSY COURSE ERRATIC..."

While in humorous vein, this PIREP alerted pilots and controllers to a bird hazard.

Your PIREP always helps someone else and becomes part of the aviation weather service. Please report anything you observe that may be of concern to other pilots.

RADAR WEATHER REPORTS (RAREPS)

Thunderstorms and general areas of precipitation can be observed by radar. Most radar stations report each hour at H+35 with intervening special reports as required. The report includes location of precipitation along with type, intensity, and intensity trend. Table 3-1 explains symbols denoting intensity and trend. Table 3-2 shows the order and content of a radar weather report.

---

*Note that all heights are referenced to MSL.

TABLE 3-1. Precipitation intensity and intensity trend

Intensity		*Intensity Trend*	
Symbol	*Intensity*	*Symbol*	*Trend*
–	Light	+	Increasing
(none)	Moderate		
+	Heavy	–	Decreasing
++	Very Heavy		
X	Intense	NC	No change
XX	Extreme		
U	Unknown	NEW	New echo

TABLE 3-2. Ordered content of a radar weather report

OKC 1934 LN 8TRW++/+ 86/40 164/60 199/115 15W L2425
MT570 AT 159/65 2 INCH HAIL RPRTD THIS CELL
MO1 NO2 ON3 PM34 QM3 RL2 SL9

OKC 1934	*LN*	*8*	*TRW++/+*	*86/40 164/60 199/115*
a.	b.	c.	d.	e.

15W	*L2425*	*MT 570 AT 159/65*
f.	g.	h.

2 INCH HAIL RPRTD THIS CELL
i.

MO1 NO2 ON3 PM34 QM3 RL2 SL9
j.

See TABLE 7-1 for corresponding rainfall rates defining intensities. Note that intensity and intensity trend is not applicable to frozen precipitation.

Refer to TABLE 3-2:

a. Location identifier and time of radar observation (Oklahoma City RAREP at 1934 GMT in this example).
b. Echo pattern (line in this example)—
   Echo pattern or configuration may be a
   1. Line (LN)—a line of precipitation echoes at least 30 miles long, at least five times as long as it is wide and at least 30% coverage within the line.
   2. Fine Line (FINE LN)—a unique *clear air* echo (usually precipitation free and cloud free) in the form of a thin or fine line on the PPI scope. It represents a strong temperature/moisture boundary such as an advancing dry cold front.
   3. Area (AREA)—a group of echoes of similar type and not classified as a line.
   4. Spiral Band Area (SPRL BAND AREA)—an area of precipitation associated with hurricane that takes on a spiral band configuration around the center.
   5. Single Cell (CELL)—a single isolated precipitation not reaching the ground.
   6. Layer (LYR)—an elevated layer of stratform precipitation not reaching the ground.
c. Coverage in tenths (8/10 in this example).
d. Type, intensity, and intensity trend of weather (thunderstorm (T), very heavy rainshowers (RW++) and increasing in intensity (/+) in this example)—See TABLE 7-1 for weather symbols used except hail is reported as "A" in a RAREP. See TABLE 3-1 for intensity and intensity trend symbols.
e. Azimuth (reference true N) and range in nautical miles (NM) of points defining the echo pattern (86/40 164/60 199/115 in this example)—See following examples for elaboration of echo patterns.
f. Dimension of echo pattern (15 NM wide in this example)—Dimension of an echo pattern is given when azimuth and range define *only* the center line of the pattern. In this example, "15W" means the line has a total width of 15 NM, 71/2 miles either side of a center line drawn from the points given i.e. "D15" means a convective echo is 15 miles in diameter around a given center point.
g. Pattern movement (line moving *from* 240 degrees at 25 knots in this example)—may also show movement of individual storms or cells "C" and movement of an area "A".
h. Maximum top and location (57,000 feet MSL on radial 159 degrees at 65 NM in this example).
i. Remarks—self-explanatory using plain language contractions.
j. Digital section—used for preparing radar summary chart.

To assist you in interpreting RAREPs, four examples are decoded into plain language:

LZK 1133 AREA 4TRW+/+ 22/100 88/170 196/180 220/115 C2425 MT 310 AT 162/110

Little Rock AR radar weather observation at 1133 GMT. An area of echoes, four-tenths coverage, containing thunderstorms and heavy rainshowers, increasing in intensity.
Area is defined by points (referenced from LZK radar site) at 22 degrees, 100 NM (nautical miles); 88 degrees, 170 NM; 196 degrees, 180 NM and 220 degrees, 115 NM. (These points plotted on a map and connected with straight lines outline the area of echoes.

Maximum top (MT) is 31,000 feet MSL located at 162 degrees and 110 NM from LZK.

JAN 1935 SPL LN 10TRWX/NC 86/40 164/60 199/115 12W C2430 MT 440 AT 159/65 D10

Jackson MS special radar report at 1935 GMT. Line of echoes, ten-tenths coverage, thunderstorm, intense rainshowers, no change in intensity. Center of the line extends from 86 degrees, 40 NM; 164 degrees, 60 NM to 199 degrees, 115 NM. The line is 12 NM wide (12W). (To display graphically, plot the center points on a map and connect the points with a straight line; since the thunderstorm line is 12 miles wide, it extends 6 miles either side of your plotted line.)
Thunderstorm cells are moving from 240 degrees at 30 knots.
Maximum top is 44,000 feet MSL centered at 159 degrees, 65 NM from JAN.
Diameter of this cell is 10 NM (D10).

MAF 1130 AREA 2S 27/80 90/125 196/50 268/100 A2410 MT U100

Midland TX radar weather report at 1130 GMT. An area, two-tenths coverage, of snow (no intensity or trend is assigned for non-liquid precipitation) Area is bounded by points 27 degrees, 80 NM; 90 degrees, 125 NM; 196 degrees, 50 NM and 268 degrees, 100 NM.
Area movement is from 240 degrees at 10 knots.
Maximum tops are 10,000 feet MSL, tops are uniform (smooth). Note that these are precipitation tops and not cloud tops.

HDO 1132 AREA 2TRW++6R-/NC 67/130 308/45 105W C2240 MT 380 AT 66/54

Hondo TX radar weather report at 1132 GMT.
An area of echoes, total coverage eight-tenths, containing two-tenths coverage of thunderstorms with very heavy rainshowers and six-tenths coverage of light rain. No change in intensity. (Suggests thunderstorms embedded in an area of light rain.). Although the pattern is an "area", only two points are given followed by "105W". This means the area lies 52 and 1/2 miles either side of the line defined by the two points - 67 degrees, 130 NM and 308 degrees, 45 NM.
Thunderstorm cells are moving from 220 degrees at 40 knots.
Maximum top is 38,000 feet at 66 degrees, 54 NM from HDO.

When a radar report is transmitted but doesn't contain any encoded weather observation, a contraction is sent which indicates the operational status of the radar. For example,

OKC 1135 PPINE means Oklahoma City OK radar at 1135 GMT detects no echoes.

TABLE 3-3 explains the contractions.

TABLE 3-3. Contractions of radar operational status.

Contraction	*Operational status*
PPINE	Equipment normal and operating in PPI (Plan Position Indicator) mode; no echoes observed.
PPIOM	Radar inoperative or out of service for preventative maintenance.
PPINA	Observations not available for reasons other than PPINE or PPIOM.
ROBEPS	Radar operating below performance standards.
ARNO	"A" scope or azimuth/range indicator inoperative.
RHINO	Radar cannot be operated in RHI (Range-height indicator) mode. Height data not available.

A radar weather report may contain remarks in addition to the coded observation. Certain types of severe storms may produce distinctive patterns on the radar scope. For example, a hook-shaped echo may be associated with a tornado. A line echo wave pattern (LEWP) in which one portion of a squall line bulges out ahead of the rest of the line may produce strong gusty winds at the bulge. A "vault" on the Range-Height Indicator scope may be associated with a severe thunderstorm producing large hail and strong gusty winds at the surface. If hail, strong winds, tornado activity, or other adverse weather is known to be associated with identified echoes on the radar scope, the location and type of phenomenon are included as a remark. Examples of remarks are, "HAIL REPORTED THIS CELL", "TORNADO ON GROUND AT 338/15" AND "HOOK ECHO 243/18". As far as indicating precipitation *not* reaching the ground, two contractions are aloft and mostly aloft respectively. That is, some or most of the precipitation is *not* reaching the ground. Bases of the precipitation will be given in hundred of feet MSL; example "PALF BASE 40" means part of the precipitation detected is evaporating at 4,000 feet MSL.

Radar weather reports also contain groups of digits, ie, MO1 NO2 ON3 PM34, etc., which are entered on a line following the RAREP. This digitized radar information (omitted from the foregoing examples) is used primarily by meteorologists and hydrologists for estimating amount of rainfall and in preparing the radar summary chart. However, this code is useful in determining more precisely where precipitation is occurring within an area and the intensity of the precipitation by using a proper grid overlay chart for the corresponding radar site. See Figure 3-2 for an example of a digital code plotted from the OKC RAREP in Table 3-2.

The digit assigned to a box represents encoded intensity levels of the precipitation as determined by a video integrator processor. See Table 7-1 for definitions of intensity levels 1-6. Thus, the term VIP LEVEL 1 simply means the precipitation intensity is weak or light, VIP LEVEL 2 is moderate, etc. Note that the *maximum* VIP LEVEL is encoded for any given box on the grid identified in the digital code. A box is identified by two letters, the first representing the row in which the box is found and the second letter representing the column. For example "MO1" identifies the box located in row M and column 0 as containing precipitation with a maximum VIP LEVEL of one (1). A code of "MO1324" indicates precipitation in four consecutive boxes in the same row. Working from left to right box MO = 1, box MP = 3, box MQ = 2 and box MR = 4.

When using hourly and special radar weather reports in preflight planning, note the location and

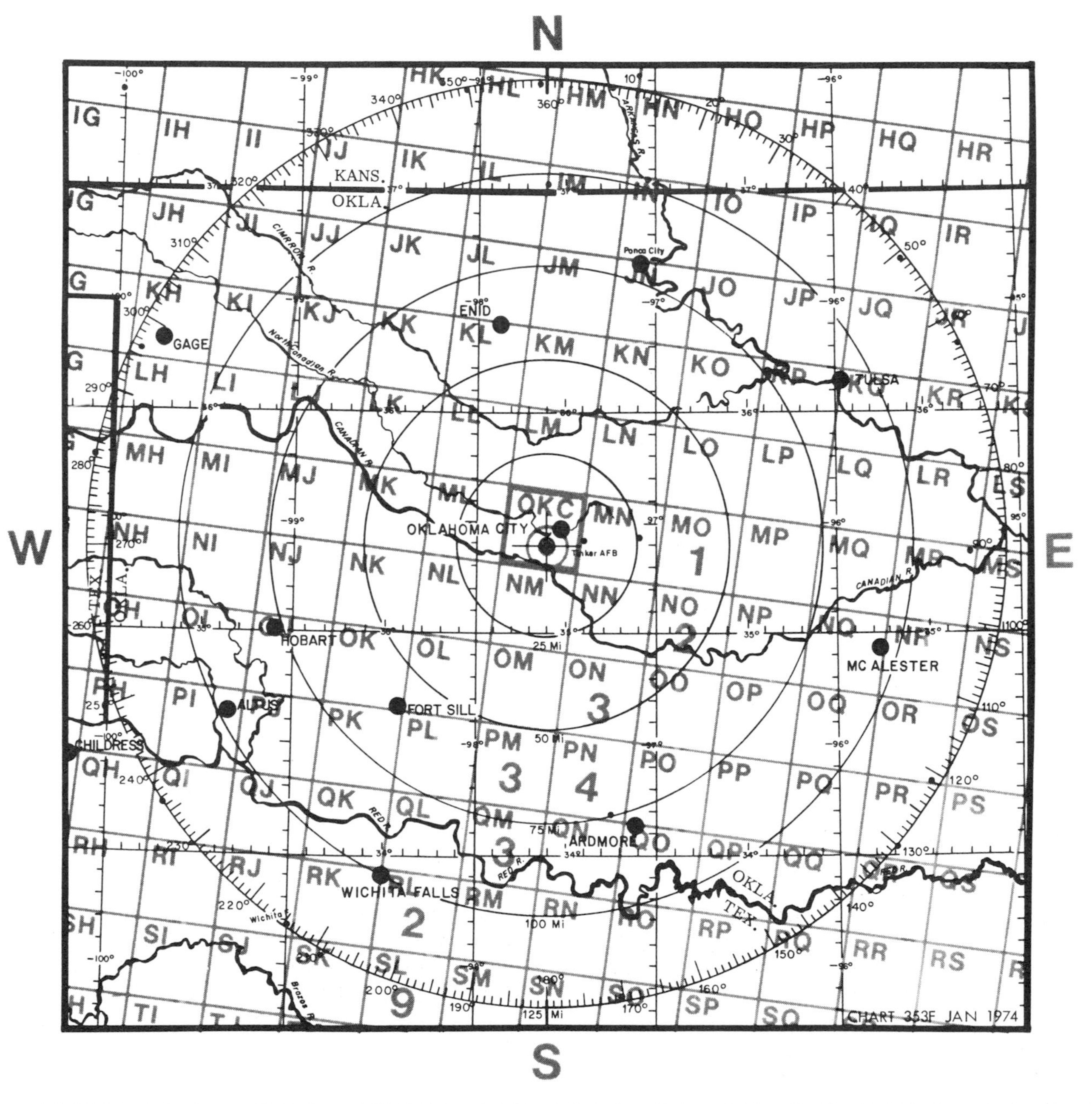

FIGURE 3-2. Digital Radar Report Plotted on a PPI Grid Overlay Chart. Data from Table 3-2. *Note*: See Table 7-1 for Intensity Level Codes 1 through 6.

The following VIP LEVEL codes are used for echoes beyond 125 nautical miles:

8 = Echoes of unknown intensity but believed to be severe from other reports.

9 = Echoes of unknown intensity but not believed to be severe.

```
SDUS KNKA 112113
SDUS24 KWBC 112035                .......                 . 25562213..  .
+ 51 Ø41                          .      .                . 24342 1.    .
                                  .      .                . 32543  .    .
                                  .      .                .  94411 .     .
                                  .      .                .  95421 .     .
                                  .      ..             . .  1522  .     .
                                  .        ........ ...1511...     .
                                  .                    .1511   .     .
                                  .                    .13      .
                                  .                     531    . 1 11
                                  .                    9531    .  2221
                                  .                    2512    ....221
                                  .                    33.     .  1221
                                                       11.2      1   .
                                                       11 .    ..2
                              .                         1       21.
                              .                        .
                              .    ....               . 1
                              ..MM.   ..             .  1   1
                                MMMMM  .         ..
                                 MMMM   .       .
                                 MMMM    .      ..
                                 MMMMM    .     .
                                  MMMMM    .    .
                                   MMMMM   .    .
                                     MMMMM .    .
                                            ..  .
                                             .....
```

FIGURE 3-3. Teletypewriter Plot of Echo Intensities for the South Central United States. *Note:* See Table 7-1 for Intensity Level Codes 1 through 6.
M = Missing.
8 = Echoes of unknown intensity (beyond 125 NM) but believed to be severe.
9 = Echoes of unknown intensity (beyond 125 NM) but not believed to be severe.

coverage of echoes, the type of weather reported, the intensity trend and especially the direction of movement.

A WORD OF CAUTION—remember that when National Weather Service radar detects objects in the atmosphere it only detects those of precipitation size or greater. It is *not* designed to detect ceilings and restrictions to visibility. An area may be blanketed with fog or low stratus but unless precipitation is also present the radar scope will be clear of echoes. Use radar reports along with PIREPs and aviation weather reports and forecasts.

RAREPs help you to plan ahead to avoid thunderstorm areas. Once airborne, however, you must depend on visual sighting or airborne radar to evade individual storms.

Another product provided for the use of the pilot in flight planning is the teletypewriter digital plot (see figure 3-3). The digital plot may be obtained through the request/reply circuit. The chart has digitized intensities plotted over a section map of several states. The numbers on the plot refer to RADAR intensity and represent the strongest return found in the area. From this plot the briefer/pilot may obtain the location of the strongest RADAR returns in the area of interest.

SATELLITE WEATHER PICTURES

Prior to the space age, weather observations were made only at distinct points within the atmosphere and complemented by pilot observations (PIREPs) of en route clouds and weather. These PIREPs give a "sense" of weather as viewed from above. However, with the advent of weather satellites a whole new dimension to weather observing and reporting has emerged. There are two types of weather satellites in use today: GOES (a geostationary satellite) and NOAA (a near polar orbiter satellite).

There are two U.S. GOES (Geostationary Operational Environmental Satellite) satellites used for picture taking. One stationed over the equator at 75 degrees west and the other at 135 degrees west. Together they cover North and South America and surrounding waters. They normally each transmit a picture of the earth, pole to pole each half hour. When disastrous weather threatens the U.S., the satellites can scan small areas rapidly so that we can receive a picture as often as every three minutes. Data from these rapid scans are used at national warning centers.

However, since the GOES satellite is stationary over the equator, the pictures poleward of about 50% degrees latitude become significantly distorted. Thus, another type of satellite is employed. The NOAA satellite is a near polar orbiter with an inclination angle (to the equator) of 98.6 degrees. In other words, this satellite orbits the earth on a track

FIGURE 3-4. GOES Visible Impagery.

which nearly crosses the North and South Poles. A high resolution picture is produced about 800 miles either side of it's track on the journey from pole to pole. The NOAA pictures are essential to weather personnel in Alaska and Canada.

NOTE: At the time of this revision, GOES East (75 degrees west) has become inoperative and GOES West has been positioned at 100 degrees west longitude in order to cover as much of the U.S. as possible. As a result, from the west coast out into the Pacific Ocean and from the east coast out into the Atlantic Ocean, the most useful satellite imagery will now be provided by the two (2) NOAA satellites currently in use. The next GOES satellite to be sent up to replace GOES East will be sometime in 1985 which will return the satellite to normal.

Basically, two types of imagery are available and when combined give a great deal of information about clouds. Through interpretation, the analyst can determine the type of cloud, the temperature of cloud tops (from this the approximate height of the cloud) and the thickness of cloud layers. From this information, the analyst gets a good idea of the associated weather.

One type of imagery is visible imagery (see figure 3-4 and 3-6). With a visible picture we are looking at clouds and the earth reflecting sunlight to the satellite sensor. The greater the reflected sunlight reach-

FIGURE 3-5. GOES Infra-red Imagery.

FIGURE 3-6. NOAA Visible Imagery.

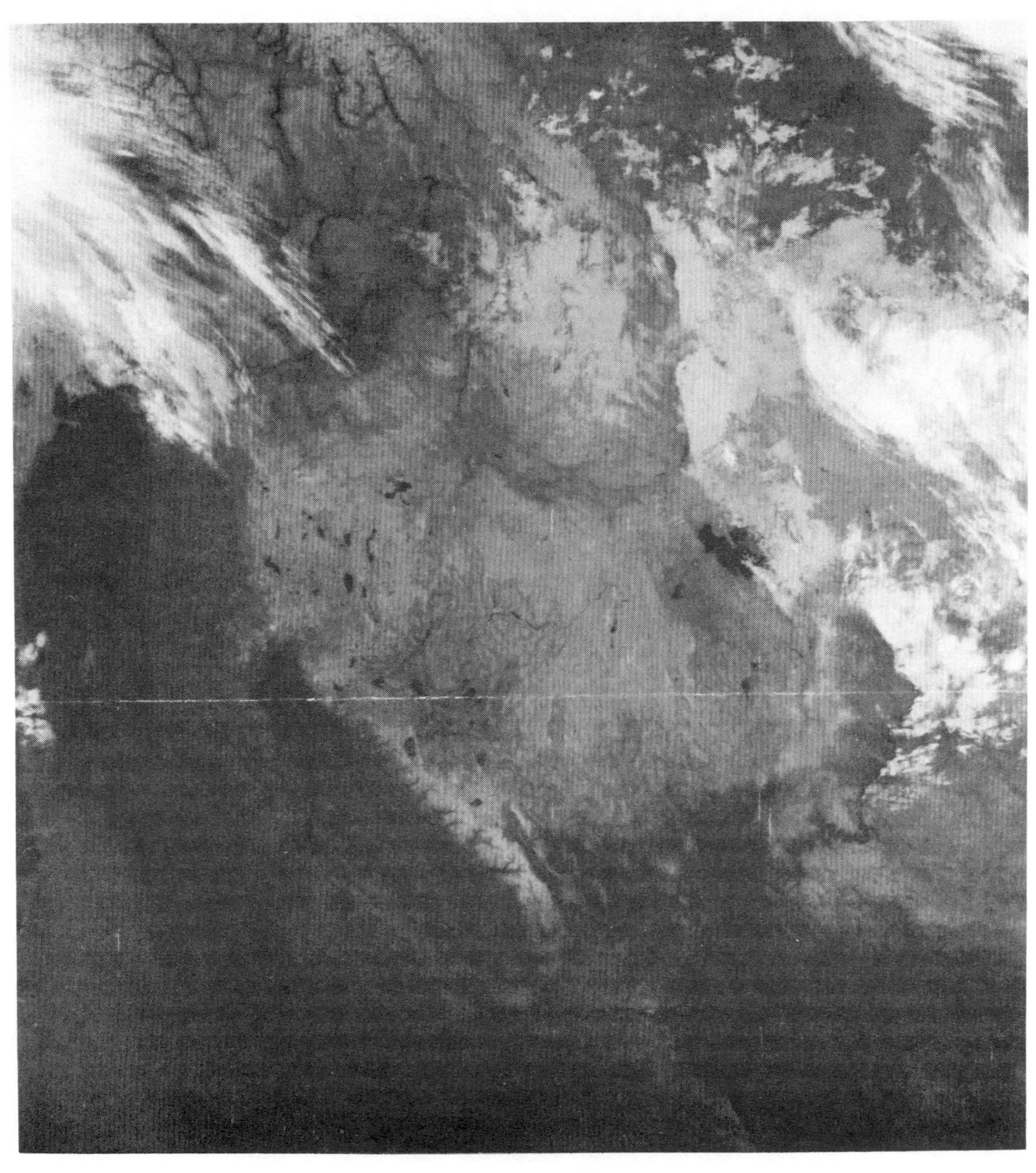

FIGURE 3-7. NOAA Infra-Imagery.

ing the sensor, the whiter the object is on the picture. The amount of reflectivity reaching the sensor depends upon the height, thickness and ability of the object to reflect sunlight. We can generalize that clouds are much more reflective than the earth and so clouds will show up white on the picture, especially thick clouds. Thus, the visible picture is primarily used to determine 1) the presence of clouds and 2) the type of cloud from shape and texture. There are no visible pictures at night from the GOES satellite. The NOAA satellite does take night visible pictures through the ability of the moon to reflect sunlight to the earth at night.

The second type of imagery is infra-red imagery (IR) (see figure 3-5 and 3-7). With an IR picture we are looking at heat radiation being emitted by the clouds and earth. The images show temperature differences between cloud tops and the ground, as well as, temperature gradations of cloud tops and temperature gradations along the earth's surface. Ordinarily, cold temperatures are displayed as light gray or white, making high clouds appear whitest, but various enhancements are sometimes used to sharply illustrate important temperature changes. IR pictures are used to determine cloud top temperatures and thus the approximate height of the cloud. IR pictures are available both day and night. From this, one can see the importance of using visible and IR imagery together when interpreting clouds.

Operationally, at WSFOs and FSSs, pictures are received once every 30 minutes. In these they can see development and dissipation of weather, such as fog and convection, all over the country. Much of this is not visible from reporting points.

Section 4
AVIATION WEATHER FORECASTS

Good flight planning considers forecast weather. This section explains the following aviation forecasts:

1. TERMINAL FORECASTS
   (a). Domestic (FT)
   (b). International (ICAO TAF)
2. AREA FORECAST (FA)
3. TWEB ROUTE FORECAST AND SYNOPSIS
4. CONVECTIVE SIGMET (WST)
5. SIGMET AND AIRMET (WS and WA)
6. WINDS AND TEMPERATURES ALOFT FORECAST (FD)
7. SPECIAL FLIGHT FORECASTS
8. CENTER WEATHER SERVICE UNIT (CSWU) PRODUCTS

Also discussed are the following general forecasts which may aid in flight planning:

1 . HURRICANE ADVISORY (WH)
2 . CONVECTIVE OUTLOOK (AC)
3 . SEVERE WEATHER WATCH BULLETIN (WW)
4 . ALERT SEVERE WEATHER WATCH MESSAGE (AWW)

TERMINAL FORECASTS (FT AND TAF)

FT (Domestic)

A terminal forecast (FT) is a description of the surface weather expected to occur at an airport. The forecast cloud heights and amounts, visibility, weather and wind relate to flight operations within 5 nautical miles of the center of the runway complex. The term vicinity (VCNTY) covers the area from 5 miles beyond the center of the runway complex out to 25 miles. For example, "TRW VCNTY" means thunderstorms are expected to occur between 5 and 25 miles from the station. Scheduled forecasts are issued by WSFOs for their respective areas (figures 1-4 and 1-4A) 3 times daily and are valid for 24 hours. Issue and valid times are according to time zones (see Section 14). The format of the FT is essentially the same as that of the SA report.

Example of a Terminal Forecast (FT)

STL 251010 C5 X 1/2S-BS 3325G35 OCNL C0 X 0S+BS. 16Z C30 BKN 3BS 3320 CHC SW-. 22Z 30 SCT 3315. 00Z CLR. 04Z VFR WND..

To aid in the discussion, we have divided the forecast into the following elements lettered "a" through "i"

STL (a.) *251010* (b.) *C5 X* (c.) *1/2* (d.) *S-BS* (e.) *3325G35* (f.) *OCNL C0 X 0S+BS.* (g.)

16Z C30 BKN 3BS 3320 CHC SW–. 22Z 20SCT 3315. 00Z CLR. (h.)

04Z VFR WND.. (i.)

a. Station identifier. "STL" identifies St. Louis, MO. The forecast is for St. Louis.
b. Date-time group. "251010" is date and valid times. The forecast is valid beginning on the 25th day of the month at 1000Z valid until 1000Z the following day.
c. Sky and ceiling. "C5 X" means ceiling 500 feet, sky obscured. The letter "C" always identifies a forecast ceiling layer. Cloud heights are always referenced to ground level. Sky cover designators and height coding are identical to those used in the SA reports.

The summation of cloud layers concept is used here as in the SA. Total sky cover at each level from a ground observers point of view is determined instead of the individual cloud layers at each level. Thus, there will not be a "SCT" layer above a "BKN" layer because total sky cover can only increase with each succeeding higher level.

d. Visibility. "1/2" means visibility 1/2 mile. Visibility is in statute miles and fractions. Absence of a visibility entry specifically implies visibility more than 6 statute miles.
e. Weather and obstructions to vision. "S-BS" means light snow and blowing snow. These elements are in symbols identical to those used in SA reports and are entered only when forecast.
f. Wind. "3325G35" means wind from 330 degrees at 25 knots gusting to 35 knots which is the same as in the SAs. Omission of a wind entry specifically implies wind less than 6 knots.
g. Remarks. "OCNL C0 X 0S+BS" means occa-

286-959 - 91 - 3 : QL 3

sional conditions of ceiling zero, sky obscured, visibility zero, heavy snow and blowing snow. Remarks may be added to more completely describe forecast weather by indicating variations from prevailing conditions. See table 4-10 for the definitions of the variability terms used in remarks. In this case the occasional conditions described above are expected to occur with a greater than 50% probability but for less than 1/2 of the time period from 1000Z to 1600Z. Thus, c. through f. in the above example are the prevailing conditions from 1000Z to 1600Z with g. specifying variations from the prevailing conditions during the same time period. Also, "LLWS" will be included in the remarks section if low level wind shear is forecast.

In the remarks section, caution must be used in interpreting the data when two (2) or more variable conditions are forecast for different time periods within the same forecast group. If a phenomenon in the remarks is forecast for only a portion of the time period of the forecast group, the time period for the phenomenon is indicated immediately after the phenomenon. Example:

00Z C30 OVC OCNL 12 OVC 02Z-04Z CHC TRW−. 06Z etc.

This forecast says prevailing conditions will be ceiling 3000 feet from 00Z to 06Z. A ceiling of 1200 feet is expected between 02Z and 04Z for less than 1/2 of the 2 hours. The chance of thunderstorms with light rainshowers is for the period 00Z to 06Z.

h. Expected changes. When changes are expected, preceding conditions are followed by a period before the time and conditions of the expected change. " 16Z C30 BKN 3BS 3320 CHC SW−. 22Z 30 SCT 3315. 00Z CLR." means by 1600 Z the prevailing conditions will be 3000 Broken, visibility 3 miles, blowing snow, wind 330 degrees at 20 knots and a 30 to 50% chance of light snow showers. By 2200Z, the prevailing conditions will change to 3,000 scattered, visibility more than 6 miles (implied) and wind 330 degrees at 15 knots. By 0000Z, prevailing conditions will become sky clear, visibility more than 6 miles and wind less than 10 knots (implied).

i. 6-hour categorical outlook. The last 6 hours of the forecast is a categorical outlook. "04Z VFR WND.." means that from 0400Z until 1000Z (the end of the forecast period) weather will be ceiling more than 3,000 feet or none and visibility greater than 5 miles (VFR) with wind 25 knots or stronger. The double period (..) signifies the end of the forecast for the specific terminal.

See table 4-1 for a list of applicable categories and remarks that may be appended to the category "VFR" to better describe the expected conditions. The cause(s) of below VFR categorical outlooks must be stated. Below VFR categories can be due to ceilings only (CIG), restrictions to visibility only (TRW F etc.). See table 4-2 for examples. The term "WND" is not included if winds (sustained or gusts) are forecast to be less than 25 knots.

TABLE 4-1. Categories

Category	*Definition*
LIFR	Low IFR-ceiling less than 500 feet and/or visibility less than 1 mile.
IFR	Ceiling 500 to less than 1,000 feet and/or visibility 1 to less than 3 miles.
MVFR	Marginal VRF-ceiling 1,000 to 3,000 feet and/or visibility 3 to 5 miles inclusive.
VFR	No ceiling or ceiling greater than 3,000 feet and visibility greater than 5 miles.

Remarks that may be appended to VFR at forecaster's discretion	*Definition*
VFR CIG ABV 100	Ceiling greater than 10,000 feet and visibility greater than 5 miles.
VFR NO CIG	Cloud coverage less than 6/10 or thin clouds and visibility greater than 5 miles.
VFR CLR	Cloud coverage less than 1/10 and visibility greater than 5 miles.

TABLE 4-2. Examples of categorical groupings

Example	*Definition*
LIFR CIG	Low IFR due to low ceiling *only*.
IFR F	IFR due to visibility restricted by fog *only*
MVFR CIG H K	Marginal VFR due *both* to low ceiling and to visibility restricted by haze and smoke.
IFR CIG R WND	IFR due *both* to low ceiling and to visibility restricted by rain, wind expected to be 25 knots or greater.

Scheduled FT Collectives

The heading of an FT collective identifies the message as an FT along with a 6-digit date-time group giving the transmission time. For example, "FT130940 means a collective transmitted on the 13th at 0940Z. A collective FT message will usually be broken down into states, ie "TX 130940" would be followed by a group of FTs for terminals in the state of Texas.

Out of Sequence FTs.

A delayed, corrected, or amended FT is identified in the message rather than in the heading. The following are a delayed FT for Binghamton NY, a corrected FT for Memphis TN, and an amended FT for Lufkin TX.

BGM FT RTD 131615 1620Z 100 SCT 250 SCT 1810. 18Z 50 SCT 100 SCT 1913 CHC C30 BKN 3TRW AFT 20Z. 03Z 100 SCT C250 BKN. 09Z VFR..

MEM FT COR 132222 2230Z 40 SCT 300 SCT CHC TRW. 02Z CLR. 16Z VFR..

LFK FT AMD 1 131410 1425Z C8 OVC 4F OVC OCNL BKN. 15Z 20 SCT 250-BKN. 19Z 40 SCT 120 SCT CHC C30 BKN 3TRW. 04Z MVFR CIG F..

Note in each forecast a time group following the valid period which is the issue time. For example, the BGM delayed forecast was issued at 1620Z and not at the scheduled issuance time of 1440Z. This changes the beginning of the valid forecast period from 1500Z to 1600Z.

An FT will not be issued unless two consecutive observations are received from the station. So a routine delayed FT (RTD) is usually due to a station not being on a 24 hour observing schedule with the first two observations of the day being received after the regularly scheduled FT issuance time. If the time is known when observations usually end for the day the phrase "NO AMDTS AFT (TIME) Z" will be appended to the *last* scheduled FT. For example, let's say Oklahoma City's last observation for the day is 03Z. To indicate that no amendments will be available after 03Z due to lack of observations, the last regularly scheduled FT might look like this:

OKC 172222 CLR 16Z VFR CLR. NO AMDTS AFT 03Z..

A new FT will not be issued until two (2) consecutive observations are received.

A corrected FT is necessary due to a typographical error in the FT.

An amended FT is necessary for a situation in which the forecast has to be revised due to significant changes in the weather. Note also that the amended forecast for LFK has the entry "AMD 1". Amended FTs for each terminal are numbered sequentially starting after each scheduled forecast.

International Civil Aviation Organization (ICAO) Terminal Forecast (TAF)

Terminal forecasts for long overwater international flights (TAF) are in an alphanumeric code. They are scheduled four times daily for 24-hour periods beginning at 0000Z, 0600Z, 1200Z and 1800Z (see Section 14 for issuance times).

Format. The TAF is a series of groups made up of digits and letters. An individual group is identified by its position in the sequence, by its alphanumeric coding or by a numerical indicator. Listed below are a few contractions used in the TAF. Some of the contractions are followed by time entries indicated by "tt" or "tttt" or by probability, "pp".

Significant weather change indicators

GRADU tttt—A gradual change occurring during a period in excess of one-half hour. "tttt" are the beginning and ending times of the expected change to the nearest hour; i.e., "GRADU 1213" means the transition will occur between 1200Z and 1300Z.

RAPID tt —A rapid change occurring in one-half hour or less. "tt" is the time to the nearest hour of the change; i.e., "RAPID 23" means the change will occur about 2300Z.

Variability terms—indicate that short time period variations from prevailing conditions are expected with the total occurrence of these variations less than 1/2 of the time period during which they are called for.

TEMPO tttt—Temporary changes from prevailing conditions of less than one hour duration in each instance. There may be more than one (1) instance for a specified time period. "tttt" are the earliest and latest times during which the temporary changes are expected; i.e., "TEMPO 0107" means the temporary changes may occur between 0100Z and 0700Z.

INTER tttt —Changes from prevailing conditions are expected to occur frequently and briefly. "tttt" are the earliest and latest times the brief changes are expected; i.e., "IN-

TER 1518" means that the brief, but frequent, changes may occur between 1500Z and 1800Z. INTER has shorter and more frequent changes than TEMPO.

Probability

PROB pp —Probability of conditions occurring. "pp" is the probability in percent; i.e., "PROB 20" means a 10 or 20% probability of the conditions occurring. "PROB 40" means a 30 to 50% inclusive probability.

Cloud and weather terms

CAVOK —No clouds below 50,000 feet or below the highest minimum sector altitude whichever is greater, and no cumulonibus. Visibility *6 miles* or greater. No precipitation, thunderstorms, shallow fog or low drifting snow.

WX NIL —No significant weather (no precipitation, thunderstorms or obstructions to vision).

SKC —Sky clear.

Following is a St. Louis MO forecast in TAF code. It is the same as the preceding FT example on page 4-1 except that it begins 2 hours later.

KSTL 1212 33025/35 0800 71SN 9//05 INTER 1215 0000 75XXSN 9//000 GRADU 1516 33020 4800 38BLSN 7SC030 PROB 40 85SNSH GRADU 2122 33015 9999 WX NIL 3SC030 RAPID 00 VRB05 9999 SKC GRADU 0304 24015/25 CAVOK Ⓛ

The forecast is broken down into the elements lettered "a" to "1" to aid in the discussion. Not included in the example but explained at the end are three optional forecast groups for "m" icing, "n" turbulence and "o" temperature.

KSTL	*1212*	*33025/35*	*0800*	*71SN*	*9//005*
a.	b.	c.	d.	e.	f.

INTER 1215 0000 75XXSN 9//000
g.

GRADU 1516 33020 4800 38BLSN 7SC030
h.

PROB 40 85SNSH
i.

GRADU 2122 33015 9999 WX NIL 3SC030
j.

RAPID 00 VRB05 9999 SKC
k.

GRADU 0304 24015/25 CAVOK Ⓛ
l.

a. Station identifier. The TAF code uses ICAO 4-letter station identifiers. In the contiguous 48 states the 3-letter identifier is prefixed with a "K"; i.e., the 3-letter identifier for Seattle is SEA while the ICAO identifier is KSEA. Elsewhere, the first two letters of the ICAO identifier tell what region the station is in. "MB" means Panama/Canal Zone (MBHO is Howard AFB); "MI" means Virgin Islands (MISX is St. Croix); "MJ" is Puerto Rico (MJSJ is San Juan); "PA" is Alaska (PACD is Cold Bay); " PH" is Hawaii (PHTO is Hilo).

b. Valid time. Valid time of the forecast follows station identifier. "1212" means a 24-hour forecast valid from 1200Z until 1200Z the following day.

c. Wind. Wind is forecast usually by a 5-digit group giving degrees in 3 digits and speed in 2 digits. When wind is expected to be 100 knots or more, the group is 6-digits with speed given is 3 digits. When speed is gusty or variable, peak speed is separated from average speed with a slash. For example, in the KSTL TAF, "33025/35" means wind 330 degrees, average speed 25 knots, peak speed 35 knots. A group "160115/130" means wind 160 degrees, 115 knots, peak speed 130 knots. "00000" means calm; "VRB" followed by speed indicates direction variable; i.e., "VRB10" means wind direction variable at 10 knots.

d. Visibility. Visibility is in meters. TABLE 4-5 is a table for converting meters to miles and fractions. "0800" means 800 meters converted from table to 1/2 mile.

e. Significant weather. Significant weather is decoded using TABLE 4-3. Groups in the table are numbered sequentially. Each number is followed by an acronym suggestive of the weather; you can soon learn to read most of the acronyms without reference to the table. Examples: "177TS", thunderstorm; "18SQ", squall; "31SA", sandstorm; "60RA", rain; "85SNSH", snow shower. "XX" freezing rain. In the KSTL forecast, "71SN" means light snow. The TAF encodes only the single most significant type of weather; the U.S. domestic FT permits encoding of multiple weather types. See TABLE 4-4 to convert weather from FT to TAF.

f. Clouds. A cloud group is a 6-character group. The first digit is coverage in octas (eighths) of the individual cloud layer only. The summation of cloud layer to determine total sky cover from a ground observers point of view is NOT used. See TABLE 4-6. The two letters identify cloud type as shown in the same table. The last

Table 4-3. TAF weather codes

Code	*Simple Definition*	*Detailed Definition*
04FU	Smoke	Visibility reduced by smoke. No visibility restriction.*
05HZ	Dust haze	Visibility reduced by haze. No visibility restriction.
06HZ	Dust haze	Visibility reduced by dust suspended in the air but wind not strong enough to be adding more dust. No visibility restriction.*
07SA	Duststorm, sandstorm, rising dust or sand	Visibility reduced by dust suspended in the air and wind strong enough to be adding more dust. No well developed dust devils, duststorm or sandstorm. Visibility 6 miles or less.
08PO	Dust devil	Basically the same as 07SA but with well developed dust devils. Visibility 6 miles or less.
10BR	Mist	Fog, ground fog or ice fog with visibility 5/8 to 6 milcs.
11MIFG	Shallow fog	Patchy shallow fog (less than 6 feet deep and coverage less than half) with visibility in the fog less than 5/8 mile.
12MIFG	Shallow fog	Shallow fog (less than 6 feet deep with more or less continuous coverage) with visibility in the fog less than 5/8 miles.
17TS	Thunderstorm	Thunderstorm at the station but with no precipitation.
18SQ	Squall	No precipitation. A sudden increase of at least 15 knots in average wind speed, sustained at 20 knots or more for at least one (1) minute. Not an easy thing to forecast!
19FC	Funnel cloud	Used to forecast a tornado, funnel cloud or waterspout at or near the station. Also not easy to forecast and likely to be overshadowed by some other more violent weather such as thunderstorms.
30SA	Duststorm, sandstorm, rising dust or sand	Duststorm or sandstorm, visibility 5/16 to less than 5/8 mile, decreasing in intensity.
31SA		Basically the same as 30SA but with no change in intensity.
32SA		Basically the same as 30SA but increasing in intensity.
33XXSA	Heavy duststorm or sandstorm	Severe duststorm or sandstorm, visibility less than 5/16 mile, decreasing in intensity.
34XXSA		Basically the same as 33XXSA but with no change in intensity.
35XXSA		Basically the same as 33XXSA but increasing in intensity.
36DRSN	Low drifting snow	Low drifting snow (less than 6 feet) with visibility in drifting snow less than 5/16 miles.
37DRSN		Low drifting snow (less than 6 feet) with visibility in drifting snow less than 5/16 miles.
38BLSN	Blowing snow	Blowing snow (more than 6 feet) with visibility 5/16 to 6 miles.
39BLSN		Blowing snow (more than 6 feet deep) with visibility 5/16 to 6 miles.
40BCFG	Fog patches	Distant fog (not at station).
41BCFG		Patchy fog at the station, visibility in the fog patches less than 5/8 miles.
42FG	Fog	Fog at the station, visibility less than 5/8 mile, sky visible, fog thinning.
43FG		Fog at the station, visibility less than 5/8 mile, sky not visible, fog thinning.
44FG		Fog at the station, visibility less than 5/8 mile, sky visible, no change in intensity.
45FG		Fog at the station, visibility less than 5/8 mile, sky not visible, no change in intensity.
46FG		Fog at the station, visibility less than 5/8 mile, sky visible, fog thickening.
47FG		Fog at the station, visibility less than 5/8 mile, sky not visible, fog thickening.

* While this may seem to be contradictory, it means that while visibility *is* restricted, the *amount* of the restriction is not limited.

NOTE: In code figures 40 through 47, "fog" includes both fog and ice fog.
See FMH No.1 (Surface Observations) for definitions of precipitation intensities.

TABLE 4-3. TAF weather codes (Cont.)

Code	*Simple Definition*	*Detailed Definition*
48FZFG	Freezing	Fog depositing rime ice, visibility less than 5/8 mile, sky visible.
49FZFG	fog	Fog depositing rime ice, visibility less than 5/8 mile, sky not visible.
50DZ		Light intermittent drizzle.
51DZ	Drizzle	Light continuous drizzle.
52DZ		Moderate intermittent drizzle.
53DZ		Moderate continuous drizzle.
54XXDZ	Heavy	Heavy intermittent drizzle.
55XXDZ	drizzle	Heavy continuous drizzle.
56XXDZ	Freezing drizzle	Light freezing drizzle.
57XXFZDZ	Heavy freezing drizzle	Moderate or heavy freezing drizzle.
58RA		Mixed rain and drizzle, light.
59RA		Mixed rain and drizzle, moderate or heavy.
60RA	Rain	Light intermittent rain.
61RA		Light continuous rain.
62RA		Moderate intermittent rain.
63RA		Moderate continuous rain.
64XXRA	Heavy	Heavy intermittent rain.
65XXRA	rain	Heavy continuous rain.
66FZRA	Freezing rain	Freezing rain or mixed freezing rain and freezing drizzle, light.
67XXFZRA	Heavy freezing rain	Freezing rain or mixed freezing rain and freezing drizzle, moderate or heavy.
68RASN	Rain and snow	Mixed rain and snow or drizzle and snow, light.
69XXRASN	Heavy rain and snow	Mixed rain and snow or drizzle and snow, moderate or heavy.
70SN		Light intermittent snow.
71SN	Snow	Light continuous snow.
72SN		Moderate intermittent snow.
73SN		Moderate continuous snow.
74XXSN	Heavy	Heavy intermittent snow.
75XXSN	snow	Heavy continuous snow.

NOTE: See FMH No.1 (Surface Observations) for definitions of precipitation intensities.

TABLE 4-3. TAF weather codes (Cont.)

Code	*Simple Definition*	*Detailed Definition*
77SG	Snow grains	Snow grains, any intensity. May be accompanied by fog or ice fog.
79PE	Ice pellets	Ice pellets, any intensity. May be mixed with some other precipitation.
80RASH	Showers	Light rain showers.
81XXSH	Heavy	Moderate or heavy rain showers.
82XXSH	showers	Violent rain showers (more than 1 inch per hour or 0.1 inch in 6 minutes).
83RASN	Showers of rain and snow	Mixed rain showers and snow showers. Intensity of both showers is light.
84XXRASN	Heavy showers of rain and snow	Mixed rain showers and snow showers. Intensity of either shower is moderate or heavy.
85SNSH	Snow showers	Light snow showers.
86XXSNSH	Heavy snow showers	Moderate or heavy snow showers.
87GR	Soft hail	Light ice pellet showers. There may also be rain or mixed rain or snow.
88GR		Moderate or heavy ice pellet showers. There may also be rain or mixed rain and snow.
89GR	Hail	Hail, not associated with a thunderstorm. There may be also rain or mixed rain and snow.
90XXGR	Heavy hail	Moderate or heavy hail, not associated with a thunderstorm. There may also be rain or mixed rain or snow.
91RA	Rain	Light rain or light rain shower at the time of the forecast *and* thunderstorms during the preceding hour but not at the time of the forecast.
92XXRA	Heavy rain	Basically the same as 91RA but the intensity of the rain or rain shower is moderate or heavy.
93GR	Hail	Basically the same as 91RA but the precipitation is light snow or snow showers, or light mixed rain and snow or rain showers and snow showers, or light ice pellets or ice pellet showers.

TABLE 4-3. TAF weather codes (Cont.)

Code	Simple Definition	Detailed Definition
94XXGR	Heavy hail	Basically the same as 93GR but the intensity of any precipitation is moderate or heavy.
95TS	Thunderstorm	Thunderstorm with rain or snow, or a mixture of rain and snow, but no hail, ice pellets or snow pellets.
96TSGR	Thunderstorm with hail	Thunderstorm with hail, ice pellets or snow pellets. There may also be rain or snow, or mixed rain and snow.
97XXTS	Heavy thunderstorm	Severe thunderstorm with rain or snow, or a mixture of rain and snow, but no hail, ice pellets or snow pellets.
98TSSA	Thunderstorm with duststorm or sandstorm	Thunderstorm with duststorm or sandstorm. There may also be some form of precipitation with the thunderstorm.
99XXTSGR	Heavy thunderstorm with hail	Basically the same as 97XXTS but in addition to everything else there is hail.

three digits are cloud height in hundreds of feet above ground level (AGL). In the KSTL TAF, "9//005" means sky obscured (9), clouds not observed (//), vertical visibility 500 feet (005). The TAF may include as many cloud groups as necessary to describe expected sky condition.

g. Expected variation from prevailing conditions. Variations from prevailing conditions are identified by the contractions INTER and TEMPO as defined earlier. In the KSTL TAF, "INTER 1215 0000 75XXSN 9//000" means intermittently from 1200Z to 1500Z (1215) visibility zero meters (0000) or zero miles, heavy snow (75XXSN), sky obscured, clouds not observed, vertical visibility zero (9//000).

h, i, j, k, and l. An expected change in prevailing conditions is indicated by the contraction GRADU and RAPID as defined earlier. In the KSTL TAF, "GRADU 1516 33020 4800 38BLSN 7SC030" means a gradual change between 1500Z and 1600Z to wind 330 degrees at 20 knots, visibility 4,800 meters or 3 miles (TABLE 4-5), blowing snow, 7/8 stratocumulus (TABLE 4-6) at 3000 feet AGL. "PROB 40 85SNSH" means there is a 30 to 50% probability that light snow showers will occur between 1600Z and 2100Z. "GRADU 2122 33015 9999 WX NIL 3SC030" means a gradual change between 2100Z and 2200Z to wind 330 degrees at 15 knots, visibility 10 kilometers or more (more than 6 miles), no significant weather, 3/8 stratocumulus at 3000 feet. "RAPID 00 VRB05 9999 SKC" means a rapid change about 0000Z to wind direction variable at 5 knots, visibility more than 6 miles, sky clear. "GRADU 0304 24015/25 CAVOK Φ" means a gradual change between 0300Z and 0400Z to wind 240 degrees at 15 knots, peak gust to 25 knots with CAVOK conditions. Φ means end of message.

m. Icing. An icing group may be included. It is a 6-digit group. The first digit is 6 identifying it as an icing group. The second digit is the type of ice accretion from TABLE 4-7 top. The next three digits are height of the base of the icing layer in hundreds of feet (AGL). The last digit is the thickness of the layer in *thousands* of feet. For example, let's decode the group "680304". "6" indicates an icing forecast; "8" indicates severe icing in cloud; "030" says the base of the icing is at 3,000 feet (AGL); and "4" specifies a layer 4,000 feet thick.

n. Turbulence. A turbulence group also may be included. It also is a 6-digit group coded the same as the icing group except a "5" identifies the group as a turbulence forecast. Type of turbulence is from TABLE 4-7 bottom. For example, decoding the group "590359", "5" indentifies a turbulence forecast; "9" specifies frequent severe turbulence in cloud (TABLE 4-7); "035" says the base of the turbulent layer is 3,500 feet (AGL); "9" specifies that the turbulence layer is 9,000 feet thick.

When either an icing layer or a turbulent layer is expected to be more than 9,000 feet thick, multiple groups are used. The top specified in one group is coincident with the base in the following group. Let's assume the forecaster expects frequent turbulence from the surface to 45,000 feet with the most hazardous turbulence at mid-levels. This could be encoded "530005 550509 592309 553209 554104". While you most likely will never see such a complex coding with this many groups, the flexible TAF code permits it.

o. Temperature. A temperature code is seldom included in a terminal forecast. However, it may be included if critical to aviation. It may be used to alert the pilot to high density altitude or possible frost when on the ground. The temperature group is identified by the digit "0". The next two (2) digits are the time to the nearest whole hour (GMT) to which the fore-

TABLE 4-4. Converting significant weather from U.S. terms to WMO terms.

When forecasting any of these in U.S. domestic code	Express the TAF code equivalents as shown in the appropriate column below	Precipitation & Intensity		
		Light	Moderate	Heavy
a	89GR			
BD or BN (vsby 5/16 to 1/2 mi)	31SA			
BD or BN (vsby 0 to 1/4 mi)	34XXSA			
BS (vsby 6 mi or less)	38BLSN			
D (vsby 6 mi or less)	06HZ			
GF (vsby 1/2 mi or less)	44FG			
H (vsby 6 mi or less)	05HZ			
F or IF (vsby 1/2 mi or less)	45FG			
F, GF or IF (vsby 5/8 to 6 mi)	10BR			
IP	79PE			
IPW	87GR			
K	04FU			
L		51DZ	53DZ	55DZ
R		61RA	63RA	64RA
RS		68RASN	68RASN	69XXRASN
RW		80RASH	80RASH	81XXSH
RWSW		83RASN	83RASN	84XXRASN
S		71SN	73SN	75XXSN
SG	77SG			
SP	87GR			
SW		85SNSH	85SNSH	86XXSNSH
ZL		56FZDZ	56FZDZ	57XXFZDZ
ZR		66FZRA	67XXFZRA	67XXFZRA
TRW− or TRW	95TS			
TRW+	95TS			INTER 81XXSH
				INTER 82XXSH*
TRW−A or TRWA	96TSGR			
T+RW	97XXTS			
T+RW+	97XXTS			INTER 81XXSH
	97XXTS			INTER 82XXSH*
T+RWA	99XXTSGR			
T+RW+A	99XXTSGR			INTER 81XXSH
	99XXTSGR			INTER 82XXSH*

* INTER 82XXSH is to be encoded in a TAF only when a violent rainshower (at least 1 inch of rain per hour or 0.10 inch in 6 minutes) is forecast.
NOTE: Conversions from TAF to FT will not be exact in some cases due to a lack of a one to one relationship.

cast temperature applies. The last two (2) digits are temperature in degrees Celsius. A minus temperature is preceded by the letter "M". Examples: "02137" means temperature at 2100Z is expected to be 37 degrees Celsius (about 99 degrees F); "012M02" means temperature at 1200Z is expected to be minus 2 degrees Celsius. A forecast may include more than one temperature group.

Example of a DOMESTIC FA

DFWH FA 041040
HAZARDS VALID UNTIL 042300
OK TX AR LA TN MS AL AND CSTL WTRS

TABLE 4-5. Visibility conversion—TAF code to miles

Meters	Miles	Meters	Miles	Meters	Miles
0000	0	1200	3/4	3000	1 7/8
0100	1/16	1400	7/8	3200	2
0200	1/8	1600	1	3600	2 1/4
0300	3/16	1800	1 1/8	4000	2 1/2
0400	1/4	2000	1 1/4	4800	3
0500	5/16	2200	1 3/8	6000	4
0600	3/8	2400	1 1/2	8000	5
0800	1/2	2600	1 5/8	9000	6
1000	5/8	2800	1 3/4	9999	>6

TABLE 4-6. TAF cloud code

Code	Cloud amount		Cloud type
0	0 (clear)	CI	Cirrus
1	1 octa or less but not zero	CC	Cirrocumulus
2	2 octas	CS	Cirrostratus
3	3 octas	AC	Altocumulus
4	4 octas	AS	Altostratus
5	5 octas	NS	Nimbostratus
6	6 octas	SC	Stratocumulus
7	7 octas or more but not 8 octas	ST	Stratus
		CU	Cumulus
8	8 octas (overcast)	CB	Cumulonimbus
9	Sky obscured or cloud amount not estimated	//	Cloud not visible due to darkness or obscuring phenomena

TABLE 4-7. TAF icing and turbulence

Figure code	Amount of ice accretion (TAF group 6)
0	No icing
1	Light icing
2	Light icing in cloud
3	Light icing in precipitation
4	Moderate icing
5	Moderate icing in cloud
6	Moderate icing in precipitation
7	Severe icing
8	Severe icing in cloud
9	Severe icing in precipitation

Figure code	Turbulence (TAF group 5)
0	None
1	Light Turbulence
2	Moderate turbulence in clear air, infrequent
3	Moderate turbulence in clear air, frequent
4	Moderate turbulence in cloud, infrequent
5	Moderate turbulence in cloud, frequent
6	Severe turbulence in clear air, infrequent
7	Severe turbulence in clear air, frequent
8	Severe turbulence in cloud, infrequent
9	Severe turbulence in cloud, frequent

FLT PRCTNS. . . TURBC. . . TN AL AND CSTL
WTRS . . . ICG. . . TN
. . . IFR. . . TX
TSTMS IMPLY PSBL SVR OR GTR TURBC
SVR ICG AND LLWS NON MSL HGTS NOTED
BY AGL OR CIG
THIS FA ISSUANCE INCORPORATES THE
FOLLOWING AIRMETS STILL IN EFFECT
. . . NONE.

DFWS FA 041040
SYNOPSIS VALID UNTIL 050500
AT 11Z RDG OF HI PRES ERN TX NWWD TO
CNTRL CO WITH HI CNTR OVR ERN TX. BY
05Z HI CNTR MOVS TO CNTRL LA.

DFWI FA 041040
ICING AND FRZLVL VALID UNTIL 042300
TN
FROM SLK TO HAT TO MEM TO ORD TO SLK
OCNL MDT RIME ICGIC ABV FRZLVL TO
100. CONDS ENDING BY 17Z. FRZLVL 80
CHA SGF LINE SLPG TO 120 S OF A IAH MAF
LINE.

DFWT FA 041040
TURBC VALID UNTIL 042300
TN AL AND CSTL WTRS
FROM SLK TO FLO TO 90S MOB TO MEI TO
BUF TO SLK
OCNL MDT TURBC 250-380 DUE TO JTSTR.
CONDS MOVG SLOLY EWD AND CONTG
BYD 23Z.

DFWC FA 041040
SGFNT CLOUD AND WX VALID UNTIL 042300
. . . OTLK 042300-050500
IFR. . . TX
FROM SAT TO PSX TO BRO TO MOV TO SAT
VSBY BLO 3F TIL 15Z.
OK AR TX LA MS AL AND CSTL WTRS
80 SCT TO CLR EXCP VSBY BLO 3F TIL 15Z
OVR PTNS S CNTRL TX. OTLK. . . VFR.
TN
CIGS 30-50 BKN 100 VSBYS OCNLY 3-5F
BCMG AGL 40-50 SCT TO CLR BY 19Z.
OTLK. . . VFR.

DOMESTIC AREA FORECAST (FA)

An area forecast (FA) is a forecast of general weather conditions over an area the size of several states. It is used to determine forecast enroute weather and to interpolate conditions at airports which do not have FTs issued. Figure 1-5, section 1, maps the FA areas. FAs are issued 3 times a day by the National Aviation Weather Advisory Unit (NAWAU) in Kansas City for each of the 6 areas in the contiguous 48 states. In Alaska, FAs are issued by the WSFOs in Anchorage, Fairbanks, and Juneau for their respective areas (figure 1-5A). The WSFO in Honolulu issues FAs for Hawaii (figure 1-5A). See Section 14 for issuance times.

Each FA consists of a 12 hour forecast plus a 6 hour outlook. All times are Greenwich Mean Time (GMT). All distances except visibility are in nautical miles. Visibility is in statute miles.

The FA is comprised of 5 sections, HAZARDS/FLIGHT PRECAUTIONS (H), SYNOPSIS (S), ICING (I), TURBULENCE (T) (AND LOW LEVEL WIND SHEAR, if applicable), and SIGNIFICANT CLOUDS AND WEATHER (C). Each section has an unique communications header which allows replacement of individual sections, due to amendments or corrections, instead of replacing the entire FA. For example (using the FA example on the previous page):

DFWH FA 041040

states that this section of the FA which deals with the hazards section (H) has been issued on the 4th day of the month at 1040Z for the Dallas-Fort Worth (DFW) forecast area.

HAZARDS/FLIGHT PRECAUTIONS (H) Section

A 12 hour forecast that identifies and locates aviation weather hazards which meet Inflight Advisory criteria and thunderstorms that are forecast to be at least scattered in area coverage. These hazards include IFR conditions, icing (ICG), turbulence (TURBC), mountain obscurations (MTN OBSCN), and thunderstorms (TSTMS). A discussion of the hazards section from the above DFW FA continues:

HAZARDS VALID UNTIL 042300

This states that the hazards listed may be valid for the 12 hour forecast time period 11Z to 23Z or for only a portion of the time period. If a specific time period for a hazard is to be stated, it will be stated in the appropriate subsequent section.

OK TX AR LA TN MS AL AND CSTL WTRS

This identifies the states and geographical area that make up the DFW forecast area. This statement will be found in all DFW FAs and does *not* outline the hazard areas.

FLT PRCTNS...TURBC...TN AL AND
CSTL WTRS...ICG...TN
...IFR...TX

This states that TURBC, ICG, and IFR conditions are forecast within the 12 hour period for the listed states, within the designated FA boundary. The forecasts are stated in subsequent sections. If no hazards are expected, "NONE EXPECTED" will be written.

TSTMS IMPLY PSBL SVR OR GTR TURBC
SVR ICG AND LLWS

is found in all FAs as a reminder of the hazards existing in all thunderstorms. Thus, these thunderstorm associated hazards are not spelled out within the body of the FA.

NON MSL HGTS NOTED BY AGL OR CIG

You will find that this statement is contained in all FAs to alert the user that heights, for the most part, are *above sea level.* All heights are in hundreds of feet. For example, 30 BKN 100 HIR TRRN OBSCD means bases of broken clouds 3,000 feet with tops 10,000 feet MSL. Terrain above 3,000 will be obscured. The tops of clouds and icing/freezing level heights are *always* MSL.

Heights *above ground level* will be denoted in either of two ways:

(1) Ceilings by definition are above ground. Therefore, the contraction "CIG" indicates above ground. For example, "CIGS GENLY BLO 10" means that ceilings are expected to be generally below 1,000 feet.

(2) The contraction "AGL" means above ground level. Therefore, "AGL" 20 SCT" means scattered clouds with bases 2,000 feet above ground level.

Thus, if the contraction "AGL" or "CIG" is not denoted, height is automatically above MSL.

THIS FA ISSUANCE INCORPORATES THE FOLLOWING AIRMETS STILL IN EFFECT ...NONE.

A statement contained in all FAs stating that any AIRMET in effect at the time of FA issuance is incorporated into the FA. The AIRMET is then cancelled. In this example, no AIRMETs were in effect when this FA was issued.

SYNOPSIS Section

A brief summary of the location and movement of fronts, pressure systems, and circulation patterns for an 18 hour period.

ICING Section

A forecast of non-thunderstorm related icing of light or greater intensity for up to 12 hours. If a trace or less of icing is expected, the remark "NO SGFNT ICING EXPCD" is used. Otherwise, the location of each icing phenomenon is specified in a separate paragraph containing (1) The affected states or areas within the designated FA boundary, (2) The VOR points outlining the *entire* area of icing and (3) the type, intensity, and heights of the icing. For example:

TN
FROM SLK TO HAT TO MEM TO ORD TO SLK
OCNL MDT RIME ICGIC ABV FRZLVL TO 100

identifies Tennessee as the only state within the DFW forecast area that is forecast to experience OCNL MDT RIME ICGIC ABV FRZLVL TO 100. The entire area of icing is enclosed by the VOR points (see figure 4-2 on page 4-38) listed and includes states covered in other FA areas. The lowest freezing level heights are specified in a separate statement in hundreds of feet MSL.

TURBULENCE/LOW LEVEL WIND SHEAR Section

This section forecasts non-thunderstorm related turbulence of moderate or greater intensity and low level wind shear for up to 12 hours. If moderate or greater turbulence is not expected, the remark "NO SGFNT TURBC EXPCD" is used. Otherwise, each location of turbulence phenonmenon is specified in a separate paragraph using the same format found in the icing section. For example:

TN AL AND CSTL WTRS
FROM SLK TO FLO TO 90S MOB TO MEI TO BUF TO SLK
OCNL MDT TURBC 250-380 DUE TO JTSTR. CONDS MOVG SLOLY EWD AND CONTG BYD 23Z.

Low level wind shear (LLWS) potential, when forecast, is included as a separate paragraph. It is omitted from the DFW FA example above since none was forecast. If LLWS had been forecast, an example of how it would appear follows:

LLWS POTENTIAL OVR WRN NY FROM 03Z-05Z DUE TO STG WMFNT.

Note: If the total area to be affected by any hazard (ICG, TURBC, MTN OBSCN, IFR, TSTMS) during the forecast period (as outlined by VOR's) is very large, it could be only a portion of this total area may be affected at any one time.

As a specific example, let us examine the above forecast of jetstream turbulence from 250-380. Figure 4-1 outlines the total areas to be affected during the 12 hour period. The forecast for CONDITIONS MOVING SLOWLY EASTWARD AND CONTINUING BEYOND 23Z tells us that the phenomenon will move slowly out of the western portion of the outlined area and into the eastern portion of the outlined area during the 12 hour period. By late in the time period, the western portion will be free of significant turbulence but the eastern portion will continue to experience significant turbulence beyond 23Z.

SIGNIFICANT CLOUD AND WEATHER Section

A 12 hour forecast, in broad terms, of clouds and weather significant to flight operations plus a 6 hour categorical outlook. Table 4-8 defines the contractions and compares them to the designators used in the FT. Surface visibility and obstructions to vision are included when forecast visibility is 5 miles or less. Precipitation, thunderstorms, and sustained winds of 30 knots or greater are always included when forecast. Table 4-9 gives expected coverage indicated by the terms "isolated", "widely scattered", "scattered", and "numerous". Table 4-10 identifies variability terms used.

TABLE 4-8. Contractions in FAs

Contraction	*FT Designator*	*Definition*
CLR	CLR	Sky clear
SCT	SCT	Scattered
BKN	BKN	Broken
OVC	OVC	Overcast
OBSCD	X	Obscured
PTLY OBSCD	−X	Partly obscured
	−	Thin
CIG	C	Ceiling

TABLE 4-9. Area coverage of showers and thunderstorms

Adjective	*Coverage*
Isolated	Single cells (no percentage)
Widely scattered	Less than 25% of area affected
Scattered	25 to 54% of area affected
Numerous	55% or more of area affected

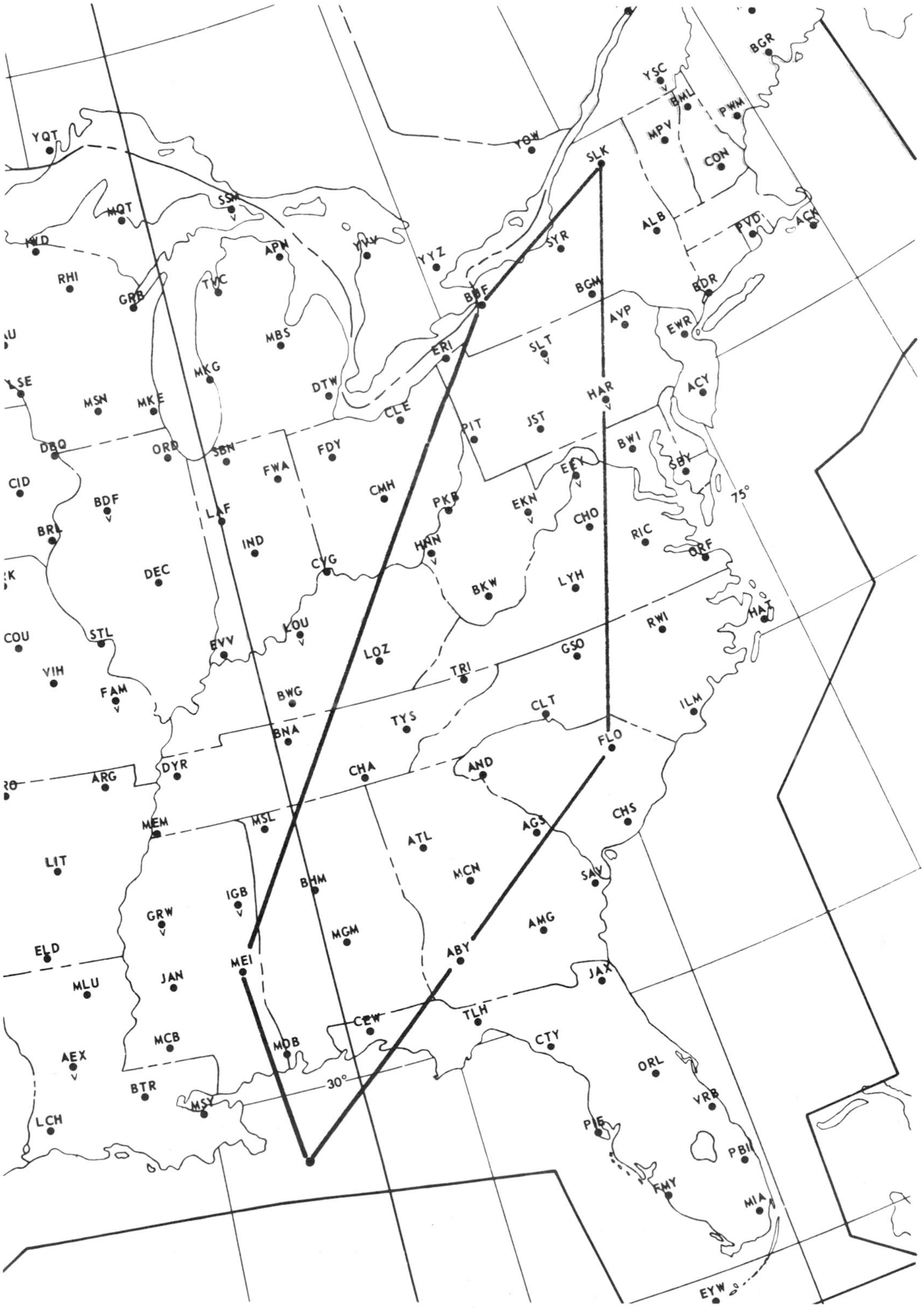

FIGURE 4-1. Area of jetstream turbulence in FA example.

TABLE 4-10. Variability terms

Term	Description
OCNL	Greater than 50% probability of the phenomenon occurring but for less than 1/2 of the forecast period
CHC	30 to 50% probability (precipitation only)
SLGT CHC	10 or 20% probability (precipitation only)

The SGFNT CLOUD AND WX section is usually several paragraphs. The breakdown may be by states, by well known geographical areas, or in reference to location and movement of a pressure system or front. Figure 4-2 is a map to assist you in identifying geographical areas. An example would be

OK AR TX LA MS AL AND CSTL WTRS
80 SCT TO CLR EXCP VSBY BLO 3F TIL
15Z OVR PTNS S CNTRL TX.

A categorical outlook, identified by "OTLK", is included for each area breakdown. For example,

OTLK. . . VFR.

Categorical outlooks of IFR and MVFR can be due to ceilings only (CIG), restrictions to visibility only (TRW F etc.), or a combination of both (CIG TRW F etc.). For example, "OTLK. . . VFR BCMG MVFR CIG F AFT 09Z" means the weather is expected to be VFR becoming MVFR (marginal VFR) due to low ceilings and visibilities restricted by fog after 0900Z. "WIND" is included in the outlook if winds, sustained or gusty, are expected to be 30 knots or greater. For definitions of each category, refer to the section on the weather depiction chart.

FLT PRCTNS from the hazards section which are described in this section, as IFR and MTN OBSCN, will be specified in a separate paragraph at the beginning of the section using the same format as the icing and turbulence sections. TSTMS, however, will *not* be included as a separate paragraph. For example,

IFR. . . TX
FROM SAT TO PSX TO BRO TO MOV TO SAT
VSBY BLO 3F TIL 15Z.

AMENDED AREA FORECASTS

Amendments to the FA are issued as needed. Only that section of the FA being revised is transmitted as an amendment. Area forecasts are also amended and updated by inflight advisories but the FA section affected will always be kept current by amending it. An amended FA is identified by "AMD", a corrected FA is identified by "COR", and a delayed FA is identified by "RTD".

TWEB ROUTE FORECASTS AND SYNOPSIS

The TWEB Route Forecast is similar to the Area Forecast (FA) except information is contained in a route format. Forecast sky cover (height and amount of cloud bases), cloud tops, visibility (including vertical visibility), weather, and obstructions to vision are described for a corridor 25 miles either side of the route. Cloud bases and tops are always MSL unless noted. Ceilings are always above ground.

The Synopsis is a brief statement of frontal and pressure systems affecting the route during the forecast valid period.

The TWEB Route Forecasts are prepared by the WSFOs for more than 300 selected short-leg and cross-country routes over the contiguous U.S. (figure 1-6 and figure 1-7, section 1). WSFOs prepare synopses for the routes in their areas. These forecasts go into the Transcribed Weather Broadcasts (TWEB) and the Pilot's Automatic Telephone Weather Answering Service (PATWAS) transcriptions described in section 1. Individual route forecasts and synopses are also available by request/reply teletypewriter through any FSS or WSO.

The TWEB Route Forecasts and Synopses are issued by the WSFOs three times per day according to time zone. See Section 14 for issuance times. The early morning and midday forecasts are valid for 12 hours and the evening forecast for 18 hours. This schedule provides 24-hour coverage with most frequent updating during the hours of greatest general aviation activity.

Example of a TWEB Synopsis:

BIS SYNS 252317. LO PRES TROF MVG ACRS ND TDA AND TNGT. HI PRES MVG SEWD FM CANADA INTO NWRN ND BY TNGT AND OVR MST OF ND BY WED MRNG.

BIS—Bismarck ND. WSFO issuing Synopsis and Route Forecasts
SYNS—Synopsis for the area covered by the Route Forecasts
25—25th day of the month
2317—Valid 23Z on the 25th to 17Z on the 26th (18 hours)
(Rest of message)
—LOW PRESSURE TROUGH MOVING ACROSS NORTH DAKOTA TODAY AND TONIGHT. HIGH PRESSURE MOVING SOUTHEASTWARD FROM CANADA INTO NORTHWESTERN NORTH DAKOTA BY TONIGHT AND OVER MOST OF NORTH DAKOTA BY WEDNESDAY MORNING.

Example of a TWEB Route Forecast:

249 TWEB 252317 GFK-MOT-ISN. GFK VCNTY CIGS AOA 5 THSD TILL 12Z OTRW OVR RTE CIGS 1 TO 3 THSDS VSBY 3 TO 5 MI IN LGT SNW WITH CONDS BRFLY LWR IN HVYR SNW SHWRS

249—Route number
TWEB—TWEB Route Forecast
25—25th day of month
2317—Valid 23Z on the 25th to 17Z on the 26th (18 hours)
GFK-MOT-ISN-Route: Grand Forks to Minot to Williston ND

(Rest of message)
—GRAND FORKS VICINITY CEILINGS AT OR ABOVE 5000 FEET UNTIL 1200Z OTHERWISE OVER ROUTE CEILINGS 1 TO 3 THOUSAND FEET VISIBILITY 3 TO 5 MILES IN LIGHT SNOW WITH CONDITIONS BRIEFLY LOWER IN HEAVIER SNOW SHOWERS.

When visibility is not stated it is implied to be greater than 6 miles.

Because of their varied accessibility and route format, these forecasts are important and useful weather information available to the pilot for flight operations and planning. You should become familiar with them and use them regularly.

INFLIGHT ADVISORIES (WST, WS, WA)

Inflight advisories are unscheduled forecasts to advise enroute aircraft of development of potentially hazardous weather. All inflight advisories in the conterminous U.S. (48 states) are issued by the National Aviation Weather Advisory Unit (NAWAU) in Kansas City. In Alaska, the three WSFOs (Anchorage, Fairbanks, and Juneau) issue inflight advisories for their respective areas. The WSFO in Honolulu issues advisories for Hawaii. All heights are referenced to MSL, except in the case of ceilings (CIG) which indicates above ground level. The advisories are of three types—CONVECTIVE SIGMET (WST), SIGMET (WS), and AIRMET (WA). All inflight advisories use the same location identifiers (either VORs or well known geographic areas) to describe the hazardous weather areas (see figures 4-2 and 4-3).

CONVECTIVE SIGMET (WST)

CONVECTIVE SIGMETs are issued in the conterminous U.S. for any of the following:

1. Severe thunderstorm due to a) surface winds greater than or equal to 50 knots or b) hail at the surface greater than or equal to 3/4 inches in diameter or c) tornadoes.
2. Embedded thunderstorms.
3. Line of thunderstorms.
4. Thunderstorms greater than or equal to VIP level 4 affecting 40% or more of an area at least 3000 square miles.

Any CONVECTIVE SIGMET implies severe or greater turbulence, severe icing, and low level wind shear. A CONVECTIVE SIGMET may be issued for any convective situation which the forecaster feels is hazardous to all categories of aircraft.

CONVECTIVE SIGMET bulletins are issued for the Eastern (E), Central (C) and Western (W) United States. The areas separate at 87 and 107 degrees west longitude with sufficient overlap to cover most cases when the phenomenon crosses the boundaries. Thus, a bulletin will usually be issued only for the area where the bulk of observed and forecast conditions are located. Bulletins are issued hourly at H+55. Special bulletins are issued at any time as required and updated at H+55. If no criteria meeting a CONVECTIVE SIGMET are observed or forecast, the message "CONVECTIVE SIGMET. . .NONE" will be issued for each area at H+55. Individial SIGMETs for each area are numbered sequentially (01-99) each day, beginning at 00Z. A continuing CONVECTIVE SIGMET phenomenon will be reissued every hour at H+55 with a new number. The text of the bulletin consists of either an observation and a forecast or just a forecast. The forecast is valid for up to 2 hours.

EXAMPLE

The following are examples of CONVECTIVE SIGMET bulletins for the Central U.S., For the Western U.S., they would be numbered 17W, 18W, and 19W while for the Eastern U.S., they would be numbered 17E, 18E, and 19E.

MKCC WST 221655
CONVECTIVE SIGMET 17C
KS OK TX
VCNTY GLD-CDS LINE
NO SGFNT TSTMS RPRTD
FCST TO 1855Z
LINE TSTMS DVLPG BY 1755Z WILL MOV EWD 30-35 KTS THRU 1855Z
HAIL TO 1 1/2 IN PSBL.

MKCC WST 221655
CONVECTIVE SIGMET 18C
SD NE IA
FROM FSD TO DSM TO GRI TO BFF TO FSD
AREA TSTMS WITH FEW EMBDD CELLS
MOVG FROM 2725 TOPS 300
FCST TO 1855Z
DSPTG AREA WILL MOV EWD 25 KNTS.

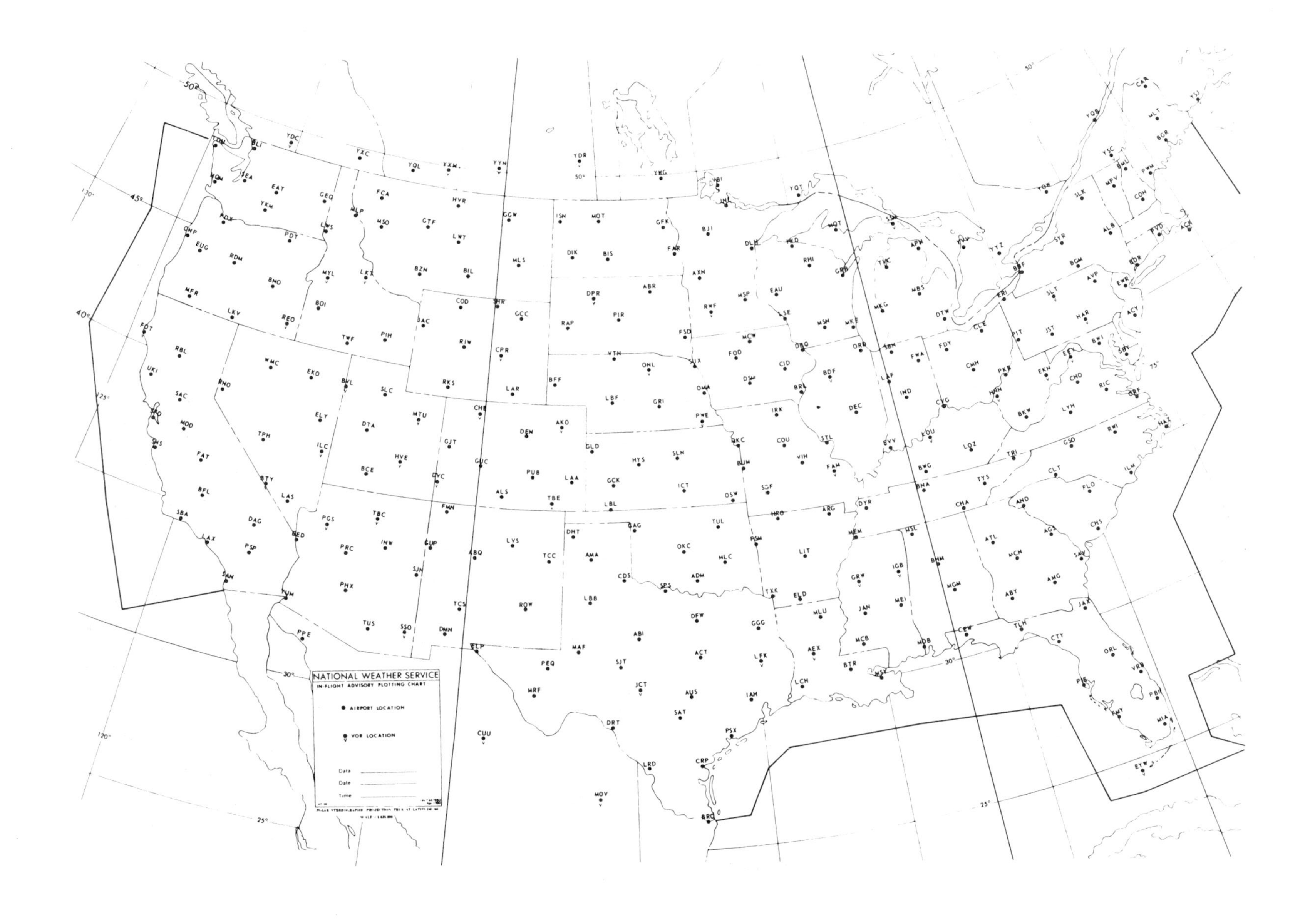

FIGURE 4-2. Inflight Weather Advisory Location Identifier (VORs)

FIGURE 4-3. Geographical areas and terrain features. Forecasts often best locate weather by reference to terrain.

MKCC WST 221755
CONVECTIVE SIGMET 19C
KS OK
FROM 30E TO 20E GAG
DVLPG LINE TSTMS 25 MI WIDE MOVG FROM 2315 TOPS 450.
HAIL TO 1 IN. . . WIND GUSTS TO 55.
FCST TO 1955Z
LINE WILL CONT INTSFYG AND MOV NEWD 25-30 KTS THRU 1955Z. HAIL TO 2 IN PSBL.

The first example is a bulletin issued at 1655Z on the 22nd day of the month. It is the 17th CONVECTIVE SIGMET of the day in the Central U.S.. Although no significant thunderstorm activity is noted at 1655Z, a line of thunderstorms is expected to develop by 1755Z near a Goodland-Childress line (see figure 4-2 for complete list of VORs) in the states of Kansas, Oklahoma, and Texas and move eastward 30-35 knots possibly producing 1 1/2 inch hail thru 1855Z.

Note that 19C is an update and re-issuance of 17C.

SIGMET (WS)/AIRMET (WA)

A SIGMET advises of weather potentially hazardous to all aircraft other than convective activity. In the conterminous U.S., items covered are:

1. Severe icing
2. Severe or extreme turbulence
3. Duststorms, sandstorms, or volcanic ash lowering visibilities to less than three (3) miles

In Alaska and Hawaii there are no CONVECTIVE SIGMETs. In these states we add:

4. Tornadoes
5. Lines of thunderstorms
6. Embedded thunderstorms
7. Hail greater than or equal to 3/4 inch diameter

An AIRMET is for weather that may be hazardous to single engine, other light aircraft, and VFR pilots. AIRMETs should be read by all pilots. The items covered are:

1. Moderate icing
2. Moderate turbulence
3. Sustained winds of 30 knots or more at the surface
4. Ceilings less than 1000 feet and/or visibility less than 3 miles affecting over 50% of the area at one time.
5. Extensive mountain obscurement

These SIGMET/AIRMET items are considered "widespread" because they must be affecting or be forecast to affect an area of at least 3000 square miles at any one time. However, if the total area to be affected during the forecast period (as outlined by VOR's) is very large, it could be that only a small portion of this total area would be affected at any one time. An example would be a 3000 square mile phenomenon forecast to move across an area totaling 25,000 square miles during the forecast period. For a specific example, see turbulence section in the Area Forecast (FA).

SIGMETs/AIRMETs are issued for 6 areas corresponding to the FA areas (figure 1-5) with a maximim forecast period of 4 hours for SIGMETs and 6 hours for AIRMETs. If conditions persist beyond the forecast period, the SIGMET/AIRMET must be updated and reissued.

A phenomenon is identified by an alphabetic designator in the series ALFA through NOVEMBER for SIGMETs and OSCAR through ZULU for AIRMETs. Issuances for the same phenomenon will be sequentially numbered. For example, ALFA 1 is the first issuance for a SIGMET phenomenon, ALFA 2 is the second issuance for the same phenomenon, etc. The first issuance of a SIGMET will be labeled UWS (Urgent Weather SIGMET). UWS will also be used in subsequent issuances at the discretion of the forecaster. For an AIRMET the first issuance will be OSCAR 1, etc. All designators in the series will be used before starting over again with ALFA and OSCAR. The alphabetic designator assigned to a phenomenon will be retained until the phenomenon ends, even when moving from one area into another. For example, the first issuance in the CHI area for phenomenon moving in from the SLC area will be SIGMET ALFA 3 if the previous two issuances, ALFA 1 and ALFA 2 had been in the SLC area. Since no two different phenomena across the country can have the same alphabetic designator at the same time, all 6 areas must use the same ALFA through NOVEMBER and OSCAR through ZULU series.

EXAMPLES:

DFWA UWS 051710
SIGMET ALFA 1 VALID UNTIL 052110
AR LA MS
FROM MEM TO 30N MEI TO BTR TO MLU TO MEM
OCNL SVR ICING ABV FRZLVL EXPCD. FRZLVL 080 E TO 120 W.
CONDS CONTG BYD 2100Z.

SFOB WS 100130
SIGMET BRAVO 2 VALID UNTIL 100530
OR WA
FROM SEA TO PDT TO EUG TO SEA
OCNL MOGR CAT BTN 280 AND 350 EXPCD DUE TO JTSTR. CONDS BGNG AFT 0200Z CONTG BYD 0530Z AND SPRDG OVR CNTRL ID BY 0400Z.

MIAP WAS 151900
AIRMET PAPA 2 VALID UNTIL 160100
GA FL
FROM SAV TO JAX TO CTY TO TLH TO SAV
MDT TURBC BLO 100 EXPCD. CONDS IPVG
AFT 160000Z.

The first example above is a SIGMET bulletin issued for the DFW area at 1710Z on the 5th and is valid until 2110Z (Note maximum forecast period of 4 hours for a SIGMET). The designator ALFA identifies the phenomenon, in this case, severe icing. This is the first issuance of the SIGMET as indicated by "UWS" and "ALFA 1". The affected states *within* the DFW area are Arkansas, Louisiana, and Mississippi. VORs (see figure 4-2) outline the *entire* area to be affected (irrespective of FA boundaries) by severe icing during the forecast period. Freezing level data and notation that conditions are expected to continue beyond 4 hours are included. See TABLE 4-10 for defintions of variability terms.

It is important to note that AIRMETs will only be issued for conditions meeting AIRMET criteria that are *not* already forecast in the area forecast (FA). SIGMETs will be issued whether or not they are forecast in the FA.

WINDS AND TEMPERATURES ALOFT FORECAST (FD)

Winds and temperatures aloft are forecast for specific locations in the contiguous U.S. as shown in figure 1-3, section 1. FD forecasts are also prepared for a network of locations in Alaska as shown in figure 1-3-A, section 1. Forecasts are made twice daily based on 00Z and 12Z data for use during specific time intervals.

Below is a sample FD message containing a heading and six FD locations. The heading always includes the time during which the FD may be used (1700-2100Z in the example) and a notation "TEMPS NEG ABV 24000". Since temperatures above 24,000 feet are always negative, the minus sign is omitted.

FD KWBC 151640
BASED ON 151200Z DATA
VALID 151800Z FOR USE 1700-2100Z TEMPS NEG ABV 24000

FT	3000	6000	9000	12000	18000	24000	30000	34000	39000
ALA			2420	2635−08	2535−18	2444−30	245945	246755	246862
AMA		2714	2725+00	2625−04	2531−15	2542−27	265842	256352	256762
DEN			2321−04	2532−08	2434−19	2441−31	235347	236056	236262
HLC		1707−01	2113−03	2219−07	2330−17	2435−30	244145	244854	245561
MKC	0507	2006+03	2215−01	2322−06	2338−17	2348−29	236143	237252	238160
STL	2113	2325+07	2332+02	2339−04	2356−16	2373−27	239440	730649	731960

FORECAST LEVELS

The line labelled "FT" shows 9 of 11 standard FD levels. The 45,000 and 53,000 foot levels are not transmitted on teletypewriter circuits but are available in the communications system. The pilot may request these levels from the FSS briefer or NWS meteorologist. Through 12,000 feet the levels are true altitude, 18,000 feet and above are pressure altitude. The FD locations are transmitted in alphabetical order.

Note that some lower level groups are omitted. No winds are forecast within 1,500 feet of station elevation. No temperatures are forecast for the 3,000 foot level or for a level within 2,500 feet of station elevation.

DECODING

A 4-digit group shows wind direction (reference true north) and windspeed. Look at the St. Louis (STL) forecast for 3,000 feet. The group 2113 means wind from 210 degrees at 13 knots. The first two digits give direction in tens of degrees and the second two speed in knots.

A 6-digit group includes forecast temperature. In the STL forecast, the coded group for 9,000 feet is 2332+02 which is wind from 230 degrees at 32 knots and temperature +2 degrees Celsius.

Encoded windspeed 100 to 99 knots have 50 added to the direction code and 100 subtracted from the speed. The STL forecast for 39,000 feet is "731960". Wind is from 230 degrees at 119 knots and temperature −60 degrees Celsius.

How do you recognize when coded direction has been increased by 50? Coded direction (in tens of degrees) range from 01 (010 degrees) to 36 (360 degrees). Thus, a coded direction of more than "36" indicates winds 100 knots or more. Coded direction with speeds of over 100 knots range from 51 through 86.

If windspeed is forecast at 200 knots or greater, the wind group is coded as 199 knots; i.e., "7799" is decoded 270 degrees at 199 knots or greater.

When the forecast speed is less than 5 knots, the coded group is "9900" and read, "LIGHT AND VARIABLE".

Examples of decoding FD winds and temperatures:

Coded	Decoded
9900+00	Wind light and variable, temperature 0 degree Celsius
2707	270 degrees at 7 knots
850552	350 degrees (85-50=35) at 105 knots (05+100=105), temperature −52 degrees Celsius

SPECIAL FLIGHT FORECAST

When planning a special category flight and scheduled forecasts are insufficient to meet your needs, you may request a special flight forecast through any FSS or WSO. Special category flights are hospital or rescue flights; experimental, photographic or test flights; record attempts; and mass flights such as air tours, air races and fly-aways from special events.

Make your request far enough in advance to allow ample time for preparing and transmitting the forecast. Advance notice of 6 hours is desirable. In making a request, give the:

1. Aircraft mission
2. Number and type of aircraft
3. Point of departure
4. Route of flight (including intermediate stops, destination, alternates)
5. Estimated time of departure
6. Time enroute
7. Flight restrictions (such as VFR, below certain altitudes, etc.)
8. Time forecast is needed

The forecast is written in plain language contractions as in the examples:

SPL FLT FCST ABQ-PHOTO MISSION-ABQ 121500Z. THIN CI CLDS AVGG LESS THAN TWO TENTHS CVR. VSBY MORE THAN 30. WNDS AND TEMPS ALF AT FLT ALTITUDE 2320+03. ABQ WSFO 121300Z.

SPL FLT OTLK MKC-RST 062100Z–062400Z. CIG 2 THSD OVC OR BTR. WNDS ALF AT FLT ALTITUDE 2320. MKC WSFO 052300Z.

CENTER WEATHER SERVICE UNIT (CWSU) PRODUCTS

Center Weather Service Unit products are issued by the CWSU meteorologist located in the ARTCCs. Coordination between the CWSU meteorologist and the nearby NWS WSFO is extremely important because both will address the same event. If time permits, coordination should take place before the CWSU meteorologist issues a product.

METEOROLOGICAL IMPACT STATEMENT (MIS)

A Meteorological Impact Statement (MIS) is an unscheduled traffic/flight operations planning forecast of conditions expected to begin generally 4 to 12 hours after issuance. This enables the impact of expected weather conditions to be included in traffic control related decisions of the near future.

A MIS will be issued when the following three (3) conditions are met:

1. if any one of the following conditions are forecast.
   a. convective SIGMET criteria
   b. moderate or greater icing and/or turbulence
   c. heavy or freezing precipitation
   d. low IFR conditions
   e. surface winds/gusts 30 knots or greater
   f. low level wind shear within 2000 feet of the surface
   g. volcanic ash, dust, or sandstorm
2. if impact occurs on air traffic flow within the ARTCC area of responsibility.
3. if forecast lead time (the time between issuance and outset of a phenomenon), in the forecaster's judgement, is sufficient to make issuance of a Center Weather Advisory (CWA) unnecessary.

An example of a MIS:

ZKC MIS 02 031800Z-040100Z
ENROUTE. . .NONE.
TERMINAL. . .STL. . .SCT OCNL BKN 015-020 BKN OCNL OVC 040-080 CHC LVL 3/4 TSTMS. WIDELY SCT TSTMS MOVG E 20 KTS. 22Z MVFR CONDS WITH BKN-OVC CIGS 008-015 CHC VSBYS 4-6 FOG/HAZE. SFC WINDS 250-280 10-15 KTS SHFTG 280-310 AFT 22Z.

This MIS from Kansas City MO ARTCC is the 2nd issuance of the day; issued at 1800Z on the 3rd and is valid until 0100Z on the 4th.

CENTER WEATHER ADVISORY (CWA)

A Center Weather Advisory (CWA) is an unscheduled inflight flow control, air traffic and air crew *advisory* for use in anticipating and avoiding adverse weather conditions in the enroute and terminal areas. The CWA is *not* a flight planning forecast but a *nowcast* for conditions beginning within the next two (2) hours. Maximum valid time of a CWA is two (2) hours, i.e. no more than 2 hours between issuance time and "valid until time". If conditions are expected to continue beyond the valid period, a statement will be included in the advisory.

A CWA may be issued for the following three (3) situations:

1. as a supplement to an *existing* inflight advisory or area forecast (FA) section for the purpose of improving or updating the definition of the phenomenon in terms of location, movement, extent, or intensity *relevant* to the ARTCC area of responsibility. This is important for the following reason. A SIGMET for severe turbulence issued by NAWAU may outline the entire ARTCC area for the total four (4) hour valid

period but may only be covering a relatively small portion of the ARTCC area at any one time during the four (4) hour period.

2. when an inflight advisory has not yet been issued but conditions meet inflight advisory criteria based on current pilot reports and the information must be disseminated sooner than NAWAU can issue the inflight advisory. In this case of an impending SIGMET, the CWA will be issued as urgent "UCWA" to allow the fastest possible dissemination.
3. when inflight advisory criteria is not met but conditions are or will shortly be adversely affecting the safe flow of air traffic within the ARTCC area of responsibility.

Format of a CWA heading:

ARTCC Designator and Phenomenon number (numbers 1 through 6 used for replaceability) /"CWA" /issuance number (2 digits) /inflight advisory alphanumeric designator (if applicable) /date and time issued /"—" /valid until time.

Examples of a CWA:

ZFW3 CWA 03 032140-2340
ISOLD SVR TSTM OVR MLU MOVG SWWD 10 KTS. TOP 610. WND GUSTS TO 55 KTS. HAIL TO 1 INCH RPRTD AT MLU. SVR TSTM CONTG BYND 2340.

ZKC1 CWA 01/ALFA 4 121528-1728
NUMEROUS RPRTS OF MDT TO SVR ICG 080-090 30 MILE RADIUS OF STL. LGT OR NEG ICG RPRTD 040-120 RMNDR OF ZKC AREA.

HURRICANE ADVISORY (WH)

When a hurricane threatens a coast line, but is located at least 300 NM off shore, an abbreviated hurricane advisory (WH) is issued to alert aviation interests. The advisory gives location of the storm center, its expected movement, and maximum winds in and near the storm center. It does not contain details of associated weather. Specific ceilings, visibilities, weather, and hazards are found in the area and terminal forecasts and inflight advisories.

An example of an abbreviated aviation hurricane advisory:

MIA WH 181010
HURCN IONE AT 1000Z CNTRD 29.4N 74.2W OR 400 NMI E OF JACKSONVILLE FL EXPCTD TO MOV N ABT 12 KT. MAX WNDS 110 KT OVR SML AREA NEAR CNTR AND HURCN WNDS WITHIN 55-75 NM.

CONVECTIVE OUTLOOK (AC)

A convective outlook (AC) describes the prospects for general thunderstorm activity during the following 24 hours. Areas in which there is a high, moderate or slight risk of severe thunderstorms are included as well as areas where thunderstorms may approach severe limits (approaching means winds greater than or equal to 35 knots but less than 50 knots and/or hail greater than or equal to 1/2 inch in diameter but less than 3/4 inch). Refer to the "Severe Weather Outlook Chart" for "risk" definitions. Forecast reasoning is also included in all ACs.

Outlooks are transmitted by the National Severe Storm Forecast Center (NSSFC) in Kansas City MO at 0800Z and 1500Z, and between February 1 and August 31 at 1930Z. Forecasts in each AC are valid until 1200Z the next day and are used for preparing and updating the Severe Weather Outlook Chart.

Use the outlook primarily for planning flights later in the day.

Severe thunderstorm criteria:

a. wind greater than or equal to 50 knots at the surface or
b. hail greater than or equal to 3/4 inch diameter at the surface or
c. tornadoes

The following is a convective outlook:

MKC AC 031500
VALID 031500—041200Z

THERE IS A MDT RISK OF SVR TSTMS THIS AFTN AND EVE PTNS ERN AL..ERN TN..ERN KY..WV..PA..NY ..VT..NH..MA..CT..NJ..DE..MD.. VA..NC..SC..GA. AREA IS TO RT OF LN FM DHN MGM HSV LOZ HTS PIT SYR MPV PSM BOS GON ACY SBY RDU AGS ABY DHN.

GEN TSTM ACTVY TO RT OF LN FM BPT MLU MEM OWB TOL..CONTD JAX CTY. ALSO TO RT OF LN FM CDC ELY BYI IDA LND RWL DEN CEZ CDC.
UPR LVL LOW NR MLI WITH TROF EXTNDG SWD INTO ERN TX EXPCD TO MOV NEWD. VRY STG UPR LVL JET FM GGG DAY PWM EXPCD TO CONT MOVG SLOLY EWD PROVIDING UPR LVL SHEAR AND DVRG FIELDS. NARROW BAND OF INSTBLTY RANGES FM MINUS 7 IN AL TO MINUS 4 IN SERN NY. RPDLY MOVG CELLS EXPCD TO MOV THRU WARM AND MOIST AIR ORIENTED FM AL TO WV TO ERN NY. ISOLD TRW PSBL THIS AFTN FM SRN ID INTO CO AS WK UPR TROF MOV ACRS AREA OF MARGINAL INSTBLY.

SEVERE WEATHER WATCH BULLETIN (WW)

A severe weather watch bulletin (WW) defines areas of possible severe thunderstorms or tornado activity. The bulletins are issued by the National Severe Storm Forecast Center at Kansas City MO. WWs are unscheduled and are issued as required.

A severe thunderstorm watch describes expected severe thunderstorms and a tornado watch states that the additional threat of tornadoes exists in the designated watch area.

In order to alert the WSFOs, WSOs, CWSUs, and FSS's, and other users, a preliminary message called the Alert Severe Weather Watch message (AWW) is sent before the main bulletin.

Example of a preliminary message:

MKC AWW 161755
WW 279 SEVERE TSTM NY PA NJ
161830Z—17000Z AXIS..70 STATUTE
MILES EITHER SIDE OF LINE..
10W MSS.20E ABE
HAIL SURFACE AND ALOFT..2 INCHES.
WIND GUSTS..65 KNOTS.
MAX TOPS TO 540. MEAN WIND VECTOR
19020. REPLACES WW 278.. OH PA NY.

The Severe Weather Watch Bulletin format:

A. Type of severe weather watch, watch area, valid time period, type of severe weather possible, watch axis, meaning of a watch, and a statement that persons be on the lookout for severe weather.
B. Other watch information..references to previous watches.
C. Phenomena, intensities, hail size, wind speeds (knots), maximum CB tops, and estimated cell movement (mean wind vector).
D. Cause of severe weather.
E. Information on updating ACs.

Example of a Severe Weather Watch Bulletin (WW)

MKC WW 161800
BULLETIN IMMEDIATE BROADCAST REQUESTED
SEVERE THUNDERSTORM WATCH NUMBER 279
NATIONAL WEATHER SERVICE KANSAS CITY MO
200 PM EDT THU JUN 16 1983

A..THE NATIONAL SEVERE STORMS FORECAST CENTER HAS ISSUED A SEVERE THUNDERSTORM WATCH FOR

EASTERN HALF OF NEW YORK
NORTHEASTERN PENNSYLVANIA
NORTHERN NEW JERSEY

FROM 230 PM EDT UNTIL 800 PM EDT THIS THURSDAY AFTERNOON AND EVENING

LARGE HAIL AND DAMAGING THUNDERSTORM WINDS ARE POSSIBLE. IN THESE AREAS. THE SEVERE THUNDERSTORM WATCH AREA IS ALONG AND 70 STATUTE MILES EITHER SIDE OF A LINE FROM 10 MILES WEST OF MASSENA NEW YORK TO 20 MILES EAST OF ALLENTOWN PENNSYLVANIA

REMEMBER....A SEVERE THUNDERSTORM WATCH MEANS CONDITIONS ARE FAVORABLE FOR SEVERE THUNDERSTORMS IN AND CLOSE TO THE WATCH AREA.
PERSONS IN THESE AREAS SHOULD BE ON THE LOOKOUT FOR THREATENING WEATHER CONDITIONS AND LISTEN FOR LATER STATEMENTS AND POSSIBLE WARNINGS.

B..OTHER WATCH INFORMATION..THIS SEVERE THUNDERSTORM WATCH REPLACES SEVERE THUNDERSTORM WATCH NUMBER 278. WATCH NUMBER 278 WILL NOT BE IN EFFECT AFTER 230 PM EDT.

C..A FEW SVR TSTMS WITH HAIL SFC AND ALF TO 2 IN. EXTRM TURBC AND SFC WIND GUSTS TO 65 KT. A FEW CBS WITH MAX TOPS TO 540. MEAN WIND VECTOR 19020.

D..TSTMS EXPCD TO INCRS RPDLY IN ZONE OF WK SFC CONVG WHERE AMS HAS LI OF MINUS 8.

E..OTR TSTMS...WW MAY BE RQD SOON FOR PTNS ERN WY NERN CO AND WRN NEB. UPDATE AC TO INCL GEN TSTM ACTVY IN SRN FL THIS AFTN TO RT OF LINE FROM FMY PBI.

Status reports are issued as needed to show progress of storms and to delineate areas no longer under the threat of severe storm activity. Cancellation bulletins are issued when it becomes evident that no severe weather will develop or that storms have subsided and are no longer severe.

When tornadoes or severe thunderstorms have developed, local WSOs and WSFOs issue local warnings.

Section 5
SURFACE ANALYSIS CHART

A surface analysis is commonly referred to as a surface weather chart. In the contiguous 48 states a computer prepared chart covering these states and adjacent areas is transmitted every three hours. Areas with facsimile receive surface weather charts at regularly scheduled intervals. Figure 5-1 is a section of a surface weather chart and Figure 5-2 illustrates the symbols depicting fronts and pressure centers. The following explains the contents of the chart.

VALID TIME

Valid time of the chart corresponds to the time of the plotted observations. A date-time group in Greenwich Mean Time tells the user when conditions portrayed on the chart were occurring.

ISOBARS

Isobars are solid lines depicting the sea level pressure pattern. They are usually spaced at 4 millibar intervals. When the pressure gradient is weak, dashed isobars are sometimes inserted at 2 millibar intervals to more clearly define the pressure pattern. Each isobar is labelled by a two-digit number. For example, 32 signifies 1032.0 mb, 00 signifies 1000.0 mb, 92 signifies 992.0 mb, and 88 signifies 988.0 mb.

PRESSURE SYSTEMS

The letter "L" denotes a low pressure center and an "H" denotes a high pressure center. The pressure at each center is indicated by a two-digit underlined number which is interpreted the same as the isobar labels.

FRONTS

The analysis shows frontal positions and types of fronts by the symbols in figure 5-2. The "pips" on the front indicate the type of front and point in the direction toward which the front is moving. Pips on either side of a front suggest little or no movement, i.e. a stationary front. Briefing offices sometimes color the symbols to facilitate use of the map.

A three-digit number entered along a front classifies the front as to type, table 5-1; intensity, table 5-2; and character, table 5-3. For example in figure 5-1, the front extending from North Dakota southwest toward Arizona is labeled "427" which means a cold front at the surface ("4" in table 5-1), weak with little or no change ("2" in table 5-2) and with waves along the front ("7" in table 5-3). The waves along the front may be weak low pressure centers which are not indicated or simply one part of the front moving faster than the other. The triangular pips also identify this front as a cold front. The pips point toward the east over the Dakotas indicating the cold front is moving to the east in this region while in Arizona the pips point toward the southeast indicating the cold front is moving to the southeast in that region.

Two short lines across a front indicate a change in classification. Note in figure 5-1 the two lines crossing the front off the coast of Georgia indicating a change from "225" to "420". In this case a warm front extends westward from the two short lines intersecting the front while a cold front extends eastward. Note that the stationary front along the Gulf Coast is undergoing frontolysis (dissipation).

TABLE 5-1. Type of front

Code Figure	Description
0	Quasi-stationary at surface
1	Quasi-stationary above surface
2	Warm front at surface
3	Warm front above surface
4	Cold front at surface
5	Cold front above surface
6	Occlusion
7	Instability line
8	Intertropical front
9	Covergence line

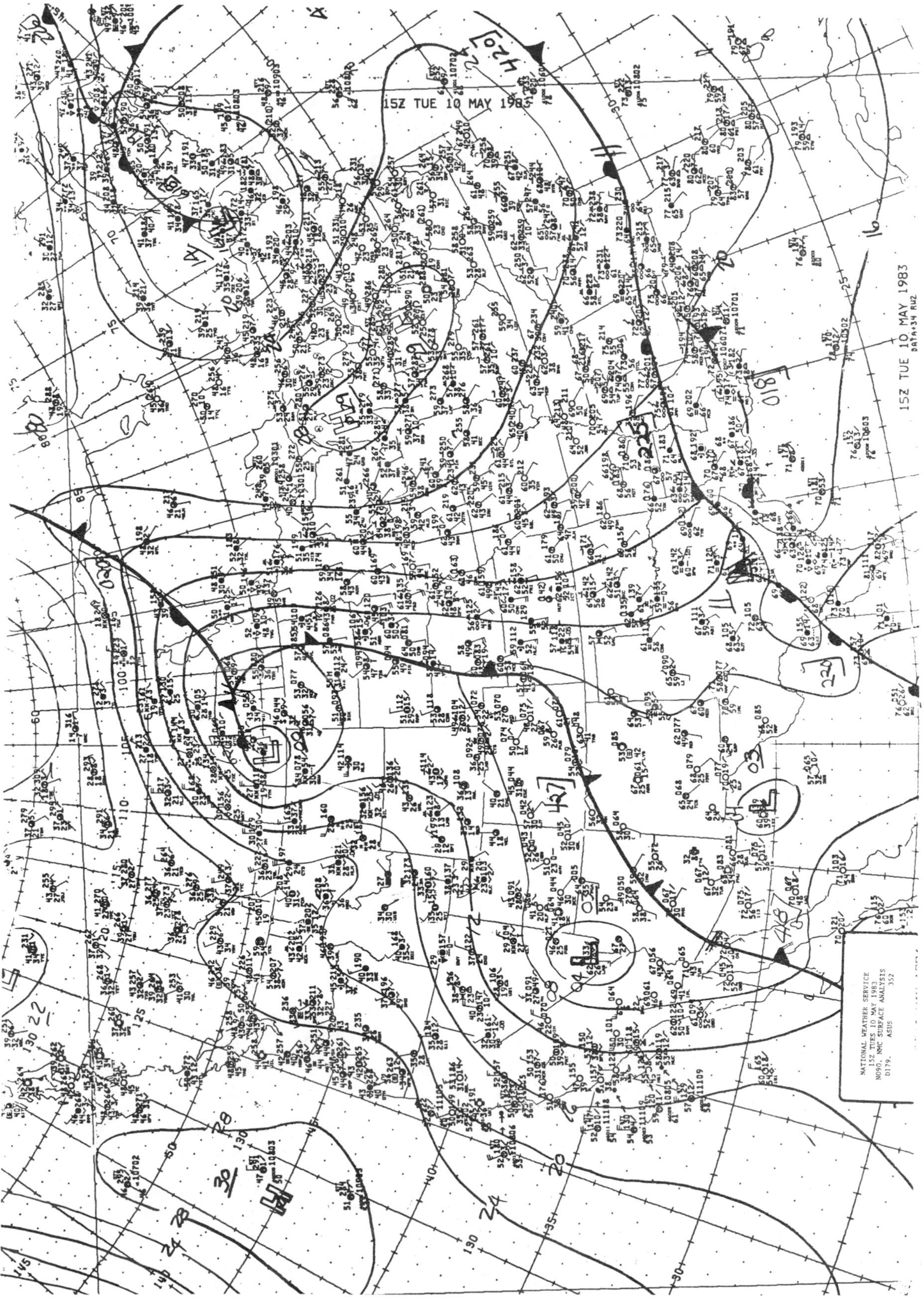

FIGURE 5-1. Surface Weather Analysis Chart.

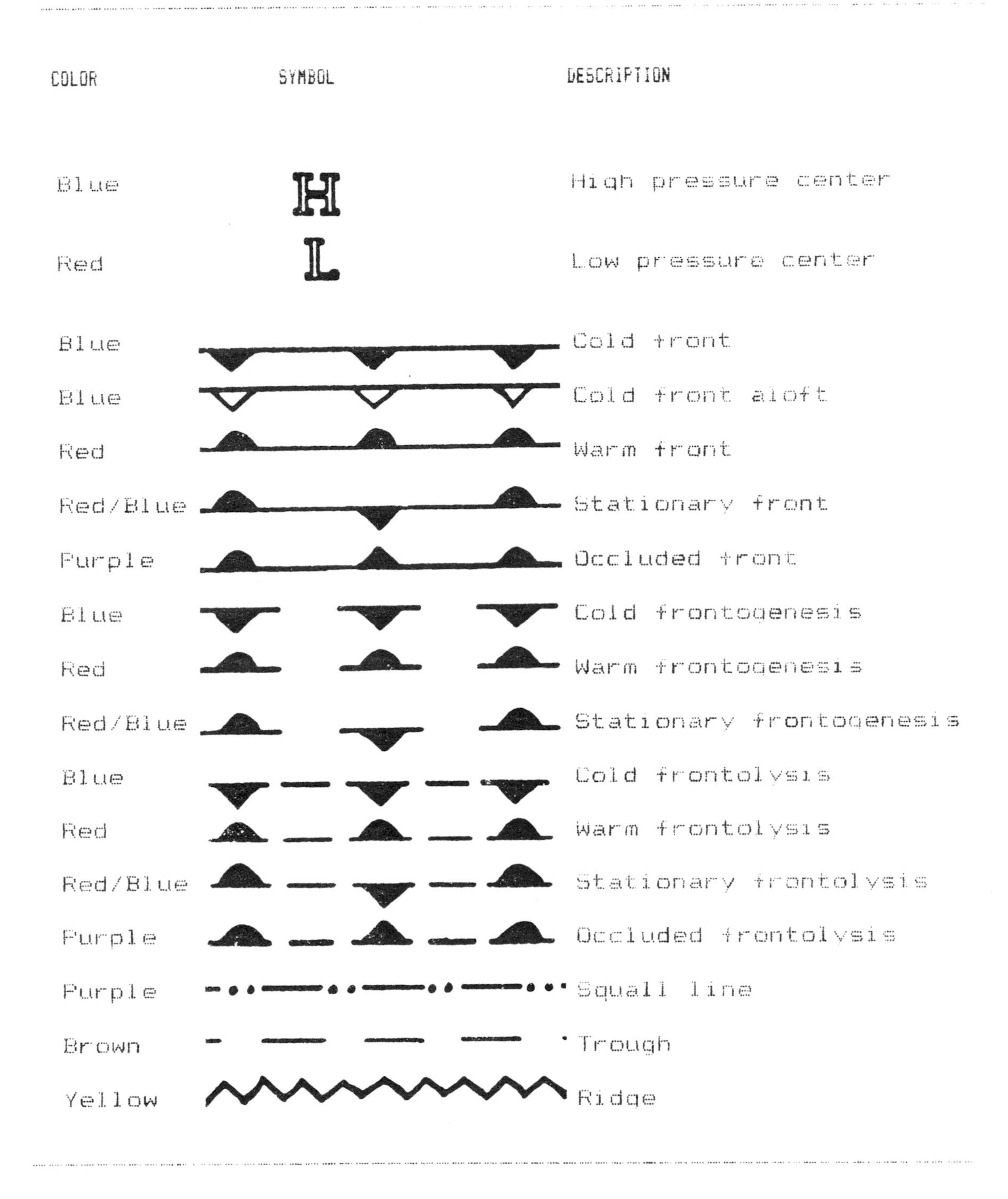

FIGURE 5-2. List of symbols on surface analyses. Colors are those suggested for on-station use.
NOTE: A trough line usually is further identified by the word "TOF".
A trough line is not a front.

Table 5-2. Intensity of front

Code Figure	Description
0	No specification
1	Weak, decreasing
2	Weak, little or no change
3	Weak, increasing
4	Moderate, decreasing
5	Moderate, little or no change
6	Moderate, increasing
7	Strong, decreasing
8	Strong, little or no change
9	Strong, increasing

Table 5-3. Character of front

Code Figure	Description
0	No specification
1	Frontal area activity, decreasing
2	Frontal area activity, little change
3	Frontal area activity, increasing
4	Intertropical
5	Forming or existence expected
6	Quasi-stationary
7	With waves
8	Diffuse
9	Position doubtful

TROUGHS AND RIDGES

A trough of low pressure with significant weather will be depicted as a thick, dashed line running through the center of the trough and identified with the word "TROF". The symbol for a ridge of high pressure is very rarely, if at all, depicted (see figure 5-2 for symbols).

OTHER INFORMATION

Figure 5-3 shows a station model which shows where the weather information is plotted. Figure 5-4 through figure 5-7 help explain the decoding of the station model.

USING THE CHART

The surface analysis provides you with a ready means of locating pressure systems and fronts. It also gives you an overview of winds, temperatures, and dew point temperatures as of chart time. When using the chart, keep in mind that weather moves and conditions change. For example, a front located over Kansas may be nearing Oklahoma by the time you see the chart. Using the surface analysis chart in conjunction with other charts such as weather depiction, radar summary, upper air, and prognostics (forecast charts) gives a more complete weather picture.

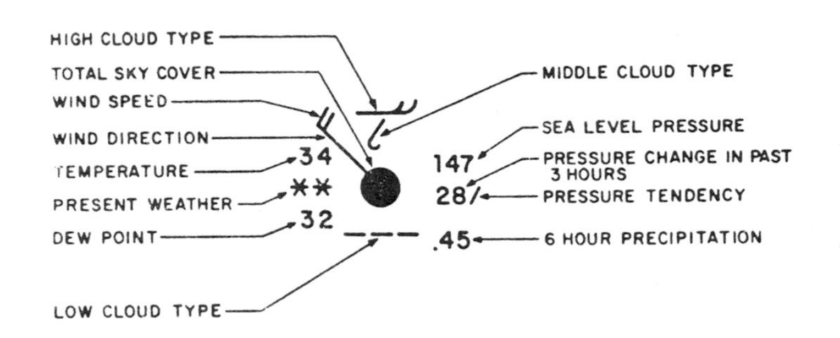

1. Total sky cover: overcast (see figure 5-4).
2. Temperature: 34 degrees F, Dew point: 32 degrees F.
3. Wind: blowing from the northwest at 20 knots relative to true north.
   Wind speeds are in knots and are indicated as:

calm 5 knots (kts) 10 kts 15 kts 50 kts 65 kts

   Circle around the station means calm. A half-flag has a value of 5 knots, a full-flag 10 knots, and a pennant 50 knots. These are used in an appropriate combination to represent wind speed; e.g. two pennants indicate wind speed of 100 knots. Wind direction stated is the direction FROM which the wind is blowing. For example:

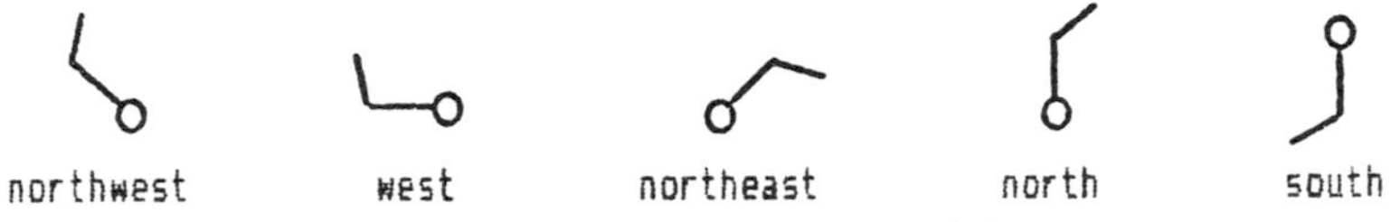

4. Present weather: continuous light snow (see figure 5-5).
5. Predominant low, mid, high cloud reported: fractostratus or fractocumulus of bad weather, altocumulus in patches, dense cirrus (see figure 5-7).
6. Sea level pressure: 1014.7 millibars (mb). Always shown as 3 digits to the nearest tenth of a millibar. For 1000 mb or greater, prefix a 10 to the 3 digits. For less than 1000 mb prefix a 9 to the 3 digits.

108 = 1010.8 mb	888 = 988.8 mb
225 = 1022.5 mb	961 = 996.1 mb
000 = 1000.0 mb	720 = 972.0 mb

7. Pressure change in past 3 hours: increased steadily or unsteadily by 2.8 mb. Actual change is in tenths of a millibar. See figure 5-6 for tendency explanation.
8. 6 hour precipitation: 45 hundredths of an inch. Amount is given to the nearest hundredth of an inch.

FIGURE 5-3. Station model and explanation.

Symbol	Total sky cover
	Sky clear
	Less than 1/10 (Few)
	1/10 to 5/10 inclusive (Scattered)
	6/10 to 9/10 inclusive (Broken)
	10/10 with breaks (BINOVC)
	10/10 (Overcast)
	Sky obscured or partially obscured

FIGURE 5-4. Sky cover symbols.

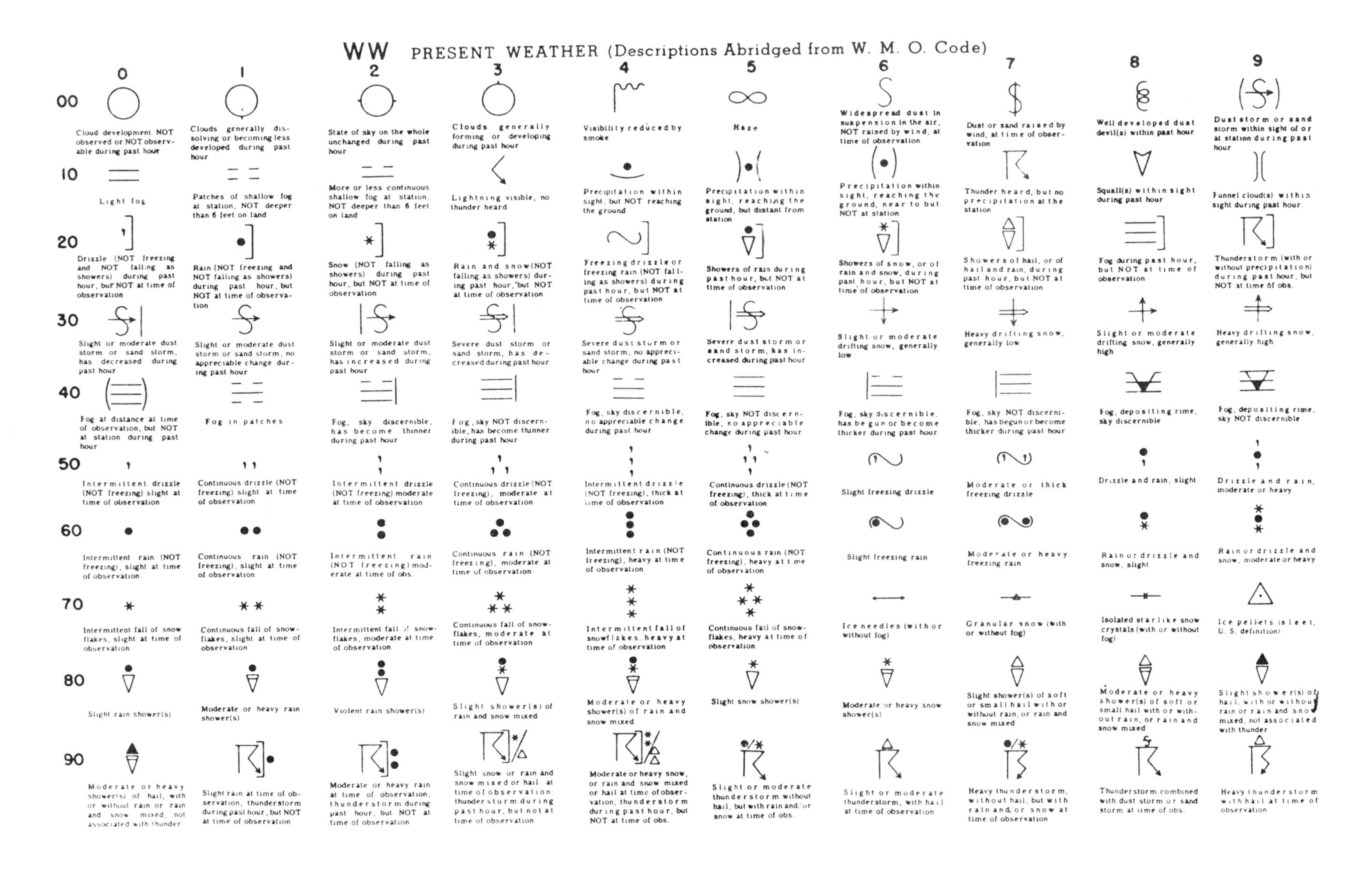

FIGURE 5-5. Present weather.

Description of Characteristic		Graphic
Primary Unqualified Requirement	Additional Requirements	
HIGHER Atmospheric pressure now higher than 3 hours ago	Increasing then decreasing	
	Increasing then steady; or - - - - - - - - - increasing then increasing more slowly	
	Increasing: Steadily / Unsteadily	
	Decreasing or steady then increasing; or - - - - - - - - - increasing then increasing more rapidly	

Description of Characteristic		Graphic
Primary Unqualified Requirement	Additional Requirements	
THE SAME Atmospheric pressure now same as 3 hours ago	Incresing then decreasing	
	Steady	
	Decreasing then increasing	
LOWER Atmospheric pressure now lower than 3 hours ago	Decreasing then increasing	
	Decreasing then steady or - - - - - - - - - decreasing then decreasing more slowly	
	Decreasing: Steadily / Unsteadily	
	Steady or increasing then decreasing; or - - - - - - - - - decreasing then decreasing more rapidly	

FIGURE 5-6. Barometer tendencies.

CLOUD ABBREVIATION		C_L	DESCRIPTION (Abridged From W.M.O. Code)		C_M	DESCRIPTION (Abridged From W.M.O. Code)		C_H	DESCRIPTION (Abridged From W.M.O. Code)
St or Fs - Stratus or Fractostratus	1		Cu of fairweather, little vertical development and seemingly flattened	1		Thin as (most of cloud layer semi-transparent)	1		Filaments of Ci, or "mares tail," scattered and not increasing
Ci - Cirrus	2		Cu of considerable development, generally towering, with or without other Cu or Sc bases all at same level	2		Thick As, greater part sufficiently dense to hide sun (or moon), or Ns	2		Dense Ci in patches or twisted sheaves, usually not increasing, sometimes like remains of Cb; or towers or tufts
Cs - Cirrostratus	3		Cb with tops lacking clear-cut outlines, but distinctly not cirriform or anvil-shaped; with or without Cu, Sc, or St	3		Thin Ac, mostly semi-transparent; cloud elements not changing much and at a single level	3		Dense Ci, often anvil-shaped, derived from or associated with Cb
Ce - Cirrocumulus	4		Sc formed by spreading out of Cu; Cu often present also	4		Thin Ac in patches; cloud elements continually changing and/or occurring at more than one level	4		Ci, often hook-shaped, gradually spreading over the sky and usually thickening as a whole
Ac - Altocumulus	5		Sc not formed by spreading out of Cu	5		Thin Ac in bands or in a layer gradually spreading over sky and usually thickening as a whole	5		Ci and Cs, often in converging bands, or Cs alone; generally overspreading and growing denser; the continuous layer not reaching 45° altitude
As - Altostratus	6		St or Fs or both, but no Fs of bad weather	6		Ac formed by the spreading out of Cu	6		Ci and Cs, often in coverging bands, or Cs alone; generally overspreading and growing denser; the continuous layer exceeding 45° altitude
Sc - Stratocumulus	7		Fs and/or Fc of bad weather (scud)	7		Double-layered Ac, or a thick layer of Ac, not increasing; or Ac with As and/or Ns	7		Veil of Cs covering the entire sky
Ns - Nimbostratus	8		Cu and Sc (not formed by spreading out of Cu) with bases at different levels	8		Ac in the form of Cu-shaped tufts or Ac with turrets	8		Cs not increasing and not covering entire sky
Cu or Fc - Cumulus or Fractocumulus; Cb - Cumulonimbus	9		Cb having a clearly fibrous (cirriform) top, often anvil-shaped, with or without Cu, Sc, St, or scud	9		Ac of a chaotic sky, usually at different levels; patches of dense Ci are usually present also	9		Cc alone or Cc with some Ci or Cs, but the Cc being the main cirriform cloud

FIGURE 5-7. Cloud abbreviation.

Section 6
WEATHER DEPICTION CHART

The weather depiction chart, figure 6-1, is computer prepared from surface aviation (SA) reports to give a broad overview of observed flying category conditions as of the valid time of the chart. The computer prepared chart is valid at the time of the plotted data. Beginning at 01Z each day, charts are transmitted at 3 hour intervals.

PLOTTED DATA

The plotted data shown for each station as required are:

Total Sky Cover

Total sky cover is shown by the station circle shaded as in table 6-1.

TABLE 6-1. Total sky cover.

Symbol	*Total sky cover*
○	Sky clear
⦶	Less than 1/10 (Few)
◔	1/10 to 5/10 inclusive (Scattered)
◕	6/10 to 9/10 inclusive (Broken)
◑	10/10 with breaks (BINOVC)
●	10/10 (Overcast)
⊗	Sky obscured or partially obscured

Cloud Height or Ceiling

Cloud height, above ground level, is entered under the station circle in hundreds of feet, the same as coded in a SA report. If total sky cover is few or scattered, the cloud height entered is the base of the lowest layer. If total sky cover is broken or greater, the cloud height entered is the ceiling. Broken or greater total sky cover without a height entry indicates thin sky cover. A partially or totally obscured sky is shown by the same sky cover symbol (X). However, a partially obscured sky without a cloud layer above is denoted by the absence of a height entry while a partially obscured sky with clouds above will have a cloud layer or ceiling height entry. A totally obscured sky always has a height entry of the ceiling (vertical visibility into the obscuration).

Weather and Obstructions to Vision

Weather and obstructions to vision symbols are entered just to the left of the station circle. Figure 5-5 explains most of the symbols used. When an SA Reports clouds topping ridges a symbol unique to the weather depiction chart is entered to the left of the station circle:

▲ denotes clouds topping ridges.

When several types of weather and/or symbols obstructions are reported at a station, only the most significant one is entered (i.e. the highest coded number in figure 5-5).

Visibility

When visibility is 6 miles or less, it is entered to the left of weather or obstructions to vision. Visibility is entered in statute miles and fractions of a mile.

Table 6-2 shows examples of plotted data.

ANALYSIS

The chart shows observed ceiling and visibility by categories as follow:

1. IFR—Ceiling less than 1,000 feet and/or visibility less than 3 miles; hatched area outlined by a smooth line.
2. MVFR (Marginal VFR)—Ceiling 1,000 feet to 3,000 feet inclusive and/or visibility 3 to 5 miles inclusive; non-hatched area outlined by a smooth line.
3. VFR—Ceiling greater than 3,000 feet or unlimited and visibility greater than 5 miles; not outlined.

The three (3) categories are also explained in the lower right portion of the chart for quick reference.

Referring to figure 6-1, the MVFR conditions in southwest Oregon are indicated in an area where

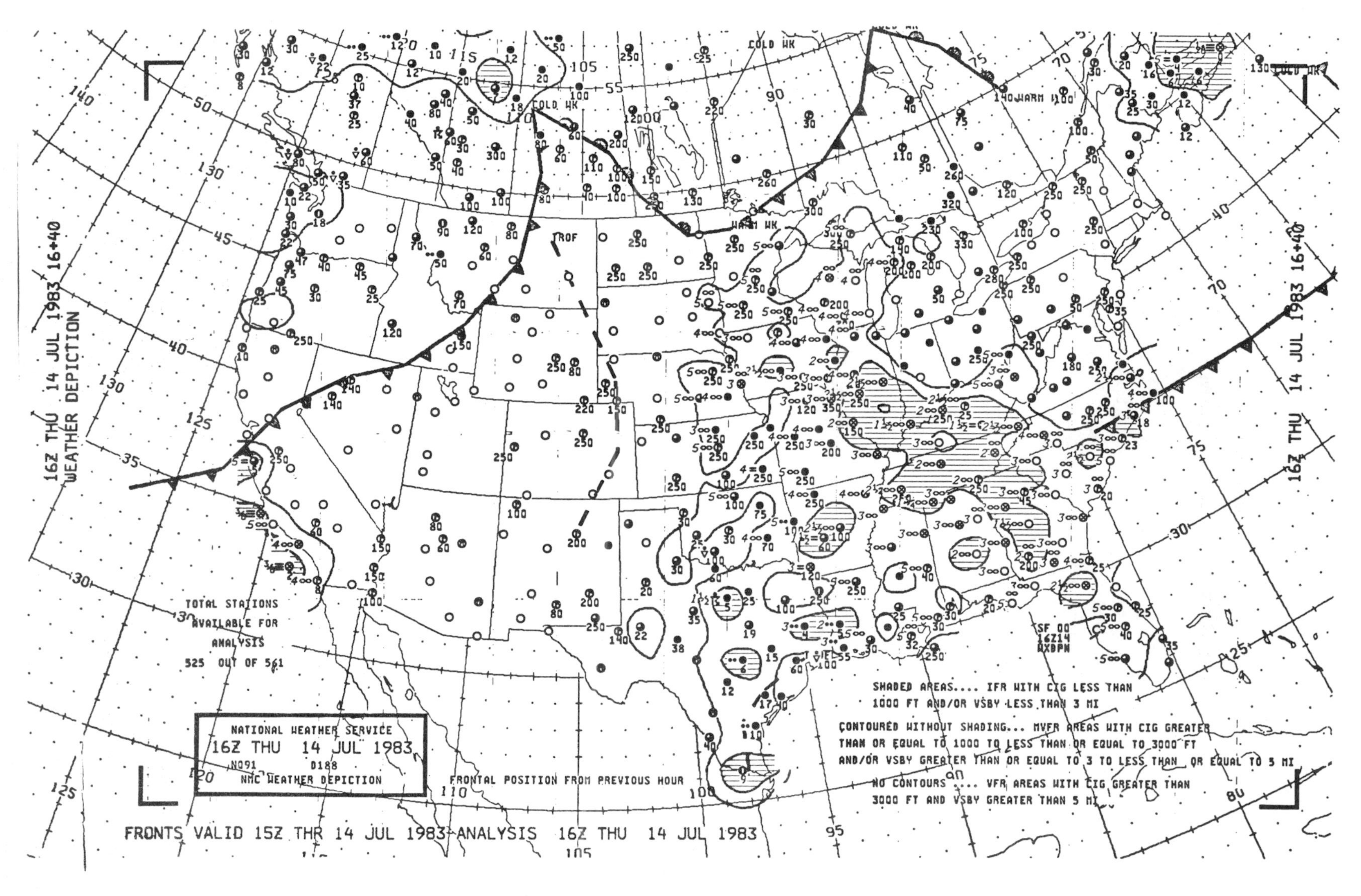

FIGURE 6-1. A Weather Depiction Chart.

TABLE 6-2. Examples of plotting on the Weather Depiction Chart

Plotted	Interpreted
8	Few clouds, base 800 feet, visibility more than 6
12	Broken sky cover, ceiling 1,200 feet, rain shower, visibility more than 6
5 ∞	Thin overcast with breaks, visibility 5 in haze
30	Scattered at 3,000 feet, clouds topping ridges, visibility more than 6
2 =	Sky clear, visibility 2, ground fog or fog
1/2	Sky partially obscured, visibility 1/2, blowing snow, no cloud layers observed
2 = 200	Sky partially obscured, visibility 2, fog, cloud layer at 20,000 feet Assume sky is partially obscured since 20,000 feet cannot be vertical visibility into fog. It is questionable if 20,000 feet is lowest scattered layer or ceiling.*
1/4 * 5	Sky obscured, ceiling 500, visibility 1/4, snow
1 12	Overcast, ceiling 1,200 feet thunderstorm, rain shower, visibility 1
(M)	Data missing

* Note: Since a partial and a total obscuration (X) is entered as total sky cover, it can be difficult to determine if a height entry is a cloud layer above a partial obscuration or vertical visibility into a total obscuration. Check the SA.

the plotted stations show only VFR conditions. Note that off the Baja California coast it is stated that the total stations analyzed for this chart are far more numerous than the number of stations actually plotted. Thus, there are stations in southwest Oregon, not plotted on the chart, that are reporting MVFR conditions.

In addition the chart shows fronts and troughs from the surface analysis for the preceding hour. These features are depicted the same as the surface chart.

USING THE CHART

The weather depiction chart is a choice place to begin your weather briefing and flight planning. From it, you can determine general weather conditions more readily than any other source. It gives you a "bird's eye" view at chart time of areas favorable and adverse weather and frontal systems associated with the weather.

The chart may not completely represent enroute conditions because of variations in terrain and weather between stations. Futhermore, weather changes and by the time the chart is available, plotted data around the stations have been superseded by SA reports. After you initially size up the general picture, your final flight planning must consider forecasts, progs, and the latest pilot, radar and surface weather reports.

Section 7
RADAR SUMMARY CHART

A radar summary chart, figure 7-1, graphically displays a collection of radar reports. Figure 1-2 depicts the National Weather Service radar network. The computer generated chart is valid at the time of the plotted radar reports, i.e., at H+35. Charts are available for 16 hours daily via NAFAX and 24 hours daily by DIFAX constructed from regularly scheduled radar observations. The chart displays the type of precipitation echoes and indicates their intensity, intensity trend, configuration, coverage, echo tops and bases, and movement. Severe weather watches are also plotted if they are in effect when the chart is valid. This section explains chart notations, symbols, and use.

ECHO TYPE, INTENSITY, AND INTENSITY TREND

Radar primarily detects particles of precipitation size within a cloud or falling from a cloud. The type of precipitation can be determined by the radar operator from the scope presentation in combination with other sources. TABLE 7-1 lists the symbols used to denote types of precipitation, intensity and intensity trend. The intensity is obtained from the Video Integrator Processor (VIP) and is indicated on the chart by *contours.* The six (6) VIP levels are combined into three (3) contours as indicated in TABLE 7-1. For example in Table 7-1, the area of precipitation between the first (outer) contour and the second contour would have an intensity of VIP level 1 and possibly 2. Whether we really have VIP level 2 in the area cannot be determined. However, we can say that the maximum intensity is definitely below VIP level 3. When determining intensity levels from the radar summary chart, it is recommended that the maximum possible intensity be used. To determine the actual maximum VIP level, you would examine the RAREP message (SD) for that particular time. The intensity trend is indicated by a symbol plotted beside the precipitation type. The absence of a trend symbol indicates no change. For example in figure 7-1, over eastern North Dakota and western Minnesota, there is an area of light to moderate rainshowers with no change in intensity from the previous observation. Actual intensity for frozen precipitation cannot be determined from the contours since the intensity levels are only correlated to liquid precipitation. Intensity trend for frozen precipitation is not reported on a RAREP and thus, is not indicated on the radar chart. The symbol "S+" on the chart means the area of snow is *new.* The symbol "S" on the chart means an area of snow is indicated by radar with no reference to intensity trend. Remember the intensity trend symbols; (−) decreasing, (no symbol) no changes, (+) increasing; refer only to liquid precipitation.

It is important to remember that intensity on the radar summary chart is shown by contours and *not* by the symbol following the type of precipitation. For example in figure 7-1, along the east central coast of Florida near Vero Beach, there is an area of light to moderate rainshowers that is either new or has increased in intensity from the previous observation.

Also note that hail possibly reaching the surface is associated with the thunderstorms and intense to extreme rainshowers in the southern half of Missouri. The actual locations of hail are indicated by a line drawn from "HAIL" to the symbol ■.

ECHO CONFIGURATION AND COVERAGE

The configuration is the arrangement of echoes. There are three designated arrangements, (1) a LINE of echoes, (2) an AREA of echoes and (3) an isolated CELL. A fine line would appear on the chart as a pencil thin line with a movement indicated. See section 3 under radar reports for definitions of the above.

Coverage is simply the area covered by echoes. All of the hatched area inside of the contours on the chart is considered to be covered by echoes. When the echoes are reported as a LINE, a line will be drawn through them on the radar chart. When there is at least 8/10 coverage in the line, it is labeled solid (SLD) at both ends of the line. In the absence of this label it can be assumed that there is less than 8/10 coverage. For example in figure 7-1, there is a solid line of thunderstorms with intense to extreme rainshowers extending from northeast Arkansas to southwest Kentucky.

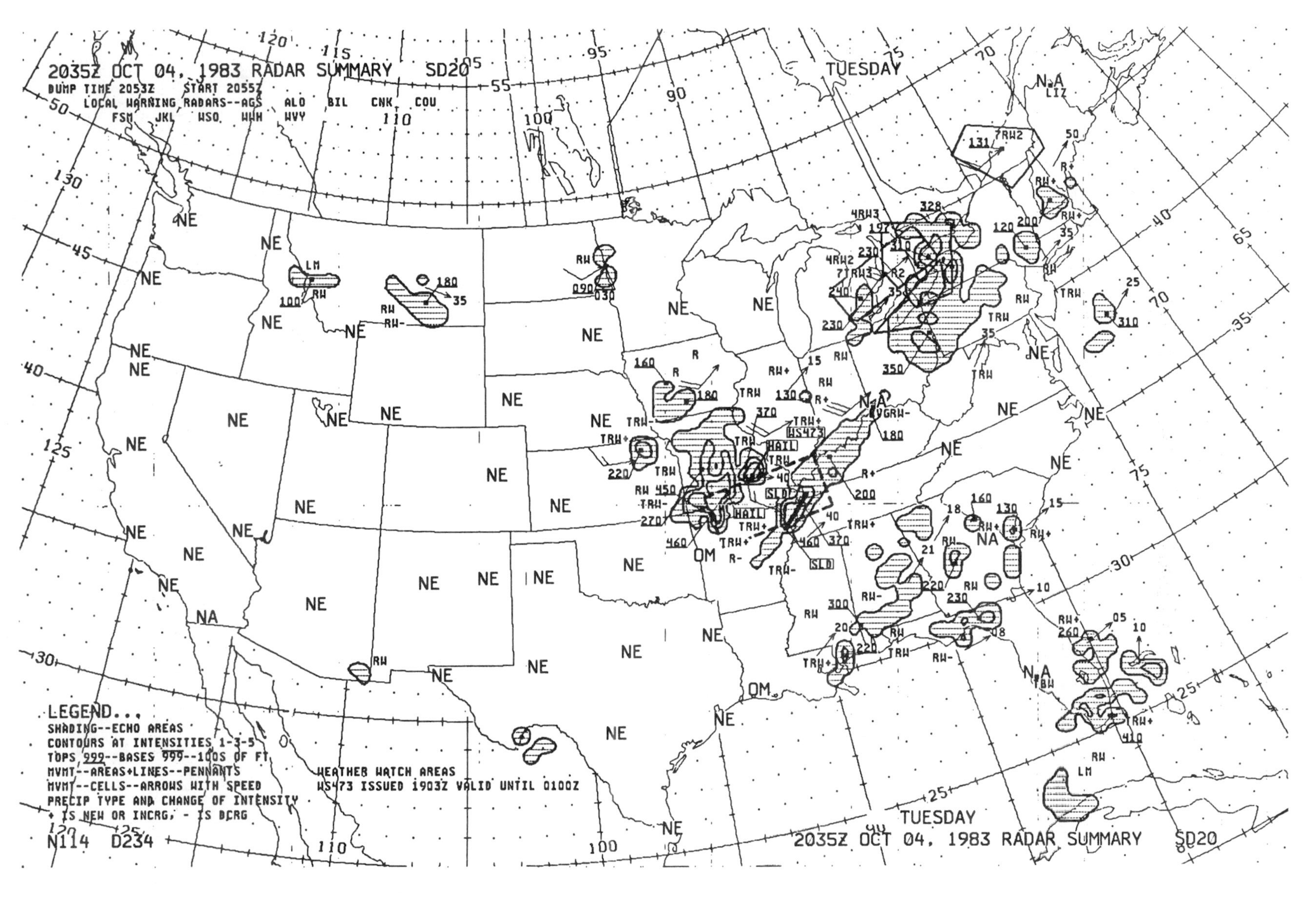

FIGURE 7-1. A Radar Summary Chart.

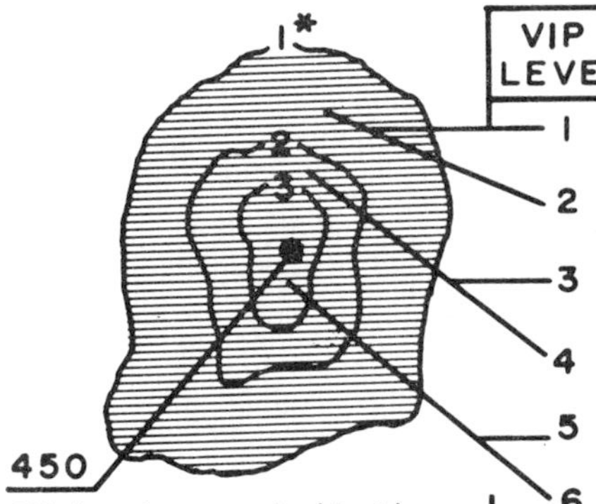

450

Highest precipitation top in area in hundreds of feet MSL.
(45,000 FEET MSL)

VIP LEVEL	ECHO INTENSITY	PRECIPITATION INTENSITY	RAINFALL RATE in/hr STRATIFORM	RAINFALL RATE in/hr CONVECTIVE
1	WEAK	LIGHT	LESS THAN 0.1	LESS THAN 0.2
2	MODERATE	MODERATE	0.1 - 0.5	0.2 - 1.1
3	STRONG	HEAVY	0.5 - 1.0	1.1 - 2.2
4	VERY STRONG	VERY HEAVY	1.0 - 2.0	2.2 - 4.5
5	INTENSE	INTENSE	2.0 - 5.0	4.5 - 7.1
6	EXTREME	EXTREME	MORE THAN 5.0	MORE THAN 7.1

* The numbers representing the intensity level do not appear on the chart. Beginning from the first contour line, bordering the area, the intensity level is 1-2; second contour is 3-4; and third contour is 5-6.

SYMBOLS USED ON CHART

SYMBOL	MEANING
R	RAIN
RW	RAIN SHOWER
HAIL	HAIL
S	SNOW
IP	ICE PELLETS
SW	SNOW SHOWER
L	DRIZZLE
T	THUNDERSTORM
ZR, ZL	FREEZING PRECIPITATION
NE	NO ECHOES OBSERVED
NA	OBSERVATIONS UNAVAILABLE
OM	OUT FOR MAINTENANCE
STC	STC ON - all precipitation may not be seen
ROBEPS	RADAR OPERATING BELOW PERFORMANCE STANDARDS
RHINO	RANGE HEIGHT INDICATOR NOT OPERATING

SYMBOL	MEANING
+	INTENSITY INCREASING OR NEW ECHO
–	INTENSITY DECREASING
NO SYMBOL	NO CHANGE IN INTENSITY
35  (arrow)	CELL MOVEMENT TO NE AT 35 KNOTS
 (arrow)	LINE OR AREA MOVEMENT TO EAST AT 20 KNOTS
LM	LITTLE MOVEMENT
MA	ECHOES MOSTLY ALOFT
PA	ECHOES PARTLY ALOFT

SYMBOL	MEANING
 (line)	LINE OF ECHOES
SLD	8/10 OR GREATER COVERAGE IN A LINE
WS999	SEVERE THUNDERSTORM WATCH
WT999	TORNADO WATCH
LEWP	LINE ECHO WAVE PATTERN
HOOK	HOOK ECHO

RAINFALL RATES SHOULD BE USED WITH CAUTION

TABLE 7-1. Key to Radar Summary Chart.

ECHO HEIGHTS

Echo heights in locations with radars designed for weather detection are obtained by use of range height indicators and are PRECIPITATION tops and bases. In those areas not served by National Weather Service radars the tops are obtained from pilot reports and are actual CLOUD tops. Usually, echo height will be missing in the western mountain regions because ARTCC radars are used.

Heights are displayed in hundreds of feet MSL and should be considered only as approximations because of radar limitations. Tops are entered above a short line while any available bases are entered below. The top height displayed is the highest in the indicated area.

Examples are:

$\frac{220}{080}$ Bases 8,000 feet, Max top 22,000 feet

$\frac{500}{\quad}$ Bases at the surface, Max top 50,000 feet

$\frac{\quad}{020}$ Bases 2,000 feet, Max top either missing or reported in another place

Absence of a figure below the line indicates that the echo base is at the surface. Radar detects tops more readily than bases because precipitation usually reaches the ground. For example in figure 7-1, over eastern North Dakota and western Minnesota the maximum precipitation top in the area is 9000 feet MSL at the location indicated by a line drawn to the symbol ■. The base of the rainshower associated with the maximum top and therefore probably over most of the area is 3000 feet MSL. This indicates dry air near the surface which is causing the rain to evaporate before reaching the ground. The level at which the precipitation is evaporating above *ground level* depends on station elevation. For example, if the ground elevation for the area just mentioned is 2500 feet MSL, the base of the precipitation is only 500 feet AGL.

ECHO MOVEMENT

Individual cell movement within a line or area is often different from that of the line or area itself. This difference is indicated by the use of different symbols, as shown in TABLE 7-1. Line or area movement is indicated by a shaft and barb combination with the shaft indicating the direction and the barbs the speed. A whole barb is 10 knots, a half barb is 5 knots, and a pennant is 50 knots. Individual cell movement is indicated by an arrow with the speed in knots entered as a number. Little movement is indentified by "LM". For example in figure 7-1, no cell movement is given over eastern North Dakota but the area movement is toward the northeast at 10 knots. Over eastern Montana, no area movement is given but the cell movement of the rainshowers is toward the east southeast at 35 knots. Over western Montana, the rainshowers show little movement.

SEVERE WEATHER WATCH AREAS

Severe weather watch areas are outlined by heavy dashed lines, usually in the form of a large rectangular box. There are two types, (1) tornado watches and (2) severe thunderstorm watches. Referring to TABLE 7-1 and figure 7-1, the type of watch and the watch number are enclosed in a small rectangle and positioned as closely as possible to the northeast corner of the watch box. For example figure 7-1, "WS 473" means a severe thunderstorm watch and is the 473rd severe weather watch issued so far in the year. The watch number is also printed at the bottom of the chart together with the issuance time and valid until time.

CANADIAN DATA

Radar data from six Canadian radar stations are plotted when available. The stations in Ontario are: Carp, Exeter, Toronto, Villeroy and Ottawa. Montreal, Quebec is also plotted. The data is displayed in AZRAN (azimuth-range) format with echo areas outlined by solid lines. Area, line and cell movements are shown in the same manner as U.S. data. An alphanumeric code associated with each echo shows, in order, area coverage, precipitation type, intensity, and intensity trend. Precipitation type and intensity trend are the same as U.S. data. For area coverage, a blank designator represents cells, a 1 equals less than 1/10 coverage, a 4 equals 1/10 to 5/10 coverage, a 7 equals 6/10 to 9/10 coverage and 10 equals 10/10 coverage. For intensity levels, 0 is very weak, 1 is weak, 2 is moderate, 3 is strong and 4 is very strong with levels 1 through 4 being comparable to the U.S. VIP LVLS 1 through 4.

For example in figure 7-1, the region in southeast Canada and covering a portion of New England constitutes a Canadian radar report "7RW2". Decoded, there is an area of moderate rainshowers with no change in intensity. 6/10 to 9/10 of the area is covered with rainshowers with area movement toward the east northeast at 20 knots. Maximum top within the area is 13,100 feet MSL.

Canadian echo top reports are converted from meters to feet and are plotted to the nearest hundreds of feet MSL. For example, west of New York State in southern Canada, the tops are decoded as follows: "197" is 19,700 feet MSL and "328" is 32,800 feet MSL.

It can sometimes be difficult to interpret the data where both American and Canadian reports are plotted, such as in the Great Lakes region in figure 7-1.

Do not confuse Canadian data with a severe weather watch box. In figure 7-1, the rectangular box in the Great Lakes region is not a severe weather watch box but a Canadian radar plot. This box is *not* outlined by heavy dashed lines as is required for a severe weather watch area. On charts transmitted after December 1983, the areas of Canadian data are plotted as light solid lines instead of heavy solid lines as shown in figure 7-1. This should lead to less confusion as to U.S. verus Canadian data.

USING THE CHART

The radar summary chart aids in preflight planning by identifying general areas and movement of precipitation and/or thunderstorms. Radar detects ONLY drops or ice particles of precipitation size, it DOES NOT detect clouds and fog. Therefore, the absence of echoes does not guarantee clear weather. Furthermore, cloud tops may be higher than precipitation tops detected by radar. The chart must be used in conjunction with other charts, reports, and forecasts.

Examine chart notations carefully. Always determine location and movement of echoes. If echoes are anticipated near your planned route, take special note of echo intensity and trend. Be sure to examine for missing radar reports (NA, OM) before briefing "no echoes present". For example, the Covington (CVG) radar report in northern Kentucky is shown as not available (NA). There could very well be echoes in southwest Ohio but too far away to be detected by the other surrounding radars.

Suppose your proposed route will take you through an area of widely scattered thunderstorms with no increase anticipated. When these storms are separated by good VFR weather, you most likely can pick your way among them, visually sighting and circumnavigating the storms. However, widespread cloudiness may conceal the thunderstorms. To avoid these embedded thunderstorms, you must either use airborne radar or detour the area. Most details on avoiding hazards of thunderstorms are given in Chapter 11, Aviation Weather.

Keep in mind that the chart is for preflight planning only and should be updated by hourly radar reports. Once airborne, you must evade individual storms from inflight observations either by visual sighting or by airborne radar or request weather radar echo information from FSS Flight Watch which has access to Radar Remote Weather Displays (RRWDS).

One more thought before this section ends. There can be an interpretation problem concerning an area of precipitation that is reported by more than one radar site. As an example, station A may be reporting RW− with cell movement to the northeast at 10 knots while station B may be reporting TRW+ with cell movement to the northeast at 30 knots for the same area. This difference in reports may be due to the different perspective and distance of the radar sites from the area of echoes. The area may be moving away from station A and approaching station B. The rule of thumb is to use that plotted data associated with the area which presents the greatest hazard to aviation, i.e. in the case above, the plotted data from station B. In figure 7-1, the area of rainshowers in eastern Montana should be briefed as having no change in intensity (RW) rather than decreasing in intensity (RW−).

Section 8

SIGNIFICANT WEATHER PROGNOSTICS

Significant weather prognostic charts, called "progs" for brevity, portray forecast weather which may influence flight planning. TABLE 8-1 explains some symbols used on these charts. Significant weather progs are issued both for domestic and international flights.

TABLE 8-1. Some standard weather symbols

Symbol	*Meaning*	*Symbol*	*Meaning*
	Moderate turbulence		Rain shower
	Severe turbulence		Snow shower
	Moderate icing		Thunderstorms
	Severe icing		Freezing rain
●	Rain		Tropical storm
*	Snow		Hurricane (typhoon)
	Drizzle		

NOTE: Character of stable precipitation is the manner in which it occurs. It may be intermittent or continuous. A single symbol denotes intermittent and a pair of symbols denotes continuous.

Examples,

Intermittent	Continuous	
●	● ●	Rain
		Drizzle
*	* *	Snow

DOMESTIC FLIGHTS

Significant weather progs are manually prepared by the forecaster for the conterminous U.S. and adjacent areas. The U.S. low level significant weather prog is designed for domestic flight planning to 24,000 feet and a U.S. high level prog is for domestic flights from 24,000 feet to 63,000 feet. Chart legends include valid time in GMT.

U.S. Low Level Significant Weather Prog

The low level prog is a four-panel chart as shown in figure 8-1. The two lower panels are 12- and 24-hour surface progs. The two upper panels are 12- and 24-hour progs of significant weather from the surface to 400 millibars (24,000 feet). The charts show conditions as they are forecast to be at the valid time of the chart. A chart is issued four (4) times daily; the 12 and 24 hour forecasts are based on the 00Z, 06Z, 12Z and 18Z synoptic data. For example, the prog in figure 8-1 is based on the 12Z 9 NOV initial data.

Surface Prog. The two surface prog panels use standard symbols for fronts and pressure centers explained in section 5. Movement of each pressure center is indicated by an arrow showing direction and a number indicating speed in knots. Isobars depicting forecast pressure patterns are included on some 24-hour surface progs.

The surface prog also outlines areas of forecast precipitation and/or thunderstorms as shown in the lower panels of figure 8-1. Smooth lines enclose areas of expected continuous or intermittent (stable) precipitation; dash-dot lines enclose areas of showers or thunderstorms (unstable precipitation). Areas of continuous or intermittent precipitation with embedded showers and thunderstorms will also be enclosed by dash-dot lines.

Note that symbols indicate precipitation type and character (see TABLE 8-1 and 8-2). If precipitation will affect half or more of an area, that area is shaded; absence of shading denotes more sparse precipitation, specifically less than half areal coverage. Look at the lower left panel of figure 8-1. At 0000Z the forecast is for continuous snow and rain affecting half or more of an area extending from north of the

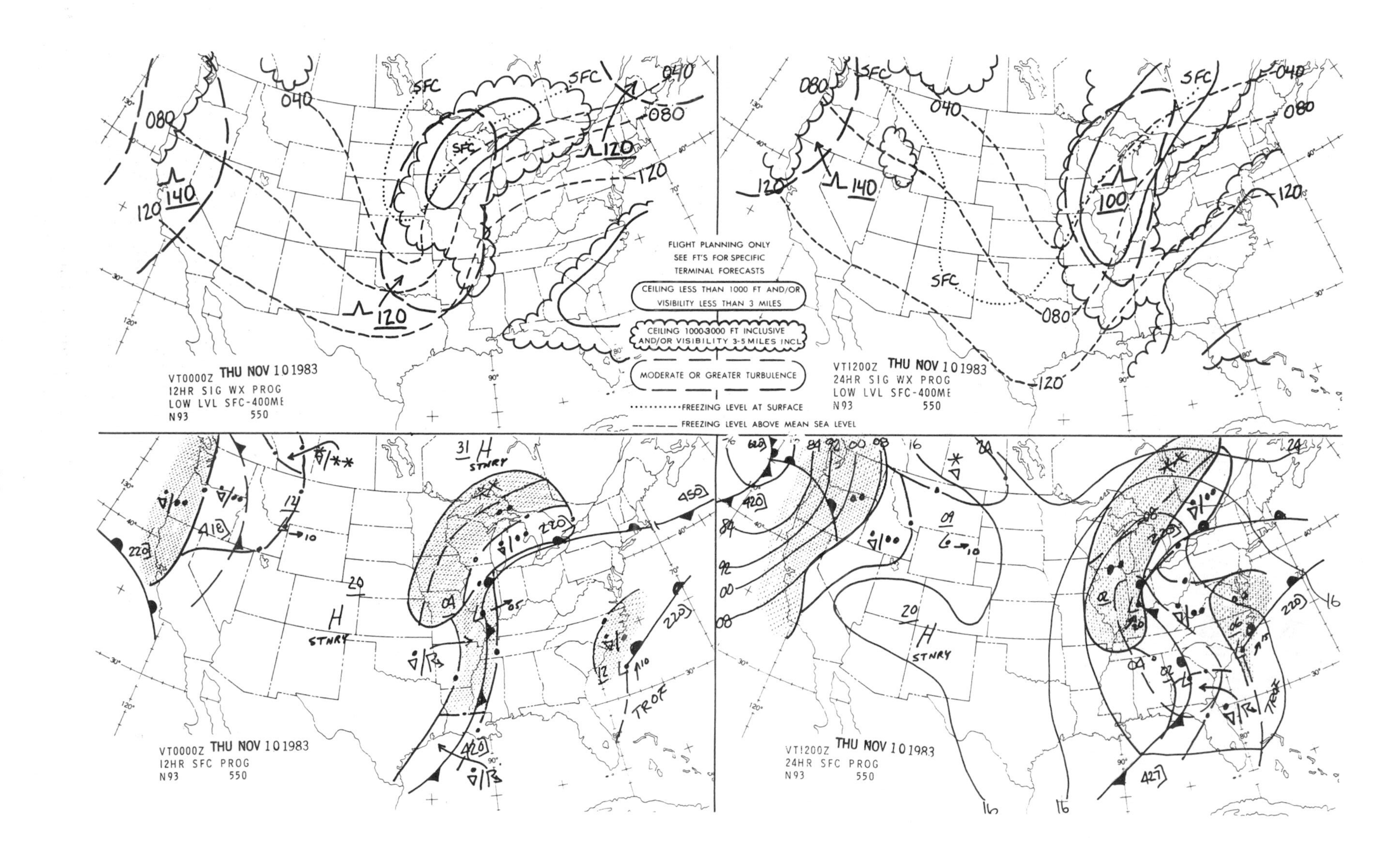

FIGURE 8-1. U.S. Low Level Significant Weather Prog (Sfc-400 mb).

TABLE 8-2. Significant weather prognostic symbols

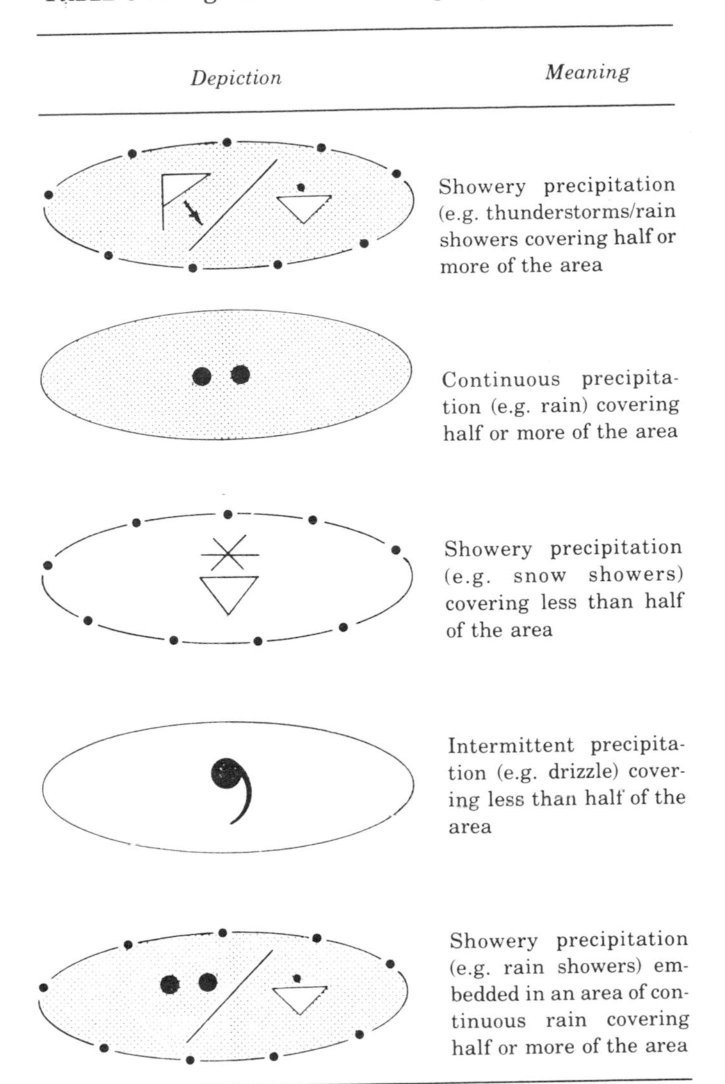

Depiction	Meaning
	Showery precipitation (e.g. thunderstorms/rain showers covering half or more of the area
	Continuous precipitation (e.g. rain) covering half or more of the area
	Showery precipitation (e.g. snow showers) covering less than half of the area
	Intermittent precipitation (e.g. drizzle) covering less than half of the area
	Showery precipitation (e.g. rain showers) embedded in an area of continuous rain covering half or more of the area

Great Lakes southwestward to Kansas. The snow is forecast over the west portion of the area with the rain-snow line represented by a dashed line. On the same prog, along the west coast and extending inland to Montana, embedded showers are forecast within an area of continuous rain. Along the coast, coverage is expected to be half or more of the area while further inland coverage is expected to be less than half of the area. From central Illinois southward to central Louisiana, showers and thunderstorms are forecast with half or more areal coverage, however, further south coverage is expected to be less than half.

Significant Weather. The upper panels of figure 8-1 depict IFR, MVFR, turbulence, and freezing levels. Note the legend near the center of the chart which explains methods of depiction.

Smooth lines enclose areas of forecast IFR weather and scalloped lines enclose areas of marginal weather (MVFR). VFR areas are not outlined. This is NOT the same manner of depiction used on the weather depiction chart to portray IFR and MVFR. Referring to figure 8-1, at 00Z an area of IFR is depicted along the east coast from North Carolina to northern Florida and is surrounded by an area of MVFR. Note that depictions are *not* extended over the open waters even though IFR conditions may exist.

Forecast areas of moderate or greater turbulence are enclosed by long-dashed lines. Thunderstorms always imply moderate or greater turbulence; thus the area of thunderstorm turbulence will not be outlined.

A symbol entered within a general area of forecast turbulence denotes intensity. Figures below and above a short line show expected base and top of the turbulent layer in hundreds of feet MSL. Absence of a figure below the line indicates turbulence from the surface upward. No figure above the line indicates turbulence extending above the upper limit of the chart. Turbulence forecast from the surface to above 24,000 feet is indicated by the notation "SFC" below the line with the upper value left blank. Referring to figure 8-1, at 00Z an area of moderate non-thunderstorm related turbulence is forecast over the far western U.S. from the surface to 14,000 feet MSL; moderate non-thunderstorm related turbulence is forecast over the middle part of the country and over eastern Maine from the surface to 12,000 feet MSL. Thunderstorm related turbulence is indicated on the lower panels by forecast areas of thunderstorms.

Freezing level height contours for the *highest* freezing level are drawn at 4,000 foot intervals. The 4,000 foot contour terminates at the 4,000 foot terrain level along the Rocky Mountains. Contours are labelled in hundreds of feet MSL. The dotted line shows where the freezing level is forecast to be at the surface and is labelled "32 F" or "SFC". An upper freezing level contour crossing the surface 32 degree line indicates multiple freezing levels due to layers of warmer air aloft. If clouds and precipitation are forecast in this area, icing hazards should be considered.

The low level significant weather prog does not specifically outline areas of icing. However, icing is implied in clouds and precipitation above the freezing level. Interpolate for freezing levels between the given contours. For example in figure 8-1, at 00Z the forecast *highest* freezing level over Oklahoma City is approximately 6,000 feet MSL.

36 and 48 Hour Surface Weather Prog

This prog is an extension of those 12 and 24 hour surface prog panels which are based on 00Z and 12Z initial synoptic data. The prog in figure 8-2 is a continuation of the 12 and 24 hour prog in figure 8-1.

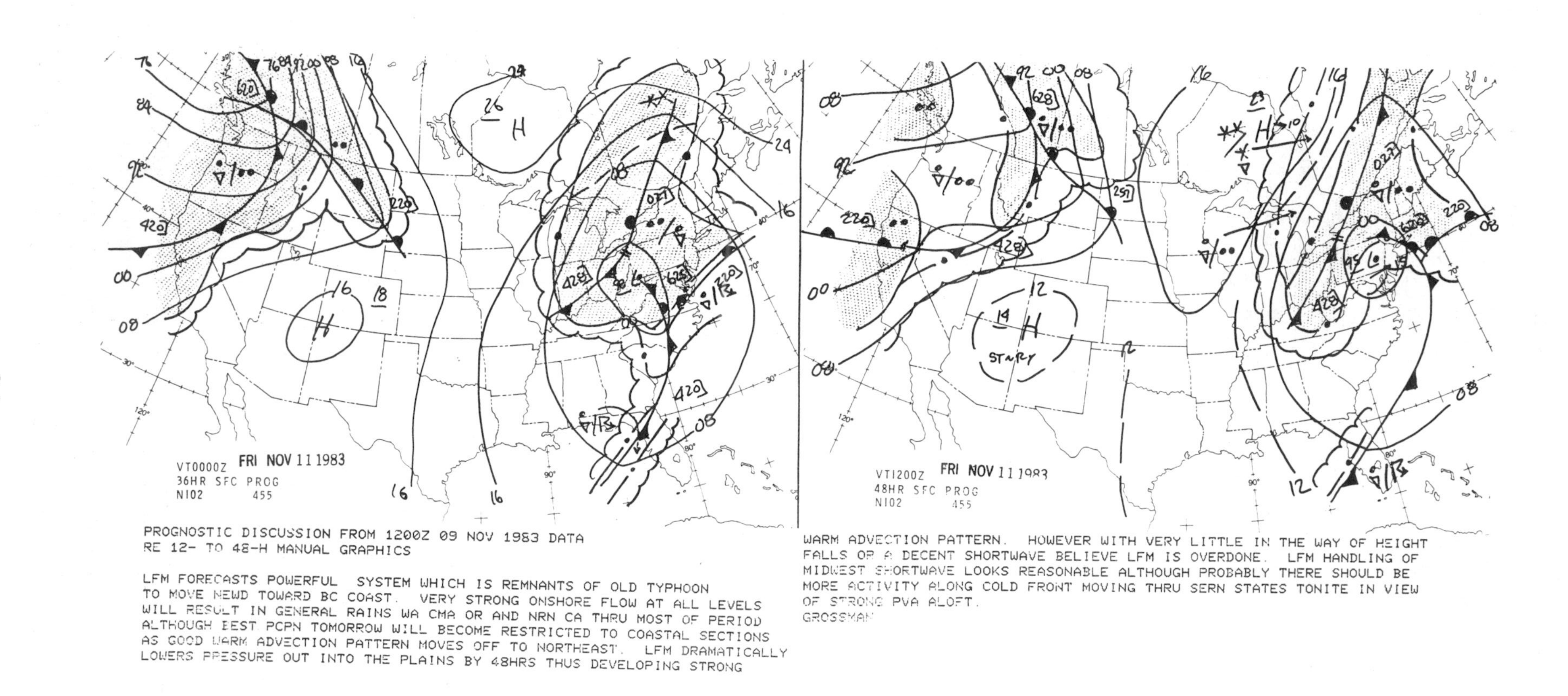

FIGURE 8-2. U.S. Low Level 36 and 48 hour Significant Weather Prog.

The depiction of data is the same as on the 12 and 24 hour surface prog with the following exceptions:

No freezing precipitation is forecast.

Scalloped lines denote area of *overcast* clouds with *no* reference to the height of cloud base.

A prognostic discussion is included to explain the forecaster's reasoning for the 12 hour through 48 hour surface progs.

Use of the Chart.

The 36 and 48 hour surface prog should only be used for outlook purposes, i.e. to get a very generalized weather picture of conditions that are in the relatively distant future.

High Level Significant Weather Progs

Manually produced forecast charts of significant weather are available for both domestic and international flights. The U.S. National Meteorological Center (NMC), near Washington, D.C., is a component of the World Area Forecast System (WAFS). NMC is designated in the WAFS as both a World Area Forecast Center and a Regional Area Forecast Center (RAFC). Its main function as a World Area Forecast Center is to prepare global forecasts in grid-point form of upper winds and upper air temperatures and to supply the forecasts to associated RAFCs. One of its main RAFC functions is to prepare and supply to users charts of forecast winds and temperatures and of forecast significant weather. This section will deal with the content of significant weather progs.

Significant weather to be depicted on the charts are the following:

a. cumulonimbus (CB) clouds meeting at least one of the following criteria:
   1. widespread cumulonimbus clouds or cumulonimbus clouds along a line with little or no space between individual clouds;
   2. cumulonimbus clouds embedded in cloud layers or concealed by haze or dust;
b. tropical cyclones;
c. severe squall lines;
d. moderate or severe turbulence (in cloud or clear air);
e. widespread sandstorm/duststorm;
f. surface positions, speed and direction of movement enroute weather phenomena;
g. tropopause heights; and
h. jetstreams.

Depiction of Thunderstorms and Cumulonimbus Clouds (CB)

Required thunderstorm activity is to be depicted by means of the abbreviation, or symbol, "CB". By definition, this symbol is to refer to the occurrence or expected occurrence of an area of widespread cumulonimbus clouds along a line with little or no space between individual clouds, or cumulonimbus clouds embedded in cloud layers or concealed by haze or dust. It does not refer to isolated or scattered (occasional) cumulonimbus clouds not embedded in cloud layers or concealed by haze or dust. The symbol "CB" automatically implies moderate or greater turbulence and icing; thus, these associated hazards will not be depicted separately.

CB data will normally be identified as ISOL EMBD CB (isolated embedded CB), OCNL EMBD CB (occasional embedded CB), ISOL CB in HAZE (isolated CB in HAZE) or OCNL CB in HAZE (occasional CB in haze). In rare instances, CB coverage above FL240 may exceed 4/8 coverage; in these instances, CB activity will be described as FRQ CB (frequent, cumulonimbus clouds with little or no separation). The meanings of these area coverage terms are: ISOL, less than 1/8; OCNL, 1/8 to 4/8; and FRQ, 5/8 to 8/8.

CB bases are considered certain to be below FL240 and will be shown as XXX. CB tops are to be expressed in hundreds of feet MSL. The area to which the forecast applies will be shown by scalloped lines. Examples:

Meaning: Occasional (1/8 to 4/8 area coverage) embedded cumulonimbus clouds with bases below FL240 and tops forecast to reach FL450.

ISOL EMBD
CB
350
XXX

Meaning: Isolated (less than 1/8 area coverage) embedded cumulonimbus clouds with bases below FL240 and tops forecast to reach FL350.

Depiction of Tropical Cyclones

Tropical storms are depicted by the symbol 6. Areas of associated cumulonimbus activity, if meeting the previously given criteria (ISOL EMBD CB, OCNL EMBD CB, ISOL CB IN HAZE, OCNL CB IN HAZE. FRQ CB), are enclosed by scalloped lines and labelled with the vertical extent. Example:

Meaning: Forecast position of a tropical cyclone with no associated thunderstorm area.

Meaning: A thunderstorm area (5/8 to 8/8 area coverage, bases below FL240, tops FL500) associated with a tropical cyclone.

Notes: 1. The names of tropical cyclones, when relevant, will be entered adjacent to the symbol.
2. Significant weather chart depicting the tropical cyclone symbol will have a statement to the effect that the latest tropical cyclone advisory, rather than the tropical cyclone's prognostic position on the chart, is to be given public dissemination.

Depiction of Severe Squall Lines

Severe squall lines are depicted within areas of CB activity by the symbol:

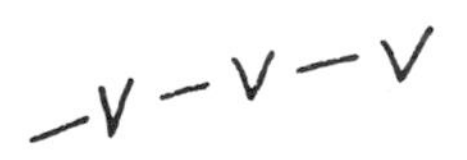

Example of depiction of severe squall line and associated CB activity:

Meaning: Forecast severe squall line with associated CB, coverage 5/8 to 8/8, bases below FL240 and tops forecast to reach FL500.

Depiction of Clear Air Turbulence

Area of forecast moderate or greater clear air turbulence (CAT) are bounded by heavy dashed lines. Clear air turbulence is interpreted as including all turbulence (including wind-shear induced and mountain-wave induced) not caused by convective activity. Areas are labelled with the appropriate symbol (⁄_ for moderate CAT; ⁄\̂_ for severe CAT) and the vertical extent in hundreds of feet MSL.

Examples:

360
280

Meaning: An area of forecast moderate CAT, vertical extent from FL280 to FL360.

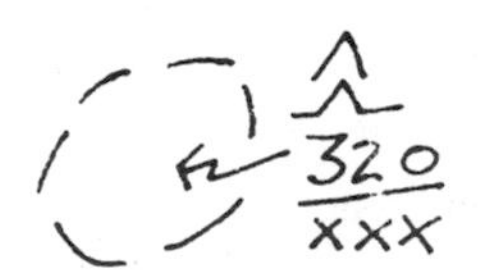

Meaning: An area of forecast severe CAT, vertical extent from below FL240 to FL320.

Note: THE SYMBOL CB IMPLIES HAIL, MODERATE OR GREATER TURBULENCE AND ICING.

Depiction of Widespread Sandstorm or Duststorm

Areas of these phenomena are enclosed by scalloped lines and labelled by symbol and vertical extent. Example:

Meaning: Widespread sandstorm or duststorm, bases below FL240 (i.e., at the surface) tops FL300.

Depiction of Inter-Tropical Convergence Zone

Areas of associated cumulonimbus activity, if meeting the previously given criteria (ISOL EMBD CB, OCNL EMBD CB, ISOL CB IN HAZE, OCNL CB IN HAZE, FRQ CB) are enclosed by scalloped lines and labelled with the vertical extent.

Example:

Meaning: Forecast position of inter-tropical convergence zone, with associated thunderstorm areas, coverage respectively frequent (5/8 to 8/8), bases below FL240, tops FL450; and occasional (1/8 to 4/8), bases below FL240, tops FL350.

Depiction of Fronts

Forecast surface position and speed (knots) and direction of movement of frontal systems associated with significant weather are depicted. Example:

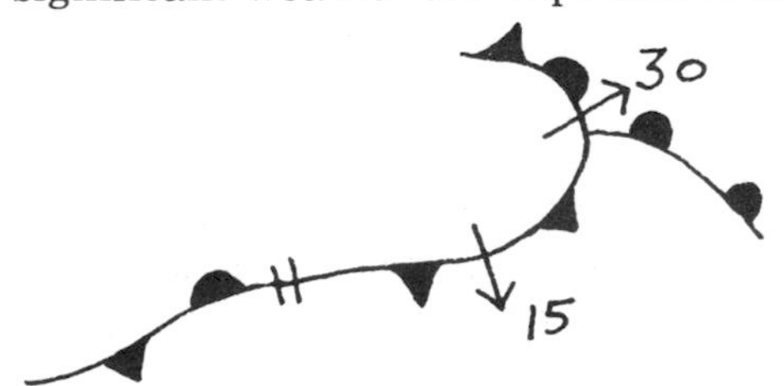

Meaning: A frontal system forecast to be at the position and with the orientation indicated at the valid time of the prognostic chart. Forecast movement related to true north and speed in knots are indicated by arrow shafts and adjacent numbers.

Depiction of Tropopause Heights

Tropopause heights are to be depicted in hundreds of feet MSL as 240, 270, 300, 340, 390, 450, and 530. Other heights may be used occasionally to define areas of very flat tropopause slope. Heights depicted are enclosed in small rectangular blocks. For example, in figure 8-3, note how the tropopause slopes from 39,000 feet in southern Minnesota to 45,000 feet in Texas.

Depiction of Jetstreams

The height and maximum wind speed of jetstreams having a core speed of 80 knots or greater are shown. Height is given as flight level (FL). Points along the jetstream at which the maximum wind speed is forecast are depicted with shafts, pennants, and feathers. Example:

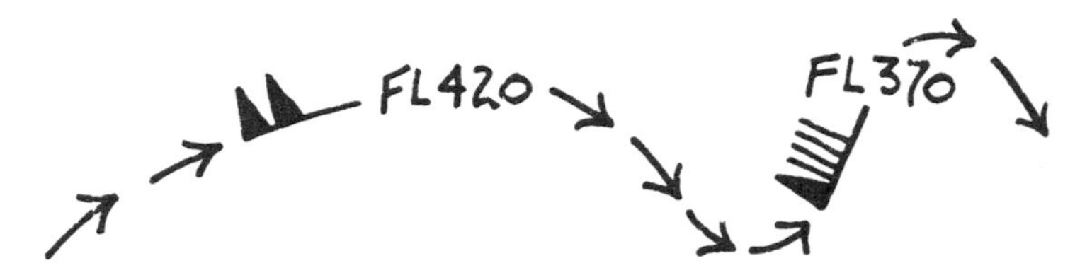

Meaning: A jetstream with forecast maximum speed of 100 knots at a height of 42,000 feet at another location and 90 knots at a height of 37,000 feet at another location. Wind speed along other portions is forecast to be less. Wind directions are indicated by the orientation of arrow shafts in relation to true north.

U.S. High Level Significant Weather Prog

The U.S. high level significant weather prog, figure 8-3, encompasses airspace from 24,000 feet to 63,000 feet pressure altitude. The prog is manually produced by the U.S. National Meteorological Center (NMC). Figure 8-3 outlines areas of forecast turbulence and cumulonimbus clouds. Table 8-3 interprets some examples of chart annotation.

Turbulence. Large-dashed lines enclose areas of probable moderate or greater turbulence not caused by convective activity. Symbols denote intensity, base, and top.

Cumulonimbus Clouds. Small-scalloped lines enclose areas of expected cumulonimbus development. The contraction "CB" denotes cumulonimbus. This symbol refers to the occurrence or expected occurrence of an area of widespread cumulonimbus clouds or cumulonimbus clouds along a line with little space between the individual clouds. It also depicts cumulonimbus clouds embedded in cloud layers or concealed by haze or dust. It does not refer to isolated or scattered cumulonimbus clouds not embedded in cloud layers or concealed by haze or dust. Cumulonimbus clouds *imply* moderate or greater turbulence and icing.

Cumulonimbus coverage and heights represent an overall average for the forecast area. When a wide variation is expected within an area, separate CB amounts and heights may be indicated.

The meanings of area coverage terms are: ISOL, less than 1/8; OCNL, 1/8 to 4/8; and FRQ, 5/8 to 8/8. CB bases are considered certain to be below 24,000 feet and are shown as XXX.

TABLE 8-3. Depiction of clouds and turbulence on a High Level Significant Weather Prog

	Depiction	*Meaning*
1.	ISOL EMBD CB 420 / XXX	ISOL embedded (less than one-eight cumulonimbus, tops 42,000 feet. Bases are below 24,000 ft.—the lower limit of the prog.
2.	OCNL EMBD CB 520 / XXX	1/8 to 4/8 coverage, embedded cumulonimbus, tops 52,000 feet, bases below 24,000 feet.
3.	FRQ CB 330 / XXX	5/8 to 8/8 coverage cumulonimbus, bases below 24,000 feet and tops 33,000 feet.
4.	330 / XXX to	Moderate to severe turbulence from below lower limit of the prog (24,000 feet) to 33,000 feet. (Consult low-level prog for turbulence forecasts below 24,000 feet.)
5.	ABV 630 / 350	Moderate turbulence from 35,000 feet to above upper limit of the prog.

NOTES:

Base and top shown by figures below and above a short line respectively.

Cumulonimbus Clouds, Examples 1, 2 and 3. Bases always below 24,000 feet and are shown by XXX. Tops above the upper limits of chart shown

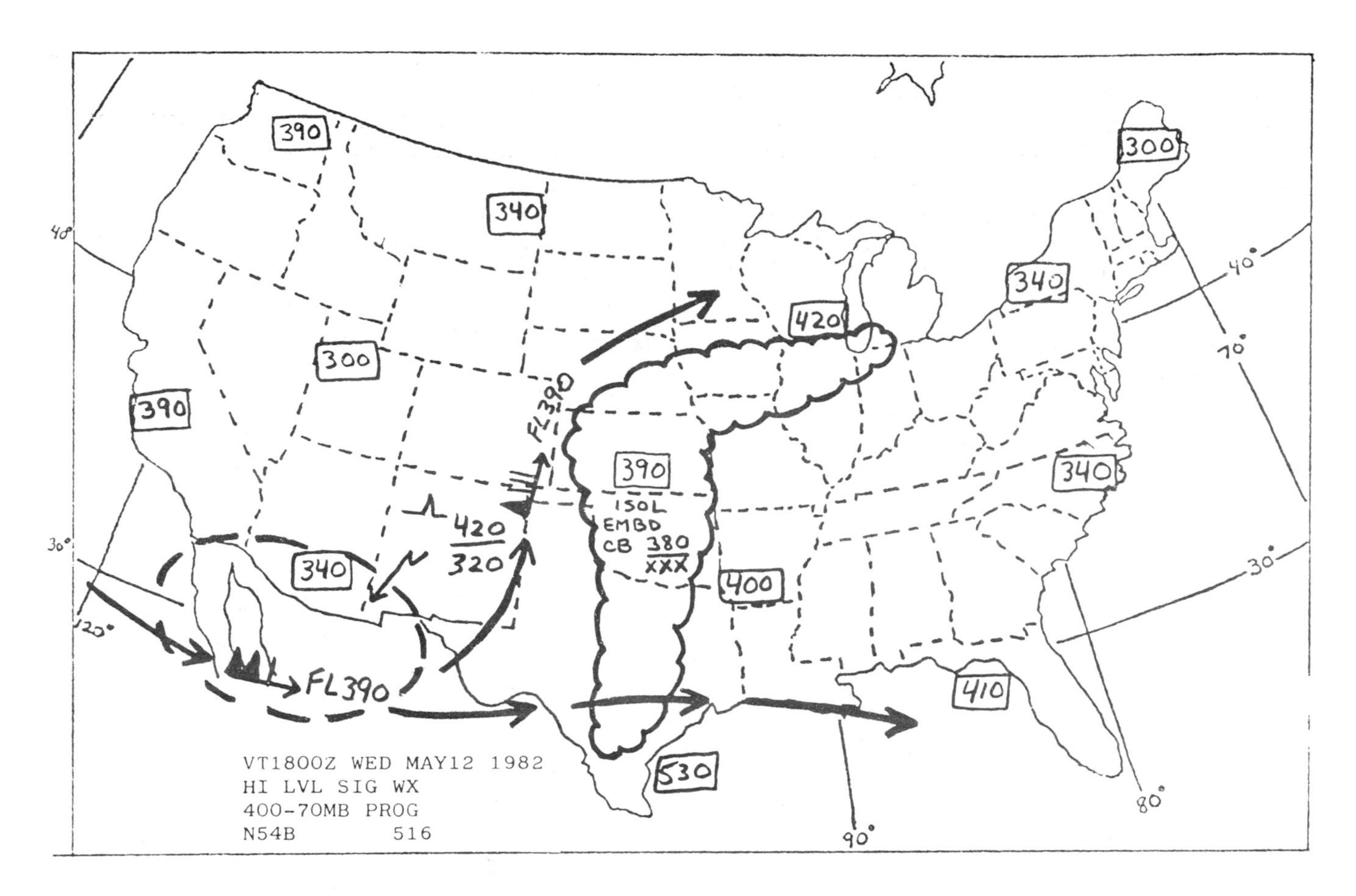

FIGURE 8-3. U.S. High Level Significant Weather Prog.

as "ABV 630" depending on the chart.

Turbulence. Bases and tops depicted the same as for cumulonimbus clouds.

International Flights

Figure 8-4 is an example of the significant weather prog for international flights. A manually produced forecast chart of high level significant weather.

Referring to figure 8-4, the legend shows NMC as a Regional Area Forecast Center (RAFC) and the originator of this significant weather prog. Significant weather is limited to the occurrence or expected occurrence of meteorological conditions considered to be of concern to aircraft operations. Significant weather progs are prepared only for flight levels from 25,000 feet to 60,000 feet. The valid time of this particular prog in figure 8-4 is 0000Z on February 9, 1984. All heights are in Flight Level (FL) and in hundreds of feet MSL.

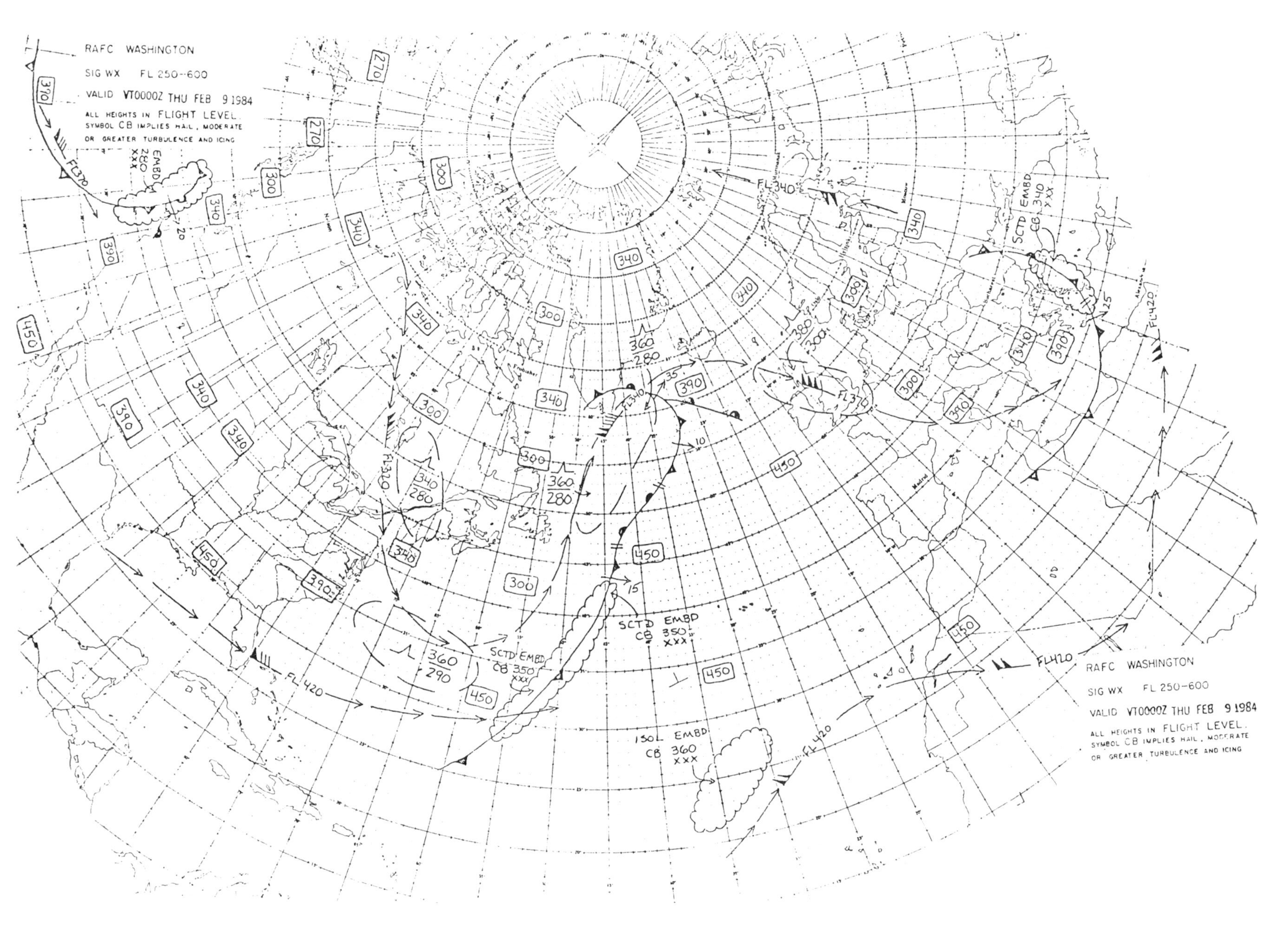

Figure 8-4. International High Level Significant Weather Prog Chart.

Section 9
WINDS AND TEMPERATURES ALOFT

Winds aloft, both forecast and observed, are computer prepared and routinely transmitted by facsimile. The forecast winds aloft charts also contain forecast temperatures aloft.

FORECAST WINDS AND TEMPERATURES ALOFT (FD)

Forecast winds and temperatures aloft charts are prepared for eight levels on eight separate panels. Those levels being 6,000; 9,000; 12,000; 18,000; 24,000; 30,000; 34,000 and 39,000 feet MSL. They are available daily as 12-hour progs valid at 1200Z and 0000Z. A legend on each panel shows the valid time and the level of the panel. Levels below 18,000 feet are true altitudes. Levels 18,000 and above are pressure altitudes of flight levels. Figure 9-1 is one panel of a winds and temperatures aloft forecast.

Temperature in whole degrees Celsius for each forecast point is entered above the station circle. Arrows with pennants and barbs similar to those used on the surface map show wind direction and speed. Wind direction is drawn to the nearest 10 degrees with the second digit of the coded direction entered at the outer end of the arrow. First you determine the general direction to the nearest 10 degrees. For example, a wind in the northwest quadrant with a digit 3 indicates 330 degrees. A calm or light and variable wind is shown by "99" entered to the lower left of the station circle.

Following are examples of plotted temperatures and winds with their interpretations:

Plotted	Interpretation
12 6	12 degrees C, wind 060 degrees at 5 knots
3 6	3 degrees C, wind 160 degrees at 25 knots

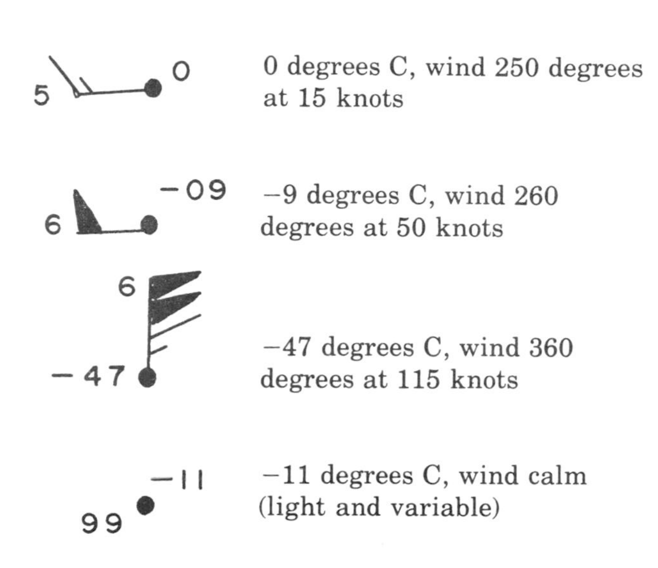

0 degrees C, wind 250 degrees at 15 knots

−9 degrees C, wind 260 degrees at 50 knots

−47 degrees C, wind 360 degrees at 115 knots

−11 degrees C, wind calm (light and variable)

OBSERVED WINDS ALOFT

Charts of observed winds for selected levels are sent twice daily on a four panel chart valid at 1200Z and 0000Z as shown in figure 9-2. Wind direction and speed at each observing station (figure 1-3) is shown by arrows the same as on the forecast charts. A calm or light and variable wind is shown as "LV" and a missing wind as "M", both plotted to the lower right of the station circle. The station circle is filled in when the reported temperature—dew point spread is 5 degrees Celsius or less. Figure 9-3 is a panel of the observed winds aloft chart. Observed temperatures are included on the upper two (2) panels (24,000 feet and 34,000 feet). A dotted bracket around the temperature means a calculated temperature.

The second standard level for a reporting station is found between 1,000 feet and 2,000 feet above the surface, depending on station elevation. To compute the second standard level, find the next thousand foot level above the station elevation and add 1,000 feet to that level. For example, the next thousand foot level above Oklahoma City OK (station elevation 1,290 feet MSL) is 2,000 feet MSL. The second standard level for Oklahoma City OK (2,000 feet + 1,000 feet) is 3,000 feet MSL or 1,710 feet AGL.

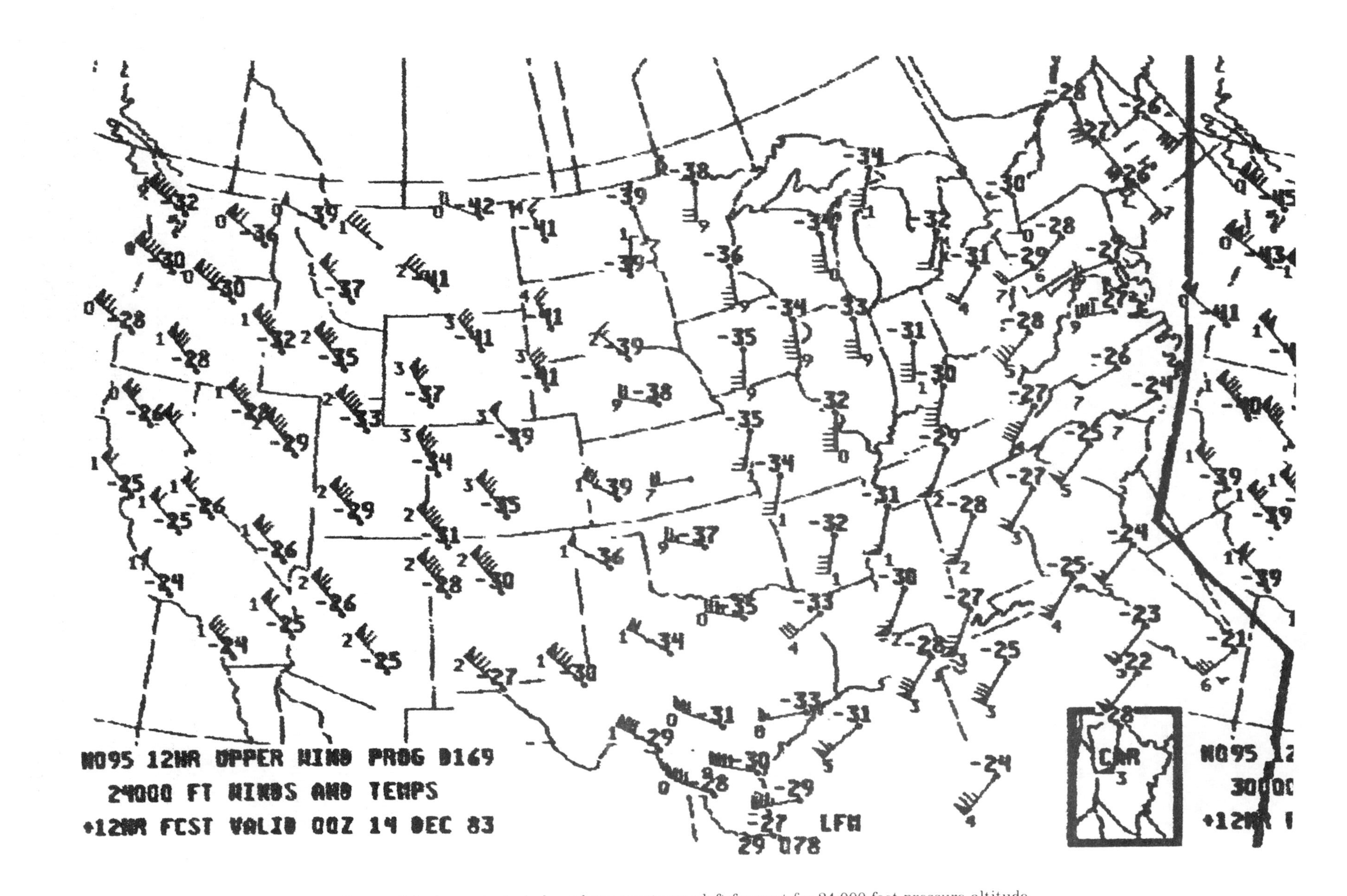

FIGURE 9-1. A panel of winds and temperatures aloft forecast for 24,000 feet pressure altitude.

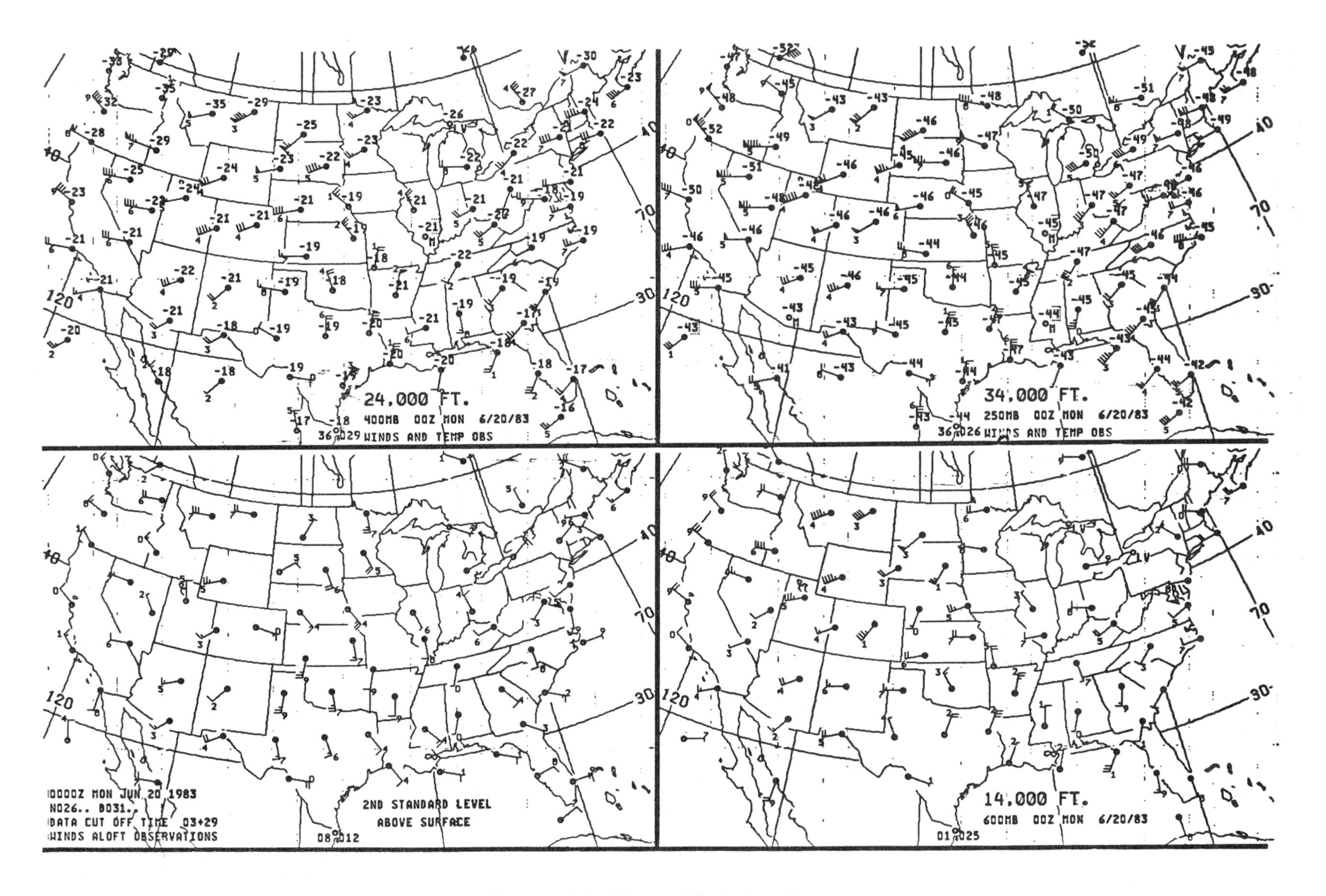

FIGURE 9-2. An Observed Winds Aloft Chart.

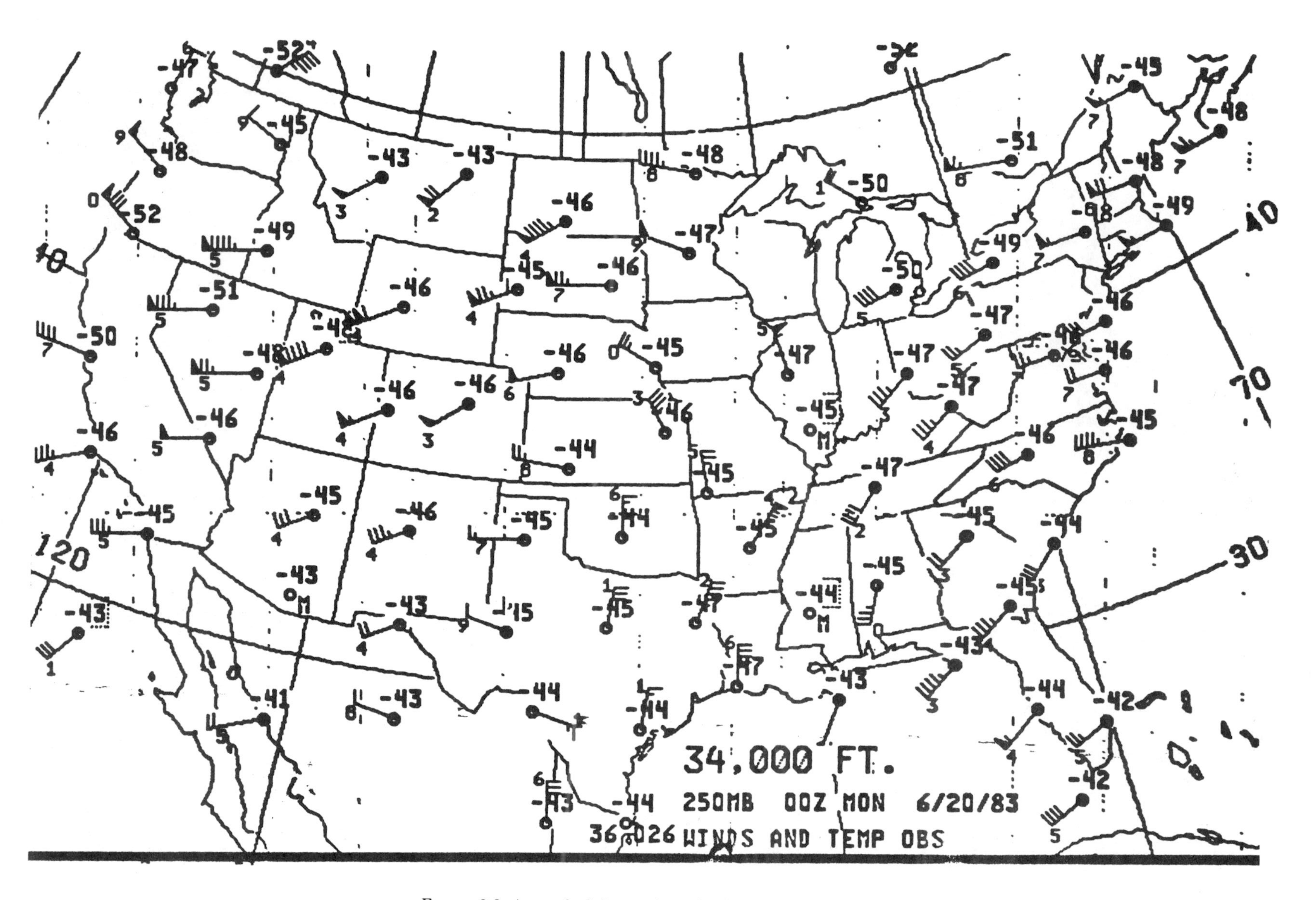

FIGURE 9-3. A panel of observed winds aloft for 34,000 feet.

Examples:

Station	Denver CO	Bismarck ND	Topeka KS	Key West FL
Station elevation:	3604 MSL	1677 MSL	879 MSL	0 MSL
Next thousand foot level above station:	4000 MSL +1000	2000 MSL +1000	1000 MSL +1000	1000 MSL +1000
Second standard level:	5000 MSL or 1396 AGL	3000 MSL or 1323 AGL	2000 MSL or 1121 AGL	2000 MSL or 2000 AGL

The 14,000 feet MSL panel is true altitude while the 24,000 and 34,000 feet MSL panels are pressure altitude.

USING THE CHARTS

The use of winds aloft charts seems obvious—to determine winds at a proposed flight altitude or to select the best altitude for a proposed flight. Temperatures also can be determined from the forecast charts. To determine winds and temperatures at a level between chart levels, interpolate. Also interpolate the data when the time period is other than the valid time of the chart.

Forecast winds are generally preferable to observed winds since they are more relevant to flight time. Although, observed winds are 5 to 8 hours old when received by facsimile and their reliability diminishes with time they can be a useful reference to check for gross errors on the 12-hour prog.

INTERNATIONAL FLIGHTS

Computer generated forecast charts of winds and temperatures aloft are available for international flights at specified levels. The U.S. National Meteorological Center (NMC), near Washington, D.C., is a component of the World Area Forecast System (WAFS). NMC is designated in the WAFS as both a World Area Forecast Center and a Regional Area Forecast Center (RAFC). Its main function as a World Area Forecast Center is to prepare global forecasts in grid-point form of upper winds and upper air temperatures and to supply the forecasts to associated RAFCs. One of NMC's main RAFC functions is to prepare and supply to users charts of forecast winds and temperatures. Figure 9-4 and 9-5 are examples of the forecast winds and temperatures aloft charts for international flights.

For example on figures 9-4 and 9-5, the lower left portion of the chart shows the originating office, NMC (Washington DC); flight level of chart (34,000 feet MSL); valid time of the chart and data base time (data from which the forecast was derived). Forecast winds are expressed in knots for spot locations with direction and speed depicted in the same manner as the U.S forecast winds and temperatures aloft chart (figure 9-1). Forecast temperatures are depicted for spot locations inside small circles; expressed in degrees Celsius. For charts with flight levels at or below FL180 (18,000 feet), temperatures are depicted as negative (−) or positive (+). On charts for flight levels (FL) above FL180, temperatures are always negative and so no sign (− or +) will be depicted.

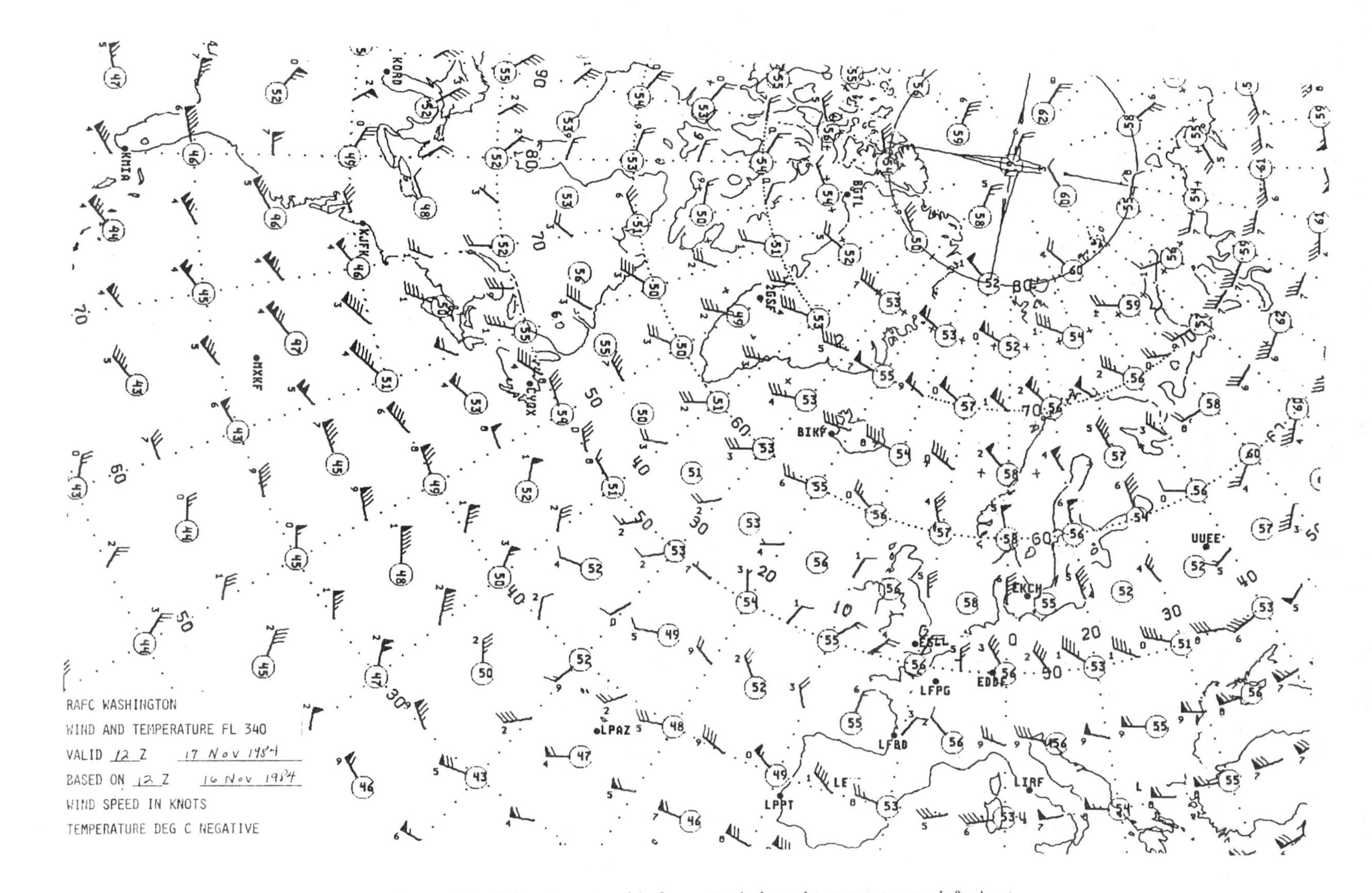

FIGURE 9-4. Polar stereographic forecast winds and temperatures aloft chart.

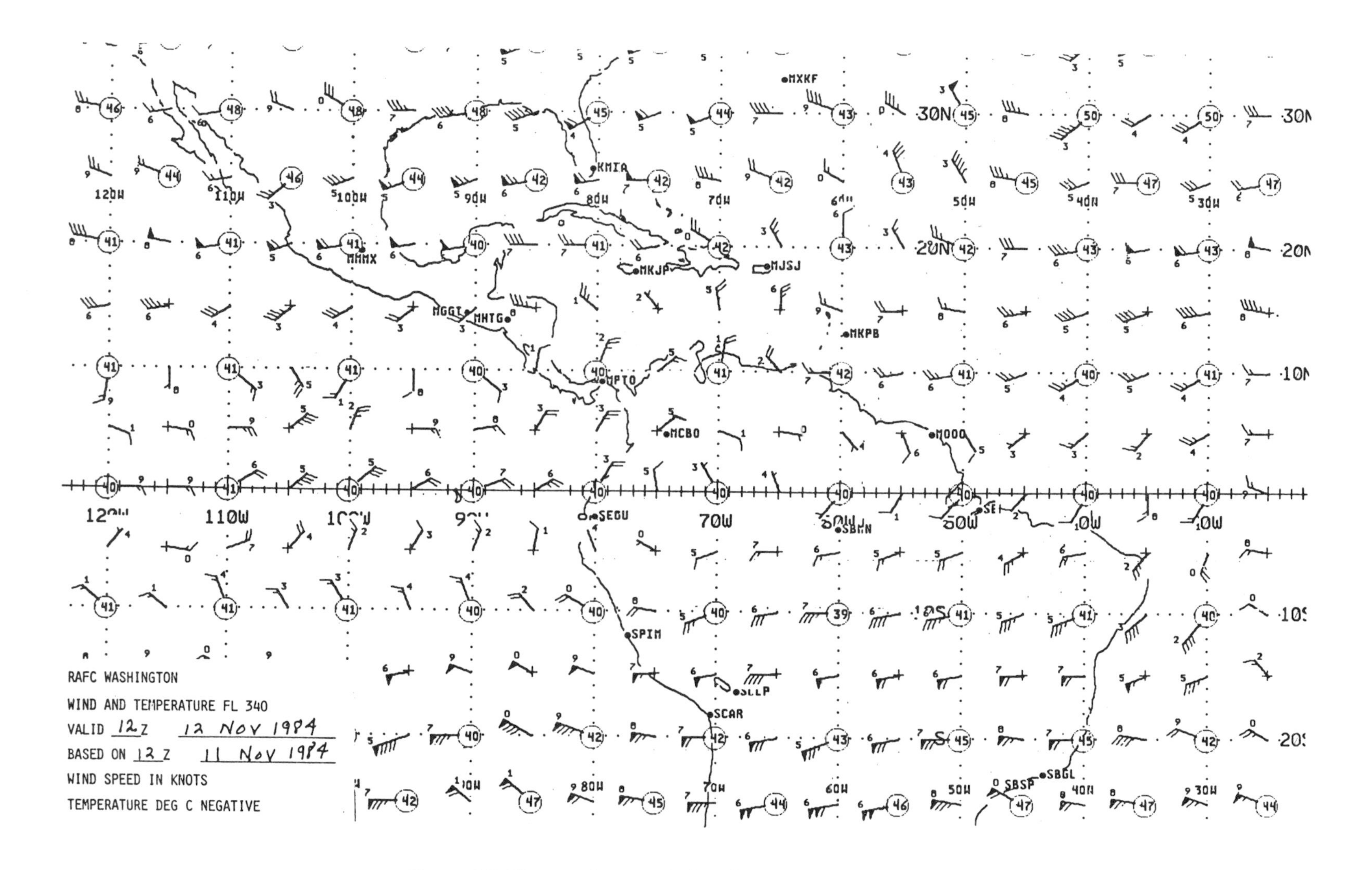

FIGURE 9-5. Mercator forecast winds and temperatures aloft chart.

Section 10

COMPOSITE MOISTURE STABILITY CHART

The Composite Moisture Stability Chart, figure 10-1, is an analysis chart using observed upper air data. It is composed of the following four panels: stability, freezing level, precipitable water, and average relative humidity. This computer generated chart is available twice daily with valid times of 12Z and 00Z. The availability of upper air data (on all the panels) for analysis is indicated by the shape of the station model. Use the legend on the precipitable water panel of figure 10-4 for the explanation. On this chart, the mandatory levels used are the surface, the 1000 mb, the 850 mbs, the 700 mb, and the 500 mb. Significant levels are the levels between the mandatory levels where significant changes in temperature and/or moisture occur when compared to below or above that level.

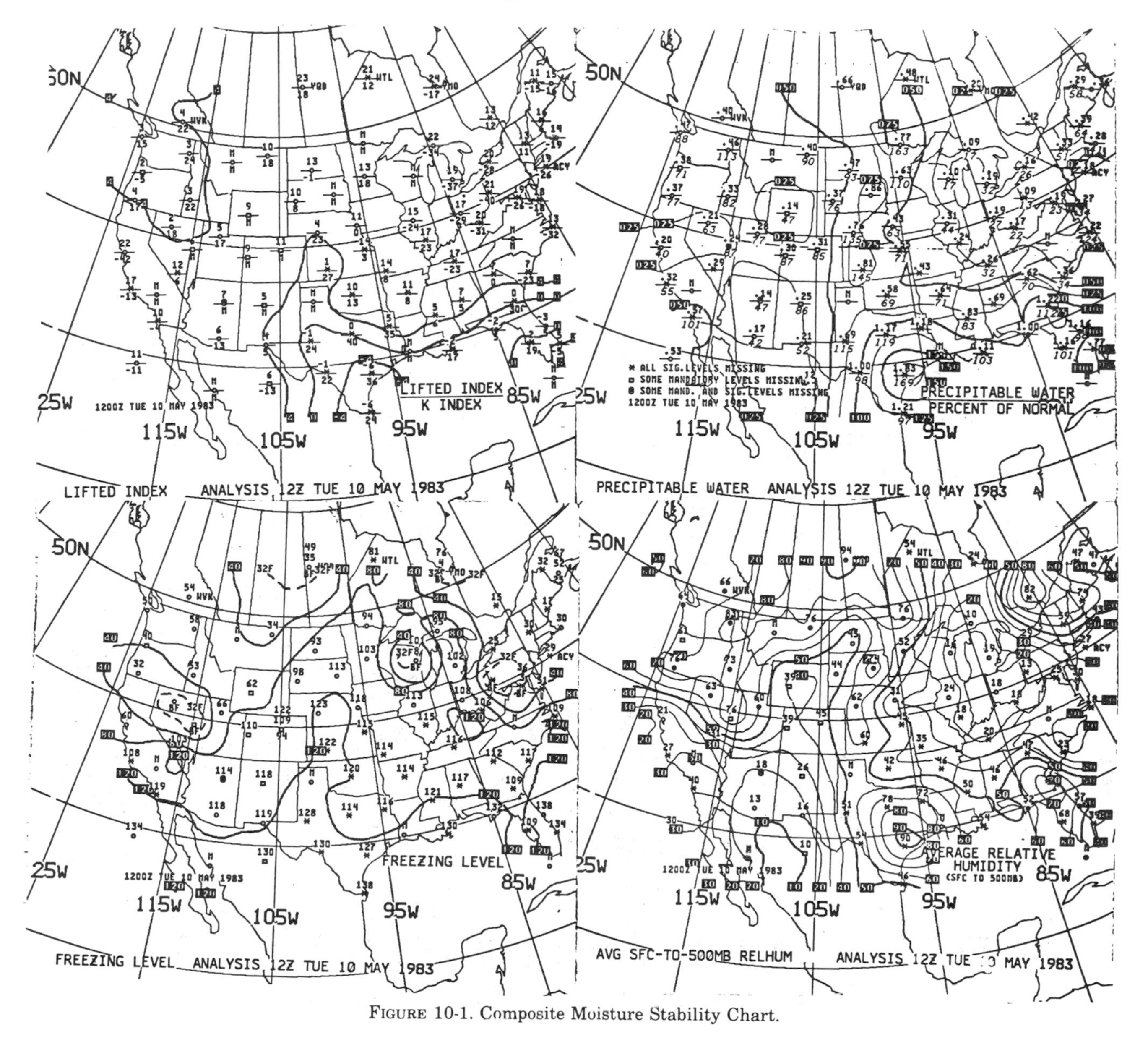

FIGURE 10-1. Composite Moisture Stability Chart.

STABILITY PANEL

The stability panel (upper left panel of chart), (figure 10-2), outlines areas of stable and unstable air. Two stability indices are computed for each upper air station, one is the *lifted index* and the other the *K index.* At each station, the lifted index is plotted above a short line and the K index below the line. An "M" indicates the value is missing.

The following explains the computation of the indices and the analysis and use of the panel. If you run into trouble with the discussion, you should review AVIATION WEATHER, chapter 6, "Stable and Unstable Air".

LIFTED INDEX (LI)

The lifted index is computed as if a parcel of air near the surface were lifted to 500 millibars. As the air is "lifted", it cools by expansion. The temperature the parcel would have at 500 millibars is then subtracted from the environmental 500 millibar temperature. The difference is the lifted index which may be positive, zero, or negative. Thus, the lifted index indicates stability at 500 mb. (18,000 feet MSL).

Positive Index

A positive index means that a parcel of a air *if lifted* would be colder than existing air at 500 millibars. The air is stable. Large positive (high) values indicate very stable air.

Zero Index

A zero index denotes that air *if lifted* to 500 millibars would attain the same temperature as the existing 500 millibar environmental temperature. The air is neutrally stable (neither stable nor unstable).

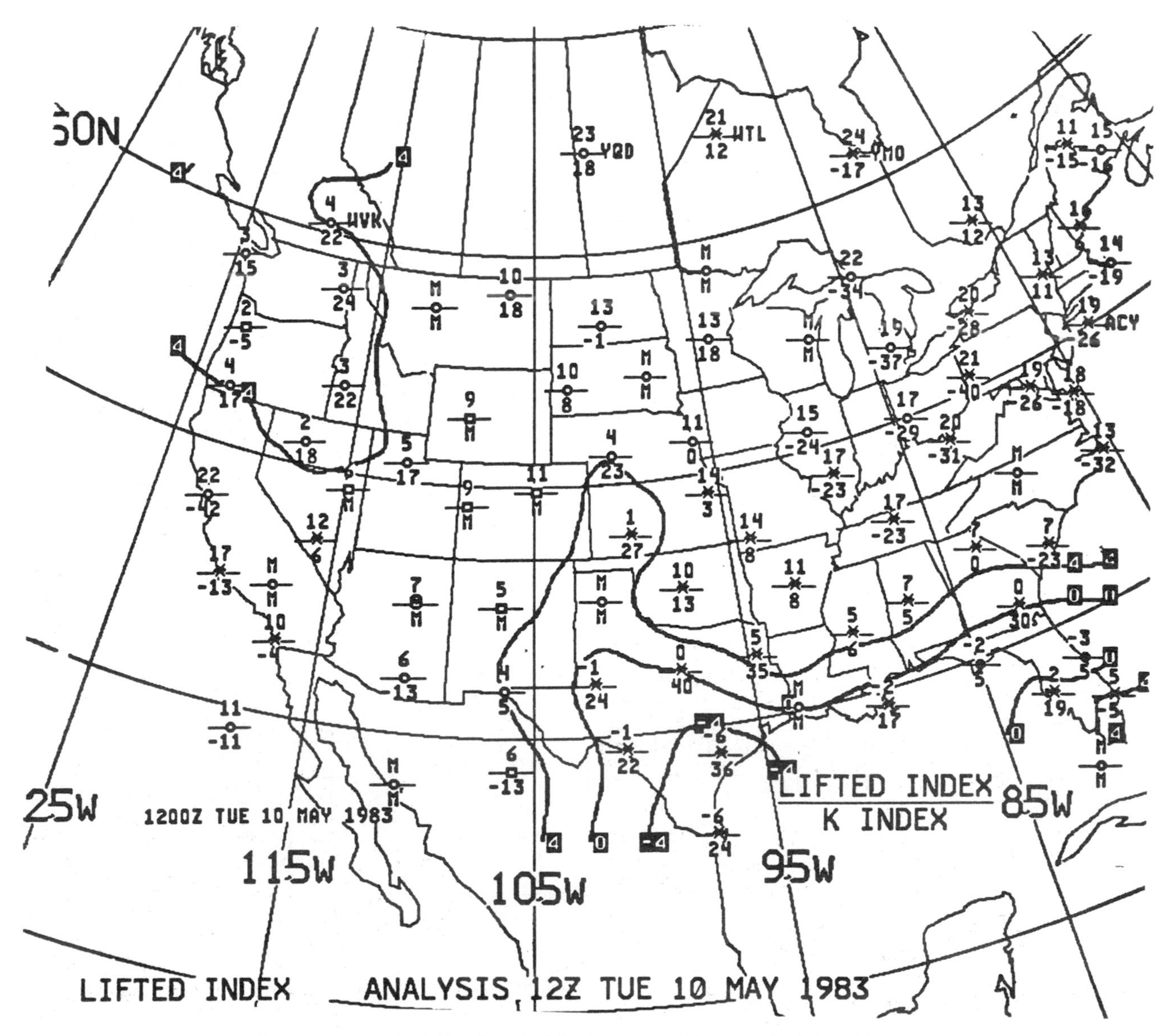

FIGURE 10-2. A Stability Panel of the Composite Moisture Stability Chart.

Negative Index

A negative index means that the low-level air *if lifted* to 500 millibars would be warmer than existing air at 500 millibars. The air is unstable and suggests the possibility of convection. Large negative (low) values indicate very unstable air.

Be aware of the following: 1) The LI assumes air near the surface will reach 500 millibars. Whether or not the near surface air will be lifted to 500 millibar level depends on what is happening below the level. It is possible to have a negative LI with no thunderstorms occurrence because the air below 500 millibars may not be lifted high enough for thunderstorms to develop. For use, the LI is more indicative of the severity of the thunderstorms, if they occur, rather than the probability of general thunderstorm occurrence (see TABLE 10-1). 2) The LI can change dramatically due to several causes, especially due to surface heating. Surface heating tends to make the LI value less positive (less stable) at 00Z as compared to 12Z.

K INDEX

The K index is primarily for the meteorologist but a discussion is included for those who are interested. The K index examines the temperature and moisture profile of the environment. Since a parcel of air is not lifted and compared to the enviroment, the K index is not truly a stability index. However, the meteorologist, looking at the environmental temperature and moisture profile, can make a good judgement as to the stability of the air. The K index is computed using three terms as follows:

$$K = \begin{array}{c}(\text{850 mb temp—500 mb temp}) \\ + \\ (\text{850 mb dew point}) - \\ (\text{700 mb temp dew point spread})\end{array}$$

The first term (850mb temp—500 mb temp) is proportional to the average lapse rate. A large temperature difference shows a steep or unstable lapse rate. The greater the difference, the more unstable the air and the higher the K value.

The second term (850 mb dew point) is a measure of low level moisture. Since the dew point is added, high moisture content at 850 millibars increases the K value.

The third term (700 mb temp dew point spread) is a measure of saturation at 700 millibars. The greater the spread, the drier the air. Since the term is subtracted, it lowers the K value. However, moist air (small spread) lowers the value less than does dry air (large spread). Thus, the greater the degree of saturation at 700 millibars, the larger is the K value.

Putting the three terms together, we see that each of the following contributes to a large or high K index:

1. An unstable lapse rate
2. High moisture content at 850 millibars, and
3. A high degree of saturation at 700 millibars.

Since the K index is not a true stability index and with the moisture variables in the equation affecting the value and meaning of the index, great caution should be exercised as to when and how the K index is to be used. Thus, the K index is used primarily by the meteorologist. However, some general use may be made of the K index but only with caution.

During the thunderstorm season, a large K index indicates conditions favorable for air mass thunderstorms. See table 10-1 for thunderstorm potential. The K index values and meanings in table 10-1 can decrease significantly for thunderstorm development associated with a synoptic scale low pressure system (non air-mass thunderstorms).

In the winter, when temperatures are very cold, the moisture terms are very small. The temperature terms completely dominate the K value computation and even fairly large values do not mean conditions are favorable for thunderstorms because of lack of moisture.

Also be aware the K values can change significantly over a short time period due to temperature and moisture advection.

STABILITY ANALYSIS

The analysis is based on the lifted index only. Station circles are blackened for LI values of zero or

TABLE 10-1. Thunderstorm Potential

LIFTED INDEX (LI)	*"K" INDEX **	*AIRMASS THUNDERSTORM PROBABILITY*
0 to −2 weak indication	<15	near 0%
of severe thunderstorms	15-20	20%
−3 to −5 moderate indication of severe thunderstorms	21-25	21-40%
indication of severe thunderstorms	26-30	41-60%
≤ -6 strong indication of severe thunderstorms	31-35	61-80%
	36-40	81-90%
	>40	near 100%

Note: See TABLE 4-9 for Areal Coverage Definitions

less. Solid lines are drawn for values of +4 and less at intervals of 4 (+4, 0, −4, −8 etc).

USING THE PANELS

As a user of stability indices, your question is "what helpful information can I obtain"? When clouds and precipitation are forecast or are occurring, the stability index is used to determine the type of clouds and precipitation. That is, stratiform clouds and steady precipitation occur with stable air while convective clouds and showery precipitation occur with unstable air. Remember that unstable air is associated with a negative lifted index and usually a high K index during the thunderstorm season (see table 10-1). Reliability of K index values decreases for non air-mass thunderstorms. Stability is also very important when considering the type, extent, and intensity of aviation weather hazards. For example, a quick estimate of areas of probable *convective* turbulence can be made by associating the areas with *unstable* air. An area of *extensive* icing would be associated with *stratiform* clouds and *steady* precipitation which are characterized by *stable* air.

It is essential to note that an unstable index does *not* automatically mean thunderstorms. Upon looking at the synoptic situtation, if thunderstorms are expected to develop in unstable air, Table 10-1 may be used with caution as stated in this section. * Caution to be used for K index values in western mountainous terrain due to elevation.

TABLE 10-2. Plotting freezing levels

Plotted	*Interpretion*
BF	Entire observation below freezing (0 degree C).
000	Surface temperature 0 degree C. Freezing level at surface and below freezing above.
120	Lowest and only freezing level 12,000 feet MSL; above freezing below 12,000 feet.
Three digits other than 000	Height of a freezing level aloft in hundreds of feet MSL, i.e.; 002, 200 feet MSL; 120, 12,000 feet MSL.
110 051 BF	Below freezing from surface to 5,100 feet; above freezing from 5,100 feet to 11,000 feet; and below freezing above 11,000 feet.
090 034 003	Lowest freezing level, 300 feet; below freezing from 300 feet to 3,400 feet; above freezing 3,400 feet to 9,000 feet; below freezing above 9,000 feet.
M	Data missing.
106 101 89 28 BF	Below freezing from surface to 2,800 ft.; above freezing from 2,800 ft. to 8,900 ft.; below freezing from 8,900 ft. to 10,100 ft.; above freezing from 10,100 ft. to 10,600 ft.; and below freezing above 10,600 ft.

FREEZING LEVEL PANEL

The freezing level panel (lower left panel of chart) is an analysis of observed freezing level data from upper air observations (see figure 10-3).

PLOTTED DATA

Table 10-2 explains plotting freezing level data. Note that more than one entry denotes multiple crossings of the zero (0) degree Celsius isotherm. See TABLE 10-3.

ANALYSIS

Solid lines are contours of the lowest freezing level and are drawn for 4000 foot intervals and labelled in hundreds of feet MSL. When a station reports more than one crossing of the zero (0) degree Celsius isotherm, the lowest crossing is used in the analysis. This is in contrast to the low level significant weather prog on which the depicted forecast freezing level aloft is the highest freezing level. A dashed line represents the 32 degree Fahrenheit isotherm at the surface and will outline an area of stations reporting "BF" (belowing freezing).

USING THE PANEL

The contour analysis shows an overall view of the lowest observed freezing level. Always plan for possible icing in clouds or precipitation above the freezing level—especially between temperatures of zero (0) Celsius and −10 Celsius.

Plotted multiple crossings of the zero (0) degree Celsius isotherm at a station always show an inversion with warm air above subfreezing temperatures (see TABLE 10-3). This situation can produce very hazardous icing when precipitation is occurring. Area forecasts show more specifically the areas of expected icing. Low level significant weather progs show anticipated changes in the freezing level.

PRECIPITABLE WATER PANEL

The precipitable water panel (upper right panel of chart) is an analysis of the water vapor content from

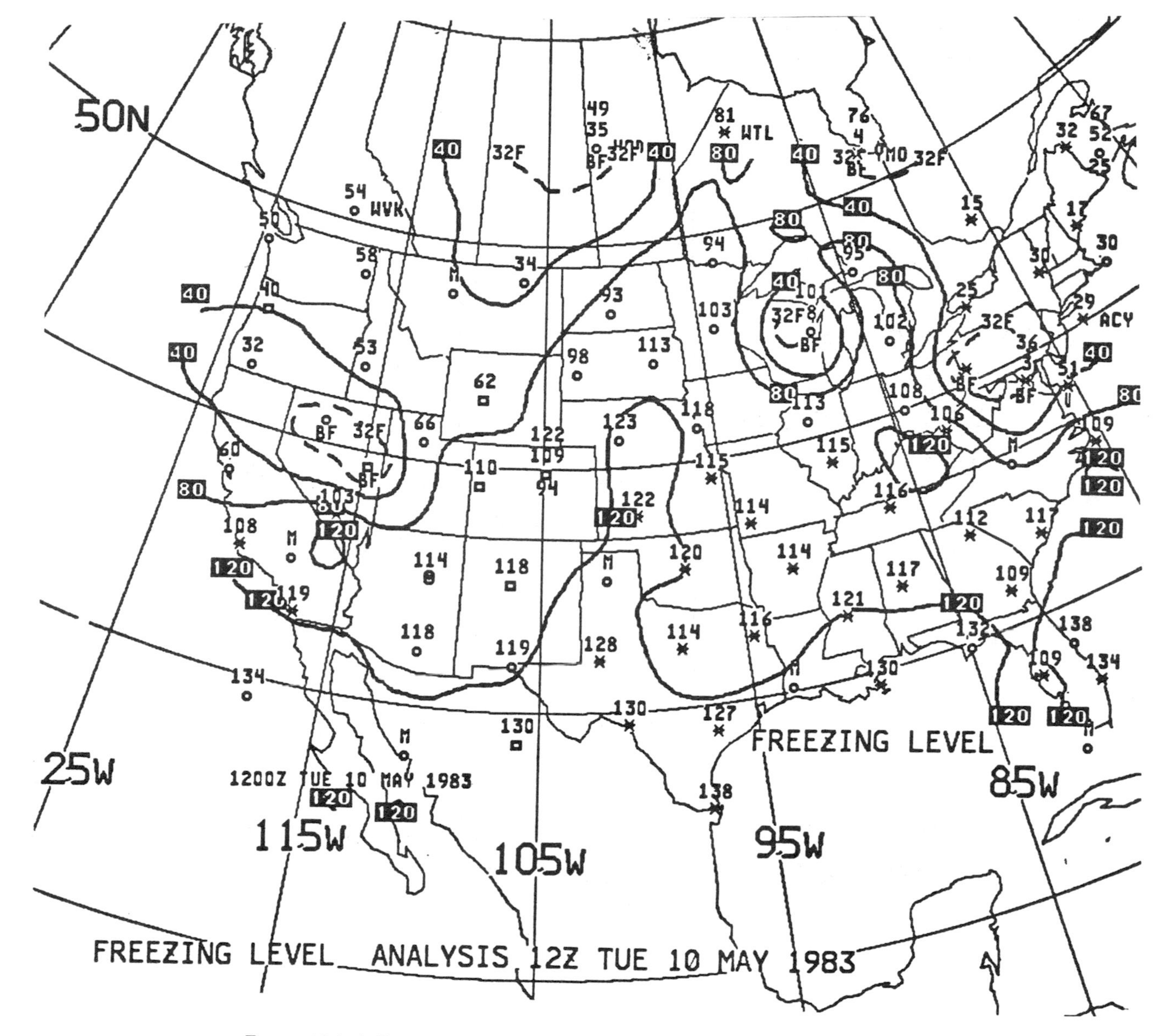

FIGURE 10-3. A Freezing Level Panel of the Composite Moisture Stability Chart.

the surface to the 500 mb level (see figure 10-4). The amount of water vapor observed is shown as precipitation water, which is the amount of liquid precipitation that would result if all the water vapor were condensed.

PLOTTED DATA

At each station precipitable water values to the nearest hundredth of an inch are plotted above a short line and the percent of normal value for the month below the line. the percent of normal value is the amount of precipitable water actually present compared to what is normally expected. As examples on figure 10-4, .58/69 at Oklahoma City OK indicates 58 hundredths of an inch of precipitable water is present which is only 69 percent of normal (below normal) for any day during this month, .81/1.45 at Dodge City KS indicates 81 hundredths of an inch of precipitable water is present which is 145 percent of normal (above normal) for any day during this month. An "M" plotted above the line indicates missing data as shown at Amarillo, TX. At Las Vegas NV the percent of normal value is not plotted which indicates insufficient climatological data to compute this value.

ANALYSIS

Stations with blackened in circles indicate precipitable water values of 1.00 inch or more. Isopleths of precipitable water are drawn and labelled for every

286-959 - 91 - 5 : QL 3

TABLE 10-3. Vertical temperature profile of plotted freezing levels at a station.

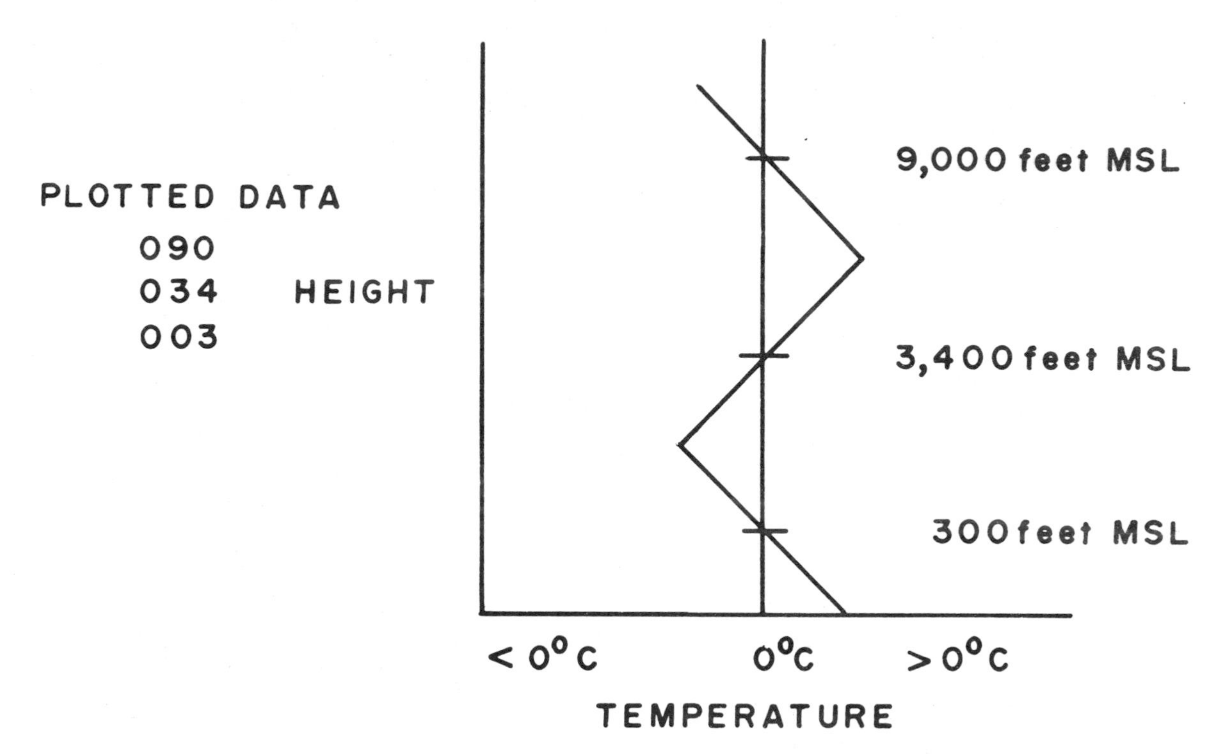

0.25 inches with the heavy isopleths drawn at 0.50 inch intervals.

USING THE PANEL

This panel is obviously used to determine water vapor content in the air between surface and 500 mb, and is especially useful to meteorologists concerned with flash flood events. By looking at the wind field upstream from your station, you can get an excellent indication of changes that will occur in moisture content, i.e. drying out or increasing moisture with time.

AVERAGE RELATIVE HUMIDITY PANEL

The average relative humidity panel (lower right panel of chart) is an analysis of the average relative humidity from the surface to 500 mb, plotted as a percentage for each reporting station (see figure 10-5). An "M" indicates the value is missing.

ANALYSIS

Station circles are blackened for humidities of 50 percent and higher. Isopleths of relative humidity (isohumes) are drawn and labelled every 10 percent with the heavy isohumes drawn for values of 10, 50 and 90 percent.

USING THE PANEL

This panel is used to determine, on the average, how saturated the air is from the surface to 500 mb. Average relative humidities of 70 percent or greater are frequently associated with areas of clouds and possible precipitation. This is because with such a

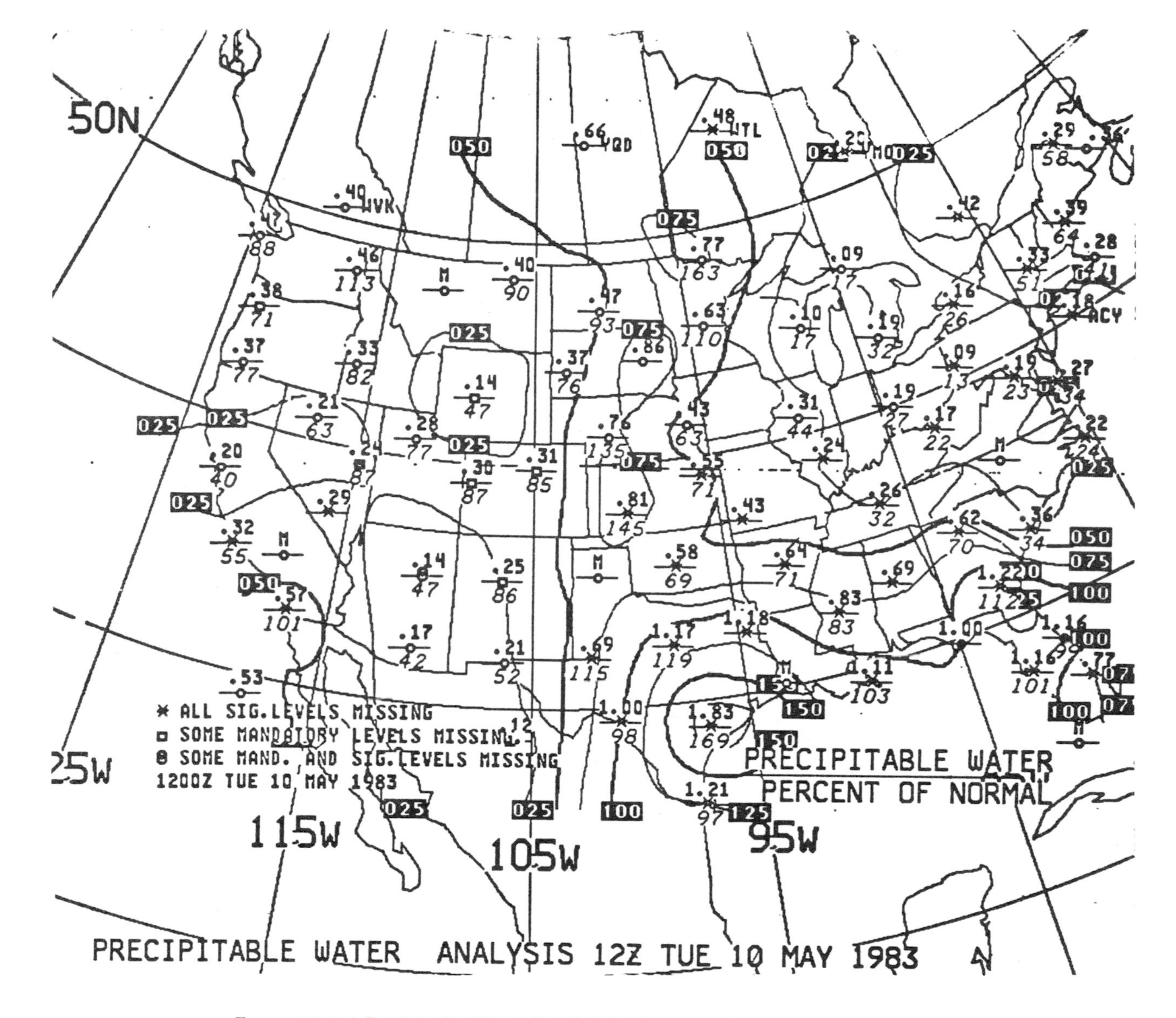

FIGURE 10-4. A Precipitable Water Panel of the Composite Moisture Stability Chart.

high average relative humidity through approximately 18,000 feet, it is likely that a specific layer(s) will have 100 percent relative humidity with clouds and possibly precipitation. It is important to remember that high values of relative humidity do not necessarily mean high values of water vapor content (precipitable water). For example in figure 10-4, the station in southwest Oregon has less water vapor content than International Falls MN (.37 and .77 respectively), but in examining figure 10-5, the average relative humidities are the same for both stations. If rain were falling at both stations, the result would likely be lighter precipitation totals for southwest Oregon.

USING THE COMPOSITE MOISTURE STABILITY CHART

Through analysis of this chart, you can determine the characteristics of a particular weather system in terms of stability, moisture, and possible aviation hazards. Even though this chart will be several hours old when received, the weather system will tend to move these characteristics with it. Thus, extrapolation techniques are an advantage to this chart although caution should be exercised due to modification of these characteristics through development, dissipation or the system moving from water to land or land to water.

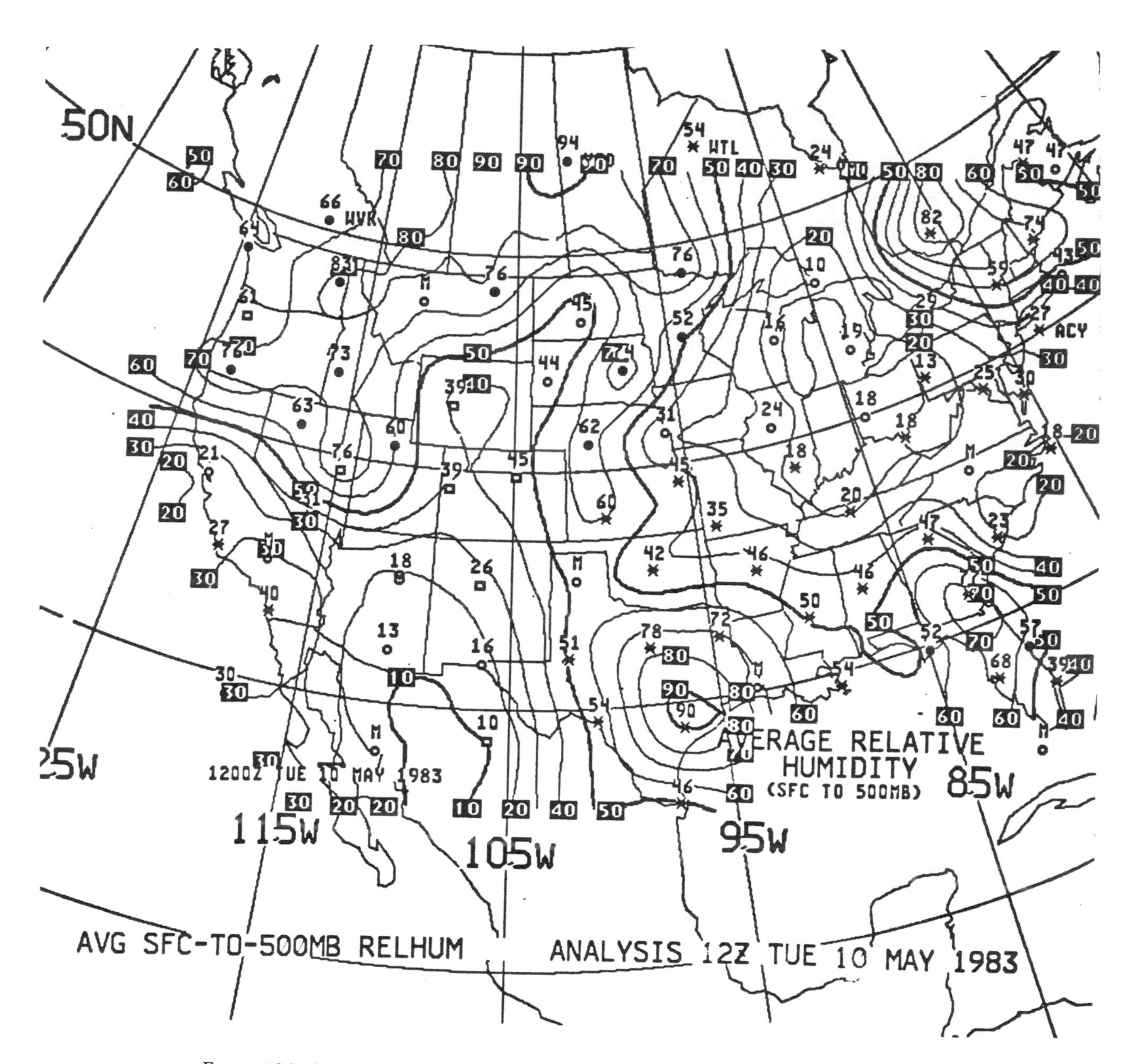

FIGURE 10-5. Average Relative Humidity Panel of the Composite Moisture Stability Chart.

Section 11
SEVERE WEATHER OUTLOOK CHART

The severe weather outlook chart, figure 11-1, is a preliminary 24-hour outlook for thunderstorm activity presented in two panels. The left-handed panel covers the 12-hour period 1200Z-0000Z. The right-hand panel covers the remaining 12 hours, 0000Z-1200Z. The manually prepared chart is issued once daily in the morning.

GENERAL THUNDERSTORMS

A line with an arrowhead delineates an area of probable general thunderstorm activity. When you face in the direction of the arrow, activity is expected to the right of the line. An area labelled APCHG indicates probable general thunderstorm activity may approach severe intensity. Approaching means winds greater than or equal to 35 knots but less than 50 knots and/or hail greater than or equal to 1/2 inch in diameter but less than 3/4 inch (surface conditions). Note in figure 11-1, from 12Z to 00Z that general thunderstorm activity is *not* forecast for the west coast, western Texas and from Tennessee northeastward to western New York.

SEVERE THUNDERSTORMS

The single-hatched area indicates possible severe thunderstorms. The following notations show possible coverage:

TABLE 11-1. Notation of Coverage

Notation	*Coverage*
SLIGHT RISK	2 to 5% coverage or 4 to 10 radar grid boxes containing severe thunderstorms per 100,000 square miles.
MODERATE RISK	6 to 10% coverage or 11 to 21 radar grid boxes containing severe thunderstorms per 100,000 square miles.
HIGH RISK	More than 10% coverage or more than 21 radar grid boxes containing severe thunderstorms per 100,000 square miles.

In figure 11-1, note the moderate risk of severe thunderstorms in the eastern Dakotas and Minnesota surrounded by a slight risk area. In the moderate risk area, severe thunderstorms are possible with severe storm coverage of 6 to 10 percent of the area.

TORNADOES

Tornado watches are plotted only if a tornado watch is in effect at chart time. The watch area is cross-hatched. No coverage is specified. Figure 11-1 shows a tornado watch in effect for eastern North Dakota and northern Minnesota at the time that the chart was issued.

USING THE CHART

The severe weather outlook is strictly for advanced planning. It alerts all interests to the possibility of future storm development. As the time of severe weather approaches, the forecaster can more specifically delineate the time, extent, and nature of the weather and issue a severe weather watch (WW).

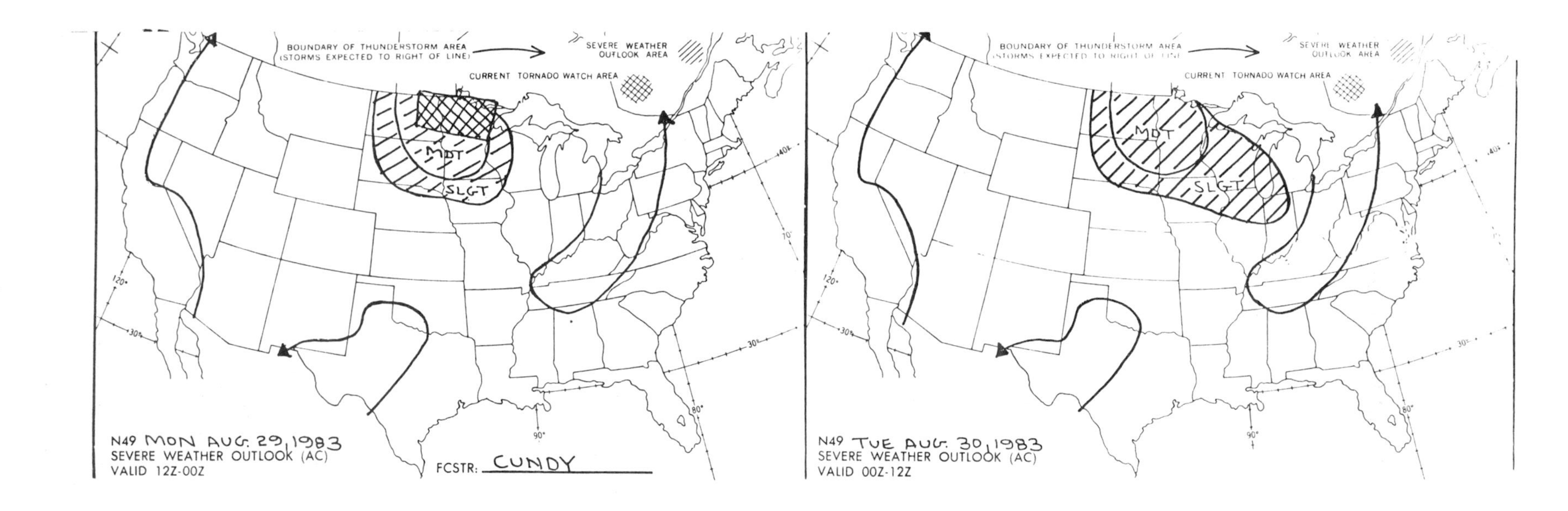

FIGURE 11-1. Severe Weather Outlook Charts.

Section 12
CONSTANT PRESSURE ANALYSIS CHARTS

Any surface of equal pressure in the atmosphere is a constant pressure surface. A constant pressure analysis chart is an upper air weather map where all the information depicted is at the specified pressure of the chart.

Twice daily, five (5) computer prepared constant pressure charts (850 mb, 700 mb, 500 mb, 300 mb and 200 mb) are transmitted by facsimile, each valid at 12Z and 00Z. Plotted at each reporting station (at the level of the specified pressure) are the observed temperature, temperature-dew point spread, wind, height of the pressure surface, as well as height changes over the previous 12 hour period. Figure 12-2 through 12-6 are sections of each constant pressure chart.

Pressure altitude (height in the standard atmosphere) for each of the five (5) pressure surfaces is shown in table 12-1. For example, 700 millibars of pressure has a pressure altitude of 10,000 feet (standard atmosphere). In the real atmosphere 700 millibars of pressure only closely approximates 10,000 feet (either above or below 10,000 feet) because the real atmosphere is seldom standard. For direct use of a constant pressure chart, assume you are planning a flight at 10,000 feet. The 700 mb chart is approximately 10,000 feet MSL. It is a source of observed temperature, temperature-dew point spread, moisture, and wind for your flight.

FIGURE 12-1. Radiosonde Data Station Plot and Decode.

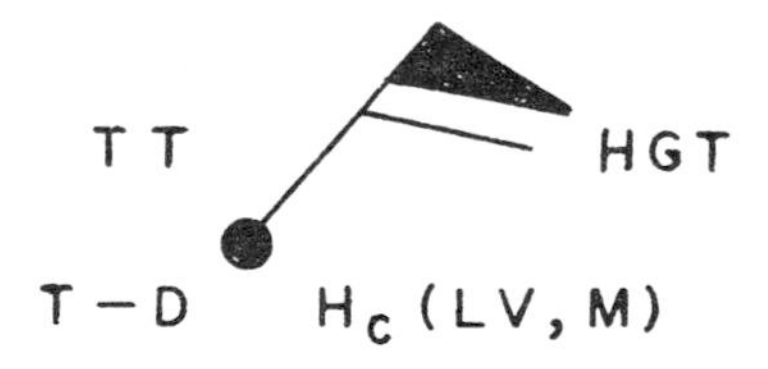

WIND—wind direction (WD) and speed (WS) plotted to the nearest 10 degrees and to the nearest 5 knots respectively.

5 knots, 10 knots, 50 knots

—if direction or speed is missing "M" is plotted in H_c space.

—if speed is less than 3 knots "LV" (light and variable) is plotted in H_c space.

HGT —height of constant pressure surface in meters. See TABLE 12-1 for decoding. If data is missing nothing is plotted in this spot.

TT —temperature to the nearest whole degree C; minus sign used if negative. Left blank if TT is missing. On the 850 mb chart, primarily in mountain regions where stations may be located above 850 mb of pressure, a bracketed temperature (and HGT) is a computed value. If two temperatures are plotted one above the other, the top temperature is used in the analysis.

T-D —temperature-dew point spread (or depression) to the nearest whole degree C. Left blank if T-D is missing. If T-D is less than or equal to 5 degrees C, the station circle is completely blackened. If T-D is greater than 29 degrees C and "X" is plotted. If TT is colder than −41 degrees C, T-D is left blank because the air is too dry at those temperatures to measure dew point.

H_c —Previous 12 hour height change plotted in tens of meters (decameters). H_c not plotted when wind is "LV" or "M". +04 means height of pressure above station has risen 40 meters, 02=meters, 11=110 meters.

PLOTTED DATA

Figure 12-1 illustrates and decodes the standard radiosonde data plot. Table 12-2 gives examples. Aircraft and satellite observations are used in analysis over areas of sparse data. A square in lieu of a station circle signifies an aircraft report. The flight level of the aircraft is plotted in hundreds of feet. Temperature and wind are at the flight level of the aircraft. The time of report is also indicated to the nearest hour GMT. For example, in figure 12-5, the aircraft report at approximately 30 degrees N and 140 degrees W is decoded as follows: flight level 34,000 feet, temperature −50 degrees Celsius, wind from 280 degrees at 10 knots, time of report to nearest hour is 2100 GMT. A star identifies satellite wind estimates made from cloud tops and is used primarily on the 850 mb chart off the west coast. These winds are representative for the 850 mb chart even though they are always labelled with a pressure altitude of 3,000 feet. See figure 12-2 for examples.

TABLE 12-1. Features of constant pressure charts-U.S.

PRESSURE (millibars)	PRESSURE ALTITUDE in feet (flight level)	PRESSURE ALTITUDE in meters	TEMPERATURE DEW POINT SPREAD	ISOTACHS	CONTOUR INTERVAL (meters)	DECODE STATION HEIGHT PLOT		EXAMPLES OF STATION HEIGHT PLOTTING	
						PREFIX TO PLOTTED VALUE	SUFFIX TO PLOTTED VALUE	PLOTTED	HEIGHT
850	5,000	1,500	YES	NO	30	1	–	530	1,530
700	10,000	3,000	YES	NO	30	2 or 3*	–	180	3,180
500	18,000	5,500	YES	NO	60	—	0	582	5,820
300	30,000	9,000	YES**	YES	120	—	0	948	9,480
200	39,000	12,000	YES**	YES	120	1	0	164	11,640

Note that pressure altitudes are rounded off to the nearest thousand for feet and to the nearest 500 for meters.

All heights are MSL.

* Prefix a "2" or "3" whichever brings the height closer to 3,000 meters.

** Omitted when air is too cold (less than −41 degrees) to measure dew point. Flight level of an aircraft is plotted in lieu of height of constant pressure surface.

TABLE 12-2. Examples of radiosonde plotted data.

	(850 MB)	(700 MB)	(500 MB)	(300 MB)	(200 MB)
Plot	22 479 / 4 LV	09 129 / 17 −03	−19 558 / X +03	−46 919 / +10	−60 191 / M
WIND	LIGHT AND VARIABLE	010° 20KTS	210° 60KTS	270° 25KTS	MISSING
TT	22°C	9° C	−19° C	−46° C	−60° C
T−D	4° C	17° C	> 29° C	not plotted	not plotted
DEW POINT	18°C	−8° C	DRY	DRY	DRY
HGT	1,479 meters	3,129 meters	5,580 meters	9,190 meters	11,910 meters
H_c	not plotted	MINUS 30 meters	PLUS 30 meters	+100 meters	not plotted

ANALYSIS

All charts contain contours and isotherms and some contain isotachs. Contours are lines of equal height, isotherms are lines of equal temperature, and isotachs are lines of equal wind speed.

Height Contours

Heights of the specified pressure for each station are analyzed through the use of solid lines called contours to give a *height* pattern. The contours depict highs, lows, troughs and ridges aloft in the same manner as isobars on the surface chart. Thus, on an upper air chart we speak of "high height centers" and "low height centers" instead of "high pressure centers" and "low pressure centers" respectively. We may compare a height analysis to a pressure analysis. A *contour* high, low, ridge or trough and the two terms are used interchangebly. For example, a high height center at 500 mb of pressure is analogous to a high pressure center at about 18,000 feet. Height and pressure analyses are just two ways of describing the same features.

Since an upper air chart is above the surface friction layer, winds for practical purposes parallel the contours. To decode contour values on the 850 mb through 300 mb chart simply add a zero to the three digit code while on the 200 mb chart you must prefix a one (1) in addition to adding a zero.

Refer to figures 12-2 and 12-3 and note the low aloft in Montana extending upward through 700 mb. Figures 12-4 through 12-6 show the low has opened into a trough at 500 mbs and above. Note that this low tilts toward the west with height.

Isotherms

Isotherms (dashed lines) drawn at 5 degrees Celsius intervals show horizontal temperature variations at chart altitude. Let's refer to the 850 mb chart (5,000 feet pressure altitude), figure 12-2 and locate an isotherm. Note the dashed line extending from North Dakota southeastward through South Carolina and labelled "+10" in the western portion of North Dakota. This is the +10 degree Celsius isotherm. North of this isotherm, temperatures at approximately 5,000 feet are below +10 degrees Celsius. The +15 degree Celsius isotherm extends across South Dakota. By inspecting isotherms, you can determine if your flight will be toward colder or warmer air. Subfreezing temperatures and a temperature-dew point spread of 5 degrees Celsius or less suggest possible icing. Note that isotherms on the 300 and 200 mb charts are the heavy dashed lines.

Isotachs

Isotachs (short, lightly dashed lines) appear only on the 300 and 200 mb charts. Isotachs are drawn at 20 knot intervals beginning with 10 knots. To aid in identifying areas of strong winds, hatching denotes wind speeds of 70 to 110 knots, a clear area within a hatched area indicates wind or 110 to 150 knots, an area of 150 to 190 knots of wind is hatched, etc. Note the alternating hatched/clear areas in figure 12-7 extending eastward from California to Florida and up the east coast. The 150 knots isotach (150 K) is over southern New Mexico and across Oklahoma and Texas.

DO NOT confuse isotherms with isotachs.

THREE DIMENSIONAL ASPECTS

As established earlier, we may treat a height contour analysis as a pressure analysis. Closely spaced contours mean strong winds as do closely spaced isobars. Wind blows clockwise around a contour high and counterclockwise around a low.

Features on synoptic surface and upper air charts are related. However, a weak surface system often loses its identity in a large scale upper air pattern while another system may be more evident on an upper chart than on the surface chart. Many times weather is closely associated with an upper air pattern than with features on the surface map.

You may have learned as a general rule to regard a surface low as a producer of bad weather and a high as a producer of good weather. Usually, this is true but an upper level low or trough usually means bad weather also. The area of cloudiness and precipitation found with an upper air low (on the east side) is usually associated with a surface low but sometimes an upper level low with clouds and precipitation will move over a rather shallow surface high with corresponding bad weather in the high. In contrast, an upper air high usually means good weather. An exception is an upper air high or ridge that has a stabilizing effect at low levels. Smoke, haze, dust, or even low stratus and fog may persist for extended periods; yet the surface map shows no cause for the restriction.

Lows generally slope to the west with ascending altitude for developing low pressure systems. Due to this slope, wind aloft with an upper system often blows across the associated surface system. Surface fronts, lows, and highs tend to move with the upper winds. For example, strong winds aloft across a surface front will cause the front to move rapidly, but if upper winds parallel a front, it moves slowly if at all.

An old, non-developing low pressure system tilts little with height. The low becomes almost vertical and is clearly evident on both surface and upper air maps. Upper winds encircle the surface low rather than blow across it. Thus, the storm moves very slowly and usually causes extensive and persistent cloudiness, precipitation, and generally adverse flying weather. The term "cold low" describes such a system and is usually identified on the surface chart

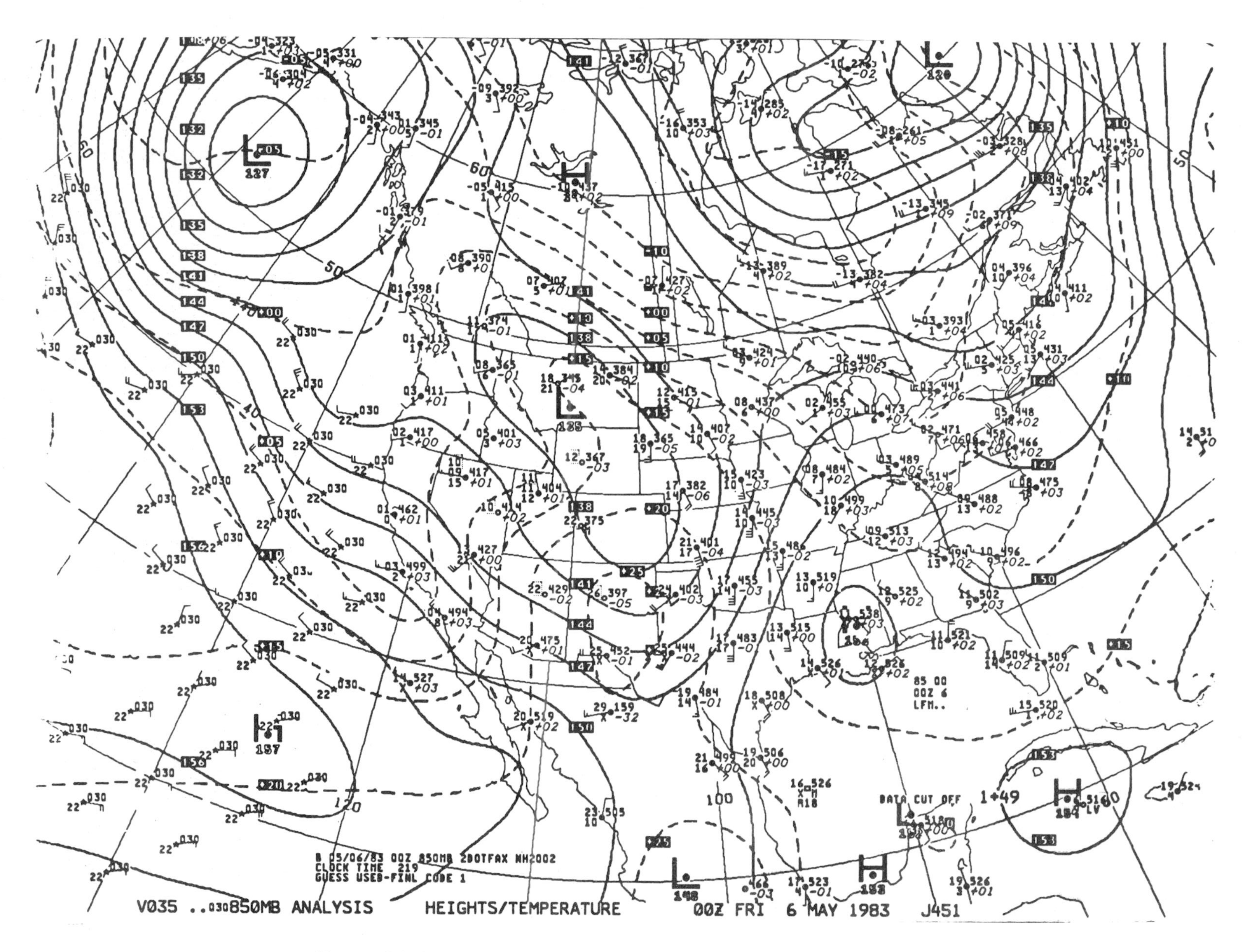

FIGURE 12-2. A section of an 850 millibar analysis, pressure altitude 5,000 feet.

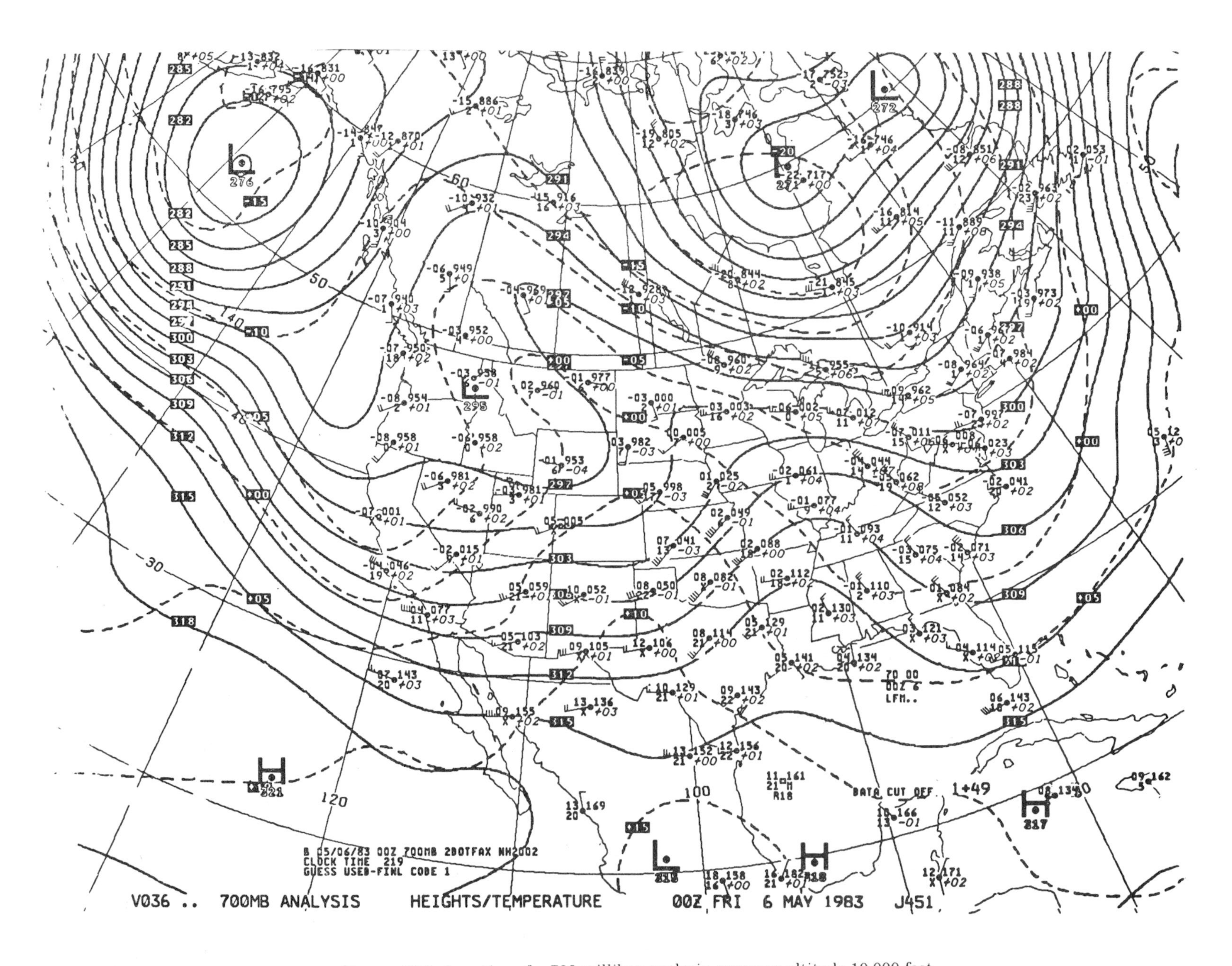

FIGURE 12-3. A section of a 700 millibar analysis, pressure altitude 10,000 feet.

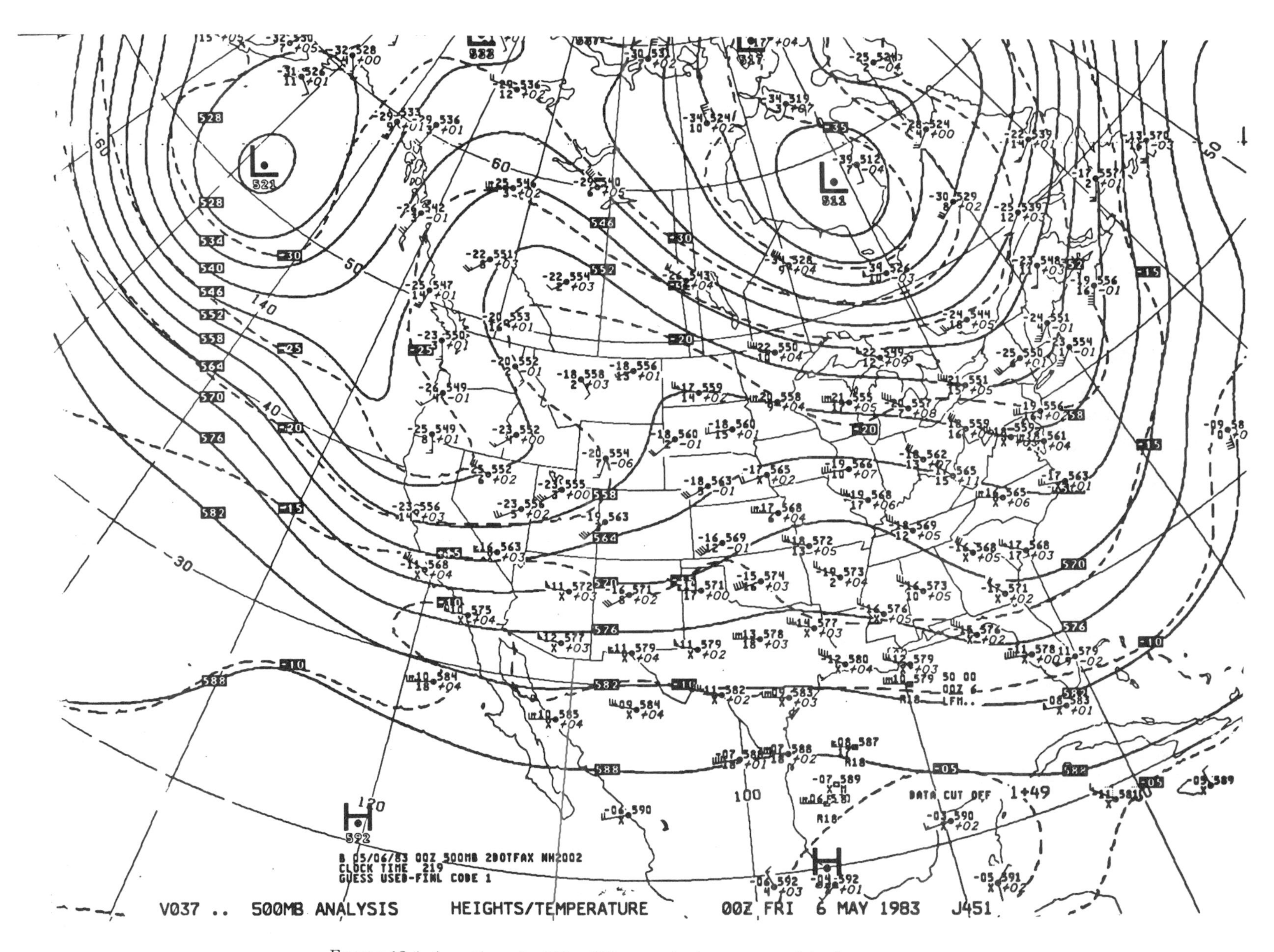

FIGURE 12-4. A section of a 500 millibar analysis, pressure altitude 18,000 feet.

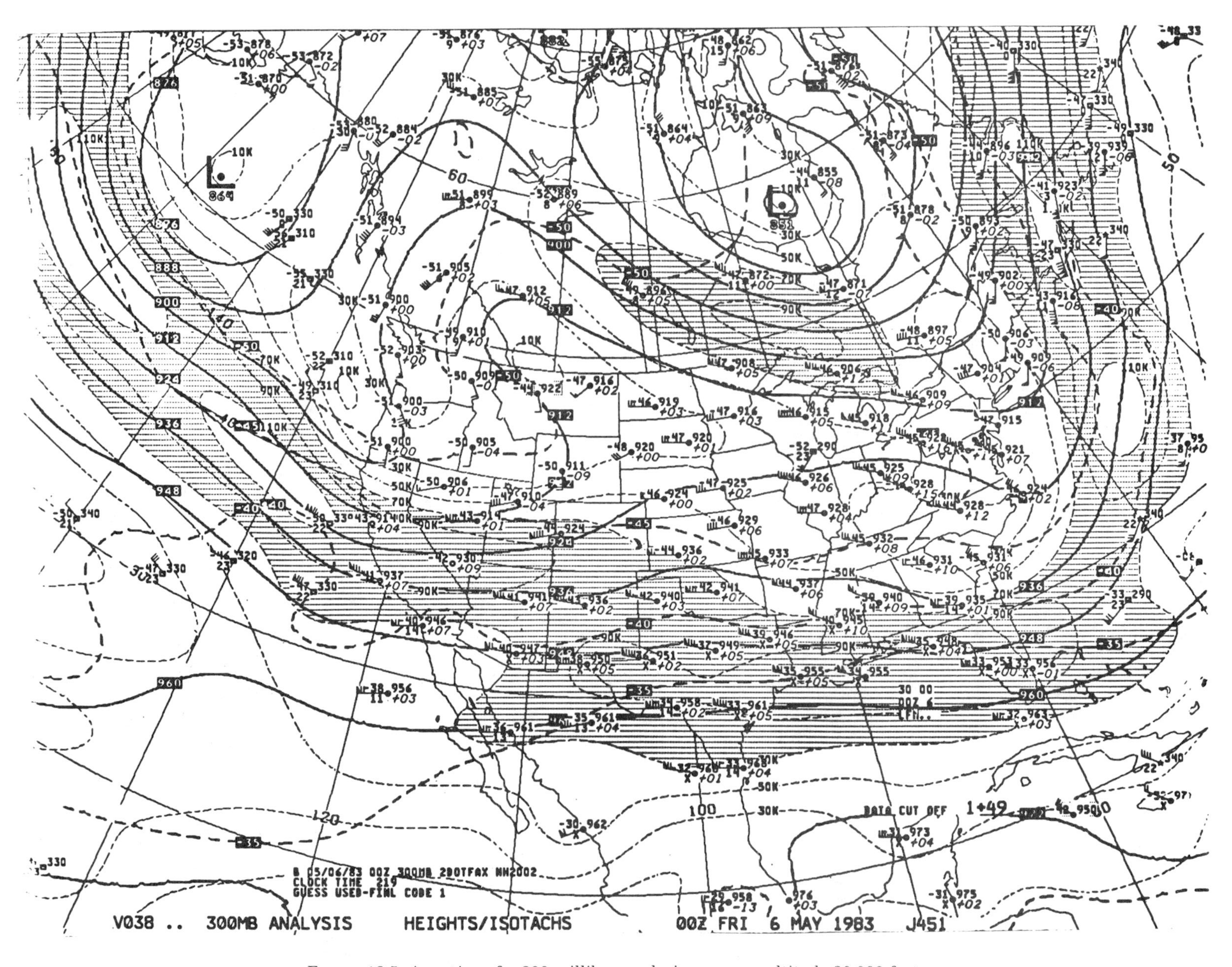

FIGURE 12-5. A section of a 300 millibar analysis, pressure altitude 30,000 feet.

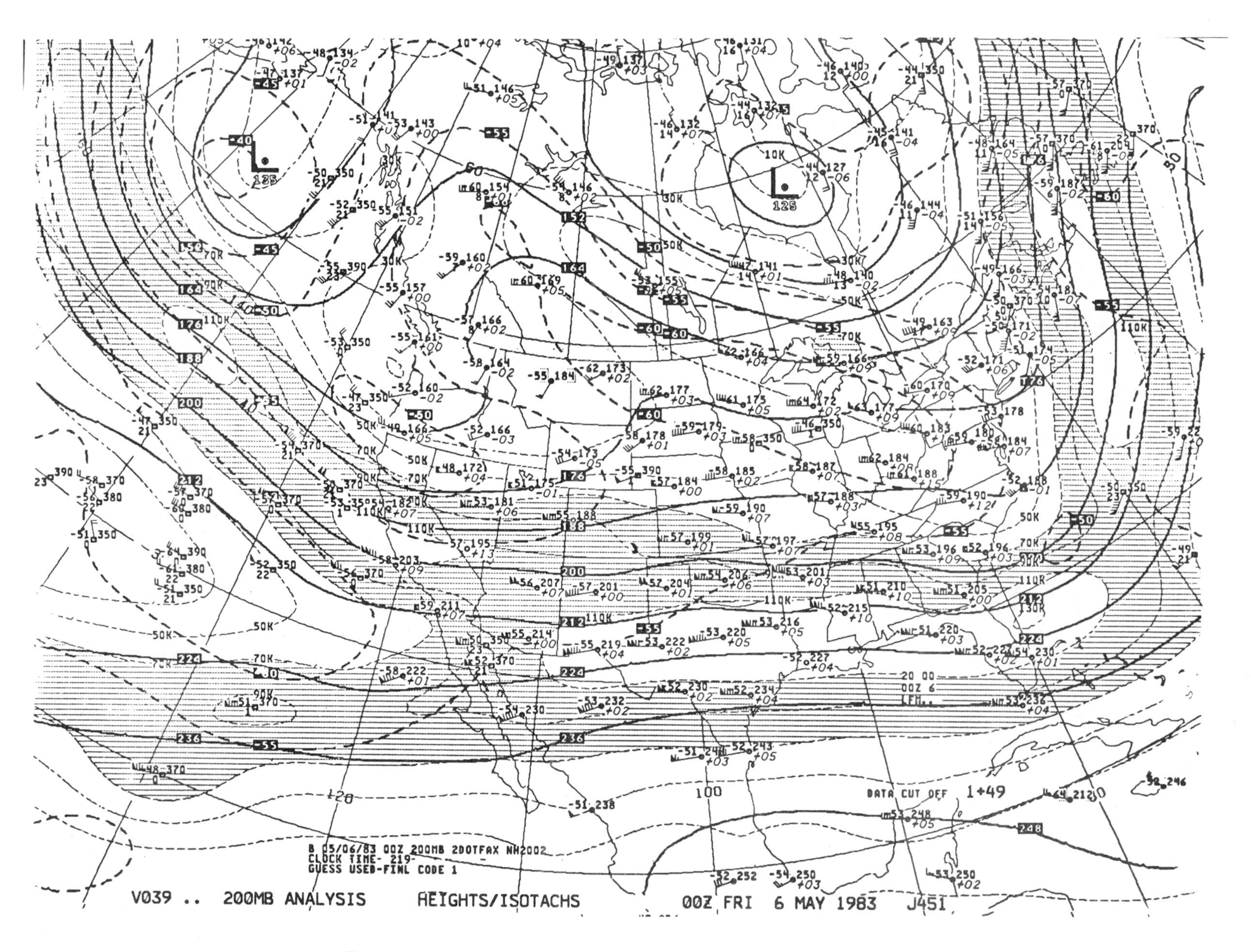

FIGURE A section of a 200 millibar analysis, pressure altitude 39,000 feet.

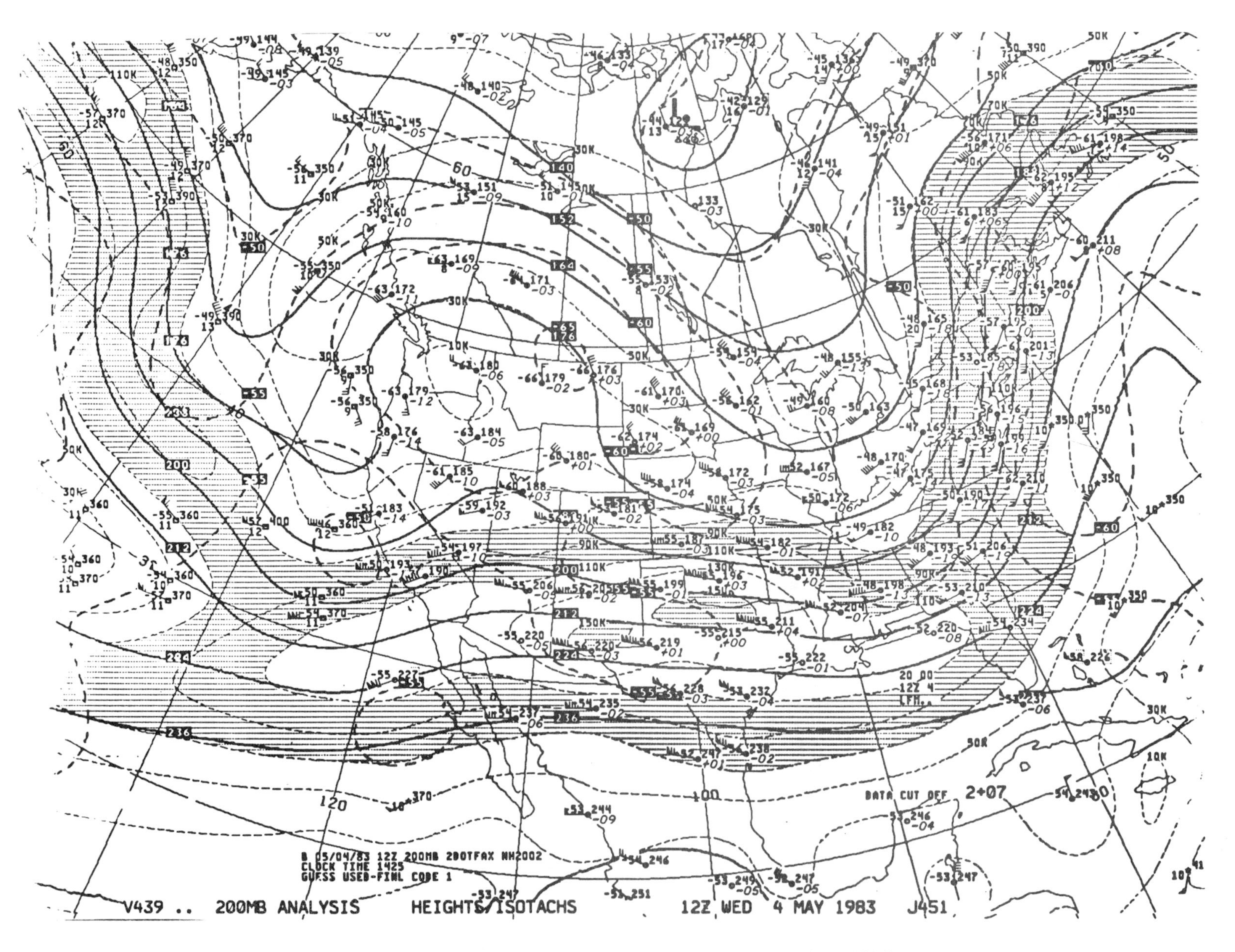

FIGURE 12-7. A section of a 200 millibar analysis, pressure altitude 39,000 feet.

as an old, occluded low with the warm air having been cut-off from the low pressure center.

In contrast to the cold low is the "thermal low". A dry, sunny region becomes quite warm from intense surface heating resulting in a surface low pressure area. The warm air is carried to high levels by convection but cloudiness is scant because of lack of moisture. The warm surface low often is "capped" by a high aloft. Unlike the cold low, the thermal low is relatively shallow with weak pressure gradients and no well defined cyclonic circulation. However, you must be alert for high density altitude, light to moderate convective turbulence, and isolated showers and thunderstorms if sufficient moisture is present. The thermal low is a semipermanent feature of the desert regions in the southwestern United States and northern Mexico during warm weather.

These are only a few rough examples of associating weather with upper air features. They point out the need to view weather in three dimensions; to get a "picture" of the atmosphere which is the first step to understanding the atmosphere and its weather.

USING THE CHARTS

From the charts you can approximate the observed temperature, wind, and temperature-dew point spread along your proposed route. Usually you can select a constant pressure chart close to your planned altitude. For altitudes about midway between two charted surfaces, interpolate between the two charts.

Determine temperature from plotted data or the pattern of isotherms. To readily delineate areas of high moisture content, station circles are shaded indicating temperature-dew point spreads of 5 degrees Celsius or less. You can get the actual spread from plotted data. A small spread alerts you to possible cloudiness, precipitation and icing. Determine windspeed for lower levels from plotted data, for the 300 and 200 millibar surfaces, determine wind speed from the isotach pattern. Wind direction parallels the contours. Using isotherms and contours, you can determine thermal advection (warming/cooling with time).

As stated earlier, constant pressure charts often show the cause of weather and its movement more clearly than does the surface map. For example, the large scale wind flow around a low aloft may spread cloudiness, low ceilings, and precipitation far more extensively than indicated by the surface map alone.

Keep in mind that constant pressure charts are observed weather.

Section 13
TROPOPAUSE DATA CHART

A four (4) panel chart containing observed tropopause data, a maximum wind prog, a vertical wind shear prog, and a high level significant weather prog is prepared for the contiguous 48 states, see figure 13-1. The high level significant weather prog was covered in Section 8. The chart is available twice daily with observed data valid at 00Z and 12Z. Progs are 18 hour forecasts valid at 18Z to 06Z.

OBSERVED TROPOPAUSE PANEL

The observed tropopause data panel shows for each upper air observing station the pressure, temperature, and wind at the tropopause. Figure 13-2 shows the panel with Alburquerque NM (ABQ) identified to aid in explaining the station model. Decode the plotted data at Alburquerque as follows: tropopause wind, 240 degrees, at 115 knots; tropopause temperature, −61 degrees Celsius; tropopause pressure, 200 millibars.

USING THE PANEL

Maximum wind occurs near the tropopause, so this panel is essentially a map of observed maximum winds. A close inspection of the map reveals a jet stream from central New Mexico across southeastern Kansas, central Missouri, and southern Illinois to West Virginia. The reason wind data are missing over Oklahoma, eastern Kansas, and Illinois is that strong winds carried the radiosonde instruments too far from observing stations to obtain reliable wind data. This area of missing wind data is actually the area of strongest winds in the jet stream.

From the map you can determine observed wind and temperature at the tropopause. You can then use constant pressure progs or the FD winds and temperatures aloft forecast to interpolate for a flight level between a constant pressure level and the tropopause.

DOMESTIC TROPOPAUSE WIND AND WIND SHEAR PROGS

Forecast parameters at the tropopause over the contiguous 48 states and some adjoining oceanic and North American areas are shown on two panels—the *tropopause winds* and the *tropopause height/vertical wind shear progs.*

TROPOPAUSE WINDS

The tropopause winds prog, figure 13-3, depicts wind direction by streamlines—solid lines. Streamlines have no dimensions and are unlabelled. They are of sufficient density to show the direction field. Winds parallel the streamlines. Direction of the streamlines basically is from west to east in mid latitudes. A high or low may be encircled by a closed streamline; you can readily determine whether it is a high or low if you know the circulation around these systems.

Wind speed is shown by isotachs at 20 knots intervals (dashed lines in figure 13-3). They are labelled in knots. Areas of wind speeds between 70 and 110 knots are hatched as are wind speeds between 150 and 190 knots. The shading criteria are the same as used on selected constant pressure analysis and progs.

TROPOPAUSE HEIGHT/VERTICAL WIND SHEAR

Tropopause height/vertical wind shear prog (figure 13-4) depicts height of the tropopause in terms of pressure altitude and vertical wind shear in knots per 1,000 feet, (see chapter 3, AVIATION WEATHER, discussion of pressure altitude). Solid lines trace intersections of the tropopause with standard constant pressure surfaces. Heights are preceded by the letter F and are in hundreds of feet.

Following is a listing of pressure and corresponding flight levels:

Millibars	*Flight Level*
500	18,000
450	21,000
400	24,000
350	27,000
300	30,000
250	34,000
200	39,000
150	45,000
100	53,000
70	63,000

Vertical wind shear is in knots per 1,000 feet depicted by dashed lines at 2 knot intervals. Wind shear is averaged through a layer from about 8,000 feet below to 4,000 feet above the tropopause.

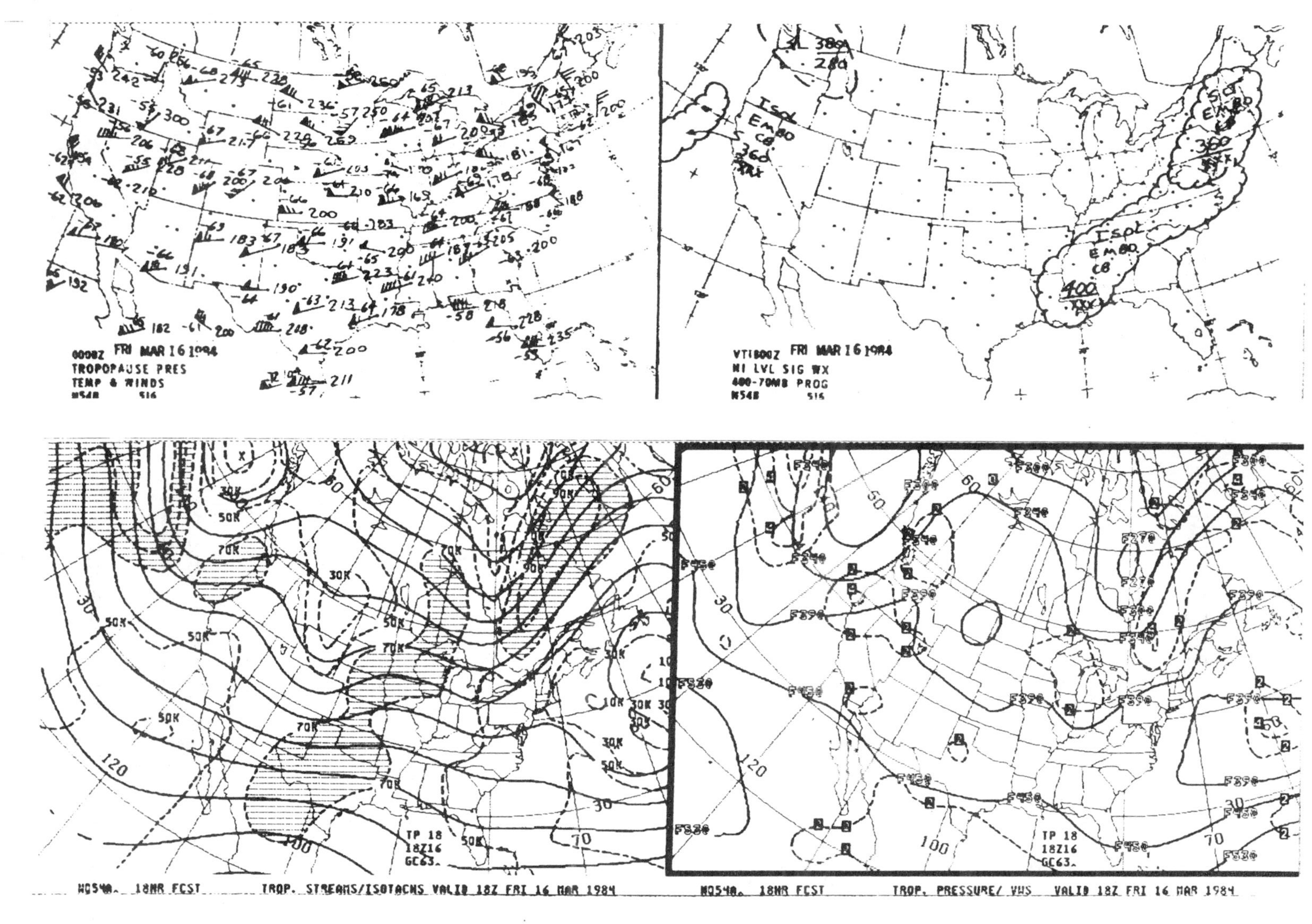

FIGURE 13-1. Tropopause data chart. Panels of observed tropopause data, maximum wind prog, vertical wind shear prog, and high level signficant weather prog.

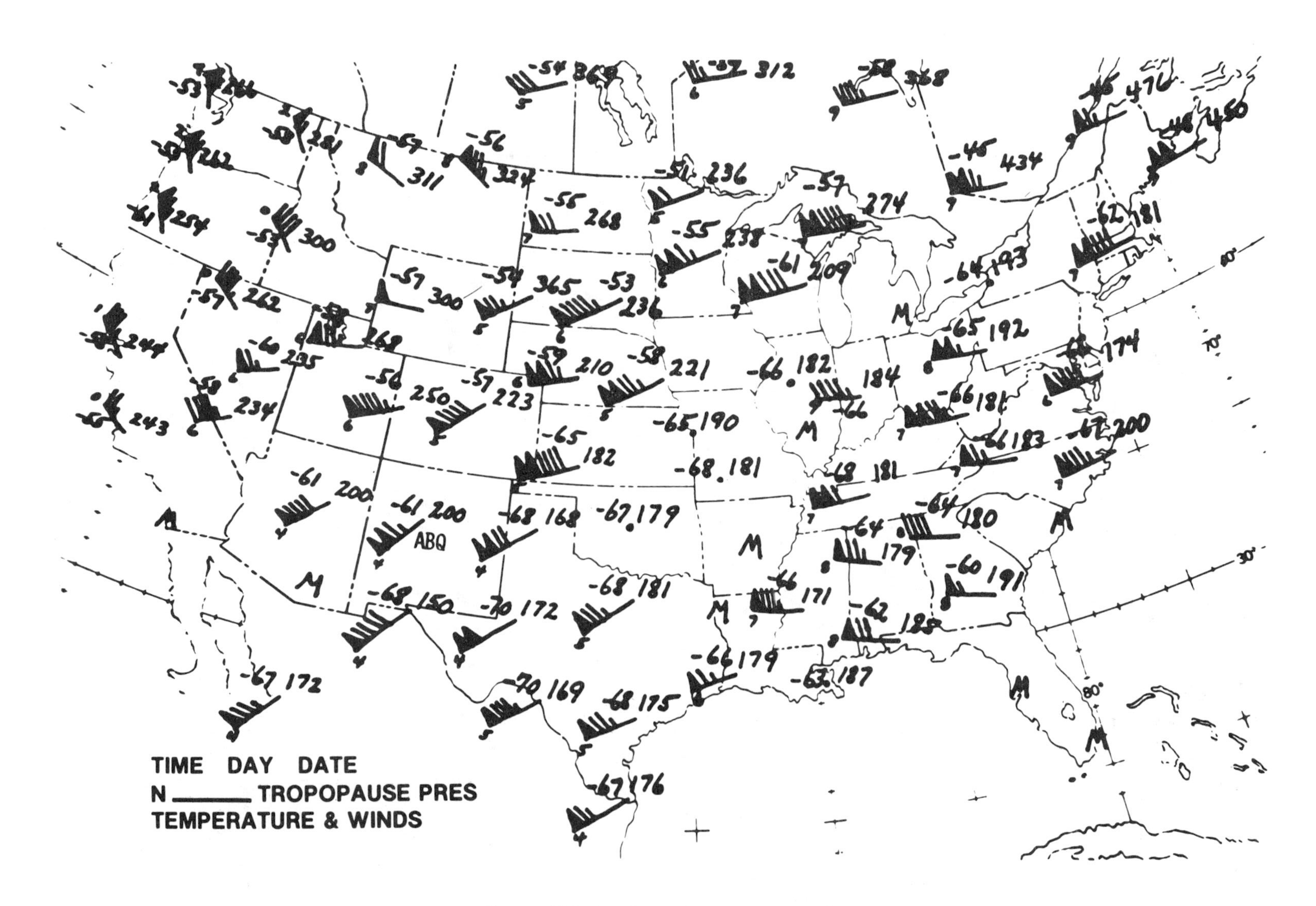

FIGURE 13-2. An observed tropopause panel.

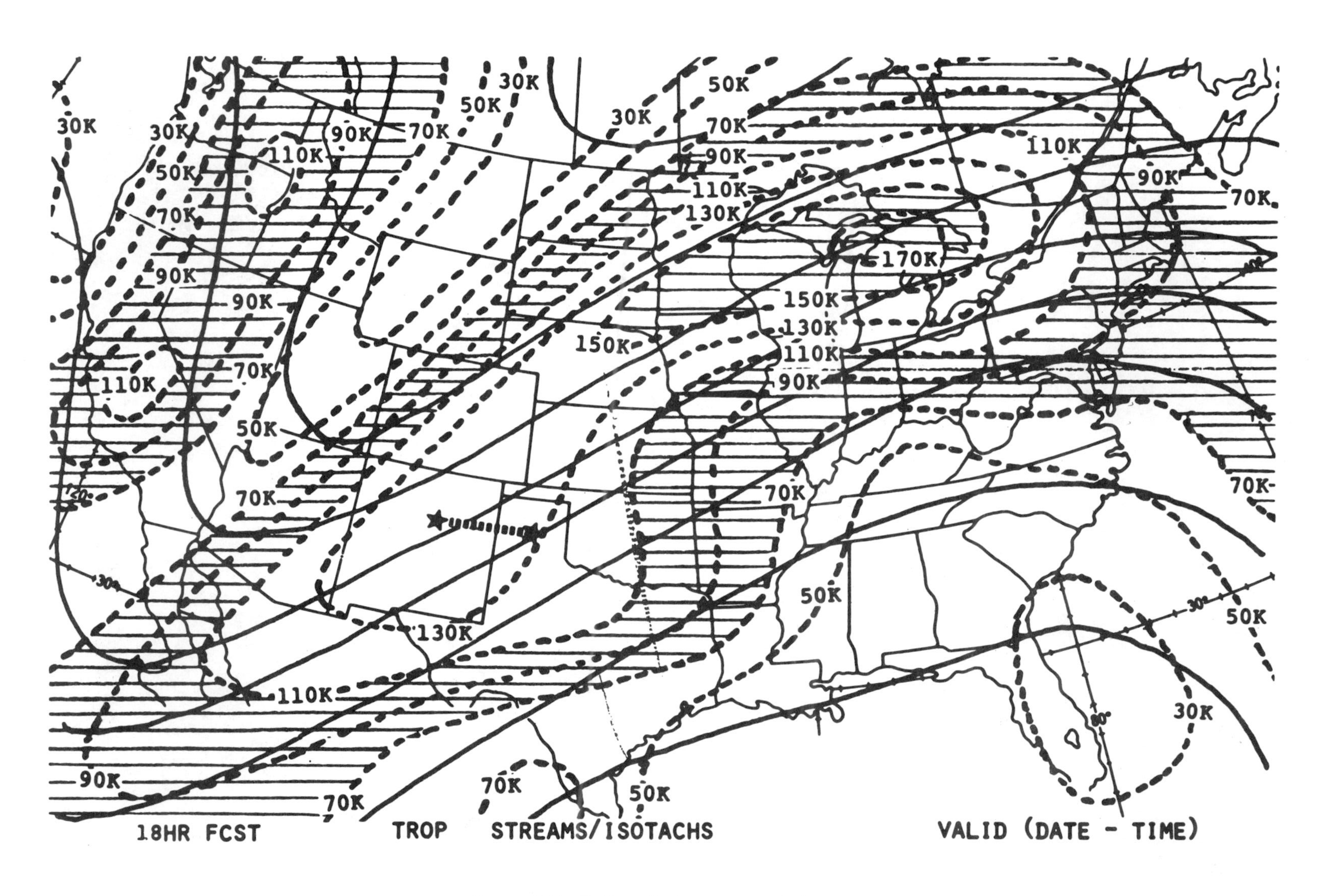

FIGURE 13-3. Section of a tropopause wind prog.

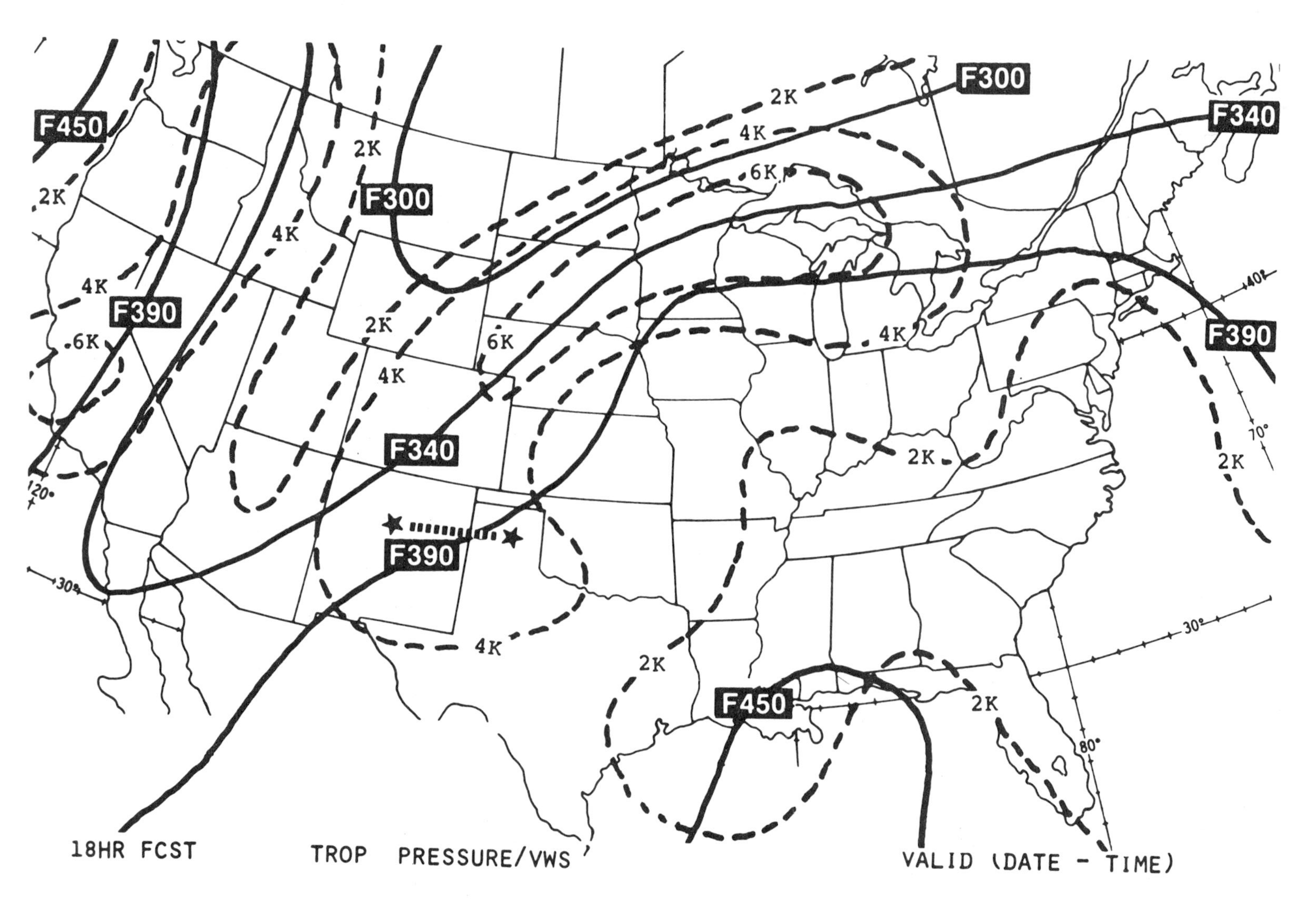

FIGURE 13-4. Section of a tropopause height/vertical wind shear prog.

USING THE PANELS

The progs are issued twice daily and may be used for a period up to plus or minus 6 hours from the valid time. The panels may be used to determine vertical and horizontal wind shears as clues to probable wind shear turbulence (see pages 14-4 and 14-5 for criteria). They also may be used to determine winds for high level flight planning.

Although neither panel depicts the jet stream, locating the jet is not difficult. It passes through isotach and vertical shear maxima. Examine figures 13-3 and 13-4, note a jet maximum from eastern Washington and Oregon extending southward and slightly westward through central California. It reappears near the southwest corner of the panels, enters the U.S. near the Arizona-New Mexico border, extends northeastward across central Nebraska, and then swings more easterly through the central Great Lakes and southern New England.

Horizontal wind shear can be determined from the spacing of isotachs. The horizontal wind shear critical for turbulence (moderate or greater) is greater than 18 knots per 150 miles (see chapter 13, AVIATION WEATHER, discussion on clear air turbulence). 150 nautical miles is equal to 2 1/2 degrees latitude.

Refer to figure 13-3 and measure 2 1/2 degrees latitude by laying a pencil along a meridian in the Atlantic. Move the pencil perpendicular to the isotach across north central Montana and you can see that the horizontal shear (the difference in wind speed) is about 40 knots along this distance. This spacing represents the wind shear critical for probable moderate or greater wind shear turbulence. The strong wind shear from southwestern Arizona to northwestern Minnesota suggests a probability of turbulence due to horizontal wind shear.

Vertical wind shear can be determined directly from the dashed lines in figure 13-4. The vertical shear critical for probable turbulence is 6 knots per 1,000 feet. You find this critical value in central California and from western Nebraska to the Great Lakes. An area of extremely high probability of moderate or greater turbulence is the three state junction of the Dakotas and Minnesota where horizontal shear is about 80 knots per 150 miles and vertical shear is in excess of 6 knots per 1,000 feet.

Wind direction and speed at the tropopause flight level may be read directly from the streamlines and isotachs. To determine wind at a flight level below and above the tropopause, first determine direction and speed at the tropopause. Wind direction changes very little within several thousand feet of the tropopause, so this direction may be used throughout the layer for which vertical wind shear is computed. Next determine wind shear and the number of thousands of feet the desired flight level differs from flight level of the tropopause. Multiply the shear by thousands of feet and subtract this value from the speed at the tropopause.

As an example, let's assume a westbound flight wants the probability of turbulence and the wind for a leg from Amarillo to Alburquerque. Note from the panels in figures 13-3 and 13-4 that horizontal wind shear is negligible. Vertical wind shear is interpolated between the 4 and 6 knot shear lines and is about 5 knots per 1,000 feet. Widespread significant turbulence (moderate or greater) is unlikely. You should also refer to the high level significant weather prog and pilot reports for further clues to turbulence.

Wind direction along the route determined from the streamline is about 230 degrees, a quartering headwind. Speed is strongest at the tropopause, so for a westbound flight, choose a flight level as far as practical above or below the tropopause. Height of the tropopause determined from figure 13-2 is flight level 39,000 feet (200 millibars). Let's assume you would like to check the wind at 43,000 feet. From figure 13-3, you determine tropopause wind speed to be on the high side of the 130 knot isotach but quite a distance from the 150 knot isotach. Let's interpolate the speed as 135 knots. The flight level, 43,000 feet, is 4,000 feet above the tropopause. Multiply the 5 knot shear by 4 and you get a difference of 20 knots. Subtract 20 knots from 135, the speed at the tropopause, and you get a speed of 115 knots. Therefore, wind at FL430 is 230 degrees at 115 knots.

Section 14

TABLES AND CONVERSION GRAPHS

This section provides graphs and tables you can use operationally in decoding weather messages during preflight and inflight planning and in transmitting pilot reports. Information included covers:

1. Icing intensities and reporting.
2. Turbulence intensities and reporting.
3. Locations of probable turbulence by intensity versus weather and terrain features.
4. Standard temperature, speed, and pressure conversions.
5. Density altitude computations.
6. Selected contractions.
7. Selected acronyms.
8. Scheduled issuance and valid times of forecast products.

The table of *Icing Intensities* classifies each intensity according to its operational effects on aircraft.

The table of *Turbulence Intensities* classifies each intensity according to its effects on aircraft control and structual integrity and on articles and occupants within the aircraft.

The table of *Locations of Probable Turbulence* lists each turbulence intensity along with terrain and weather features conducive to turbulence of that intensity.

The graph for *Density Altitudes Computations* provides a means of computing density altitude, either on the ground or aloft, using the aircraft altimeter and outside air temperature.

Contractions are used extensively in surface, radar, and pilot reports and in forecasts. Most of them are known from common usage or can be deciphered phonetically. The list of *Selected Contractions* contains only those most likely to give you difficulty. Acronyms used in this manual are defined in the list of *Acronyms*.

TABLE 14-1. Icing intensities, airframe ice accumulation and pilot report

Intensity	*Airframe ice accumulation*	*Pilot report*
Trace	Ice becomes perceptible. Rate of accumulation slightly greater than rate of sublimation. It is not hazardous even though deicing/anti-icing equipment is not used unless encountered for an extended period of time—over one hour.	Aircraft identification, location, (GMT), intensity and type of icing *, altitude/FL, aircraft type, IAS
Light	The rate of accumulation may create a problem if flight is prolonged in this environment (over one hour). Occasional use of deicing/anti-icing equipment removes/prevents accumulation. It does not present a problem if the deicing/anti-icing equipment is used.	
Moderate	The rate of accumulation is such that even short encounters become potentially hazardous and use of deicing/anti-icing equipment or diversion is necessary.	Example of pilot's transmission: Holding at Westminister VOR 1232. Light Rime Icing. Altitude six thousand, Jetstar IAS 200 kt
Severe	The rate of accumulation is such that deicing/anti-icing equipment fails to reduce or control the hazard. Immediate diversion is necessary.	

* Icing may be rime, clear or mixed.

Rime ice: Rough milky opaque ice formed by the instanteous freezing of small super cooled water droplets.

Clear ice: A glossy, clear or translucent ice formed by the relatively slow freezing of large supercooled water droplets.

Mixed ice: A combination of rime and clear ice.

Locations of Probable Turbulence by Intensities Versus Weather and Terrain Features

LIGHT TURBULENCE

1. In hilly and mountainous areas even with light winds.
2. In and near small cumulus clouds.
3. In clear-air convective currents over heated surfaces.

Table 14-2. Turbulence reporting criteria

Intensity	*Aircraft reaction*	*Reaction inside aircraft*	*Reporting term-definition*
Light	Turbulence that momentarily causes slight, erratic changes in altitude and/or attitude (pitch, roll, yaw). Report as *Light Turbulence;** or Turbulence that causes slight, rapid and somewhat what rhythmic bumpiness without appreciable changes in altitude or attitude. Report as *Light Chop.*	Occupants may feel a slight strain against seat belts or shoulder straps. Unsecured objects may be displaced slightly. Food service may be conducted and little or no difficulty is encountered in walking.	Occasional—Less than 1/2 of the time. Intermittent—1/3 to 2/3. Continous—More than 2/3.
Moderate	Turbulence that is similar to Light Turbulence but of greater intensity. Changes in altitude and/or attitude occur but the aircraft remains in positive control at all times. It usually causes variations in indicated airspeed. Report as *Moderate Turbulence;** or Turbulence that is similar to Light Chop but of greater intensity. It causes rapid bumps or jolts without appreciable changes in aircraft altitude or attitude. Report as *Moderate Chop.*	Occupants feel definite strains against seat belts or shoulder straps. Unsecured objects are dislodged. Food service and walking are difficult.	NOTE 1. Pilots should report location(s), time (GMT), intensity, whether in or near clouds, altitude, type of aircraft and, when applicable, duration of turbulence. 2. Duration may be based on time between two locations or over a single location. All locations should be readily identifiable.
Severe	Turbulence that causes large, abrupt changes in altitude and/or attitude. It usually causes large variations in indicated airspeed. Aircraft may be momentarily out of control. Report as *Severe Turbulence.**	Occupants are forced violently against seat belts or shoulder straps. Unsecured objects are tossed about. Food service and walking are impossible.	EXAMPLES: a. Over Omaha, 1232Z, Moderate Turbulence, in cloud, Flight Level 310, B707. b. From 50 miles south of Alburquerque to 30 miles north of Phoenix, 1210Z to 1250Z, occasional Moderate Chop, Flight Level 330, DC8.
Extreme	Turbulence in which the aircraft is violently tossed about and is practically impossible to control. It may cause structural damage. Report as *Extreme Turbulence.**		

* High level turbulence (normally above 15,000 feet AGL) not associated with cumuliform cloudiness, including thunderstorms, should be reported as CAT (clear air turbulence) preceded by the appropriate intensity, or light or moderate chop.

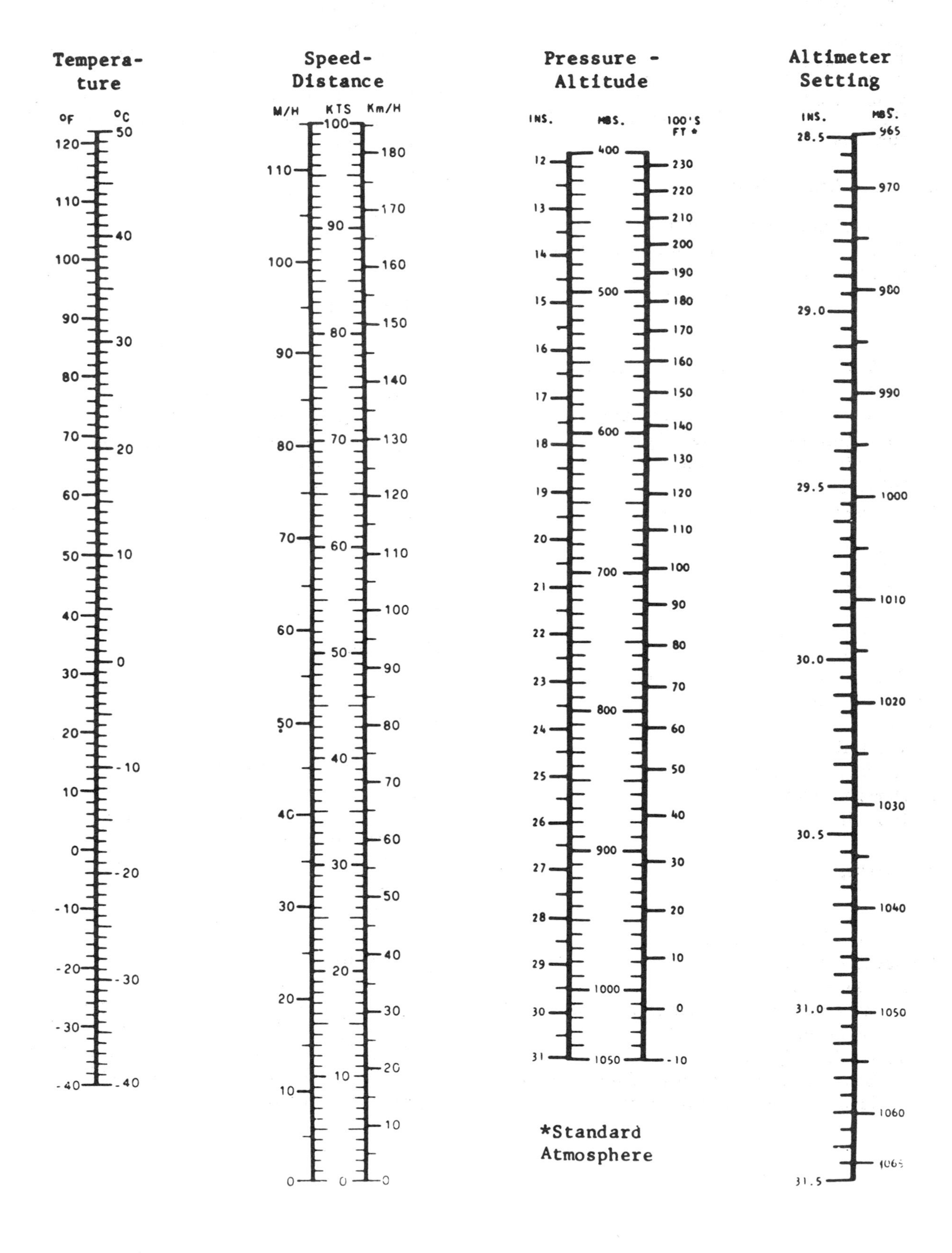
Tempera-
ture
°F
°C
Speed-
Distance
M/H
KTS
Km/H
Pressure -
Altitude
INS.
MBS.
100'S
FT *
Altimeter
Setting
INS.
MBS.
*Standard
Atmosphere

4. With weak wind shears in the vicinity of:
   a. Troughs aloft.
   b. Lows aloft.
   c. Jet streams.
   d. The tropopause.
5. In the lower 5,000 feet of the atmosphere:
   a. When winds are near 15 knots.
   b. Where the air is colder than the underlying surfaces.

MODERATE TURBULENCE

1. In mountainous areas with a wind component of 25 to 50 knots perpendicular to and near the level of the ridge:
   a. At all levels from the surface to 5,000 feet above the tropopause with preference for altitudes:
      (1) Within 5,000 feet of the ridge level.
      (2) At the base of relatively stable layers below the base of the tropopause.
      (3) Within the tropopause layer.
   b . Extending outward on the lee of the ridge for 150 to 300 miles.
2. In and near thunderstorms in the dissipating stage.
3. In and near other towering cumuliform clouds.
4. In the lower 5,000 feet of the tropopause:
   a. When surface winds are 30 knots or more.
   b. Where heating of the underlying surface is unusually strong.
   c. Where there is an invasion of very cold air.
5. In fronts aloft.
6. Where:
   a. Vertical wind shears exceed 6 knots per 1,000 feet, and/or
   b. Horizontal wind shears exceed 18 knots per 150 miles.

SEVERE TURBULENCE

1. In mountainous areas with a wind component exceeding 50 knots perpendicular to and near the level of the ridge:
   a. In 5,000—foot layers:
      (1) At and below the ridge level in rotor clouds or rotor action.
      (2) At the tropopause.
      (3) Sometimes at the base of other stable layers below the tropopause.
   b. Extending outward on the lee of the ridge for 50 to 150 miles.
2. In and near growing and mature thunderstorms.
3. Occasionally in other towering cumuliform clouds.
4. 50 to 100 miles on the cold side of the center of the jet stream, in troughs aloft, and in lows aloft where:
   a. Vertical wind shears exceed 6 knots per 1,000 feet, and
   b. Horizontal winds shears exceed 40 knots per 150 miles.

EXTREME TURBULENCE

1. In mountain wave situations, in and below the level of well-developed rotor clouds. Sometimes it extends to the ground.
2. In severe thunderstorms (most frequently in organized squall lines) indicated by:
   a. Large hailstones (3/4 inch or more in diameter).
   b. Strong radar echoes, or
   c. Almost continuous lightning.

DENSITY ALTITUDE COMPUTATION

Use this graph to find density either on the ground or aloft. Set your altimeter at 29.92 inches; it now indicates pressure altitude. Read outside air temperature. Enter the graph at your pressure altitude and move horizontally to the temperature. Read density altitude from the sloping lines.

Example 1.

Find density altitude in flight. Pressure altitude is 9,500 feet; and temperature, −8 degrees C. Find 9,500 feet on the left of the graph and move across to −8 degrees C. Density altitude is 9,000 feet (marked "1" on the graph).

Example 2.

Find density altitude for take-off. Pressure altitude is 4,950 feet; and temperature 97 degrees F. Enter the graph at 4,950 feet and move across to 97 degrees F. Density altitude is 8,200 feet (marked "2" on graph). Note that in the warm air, density altitude is considerably higher than pressure altitude.

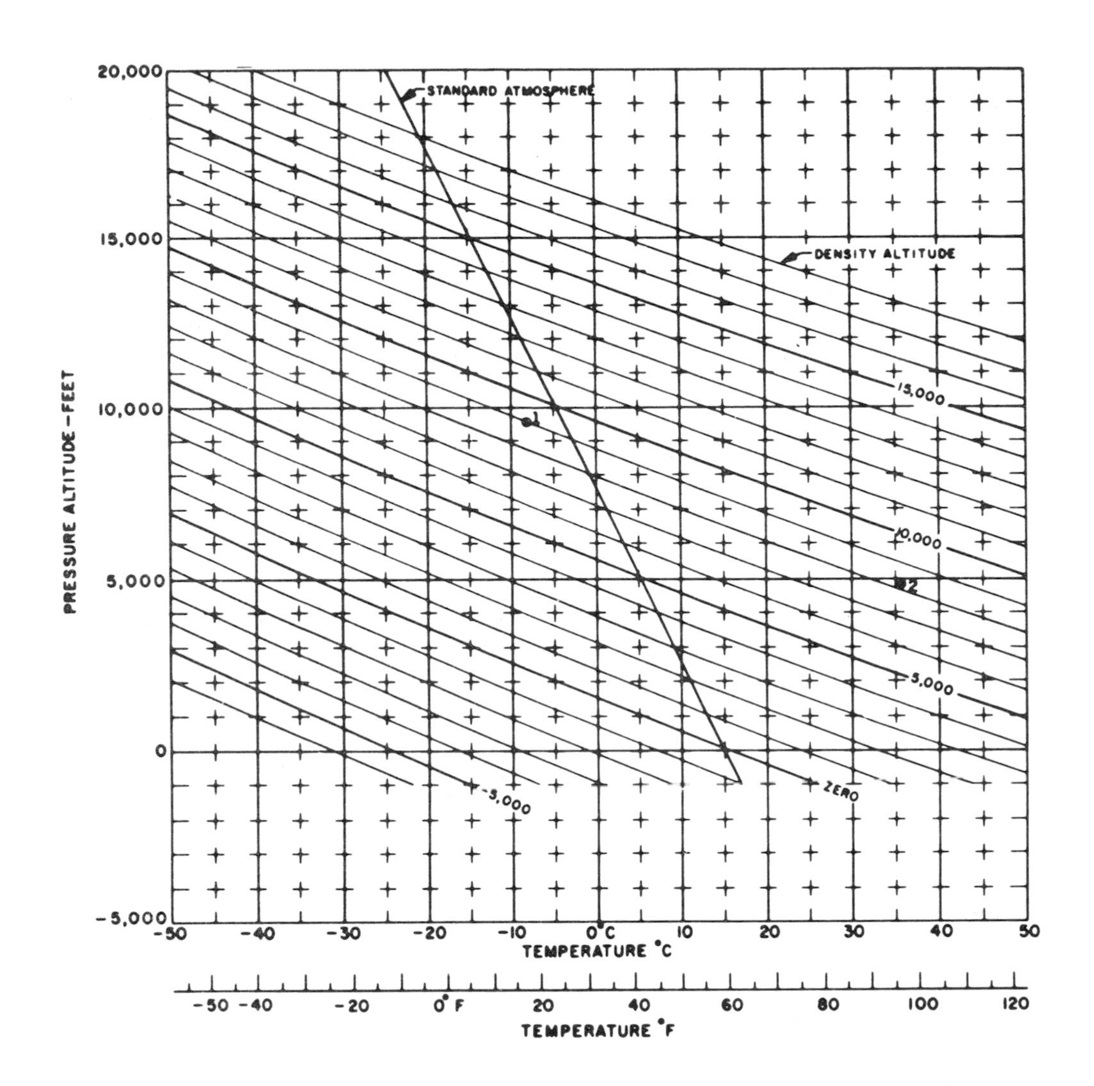

SELECTED CONTRACTIONS

A

ACLD	above clouds
ACSL	standing lenticular altocumulus
ACYC	anticyclonic
AFDK	after dark
ALQDS	all quadrants
AC	altocumulus
ACCAS	altocumulus castellanus
AS	altostratus
AOA	at or above
AOB	at or below

B

BCKG	backing
BFDK	before dark
BINOVC	breaks in overcast
BL	between layers
BLZD	blizzard
BOVC	base of overcast

C

CBMAM	cumulonimbus mamma
CC	cirrocumulus
CCSL	standing lenticular cirrocumulus
CFP	cold frontal passage
CI	cirrus
CLRS	clear and smooth
CRLCN	circulation
CS	cirrostratus
CU	cumulus
CUFRA	cumulus fractus
CYC	cyclonic

D

DFUS	diffuse
DNSLP	downslope
DP	deep
DTRT	deteriorate
DURGC	during climb

DURGD	during descent
DWNDFTS	downdraft

E

EMBDD	embedded

F

FNTGNS	frontogenesis (front forming)
FNTLYS	frontolysis (front decaying)
FROPA	frontal passage

G

GFDEP	ground fog estimated—feet deep

H

HDEP	haze layer estimated—feet deep
HLSTO	hailstones
HLYR	haze layer aloft

I

ICG	icing
ICGIC	icing in clouds
ICGICIP	icing in clouds and in precipitation
ICGIP	icing in precipitation
INTMT	intermittent
INVRN	inversion
IPV	improve
ISOLD	isolated

K

KDEP	smoke layer estimated—feet deep
KLYR	smoke layer aloft
KOCTY	smoke over city

L

LLWS	low-level wind shear
LTG, LTNG	lightning
LTGCC	lightning cloud-to cloud
LTCCCCG	lightning cloud-to-cloud, cloud-to-ground
LTGCW	lightning cloud-to-water
LTGIC	lightning in cloud
LTNG, LTG	lightning

M

MEGG	merging
MLTLVL	melting level
MNLD	mainland
MOGR	moderate or greater
MRGL	marginal
MSTR	moisture

N

NCWX	no change in weather
NPRS	non persistent
NRW	narrow
NS	nimbostratus

O

OAOI	on and off instruments
OAT	outside air temperature
OCFNT	occluded front
OCLD	occlude
OFP	occluded frontal passage
OFSHR	off shore
OI	on instruments
OMTNS	over mountains
ONSHR	on shore
OTAS	on top and smooth
OVRNG	overrunning

P

PDW	priority delayed weather
PRESFR	pressure falling rapidly
PRESRR	pressure rising rapidly
PRIND	present indications are
PRST	persist

Q

QSTNRY	quasistationary
QUAD	quadrant

R

RGD	ragged
RTD	routine delayed weather

S

SC	stratocumulus
SKC, CLR	sky clear
SNOINCR	snow depth increase in past hour
SNRS, SR	sunrise
SNST, SS	sunset
SNWFL	snowfall
SQAL	squall
SQLN	squall line
SR, SNRS	sunrise
SS, SNST	sunset
ST	stratus
STFRA	stratus fractus
STFRM	stratiform
STM	storm

T

TCU	towering cumulus
TOVC	top of overcast
TROP	tropopause
TWRG	towering

U

UDDF	up and down drafts
UPDFTS	updrafts
UPSLP	upslope

V

VLNT	violent
VR	veer

W

WDSPRD	widespread
WFP	warm frontal passage
WK	weak
WRMFNT	warm front
WSHFT	wind shift
WV	wave

ACRONYMS

AC —Convective Outlook Bulletin; identifies a forecast of probable convective storms.

AIRMET —Airman's Meteorological Information; an inflight advisory forecast of conditions possible hazardous to light aircraft or inexperienced pilots.

ARTCC —Air Route Traffic Control Center, FAA.

CWSU —Center Weather Service Unit, NWS and FAA.

EFAS —Enroute Flight Advisory Service (Flight Watch), FAA.

FA —Area Forecast; identifies a forecast of general aviation weather over a relatively large area.

FD —Winds and Temperatures Aloft Forecast; a forecast identifier.

FSS —Flight Service Station, FAA.

FT —Terminal Forecast; identifies a forecast in the U.S. forecast code.

GOES —Geostationary Operational Environmental Satellite.

HIWAS —Hazardous In-flight Weather Advisory Service, FAA

ICAO —International Civil Aviation Organization.

IFSS —International Flight Service Station, FAA.

LAWRS —Limited Aviation Weather Reporting Station; usually a control tower; reports fewer weather elements than a complete SA.

NAWAU —National Aviation Weather Advisory Unit, NWS.

NESDIS —National Environmental Satellite Data and Information Service.

NHC —National Hurricane Center, NWS.

NMC —National Meteorological Center, NWS.

NOAA —National Oceanic and Atmospheric Administration, Department of Commerce.

NOTAM —Notice to Airmen.

NSSFC —National Severe Storms Forecast Center, NWS.

NWS —National Weather Service, National Oceanic and Atmospheric Administration, Department of Commerce.

PATWAS —Pilot's Automatic Telephone Weather Answering Service; a self-briefing service.

PIREP —Pilot Weather Report.

RAREP —Radar Weather Report.

SA —Surface Aviation Weather Report; a message identifier.

SAWRS —Supplemental Aviation Weather Reporting Station; usually an airline office at a terminal not having NWS or FAA facilities.

SFSS —Satellite Field Service Station.

SIGMET —Significant Meteorological Information; an inflight advisory forecast of weather hazardous to all aircraft.

TAF —Terminal Aviation Forecast; identifies a terminal forecast in the ICAO code.

TWEB —Transcribed Weather Broadcast; a self-briefing radio broadcast service.

UA —Pilot Report (PIREP); a message identifier.

WA —AIRMET valid for a specified period, a message identifier.

WS —SIGMET valid for a specified period, a message identifier.

WSFO —Weather Service Forecast Office, NWS.

WSO —Weather Service Office, NWS.

WST —Convective SIGMET, a message identifier.

WW —Severe Weather Watch; identifies a forecast of probable severe thunderstorms or tornadoes.

TABLE 14-3. Scheduled issuance and valid times of forecast products

FORECAST PRODUCTS	*TIME ZONE*	*AREA*	*ISSUANCE TIME*	*VALID PERIOD*
Terminal Forecast (FT)	Pacific Mountain	—	0940Z	10-10Z
			1540Z	16-16Z
			2240Z	23-23Z
	Central Eastern	—	0940Z	10-10Z
			1440Z	15-15Z
			2140Z	22-22Z
		Anchorage Fairbanks	0440Z	05-05Z
			1140Z	12-12Z
			1640Z	17-17Z
			2140Z	22-22Z
		Juneau	0340Z	04-04Z
			1040Z	11-11Z
			1440Z	15-15Z
			2040Z	21-21Z
		Honolulu	0540Z	06-06Z
			1140Z	12-12Z
			1740Z	18-18Z
			2340Z	00-00Z
ICAO Terminal		All	2340Z	00-00Z
			0540Z	06-06Z
			1140Z	12-12Z
			1740Z	18-18Z
Area Forecast (FA)		Boston Miami	0840Z	09-03Z
			1740Z	18-12Z
			2340Z	00-18Z
		Chicago Dallas-Fort Worth	1040Z	11-05Z
			1840Z	19-13Z
			0040Z	01-19Z
		San Francisco Salt Lake City	1140Z	12-06Z
			1940Z	20-14Z
			0140Z	02-20Z
		Anchorage Fairbanks	0640Z	07-01Z
			1540Z	16-10Z
			2340Z	00-18Z
		Juneau	0640Z	07-01Z
			1340Z	14-08Z
			2240Z	23-17Z
		Honolulu	0340Z	04-22Z
			0940Z	10-04Z
			1540Z	16-10Z
			2140Z	22-16Z
Transcribed Weather Broadcast (TWEB)	Pacific Mountain		1140Z	12-00Z
			1840Z	19-07Z
			2340Z	00-18Z
	Central Eastern		1040Z	11-23Z
			1740Z	18-06Z
			2240Z	23-17Z
		Alaska Hawaii	None	
Inflight		All	Not Scheduled	See Section 4

U.S. GOVERNMENT PRINTING OFFICE : 1991 - 286-959 : QL 3

APPENDIX B

OBJECTIVES FOR AVIATION WEATHER SERVICES

NOTE: Many of the following objectives are also answered in AVIATION METEOROLOGY UNSCRAMBLED.

1. Define weather observation.
2. State the primary function of the following: RRWDS, LLWAS, NOAA, NESDIS, NMC, NHC, NSSFC, NAWAU, WSFO, WSO, TWEB, PATWAS, HIWAS, EFAS, ATCC, FSS, ARTCC, CWSU, CFWSU, ATIS, AM WEATHER, AFOS, and TERMINAL CONTROL FACILITY.
3. List the facts about your flight that you should give when requesting a weather briefing.
4. List and describe the three types of weather briefings.
5. Who supplies request/reply service? What information is available?
6. Define weather report. Name and describe the two basic types of reports.
7. List and give the meaning of the sky cover designators.
8. Define transparent, thin, and ceiling.
9. List and describe the ceiling designators.
10. Describe the use of the letter V when it is appended to ceiling heights.
11. Define prevailing, vertical, and slant range visibilities.
12. Describe the use of the letter V when suffixed to prevailing visibility.
13. List and give the meaning of weather symbols.
14. State the intensities of precipitation.
15. Define severe thunderstorm.
16. Obstructions to visibility are reported for only what visibility values?
17. Describe the two types of remarks concerning either a surface based obscuration or an obscuring phenomenon aloft.
18. How is sea level pressure recorded? What is the usual value range?
19. How is wind recorded?
20. Define gust, and squall.
21. What is the use of E when it placed before the wind group?
22. How are altimeter settings recorded?
23. Define runway visibility and runway visual range.
24. Under what conditions are PIREPs omitted from weather reports?
25. What stations report RADATs? When is icing data included?
26. Read surface aviation reports.
27. State the function of AMOS, AUTOB, RAMOS, and read/interpret their reports.
28. Read/interpret pilot and radar reports, and satellite pictures.
29. Read/interpret domestic (FT) forecasts.
30. How/when is VCNTY used?
31. Define area forecast. How are they used and how often are they issued?

32. When are surface visibility and obstructions to vision included in the significant cloud and weather section of area forecasts? What is the meaning of WIND?
33. Read/interpret area forecasts.
34. Compare TWEB Route Forecasts and Area Forecasts.
35. Who prepares TWEB Route Forecasts? How wide (miles) are such forecasts? How are these forecasts received?
36. Read/interpret TWEB Route Forecasts. At least how much visibility is available when visibility is not stated?
37. Define inflight advisory.
38. For what meteorological phenomena are CONVECTIVE SIGMENTS issued?
39. What do all CONVECTIVE SIGMETS imply and for what categories of aircraft are they valid?
40. When are bulletins and special bulletins issued (CONVECTIVE SIGMETS)?
41. For what types of aircraft are SIGMETs and AIRMETs issued?
42. Read/interpret winds and temperatures aloft forecasts. For which levels (low) are winds and temperatures omitted?
43. When (and from whom) may a special flight forecast be requested?
44. List some examples (types) of special flight forecasts.
45. Read/interpret special flight forecasts.
46. What kind of forecast is an MIS? Under what conditions is an MIS issued?
47. Read/interpret MIS forecasts.
48. What is the purpose of a CWA? Under what conditions is a CWA issued?
49. Read/interpret CWAs.
50. When is a hurricane advisory issued?
51. Read/interpret hurricane advisories.
52. What is the purpose of the convective outlook?
53. Read/interpret convective outlooks.
54. What is the purpose of a Severe Weather Watch bulletin and the Alert Severe Weather Watch message?
55. Read/interpret WWs and AWWs.
56. Read/interpret surface charts. How often are they transmitted?
57. Read/interpret weather depiction charts. How often are they transmitted? What is the data source for these charts?
58. What is the best use of the weather depiction chart?
59. What does weather radar primarily detect?
60. State the types of designated arrangements for the radar summary chart.
61. How are echo heights determined on a radar summary chart?
62. What are the two types of weather watch areas on radar summary charts?
63. What is the primary use of the radar summary chart?
64. Read/interpret radar summary charts.
65. How often are U.S. Low Level Significant Weather Progs issued?

66. Read/interpret U.S. Low Level and High Level Significant Weather Progs.
67. How often are winds and temperatures aloft issued?
68. How are light and variable winds and missing winds indicated on winds and temperatures aloft charts? How is a temperature-dew point spread of five degrees Celsius or less indicated?
69. Read/interpret winds and temperatures aloft charts.
70. How are significant levels determined on the composite moisture stability chart?
71. How is the lifted index computed? Compare positive, zero, and negative index values.
72. What factors contribute to a high K index? Describe uses of the K index.
73. State air mass thunderstorm possibilities for various values of the lifted index and the K index.
74. Describe how the lifted index and the K index indicate various types of clouds and precipitation, turbulence, and icing.
75. State the most appropriate use of the freezing level panel of the composite moisture stablility chart.
76. What kind of analysis is the precipitable water panel? How are values of 1.00 inch or more indicated? What is the best use of the panel?
77. State how the average relative humidity panel is analyzed and used.
78. Read/interpret composite moisture stability charts. How are such charts best used?
79. Define APCHG as used on the severe weather outlook chart. What are the meanings of the notations SLIGHT RISK, MODERATE RISK, and HIGH RISK?
80. Read/interpret severe weather outlook charts. How are such charts best used?
81. How often and for what levels are constant pressure analysis charts issued?
82. How are strong winds depicted on constant pressure charts?
83. Describe several three dimensional aspects of constant pressure charts.
84. Describe several uses of constant pressure charts.
85. How can the jet stream be located on the tropopause data chart?
86. How can horizontal and vertical wind shear be determined from the tropopause data chart?
87. How can wind at a flight level below and above the tropopause be determined?
88. Read/interpret tropopause data charts.

APPENDIX C

SUPPLEMENTAL QUESTIONS FROM FAA QUESTION BOOKS

This appendix contains two parts. Part A contains questions over the seventeen chapters of the text and Part B contains questions over AVIATION WEATHER SERVICES. All of the questions are from FAA question books, private through ATP.

The questions in Part A are grouped by the chapter which contains the answers to those questions. The questions are listed in order, private first (1000 numbers), commercial second (2000 numbers), instrument third (7000 numbers), instructor fourth (3000 numbers), and ATP last (6000 numbers). Duplicate questions are listed only once; the one listed is the one on the lowest level exam. For example, a question on both the private and instrument exams is listed only once, the private one, so a person studying for an instrument rating should know how to answer all questions at and below the instrument level. Of course, questions which are the same as the ones at the end of some chapter are also not listed below. Many of the questions below are still close to being the same as ones at the end of some chapter, but they are listed anyhow. A complete listing of the questions from each exam is in the instructor supplements.

Part B contains all of the FAA questions over AVIATION WEATHER SERVICES, PRIVATE THROUGH ATP, in that order; duplicate questions have not been eliminated. Although much of AC 00-30A and AC 00-50A is contained in the text, they are duplicated in their entirety in this appendix, pages 15-34, for ease of reference.

Finally, it is unlikely that there will be any significant change in the questions that follow over the next couple of years. Most of the questions in current question books were in previous ones, although possibly with a slightly different number (6014 rather than 6018, for example).

PART A

CHAPTER I

7147. The average height of the troposphere in the mid-latitudes is

A—20,000 feet.
B—25,000 feet.
C—37,000 feet.

3182. In what part of the atmosphere does most weather occur?

A— Tropopause.
B— Troposphere.
C— Stratosphere.

6258. What is a feature of air movement in a high pressure area?

A— Ascending from the surface high to lower pressure at higher altitudes.
B— Descending to the surface and then outward.
C— Moving outward from the high at high altitudes and into the high at the surface.

6264. What term describes an elongated area of low pressure?

A— Trough.
B— Ridge.
C— Hurricane or typhoon.

CHAPTER II

2303. What is the standard temperature at 20,000 feet?

A— - 15° C.
B— - 20° C.
C— - 25° C.

2381. Which feature is associated with the tropopause?

A— Constant height above the Earth.
B— Abrupt change in temperature lapse rate.
C— Absolute upper limit of cloud formation.

7089. The primary cause of all changes in the Earth's weather is

A—variation of solar energy received by the Earth's regions.
B—changes in air pressure over the Earth's surface.
C—movement of the air masses.

7090. A characteristic of the stratosphere is

A—an overall decrease of temperature with an increase in altitude.
B—a relatively even base altitude of approximately 35,000 feet.
C—relatively small changes in temperature with an increase in altitude.

7106. If the air temperature is +8° C at an elevation of 1,350 feet and a standard (average) temperature lapse rate exists, what will be the approximate freezing level?

A—3,350 feet MSL.
B—5,350 feet MSL.
C—9,350 feet MSL.

7117. Unsaturated air flowing upslope will cool at the rate of approximately (dry adiabatic lapse rate)

A—3° C per 1,000 feet.
B—2° C per 1,000 feet.
C—2.5° C per 1,000 feet.

3181. The layer of the atmosphere typified by relatively small changes in temperature with increased height, except for a warming trend near the top, is the

A— tropopause.
B— troposphere.
C— stratosphere.

3184. The tropopause is the dividing line between the

A— troposphere and ionosphere.
B— ionosphere and stratosphere.
C— troposphere and stratosphere.

3185. Which is the primary driving force of weather on the Earth?

A— The Sun.
B— Coriolis.
C— Rotation of the Earth.

3186. The average lapse rate in the troposphere is

A— 2.0° C per 1,000 feet.
B— 3.0° C per 1,000 feet.
C— 5.4° C per 1,000 feet.

6250. What is a characteristic of the troposphere?

A— It contains all the moisture of the atmosphere.
B— There is an overall decrease of temperature with an increase of altitude.
C— The average altitude of the top of the troposphere is about 6 miles.

6254. When does minimum temperature normally occur during a 24–hour period?

A— After sunrise.
B— About 1 hour before sunrise.
C— At midnight.

6267. Where is a common location for an inversion?

A— At the tropopause.
B— In the stratosphere.
C— At the base of cumulus clouds.

6269. Which term applies when the temperature of the air changes by compression or expansion with no heat added or removed?

A— Katabatic.
B— Advection.
C— Adiabatic.

6285. Which process causes adiabatic cooling?

A— Expansion of air as it rises.
B— Movement of air over a colder surface.
C— Release of latent heat during the vaporization process.

CHAPTER IV

2310. What causes wind?

A— The Earth's rotation.
B— Airmass modification.
C— Pressure differences.

2315. What prevents air from flowing directly from high-pressure areas to low-pressure areas?

A— Coriolis force.
B— Surface friction.
C— Pressure gradient force.

2319. When flying into a low-pressure area in the Northern Hemisphere, the wind direction and velocity will be from the

A— left and decreasing.
B— left and increasing.
C— right and decreasing.

7100. What relationship exists between the winds at 2,000 feet above the surface and the surface winds?

A—The winds at 2,000 feet and the surface winds flow in the same direction, but the surface winds are weaker due to friction.
B—The winds at 2000 feet tend to parallel the isobars while the surface winds cross the isobars at an angle toward lower pressure and are weaker.
C— The surface winds tend to veer to the right of the winds at 2,000 feet and are usually weaker.

3196. The windflow around a low pressure is

A— cyclonic.
B— adiabatic.
C— anticyclonic.

6256. At lower levels of the atmosphere, friction causes the wind to flow across isobars into a low because the friction

A— decreases windspeed and Coriolis force.
B— decreases pressure gradient force.
C— creates air turbulence and raises atmospheric pressure.

6257. Which type wind flows downslope becoming warmer and dryer?

A— Land breeze.
B— Valley wind.
C— Katabatic wind.

6274. Isobars on a surface weather chart represent lines of equal pressure

A— at the surface.
B— reduced to sea level.
C— at a given atmospheric pressure altitude.

6275. At which location does Coriolis force have the least effect on wind direction?

A— At the poles.
B— Middle latitudes (30° to 60°).
C— At the Equator.

6276. How does Coriolis force affect wind direction in the Southern Hemisphere?

A— Causes clockwise rotation around a low.
B— Causes wind to flow out of a low toward a high.
C— Has exactly the same effect as in the Northern Hemisphere.

6277. Which weather condition is defined as an anticyclone?

A— Calm.
B— High pressure area.
C— COL.

CHAPTER V

1387. If a pilot changes the altimeter setting from 30.11 to 29.96, what is the approximate change in indication?

A—Altimeter will indicate .15" Hg higher.
B—Altimeter will indicate 150 feet higher.
C—Altimeter will indicate 150 feet lower.

1390. If a flight is made from an area of low pressure into an area of high pressure without the altimeter setting being adjusted, the altimeter will indicate

A—the actual altitude above sea level.
B—higher than the actual altitude above sea level.
C—lower than the actual altitude above sea level.

1391. If a flight is made from an area of high pressure into an area of lower pressure without the altimeter setting being adjusted, the altimeter will indicate

A—lower than the actual altitude above sea level.
B—higher than the actual altitude above sea level.
C—the actual altitude above sea level.

1392. Under what condition will true altitude be lower than indicated altitude?

A—In colder than standard air temperature.
B—In warmer than standard air temperature.
C—When density altitude is higher than indicated altitude.

1393. Which condition would cause the altimeter to indicate a lower altitude than true altitude?

A—Air temperature lower than standard.
B—Atmospheric pressure lower than standard.
C—Air temperature warmer than standard.

1394. Which factor would tend to increase the density altitude at a given airport?

A—An increase in barometric pressure.
B—An increase in ambient temperature.
C—A decrease in relative humidity.

3191. An altimeter indicates 1,850 feet MSL when set to 30.18. What is the approximate pressure altitude?

A— 1,590 feet.
B— 1,824 feet.
C— 2,110 feet.

6263. What is corrected altitude (approximate true altitude)?

A— Pressure altitude corrected for instrument error.
B— Indicated altitude corrected for temperature variation from standard.
C— Density altitude corrected for temperature variation from standard.

6273. Which pressure is defined as station pressure?

A— Altimeter setting.
B— Actual pressure at field elevation.
C— Station barometric pressure reduced to sea level.

CHAPTER VI

3203. The ratio of the existing water vapor in the air, as compared to the maximum amount that could exist at a given temperature, is called

A— the dewpoint.
B— saturation point.
C— relative humidity.

3204. What is the process by which ice can form on a surface directly from water vapor on a cold, clear night?

A— Sublimation.
B— Condensation.
C— Supersaturation.

6292. What minimum thickness of cloud layer is indicated if precipitation is reported as light or greater intensity?

A— 4,000 feet thick.
B— 2,000 feet thick.
C— A thickness which allows the cloud tops to be higher than the freezing level.

CHAPTER VII

1419. Crests of standing mountain waves may be marked by stationary, lens-shaped clouds known as

A—mammatocumulus clouds.
B—standing lenticular clouds.
C—roll clouds.

2339. The presence of standing lenticular altocumulus clouds is a good indication of

A— an approaching storm.
B— very strong turbulence.
C— heavy icing conditions.

7127. What are the four families of clouds?

A—Stratus, cumulus, nimbus, and cirrus.
B—Clouds formed by: updrafts, fronts, cooling layers of air, and precipitation into warm air.
C—High, middle, low, and those with extensive vertical development.

7142. Fair weather cumulus clouds often indicate

A—turbulence at and below the cloud level.
B—poor visibility.
C—smooth flying conditions.

7150. A high cloud is composed mostly of

A—ozone.
B—condensation nuclei.
C—ice crystals.

6289. Which type clouds are indicative of very strong turbulence?

A— Nimbostratus.
B— Standing lenticular.
C— Cirrocumulus.

CHAPTER VIII

1395. The development of thermals depends upon

A—a counterclockwise circulation of air.
B—temperature inversions.
C—solar heating.

1397. What condition does a rising barometer indicate for balloon operations?

A—Decreasing clouds and wind.
B—Chances of thunderstorms.
C—Approaching frontal activity.

1408. If an unstable air mass is forced upward, what type clouds can be expected?

A—Stratus clouds with little vertical development.
B—Stratus clouds with considerable associated turbulence.
C—Clouds with considerable vertical development and associated turbulence.

1409. What feature is associated with a temperature inversion?

A—A stable layer of air.
B—An unstable layer of air.
C—Chinook winds on mountain slopes.

1410. What is the approximate base of the cumulus clouds if the temperature at 1,000 feet MSL is 70 °F and the dewpoint is 48 °F?

A—4,000 feet MSL.
B—6,000 feet MSL.
C—8,000 feet MSL.

1411. At approximately what altitude above the surface would the pilot expect the base of cumuliform clouds if the surface air temperature is 82 °F and the dewpoint is 38 °F?

A—9,000 feet AGL.
B—10,000 feet AGL.
C—11,000 feet AGL.

1412. What early morning weather observations indicate the possibility of good weather conditions for balloon flight most of the day?

A—Clear skies and surface winds, 5 knots or less.
B—Low moving, scattered cumulus clouds and surface winds, 5 knots or less.
C—Overcast with stratus clouds and surface winds, 7 knots or less.

1451. Convective circulation patterns associated with sea breezes are caused by

A—warm, dense air moving inland from over the water.
B—water absorbing and radiating heat faster than the land.
C—cool, dense air moving inland from over the water.

1452. What are some visual cues that identify a sea breeze front?

A—No restrictions to visibilities on either side.
B—A difference in visibility between land and sea.
C—Low stratus clouds on the landward side.

2330. What determines the structure or type of clouds which will form as a result of air being forced to ascend?

A— The method by which the air is lifted.
B— The stability of the air before lifting occurs.
C— The relative humidity of the air after lifting occurs.

2332. What are the characteristics of stable air?

A— Good visibility; steady precipitation; stratus clouds.
B— Poor visibility; steady precipitation; stratus clouds.
C— Poor visibility; intermittent precipitation; cumulus clouds.

2334. From which measurement of the atmosphere can stability be determined?

A— Atmospheric pressure.
B— The ambient lapse rate.
C— The dry adiabatic lapse rate.

2335. What type weather can one expect from moist, unstable air and very warm surface temperature?

A— Fog and low stratus clouds.
B— Continuous heavy precipitation.
C— Strong updrafts and cumulonimbus clouds.

2341. Which combination of weather-producing variables would likely result in cumuliform-type clouds, good visibility, and showery rain?

A— Stable, moist air and orographic lifting.
B— Unstable, moist air and orographic lifting.
C— Unstable, moist air and no lifting mechanism.

2345. Which is a characteristic of stable air?

A— Cumuliform clouds.
B— Excellent visibility.
C— Restricted visibility.

2386. Select the true statement concerning thermals.

A— Thermals are unaffected by winds aloft.
B— Strong thermals have proportionately increased sink in the air between them.
C— A thermal invariably remains directly above the surface area from which it developed.

2387. A thermal column is rising from an asphalt parking lot and the wind is from the south at 12 knots. Which statement would be true?

A— As altitude is gained, the best lift will be found directly above the parking lot.
B— As altitude is gained, the center of the thermal will be found farther north of the parking lot.
C— The slowest rate of sink would be close to the thermal and the fastest rate of sink farther from it.

2388. Which is true regarding the development of convective circulation?

A— Cool air must sink to force the warm air upward.
B— Warm air is less dense and rises on its own accord.
C— Warmer air covers a larger surface area than the cool air; therefore, the warmer air is less dense and rises.

2389. Which is generally true when comparing the rate of vertical motion of updrafts with that of downdrafts associated with thermals?

A— Updrafts and downdrafts move vertically at the same rate.
B— Downdrafts have a slower rate of vertical motion than do updrafts.
C— Updrafts have a slower rate of vertical motion than do downdrafts.

2390. Which thermal index would predict the best probability of good soaring conditions?

A— 10. [-10]
B— 5. [-5]
C— + 5.

2391. Which is true regarding the effect of fronts on soaring conditions?

A— A slow moving front provides the strongest lift.
B— Good soaring conditions usually exist after passage of a warm front.
C— Frequently, the air behind a cold front provides excellent soaring for several days.

2392. Convective circulation patterns associated with sea breezes are caused by

A— water absorbing and radiating heat faster than the land.
B— land absorbing and radiating heat faster than the water.
C— cool and less dense air moving inland from over the water, causing it to rise.

2393. The conditions most favorable to wave formation over mountainous areas are a layer of

A— stable air at mountaintop altitude and a wind of at least 20 knots blowing across the ridge.
B— unstable air at mountaintop altitude and a wind of at least 20 knots blowing across the ridge.
C— moist, unstable air at mountaintop altitude and a wind of less than 5 knots blowing across the ridge.

2394. When soaring in the vicinity of mou... ranges, the greatest potential danger from vertical and rotor-type currents will usually be encountered on the

A— leeward side when flying with the wind.
B— leeward side when flying into the wind.
C— windward side when flying into the wind.

2396. (Refer to figure 6.) The Pseudo-Adiabatic Chart and the soundings taken at 1400 GMT. Between what altitudes could optimum thermalling be expected at the time of the sounding?

A— From 2,500 to 6,000 feet.
B— From 6,000 to 10,000 feet.
C— From 13,000 to 15,000 feet.

2397. (Refer to figure 6.) The soundings taken at 0900 GMT from 2,500 feet to 15,000 feet, as shown on the Pseudo-Adiabatic Chart. What minimum surface temperature is required for instability to occur and for good thermals to develop from the surface to 15,000 feet MSL?

A— 58° F.
B— 68° F.
C— 80° F.

3212. At approximately what altitude above the surface would you expect the base of cumuliform clouds if the surface air temperature is 33 °C and the dewpoint is 15 °C?

A— 4,100 feet AGL.
B— 6,000 feet AGL.
C— 7,200 feet AGL.

3259. One method for locating thermals is to

A— fly an ever increasing circular path.
B— look for diverging streamers of dust or smoke.
C— look for converging streamers of dust or smoke.

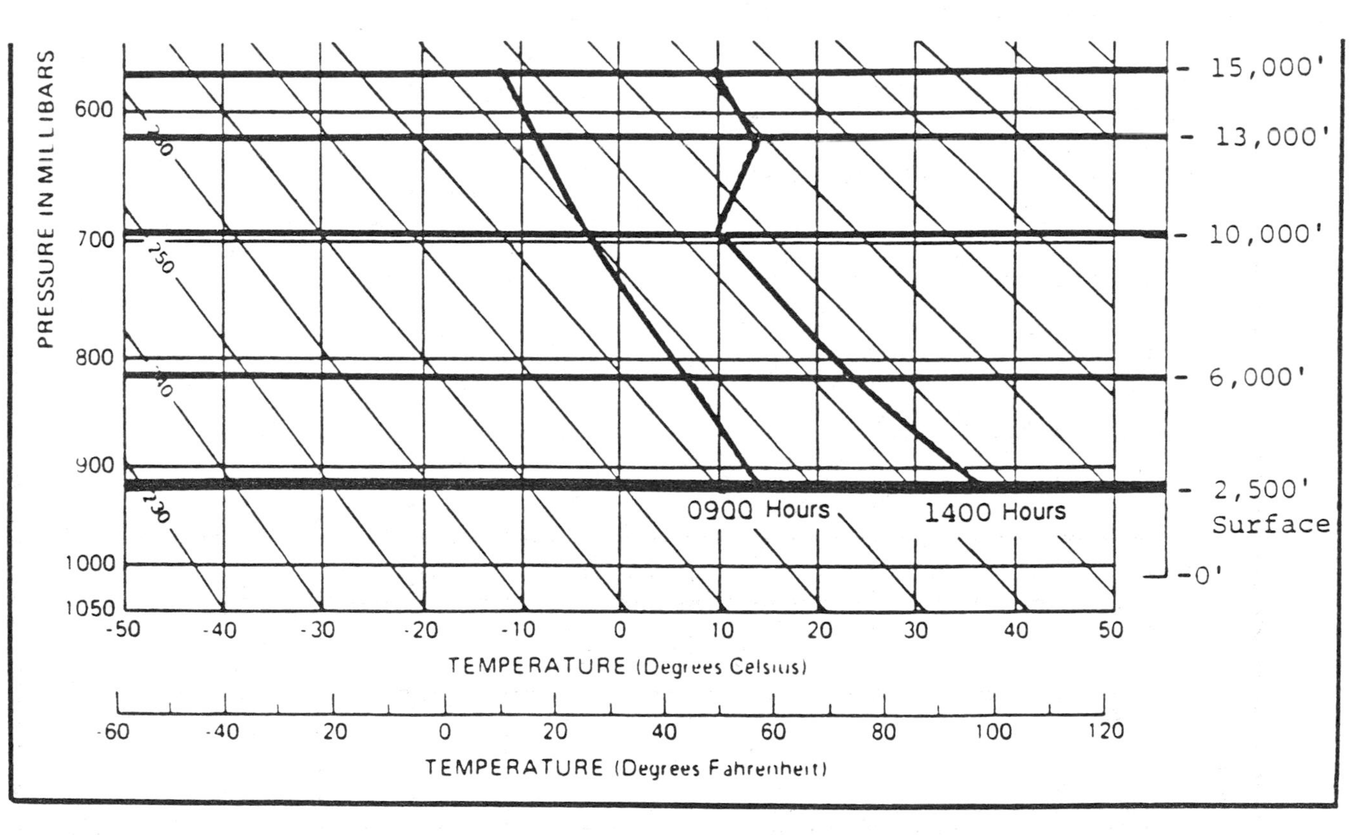

FIGURE 6. —Pseudo-Adiabatic Chart.

3210. The airport elevation is 1,294 feet MSL. At what elevation above mean sea level would you expect cumuliform clouds if the surface air temperature is 26 °C and the dewpoint is 21 °C?

A— 2,000 feet MSL.
B— 3,136 feet MSL.
C— 3,294 feet MSL.

3260. An important precaution when soaring in a dust devil is to

A— avoid the eye of the vortex because of extreme turbulence.
B— avoid steep turns on the upwind side to prevent being blown into the vortex.
C— avoid the clear area at the outside edge of the dust because of severe downdrafts.

3261. One of the best visual indications of a thermal is a

A— smooth cumulus cloud with a concave base.
B— broken to overcast sky with cumulus clouds.
C— fragmented cumulus cloud with a concave base.

3262. The most favorable type thermals for cross-country soaring may be found

A— under mountain waves.
B— along thermal streets.
C— just ahead of a warm front.

3265. Under what condition can enough lift be found for soaring under stable weather conditions?

A— Over steep escarpments or cliffs.
B— In mountain waves that form on the upwind side of the mountains.
C— On the upwind side of hills or ridges with moderate winds present.

3267. (Refer to figure 2.) At the 0900 sounding and the line plotted from the surface to 10,000 feet, what temperature must exist at the surface for instability to take place between these altitudes?

A— Any temperature more than 68 °F.
B— Any temperature less than 68 °F.
C— Any temperature between 43 °F and 68 °F.

3268. (Refer to figure 2.) According to the soundings taken at 1400, is the atmosphere stable or unstable and at what altitudes?

A— Stable from 6,000 to 10,000 feet.
B— Stable from 10,000 to 13,000 feet.
C— Unstable from 10,000 to 13,000 feet.

3269. (Refer to figure 2.) Using the 1400 soundings, does an inversion exist and, if so, at what altitudes?

A— No; there is no inversion shown.
B— Yes; between 10,000 and 13,000 feet.
C— Yes; between 13,000 and 15,000 feet.

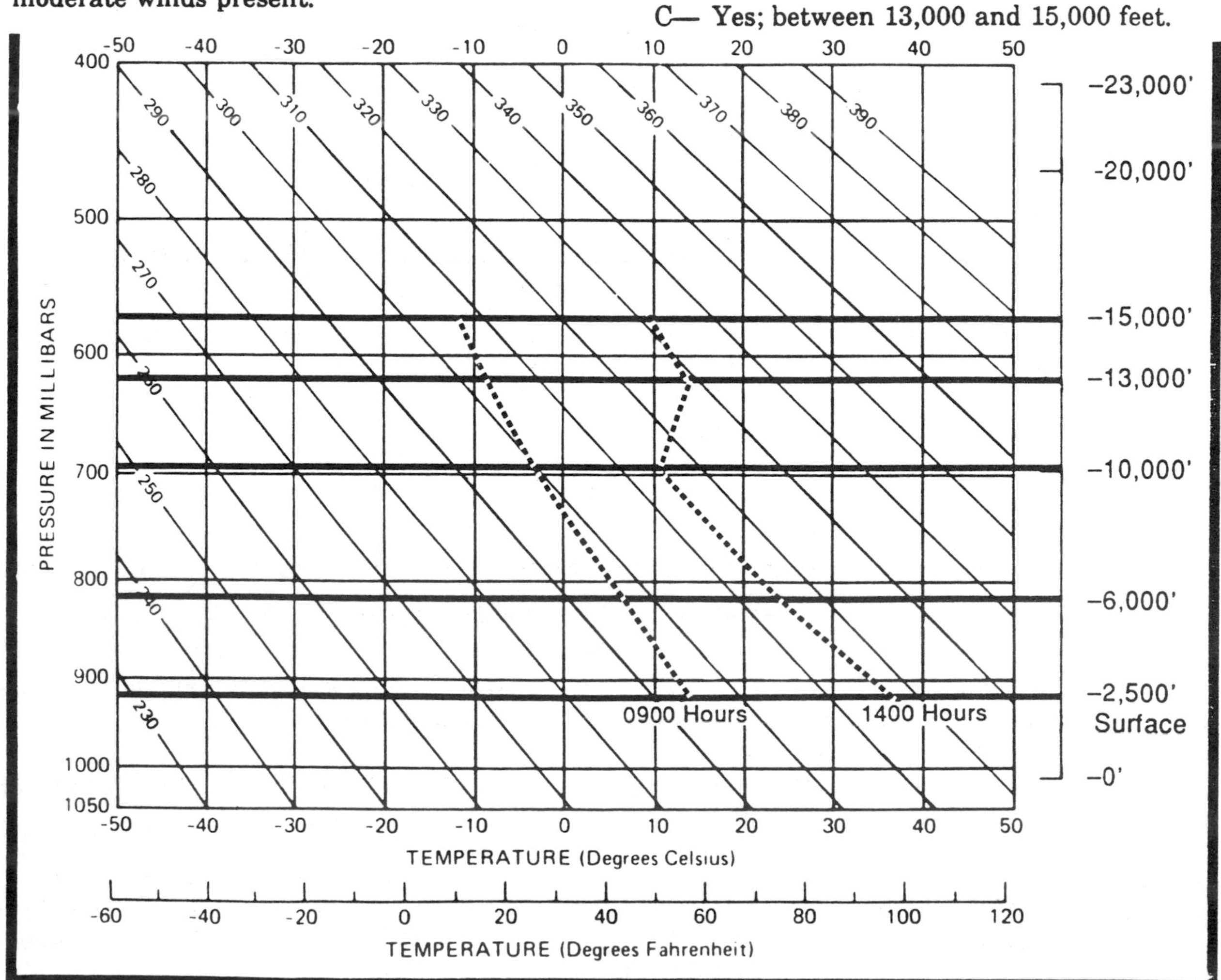

FIGURE 2.—Pseudo-Adiabatic Chart.

6259. Where is the usual location of a thermal low?

A— Over the arctic region.
B— Over the eye of a hurricane.
C— Over the surface of a dry, sunny region.

6284. What weather condition occurs at the altitude where the dewpoint lapse rate and the dry adiabatic lapse rate converge?

A— Cloud bases form.
B— Precipitation starts.
C— Stable air changes to unstable air.

6286. When saturated air moves downhill, its temperature increases

A— at a faster rate than dry air because of the release of latent heat.
B— at a slower rate than dry air because vaporization uses heat.
C— at a slower rate than dry air because condensation releases heat.

6287. Which condition is present when a local parcel of air is stable?

A— The parcel of air resists convection.
B— The parcel of air cannot be forced uphill.
C— As the parcel of air moves upward, its temperature becomes warmer than the surrounding air.

CHAPTER IX

1413. What are characteristics of a moist, unstable air mass?

A—Cumuliform clouds and showery precipitation.
B—Poor visibility and smooth air.
C—Stratiform clouds and continuous precipitation.

1415. A stable air mass is most likely to have which characteristic?

A—Showery precipitation.
B—Turbulent air.
C—Smooth air.

2336. Which would increase the stability of an airmass?

A— Warming from below.
B— Cooling from below.
C— Decrease in water vapor.

2344. When an airmass is stable, which of these conditions are most likely to exist?

A— Numerous towering cumulus and cumulonimbus clouds.
B— Moderate to severe turbulence at the lower levels.
C— Smoke, dust, haze, etc., concentrated at the lower levels with resulting poor visibility.

2346. Which is a characteristic typical of a stable airmass?

A— Cumuliform clouds.
B— Showery precipitation.
C— Continuous precipitation.

2348. Which are characteristics of a cold airmass moving over a warm surface?

A— Cumuliform clouds, turbulence, and poor visibility.
B— Cumuliform clouds, turbulence, and good visibility.
C— Stratiform clouds, smooth air, and poor visibility.

7151. An airmass is a body of air that

A—has similar cloud formations associated with it.
B—creates a wind shift as it moves across the Earth's surface.
C—covers an extensive area and has fairly uniform properties of temperature and moisture.

3226. A moist, cold air mass that is being warmed from below is characterized, in part, by

A— instability and showers.
B— stability, fog, and drizzle.
C— instability and continuous precipitation.

6283. What is indicated about an air mass if the temperature remains unchanged or decreases slightly as altitude is increased?

A— The air is unstable.
B— A temperature inversion exists.
C— The air is stable.

CHAPTER X

1422. The boundary between two different air masses is referred to as a

A—frontolysis.
B—frontogenesis.
C—front.

1423. One of the most easily recognized discontinuities across a front is

A—temperature.
B—cloud formation.
C—precipitation.

1425. Steady precipitation preceding a front is an indication of

A—stratiform clouds with moderate turbulence.
B—cumuliform clouds with little or no turbulence.
C—stratiform clouds with little or no turbulence.

2363. The most severe weather conditions, such as destructive winds, heavy hail, and tornadoes, are generally associated with

A— fast-moving warm fronts.
B— slow-moving warm fronts.
C— squall lines and steady-state thunderstorms.

7130. Where do squall lines most often develop?

A—In an occluded front.
B—In a cold air mass.
C—Ahead of a cold front.

7134. What is indicated by the term "embedded thunderstorms"?

A—Severe thunderstorms are embedded within a squall line.
B—Thunderstorms are predicted to develop in a stable air mass.
C—Thunderstorms are obscured by massive cloud layers and cannot be seen.

3227. If a wave were to form on a stationary front running east and west across the United States, that portion east of the wave would normally

A— remain stationary with that portion west of the wave becoming a cold front.
B— become a warm front and that portion west of the wave would become a cold front.
C— become a cold front and that portion west of the wave would become a warm front.

6290. What is a feature of a stationary front?

A— The warm front surface moves about half the speed of the cold front surface.
B— Weather conditions are a combination of strong cold front and strong warm front weather.
C— Surface winds tend to flow parallel to the frontal zone.

6291. Which event usually occurs after an aircraft passes through a front into the colder air?

A— Temperature/dewpoint spread decreases.
B— Wind direction shifts to the left.
C— Atmospheric pressure increases.

6312. What type weather change is to be expected in an area where frontolysis is reported?

A— The frontal weather is becoming stronger.
B— The front is dissipating.
C— The front is moving at a faster speed.

6313. Which weather condition is an example of a nonfrontal instability band?

A— Squall line.
B— Advective fog.
C— Frontogenesis.

6314. Which atmospheric factor causes rapid movement of surface fronts?

A— Upper winds blowing across the front.
B— Upper low located directly over the surface low.
C— The cold front overtaking and lifting the warm front.

6315. In which meteorological conditions can frontal waves and low pressure areas form?

A— Warm fronts or occluded fronts.
B— Slow-moving cold fronts or stationary fronts.
C— Cold front occlusions.

6316. What weather difference is found on each side of a "dry line"?

A— Extreme temperature difference.
B— Dewpoint difference.
C— Stratus versus cumulus clouds.

6322. Which type precipitation is an indication that supercooled water is present?

A— Wet snow.
B— Freezing rain.
C— Ice pellets.

6323. Which type of icing is associated with the smallest size of water droplet similar to that found in low-level stratus clouds?

A— Clear ice.
B— Frost ice.
C— Rime ice.

CHAPTER XI

1427. Possible mountain wave turbulence could be anticipated when winds of 40 knots or greater blow

A—across a mountain ridge, and the air is stable.
B—down a mountain valley, and the air is unstable.
C—parallel to a mountain peak, and the air is stable.

1428. Where does wind shear occur?

A—Only at higher altitudes.
B—Only at lower altitudes.
C—At all altitudes.

1429. Hazardous wind shear is commonly encountered near the ground during periods

A—when the wind velocity is stronger than 35 knots.
B—of strong low-level temperature inversion.
C—following frontal passage.

1430. A pilot can expect a wind shear zone in a temperature inversion whenever the windspeed at 2,000 to 4,000 feet above the surface is at least

A—10 knots.
B—15 knots.
C—25 knots.

1444. Upon encountering severe turbulence, which flight condition should the pilot attempt to maintain?

A—Constant altitude.
B—Constant airspeed (V_A).
C—Level flight attitude.

2351. What is an important characteristic of wind shear?

A— It exists in a horizontal direction only, and is normally found near a jetstream.
B— It occurs primarily at the lower levels and is usually associated with mountain waves.
C— It can be present at any level and can exist in both a horizontal and vertical direction.

2352. Hazardous wind shear is commonly encountered during periods

A— of high pressure.
B— when the wind velocity is stronger than 35 knots.
C— of temperature inversion, and near thunderstorms.

2355. When climbing or descending through an inversion or wind-shear zone, the pilot should be alert for which changes in airplane performance?

A— A sudden decrease in power.
B— A sudden change in airspeed.
C— A fast rate of climb and a slow rate of descent.

2356. Convective currents are most active on warm summer afternoons when winds are

A— light.
B— moderate.
C— strong.

2357. When flying low over hilly terrain, ridges, or mountain ranges the greatest potential danger from turbulent air currents will usually be encountered on the

A— leeward side when flying with the wind.
B— leeward side when flying into the wind.
C— windward side when flying into the wind.

2382. A common location of clear air turbulence is

A— in an upper trough on the polar side of a jetstream.
B— near a ridge aloft on the equatorial side of a high-pressure flow.
C— south of an east/west oriented high-pressure ridge in its dissipating stage.

2383. The jetstream and associated clear air turbulence can sometimes be visually identified in flight by

A— dust or haze at flight level.
B— long streaks of cirrus clouds.
C— a constant outside air temperature.

2384. During the winter months in the middle latitudes, the jetstream shifts toward the

A— north and speed decreases.
B— south and speed increases.
C— north and speed increases.

2385. The strength and location of the jetstream is normally

A— weaker and farther north in the summer.
B— stronger and farther north in the winter.
C— stronger and farther north in the summer.

7148. A jetstream is defined as wind of

A—30 knots or greater.
B—40 knots or greater.
C—50 knots or greater.

3183. The height of the tropopause over the contiguous 48 states is

A— relatively constant.
B— lower in the winter.
C— lower in the summer.

3231. Low-level wind shear, which results in a sudden change of wind direction, may occur

A— after a warm front has passed.
B— when surface winds are light and variable.
C— when there is a low-level temperature inversion with strong winds above the inversion.

3232. Which condition could be expected if a strong temperature inversion exists near the surface?

A— Strong, steady downdrafts and an increase in OAT.
B— A wind shear with the possibility of a sudden loss of airspeed.
C— An OAT increase or decrease with a constant wind condition.

6227. What action is appropriate when encountering the first ripple of reported clear air turbulence?

A— Extend flaps to decrease wing loading.
B— Extend gear to provide more drag and increase stability.
C— Adjust airspeed to that recommended for rough air.

6228. If severe turbulence is encountered, which procedure is recommended?

A— Maintain a constant altitude.
B— Maintain a constant attitude.
C— Maintain constant airspeed and altitude.

6232. Which INITIAL cockpit indications should a pilot be aware of when a headwind shears to a calm wind?

A— Indicated airspeed decreases, aircraft pitches up, and altitude decreases.
B— Indicated airspeed increases, aircraft pitches down, and altitude increases.
C— Indicated airspeed decreases, aircraft pitches down, and altitude decreases.

6233. Which condition would INITIALLY cause the indicated airspeed and pitch to increase and the sink rate to decrease?

A— Sudden decrease in a headwind component.
B— Tailwind which suddenly increases in velocity.
C— Sudden increase in a headwind component.

6234. Which INITIAL cockpit indications should a pilot be aware of when a constant tailwind shears to a calm wind?

A— Altitude increases; pitch and indicated airspeed decrease.
B— Altitude, pitch, and indicated airspeed decrease.
C— Altitude decreases; pitch and indicated airspeed increase.

6236 Which wind-shear condition results in a loss of airspeed?

A— Decreasing headwind or tailwind.
B— Decreasing headwind and increasing tailwind.
C— Increasing headwind and decreasing tailwind.

6237. Which wind shear condition results in an increase in airspeed?

A— Increasing tailwind and decreasing headwind.
B— Increasing tailwind and headwind.
C— Decreasing tailwind and increasing headwind.

6240. Which airplane performance characteristics should be recognized during takeoff when encountering a tailwind shear that increases in intensity?

A— Loss of, or diminished, airspeed performance.
B— Decreased takeoff distance.
C— Increased climb performance immediately after takeoff.

6241. Thrust is being managed to maintain desired indicated airspeed and the glide slope is being flown. Which characteristics should be observed when a tailwind shears to a constant headwind?

A— PITCH ATTITUDE: Increases.
REQUIRED THRUST: Reduced, then increased.
VERTICAL SPEED: Increases.
INDICATED AIRSPEED: Decreases, then increases to approach speed.
B— PITCH ATTITUDE: Increases.
REQUIRED THRUST: Increased, then reduced.
VERTICAL SPEED: Decreases.
INDICATED AIRSPEED: Decreases, then increases to approach speed.
C— PITCH ATTITUDE: Decreases.
REQUIRED THRUST: Increased, then reduced.
VERTICAL SPEED: Decreases.
INDICATED AIRSPEED: Decreases, then increases to approach speed.

6266. What information from the control tower is indicated by the following transmission?

"SOUTH BOUNDARY WIND ONE SIX ZERO AT TWO FIVE, WEST BOUNDARY WIND TWO FOUR ZERO AT THREE FIVE."

A— A downburst is located at the center of the airport.
B— Wake turbulence exists on the west side of the active runway.
C— There is a possibility of wind shear over or near the airport.

6317. Under what conditions would clear air turbulence (CAT) most likely be encountered?

A— When constant pressure charts show 20-knot isotachs less than 60 NM apart.
B— When constant pressure charts show 60-knot isotachs less than 20 NM apart.
C— When a sharp trough is moving at a speed less than 20 knots.

6318. What action is recommended when encountering turbulence due to a wind shift associated with a sharp pressure trough?

A— Establish a course across the trough.
B— Climb or descend to a smoother level.
C— Increase speed to get out of the trough as soon as possible.

6319. In comparison to an approach in a moderate headwind, which is an indication of a possible wind shear due to a decreasing headwind when descending on the glide slope?

A— Less power is required.
B— Higher pitch attitude is required.
C— Lower descent rate is required.

6324. Which is a necessary condition for the occurrence of a low-level temperature inversion wind shear?

A— The temperature differential between the cold and warm layers must be at least 10 °C.
B— A calm or light wind near the surface and a relatively strong wind just above the inversion.
C— A wind direction difference of at least 30° between the wind near the surface and the wind just above the inversion.

6325. What is the lowest cloud in the stationary group associated with a mountain wave?

A— Rotor cloud.
B— Standing lenticular.
C— Low stratus.

6326. Where is the normal location of the jetstream relative to surface lows and fronts?

A— The jetstream is located north of the surface systems.
B— The jetstream is located south of the low and warm front.
C— The jetstream is located over the low and crosses both the warm front and the cold front.

6327. Which type frontal system is normally crossed by the jetstream?

A— Cold front and warm front.
B— Warm front.
C— Occluded front.

6328. Which type clouds may be associated with the jetstream?

A— Cumulonimbus cloud line where the jetstream crosses the cold front.
B— Cirrus clouds on the equatorial side of the jetstream.
C— Cirrostratus cloud band on the polar side and under the jetstream.

6329. Which action is recommended if jetstream turbulence is encountered with a direct headwind or tailwind?

A— Increase airspeed to get out of the area quickly.
B— Change course to fly on the polar side of the jetstream.
C— Change altitude or course to avoid a possible elongated turbulent area.

6330. Which action is recommended regarding an altitude change to get out of jetstream turbulence?

A— Descend if ambient temperature is falling.
B— Descend if ambient temperature is rising.
C— Maintain altitude if ambient temperature is not changing.

6331. Clear air turbulence associated with a mountain wave may extend as far as

A— 1,000 miles or more downstream of the mountain.
B— 5,000 feet above the tropopause.
C— 100 miles or more upwind of the mountain.

6334. Turbulence encountered above 15,000 feet AGL, not associated with cloud formations, should be reported as

A— convective turbulence.
B— high altitude turbulence.
C— clear air turbulence.

6335. A strong wind shear can be expected

A— on the low–pressure side of a 100–knot jetstream core.
B— where the 20–knot isotachs are spaced 100 NM or closer together.
C— if the 5 °C isotherms are spaced 100 NM or closer together.

6336. What is a likely location of clear air turbulences?

A— In an upper trough on the polar side of a jetstream.
B— Near a ridge aloft on the equatorial side of a high pressure flow.
C— Downstream of the equatorial side of a jetstream.

6337. Where do the maximum winds associated with the jetstream usually occur?

A— In the vicinity of breaks in the tropopause on the polar side of the jet core.
B— Below the jet core where a long straight stretch of the jetstream is located.
C— On the equatorial side of the jetstream where moisture has formed cirriform clouds.

6338. Which type jetstream can be expected to cause the greater turbulence?

A— A straight jetstream associated with a high pressure ridge.
B— A jetstream associated with a wide isotherm spacing.
C— A curving jetstream associated with a deep low pressure trough.

6339. What weather feature occurs at altitude levels near the tropopause?

A— Maximum winds and narrow wind shear zones.
B— Abrupt temperature increase above the tropopause.
C— Thin layers of cirrus (ice crystal) clouds at the tropopause level.

6340. Where are jetstreams normally located?

A— In areas of strong low pressure systems in the stratosphere.
B— At the tropopause where intensified temperature gradients are located.
C— In a single continuous band, encircling the Earth, where there is a break between the equatorial and polar tropopause.

Subject: RULES OF THUMB FOR AVOIDING OR MINIMIZING ENCOUNTERS WITH CLEAR AIR TURBULENCE

Date: 11/21/88
Initiated by: AFS-400
AC No: 00-30A
Change:

1. PURPOSE. This advisory circular (AC) describes to pilots, aircrew members, dispatchers, and other operations personnel the various types of clear air turbulence (CAT) and some of the weather patterns associated with it, and provides the "Rules of Thumb" for avoiding or minimizing encounters with CAT.

2. CANCELLATION. This revision supersedes AC 00-30, Rules of Thumb for Avoiding or Minimizing Encounters With Clear Air Turbulence, dated March 5, 1970.

3. BACKGROUND.

a. In 1966, the National Committee for Clear Air Turbulence officially defined CAT as "all turbulence in the free atmosphere of interest in aerospace operations that is not in or adjacent to visible convective activity (this includes turbulence found in cirrus clouds not in or adjacent to visible convective activity)." Over time, less formal definitions of CAT have evolved. Advisory Circular 00-45C, Aviation Weather Services, defines CAT as "high level turbulence (normally above 15,000 feet AGL) not associated with cumuliform cloudiness, including thunderstorms." The Airman's Information Manual expands on the basic AC 00-45C CAT definition as "turbulence encountered in air where no clouds are present. This term is commonly applied to high-level turbulence associated with windshear. Thus, clear air turbulence or CAT has been defined in several ways, but the most comprehensive definition is: "turbulence encountered outside of convective clouds." This includes turbulence in cirrus clouds, within and in the vicinity of standing lenticular clouds and, in some cases, in clear air in the vicinity of thunderstorms. Generally, though, CAT definitions exclude turbulence caused by thunderstorms, low-altitude temperature inversions, thermals, or local terrain features.

b. CAT was recognized as a problem with the advent of multiengine jet aircraft in the 1950's. CAT is especially troublesome because it is often encountered unexpectedly and frequently without visual clues to warn pilots of the hazard.

4. DISCUSSION.

a. One of the principal areas where CAT is found is in the vicinity of the jetstream or jetstreams. A jetstream is a river of high-altitude wind with a speed of 50 knots, or greater, following the planetary atmospheric pattern. There are, in fact, three jetstreams: the polar front jetstream, the subtropical jetstream, and the polar night jetstream. The polar front jetstream as its name implies, is associated with the polar front or the division between the cold polar and warm tropical air masses. The mean latitude of the jetstream core varies from 25° north latitude during the winter months to 42° north latitude during the summer months

(1) The polar front jetstream is the center of the planetary wave pattern and as such meanders over a large portion of the hemisphere throughout the year, particularly during the winter months when it is most intense. Although the polar front jetstream varies in altitude, the core is most commonly found around 30,000 feet and it is generally best depicted on the 300 millibar constant pressure map.

(2) The subtropical jetstream is a very persistent circumpolar jetstream found on the northern periphery of the tropical latitudes between 20° and 30° north latitude. It normally forms three waves around the globe with crests over the eastern coasts of Asia and North America and the Near East. Like the polar front jetstream, the subtropical jetstream is most active during the winter months and often intrudes well into the southeastern United States. It is generally higher than the polar front jetstream with the core between 35,000 and 45,000 feet.

(3) The polar night jetstream is found in the stratosphere in the vicinity of the Arctic Circle during the winter months and does not have a significant affect on air travel over the United States and southern Canada.

b. CAT associated with a jetstream is most commonly found in the vicinity of the tropopause and upper fronts. The tropopause is actually an upper front separating the troposphere from the stratosphere. Analyses of the tropopause are issued by the National Weather Service on a scheduled basis. In the absence of other information, the tropopause will generally have a temperature of between -55°C. and -65°C. In some cases, it will be at the top of a cirrus cloud layer. Clouds are very seldom found above the tropopause in the dry stratosphere, except in the summertime when occasionally large thunderstorms will poke through the tropopause and spread anvil clouds in the stratosphere. CAT is most frequently found on the poleward side of the jetstream (the left side facing downwind). It is additionally common in the vicinity of a jetstream maxima (an area of stronger winds that moves along the jetstream).

c. There are several patterns of upper-level winds that are associated with CAT. One of these is a deep, upper trough. The CAT is found most frequently at and just upwind of the base of the trough, particularly just downwind of an area of strong temperature advection. Another area of the trough in which to suspect CAT is along the centerline of a trough where there is a strong horizontal windshear between the northerly and southerly flows. CAT is also found in the back side of a trough in the vicinity of a wind maxima as the maxima passes through.

d. One noteworthy generator of CAT is the confluence of two jetstreams. On occasion, the polar front jetstream will dip south and pass under the subtropical jetstream. The area of windshear between the two jetstreams in the area of confluence and immediately downstream is frequently turbulent.

e. CAT is very difficult to predict accurately, due in part to the fact that CAT is spotty in both dimensions and time. Common dimensions of a turbulent area associated with a jetstream are on the order of 100 to 300 miles long, elongated in the direction of the wind, 50 to 100 miles wide, and 2,000 to 5,000 feet deep. These areas may persist from 30 minutes to a day. In spite of the difficulty forecasting CAT, there are rules that have been developed to indicate those areas where CAT formation is likely.

f. The threshold windspeed in the jetstream for CAT is generally considered to be 110 knots. Windspeed in jetstreams can be much stronger than 110 knots and the probability of encountering CAT increases with the windspeed and the windshear it generates. It is not the windspeed itself that causes CAT; it is the windshear or difference in windspeed from one point to another that causes the wave motion or overturning in the atmosphere that is turbulence to an aircraft. Windshear occurs in all directions, but for convenience it is measured along vertical and horizontal axes, thus becoming horizontal and vertical windshear. Moderate CAT is considered likely when the vertical windshear is 5 knots per 1,000 feet, or greater, and/or the horizontal windshear is 20 knots per 150 nautical miles, or greater. Severe CAT is considered likely when the vertical windshear is 6 knots per 1,000 feet and/or the horizontal windshear is 40 knots per 150 miles or greater.

g. Until practical airborne detectors are developed, pilots are urged to use the "Rules of Thumb to Assist in Avoiding or Minimizing Encounters With Clear Air Turbulence" contained in appendix 1. The majority of these guidelines were developed initially by the International Civil Aviation Organization's (ICAO) Sixth Air-Navigation Conference of April/May 1969, but have been expanded based on recommendations from the Department of Defense, the National Transportation Safety Board, and the Federal Aviation Administration.

5. RECOMMENDATION. All pilots and other personnel concerned with flight planning should carefully consider the hazards associated with flight through areas where pilot reports or aviation weather forecasts indicate the presence of CAT including mountain wave turbulence. The "Rules of Thumb" in appendix 1 are intended to assist pilots in avoiding potentially hazardous CAT during flight.

D.C. Beaudette
Acting Director, Flight Standards Service

APPENDIX 1. RULES OF THUMB TO ASSIST IN AVOIDING OR MINIMIZING ENCOUNTERS WITH CLEAR AIR TURBULENCE (CAT)

Note: The following "Rules of Thumb" apply primarily to the westerly jetstreams.

1. Jetstreams stronger than 110 knots (at the core) are apt to have areas of significant turbulence near them in the sloping tropopause above the core, in the jetstream front below the core, and on the low-pressure side of the core.

2. Windshear and its accompanying CAT in jetstreams is more intense above and to the lee of mountain ranges. CAT should be anticipated whenever the flightpath traverses a strong jetstream in the vicinity of mountainous terrain.

3. Both vertical and horizontal windshear are, of course, greatly intensified in mountain wave conditions. Therefore, when the flightpath traverses a mountain wave type of flow, it is desirable to fly at turbulence-penetration speed and avoid flight over areas where the terrain drops abruptly, even though there may be no lenticular clouds to identify the condition.

4. On charts for standard isobaric surfaces, such as 300 millibars, if 20-knot isotachs are spaced closer together than 150 nautical miles (2-1/2 degrees latitude), there is sufficient horizontal shear for CAT. This area is normally on the poleward (low-pressure) side of the jetstream axis, but in unusual cases may occur on the equatorial side.

5. Turbulence is also related to vertical shear. From the tropopause height/vertical windshear chart, determine the vertical shear in knots-per-thousand feet. If it is greater than 5 knots per 1,000 feet, turbulence is likely.

6. Curving jetstreams are more apt to have turbulent edges than straight ones, especially jetstreams which curve around a deep pressure trough.

7. Wind-shift areas associated with pressure troughs and ridges are frequently turbulent. The magnitude of the windshear is the important factor.

8. If jetstream turbulence is encountered with direct tailwinds or headwinds, a change of flight level or course should be initiated since these turbulent areas are elongated with the wind and are shallow and narrow.

9. If jetstream turbulence is encountered in a crosswind, it is not so important to change course or flight level since the rough areas are narrow across the wind.

10. If turbulence is encountered in an abrupt wind shift associated with a sharp pressure trough line, establish a course across the trough rather than parallel to it.

11. If turbulence is expected because of penetration of a sloping tropopause, watch the temperature gauge. The point of coldest temperature along the flightpath will be the tropopause penetration. Turbulence will be most pronounced in the temperature-change zone on the stratospheric (upper) side of the sloping tropopause.

12. If possible, when crossing the jet, climb with a rising temperature and descend with a dropping temperature.

13. Weather satellite pictures are useful in identifying jetstreams associated with cirrus cloud bands. CAT is normally expected in the vicinity of jetstreams. Satellite imagery showing "wave-like" or "herringbone" cloud patterns are often associated with mountain wave turbulence. Pilots should avail themselves of briefings on satellite data whenever possible.

14. Last, but not least, monitor your radio--pilot reports can be invaluable and if you get caught by "the CAT," file a PIREP!

Note: In this country, civil forecasts of areas of CAT are made by the National Weather Service and disseminated as follows: (1) in area forecasts every 8 hours (every 6 hours in Hawaii); (2) on high-level significant weather facsimile charts available every 6 hours, and (3) on a nonscheduled basis as in-flight advisories (SIGMETS). SIGMETS are issued when severe or extreme CAT is forecast or has been reported. This information is available to pilots through the en route advisory service (flight watch), in SIGMET alerts broadcast on air route traffic control center frequencies, and over the hazardous in-flight weather advisory service (HIWAS).

AC 00-50A

DATE 1/23/79

ADVISORY CIRCULAR

DEPARTMENT OF TRANSPORTATION
Federal Aviation Administration
Washington, D.C.

Subject: LOW LEVEL WIND SHEAR

1. PURPOSE. This advisory circular is intended to provide guidance for recognizing the meteorological situations that produce the phenomenon widely known as low level wind shear. It describes both preflight and in-flight procedures for detecting and predicting this phenomenon as well as pilot techniques that minimize its effects when inadvertently encountered on takeoff or landing.

2, CANCELLATION. AC 00-50, dated April 8, 1976, is canceled.

3. BACKGROUND.

a. Wind shear is best described as a change in wind direction and/or speed in a very short distance in the atmosphere. Under certain conditions, the atmosphere is capable of producing some dramatic shears very close to the ground; for example, wind direction changes of 180 degrees and speed changes of 50 knots or more within 200 feet of the ground have been observed. It has been said that wind cannot affect an aircraft once it is flying except for drift and groundspeed. However, studies have shown that this is not true if the wind changes faster than the aircraft mass can be accelerated or decelerated.

b. The most prominent meteorological phenomena that cause significant low level wind shear problems are thunderstorms and certain frontal systems at or near the airport.

c. Appendix 1 contains a bibliography of FAA publications on wind shear.

4. METEOROLOGY.

a. Thunderstorms. The winds around a thunderstorm are complex (Figure 1). Wind shear can be found on all sides of a thunderstorm cell and in the downdraft directly under the cell. The wind shift line or gust front associated with thunderstorms can precede the actual storm by 15 nautical miles or more. Consequently, if a thunderstorm is near an airport of intended takeoff or landing, low level wind shear hazards may exist.

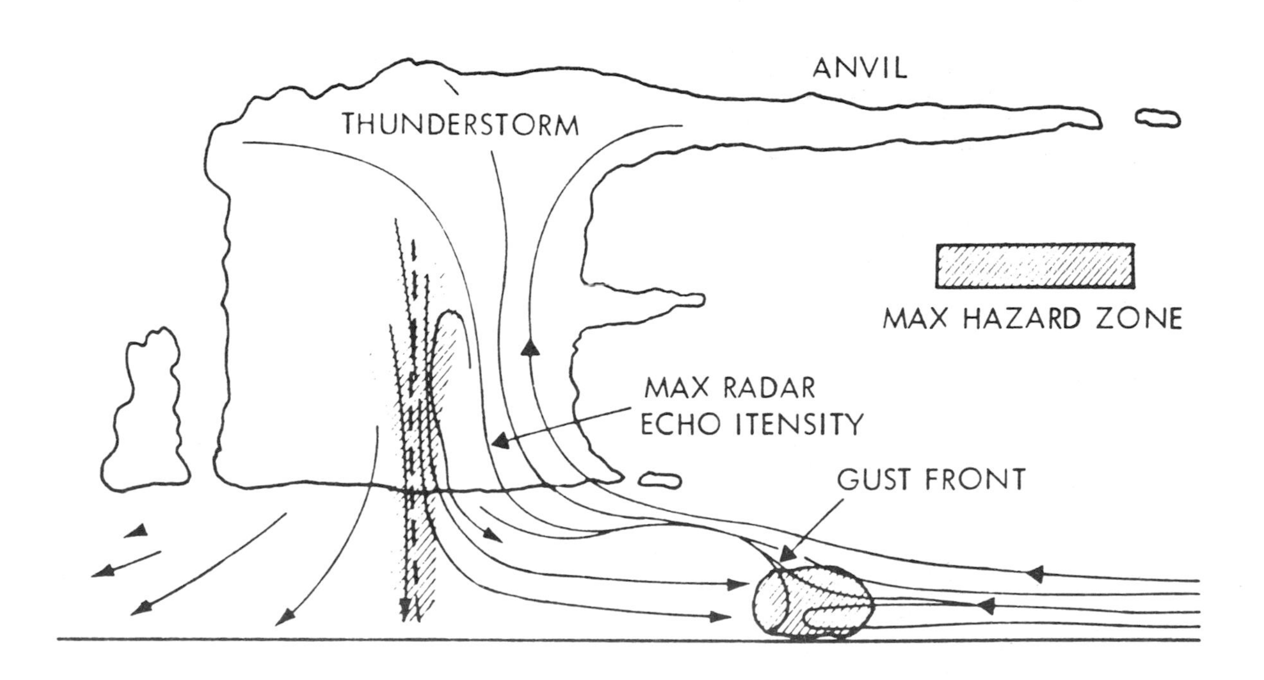

FIGURE 1. THUNDERSTORM HAZARD ZONES

b. Fronts. The winds can be significantly different in the two air masses which meet to form a front. While the direction of the winds above and below a front can be accurately determined, existing procedures do not provide precise, current measurements of the height of the front above the airport. The following is a method for determining the approximate height of the wind shear associated with a front.

(1) Wind shear occurs with a cold front just after the front passes the airport and for a short period thereafter. If the front is moving 30 knots or more, the frontal surface will usually be 5,000 feet above the airport about three hours after the frontal passage.

(2) With a warm front, the most critical period is before the front passes the airport. Warm front shear may exist below 5,000 feet for approximately six hours. The problem ceases to exist after the front passes the airport. Data compiled on wind shear indicates that the amount of shear in warm fronts is much greater than that found in cold fronts.

(3) Turbulence may or may not exist in wind shear conditions. If the surface wind under the front is strong and gusty, there will be some turbulence associated with wind shear.

c. Strong Surface Winds. The combination of strong winds and small hills or large buildings that lie upwind of the approach or departure path can produce localized areas of shear. Observing the local terrain and requesting pilot reports of conditions near the runway are the best means for anticipating wind shear from this source. This type of shear can be particularly hazardous to light airplanes.

d. Sea Breeze Fronts. The presence of large bodies of water can create local airflows due to the differences in temperature between the land and water. Changes in wind velocity and direction can occur in relatively short distances in the vicinity of airports situated near large lakes, bays or oceans.

e. Mountain Waves. These weather phenomena often create low level wind shear at airports that lie downwind of the wave. Altocumulus standing lenticular (ACSL) clouds usually depict the presence of mountain waves, and they are clues that shear should be anticipated.

5. DETECTING WIND SHEAR. Airplanes may not be capable of safely penetrating all intensities of low level wind shear. Pilots should, therefore, learn to detect, predict, and avoid severe wind shear conditions. Severe wind shear does not strike without warning. It can be detected by the following methods:

a. Analyze the weather during preflight.

(1) If thunderstorms are observed or forecast at or near the airport, be alert for the possibility of wind shear in the departure or arrival areas.

(2) Check the surface weather charts for frontal activity. Determine the surface temperature difference immediately across the front and the speed at which the front is moving. A 10°F [5°C] or greater temperature differential, and/or a frontal speed of 30 knots or more, is an indication of the possible existence of significant low level wind shear.

b. Be aware of pilot reports (PIREPS) of wind shear. Part 1 of the Airman's Information Manual recommends that pilots report any wind shear encounter to Air Traffic Control. This report should be in specific terms and include the loss/gain of airspeed due to the shear and the altitude(s) at which it was encountered. For example: "Denver tower, Cessna 1234 encountered wind shear, loss of 20 knots at 400 feet." This simple report is extremely important so that the pilot of the next airplane in sequence can determine the safety of transiting the same location. Reported shear that causes airspeed losses in excess of 15 to 20 knots should be avoided. Reported shears associated with a thunderstorm should also be avoided due to the speed which some storms move across the ground. The storm movement can cause one aircraft to encounter an airspeed increase which may appear harmless where the next aircraft can encounter a severe airspeed loss.

c. Assume that severe wind shear is present when the following conditions exist in combination.

(1) Extreme variations in wind velocity and direction in a relatively short time span.

(2) Evidence of a gust front such as blowing dust on the airport surface.

(3) Surface temperature in excess of 80° F.

(4) Dew point spread of 40° F or more.

(5) Virga (precipitation that falls from the bases of high altitude cumulus clouds but evaporates before reaching the ground).

d. Examine the approach or takeoff area with the airplane's radar set to determine if thunderstorm cells are in the vicinity of the airport. A departure or approach should not be flown through or under a thunderstorm cell.

e. Use the airplane instruments to detect wind shear.

(1) Pilots flying airplanes equipped with inertial navigation system (INS) should compare the winds at the initial approach altitude (1500-2000' above ground level (AGL)) with the reported runway surface winds to see if there is a wind shear situation between the airplane and the runway.

(2) If frontal activity does exist, note the surface wind direction to determine the location of the front with respect to the airport. If the airplane will traverse the front, compare the surface wind direction and speed with the wind direction and speed above the front to determine the potential wind shear during climbout or approach.

(3) Pilots flying airplanes equipped with a device which reads out groundspeed should compare the airplane's groundspeed with its airspeed. Any rapid changes in the relationship between airspeed and groundspeed represents a wind shear. Some operators have adopted the procedure of not allowing their aircraft to slow below a precomputed minimum groundspeed on approach. The minimum is computed by subtracting the surface headwind component from the true airspeed on approach.

(4) Pilots flying airplanes which do not have INS or groundspeed readouts should closely monitor their airplane's performance when wind shear is suspected. When the rate of descent on an ILS approach differs from the nominal values for the aircraft, the pilot should beware of a potential wind shear situation. Since rate of descent on the glide slope is directly related to groundspeed, a high descent rate would indicate a strong tailwind; conversely, a low descent rate denotes a strong headwind. The power needed to hold the glide slope also will be different from typical, no-shear conditions. Less power than normal will be needed to maintain the glide slope when a tailwind is present and more power is needed for a strong headwind. Aircraft pitch attitude is also an important indicator. A pitch attitude which is higher than normal is a good indicator of a strong headwind and vice versa. By observing the aircraft's approach parameters - rate of descent, power, and pitch attitude - the pilot can obtain a feel for the wind he is encountering. Being aware of the wind-correction angle needed to keep the localizer needle centered provides the pilot with an indication of wind direction. Comparing wind direction and velocity at the initial phases of the approach with the reported surface winds provides an excellent clue to the presence of shear before the phenomenon is actually encountered.

f. Utilize the Low Level Wind Shear Alert System (LLWSAS) at airports where it is available. LLWSAS consists of five or six anemometers around the periphery of the airport, which have their readouts automatically compared with the center field anemometer. If a wind vector difference of 15 knots or more exists between the center field anemometer and any peripheral anemometer, the tower will let the pilot know the winds from both locations. The pilot then may assess the potential for wind shear. An example of a severe wind shear alert would be the following: "Center field wind is 230 degrees at 7 knots; wind at the north end of Runway 35 is 180 degrees at 60 knots." In this case, a pilot departing on runway 35 would be taking off into an increasing tailwind condition that would result in significant losses of airspeed and, consequently, altitude.

6. AIRPLANE PERFORMANCE IN WIND SHEAR. The following information provides a basis for understanding the operational procedures recommended in this circular.

a. Power Compensation. Serious consequences may result on an approach when wind shear is encountered close to the ground after power adjustments have been already made to compensate for wind. Figures 2 and 3 illustrate the situations when power is applied or reduced to compensate for the change in aircraft performance caused by wind shear.

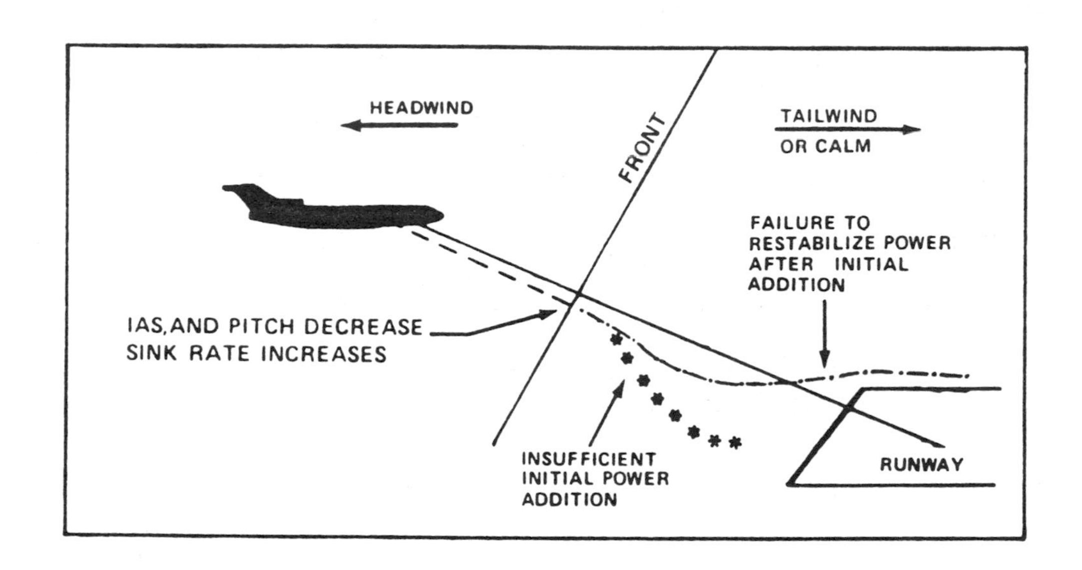

FIGURE 2. HEADWIND SHEARING TO TAILWIND OR CALM

(1) Consider an aircraft flying a 3° ILS on a stabilized approach at 140 knots indicated airspeed (IAS) with a 20-knot headwind. Assume that the aircraft encounters an instantaneous wind shear where the 20-knot headwind shears away completely. At that instant, several things will happen; the airspeed will drop from 140 to 120 knots, the nose will begin to pitch down, and the aircraft will begin to drop below the glide slope. The aircraft will then be both slow and low in a "power deficient" state. The pilot may then pull the nose up to a point even higher than before the shear in an effort to recapture the glide slope. This will aggravate the airspeed situation even further until the pilot advances the throttles and sufficient time elapses at the higher power setting for the engines to replenish the power deficiency. If the aircraft reaches the ground before the power deficiency is corrected, the landing will be short, slow, and hard. However, if there is sufficient time to regain the proper airspeed and glide slope before reaching the ground,

then the "double reverse" problem arises. This is because the throttles are set too high for a stabilized approach in a no-wind condition. So, as soon as the power deficiency is replenished, the throttles should be pulled back even further than they were before the shear (because power required for a 3° ILS in no wind is less than for a 20-knot headwind). If the pilot does not quickly retard the throttles, the aircraft will soon have an excess of power; i.e., it will be high and fast and may not be able to stop in the available runway length (Figure 2).

(2) When on approach in a tailwind condition that shears into a calm wind or headwind, the reverse of the previous statements is true. Initially, the IAS and pitch will increase and the aircraft will balloon above the glide slope. Power should initially be reduced to correct this condition or the approach may be high and fast with a danger of overshooting. However, after the initial power reduction is made and the aircraft is back on speed and glide slope, the "double reverse" again comes into play. An appropriate power increase will be necessary to restabilize in the headwind. If this power increase is not accomplished promptly, a high sink rate can develop and the landing may be short and hard (Figure 3). The double reverse problem arises primarily in downdraft and frontal passage shears. Other shears may require a consistent correction throughout the shear.

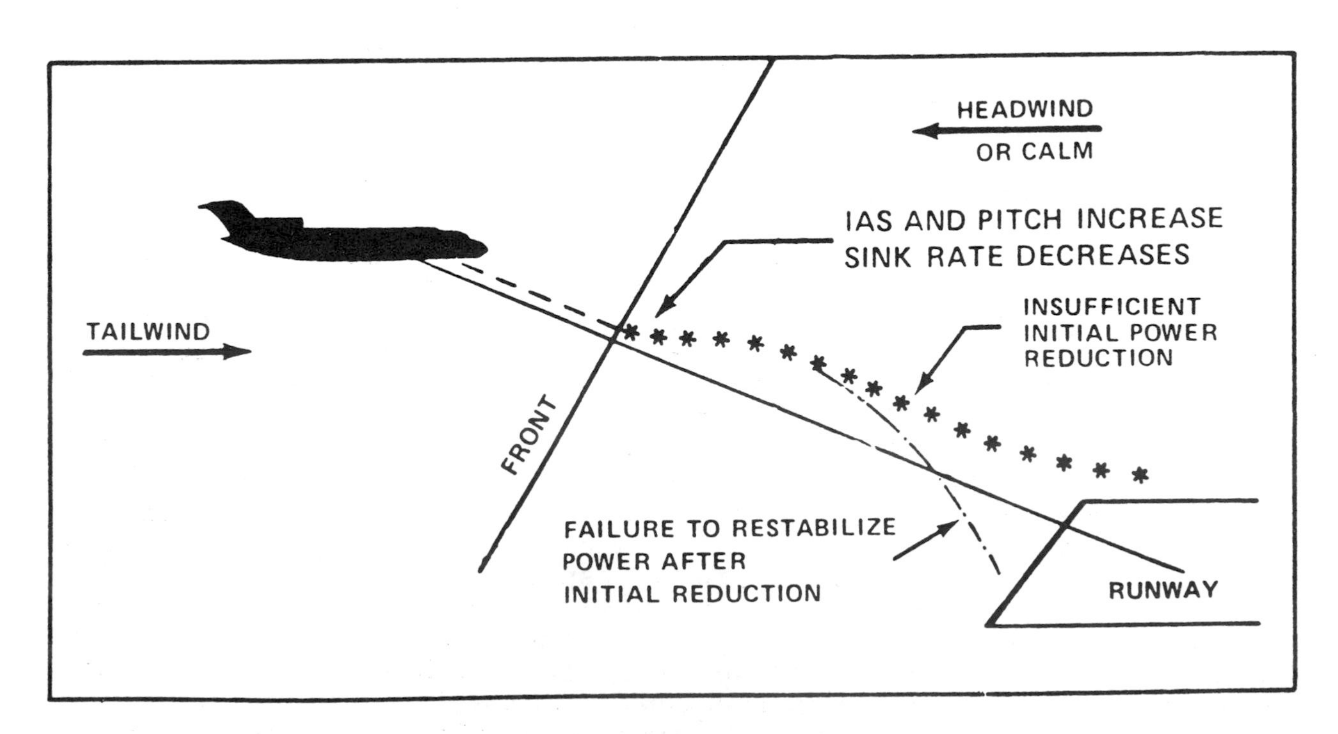

FIGURE 3. TAILWIND SHEARING TO HEADWIND OR CALM

(3) The classic thunderstorm "downburst cell" accident is illustrated in Figure 4. There is a strong downdraft in the center of the cell. There is often heavy rain in this vertical flow of air. As the vertical air flow nears the ground it turns 90 degrees and becomes a strong horizontal wind, flowing radially outward from the center. Point A in Figure 4 represents an aircraft which has not entered the cell's flow field. The aircraft is on speed and on glide slope. At Point B the aircraft encounters an increasing headwind. Its airspeed increases, and it balloons above the glide slope. Heavy rain may begin shortly. At Point C the "moment of truth" occurs. If the pilot does not fully appreciate the situation, he may attempt to regain the glide slope and lose excess airspeed by reducing power and pushing the nose down. Then in the short span of time between Points C and D the headwind ceases, a strong downdraft is entered and a tailwind begins increasing. The engines spool down, the airspeed drops below V_{ref} , and the sink rate becomes excessive. A missed approach initiated from this condition may not be successful. Note that a missed approach initiated at Point C (or sooner) would probably be successful since the aircraft is fast and high at this point. Note also that the pilot of an aircraft equipped with a groundspeed readout would see the telltale signs of a downburst cell shortly after Point B; i.e., rapidly increasing airspeed with decreasing groundspeed.

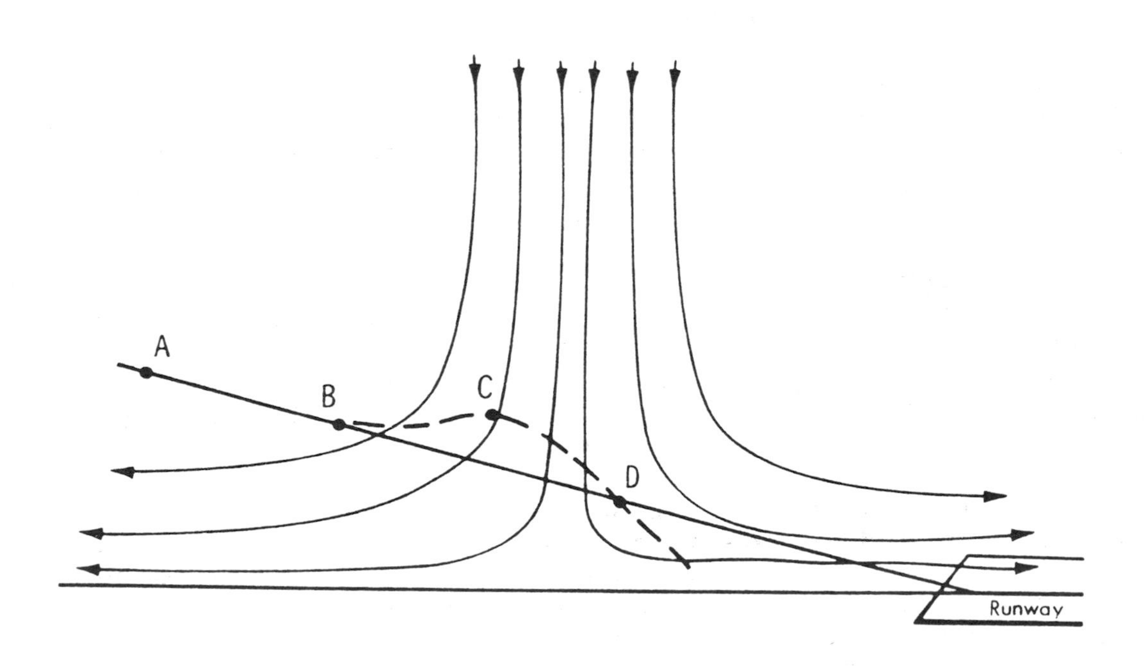

FIGURE 4. DOWNDRAFT SHEAR

b. Angle of Attack in a Downdraft. Downdrafts of falling air in a thunderstorm (sometimes called a "downburst") have gained attention in the last few years due to their role in wind shear accidents. When an airplane flies into a downdraft, the relative wind shifts so as to come down from above the horizon. This decreases angle of attack, which in turn decreases lift, and the airplane starts to sink rapidly. In order to regain the angle of attack necessary to support the weight of the airplane, the pitch attitude must be significantly increased. Such a pitch attitude may seem uncomfortably high to a pilot. However, a normal pitch attitude will result in a continued sink rate. The wing produces lift based on angle of attack - not pitch attitude. Caution should be observed when a pilot has traversed a downdraft and has pitched up sufficiently to stop the sink rate. If that pilot does not lower the nose of the airplane quickly when it exits the downdraft, the angle of attack will become too large and may approach the stall angle of attack. For these reasons, a flight director which senses angle of attack will be preferable to a flight director which calls for a fixed pitch attitude in a downdraft. However, even an angle of attack based flight director may become ineffective if it has an arbitrary pitch up command limit which is set too low (with respect to the downdraft).

c. Climb Performance. In the takeoff and landing configurations, jet transports climb best at speeds near V_2 and V_{ref} (reference speed with landing flaps), respectively. Retracting gear and flaps will even further improve climb performance. However, jet transport airplane manufacturers have pointed out that their airplanes still have substantial climb performance (generally in excess of 1000 fpm) at speeds down to stall warning or stickshaker speed, V_{ss}.

d. Energy Trade. There are only two ways an aircraft can correct for a wind shear. There can be an energy trade or a thrust change. Historically, most pilots have opted for a thrust change since they had no idea how much an energy trade would benefit them. Further information on the energy of flight, therefore, is warranted.

(1) The energy of motion (kinetic energy) is equal to $1/2\ MV^2$ where M is the mass of the airplane and V is the velocity. Kinetic energy is directly convertible to energy of vertical displacement (potential energy). More simply put, airspeed can be traded for altitude or vice versa. It is important to note that adding 10 percent to the speed of the airplane results in a 21 percent increase in kinetic energy because of the velocity being squared. This, of course, explains the concern over stopping an aircraft on the available runway when additional speed is added.

(2) The following table shows the altitude conversion capability of trading 10 or 20 knots of speed for altitude at various initial speeds. Independent of its mass, the capability of the aircraft to trade airspeed for altitude increases as its initial speed increases.

10 Knot Change From - To	Equivalent Altitude, Ft.	20 Knot Change From - To	Equivalent Altitude, Ft.
150-140	128	150-130	247
140-130	119	140-120	230
130-120	111	130-110	212
120-110	102	120-100	195
110-100	93	110-90	177

e. Trading Altitude for Speed. A pilot caught in low level wind shear who finds he is slower than the normal airspeed (even though he has gone to max power) could lower the nose and regain speed by trading away altitude. (This is trading potential energy for kinetic energy.) However, data shows that the penalty for doing this is severe; i.e., a large sink rate is built up and a great deal of altitude is lost for a relatively small increase in airspeed. Therefore, at low altitudes this alternative becomes undesirable. It is preferable to maintain the lower airspeed and rely on the airplane's climb performance at these lower speeds than to push the nose over and risk ground contact. Flight directors which attempt to maintain a given speed (such as V_2 + 10, etc.) will automatically call for trading altitude for airspeed if the airplane is below the proper airspeed. Cases have been observed in simulators where following such a flight director will result in the pilot flying the airplane into the ground. It is the pilot - not the flight director - who should decide if trading altitude for speed is desirable.

f. Trading Speed for Altitude. Conversely, a pilot caught in low level wind shear may pull the nose up and trade speed for altitude; i.e., trade kinetic energy for potential energy. If the speed is above V_2 or V_{ref} (as applicable), then this trade may well be desirable. If at or below V_2 or V_{ref}, such a trade should be attempted only in extreme circumstances. In doing so, the pilot is achieving a temporary increase in climb performance. After he has traded away all the airspeed he desires to trade, he will then be left with a permanent decrease in climb performance. In addition, if ground contact is still inevitable after the trade, there may be no airspeed margin left with which to flare in order to soften the impact. Wind shear simulations have shown, however, that in many cases trading airspeed for altitude (down to V_{ss}) prevented an accident, whereas maintaining V_{ref} resulted in ground impact.

g. Adding Speed for Wind Shear. The possibility of having to trade speed for altitude in wind shear makes it attractive to carry a great deal of extra speed. However, on landing, if the airspeed margin is not used up in the shear and the airplane touches down at an excessive speed, the airplane may not be able to stop on the available runway. It is generally agreed that if a speed margin in excess of 20 knots above V_{ref} appears to be required, the approach should not be attempted or continued.

h. Difficulties of Flying Near V_{SS}. Paragraph f stated that in simulations, wind shear "accidents" had been prevented by trading speed for altitude all the way down to V_{SS}. There are difficulties associated with flying at or near V_{SS} which should be recognized. These include:

(1) The pilot often does not know V_{SS}.

(2) The stickshaker mechanism may be miscalibrated (especially on older aircraft).

(3) The downdraft velocity may vary, which requires a change in pitch attitude to hold speed.

(4) It is hard to fly a precise airspeed in turbulence, which is often associated with wind shear.

(5) Turbulence might abruptly decrease the airspeed from V_{SS} to V_S.

(6) Pilots have historically had little training in maintaining flight at or near V_{SS}.

7. PROCEDURES FOR COPING WITH WIND SHEAR. The most important elements for the flightcrew in coping with a wind shear environment are the crew's awareness of an impending wind shear encounter and the crew's decision to avoid an encounter or to immediately respond if an encounter occurs.

a. Takeoff. If wind shear is expected on takeoff, the PIREPS and weather should be evaluated to determine if the phenomena can be safely traversed within the capability of the airplane. This is a judgment on the part of the pilot based on many factors. Wind shear is not something to be avoided at all costs, but rather to be assessed and avoided if severe. Some rules of thumb for coping with wind shear on takeoff follow:

(1) An increasing headwind or decreasing tailwind will cause an increase in indicated airspeed. If the wind shear is great enough, the aircraft will initially pitch up due to the increase in lift. The pilot should not trim the airplane at the initial high pitch attitude. After encountering the shear, if the wind remains constant, aircraft groundspeed will gradually decrease and indicated airspeed will return to its original value. This situation would normally lead to increased aircraft performance so it should not cause a problem if the pilot is aware of how this shear affects the aircraft.

(2) The worst situation on departure occurs when the aircraft encounters a rapidly increasing tailwind, decreasing headwind, and/or downdraft. Taking off under these circumstances would lead to a decreased performance condition. An increasing tailwind or decreasing headwind, when encountered, will cause a decrease in indicated airspeed. The aircraft will initially pitch down due to the decreased lift in proportion to the airspeed loss. After encountering the shear, if the wind remains constant, aircraft groundspeed will gradually increase and indicated airspeed will return to its original value.

(3) When the presence of severe wind shear is suspected for departure, the pilot should delay takeoff until conditions are more favorable.

(4) If the pilot judges the takeoff wind shear condition to be safe for departure, he should select the safest runway available considering runway length, wind directions, speed, and location of storm areas or frontal areas. He should execute a maximum power takeoff using the minimum acceptable flap position. After rotation, the pilot should maintain an airplane body angle which will result in an acceleration to V_2+25. This speed and takeoff flaps should be held through 1,000 feet AGL. Above 1,000 feet the normal noise abatement profile should be flown. If preflight planning shows that the airplane is runway length limited, or obstruction clearance is a problem, taking off into even a light shear using the V_2+25 procedure should not be attempted. This is because too much of the thrust available for climb is used for acceleration, resulting in the V_2+25 flight path falling below the engine-out flight path at V_2. This would give insufficient clearance for an obstacle in close proximity to the departure end of the runway.

(5) If severe wind shear is encountered on takeoff, the pilot should immediately confirm that maximum rated thrust is applied and trade the airspeed above V_2 (if any) for an increased rate of climb. Depending on the airplane's gross weight, pitch attitudes of 15 to 22 degrees are to be expected during this energy trade, especially if a downdraft is present. A sudden decrease in headwind will cause a loss in airspeed equal to the amount of wind shear. At this point, the pilot should quickly evaluate his airplane's performance in the shear. He/she should monitor airspeed and vertical velocity to ensure that an excessive rate of descent does not develop. If it becomes apparent that an unacceptable rate of descent cannot be prevented at V_2 speed or ground contact appears to be certain at the current descent rate, the pilot should gradually increase the airplane's pitch attitude to temporarily trade airspeed for climb capability to prevent further altitude loss. The trade should be terminated when stickshaker is encountered. The airplane should be held in an attitude that will maintain an airspeed just above the airspeed where the stickshaker was initially encountered. A general rule is to reduce pitch attitude very slightly when stickshaker is encountered. Further pitch reductions in the shear could result in a large descent rate. As the airplane departs the shear, the pilot should reduce the pitch attitude and establish a normal climb. In several recent wind shear accidents, the National Transportation Safety Board (NTSB) has found that the full performance capability of the airplane was not used following a severe wind shear encounter. Post accident studies have shown that, under similar circumstances, had flight techniques of an emergency nature (such as those outlined above) been used immediately, the airplane could have remained airborne and the accident averted.

b. Approach to Landing. Considerations involved in flying an approach and landing or go-around at an airport where wind shear is a factor are similar to those discussed for takeoff.

(1) When wind shear weather analysis, PIREPS, or an analysis of airplane performance indicates that a loss of airspeed will be experienced on an approach, the pilot should add to the V_{ref} speed as much airspeed as he expects to lose up to a maximum of $V_{ref}+20$. If the expected loss of airspeed exceeds 20 knots the approach should not be attempted unless the airplane is specially instrumented and the pilots are specially trained. The pilot should fly a stablized approach on a normal glidepath (using an electronic glidepath and the autopilot when available). In the shear when airspeed loss is encountered, a prompt and vigorous application of thrust is essential, keeping in mind that if airspeed has been previously added for the approach, the thrust application should be aimed at preventing airspeed loss below V_{ref}. An equally prompt and vigorous

reduction in thrust is necessary once the shear has been traversed and normal target speed and glidepath are reestablished to prevent exceeding desired values. Early recognition of the need for thrust is essential. Along with the thrust addition is a need for a noseup rotation to minimize departure below the glidepath. If the airplane is below 500 feet AGL and the approach becomes unstable, a go-around should be initiated immediately. Airspeed fluctations, sink rate, and glide slope deviation should be assessed as part of this decision.

(2) A pilot's chances of safely negotiating wind shear are better if he/she remains on instruments. Visual references through a rain-splattered windshield and reduced visibility may be inadequate to provide him/her with cues that would indicate deviation from the desired flightpath. At least one pilot should, therefore, maintain a continuous instrument scan until a safe landing is assured.

(3) Some autothrottle systems may not effectively respond to airspeed changes in a shear. Accordingly, the thrust should be monitored closely if autothrottles are used. Pilots should be alert to override the autothrottles if the response to increased thrust commands is too slow. Conversely, thrust levels should not be allowed to get too low during the late stages of an approach as this will increase the time needed to accelerate the engines.

(4) Should a go-around be required the pilot should initiate a normal go-around procedure, evaluate the performance of his airplane in the shear, and follow the procedures outlined in the takeoff section of this circular as applicable.

8. SUMMARY. The following summarizes the critical steps in coping with low level wind shear.

a. Be Prepared. Use all available forecasts and current weather information to anticipate wind shear. Also, make your own observations of thunderstorms, gust fronts and telltale indicators of wind direction and velocity available to pilots.

b. Giving and Requesting PIREPS on wind shear are essential. Request them and report anything you encounter. PIREPS should include:

(1) Location of shear encounter.

(2) Altitude of shear encounter.

(3) Airspeed changes experienced, with a clear statement of:

(i) the number of knots involved;

(ii) whether it was a gain or a loss of airspeed.

(4) Type of aircraft encountering the shear.

c. Avoid Known Areas of Severe Shear. When the weather and pilot reports indicate that severe wind shear is likely, delay your takeoff or approach.

d. Know Your Aircraft. Monitor the aircraft's power and flight parameters to detect the onset of a shear encounter. Know the performance limits of your particular aircraft so that they can be called upon in such an emergency situation.

e. Act Promptly. Do not allow a high sink rate to develop when attempting to recapture a glide slope or to maintain a given airspeed. When it appears that a shear encounter will result in a substantial rate of descent, promptly apply full power and arrest the descent with a noseup pitch attitude.

J. A. FERRARESE
Acting Director
Flight Standards Service

CHAPTER XII

1403. The presence of ice pellets at the surface is evidence that there

A—are thunderstorms in the area.
B—has been cold front passage.
C—is freezing rain at a higher altitude.

1431. One in-flight condition necessary for structural icing to form is

A—small temperature/dewpoint spread.
B—stratiform clouds.
C—visible moisture.

1433. Why is frost considered hazardous to flight?

A—Frost changes the basic aerodynamic shape of the airfoils, thereby decreasing lift.
B—Frost slows the airflow over the airfoils, thereby decreasing control effectiveness.
C—Frost spoils the smooth flow of air over the wings, thereby decreasing lifting capability.

2324. Ice pellets encountered during flight normally are evidence that

A— a warm front has passed.
B— there are thunderstorms in the area.
C— there exists a layer of warmer air above.

2360. Which situation would most likely result in freezing precipitation?

A— Rain falling from air which has a temperature of 32° F or less into air having a temperature of more than 32° F.
B— Rain falling from air which has a temperature of 0° C or less into air having a temperature of 0° C or more.
C— Rain falling from air which has a temperature of more than 32° F into air having a temperature of 32° F or less.

7095. What temperature condition is indicated if wet snow is encountered at your flight altitude?

A—The temperature is above freezing at your altitude.
B—The temperature is below freezing at your altitude.
C—You are flying from a warm air mass into a cold air mass.

7126. Which family of clouds is least likely to contribute to structural icing on an aircraft?

A—Low clouds.
B—High clouds.
C—Clouds with extensive vertical development.

7154. Which precipitation type normally indicates freezing rain at higher altitudes?

A—Snow.
B—Hail.
C—Ice pellets.

6280. When will frost most likely form on aircraft surfaces?

A— On clear nights with stable air and light winds.
B— On overcast nights with freezing drizzle precipitation.
C— On clear nights with convective action and a small temperature/dewpoint spread.

6282. What is a feature of supercooled water?

A— The water drop sublimates to an ice particle upon impact.
B— The unstable water drop freezes upon striking an exposed object.
C— The temperature of the water drop remains at 0 °C until it impacts a part of the airframe, then clear ice accumulates.

6320. What condition is necessary for the formation of structural icing in flight?

A— Supercooled water drops.
B— Water vapor.
C— Visible water.

CHAPTER XIII

1420. What clouds have the greatest turbulence?

A—Towering cumulus.
B—Cumulonimbus.
C—Nimbostratus.

1421. What cloud types would indicate convective turbulence?

A—Cirrus clouds.
B—Nimbostratus clouds.
C—Towering cumulus clouds.

1435. The conditions necessary for the formation of cumulonimbus clouds are a lifting action and

A—unstable air containing an excess of condensation nuclei.
B—unstable, moist air.
C—either stable or unstable air.

1437. Which weather phenomenon signals the beginning of the mature stage of a thunderstorm?

A—The appearance of an anvil top.
B—Precipitation beginning to fall.
C—Maximum growth rate of the clouds.

1438. What conditions are necessary for the formation of thunderstorms?

A—High humidity, lifting force, and unstable conditions.
B—High humidity, high temperature, and cumulus clouds.
C—High humidity, lifting force, and cumulus clouds.

1439. During the life cycle of a thunderstorm, which stage is characterized predominately by downdrafts?

A—Cumulus.
B—Dissipating.
C—Mature.

1440. Thunderstorms reach their greatest intensity during the

A—mature stage.
B—downdraft stage.
C—cumulus stage.

1441. Thunderstorms which generally produce the most intense hazard to aircraft are

A—squall line thunderstorms.
B—steady-state thunderstorms.
C—warm front thunderstorms.

1442. A nonfrontal, narrow band of active thunderstorms that often develop ahead of a cold front is a known as a

A—prefrontal system.
B—squall line.
C—shear line.

1443. If there is thunderstorm activity in the vicinity of an airport at which you plan to land, which hazardous atmospheric phenomenon might be expected on the landing approach?

A—St. Elmo's fire.
B—Wind shear turbulence.
C—Tornadoes.

2365. If airborne radar is indicating an extremely intense thunderstorm echo, this thunderstorm should be avoided by a distance of at least

A— 20 miles.
B— 25 miles.
C— 30 miles.

2366. Which statement is true regarding squall lines?

A— They are always associated with cold fronts.
B— They are slow in forming, but rapid in movement.
C— They are nonfrontal and often contain severe, steady-state thunderstorms.

2367. Which statement is true concerning squall lines?

A— They form slowly, but move rapidly.
B— They are associated with frontal systems only.
C— They offer the most intense weather hazards to aircraft.

2368. Select the true statement pertaining to the life cycle of a thunderstorm.

A— Updrafts continue to develop throughout the dissipating stage of a thunderstorm.
B— The beginning of rain at the Earth's surface indicates the mature stage of the thunderstorm.
C— The beginning of rain at the Earth's surface indicates the dissipating stage of the thunderstorm.

2372. During the life cycle of a thunderstorm, which stage is characterized predominately by downdrafts?

A— Mature.
B— Developing.
C— Dissipating.

2373. What minimum distance should exist between intense radar echoes before any attempt is made to fly between these thunderstorms?

A— 20 miles.
B— 30 miles.
C— 40 miles.

7152. What enhances the growth rate of precipitation?

A—Advective action.
B—Upward currents.
C—Cyclonic movement.

3240. Which type of cloud is associated with violent turbulence and a tendency toward the production of funnel clouds?

A— Cumulonimbus mamma.
B— Standing lenticular.
C— Altocumulus castellanus.

3245. Hail will most likely be encountered

A— beneath the anvil cloud of a large cumulonimbus.
B— during the dissipating stage of the cumulonimbus.
C— above the cumulonimbus cloud well above the freezing level.

3253. Which statement is true regarding the effect of fronts on soaring conditions?

A— A slow-moving front provides the strongest lift.
B— Excellent soaring conditions usually exist in the cold air ahead of a warm front.
C— Frequently the air behind a cold front provides excellent soaring for several days.

3254. The conditions most favorable to wave formation over mountainous areas are a layer of

A— unstable air at mountaintop altitude and a wind of at least 20 MPH blowing across the ridge.
B— stable air at mountaintop altitude and a wind of at least 20 MPH blowing across the ridge.
C— moist, unstable air at mountaintop altitude and a wind of less than 5 MPH blowing across the ridge.

3256. Select the true statement concerning thermals.

A— Strong thermals have proportionately increased sink in the air between them.
B— Thermals will not develop unless the Sun's rays strike the Earth at a vertical angle.
C— A thermal invariably remains directly above the surface area from which it developed.

6281. What is the result when water vapor changes to the liquid state while being lifted in a thunderstorm?

A— Latent heat is released to the atmosphere.
B— Latent heat is transformed into pure energy.
C— Latent heat is absorbed from the surrounding air by the water droplet.

6288. Convective clouds which penetrate a stratus layer can produce which threat to instrument flight?

A— Freezing rain.
B— Clear air turbulence.
C— Embedded thunderstorms.

6298. What is indicated by the term "embedded thunderstorms"?

A— Severe thunderstorms are embedded in a squall line.
B— Thunderstorms are predicted to develop in a stable air mass.
C— Thunderstorms are obscured by other types of clouds.

6301. Atmospheric pressure changes due to a thunderstorm will be at the lowest value

A— during the downdraft and heavy rain showers.
B— when the thunderstorm is approaching.
C— immediately after the rain showers have stopped.

6302. Why are downdrafts in a mature thunderstorm hazardous? They

A— are kept cool by cold rain which tends to accelerate the downward velocity.
B— converge toward a central location under the storm after striking the surface.
C— become warmer than the surrounding air and reverse into an updraft before reaching the surface.

6303. What is a difference between an air mass thunderstorm and a steady–state thunderstorm?

A— Air mass thunderstorms produce precipitation which falls outside of the updraft.
B— Air mass thunderstorm downdrafts and precipitation retard and reverse the updrafts.
C— Steady–state thunderstorms are associated with local surface heating.

6304. Which type storms are most likely to produce funnel clouds or tornadoes?

A— Air mass thunderstorms.
B— Cold front or squall line thunderstorms.
C— Storms associated with icing and supercooled water.

6309. Which type cloud is associated with violent turbulence and a tendency toward the production of funnel clouds?

A— Cumulonimbus mamma.
B— Standing lenticular.
C— Stratocumulus.

6310. A clear area in a line of thunderstorm echoes on a radar scope indicates

A— the absence of clouds in the area.
B— an area of no convective turbulence.
C— an area where precipitation drops are not detected.

CHAPTER XIV

1385. Which weather conditions should be expected beneath a low-level temperature inversion layer when the relative humidity is high?

A—Smooth air, poor visibility, fog, haze, or low clouds.
B—Light wind shear, poor visibility, haze, and light rain.
C—Turbulent air, poor visibility, fog, low stratus type clouds, and showery precipitation.

1446. If the temperature/dewpoint spread is small and decreasing, and the temperature is 62 °F, what type weather is most likely to develop?

A—Freezing precipitation.
B—Thunderstorms.
C—Fog or low clouds.

2350. Fog produced by frontal activity is a result of saturation due to

A— nocturnal cooling.
B— adiabatic cooling.
C— evaporation of precipitation.

2374. Which in-flight hazard is most commonly associated with warm fronts?

A— Advection fog.
B— Radiation fog.
C— Precipitation-induced fog.

2376. A situation most conducive to the formation of advection fog is

A— a light breeze moving colder air over a water surface.
B— an airmass moving inland from the coastline during the winter.
C— a warm, moist airmass settling over a cool surface under no-wind conditions.

2377. Advection fog has drifted over a coastal airport during the day. What may tend to dissipate or lift this fog into low stratus clouds?

A— Nighttime cooling.
B— Surface radiation.
C— Wind 15 knots or stronger.

2378. What lifts advection fog into low stratus clouds?

A— Nighttime cooling.
B— Dryness of the underlying land mass.
C— Surface winds of approximately 15 knots or stronger.

2379. In what ways do advection fog, radiation fog, and steam fog differ in their formation or location?

A— Radiation fog is restricted to land areas; advection fog is most common along coastal areas; steam fog forms over a water surface.
B— Advection fog deepens as windspeed increases up to 20 knots; steam fog requires calm or very light wind; radiation fog forms when the ground or water cools the air by radiation.
C— Steam fog forms from moist air moving over a colder surface; advection fog requires cold air over a warmer surface; radiation fog is produced by radiational cooling of the ground.

2380. With respect to advection fog, which statement is true?

A— It is slow to develop, and dissipates quite rapidly.
B— It forms almost exclusively at night or near daybreak.
C— It can appear suddenly during day or night and it is more persistent than radiation fog.

7155. Which weather condition can be expected when moist air flows from a relatively warm surface to a colder surface?

A—Increased visibility.
B—Convective turbulence due to surface heating.
C—Fog.

7156. Fog is usually prevalent in industrial areas because of

A—atmospheric stabilization around cities.
B—an abundance of condensation nuclei from combustion products.
C—increased temperatures due to industrial heating.

7162. Which conditions are favorable for the formation of radiation fog?

A—Moist air moving over colder ground or water.
B—Cloudy sky and a light wind moving saturated, warm air over a cool surface.
C—Clear sky, little or no wind, small temperature/dewpoint spread, and over a land surface.

3206. When warm air moves over a cold lake, what weather phenomenon is likely to occur on the leeward side of the lake?

A— Fog.
B— Showers.
C— Cloudiness.

3246. One condition necessary for the formation of fog is

A— calm air.
B— visible moisture.
C— high relative humidity.

6252. What characterizes a ground-based inversion?

A— Convection currents at the surface.
B— Cold temperatures.
C— Poor visibility.

6293. Which condition produces weather on the lee side of a large lake?

A— Warm air flowing over a colder lake may produce fog.
B— Cold air flowing over a warmer lake may produce advection fog.
C— Warm air flowing over a cool lake may produce rain showers.

6305. When advection fog has developed, what may tend to dissipate or lift the fog into low stratus clouds?

A— Temperature inversion.
B— Wind stronger than 15 knots.
C— Surface radiation.

6306. Which conditions are necessary for the formation of upslope fog?

A— Moist, stable air being moved over gradually rising ground by a wind.
B— A clear sky, little or no wind, and 100 percent relative humidity.
C— Rain falling through stratus clouds and a 10- to 25-knot wind moving the precipitation up the slope.

6307. How are haze layers cleared or dispersed? By

A— convective mixing in cool night air.
B— wind or the movement of air.
C— evaporation similar to the clearing of fog.

CHAPTER XV

1458. Which type weather briefing should a pilot request, when filing a flight plan, if no preliminary weather information has been received?

A—Outlook briefing.
B—Abbreviated briefing.
C—Standard briefing.

1460. To *update* a previous weather briefing, a pilot should request

A—an abbreviated briefing.
B—a standard briefing.
C—an outlook briefing.

2358. During an approach, the most important and most easily recognized means of being alerted to possible wind shear is monitoring the

A— amount of trim required to relieve control pressures.
B— heading changes necessary to remain on the runway centerline.
C— power and vertical velocity required to remain on the proper glidepath.

7128. Where can wind shear associated with a thunderstorm be found? (Choose the most complete answer.)

A—In front of the thunderstorm cell (anvil side) and on the right side of the cell.
B—In front of the thunderstorm cell and directly under the cell.
C—On all sides of the thunderstorm cell and directly under the cell.

7133. Which is a characteristic of low-level wind shear as it relates to frontal activity?

A—With a warm front, the most critical period is before the front passes the airport.
B—With a cold front, the most critical period is just before the front passes the airport.
C—Turbulence will always exist in wind-shear conditions.

7139. Which procedure is recommended if a pilot should unintentionally penetrate embedded thunderstorm activity?

A—The pilot should reverse aircraft heading or proceed toward an area of known VFR conditions.
B—Reduce airspeed to maneuvering speed and maintain a constant altitude.
C—Set power for recommended turbulence penetration airspeed and attempt to maintain a level-flight attitude.

7143. What is an important characteristic of wind shear?

A—It is an atmospheric condition that is associated exclusively with zones of convergence.
B—The Coriolis phenomenon in both high- and low-level airmasses is the principal generating force.
C—It is an atmospheric condition that may be associated with a low-level temperature inversion, a jetstream, or a frontal zone.

7244. What is the expected duration of an individual microburst?

A—Two minutes with maximum winds lasting approximately 1 minute.
B—One microburst may continue for as long as 2 to 4 hours.
C—seldom longer than 15 minutes from the time the burst strikes the ground until dissipation.

7245. Maximum downdrafts in a microburst encounter may be as strong as

A—8,000 feet per minute.
B—7,000 feet per minute.
C—6,000 feet per minute.

7246. An aircraft that encounters a headwind of 45 knots, within a microburst, may expect a total shear across the microburst of

A—40 knots.
B—80 knots.
C—90 knots.

7247. (Refer to figure 12.) If involved in a microburst encounter, which aircraft positions will the most severe downdraft occur?

A—4 and 5.
B—2 and 3.
C—3 and 4.

7248. (Refer to figure 12.) When penetrating a microburst, which aircraft will experience an increase in performance without a change in pitch or power?

A—3
B—2.
C—1.

7249. (Refer to figure 12.) The aircraft in position 3 will experience which effect in a microburst encounter?

A—Decreasing headwind.
B—Increasing tailwind.
C—Strong downdraft.

7250. (Refer to figure 12.) What effect will a microburst encounter have upon the aircraft in position 4?

A—Strong tailwind.
B—Strong updraft.
C—Significant performance increase.

7251. (Refer to figure 12.) How will the aircraft in position 4 be affected by a microburst encounter?

A—Performance increasing with a tailwind and updraft.
B—Performance decreasing with a tailwind and downdraft.
C—Performance decreasing with a headwind and downdraft.

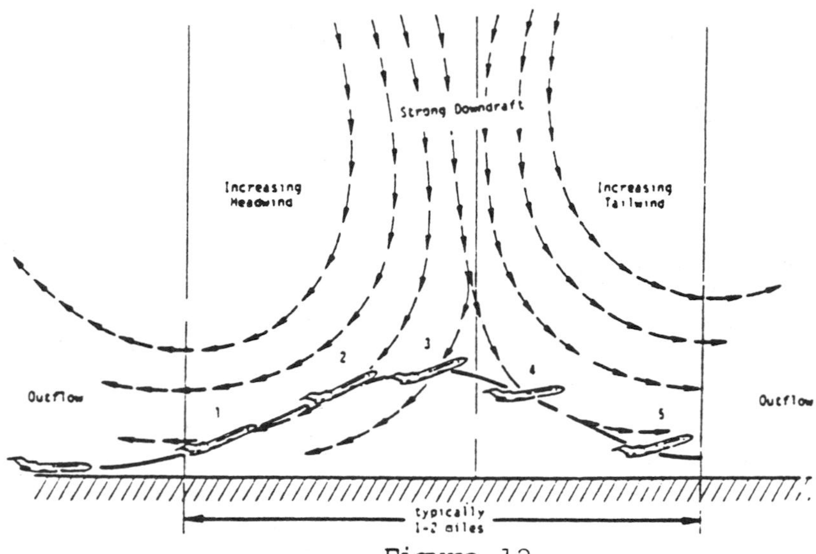

Figure 12.

6238. Which is a definition of "severe wind shear"?

A— Any rapid change of horizontal wind shear in excess of 25 knots; vertical shear excepted.
B— Any rapid change in wind direction or velocity which causes airspeed changes greater than 15 knots or vertical speed changes greater than 500 feet per minute.
C— Any change of airspeed greater than 20 knots which is sustained for more than 20 seconds or vertical speed changes in excess of 100 feet per minute.

6239. Doppler wind measurements indicate that the windspeed change a pilot may expect when flying through the peak intensity of a microburst is approximately

A— 15 knots.
B— 25 knots.
C— 45 knots.

6311. When flying over the top of a severe thunderstorm, the cloud should be overflown by at least

A— 1,000 feet for each 10 knots windspeed.
B— 2,500 feet.
C— 500 feet above any moderate to severe turbulence layer.

CHAPTER XVII

2400. Flight Service Stations in the conterminous 48 United States having voice capability on VOR's or radiobeacons (NDB's) broadcast

A— AIRMET's and SIGMET's at 15 minutes past the hour and each 15 minutes thereafter as long as they are in effect.
B— AIRMET's and Non-convective SIGMET's at 15 minutes and 45 minutes past the hour for the first hour after issuance.
C— hourly weather reports at 15 and 45 minutes past each hour for those reporting stations within approximately 150 NM of the broadcast stations.

7076. What minimum weather conditions must be forecast for your ETA at an airport that has only a VOR approach with standard alternate minimums, for the airport to be listed as an alternate on the IFR flight plan?

A—800-foot ceiling and 1 statute mile visibility.
B—800-foot ceiling and 2 statute miles visibility.
C—1,000-foot ceiling and visibility to allow descent from MEA, approach, and landing under basic VFR.

7077. Is an alternate airport required for an IFR flight to ATL (Atlanta Hartsfield) if the proposed ETA is 1930Z?

ATL FT AMD 1 161615 1630Z C20 BKN 6RW-VRBL 20 SCT C40 BKN OCNL 4RW- 3315. 18Z 40 SCT C100 BKN 3110 OCNL C20 BKN 6RW-CHC C10 OVC 3TRW 3310G20 AFT 21Z. 09Z MVFR CIG R..

A—An alternate is required because the ceiling could fall below 2,000 feet within 2 hours before to 2 hours after the ETA.
B—An alternate is not required because the ceiling and visibility are forecast to remain at or above 1,000 feet and 3 miles, respectively.
C—An alternate is not required because the ceiling and visibility are forecast to be at or above 2,000 feet and 3 miles within 1 hour before to 1 hour after the ETA.

7078. What minimum conditions must exist at the destination airport to avoid listing an alternate airport on an IFR flight plan when a standard instrument approach procedure is available?

A—From 2 hours before to 2 hours after ETA, forecast ceiling 2,000, and visibility 2-1/2 miles.
B—From 2 hours before to 2 hours after ETA, forecast ceiling 3,000, and visibility 3 miles.
C—From 1 hour before to 1 hour after ETA, forecast ceiling 2,000, and visibility 3 miles.

7079. Under what condition are you not required to list an alternate airport on an IFR flight plan for an IFR flight in a helicopter?

A—When the ceiling is forecast to be at least 1,000 feet above the lowest of the MEA, MOCA, or initial approach altitude within 2 hours of your ETA at the destination airport.
B—When the weather reports or forecasts indicate the ceiling and visibility will be at least 2,000 feet and 3 miles for 1 hour before to 1 hour after your ETA at the destination airport.
C—When the ceiling is forecast to be at least 1,000 feet above the lowest of the MEA, MOCA, or initial approach altitude and the visibility is 2 miles more than the minimum landing visibility within 2 hours of your ETA at the destination airport.

7080. What standard minimums are required to list an airport as an alternate on an IFR flight plan if the airport has VOR approach only?

A—Ceiling and visibility at ETA, 800 feet and 2 miles, respectively.
B—Ceiling and visibility from 2 hours before until 2 hours after ETA, 800 feet and 2 miles, respectively.
C—Ceiling and visibility at ETA, 600 feet and 2 miles, respectively.

7081. What are the minimum weather conditions that must be forecast to list an airport as an alternate when the airport has no approved instrument approach procedure?

A—The ceiling and visibility at ETA, 2,000 feet and 3 miles, respectively.
B—The ceiling and visibility from 2 hours before until 2 hours after ETA, 2,000 feet and 3 miles, respectively.
C—The ceiling and visibility at ETA must allow descent from MEA, approach, and landing, under basic VFR.

7082. What minimum weather conditions must be forecast for your ETA at an airport that has a precision approach procedure, with standard alternate minimums, in order to list it as an alternate for the IFR flight?

A—600-foot ceiling and 2 statute miles visibility at your ETA.
B—600-foot ceiling and 2 statute miles visibility from 2 hours before to 2 hours after your ETA.
C—800-foot ceiling and 2 statute miles visibility at your ETA.

6255. Which area or areas of the Northern Hemisphere experience a generally east to west movement of weather systems?

A— Arctic only.
B— Arctic and subtropical.
C— Subtropical only.

6322. Which weather condition is associated with the "Intertropical Convergence Zone" near the Equator?

1—Permanent low-pressure area at the surface.
2—Air rising, frequent thunderstorms, and heavy rains.
3—Development of tropical cyclones which may grow into hurricanes or typhoons.

6323. Which weather condition is present when the tropical storm is upgraded to a hurricane?

1—Highest windspeed, 100 knots or more.
2—A clear area or hurricane eye has formed.
3—Sustained winds of 65 knots or more.

6324. What is the general direction of movement of a hurricane located in the Caribbean or Gulf of Mexico region?

1—Northwesterly curving to northeasterly.
2—Westerly, until encountering land, then easterly.
3—Counterclockwise over open water, then dissipating outward over land.

6332. Summer thunderstorms in the Arctic region will generally move

A— northeast to southwest in polar easterlies.
B— southwest to northeast with the jetstream flow.
C— directly north to south with the low-level polar airflow.

6333. Which arctic flying hazard is caused when a cloud layer of uniform thickness overlies a snow or ice covered surface?

A— Ice fog.
B— Whiteout.
C— Blowing snow.

PART B

1426. What is indicated when a current ***SIGMET*** forecasts embedded thunderstorms?

A—Severe thunderstorms are embedded within a squall line.
B—Thunderstorms are dissipating.
C—Thunderstorms are obscured by massive cloud layers.

1454. Individual forecasts for specific routes of flight can be obtained from which weather service?

A—Transcribed Weather Broadcasts (TWEB's).
B—Terminal Forecasts.
C—Area Forecasts.

1455. Transcribed Weather Broadcasts (TWEB's) may be monitored by tuning the appropriate radio receiver to certain

A—airport advisory frequencies.
B—VOR and NDB frequencies.
C—FSS frequencies only.

1463. (Refer to figure 12.) Which of the reporting stations have VFR weather?

A—All.
B—INK, BOI, and JFK.
C—INK, BOI, and LAX.

1464. ***Ceiling*** is defined as the height above the Earth's surface of the

A—lowest reported obscuration and the highest layer of clouds reported as overcast.
B—lowest layer of clouds reported as broken or overcast and not classified as thin.
C—lowest layer of clouds reported as scattered, broken, or thin.

1465. (Refer to figure 12.) The wind direction and velocity at ***JFK*** is from

A—180° at 4 knots.
B—180° at 40 knots.
C—300° at 6 knots.

1466. (Refer to figure 12.) What are the wind conditions at Wink, Texas (***INK***)?

A—Calm.
B—010° at 2 knots gusting to 18 knots.
C—110° at 12 knots gusting to 18 knots.

1467. (Refer to figure 12.) The remarks section for ***MDW*** has RF2 and RB12 listed. These two entries mean

A—rain and fog have reduced visibility to 2 miles and rain began at 1812Z.
B—rain and fog are obscuring two-tenths of the sky and rain began at 1812Z.
C—freezing rain has reduced visibility to 2 miles and the barometer has risen .12" Hg.

1468. (Refer to figure 12.) What are the current conditions depicted for Chicago Midway Airport (***MDW***)?

A—Sky partially obscured, measured ceiling 700 overcast, visibility 1-1/2, heavy rain, fog.
B—Thin overcast, measured 700 ceiling overcast, visibility 1-1/2, heavy rain, fog.
C—Sky partially obscured, measured ceiling 700 overcast, visibility 11, occasionally 2, with rain and heavy fog.

1469. (Refer to figure 13.) According to the briefing, the most ideal time to launch balloons is

A—as soon after 1300Z as possible.
B—at 1500Z when the ground will be partially shaded.
C—at 2000Z when there is enough wind for cross-country.

1470. (Refer to figure 13.) According to the briefing, good balloon weather will begin to deteriorate

A—soon after 1300Z as the wind starts to increase.
B—about 1500Z when the lower scattered clouds begin to form.
C—at 2000Z due to sharp increase in wind conditions.

1471. (Refer to figure 13.) What effect do the clouds mentioned in the briefing have on soaring conditions?

A—All thermals stop at the base of the clouds.
B—Thermals persist to the tops of the clouds at 25,000 feet.
C—The scattered clouds indicate thermals at least to the tops of the lower clouds.

1472. (Refer to figure 13.) At what time will thermals begin to form?

A—Between 1300Z and 1500Z while the sky is clear.
B—By 1500Z (midmorning) when scattered clouds begin to form.
C—About 2000Z (early afternoon) when the wind begins to increase.

1473. (Refer to figure 14.) The base and tops of the overcast layer reported by a pilot are

A—1,200 feet MSL and 5,500 feet MSL.
B—5,500 feet AGL and 7,200 feet MSL.
C—7,200 feet MSL and 8,900 feet MSL.

1474. (Refer to figure 14.) The wind and temperature at 12,000 feet MSL as reported by a pilot are

A—009° at 121 MPH and 90 °F.
B—010° at 20 knots and 1 °F.
C—090° at 21 knots and -9 °C.

1475. (Refer to figure 14.) If the terrain elevation is 614 feet MSL, what is the height above ground level of the base of the ceiling?

A—586 feet AGL.
B—1,200 feet AGL.
C—6,586 feet AGL.

1476. (Refer to figure 14.) The intensity of turbulence reported by a pilot is

A—light.
B—moderate.
C—severe.

1477. (Refer to figure 14.) The intensity and type of icing reported by a pilot is

A—light rime.
B—light to moderate clear.
C—moderate rime.

1478. Which weather reports and forecasts are most important for local area balloon operations?

A—Area Forecasts, Winds Aloft Forecasts, and Radar Summary Charts.
B—Terminal Forecasts, Winds Aloft Forecasts, and Surface Analysis Charts.
C—Terminal Forecasts, Winds Aloft Forecasts, and Surface Aviation Weather Reports.

1479. From which primary source should information be obtained regarding expected weather at the estimated time of arrival at your destination?

A—Low-Level Prognostic Chart.
B—Weather Depiction Chart.
C—Terminal Forecast.

1480. (Refer to figure 15.) What ceiling is forecast for ***GAG*** between 1600Z and 0100Z?

A—6,000 scattered, chance 10,000 broken.
B—10,000 scattered, chance 2,500 scattered.
C—10,000 broken, chance 5,000 broken.

1481. (Refer to figure 15.) What wind conditions are expected at ***HBR*** at 1600Z?

A—Calm.
B—115° at 15 knots.
C—300° at 10 knots.

1482. (Refer to figure 15.) What is the outlook for weather conditions at ***MLC***?

A—Ceilings 2,000 - 3,000 feet with southerly winds.
B—Ceiling 700 feet, sky obscured, visibility 1/2 mile in the thundershowers.
C—Marginal VFR, low ceilings, and thundershowers.

1483. (Refer to figure 15.) According to the Terminal Forecast for ***OKC***, the cold front should pass between

A—1800Z and 2100Z.
B—2100Z and 0200Z.
C—1515Z and 1800Z.

1484. (Refer to figure 15.) The wind condition in the Terminal Forecast 6-hour categorical outlook for ***PNC*** is for

A—velocities of 25 knots or stronger.
B—a wind shift from south to northwest.
C—the wind to change from a gusty condition to calm.

1485. (Refer to figure 15.) When is the wind forecast to shift at ***TUL***?

A—1500Z.
B—By 2300Z.
C—Between 2300Z and 0900Z the next day.

1486. (Refer to figure 15.) According to the Terminal Forecast for ***OKC***, the cold front should pass

A—by 2100Z.
B—between 1515Z and 1800Z.
C—after 0900Z the next day.

1487. (Refer to figure 15.) What is the outlook for weather conditions at ***TUL***?

A—VFR and windy.
B—Chance of ceilings 3,000 feet broken, visibility 5 miles in thundershowers.
C—VFR with winds 320° at 15 knots with gusts to 25 knots.

1488. To best determine general forecast weather conditions over several states, the pilot should refer to

A—Area Forecasts.
B—prognostic charts.
C—weather maps.

1489. (Refer to figure 16.) What is the forecast ceiling and visibility for the entire area from 2300Z through 0500Z?

A—500 feet to less than 1,000 feet and 1 mile to less than 3 miles.
B—1,000 to 3,000 feet and 3 to 5 miles.
C—3,000 feet or greater and 3 to 5 miles or greater.

1490. To determine the freezing level and areas of probable icing aloft, the pilot should refer to the

A—Radar Summary Chart.
B—Weather Depiction Chart.
C—Area Forecast.

1491. The section of the Area Forecast entitled SIG CLDS AND WX contains a summary of

A—cloudiness and weather significant to flight operations broken down by states or other geographical areas.
B—forecast sky cover, cloud tops, visibility, and obstructions to vision along specific routes.
C—AIRMET's and SIGMET's still in effect at the time of issue.

1492. (Refer to figure 16.) What hazards are forecast in the Area Forecast for ***TN***, ***AL***, and the coastal waters?

A—Thunderstorms with severe or greater turbulence, severe icing, and low-level wind shear.
B—Moderate rime icing above the freezing level to 10,000 feet.
C—Moderate turbulence from 25,000 to 38,000 feet.

1493. (Refer to figure 16.) What type obstructions to vision, if any, are forecast for the entire area from 2300Z until 0500Z the next day?

A—None of any significance, VFR is forecast.
B—Visibility 3 to 5 miles in fog.
C—Visibility below 3 miles in fog over south-central Texas.

1494. (Refer to figure 16.) What sky condition and type obstructions to vision are forecast for all the area except ***TN*** from 1040Z until 2300Z?

A—Ceilings 3,000 to 5,000 feet broken, visibility 3 to 5 miles in fog.
B—8,000 feet scattered to clear except visibility below 3 miles in fog until 1500Z over south-central Texas.
C—Generally ceilings 3,000 to 8,000 feet to clear with visibilities sometimes below 3 miles in fog.

1495. To obtain a continuous transcribed weather briefing, including winds aloft and route forecasts for a cross-country flight, a pilot should monitor a

A—Transcribed Weather Broadcast (TWEB) on a low-frequency radio receiver.
B—VHF radio receiver tuned to an Automatic Terminal Information Service (ATIS) frequency.
C—regularly scheduled weather broadcast on a VOR frequency.

1496. What information is contained in a convective SIGMET?

A—Tornadoes, embedded thunderstorms, and hail 3/4 inch or greater in diameter.
B—Severe icing, severe turbulence, or widespread dust storms lowering visibility to less than 3 miles.
C—Surface winds greater than 40 knots or thunderstorms equal to or greater than video integrator processor (VIP) levels.

1497. SIGMET's are issued as a warning of weather conditions hazardous to which aircraft?

A—Small aircraft only.
B—Large aircraft only.
C—All aircraft.

1498. AIRMET's are issued as a warning of weather conditions particularly hazardous to which aircraft?

A—Small single-engine aircraft.
B—Large multiengine aircraft.
C—All aircraft.

1499. Which in-flight advisory would contain information on severe icing?

A—Convective SIGMET.
B—SIGMET.
C—AIRMET.

1500. (Refer to figure 17.) What wind is forecast for ***STL*** at 6,000 feet?

A—210° at 13 knots.
B—230° at 25 knots.
C—232° at 5 knots.

1501. (Refer to figure 17.) What wind is forecast for ***STL*** at 18,000 feet?

A—230° at 56 knots.
B—235° at 06 knots.
C—235° at 06 gusting to 16 knots.

1502. (Refer to figure 17.) Determine the wind and temperature aloft forecast for ***DEN*** at 30,000 feet.

A—023° at 53 knots, temperature 47 °C.
B—230° at 53 knots, temperature –47 °C.
C—235° at 34 knots, temperature –7 °C.

1503. (Refer to figure 17.) Determine the wind and temperature aloft forecast for 3,000 feet at ***MKC***.

A—050° at 7 knots, temperature missing.
B—360° at 5 knots, temperature –7 °C.
C—360° at 50 knots, temperature +7 °C.

1504. (Refer to figure 17.) What wind is forecast for ***STL*** at 34,000 feet?

A—007° at 30 knots.
B—073° at 6 knots.
C—230° at 106 knots.

1505. What values are used for Winds and Temperatures Aloft Forecasts?

A—Magnetic direction and knots.
B—Magnetic direction and miles per hour.
C—True direction and knots.

1506. When the term ***light and variable*** is used in reference to a Winds and Temperatures Aloft Forecast, the coded group and windspeed is

A—0000 and less than 7 knots.
B—9900 and less than 5 knots.
C—9999 and less than 10 knots.

1507. (Refer to figure 18.) What is the status of the front that extends from New Mexico to Indiana?

A—Stationary.
B—Occluded.
C—Retreating.

1508. (Refer to figure 18.) The IFR weather in eastern Texas is due to

A—intermittent rain.
B—fog.
C—dust devils.

1509. (Refer to figure 18.) Of what value is the Weather Depiction Chart to the pilot?

A—For determining general weather conditions on which to base flight planning.
B—For a forecast of cloud coverage, visibilities, and frontal activity.
C—For determining frontal trends and air mass characteristics.

1510. (Refer to figure 18.) What is the ceiling in southeast New Mexico?

A—400 feet.
B—4,000 feet.
C—6,000 feet.

1511. (Refer to figure 18.) What weather phenomenon is causing IFR conditions along the coast of Oregon and California?

A—Squall line activity.
B—Low ceilings.
C—Heavy rain showers.

1512. (Refer to figure 18.) According to the Weather Depiction Chart, the weather for a flight from central Arkansas to southeast Alabama is

A—broken clouds at 2,500 feet.
B—visibility from 3 to 5 miles.
C—broken to scattered clouds at 25,000 feet.

1513. Radar weather reports are of special interest to pilots because they indicate

A—large areas of low ceilings and fog.
B—location of precipitation along with type, intensity, and trend.
C—location of broken to overcast clouds.

1514. What information is provided by the Radar Summary Chart that is not shown on other weather charts?

A—Lines and cells of hazardous thunderstorms.
B—Ceilings and precipitation between reporting stations.
C—Types of precipitation between reporting stations.

1515. (Refer to figure 19, area A.) What is the direction and speed of movement of the radar return?

A—020° at 20 knots.
B—East at 15 knots.
C—Northeast at 22 knots.

1516. (Refer to figure 19, area C.) What type of weather is occurring in the radar return?

A—Continuous rain.
B—Thunderstorms and rain showers.
C—Rain showers increasing in intensity.

1517. (Refer to figure 19, area D.) What is the direction and speed of movement of the radar return?

A—Southeast at 30 knots.
B—Northeast at 20 knots.
C—West at 30 knots.

1518. (Refer to figure 19, area D.) The maximum cloud top is

A—2,000 feet.
B—20,000 feet.
C—30,000 feet.

1519. (Refer to figure 19, area B.) What does the dashed line enclose?

A—Areas of heavy rain.
B—Severe weather watch area.
C—Areas of hail 1/4 inch in diameter.

1520. (Refer to figure 20.) How are Significant Weather Prognostic Charts used by a pilot?

A—For overall planning.
B—For determining areas to avoid (freezing levels and turbulence).
C—For analyzing frontal activity and cloud coverage.

1521. (Refer to figure 20.) Interpret the weather symbol depicted in southern California on the 12-hour Significant Weather Prognostic Chart.

A—Moderate turbulence, surface to 18,000 feet.
B—Thunderstorm tops at 18,000 feet.
C—Base of clear air turbulence, 18,000 feet.

1522. (Refer to figure 20.) What weather is forecast for the Gulf Coast area just ahead of the cold front during the first 12 hours?

A—Ceiling 1,000 to 3,000 feet and/or visibility 3 to 5 miles.
B—IFR with moderate or greater turbulence over the coastal areas.
C—Thunderstorm cells moving northeastward ahead of the front.

1523. (Refer to figure 20.) The band of weather associated with the cold front in the western states is forecast to move

A—east at 30 knots.
B—northeast at 12 knots.
C—southeast at 30 knots.

1524. (Refer to figure 20.) At what altitude is the freezing level over southeastern Oklahoma on the 24-hour Significant Weather Prognostic Chart?

A—4,000 feet.
B—8,000 feet.
C—10,000 feet.

1525. In addition to the standard briefing, what additional information should be asked of the weather briefer in order to evaluate soaring conditions?

A—The upper soundings to determine the thermal index at all soaring levels.
B—Dry adiabatic rate of cooling to determine the height of cloud bases.
C—Moist adiabatic rate of cooling to determine the height of cloud tops.

1526. When telephoning a weather briefing facility for preflight weather information, pilots should

A—identify themselves as pilots.
B—tell the number of hours they have flown within the preceding 90 days.
C—state the number of occupants on board and the color of the aircraft.

1527. When telephoning a weather briefing facility for preflight weather information, pilots should state

A—the full name and address of the pilot in command.
B—the intended route, destination, and type of aircraft.
C—the radio frequencies to be used.

1528. When telephoning a weather briefing facility for preflight weather information, pilots should state

A—the full name and address of the formation commander.
B—that they possess a current pilot certificate.
C—whether they intend to fly VFR only.

INK SA 1854 CLR 15 106/77/63/1112G18/000
BOI SA 1854 150 SCT 30 181/62/42/1304/015
LAX SA 1852 7 SCT 250 SCT 6HK 129/60/59/2504/991
MDW RS 1856 -X M7 OVC 11/2R-F 990/63/61/3205/980/RF2 RB12
JFK RS 1853 W5 X 1/2F 180/68/64/1804/006/R04RVR22V30 TWR VSBY 1/4

FIGURE 12.—Surface Aviation Weather Reports.

This is a telephone weather briefing from the Dallas FSS for a local operation of gliders and lighter-than-air at Caddo Mills, Texas (about 30 miles east of Dallas). The briefing is at 13Z.

"There are no adverse conditions reported or forecast for today."

'A weak low pressure over the Texas Panhandle and eastern New Mexico is causing a weak southerly flow over the area.'

'Current weather here at Dallas is clear, visibility 12 miles, temperature 70, dew point 48, wind south 5 knots, altimeter 29 point 78.'

'By 15Z we should have a few scattered clouds at 5 thousand AGL, with higher scattered cirrus at 25 thousand MSL. After 20Z the wind should pick up to about 15 knots from the south.'

'The winds aloft are: 3 thousand 170 at 7, temperature 20; 6 thousand 200 at 18, temperature 14; 9 thousand 210 at 22, temperature 8; 12 thousand 225 at 27, temperature 0; 18 thousand 240 at 30, temperature -7."

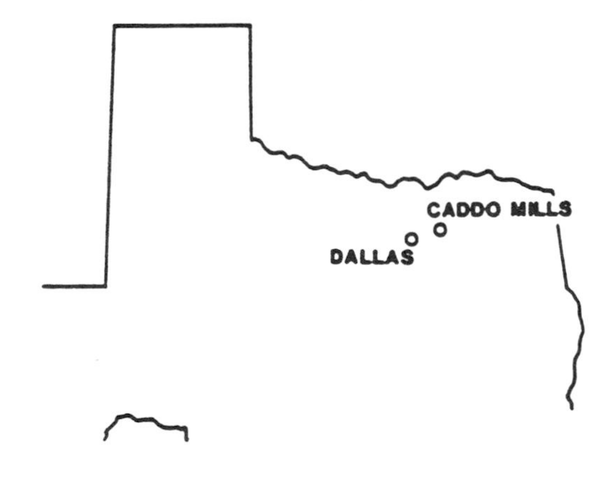

FIGURE 13.—Telephone Weather Briefing.

```
UA /OV OKC-TUL /TM 18ØØ /FL 12Ø /TP BE9Ø /SK Ø12 BKN Ø55 /
/Ø72 OVC Ø89 /CLR ABV /TA -9/WV Ø921/TB MDT Ø55-Ø72 /ICG LGT-MDT
CLR Ø72-Ø89.
```

FIGURE 14.—Pilot Weather Report.

```
OK FT Ø11447

GAG FT Ø11515 1ØØ SCT 25Ø SCT 261Ø. 16Z 6Ø SCT C1ØØ BKN 3315G22 CHC C5Ø BKN
   5TRW. Ø1Z 25Ø SCT 3515G25. Ø9Z VFR WIND..

HBR FT Ø11515 C12Ø BKN 25Ø BKN 3Ø1Ø. 17Z 1ØØ SCT C25Ø BKN 3215G25 CHC C3Ø BKN
   3TRW. ØØZ 25Ø SCT 3515G25. Ø92 VFR WIND..

MLC FT Ø11515 C2Ø BKN 1815 BKN OCNL SCT. 2ØZ C3Ø BKN 1815G22 CHC C2Ø BKN
   1TRW. Ø3Z C3Ø BKN 2Ø15 CHC C7 X 1/2TRW+G4Ø. Ø9Z MVFR CIG TRW..

OKC FT Ø11515 C12 BKN 14Ø BKN 1815G28 LWR BKN V SCT. 18Z C3Ø BKN 25Ø BKN
   2315G25 LWR BKN OCNL SCT CHC C7 X 1/2TRW+G4Ø. 21Z CFP 1ØØ SCT C25Ø BKN
   3315G25 CHC C3Ø BKN 5TRW-. Ø2Z 1ØØ SCT 25Ø SCT 3515G25. Ø9Z VFR WIND..

PNC FT Ø11515 C1ØØ BKN 25Ø BKN 181Ø. 16Z CFP 2Ø SCT C1ØØ BKN 3115 SCT V BKN. ØØZ
   25Ø SCT 3515G25. Ø9Z VFR WIND..

TUL FT Ø11515 C2Ø BKN 1915G22. 19Z C3Ø BKN 1815G25 CHC 3TRW. 23Z CFP C1ØØ BKN
   25Ø BKN 3215G25 CHC C3Ø BKN 5TRW. Ø9Z VFR WIND..
```

FIGURE 15.—Terminal Forecast.

DFWH FA 041040
HAZARDS VALID UNTIL 042300
OK TX AR LA TN MS AL AND CSTL WTRS
FLT PRCTNS...TURBC...TN AL AND CSTL WTRS
...ICG...TN
...IFR...TX
TSTMS IMPLY PSBL SVR OR GTR TURBC SVR ICG AND LLWS
NON MSL HGTS NOTED BY AGL OR CIG
THIS FA ISSUANCE INCORPORATES THE FOLLOWING AIRMETS STILL IN EFFECT...NONE.

DFWS FA 041040
SYNOPSIS VALID UNTIL 050500
AT 11Z RDG OF HI PRES ERN TX NWWD TO CNTRL CO WITH HI CNTR OVR ERN TX. BY 05Z HI CNTR MOVS TO CNTRL LA.

DFWI FA 041040
ICING AND FRZLVL VALID UNTIL 042300
TN
FROM SLK TO HAT TO MEM TO ORD TO SLK
OCNL MDT RIME ICGIC ABV FRZLVL TO 100. CONDS ENDING BY 17Z.
FRZLVL 80 CHA SGF LINE SLPG TO 120 S OF A IAH MAF LINE.

DFWT FA 041040
TURBC VALID UNTIL 042300
TN AL AND CSTL WTRS
FROM SLK TO FLO TO 90S MOB TO MEI TO BUF TO SLK
OCNL MDT TURBC 250-380 DUE TO JTSTR. CONDS MOVG SLOLY EWD AND CONTG BYD 23Z.

DFWC FA 041040
SGFNT CLOUD AND WX VALID UNTIL 042300...OTLK 042300-050500
IFR...TX
FROM SAT TO FSX TO BRO TO MOV TO SAT
VSBY BLO 3F TIL 15Z.
OK AR TX LA MS AL AND CSTL WTRS
80 SCT TO CLR EXCP VSBY BLO 3F TIL 15Z OVR PTNS S CNTRL TX.
OTLK...VFR.
TN
CIGS 30-50 BKN 100 VSBYS OCNLY 3-5F BCMG AGL 40-50 SCT TO CLR BY 19Z. OTLK...VFR.

FIGURE 16.—Area Forecast.

FD WBC 151745
BASED ON 151200Z DATA
VALID 1600Z FOR USE 1800-0300Z. TEMPS NEG ABV 24000

FT	3000	6000	9000	12000	18000	24000	30000	34000	39000
ALS			2420	2635-08	2535-18	2444-30	245945	246755	246862
AMA		2714	2725+00	2625-04	2531-15	2542-27	265842	256352	256762
DEN			2321-04	2532-08	2434-19	2441-31	235347	236056	236262
HLC		1707-01	2113-03	2219-07	2330-17	2435-30	244145	244854	245561
MKC	0507	2006+03	2215-01	2322-06	2338-17	2348-29	236143	237252	238160
STL	2113	2325+07	2332+02	2339-04	2356-16	2373-27	239440	730649	731960

FIGURE 17.—Winds and Temperatures Aloft Forecast.

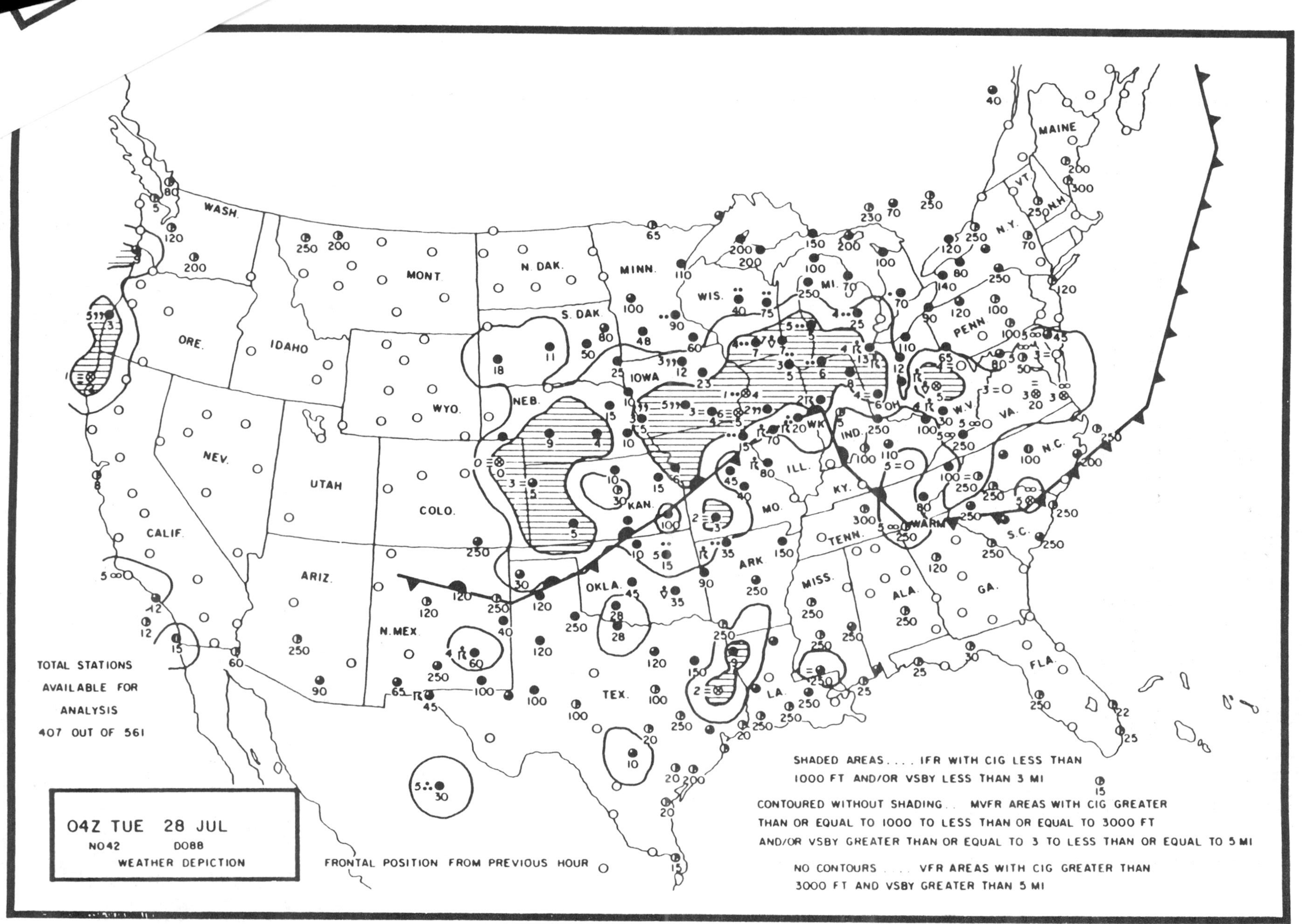

FIGURE 18.—Weather Depiction Chart.

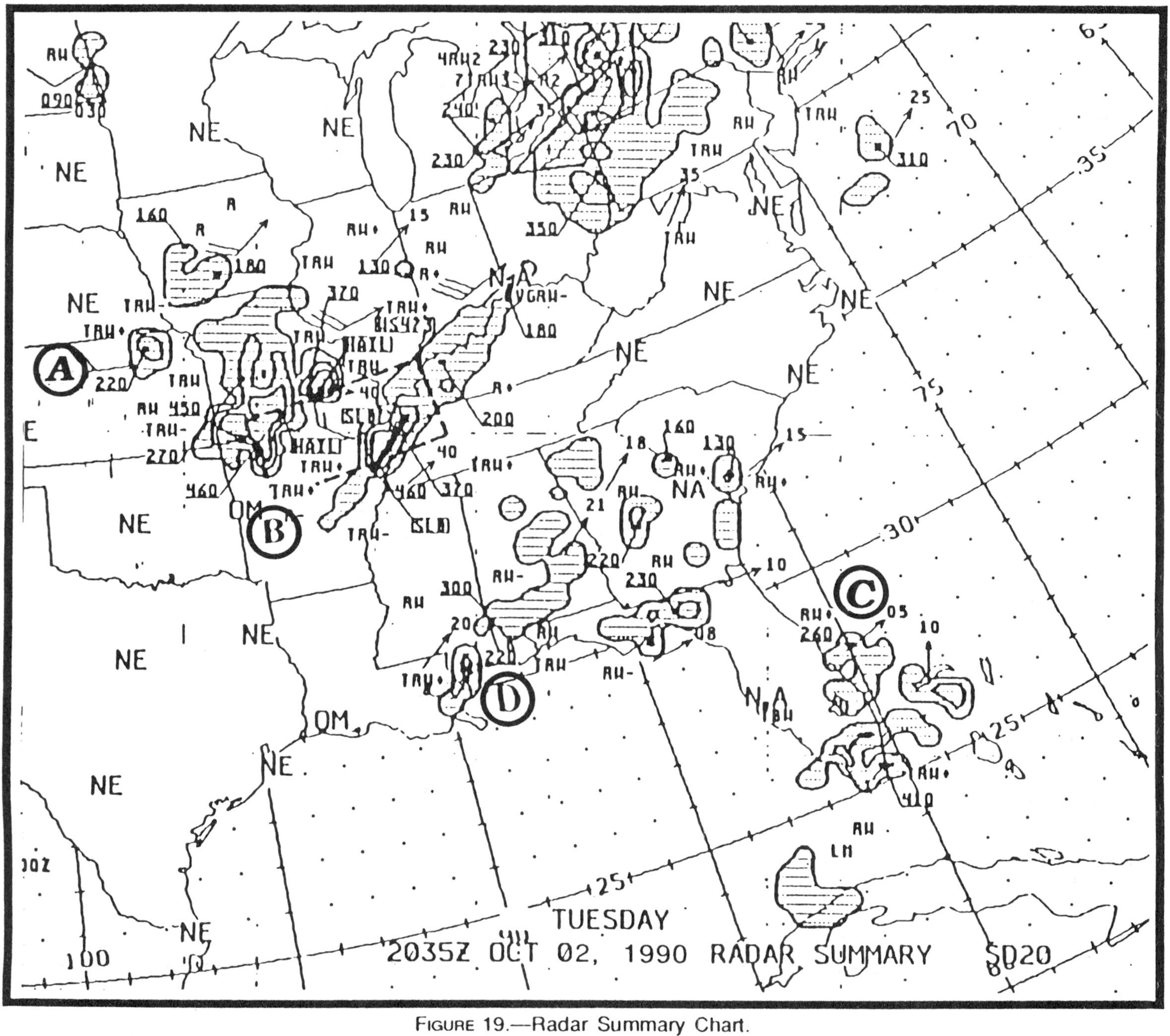

FIGURE 19.—Radar Summary Chart.

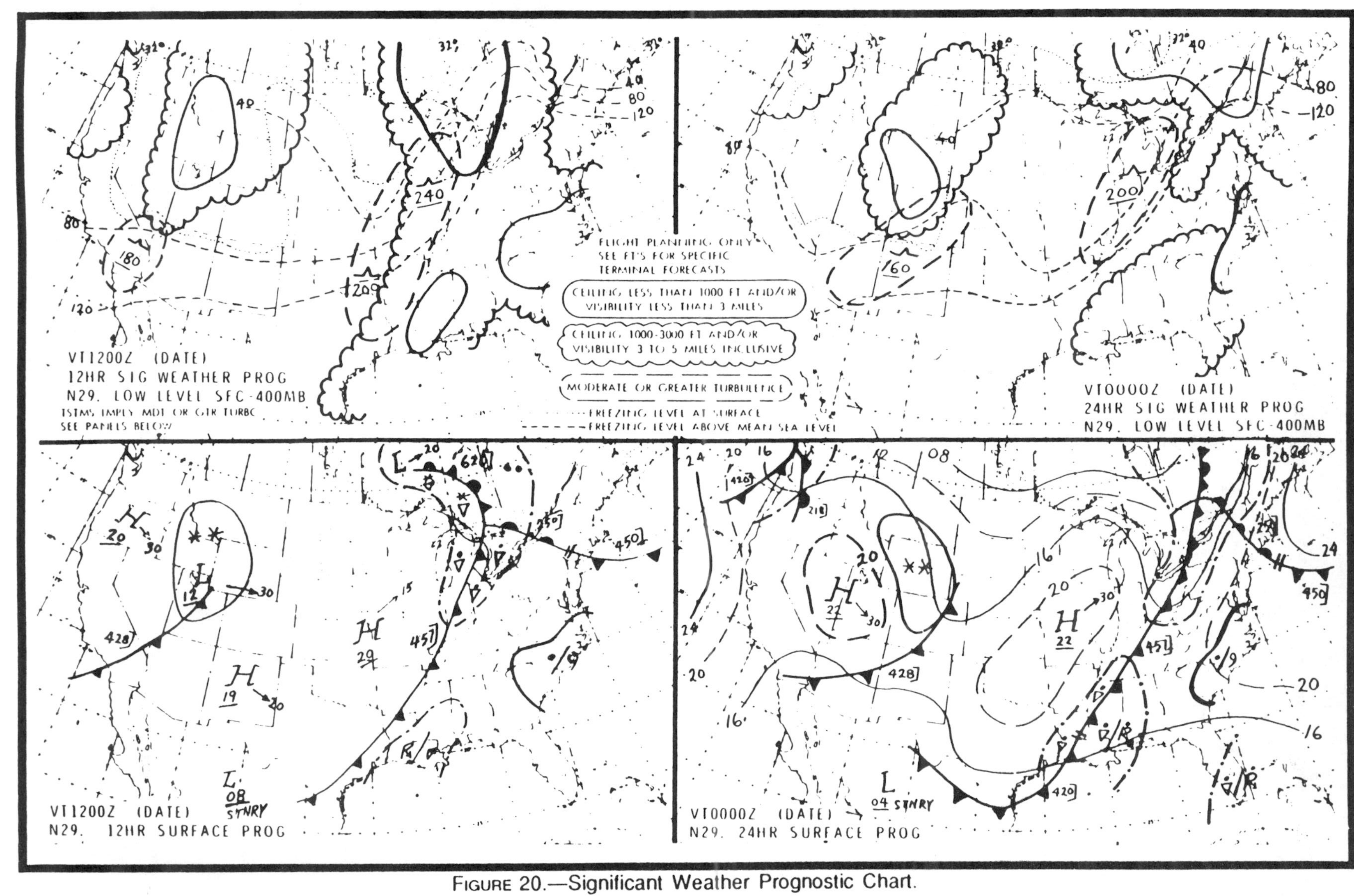

FIGURE 20.—Significant Weather Prognostic Chart.

2331. Refer to the excerpt from a surface weather report:

ABC ...194/89/45/2115/993...

At approximately what altitude AGL should bases of convective-type cumuliform clouds be expected? (Use most accurate method.)

A— 4,400 feet.
B— 8,900 feet.
C— 10,000 feet.

2398. During preflight preparation, weather report forecasts which are not routinely available at the local service outlet (FSS or WSFO) can best be obtained by means of the

A— request/reply service.
B— air route traffic control center.
C— pilot's automatic telephone answering service.

2399. The most current en route and destination flight information for an instrument flight should be obtained from

A— the FSS.
B— the ATIS broadcast.
C— Notices to Airmen (Class II).

2401. Transcribed Weather Broadcasts (TWEB's) may be monitored by tuning the appropriate radio receiver to certain

A— NDB frequencies only.
B— VOR and NDB frequencies.
C— FSS communications frequencies.

2402. The remarks section of the hourly aviation weather report contains the following coded information:

RADAT 87045

What is the meaning of this information?

A— Radar echoes with tops at 45,000 feet were observed on the 087 radial of the VORTAC.
B— A pilot reported thunderstorms 87 DME miles distance on the 045 radial of the VORTAC.
C— Relative humidity was 87 percent and the freezing level (0° C) was at 4,500 feet MSL.

2403. What is meant by the entry in the remarks section of this Surface Aviation Weather Report for "BOI"?

BOI SP 1854 -X M7 OVC 1 1/2R+F 990/63/61/ 3205/980/RF2 RB12

A— Rain and fog obscuring two-tenths of the sky; rain began at 1912.
B— Rain and fog obscuring two-tenths of the sky; rain began at 1812.
C— Runway fog, visibility 2 miles; base of the rain clouds 1,200 feet.

2404. The station originating the following weather report has a field elevation of 3,500 feet MSL. If the sky cover is one continuous layer, what is its thickness?

M5 OVC 1/2HK 173/73/72/0000/00/2/OVC 75

A— 2,500 feet.
B— 3,500 feet.
C— 4,000 feet.

2405. What wind conditions would you anticipate when squalls are reported at your destination?

A— Rapid variations in windspeed of 15 knots or more between peaks and lulls.
B— Peak gusts of at least 35 knots combined with a change in wind direction of 30° or more.
C— Sudden increases in windspeed of at least 15 knots to a sustained speed of 20 knots or more for at least 1 minute.

2406. What significant cloud coverage is reported by a pilot in this SA?

MOB...M9 OVC 2LF 131/44/43/3212/991/UA/OV 15NW MOB 1355/SK OVC 025/045 OVC 090

A— Three separate overcast layers exist with bases at 2,500, 7,500, and 13,500 feet.
B— The top of lower overcast is 2,500 feet; base and top of second overcast layer is 4,500 and 9,000 feet, respectively.
C— The base of second overcast layer is 2,500 feet; top of second overcast layer is 7,500 feet; base of third layer is 13,500 feet.

2407. To best determine observed weather conditions between weather reporting stations, the pilot should refer to

A— pilot reports.
B— Area Forecasts.
C— prognostic charts.

2408. Which is true concerning this radar weather report for OKC?

OKC 1934 LN 8TRW+/+ 86/40 164/60 199/115 15W 2425 MT 570 AT 159/65 2 INCH HAIL RPRTD THIS ECHO

A— There are three cells with tops at 11,500, 40,000, and 60,000 feet.
B— The line of cells is moving 080° with winds reported up to 40 knots.
C— The maximum top of the cells is 57,000 feet located 65 NM south-southeast of the station.

2409. What is the meaning of the term "MVFR", as used in the categorical outlook portion of Terminal and Area Forecasts?

A— A ceiling less than 1,000 feet, and/or visibility less than 3 miles.
B— A ceiling of 1,000 to 3,000 feet, and/or visibility of 3 to 5 miles.
C— A ceiling of 3,000 to 5,000 feet, and visibility of 5 to 7 miles.

2410. The contraction "WND" in the 6-hour categorical outlook in the Terminal Forecast means that the wind during that period is forecast to be

A— 15 to 20 knots.
B— less than 25 knots.
C— 25 knots or stronger.

2411. Which statement pertaining to a Terminal Forecast is true?

A— The term "WND" in the categorical outlook implies surface winds are forecast to be 10 knots or greater.
B— The term "CHC TRW VCNTY" in the remarks section pertains to an area within a 5-mile radius of the airport.
C— The term "VFR CIGS ABV 100" in the categorical outlook implies ceilings above 10,000 feet and visibility more than 5 miles.

2412. The absence of a visibility entry in a Terminal Forecast specifically implies that the surface visibility is expected to be more than

A— 3 miles.
B— 6 miles.
C— 10 miles.

2413. Terminal Forecasts are issued how many times a day and cover what period of time?

A— Three times daily and are valid for 24 hours including a 6-hour categorical outlook.
B— Four times daily and are valid for 18 hours including a 4-hour categorical outlook.
C— Six times daily and are valid for 12 hours with an additional 6-hour categorical outlook.

2414. Which information is contained in the "HAZARDS" section of the Area Forecast?

A— A summary of general weather conditions for the entire region covered in the Area Forecast.
B— A brief list of weather phenomena that meet "AIRMET" and/or "SIGMET" criteria and the location of each.
C— A brief summary of significant weather and clouds that do not meet "AIRMET" and/or "SIGMET" criteria.

2415. The section of the Area Forecast entitled "SGFNT CLOUD AND WX" contains a

A— summary of forecast sky cover, cloud tops, visibility, and obstructions to vision along specific routes.
B— summary of only those weather systems producing liquid or frozen precipitation, fog, thunderstorms, or IFR ceilings.
C— summary of sky condition, cloud heights, visibility, weather and/or obstructions to visibility, and surface winds of 30 knots or more.

2416. In the "Hazards and Flight Precautions" section of an Area Forecast, what is indicated by the forecast term — "FLT PRCTNS...IFR...TX AR LA MS TN AL AND CSTL WTRS"?

A— IFR conditions which meet in-flight advisory criteria are forecast for the states listed.
B— Each state and geographic area listed is reporting ceilings and visibilities below VFR minimums.
C— IFR conditions, turbulence, and icing are all forecast within the valid period for the listed states.

2417. In the Area Forecast, what method is used to describe the location of each icing phenomenon?

A— VOR points outline the affected area(s) within the designated FA boundary, but not beyond the FA boundary.
B— State names and portions of states, such as northwest and south-central, are used to outline each affected area.
C— VOR points are used to outline the area of icing, including VOR points outside the designated FA boundary, if necessary.

2418. What single reference contains information regarding expected frontal movement, turbulence, and icing conditions for a specific area?

A— Area Forecast.
B— Surface Analysis Chart.
C— Weather Depiction Chart.

2419. The National Aviation Weather Advisory Unit prepares "FA's" for the contiguous United States

A— twice each day.
B— three times each day.
C— every 6 hours unless significant changes in weather require it more often.

2420. Which forecast provides specific information concerning expected sky cover, cloud tops, visibility, weather, and obstructions to vision in a route format?

A— Area Forecast.
B— Terminal Forecast.
C— Transcribed Weather Broadcast.

2421. To obtain a continuous transcribed weather briefing including winds aloft and route forecasts for a cross-country flight, a pilot could monitor

A— a TWEB on a low-frequency radio receiver.
B— the regularly scheduled weather broadcast on a VOR frequency.
C— a high-frequency radio receiver tuned to En Route Flight Advisory Service.

2422. "SIGMET's" are issued as a warning of weather conditions which are hazardous

A— to all aircraft.
B— particularly to heavy aircraft.
C— particularly to light airplanes.

2423. Which correctly describes the purpose of convective SIGMET's (WST)?

A— They consist of an hourly observation of tornadoes, significant thunderstorm activity, and large hailstone activity.
B— They contain both an observation and a forecast of all thunderstorm and hailstone activity. The forecast is valid for 1 hour only.
C— They consist of either an observation and a forecast or just a forecast for tornadoes, significant thunderstorm activity, or hail greater than or equal to 3/4 inch in diameter.

2424. What values are used for Winds Aloft Forecasts?

A— True direction and MPH.
B— True direction and knots.
C— Magnetic direction and knots.

2425. On a Surface Analysis Chart, the solid lines that depict sea level pressure patterns are called

A— isobars.
B— isotachs.
C— isotherms.

2426. Dashed lines on a Surface Analysis Chart, if depicted, indicate that the pressure gradient is

A— weak.
B— strong.
C— unstable.

2427. The chart that provides a ready means of locating observed frontal positions and pressure centers is the

A— surface analysis.
B— constant pressure.
C— weather depiction.

2428. On a Surface Analysis Chart, close spacing of the isobars indicates

A— weak pressure gradient.
B— strong pressure gradient.
C— strong temperature gradient.

2429. The Surface Analysis Chart depicts

A— frontal locations and expected movement, pressure centers, cloud coverage, and obstructions to vision at the time of chart transmission.
B— actual frontal positions, pressure patterns, temperature, dewpoint, wind, weather, and obstructions to vision at the valid time of the chart.
C— actual pressure distribution, frontal systems, cloud heights and coverage, temperature, dewpoint, and wind at the time shown on the chart.

2430. Which provides a graphic display of both VFR and IFR weather?

A— Surface Weather Map.
B— Radar Summary Chart.
C— Weather Depiction Chart.

2431. When total sky cover is few or scattered, the height shown on the weather depiction chart is the

A— top of the lowest layer.
B— base of the lowest layer.
C— base of the highest layer.

2432. What information is provided by the Radar Summary Chart that is not shown on other weather charts?

A— Lines and cells of hazardous thunderstorms.
B— Ceilings and precipitation between reporting stations.
C— Areas of cloud cover and icing levels within the ATP clouds.

2433. Which weather chart depicts conditions forecast to exist at a specific time in the future?

A— Freezing Level.
B— Weather Depiction.
C— 12-Hour Significant Weather Prog.

2434. What weather phenomenon is implied within an area enclosed by small scalloped lines on a "HI LVL SIG WX PROG" chart?

A— Cirriform clouds, light to moderate turbulence, and icing.
B— Cumulonimbus clouds, icing, and moderate or greater turbulence.
C— Cumuliform or standing lenticular clouds, moderate to severe turbulence, and icing.

2435. The U.S. High Level Significant Weather Prognostic Chart forecasts significant weather for what airspace?

A— 18,000 feet to 45,000 feet.
B— 24,000 feet to 45,000 feet.
C— 24,000 feet to 63,000 feet.

2436. What is the upper limit of the Low-Level Significant Weather Prognostic Chart?

A— 30,000 feet.
B— 24,000 feet.
C— 18,000 feet.

2437. (Refer to figure 7.) According to the lifted index and K index shown on the Stability Chart, which area of the United States would have the least satisfactory conditions for thermal soaring on the day of the soundings?

A— Southeastern.
B— North-central.
C— Western seaboard.

2438. A freezing level panel of the "Composite Moisture Stability" chart is an analysis of

A— forecast freezing level data from surface observations.
B— forecast freezing level data from upper air observations.
C— observed freezing level data from upper air observations.

2439. The difference found by subtracting the temperature of a parcel of air theoretically lifted from the surface to 500 MB and the existing temperature at 500 MB, is called the

A— lifted index.
B— negative index.
C— positive index.

2440. Hatching on a constant pressure analysis chart indicates

A— hurricane eye.
B— windspeed 70 knots to 110 knots.
C— windspeed 110 knots to 150 knots.

2441. What flight planning information can a pilot derive from Constant Pressure Charts?

A— Winds and temperatures aloft.
B— Clear air turbulence and icing conditions.
C— Frontal systems and obstructions to vision aloft.

2442. From which of the following can the observed temperature, wind, and temperature/dewpoint spread be determined at a specified altitude?

A— Stability Charts.
B— Winds Aloft Forecasts.
C— Constant Pressure Charts.

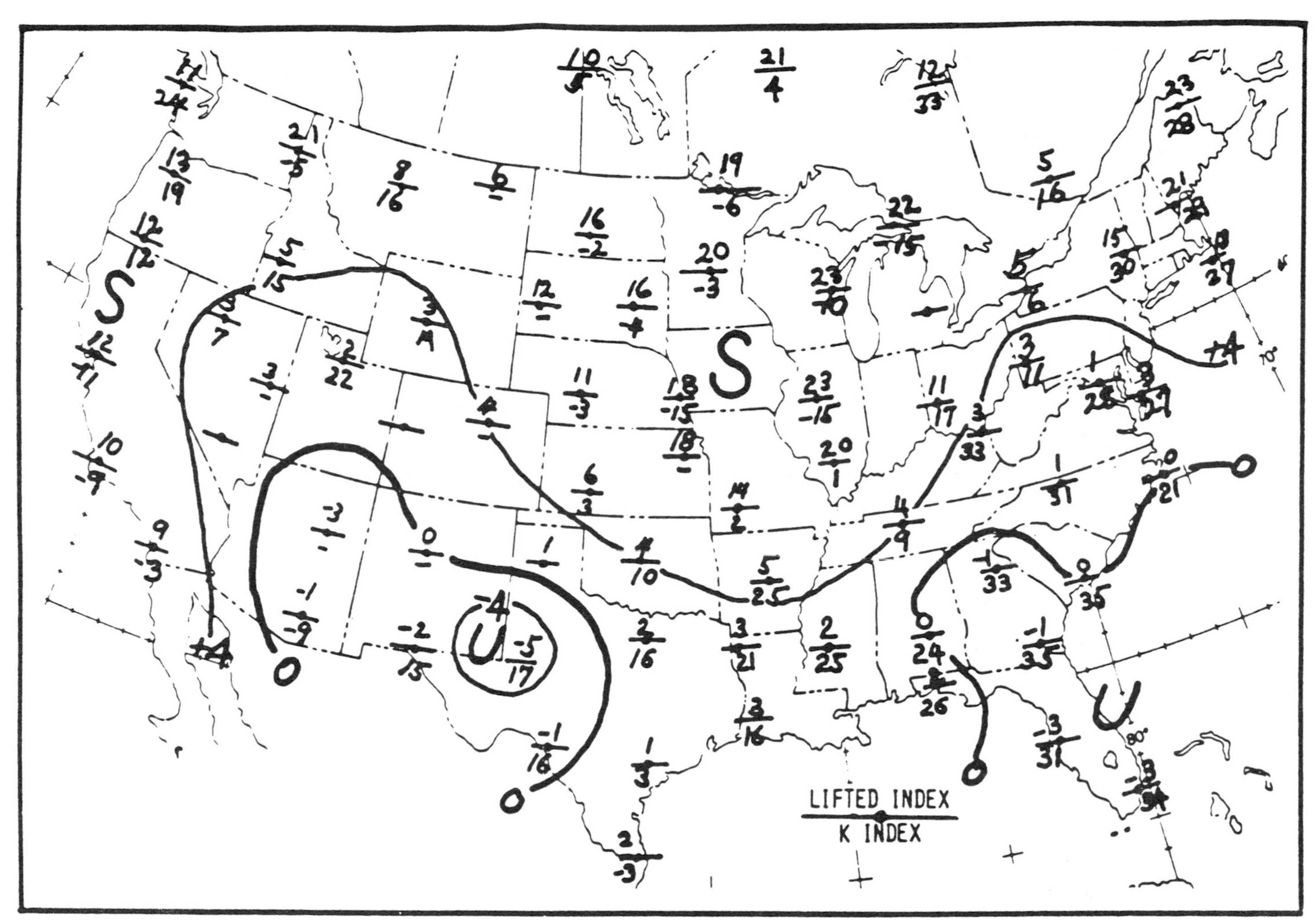

FIGURE 7.—Stability Chart

7135. If squalls are reported at your destination, what wind conditions should you anticipate?

A—Sudden increases in windspeed of at least 15 knots to a peak of 20 knots or more, lasting for at least 1 minute.
B—Peak gusts of at least 35 knots, for a sustained period of 1 minute or longer.
C—Rapid variation in wind direction of at least 20° and changes in speed of at least 10 knots between peaks and lulls.

7163. The body of an FT (Terminal Forecast) covers a geographical area within

A—a 5-mile radius of the center of a runway complex.
B—25 miles of the center of an airport.
C—10 miles of the station originating the FT.

7165. What wind direction and speed is represented by the entry 9900+00 for 9,000 feet, on an FD (winds and temperatures aloft forecast)?

A—Light and variable; less than 5 knots.
B—Vortex winds exceeding 200 knots.
C—Light and variable; less than 10 knots.

7166. What conclusion(s) can be drawn from a 500-millibar Constant Pressure Chart for a planned flight at FL180?

A—Winds aloft at FL180 generally flow across the height contours.
B—Observed temperature, wind, and temperature/dewpoint spread along the proposed route can be approximated.
C—Upper highs, lows, troughs, and ridges will be depicted by the use of lines of equal pressure.

7167. What important information is provided by the Radar Summary Chart that is not shown on other weather charts?

A—Lines and cells of hazardous thunderstorms.
B—Types of precipitation between reporting stations.
C—Areas of cloud cover and icing levels within the clouds.

7168. The section of the Area Forecast entitled "significant clouds and weather" contains a

A—summary of cloudiness and weather significant to flight operations broken down by states or other geographical areas.
B—summary of forecast sky cover, cloud tops, visibility, and obstructions to vision along specific routes.
C—statement of AIRMETS and SIGMETS still in effect at the time of issue.

7169. Which primary source should be used to obtain forecast weather information at your destination for the planned ETA?

A—Area Forecast.
B—Radar Summary and Weather Depiction Charts.
C—Terminal Forecast.

7170. Omission of a wind entry in a Terminal Forecast specifically implies that the wind is expected to be less than

A—5 knots.
B—6 knots.
C—8 knots.

7171. The absence of a visibility entry in a Terminal Forecast specifically implies that the surface visibility

A—exceeds 10 miles.
B—exceeds 6 miles.
C—is at least 15 miles in all directions from the center of the runway complex.

7172. The word WIND in the categorical outlook in the Terminal Forecast means that the wind during that period is forecast to be

A—10 knots or stronger.
B—less than 25 knots.
C—25 knots or stronger.

7173. What expected windspeed is specifically implied at 2200Z by this Terminal Forecast for Memphis?

MEM 251010 C5 X 1/2F 1710 OCNL C0 X 1/2F. 16Z C25 BKN 11/2F 1720. 22Z 20 SCT. 00Z CLR.

A—Less than 6 knots.
B—Less than 10 knots.
C—Calm and variable.

7174. SIGMETS are issued as a warning of weather conditions potentially hazardous

A—particularly to light aircraft.
B—to all aircraft.
C—only to light aircraft operations.

7175. What significant cloud coverage is reported by a pilot in this SA?

MOB...M9 OVC 2LF 131/44/43/3212/991/UA/OV 15NW MOB 1355/SK OVC 025/045 OVC 090

A—Three separate overcast layers exist with bases at 2,500, 7,500, and 13,500 feet.
B—The top of lower overcast is 2,500 feet; base and top of second overcast layer is 4,500 and 9,000 feet, respectively.
C—The base of second overcast layer is 2,500 feet; top of second overcast layer is 7,500 feet; base of third layer is 13,500 feet.

7176. Which meteorological condition is issued in the form of a SIGMET (WS)?

A—Widespread sand or duststorms lowering visibility to less than 3 miles.
B—Moderate icing.
C—Sustained winds of 30 knots or greater at the surface.

7177. A pilot planning to depart at 1100Z on an IFR flight is particularly concerned about the hazard of icing. What sources reflect the most accurate information on icing conditions (current and forecast) at the time of departure?

A—Low Level Sig Weather Prog Chart, RADATS, and the Area Forecast.
B—The Area Forecast, and the Freezing Level Chart.
C—PIREPS, AIRMETS, and SIGMETS.

7178. Which forecast provides specific information concerning expected sky cover, cloud tops, visibility, weather and obstructions to vision in a route format?

A—DFW FA 131240.
B—MEM FT 132222.
C—249 TWEB 252317.

7179. The section of the Area Forecast entitled "Hazards/Flight Precautions" contains

A—a 12-hour forecast that identifies and locates aviation weather hazards.
B—a statement listing those AIRMETS still in effect.
C—a summary of general weather conditions over several states.

7180. What is the maximum forecast period for AIRMETS ?

A—Two hours.
B—Four hours.
C—Six hours.

7181. When is the temperature at one of the forecast altitudes omitted at a specific location or station in the Winds and Temperatures Aloft Forecast (FD)?

A—When the temperature is standard for that altitude.
B—For the 3,000-foot altitude (level) or when the level is within 2,500 feet of station elevation.
C—Only when the winds are omitted for that altitude (level).

7182. When is the wind-group at one of the forecast altitudes omitted at a specific location or station in the Winds and Temperatures Aloft Forecast (FD)? When the wind

A—is less than 5 knots.
B—is less than 10 knots.
C—at the altitude is within 1,500 feet of the station elevation.

7183. Decode the excerpt from the Winds and Temperature Aloft Forecast (FD) for OKC at 39,000 feet.

FT	3000	6000	39000
OKC			830558

A—Wind 130° at 50 knots, temperature -58° C.
B—Wind 330° at 105 knots, temperature -58° C.
C—Wind 330° at 205 knots, temperature -58° C.

7184. Which values are used for winds aloft forecasts?

A—Magnetic direction and knots.
B—Magnetic direction and MPH.
C—True direction and knots.

7185. (Refer to figure 2.) What approximate wind direction, speed, and temperature (relative to ISA) should a pilot expect when planning for a flight over PSB at FL270?

A—260° magnetic at 93 knots; ISA +7° C.
B—280° true at 113 knots; ISA +3° C.
C—255° true at 93 knots; ISA +6° C.

7186. (Refer to figure 2.) What approximate wind direction, speed, and temperature (relative to ISA) should a pilot expect when planning for a flight over ALB at FL270?

A—270° magnetic at 97 knots; ISA -4° C.
B—260° true at 110 knots; ISA +5° C.
C—275° true at 97 knots; ISA +4° C.

7187. (Refer to figure 2.) What approximate wind direction, speed, and temperature (relative to ISA) should a pilot expect when planning for a flight over EMI at FL270?

A—265° true; 100 knots; ISA +3° C.
B—270° true; 110 knots; ISA +5° C.
C—260° magnetic; 100 knots; ISA -5° C.

7188. What flight planning information can a pilot derive from constant pressure charts?

A—Clear air turbulence and icing conditions.
B—Levels of widespread cloud coverage.
C—Winds and temperatures aloft.

7189. The station originating the following weather report has a field elevation of 1,800 feet. MSL. If the sky cover is one continuous layer, what is its thickness?

W8X1FK 174/74/73/0000/004/OVC/35

A—900 feet.
B—1,700 feet.
C—2,700 feet.

7190. (Refer to figure 8.) The Severe Weather Outlook Chart depicts

A—areas of probable severe thunderstorms by the use of single hatched areas on the chart.
B—areas of forecast, severe or extreme turbulence, and areas of severe icing for the next 24 hours.
C—areas of general (excluding severe) thunderstorm activity by the use of hatching on the chart.

7191. Interpret the Pilot Weather Report (PIREP).

UA/OV 20S ATL 1620 FL050/TP BE 18/IC MDT RIME ICE

A—20 NM south of Atlanta at 1620Z, a pilot flying at 5,000 feet in a Beech 18 reported moderate rime ice.
B—20 minutes after the hour, snow began at Atlanta, wind 160° at 20 knots, a Beech 18 reported moderate rime ice at 5,000 feet.
C—Snow encountered at 2,000 feet over Atlanta at 1620Z; a Beech 18 encountered rime ice at 5,000 feet.

7192. A station is forecasting wind and temperature aloft at FL390 to be 300° at 200 knots; temperature -54° C. How would this data be encoded in the FD?

A—300054.
B—809954.
C—309954.

7194. What is the single source reference that contains information regarding frontal movement, turbulence, and icing conditions for a specific area?

A—Terminal forecast (FT).
B—Weather Depiction Chart.
C—Area forecast (AF).

7195. A ceiling is defined as the height of the

A—highest layer of clouds or obscuring phenomena aloft that covers over 6/10 of the sky.
B—lowest layer of clouds that contributed to the overall overcast.
C—lowest layer of clouds or obscuring phenomena aloft that is reported as broken or overcast.

7196. The reporting station originating this Surface Aviation Weather Report has a field elevation of 1,000 feet. If the reported sky cover is one continuous layer, what is its thickness?

MDW RS 1856 M7 OVC11/2R+F 990/63/61/
3205/980/... UA.../SK OVC 65

A—4,800 feet.
B—5,000 feet.
C—5,800 feet.

7197. What is the significance of the "F2" in the remarks portion of this Surface Aviation Weather Report for CLE?

CLE SP 1350 -X E80 BKN 150 OVC 1GF
169/67/67/2105/003/R23LVV11/2 F2

A—The restriction to visibility is caused by fog and the prevailing visibility is 2 statute miles.
B—The partial obscuration is caused by fog and the visibility value is variable, 1-1/2 to 2 statute miles.
C—Fog is obscuring 2/10 of the sky.

7198. What is meant by the entry in the remarks section of this Surface Aviation Weather Report for BOI?

BOI SA 1854 -X M7 OVC1 1/2R+F
990/63/61/3205/980/RF2 RB12

A—Runway fog, visibility 2 miles; base of the rain clouds 1,200 feet.
B—Rain and fog obscuring 2/10 of the sky; rain began 12 minutes before the hour.
C—Rain and fog obscuring 2/10 of the sky; rain began at 1812.

7199. (Refer tc figure 3.) The Weather Depiction Chart indicates that northern Illinois and southern Wisconsin are reporting

A—marginal VFR conditions due to reduced visibility in drizzle and fog.
B—low IFR conditions due to ceilings below 500 feet with drizzle.
C—IFR conditions due to overcast ceilings less than 1,000 feet with reduced visibilities in rain and rain showers.

7200. (Refer to figure 3.) The Weather Depiction Chart indicates the heaviest precipitation along the front is occurring in

A—Missouri.
B—Illinois.
C—Kansas.

7201. (Refer to figure 3.) The Weather Depiction Chart indicates that the coastal sections of Texas and Louisiana are reporting

A—all ceilings at or above 20,000 feet with visibilities of 20 miles or more.
B—marginal VFR conditions due to broken ceilings of 2,000 feet.
C—VFR conditions with scattered clouds at 2,000 feet and higher cirroform.

7202. The Surface Analysis Chart depicts

A—actual pressure systems, frontal locations, cloud tops, and precipitation at the time shown on the chart.
B—frontal locations and expected movement, pressure centers, cloud coverage, and obstructions to vision at the time of chart transmission.
C—actual frontal positions, pressure patterns, temperature, dewpoint, wind, weather, and obstructions to vision at the valid time of the chart.

7203. A pilot reporting turbulence that momentarily causes slight, erratic changes in altitude and/or attitude should report it as

A—light turbulence.
B—moderate turbulence.
C—light chop.

7204. The Low Level Significant Weather Prog Chart depicts weather conditions

A—that are forecast to exist at a valid time shown on the chart.
B—as they existed at the time the chart was prepared.
C—that existed at the time shown on the chart which is about 3 hours before the chart is received.

7205. Which meteorological conditions are depicted by a prognostic chart?

A—Conditions existing at the time of the observation.
B—Interpretation of weather conditions for geographical areas between reporting stations.
C—Conditions forecast to exist at a specific time shown on the chart.

--- and Figure 4

7206. (Refer to figure 5.) What is the meaning of the symbol depicted as used on the U.S. Low Level Significant Weather Prog Chart?

A—Showery precipitation (e.g., rain showers) embedded in an area of continuous rain covering half or more of the area.
B—Continuous precipitation (e.g., rain) covering half or more of the area.
C—Showery precipitation (e.g., thunderstorms/rain showers) covering half or more of the area.

7207. A prognostic chart depicts the conditions

A—existing at the surface during the past 6 hours.
B—which presently exist from the 1,000-millibar through the 700-millibar level.
C—forecast to exist at a specific time in the future.

7208. What information is provided by a Convective Outlook (AC)?

A—It describes areas of probable severe icing and severe or extreme turbulence during the next 24 hours.
B—It provides prospects of both general and severe thunderstorm activity during the following 24 hours.
C—It indicates areas of probable convective turbulence and the extent of instability in the upper atmosphere (above 500 mb).

7209. (Refer to figure 5.) A planned low-altitude flight from central Oklahoma to western Tennessee at 1200Z is likely to encounter

A—continuous or intermittent rain or rain showers, moderate turbulence, and freezing temperatures below 8,000 feet.
B—continuous or showery rain over half or more of the area, moderate turbulence, and freezing temperatures above 10,000 feet.
C—showery precipitation covering less than half the area, no turbulence below 18,000 feet, and freezing temperatures above 12,000 feet.

7210. (Refer to figure 5.) The 12-Hour Sig Weather Prognosis Chart indicates that West Virginia will likely experience

A—continuous or showery precipitation covering half or more of the area.
B—thunderstorms and rain showers covering half or more of the area.
C—continuous rain covering less than half of the area.

7211. (Refer to figure 5.) The 12-Hour Significant Weather Prognosis Chart indicates that eastern Kentucky and eastern Tennessee can expect probable ceilings

A—less than 1,000 feet and/or visibility less than 3 miles.
B—less than 1,000 feet and/or visibility less than 3 miles, and moderate turbulence below 10,000 feet MSL.
C—less than 1,000 feet and/or visibility less than 3 miles, and moderate turbulence above 10,000 feet MSL.

7212. (Refer to figure 5.) The chart symbols over southern California on the 12-Hour Significant Weather Prognosis Chart indicate

A—expected top of moderate turbulent layer to be 12,000 feet MSL.
B—expected base of moderate turbulent layer to be 12,000 feet MSL.
C—light turbulence expected above 12,000 feet MSL.

7213. Interpret this Pilot Weather Report (PIREP).

UA/OVR MRB FL060/SK INTMTLY BL/TB MDT/RM R TURBC INCRS WWD.

A—Ceiling 6,000 intermittently below moderate thundershowers; turbulence increasing westward.
B—Flight level 60,000, intermittently below clouds; moderate rain, turbulence increasing with the wind.
C—At 6,000 feet; intermittently between layers; moderate turbulence; moderate rain; turbulence increasing westward.

7214. (Refer to figure 6.) What weather conditions are depicted within the area indicated by arrow E?

A—Frequent embedded thunderstorms, less than one-eighth coverage, tops at FL370.
B—Frequent lightning in thunderstorms at FL370.
C—Frequent cumulonimbus, five-eighths to eight-eighths coverage, bases below 24,000 feet MSL and tops at 37,000 feet MSL.

7215. (Refer to figure 6.) What weather conditions are depicted within the area indicated by arrow D?

A—Existing isolated cumulonimbus, tops above 41,000 feet and less than one-fifth coverage.
B—Forecast isolated embedded cumulonimbus, bases below 24,000 feet MSL, tops at 41,000 feet MSL and less than one-eighth coverage.
C—Forecast isolated thunderstorms, tops at FL410, less than two-fifths coverage.

7216. (Refer to figure 6.) What weather conditions are depicted within the area indicated by arrow C?

A—Severe clear air turbulence forecast within the area outlined by dashes from 32,000 feet MSL to below the lower limit of the chart.
B—Moderate turbulence at FL320 within the area outlined by dashes.
C—Moderate to severe clear air turbulence has been reported at FL320.

7217. (Refer to figure 6.) What weather conditions are depicted within the area indicated by arrow B?

A—Light to moderate turbulence at and above 33,000 feet MSL.
B—Moderate to severe turbulence from below 24,000 feet MSL to 33,000 feet MSL.
C—Moderate to severe clear air turbulence is forecast to exist at FL330.

7218. (Refer to figure 6.) What information is indicated by arrow A?

A—The height of the tropopause in meters above sea level.
B—The height of the existing layer of CAT.
C—The height of the tropopause in hundreds of feet above MSL.

7219. Which weather forecast describes prospects for an area coverage of both severe and general thunderstorms during the following 24 hours?

A—Terminal Forecast.
B—Convective Outlook.
C—Severe Weather Watch Bulletin.

7221. From which primary source should you obtain information regarding the weather expected to exist at your destination at your estimated time of arrival?

A—Weather Depiction Chart.
B—Radar Summary and Weather Depiction Chart.
C—Terminal Forecast (FT).

7222. (Refer to figure 6.) What weather conditions are depicted within the area indicated by arrow F?

A—Two-eighths to six-eighths coverage, occasional embedded thunderstorms, tops at FL510.
B—One-eighth to four-eighths coverage, occasional embedded thunderstorms, maximum tops at 51,000 feet MSL.
C—Occasionally embedded cumulonimbus, bases from 18,000 feet to 51,000 feet.

7223. (Refer to figure 7.) What weather conditions are depicted in the area indicated by arrow A on the Radar Summary Chart?

A—Moderate to strong echoes; echo tops 30,000 feet MSL; line movement toward the northwest.
B—Weak to moderate echoes; average echo bases 30,000 feet MSL; cell movement toward the southeast; rain showers with thunder.
C—Strong to very strong echoes; echo tops 30,000 feet MSL; thunderstorms and rain showers.

7224. (Refer to figure 7.) What weather conditions are depicted in the area indicated by arrow D on the Radar Summary Chart?

A—Echo tops 4,100 feet MSL; strong to very strong echoes within the smallest contour; area movement toward the northeast at 50 knots.
B—Intense to extreme echoes within the smallest contour; echo tops 29,000 feet MSL; cell movement toward the northeast at 50 knots.
C—Strong to very strong echoes within the smallest contour; echo bases 29,000 feet MSL; cell in northeast Nebraska moving northeast at 50 knots.

7225. (Refer to figure 7.) What weather conditions are depicted in the area indicated by arrow C on the Radar Summary Chart?

A—Average echo bases 2,800 feet MSL; thundershowers; intense to extreme echo intensity.
B—Cell movement toward the northwest at 20 knots; intense echoes; echo bases 28,000 feet MSL.
C—Area movement toward the northeast at 20 knots; strong to very strong echoes; echo tops 28,000 feet MSL.

7226. (Refer to figure 7.) What weather conditions are depicted in the area indicated by arrow B on the Radar Summary Chart?

A—Weak echoes; heavy rain showers; area movement toward the southeast.
B—Weak to moderate echoes; rain showers increasing in intensity.
C—Strong echoes; moderate rain showers; no cell movement.

7227. (Refer to figure 7.) What weather conditions are depicted in the area indicated by arrow E on the Radar Summary Chart?

A—Highest echo tops 30,000 feet MSL; weak to moderate echoes; thunderstorms and rain showers; cell movement toward northwest at 15 knots.
B—Echo bases 29,000 to 30,000 feet MSL; strong echoes; rain showers increasing in intensity; area movement toward northwest at 15 knots.
C—Thundershowers decreasing in intensity; area movement toward northwest at 15 knots; echo bases 30,000 feet MSL.

7228. For most effective use of the Radar Summary Chart during preflight planning, a pilot should

A—consult the chart to determine more accurate measurements of freezing levels, cloud cover, and wind conditions between reporting stations.
B—compare it with the charts, reports, and forecasts of a three dimensional picture of clouds and precipitation.
C—utilize the chart as the only source of information regarding storms and hazardous conditions existing between reporting stations.

7229. (Refer to figure 7.) What weather conditions are depicted in the area indicated by arrow G on the Radar Summary Chart?

A—Echo bases 10,000 feet MSL; cell movement toward northeast at 15 knots; weak to moderate echoes; rain.
B—Area movement toward northeast at 15 knots; rain decreasing in intensity; echo bases 1,000 feet MSL; strong echoes.
C—Strong to very strong echoes; area movement toward northeast at 15 knots; echo tops 10,000 feet MSL; light rain.

7230. (Refer to figure 7.) What weather conditions are depicted in the area indicated by arrow F on the Radar Summary Chart?

A—Line of echoes; thunderstorms; highest echo tops 46,000 feet MSL; no line movement indicated.
B—Echo bases vary from 15,000 feet to 46,000 feet MSL; thunderstorms increasing in intensity; line of echoes moving rapidly toward the north.
C—Line of severe thunderstorms moving from south to north; echo bases vary from 4,400 feet to 4,600 feet MSL; extreme echoes.

7231. Hazardous wind shear is commonly encountered near the ground

A—during periods when the wind velocity is stronger than 35 knots.
B—during periods when the wind velocity is stronger than 35 knots and near mountain valleys.
C—during periods of strong temperature inversion and near thunderstorms.

7232. (Refer to figure 8.) The Severe Weather Outlook Chart which is used primarily for advance planning, provides what information?

A—An 18-hour categorical outlook with a 48-hour valid time for severe weather watch, thunderstorm lines, and areas of expected tornado activity.
B—A preliminary 12-hour outlook for severe thunderstorm activity and probable convective turbulence.
C—A preliminary 24-hour severe weather outlook for general and severe thunderstorm activity, tornadoes, and watch areas.

7233. (Refer to figure 8.) What is the significance of the annotations "MDT" and "SLGT" on the 00Z -12Z panel of the AC?

A—Moderate risk area, surrounded by a slight risk area, of possible severe turbulence.
B—Slight to moderate chance of low-level wind shear.
C—Moderate risk of severe thunderstorms surrounded by a slight risk area.

7234. If you hear a SIGMET alert, how can you obtain the information in the SIGMET?

A—ATC will announce the hazard and advise you when to listen to an FSS broadcast.
B—Contact a weather watch station.
C—Contact the nearest FSS and ascertain whether the advisory is pertinent to your flight.

7235. (Refer to figure 9, arrow A.) The symbol on the TROP WIND SHEAR PROG represents the

A—wind direction at the tropopause (300°).
B—flight level of the tropopause.
C—height of maximum wind shear (30,000 feet).

7236. (Refer to figure 10.) What is the pressure in millibars on the tropopause pressure, temperature, and winds chart for ABQ?

A—150.
B—172.
C—200.

7237. (Refer to figure 10.) What is the approximate wind direction and velocity at ABQ?

A—220°/200 knots.
B—060°/61 knots.
C—240°/115 knots.

7238. (Refer to figure 9, arrow B.) The reference to 6K indicates

A—vertical wind shear per 1,000 feet.
B—wind velocity at 34,000 feet.
C—vertical wind velocity at 34,000 feet.

7239. (Refer to figure 11, arrow C.) What is the approximate wind direction and velocity at 34,000 feet?

A—290°/50 knots.
B—330°/50 knots.
C—090°/48 knots.

7240. (Refer to figure 11, arrow A) The wind direction and velocity on the Tropopause Wind Prog Chart is indicated from the

A—northeast at 35 knots.
B—northwest at 47 knots.
C—southwest at 35 knots.

7241. (Refer to figure 8.) The crosshatched area on the severe weather outlook chart indicates

A—moderate risk of thunderstorms.
B—forecast risk of tornadoes.
C—forecast risk of heavy thunderstorms.

7242. (Refer to figure 11, arrow A.) What is the approximate wind direction and velocity at CVG at 34,000 feet?

A—040°/35 knots.
B—097°/40 knots.
C—230°/35 knots.

7243. (Refer to figure 11, arrow B.) What is the approximate wind direction and velocity at BOI?

A—270°/55 knots.
B—250°/95 knots.
C—270°/95 knots.

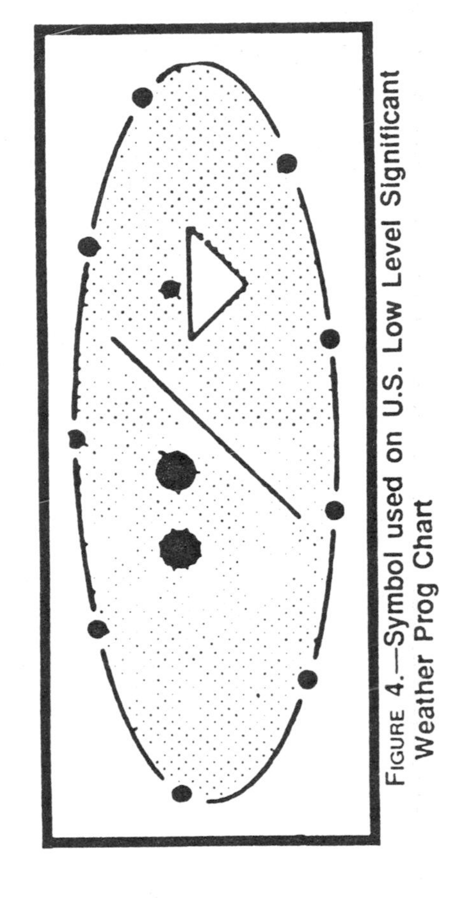

FIGURE 4.—Symbol used on U.S. Low Level Significant Weather Prog Chart

VALID 141200Z FOR USE 0900-1500Z. TEMPS NEG ABV 24000

FT	3000	6000	9000	12000	18000	24000	30000	34000	39000
EMI	2807	2715-07	2728-10	2842-13	2867-21	2891-30	751041	771150	780855
ALB	0210	9900-07	2714-09	2728-12	2656-19	2777-28	781842	760150	269658
PSB		1509+04	2119+01	2233-04	2262-14	2368-26	781939	760850	780456
STL	2308	2613+02	2422-03	2431-08	2446-19	2461-30	760142	782650	760559

FIGURE 2.—Winds and Temperatures Aloft Forecast

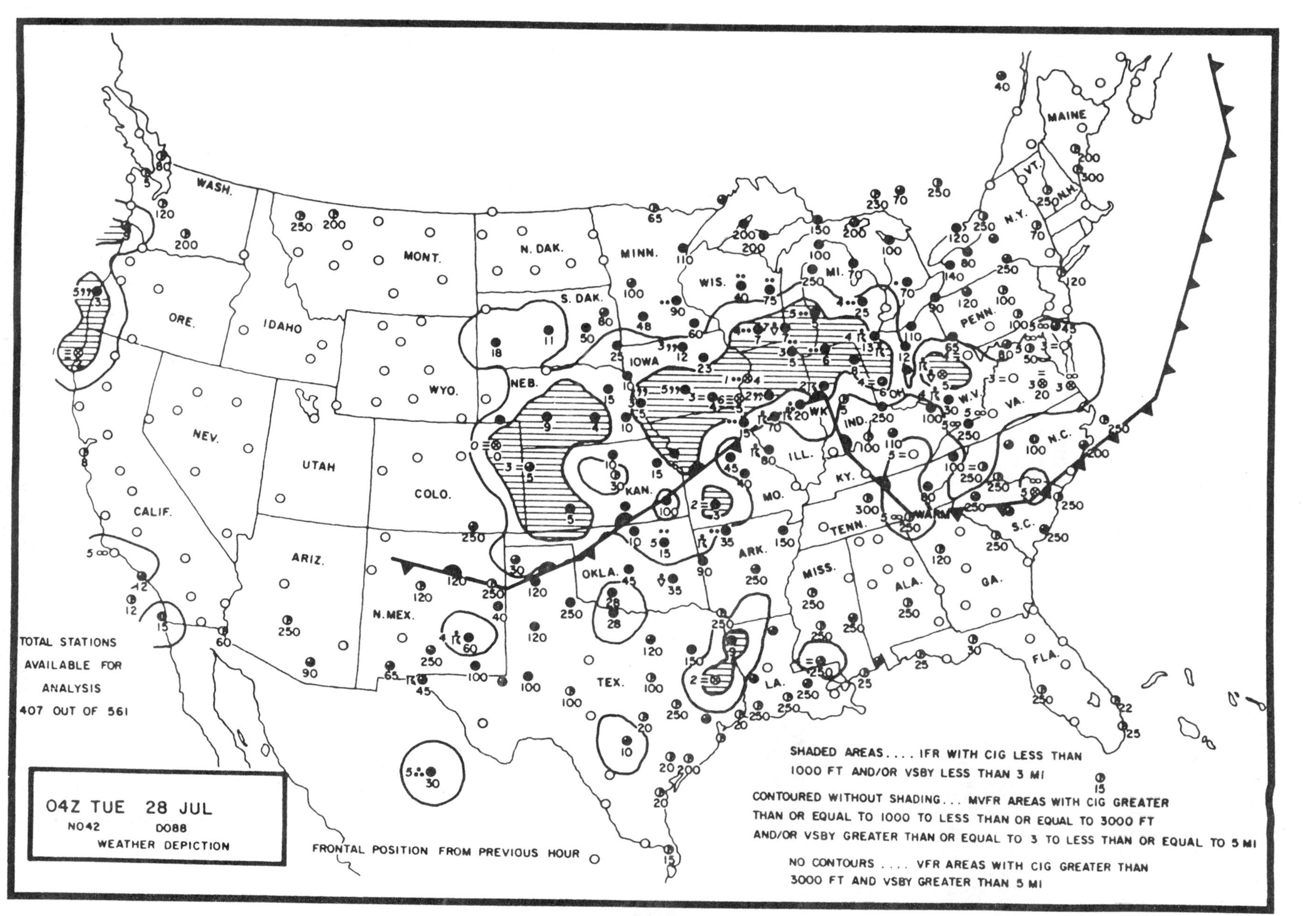

FIGURE 3.—Weather Depiction Chart

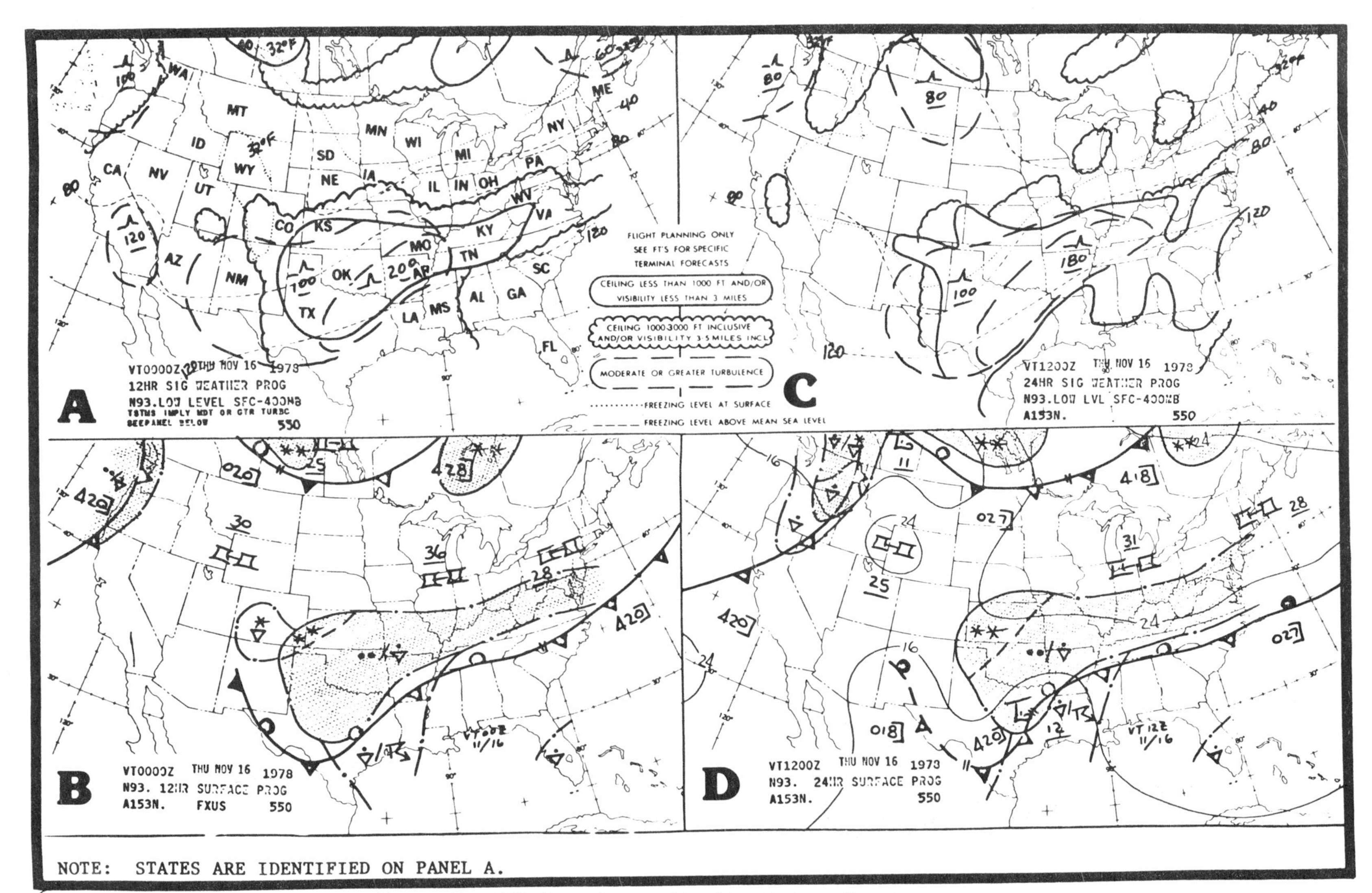

FIGURE 5.—U.S. Low Level Significant Prog Chart

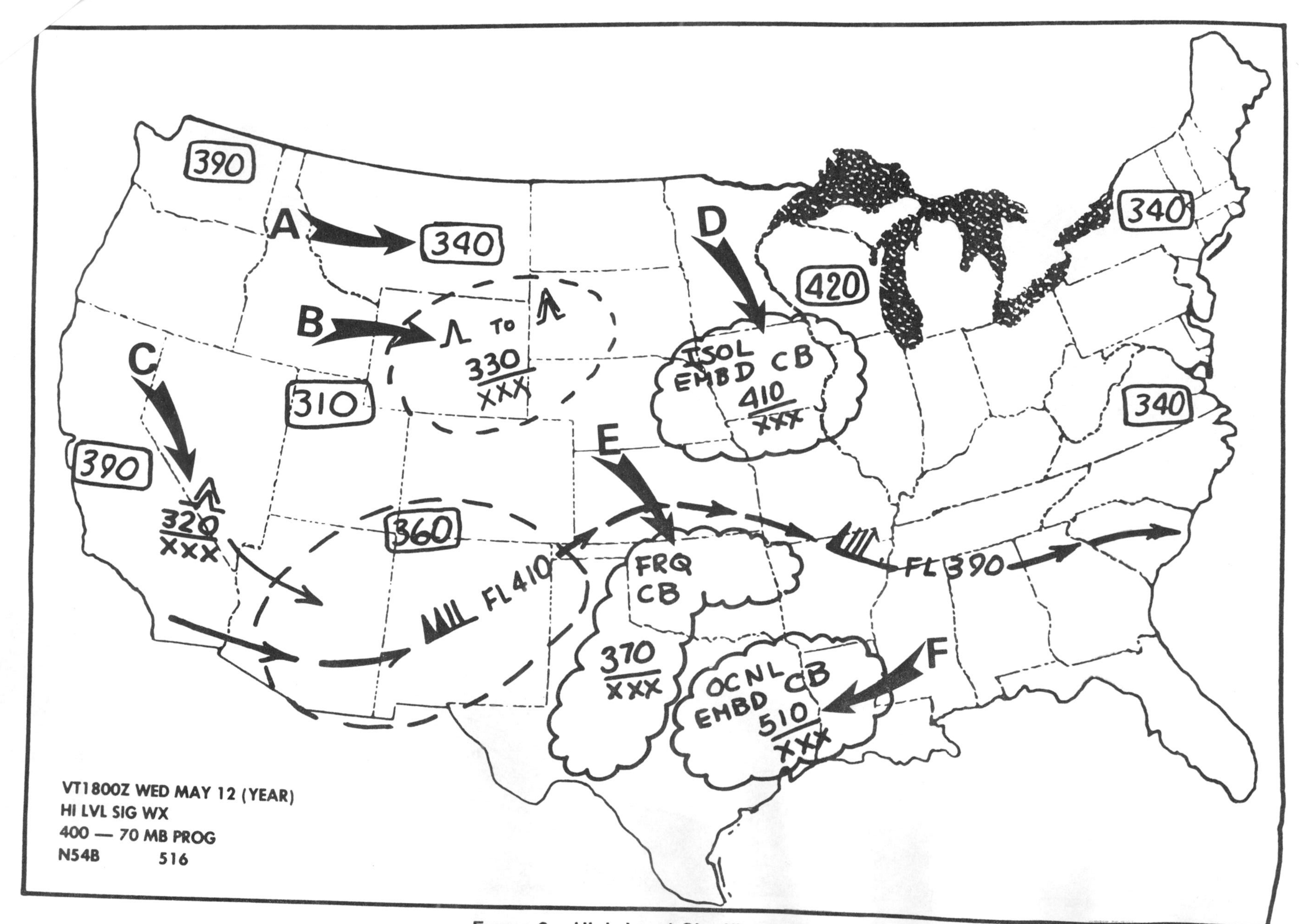

FIGURE 6.—High Level Significant Prog Chart

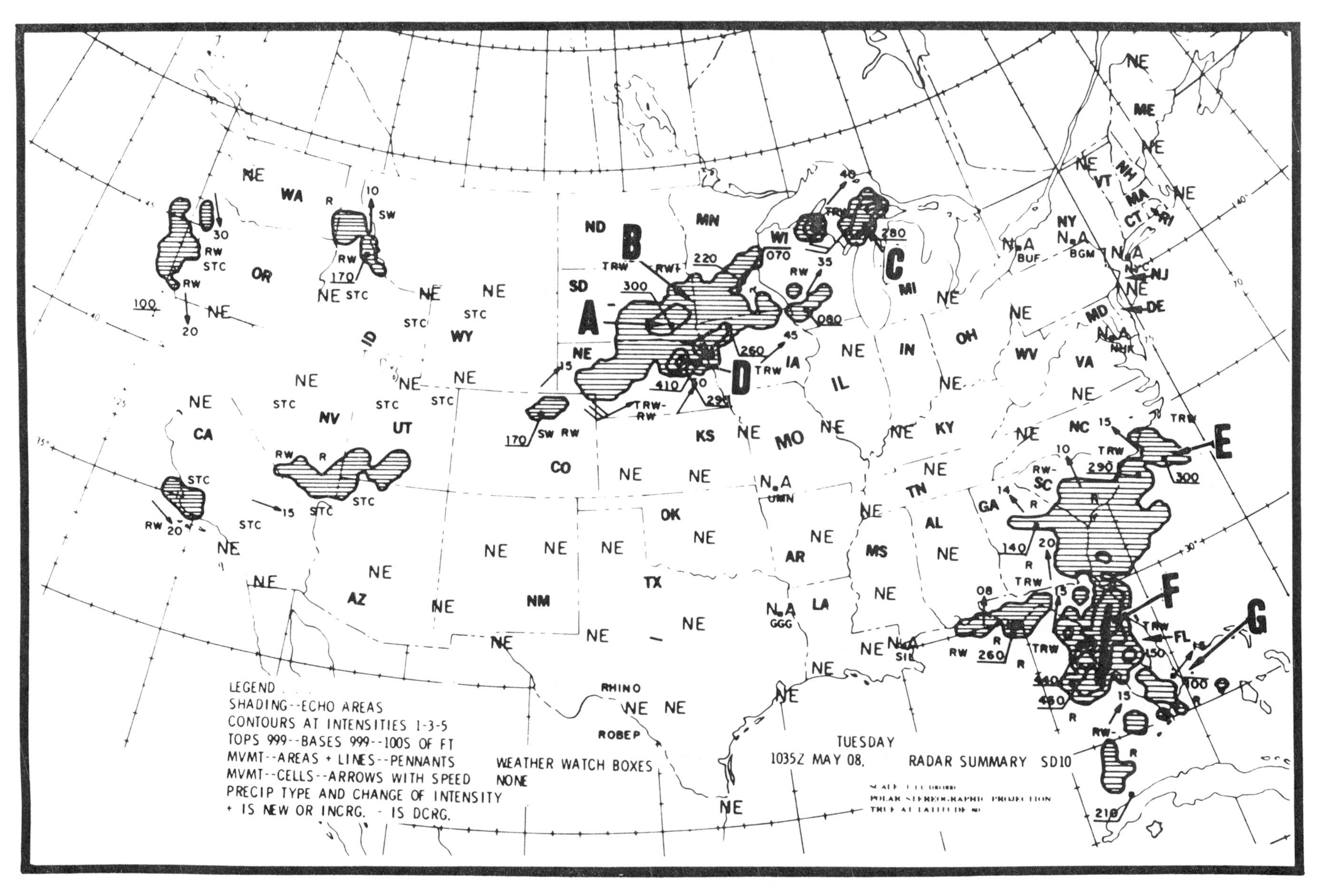

FIGURE 7.—Radar Summary Chart

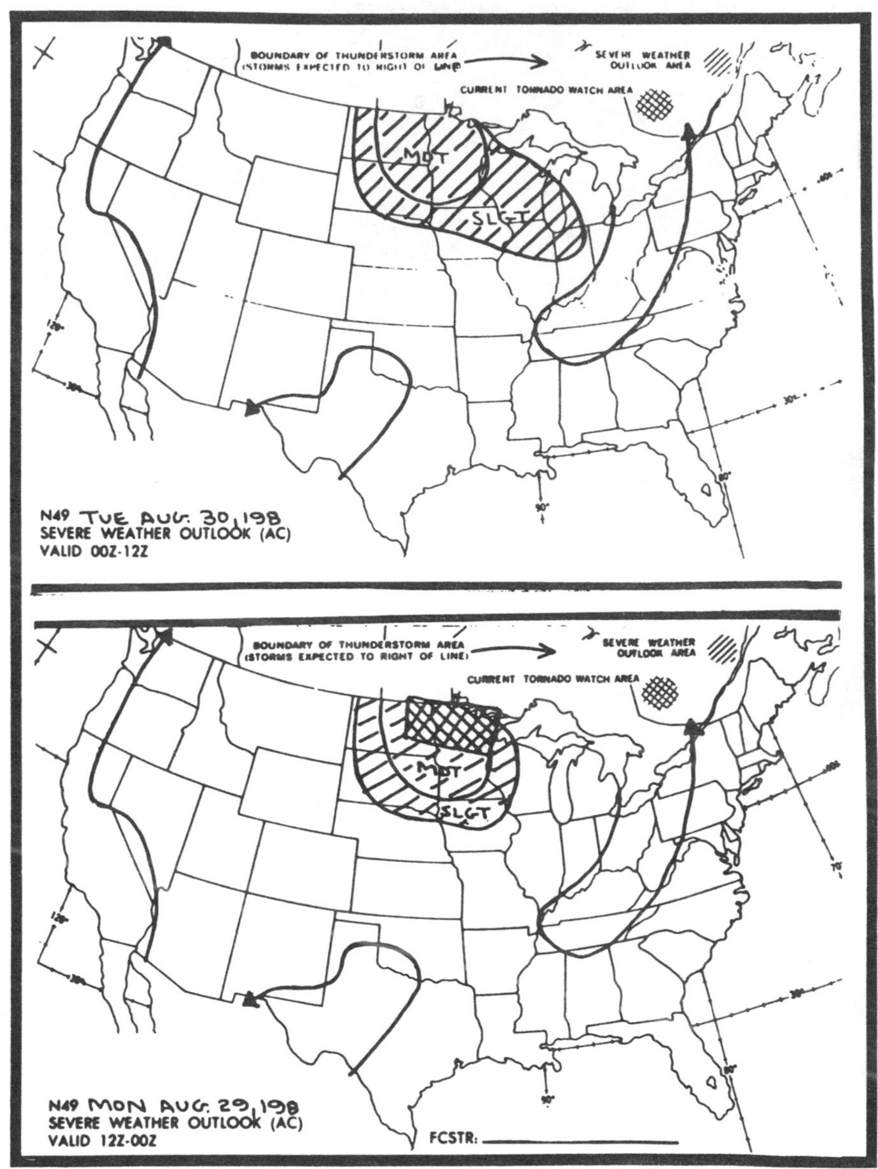

FIGURE 8.—Severe Weather Outlook Chart

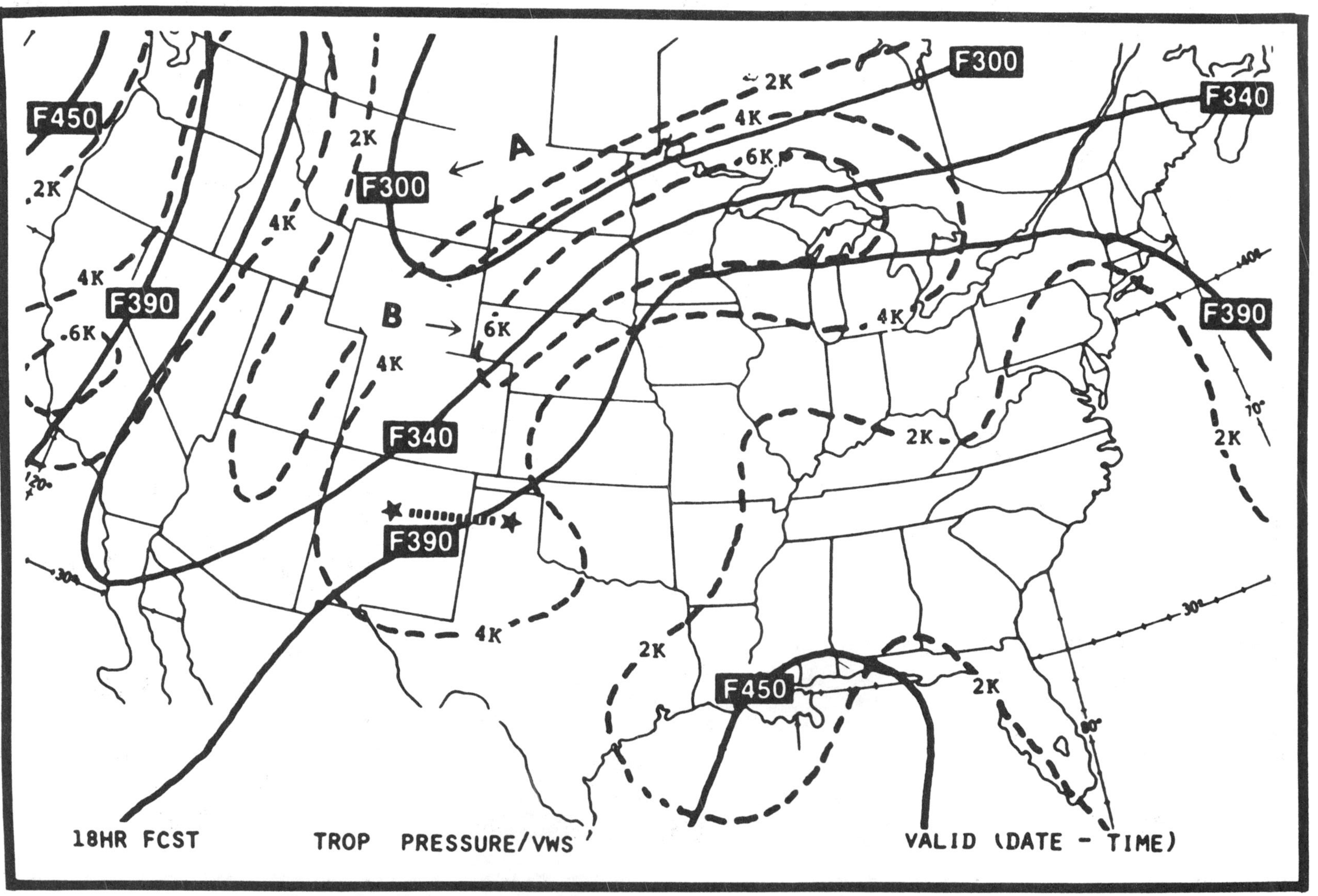

FIGURE 9.—Tropopause Height/Vertical Wind Shear Prog Chart

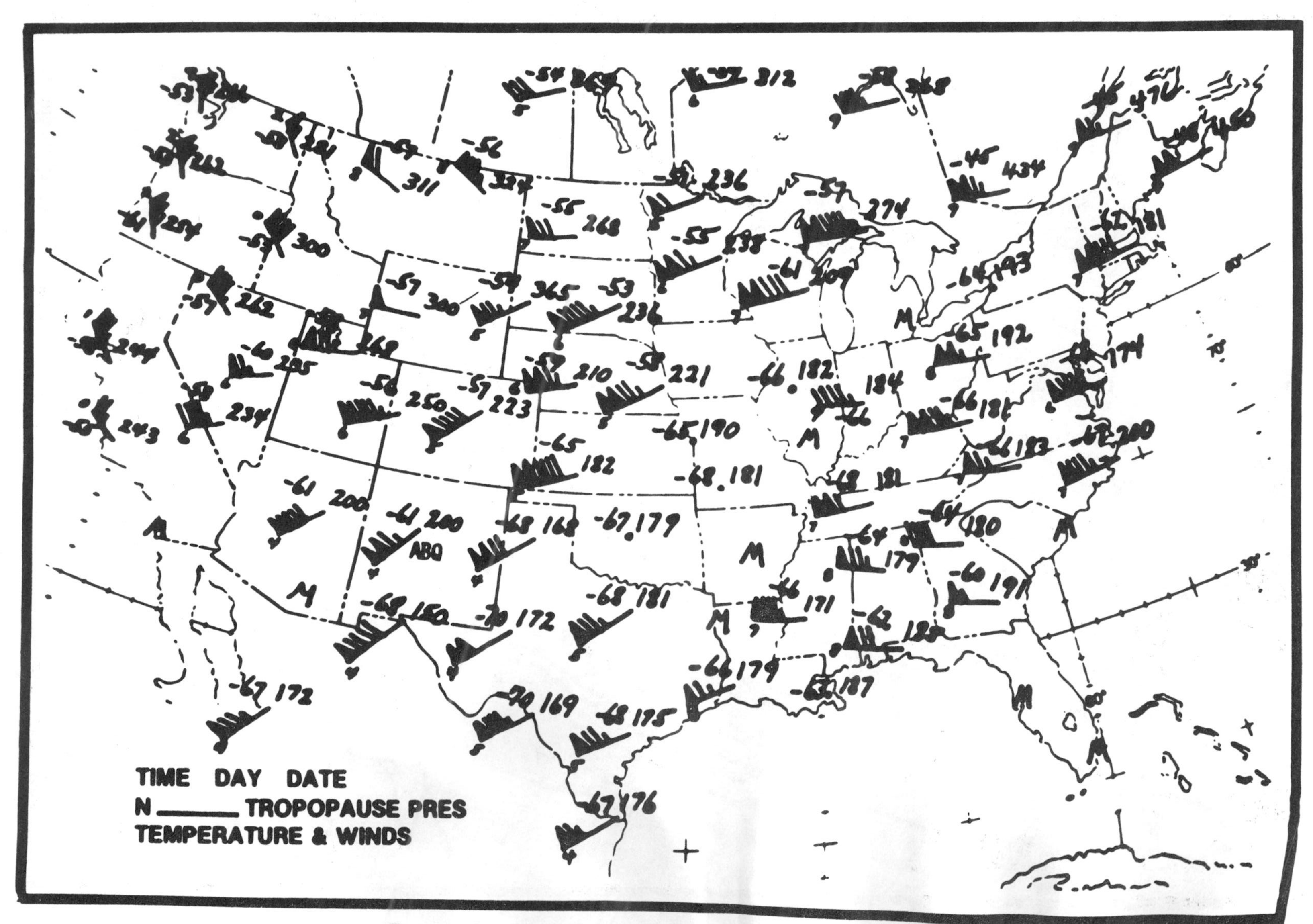

FIGURE 10.—Tropopause Pressure Temperatures and Winds

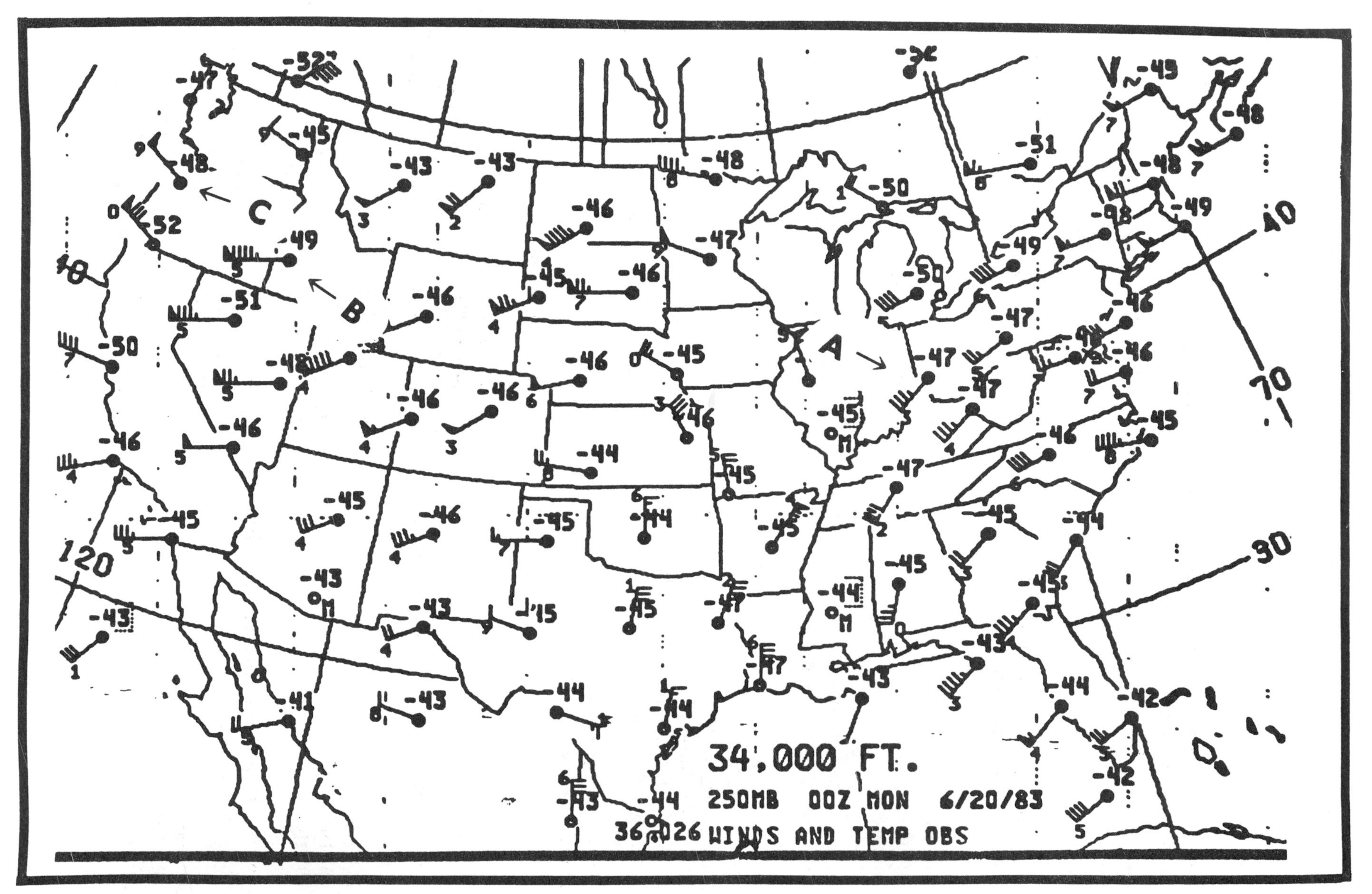

FIGURE 11.—Observed Winds Aloft for 34,000 Feet

3271. (Refer to figure 3.) Which station is reporting the wind as calm?

A— DAL.
B— FTW.
C— TYR.

3272. (Refer to figure 3.) What is the reported duration of the rain at the time of the observation at AUS?

A— 25 minutes.
B— 26 minutes.
C— 36 minutes.

3273. (Refer to figure 3.) What does the LB26E40 mean at the end of the report for AMA?

A— Drizzle began at 26 past the hour and ended at 40 past the hour.
B— Lightning began at 1726 with 40 percent coverage of radar echoes.
C— There are large buildups to the east with wind gusting 26 to 40 knots.

3274. (Refer to figure 3.) Which station is reporting the lowest visibility?

A— AUS.
B— FTW.
C— TYR.

3275. (Refer to figure 3.) The altimeter setting at AUS is

A— 1016.9 mb.
B— 30.05" Hg.
C— 31.69" Hg.

3276. (Refer to figure 3.) In the report for BKO, what is the reported ceiling?

A— 2,000 feet.
B— 13,000 feet.
C— 25,000 feet.

3277. GIVEN:

RADAT 91M024085105/1.

The relative humidity was 91 percent at

A— all levels.
B— 2,400 feet MSL.
C— 8,500 feet MSL.

3278. GIVEN:

RADAT 88M018055097/2.

How many crossings of the zero degree Celsius isotherm have occurred?

A— 5.
B— 4.
C— 3.

3279. GIVEN:

OUN AMOS 36/26/3618/007 PK WIND 27 024.

The 024 indicates

A— the peak wind occurred 24 minutes past the hour.
B— 24 hundredths of an inch of liquid precipitation since the last observation.
C— a drop of 24 hundredths of an inch of mercury in the altimeter setting in the last hour.

3280. Interpret the following radar weather report:

LIT 1133 AREA 4TRW 22/100 88/170 196/180 220/115 C2425 MT 310 AT 162/110

A— There are four cells with tops at 10,000 feet, 17,000 feet, and 11,500 feet.
B— The maximum top of the cells is located 162° and 110 NM from the station (LIT).
C— The visibility is 4 miles in thunderstorms and the intensity of thunderstorms remains unchanged.

3281. Which statement is true concerning this radar weather report for OKC?

OKC 1934 LN 8TRW / 86/ 40 164/60 199/115 15W 2425 MT 570 AT 159/65 2 INCH HAIL RPRTD THIS ECHO.

A— The visibility is 8 miles in rain showers.
B— There are three cells with tops at 11,500, 40,000, and 60,000 feet.
C— The maximum top of the cells is 57,000 feet located 65 NM south-southeast of the station.

3282. (Refer to figure 4.) Which is a true statement?

A— It is clear above 8,500 feet at DAL.
B— There are moderate buildups west of ABI.
C— The pilot reported a north wind at 30 knots at BRO.

3283. (Refer to figure 4.) Turbulence was reported west of

A— AUS.
B— ABI.
C— BRO.

3284. (Refer to figure 4.) Which is the true statement?

A— Thunderstorms were reported north of BRO.
B— Moderate turbulence was reported by a pilot east of ABI.
C— The base of the overcast at AUS was reported to be 13,000 feet.

3285. (Refer to figure 4.) The lowest cloud base reported is

A— 500 feet at BRO.
B— 4,500 feet at DAL.
C— 5,000 feet at BRO.

3286. (Refer to figure 5.) What is the visibility forecast for BRO?

A— 1 mile.
B— 3 miles.
C— Greater than 6 miles.

3287. (Refer to figure 5.) What type of weather can be expected after 0400Z at DAL?

A— At least a ceiling of 1,000 feet and visibility of 3 miles.
B— Ceiling of more than 3,000 feet and visibility greater than 5 miles.
C— Ceiling greater than 5,000 feet and visibility greater than 5 miles.

3288. (Refer to figure 5.) Which of the stations are forecasting the wind to be less than 6 knots for the entire forecast period?

A— ABI.
B— ACT.
C— AUS and ACT.

3289. (Refer to figure 5.) The valid time for the forecasts is from

A— 0940Z on the 30th until 0940Z the following day.
B— 1010Z on the 30th until 1000Z the following day.
C— 1000Z on the 30th until 1000Z on the following day.

3290. (Refer to figure 5.) What is the lowest ceiling forecast for ABI?

A— 1,000 feet.
B— 1,400 feet.
C— 10,000 feet.

3291. To determine the freezing level and areas of probable icing aloft, you should refer to the

A— Area Forecast.
B— Weather Depiction Chart.
C— Surface Analysis Weather Chart.

3292. For a brief summary of the location and movement of fronts, pressure systems, and circulation patterns, the pilot should refer to the

A— Area Forecast.
B— Stability Chart.
C— Radar Summary Chart.

3293. What is the meaning of the term MVFR, as used in the categorical outlook portion of Terminal and Area Forecasts?

A— A ceiling less than 1,000 feet and/or visibility less than 3 miles.
B— A ceiling less than 1,000 feet and/or visibility less than 1 mile.
C— A ceiling of 1,000 to 3,000 feet and/or visibility of 3 to 5 miles.

3294. An Area Forecast is valid for

A— 12 hours with an additional 6 hours categorical outlook.
B— 12 hours with an additional 12 hours categorical outlook.
C— 18 hours with an additional 12 hours categorical outlook.

3295. (Refer to figure 6.) What is the forecast for northwestern Alabama after 2300Z?

A— IFR with widely scattered thunderstorms.
B— Ceilings of 1,000 to 3,000 feet and/or visibility of 3 to 5 miles with possible rain showers.
C— Ceilings below 1,000 feet and visibility restricted to 3 miles by light rain and fog.

3296. (Refer to figure 6.) This forecast is valid for

A— 24 hours with an additional 12-hour outlook.
B— 18 hours with an additional 12-hour outlook.
C— 12 hours with an additional 6-hour outlook.

3297. (Refer to figure 6.) What is the forecast visibility for south-central Texas for the period ending 2300Z?

A— 3 miles.
B— More than 6 miles.
C— This visibility is not forecast.

3298. (Refer to figure 6.) The lowest layer of clouds forecast for Oklahoma is

A— 100 feet AGL.
B— below 1,000 feet MSL.
C— below 1,000 feet AGL.

3299. What information would be covered in an AIRMET?

A— Severe turbulence.
— Extensive mountain obscurement.
Hail of 3/4 inch or greater diameter.

3300. Which in-flight advisory would contain information on severe icing?

A— PIREP.
B— SIGMET.
C— Convective SIGMET.

3301. What information is contained in a convective SIGMET in the conterminous United States?

A— Moderate thunderstorms and surface winds greater than 40 knots.
B— Tornadoes, embedded thunderstorms, and hail 3/4 inch or greater in diameter.
C— Severe icing, severe turbulence, or widespread dust storms lowering visibility to less than 3 miles.

3302. (Refer to figure 7.) What is the temperature for 6,000 feet at AMA?

A— 8 °C.
B— The temperature is standard for that altitude.
C— No temperatures are forecast for levels within 2,500 feet of station elevation.

3303. (Refer to figure 7.) Why is there no wind forecast for 3,000 and 6,000 feet at ABQ?

A— Wind which is expected to be light and variable is omitted.
B— No winds are forecast within 1,500 feet of station elevation.
C— No winds are forecast within 3,000 feet of station elevation.

3304. (Refer to figure 7.) What is the forecast wind for 12,000 feet at AMA?

A— Calm.
B— Light and variable.
C— 090° in excess of 50 knots.

3305. (Refer to figure 7.) What is the temperature at 30,000 feet for DAL?

A— +40 °C.
B— –40 °C.
C— –40 °F.

3306. (Refer to figure 7.) What is the forecast wind at 30,000 feet for BRO?

A— 230° at 47 knots.
B— 130° at 147 knots.
C— 280° at 147 knots.

3307. (Refer to figure 7.) What is the forecast wind at 3,000 feet for DAL?

A— 100° at 8 knots.
B— 010° at 8 knots.
C— 080° at 10 knots.

3308. By referring to the isobars on a Surface Analysis Weather Chart, what can a person determine?

A— Pressure gradient.
B— Temperature changes.
C— Areas of precipitation.

3309. The intensity trend of a front (as of chart time) is best determined by referring to a

A— Surface Analysis.
B— Radar Summary Chart.
C— Weather Depiction Chart.

3310. (Refer to figure 8.) What does this symbol mean on a Surface Analysis Weather Chart?

A— Squall line.
B— Occluded front.
C— High-pressure ridge.

3311. (Refer to figure 9.) Which symbol used on a Surface Analysis Weather Chart represents a dissipating warm front?

A— 1.
B— 2.
C— 3.

3312. (Refer to figure 10.) On a Weather Depiction Chart, what does this information mean?

A— Visibility 5 miles, sky obscured.
B— Visibility 5 miles, haze, overcast, ceiling 3,500 feet.
C— Visibility 3 to 5 miles, sky obscured, ceiling 5,000 feet.

3313. On a Weather Depiction Chart, what weather conditions would be contained in an unshaded area that is enclosed by a smooth line?

A— Ceiling less than 1,000 feet and/or visibility less than 3 miles.
B— Ceiling between 5,000 and 7,000 feet and/or visibility greater than 5 miles.
C— Ceiling between 1,000 and 3,000 feet and/or visibility between 3 and 5 miles.

3314. A Weather Depiction Chart is useful to a pilot in determining

A— the temperature and dewpoint at selected stations.
B— the forecast areas of cloud cover and precipitation.
C— areas where weather conditions were reported above or below VFR minimums.

3315. (Refer to figure 11.) On a Weather Depiction Chart, what does this information mean?

A— Visibility one-half mile, 200 feet overcast, smoke.
B— Visibility 2 miles, sky obscured, haze, ceiling 2,000 feet.
C— Visibility 2 miles, sky obscured, fog, cloud layer at 20,000 feet.

3316. (Refer to figure 12.) The Weather Depiction Chart indicates that the coastal sections of Texas and Louisiana are reporting

A— marginal VFR conditions due to broken ceilings of 2,000 feet.
B— VFR conditions with scattered clouds at 2,000 feet and higher cirriform.
C— all ceilings at or above 20,000 feet with visibilities of 20 miles or more.

3317. (Refer to figure 12.) The Weather Depiction Chart indicates the heaviest precipitation along the front is occurring in

A— Kansas.
B— Missouri.
C— Illinois.

3318. (Refer to figure 12.) The Weather Depiction Chart indicates that northern Illinois and southern Wisconsin are reporting

A— low IFR conditions due to ceilings below 500 feet with drizzle.
B— marginal VFR conditions due to reduced visibility in drizzle and fog.
C— IFR conditions due to overcast ceilings less than 1,000 feet with reduced visibilities in rain and rain showers.

3319. (Refer to figure 12.) The Weather Depiction Chart indicates that most of Virginia is reporting

A— marginal VFR conditions due to extensive low ceilings.
B— IFR conditions due to very low visibilities and frontal buildups.
C— marginal VFR conditions due to reduced visibilities in fog and haze.

3320. (Refer to figure 12.) What restrictions to visibility are depicted in western Iowa?

A— Drizzle.
B— Fog, rain, and haze.
C— Drizzle, fog, and rain.

3321. (Refer to figure 13.) What is the direction and speed of movement of the line that extends from southwestern Nebraska to east-central Minnesota?

A— Northeast at 50 knots.
B— Southeast at 22 knots.
C— Northeast at 30 knots.

3322. (Refer to figure 13.) What is the VIP level of area A?

A— 2.
B— 3.
C— 5.

3323. (Refer to figure 13.) What does the 280 in area C mean?

A— The base of the clouds is 2,800 feet MSL.
B— Coverage of precipitation is 28.0 percent.
C— The highest top of precipitation is 28,000 feet MSL.

3324. (Refer to figure 13.) What is the VIP level of the black area in area F?

A— 3.
B— 4.
C— 6.

3325. A Radar Summary Chart can be very helpful to a pilot because it graphically displays

A— the intensity and movement of precipitation.
B— ceilings and precipitation between reporting stations.
C— areas of clouds, ceiling heights, and intensity of freezing precipitation.

3326. (Refer to figure 14.) Which area(s) should have the lowest ceilings at 1800Z?

A— The area just ahead of the cold front.
B— The area extending from northern Kansas to western Wisconsin.
C— The areas where precipitation is expected to occur, east of the cold front and west of the warm front.

3327. (Refer to figure 14.) Where is snow expected at 1800Z?

A— Northern Oregon and Washington.
B— In the central Great Lakes area.
C— From northwest Kansas to the Great Lakes and from northwest Colorado northward to Canada.

3328. (Refer to figure 14.) What type precipitation is expected in eastern Arkansas at 1800Z?

A— Rain showers over the entire area.
B— Continuous rain over the entire area.
C— Rain showers and thunderstorms affecting .5 or more of the area.

3329. (Refer to figure 14.) At what altitude is the freezing level in central Oklahoma as forecast on the 24-hour Significant Weather Prog?

A— 4,000 feet MSL.
B— 5,000 feet MSL.
C— 6,000 feet MSL.

3330. (Refer to figure 14.) At what altitude is the freezing level in central Oklahoma as forecast on the 12-hour Significant Weather Prog?

A— On the surface.
B— 4,000 feet MSL.
C— 8,000 feet MSL.

3331. Which weather chart depicts the conditions forecast to exist at a specific time in the future?

A— Prognostic.
B— Surface Analysis.
C— Weather Depiction.

3332. Which is an operational consideration concerning U.S. Low-Level Significant Weather Prognostic Charts?

A— The charts are designed for use in domestic flight planning to 24,000 feet.
B— This is a four-panel chart that forecasts the weather for a period of 48 hours.
C— The valid time of the charts corresponds to the time of the plotted observations and they are not forecasts.

3333. (Refer to figure 15.) This High-Level Significant Weather Prognostic Chart encompasses airspace from

A— FL240 to FL630.
B— 20,000 feet MSL to FL610.
C— 10,000 feet MSL to 51,000 feet MSL.

3334. (Refer to figure 15.) The 340 in a rectangle, at area A, means the forecast

A— height of the tropopause in millibars.
B— highest altitude for turbulence in that area.
C— height of the tropopause in hundreds of feet MSL.

3335. (Refer to figure 15.) What is the type and intensity of the forecast turbulence in area B?

A— Clear air turbulence, light to severe in intensity.
B— Convective turbulence, light to moderate in intensity.
C— Clear air turbulence, moderate to severe in intensity.

3336. (Refer to figure 15.) What is the forecast extent of the turbulence in area C?

A— From below FL240 to FL320.
B— From FL180 to 36,000 feet MSL.
C— From the surface to 32,000 feet MSL.

3337. (Refer to figure 15.) What is the forecast coverage of thunderstorms in area D?

A— 1/8 to 4/8.
B— 1/2 or less.
C— Less than 1/8.

3338. (Refer to figure 15.) In area F, what is the significance of the numbers 510 over the 3 X's?

A— Cloud bases are forecast to be from 18,000 feet MSL to 51,000 feet MSL.
B— Cloud bases are forecast to be below FL240 and the tops are forecast to reach FL510.
C— Turbulence is forecast to be severe from 24,000 feet MSL to 51,000 feet MSL.

3339. (Refer to figure 16.) What percent coverage of severe thunderstorms is forecast to occur in the area of moderate risk in the north-central United States?

A— 6 to 10 percent.
B— 10 to 50 percent.
C— 50 to 90 percent.

3340. If an area on a Severe Weather Outlook Chart is labeled APCHG, this indicates

A— possible tornadoes.
B— thunderstorm activity may approach extreme intensity.
C— winds greater than or equal to 35 knots but less than 50 knots.

3341. (Refer to figure 16.) A crosshatched area on the Severe Weather Outlook Chart indicates a

A— current tornado watch area.
B— severe weather outlook area.
C— forecast severe thunderstorm watch area.

3342. (Refer to figure 17.) What are the probable weather conditions in the area indicated by arrow B on the Stability Chart?

A— Neutral stability; showery precipitation.
B— Stable air; stratified cloudiness and steady precipitation.
C— Moderate humidity and unstable air; scattered shower activity.

3343. (Refer to figure 17.) What are the probable weather conditions in the area indicated by arrow C on the Stability Chart?

A— Unstable air; instability, showers, and thunderstorms.
B— Neutral stability; stratus clouds and light precipitation.
C— Moderately saturated air; steady precipitation and light turbulence.

3344. (Refer to figure 17.) What are the probable weather conditions in the area indicated by arrow D on the Stability Chart?

A— Stable air; predominately fair.
B— High relative humidity; showers and thunderstorms.
C— Marginally unstable air; moderate turbulence and possible thunderstorms.

3345. (Refer to figure 17.) Which symbol on the Stability Chart signifies very stable air and no precipitation?

A— $\frac{23}{-15}$
B— $\frac{-3}{34}$
C— $\frac{0}{35}$

3346. From which of the following can the observed temperature, wind, and temperature/dewpoint spread be determined at specified flight levels?

A— Stability Charts.
B— Winds Aloft Forecasts.
C— Constant Pressure Charts.

3347. When using a Constant Pressure Analysis Chart for planning a flight at 10,000 feet MSL, which analysis should the pilot refer to?

A— 850-millibar.
B— 700-millibar.
C— 500-millibar.

SELECTED SURFACE AVIATION WEATHER REPORTS

SA 301701
FTW SA 1654 -X M6 OVC 11/2HK 69/63/0904/010/ F2 TWR VSBY 21/2 CIG RGD
DFW SA 1646 M8 OVC 4F 193/68/64/1105/0111
DAL SA 1649 M9 OVC 4FH 72/63/0000/010
TYR SA 1650 -X 15 SCT 3F 76/66/0804/999/F2
ABI SA 1651 30 SCT E200 OVC 25 168/74/51/1614/008 CB W-N
AMA SA 1651 M5 OVC 4F 217/51/48/0516/023/LB26E40
AUS RS 1651 M10 BKN 24 BKN 160 OVC 4R-H 169/70/63/1208/005/RB25
BKO SA 1655 20 SCT 130 SCT E250 OVC 7 149/85/67/1515G20/997/RB19E25
TCU ALQDS

FIGURE 3.—Surface Aviation Weather Reports.

PILOT REPORTS

UA /OV DAL 090015 1725 FL090 /TP BE55 /SK 045 OVC 085 CLR ABV
UA /OV ABI 270015 1845 FL120 /TP C411 /TB MDT BLO-100
UA /OV AUS 315020 1715 FL150 /TP C500 /SK OVC 130 /IC TRACE RIME
DURGC WBND 090-130
UA /OV BRO 360030 1700 FL045 /TP PA31 /SK 050 OVC /RM LRG TSTM
50N BRO 20 WIDE

FIGURE 4.—Pilot Reports.

SELECTED TERMINAL FORECASTS

FT 300940
ABI 301010 C250 BKN 1413. 12Z 20 SCT C100 BKN 1615 CHC C14 OVC 1TRW. 18Z 30 SCT
C100 BKN 1716 CHC C10 OVC 1TRW. 02Z 80 SCT C250 BKN 04Z VFR . .

ACT 301010 C1 X 1/2F VRBL C7 BKN 3F. 17Z C20 BKN CHC C12 OVC 2TRW. 04Z VFR CIG
ABV 10 THSD BCMG IFR CIG TRWF . .

AMA 301010 C8 OVC 3R-F 0314 CHC C5 OVC 1TRW-F. 17Z C18 BKN 0418 CHC C8 OVC 1 TRWF.
20Z C30 BKN 1417 CHC C10 OVC 1TRW+F. 04Z MVFR CIG . .

AUS 301010 C5 OVC 2F. 14Z C14 OVC 1510 CHC TRW. 20Z C20 OVC 1510 CHC C5 OVC 2TRW
G25. 02Z C12 OVC 1010 CHC TRW. 04Z MVFR CIG TRW . .

BRO 301010 40 SCT C150 OVC 1210. 14Z C30 BKN 1512 CHC RW/TRW. 02Z C12 OVC 1010
CHC RW/TRW. 04Z MVFR CIG TRW . .

DAL 301010 80 SCT C250 4F VRBL C4 OVC 1F. 16Z C18 BKN 0512 SLGT CHC C12 OVC 2TRWF.
19Z 30 SCT C80 BKN 0612 SCT CHC C10 OVC 1TRWF. 04Z VFR . .

FIGURE 5.—Terminal Forecasts.

AREA FORECAST EXCERPT

DFWC FA 231040
SGFNT CLOUD AND WX VALID UNTIL 232300...OTLK 232300-240500

.
IFR...OK TX AR LA TN MS AL
FROM YDR TO YQT TO YYZ TO FWA TO CHA TO 60SE MSY TO LCH TO TXK TO LBB TO SLN TO ISN TO YDR
OCNL CIGS BLO 10 OVC AND VSBYS BLO 3SF NRN AND WRN PTN AND RF ELSW. CONDS SLOLY DMSHG FROM SW BUT CONTG BYD 23Z.

.
OK
CIGS 10–15 OVC LYRD TO 120. OCNL CIGS BLO 10 OVC AND VSBYS BLO 3R–S–F. 19Z CIGS 20 BKN–OVC 80 BKN 120. 22Z AGL 20–30 SCT–BKN 60. OTLK...VFR.

.
NWRN TX
CIGS 20–30 BKN 80. OCNL CIGS BLO 10 TIL 16Z ERN PNHDL.
18Z AGL 30–50 SCT. OTLK...VFR.

.
SWRN TX
NO CIGS BLO 120. OTLK...VFR.

.
NCNTRL AND NERN TX
AGL 15–20 SCT–BKN 50 BKN–OVC 120. ISOLD RW–. 18Z AGL 20–30 SCT 80 BKN 120. OTLK...VFR.

.
SCNTRL TX SERN TX AND CSTL WTRS
CSTL PLAIN SERN AND CSTL WTRS..CIGS 15–20 BKN–OVC 40 OVC 15. SCT TRW WITH CB TOPS 450 EXTRM SERN AND CSTL WTRS TIL 15Z. 16Z CIGS 30 BKN–SCT 80 BKN 120. OTLK...VFR.
RMNDR SCNTRL TX...CIGS 30–50 BKN 80 BKN 121. 16Z 80 BKN 120. OTLK...VFR.

.
AR LA AND CSTL WTRS
CIGS 10–15 OVC LYJD TO 180. OCNL CIGS BLO 10 OVC AND VSBYS BLO 3R–F. SCT TRW WITH CB TOPS 450 SLOLY ENDG FROM W. 18Z CIGS 20 BKN 60 BKN–OVC 120. ISOLD TRW– WITH CB TOPS 350 SERN LA. 21Z AGL 30–50 SCT 80 BKN 120 LA AND CIGS 20–30 BKN–SCT 80 BKN 120 AR. OTLK...VFR.

.
WRN TWO THIRDS TN MS AND CSTL WTRS
CIGS 10–15 OVC LYRD TO 150. OCNL CIGS BLO 10 OVC AND VSBYS BLO 3R–F. SCT TRW WITH CB TOPS 450. OTLK...MVFR CIG RW.

.
ERN THIRD IN
AGL 20 SCT 30–50 BKN–OVC LYRD TO 120. 15Z CIGS 20–30 BKN 50 OVC LYRD TO 150. WDLY SCT TRW WITH CB TOPS 450. OTLK...MVFR CIG TRW.

.
AL
NWRN HALF...CIGS 10 OVC LYRD TO 150. WDLY SCT RW– WITH OCNL CIGS BLO 10 OVC AND VSBYS BLO 3R–F TIL 17Z. WDLY SCT TRW WITH CB TOPS 450 AFT 13Z. OTLK...MVFR CIG TRW.
SERN HALF AND CSTL WTRS...CIGS 15–20 BKN–OVC LYRD TO 150 WITH WDLY SCT RW– SWRN PIN. LCL CIGS BLO 10 OVC AND VSBYS BLO 3R–F TIL 16Z. WDLY SCT TRW– WITH CB TOPS 450 AFT 16Z. OTLK...MVFR CIG TRW.
...

FIGURE 6.—Area Forecast.

WINDS AND TEMPERATURES ALOFT FORECASTS

FDUS1 KWBC 301640
DATA BASED ON 301200Z

VALID 301800Z   FOR USE 1700-2100Z. TEMPS NEG ABV 24000

FT	3000	6000	9000	12000	18000	24000	30000	34000	39000
ABI		1409+11	1611+06	1609+01	1608-13	1815-25	192440	192550	191960
ABQ			1307+08	1608+01	2005-13	2508-26	271642	282151	292460
AMA		0408	0905+05	9900+00	2108-13	2214-25	232141	232351	232261
BRO	9900	2211+14	2314+09	2318+04	2428-10	2336-23	234739	235448	246358
DAL	1008	1506+11	1705+07	2006+01	2210-13	2217-24	222740	223049	232960
ICT	3514	3412+06	3210+03	2910-01	2618-13	2526-26	253542	254052	254463

FIGURE 7.—Winds and Temperatures Aloft Forecasts.

FIGURE 8.—Surface Analysis Chart Symbol.

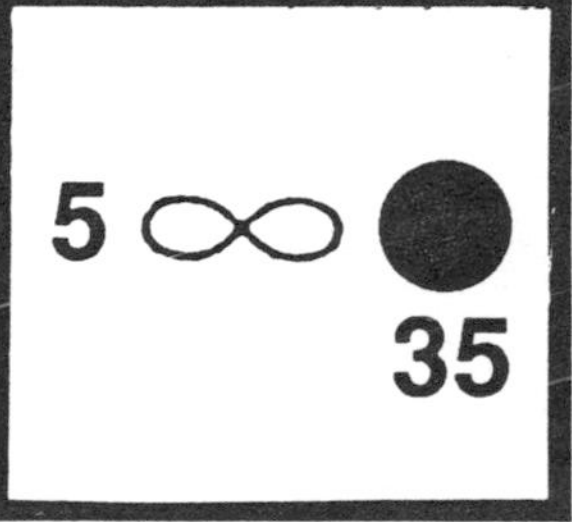

FIGURE 10.—Weather Depiction Chart Symbol.

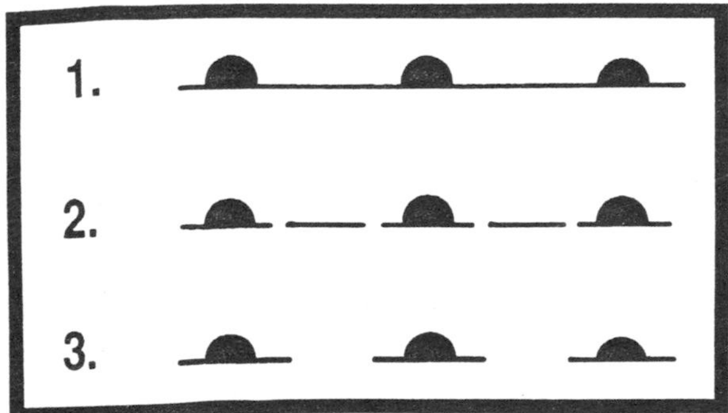

FIGURE 9.—Surface Analysis Chart Symbols.

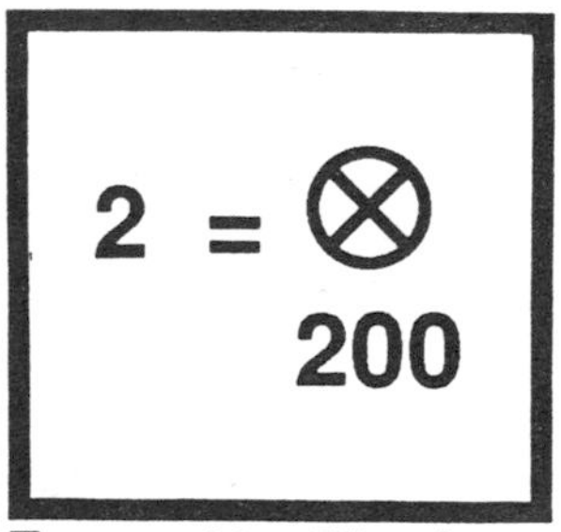

2 = ⊗
200

FIGURE 11.—Weather Depiction Chart Symbol.

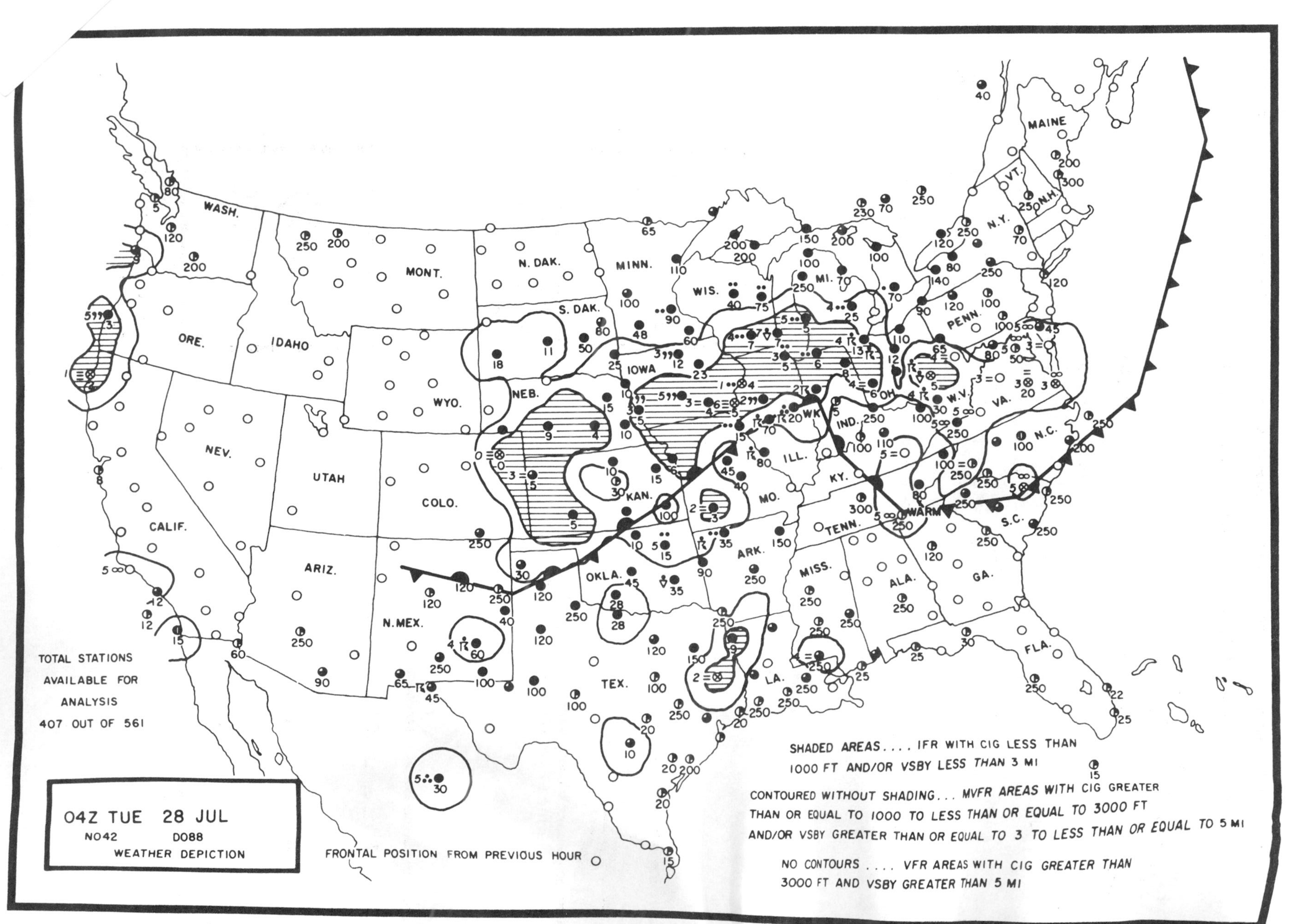

FIGURE 12.—Weather Depiction Chart.

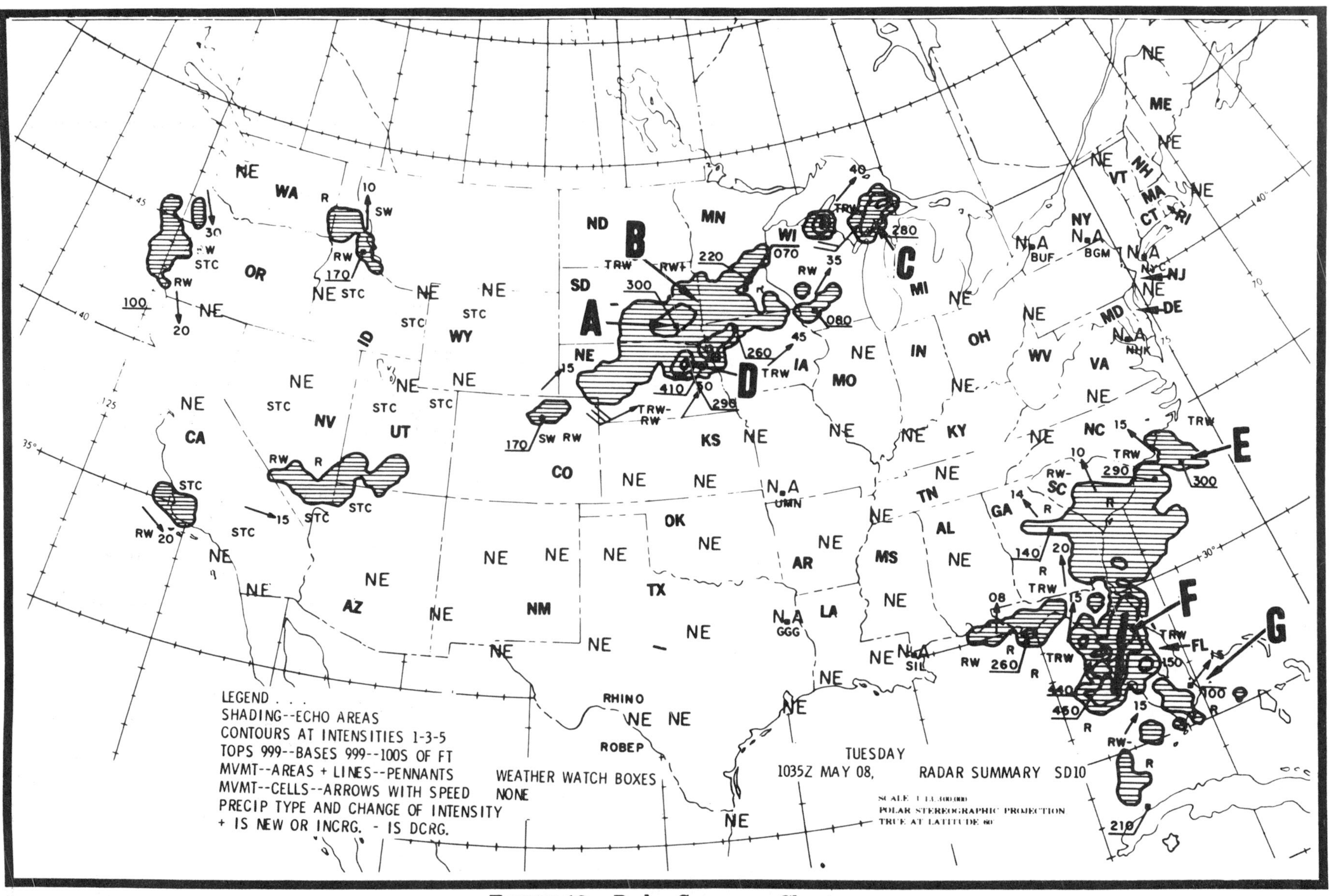

FIGURE 13.—Radar Summary Chart.

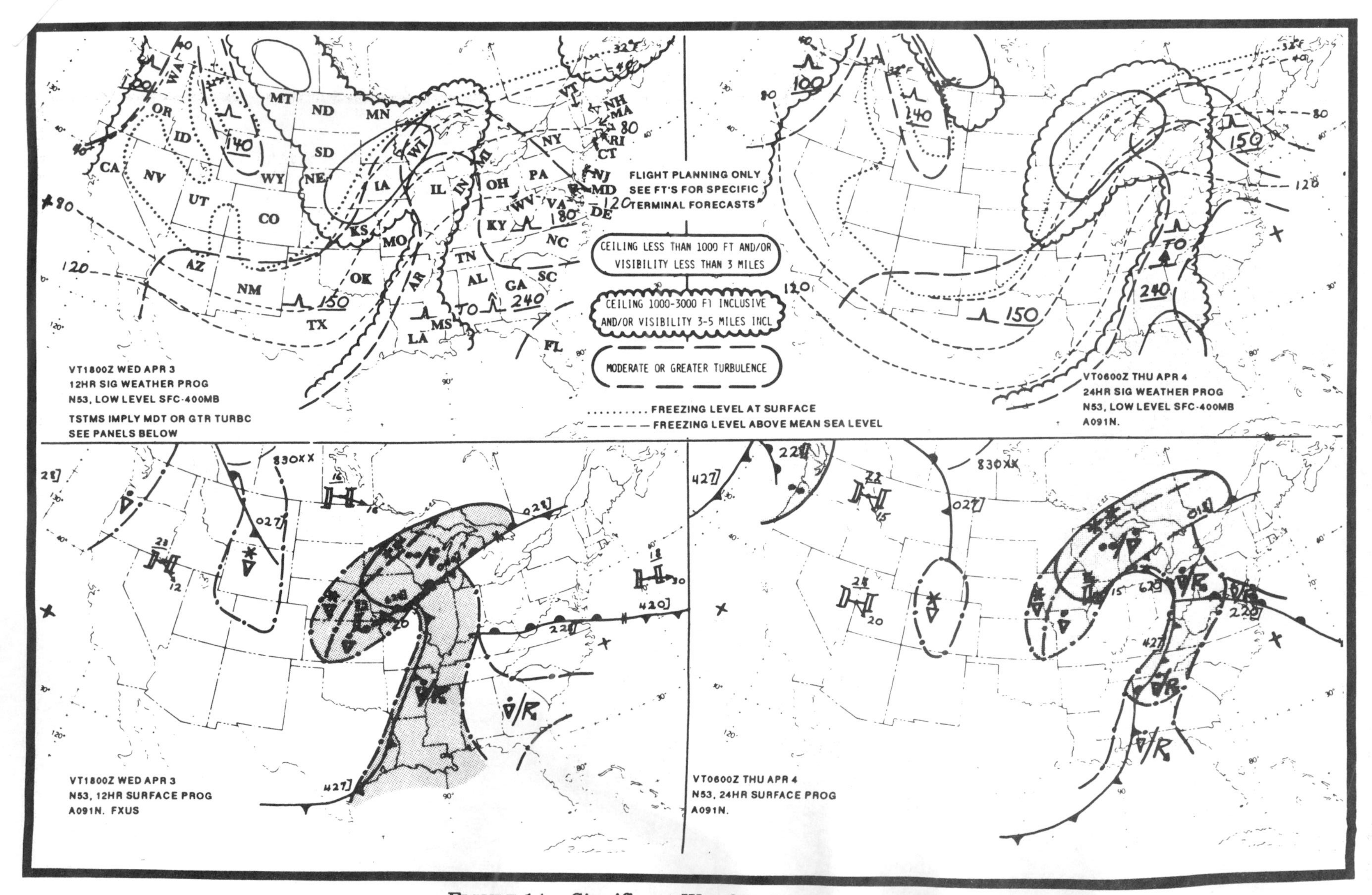

FIGURE 14.—Significant Weather Prognostic Chart.

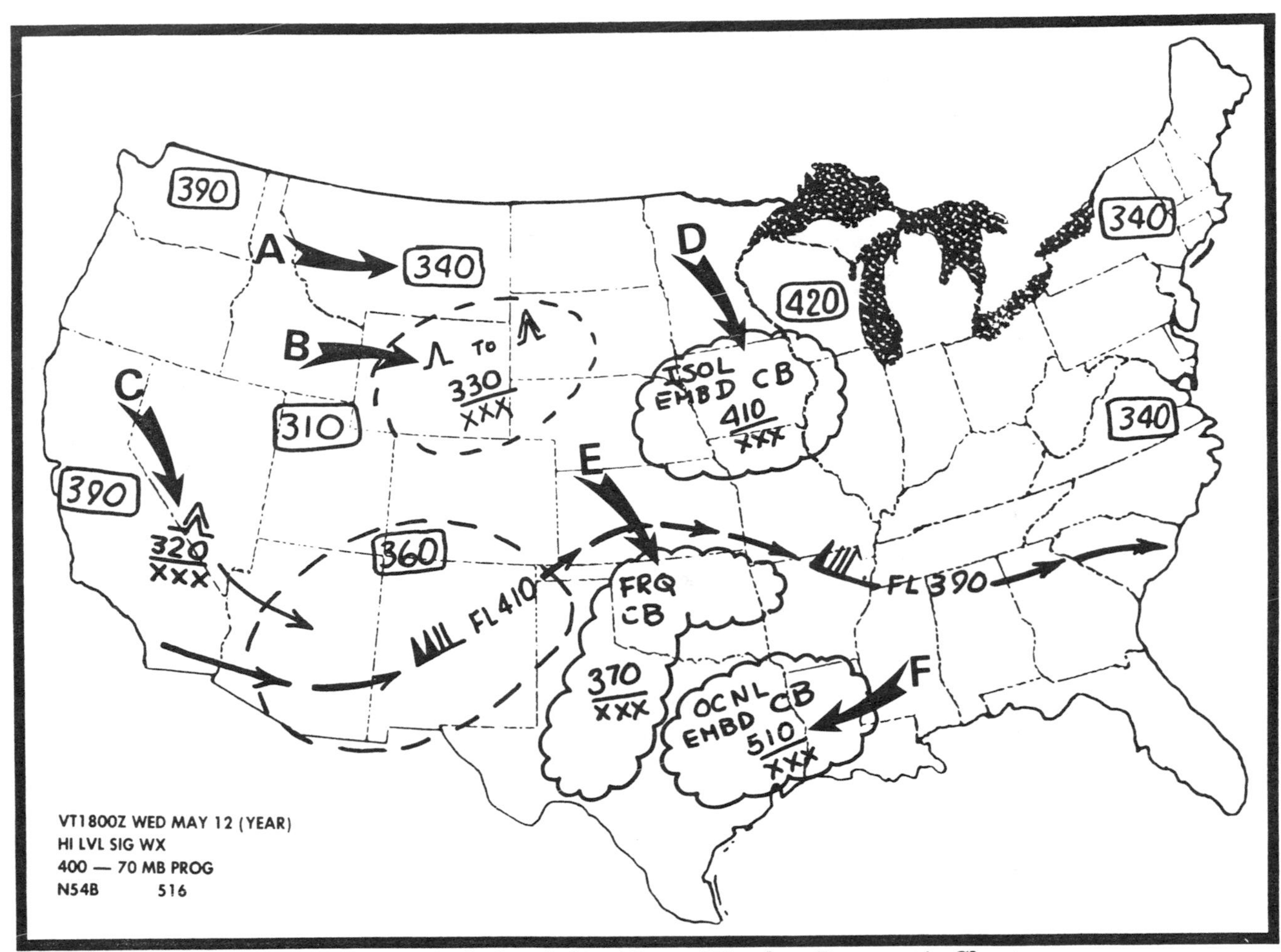

FIGURE 15.—High-Level Significant Weather Prognostic Chart.

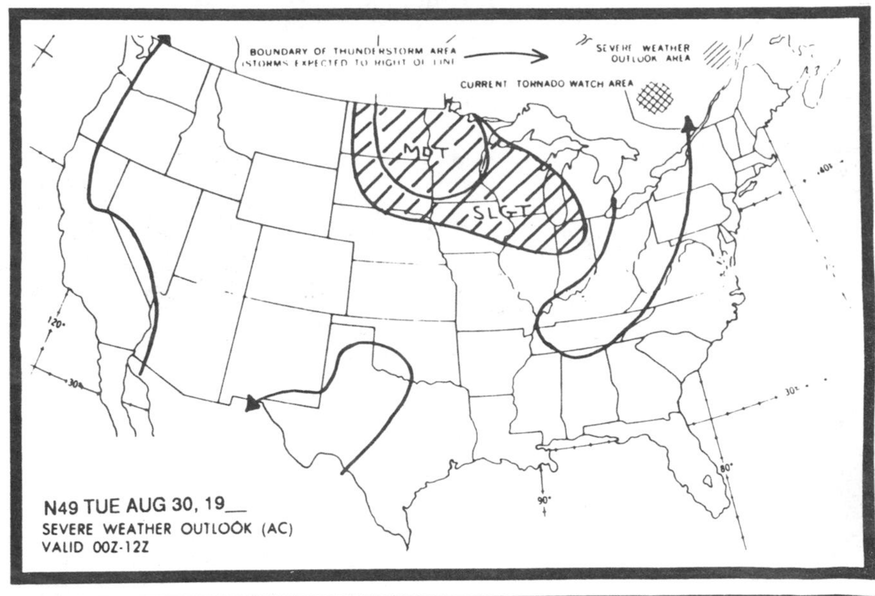

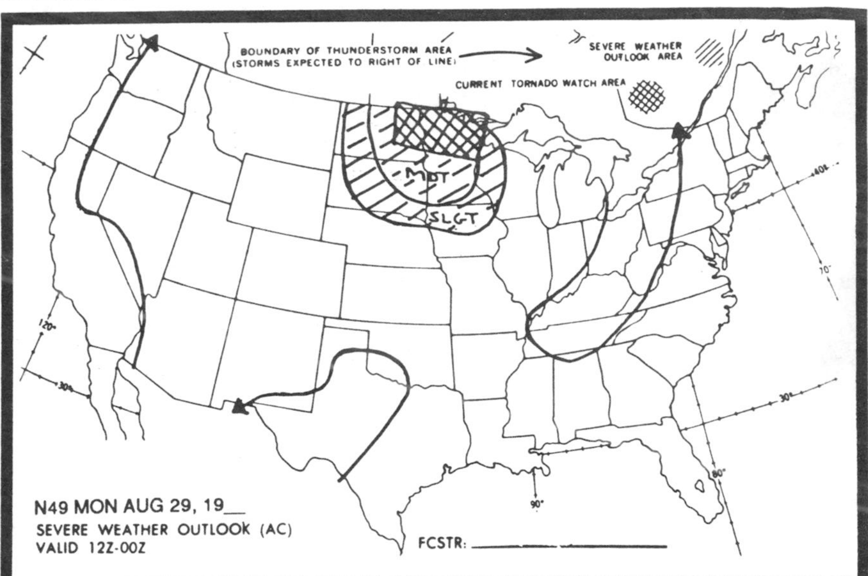

FIGURE 16.—Severe Weather Outlook Chart.

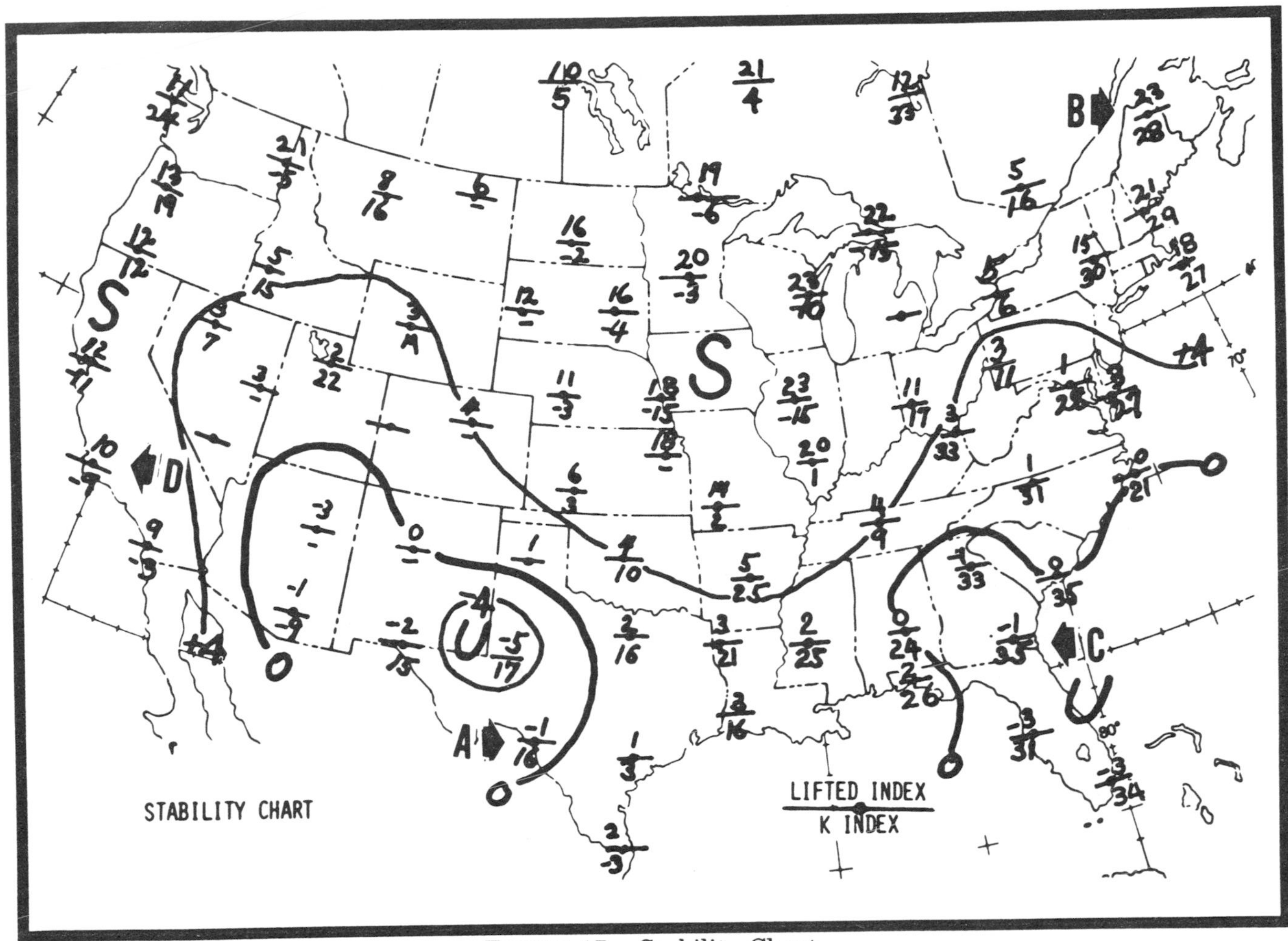

FIGURE 17.—Stability Chart.

6343. Which primary source contains information regarding the expected weather at the destination airport and at the ETA?

A— Low-Level Prog Chart.
B— Radar Summary and Weather Depiction Charts.
C— Terminal Forecast.

6344. The body of a Terminal Forecast covers a geographical area within

A— a 5-mile radius of the center of a runway complex.
B— 25 miles of the center of an airport.
C— 10 miles of the station originating the forecast.

6345. The absence of a visibility entry in a Terminal Forecast specifically implies that the surface visibility

A— is at least 1 statute mile above the minimum visibility requirement for an approach to the primary instrument runway.
B— exceeds 6 statute miles.
C— is at least 15 statute miles in all directions from the center of the runway complex.

6346. What sources reflect the most accurate information on current and forecast icing conditions?

A— Low-Level Sig Weather Prog Chart, RADAT's, and the Area Forecast.
B— PIREP's, Area Forecast, and the Freezing Level Chart.
C— PIREP's, AIRMET's, and SIGMET's.

6347. What weather is predicted by the term "TRW VICINITY" in a Terminal Forecast?

A— Thunderstorms are expected between 5 and 25 miles of the runway complex.
B— Rain showers may occur over the station and within 50 miles of the station.
C— Scattered thundershowers are predicted within the Terminal Control Area.

6348. If squalls are reported at the destination airport, what wind conditions existed at the time?

A— Sudden increases in windspeed of at least 15 knots, lasting for at least 1 minute.
B— Peak gusts of at least 35 knots for a sustained period of 1 minute or longer.
C— Rapid variation in wind direction of at least 20° and changes in speed of at least 10 knots between peaks and lulls.

6349. Which type of weather can only be directly observed during flight and then reported in a PIREP?

A— Structural icing.
B— Jetstream type winds.
C— Level of the tropopause.

6350. Forecast winds and temperatures aloft for an international flight may be obtained by consulting

A— Area Forecasts published by the departure location host country.
B— the current International Weather Depiction Chart appropriate to the route.
C— Wind and Temperature Aloft Charts prepared by a Regional Area Forecast Center (RAFC).

6351. How will an area of thunderstorm activity, that may grow to severe intensity, be indicated on the Severe Weather Outlook Chart?

A— SLGT within cross-hatched areas.
B— APCHG within any area.
C— SVR within any area.

6352. For international flights, a High-Level Significant Prognostic Chart is prepared for use at

A— any flight level above 290.
B— FL 250 to FL 600.
C— FL 180 to FL 600.

6353. The Low–Level Prognostic Chart depicts weather conditions

A— that are forecast to exist at a specific time shown on the chart.
B— as they existed at the time the chart was prepared.
C— that are forecast to exist 6 hours after the chart was prepared.

6354. A station is forecasting wind and temperature aloft to be 280° at 205 knots; temperature –51 °C at FL 390. How would this data be encoded in the FD?

A— 7800–51.
B— 789951.
C— 280051.

6355. At what time are current AIRMET's broadcast in their entirety by the AFSS?

A— 15 minutes after the hour only.
B— Every 15 minutes until the AIRMET is canceled.
C— 15 and 45 minutes after the hour during the first hour after issuance.

6356. If a SIGMET alert is announced, how can information contained in the SIGMET be obtained?

A— ATC will announce the hazard and advise when information will be provided in the FSS broadcast.
B— By contacting a weather watch station.
C— By contacting the nearest AFSS.

6357. What type service should normally be expected from an En Route Flight Advisory Service?

A— Weather advisories pertinent to the type of flight, intended route of flight, and altitude.
B— Severe weather information, changes in flight plans, and receipt of position reports.
C— Radar vectors for traffic separation, route weather advisories, and altimeter settings.

6360. Below FL 180, en route weather advisories should be obtained from an FSS on

A— 122.1 MHz.
B— 122.0 MHz.
C— 123.6 MHz.

6361. What type turbulence should be reported when it causes slight, rapid, and somewhat rhythmic bumpiness without appreciable changes in attitude or altitude, less than one–third of the time?

A— Occasional light chop.
B— Moderate turbulence.
C— Moderate chop.

6362. What type turbulence should be reported when it causes changes in altitude and/or attitude more than two–thirds of the time, with the aircraft remaining in positive control at all times?

A— Continuous severe chop.
B— Continuous moderate turbulence.
C— Intermittent moderate turbulence.

6363. What type turbulence should be reported when it momentarily causes slight, erratic changes in altitude and/or attitude, one–third to two–thirds of the time?

A— Occasional light chop.
B— Moderate chop.
C— Intermittent light turbulence.

6364. What conditions are indicated on a Weather Depiction Chart?

A— Actual sky cover, visibility restrictions, and type of precipitation at reporting stations.
B— Forecast ceilings and visibilities over a large geographic area.
C— Actual en route weather conditions between reporting stations.

6365. (Refer to figure 145.) What was the local Central Standard Time of the surface report at Austin (AUS)?

A— 11:53 a.m.
B— 5:53 p.m.
C— 10:53 p.m.

6366. (Refer to figure 145.) What type of report is listed for Dalhart (DHT)?

A— A report made by an automatic weather reporting system.
B— A special report concerning very low station pressure.
C— A record of a special report about a significant weather change.

6367. (Refer to figure 145.) What method was used to obtain the SP report at Marfa (MRF)?

A— Staffed AMOS station.
B— Automatic weather observing station (AMOS).
C— A military station observation of temperature, dewpoint, wind, and station pressure only.

6368. (Refer to figure 145.) What condition is reported at Childress (CDS)?

A— Distant heavy rain showers.
B— Heavy rain showers began 20 minutes after the hour.
C— The ceiling is solid overcast at an estimated 1,800 feet above sea level.

6369. (Refer to figure 145.) What condition is reported at Dallas (DAL)?

A— The station pressure is 1008.7 millibars.
B— Temperature/dewpoint spread is 16 °C.
C— Altimeter setting is 30.07.

6370. (Refer to figure 145.) The pilot report at Fort Worth (FTW) indicates

A— several overcast layers including one above 9,500 feet.
B— a clear layer between 3,600 feet and 6,000 feet.
C— the base of an overcast layer at 7,500 feet.

6371. (Refer to figure 145.) The SP report at Galveston (GLS) indicates which condition?

A— Wind 170° magnetic at 5 knots.
B— No precipitation since last synoptic report.
C— Sea level pressure 1000.7 millibars.

6372. (Refer to figure 145.) What weather improvement was reported at Lubbock (LBB) between 1750 and 1818 UTC?

A— The rain showers stopped.
B— The ceiling improved by 1,800 feet.
C— Visibility improved.

6373. (Refer to figure 145.) What weather condition is indicated by the report at Midland (MAF)?

A— Rain of unknown intensity was observed in the 090 to 180 quadrant.
B— The ceiling was at 25,000 feet MSL.
C— Wind was 020° magnetic at 20 knots.

6374. (Refer to figure 146.) What change took place at Wichita Falls (SPS) between 1757 and 1820 UTC?

A— The rain became heavier.
B— Atmospheric pressure increased.
C— The ceiling lowered.

6375. (Refer to figure 146.) What was the ceiling at Fort Smith (FSM)?

A— 8,000 feet AGL.
B— 2,500 feet AGL.
C— 2,900 feet MSL.

6376. (Refer to figure 146.) What change had taken place between 1755 and 1825 UTC at Harrison (HRO)?

A— Wind shifted from south to north-northwest.
B— Thundershowers began at 25 minutes after the hour.
C— Visibility reduced to IFR conditions.

6377. (Refer to figure 147.) The categorical outlook for Austin (AUS) indicates

A— marginal VFR due to ceilings and thunderstorms.
B— a chance of 1,000 feet AGL ceilings.
C— ceiling 1,400 broken and thundershowers.

6378. (Refer to figure 147.) At which time is IFR weather first predicted at Lubbock (LBB)?

A— 1500Z.
B— 1700Z.
C— 0900Z.

6379. (Refer to figure 147.) What type conditions can be expected for a flight scheduled to land at San Angelo (SJT) at 1500Z?

A— Chance of 1 nautical mile visibility.
B— Occasional ceilings 800 feet in thunderstorms.
C— IFR conditions due to low ceilings, rain, and fog.

6380. TWEB Route Forecasts provide predicted weather for

A— a corridor 25 miles either side of a numbered cross–country route.
B— a 50–mile radius of the takeoff and landing airports.
C— any route of flight specified by the requesting pilot.

6381. How can the pilot obtain TWEB Route Forecast information?

A— From the TEL TWEB and Telephone Voice Response Systems (VRS).
B— From the ATIS and Pilots Automatic Telephone Weather.
C— From ARTCC and Automated Flight Service Station briefings.

6382. What information is provided by this TWEB Route Forecast excerpt?

249 TWEB 252317 GFK––MOI––ISN, GFK VCNTY CIGS AOA 5 THSD TILL 12Z...

A— Grand Forks (GFK) ceilings at or above 5,000 feet MSL.
B— Route No. 249, from GFK to MOI to ISN.
C— Ceilings within a 50–mile radius of Grand Forks (GFK) are 5,000 feet AGL.

6383. (Refer to figure 148.) Which system in the Convective Sigmet listing has the potential of producing the most severe storm?

A— The storms in Texas and Oklahoma.
B— The storms in Colorado, Kansas, and Oklahoma.
C— The isolated storm 50 miles northeast of Memphis (MEM).

6384. (Refer to figure 148.) What time period is covered by the outlook section of the Convective Sigmet?

A— 24 hours after the valid time.
B— 2 to 6 hours after the valid time.
C— No more than 2 hours after the valid time.

6385. Which type weather conditions are covered in the Convective Sigmet?

A— Embedded thunderstorms, lines of thunderstorms, and thunderstorms with 3/4–inch hail or tornadoes.
B— Cumulonimbus clouds with tops above the tropopause and thunderstorms with 1/2–inch hail or funnel clouds.
C— Any thunderstorm with a severity level of VIP 2 or more.

6386. (Refer to figure 149.) What approximate wind direction, speed, and temperature (relative to ISA) are expected for a flight over OKC at FL 370?

A— 265° true; 27 knots; ISA +1 °C.
B— 260° true; 27 knots; ISA +6 °C.
C— 260° magnetic; 27 knots; ISA 095 °C.

6387. (Refer to figure 149.) What approximate wind direction, speed, and temperature (relative to ISA) are expected for a flight over TUS at FL 270?

A— 347° magnetic; 5 knots; ISA –10 °C.
B— 350° true; 5 knots; ISA +5 °C.
C— 010° true; 5 knots; ISA +13 °C.

6388. (Refer to figure 149.) What will be the wind and temperature trend for an SAT–ELP–TUS flight at 16,000 feet?

A— Temperature decrease slightly.
B— Windspeed decrease.
C— Wind direction shift from southwest to east.

6389. (Refer to figure 149.) What will be the wind and temperature trend for a STL–MEM–MSY flight at FL 330?

A— Windspeed decrease.
B— Wind shift from west to north.
C— Temperature increase 5 °C.

6390. (Refer to figure 149.) What will be the wind and temperature trend for a DEN–ICT–OKC flight at 11,000 feet?

A— Temperature decrease.
B— Windspeed increase slightly.
C— Wind shift from calm to a westerly direction.

6391. (Refer to figure 149.) What will be the wind and temperature trend for a DSM–LIT–SHV flight at 12,000 feet?

A— Windspeed decrease.
B— Temperature decrease.
C— Wind direction shift from northwest to southeast.

6392. (Refer to figure 149.) What is the forecast temperature at ATL for the 3,000–foot level?

A— +6 °C.
B— +6 °F.
C— Not reported.

6393. (Refer to figure 149.) What approximate wind direction, speed, and temperature (relative to ISA) are expected for a flight over MKC at FL 260?

A— 260° true; 43 knots; ISA +10 °C.
B— 260° true; 45 knots; ISA –10 °C.
C— 260° magnetic; 42 knots; ISA +9 °C.

6394. What wind direction and speed aloft are forecast by this FD report for FL 390 — "750649"?

A— 350° at 64 knots.
B— 250° at 106 knots.
C— 150° at 6 knots.

6395. What wind direction and speed aloft are forecast by this FD report for FL 390 — "731960"?

A— 230° at 119 knots.
B— 131° at 96 knots.
C— 073° at 196 knots.

6396. (Refer to figure 150.) The Weather Depiction Chart indicates the heaviest precipitation along the front is occurring in

A— Missouri.
B— Illinois.
C— Kansas.

6397. (Refer to figure 150.) The Weather Depiction Chart indicates that the coastal sections of Texas and Louisiana are reporting

A— all ceilings at or above 20,000 feet with visibilities of 20 miles or more.
B— marginal VFR conditions due to broken ceilings of 2,000 feet.
C— VFR conditions with scattered clouds at 2,000 feet and higher cirriform.

6398. What is indicated on the Weather Depiction Chart by a continuous smooth line enclosing a hatched geographic area?

A— The entire area has ceilings less than 1,000 feet and/or visibility less than 3 miles.
B— More than 50 percent of the area enclosed by the smooth line is predicted to have IFR conditions.
C— Reporting stations within the enclosed area are all showing IFR conditions at the time of the report.

6399. (Refer to figure 151.) The 12–Hour Significant Weather Prognostic Chart indicates that West Virginia will likely experience

A— continuous or showery precipitation covering half or more of the area.
B— thunderstorms and rain showers covering half or more of the area.
C— continuous rain covering less than half of the area.

6400. (Refer to figure 151.) The 12–Hour Significant Weather Prognostic Chart indicates that eastern Kentucky and eastern Tennessee can expect probable ceilings

A— less than 1,000 feet and/or visibility less than 3 miles.
B— less than 1,000 feet and/or visibility less than 3 miles, and moderate turbulence below 10,000 feet MSL.
C— less than 1,000 feet and/or visibility less than 3 miles, and moderate turbulence above 10,000 feet MSL.

6401. (Refer to figure 151.) The chart symbols over southern California on the 12–Hour Significant Weather Prognostic Chart indicate

A— expected top of moderate turbulent layer to be 12,000 feet MSL.
B— expected base of moderate turbulent layer to be 12,000 feet MSL.
C— light turbulence expected above 12,000 feet MSL.

6402. (Refer to figure 151.) A planned low-altitude flight from central Oklahoma to western Tennessee at 1200Z is likely to encounter

A— continuous or intermittent rain or rain showers, moderate turbulence, and freezing temperatures below 8,000 feet.
B— continuous or showery rain over half or more of the area, moderate turbulence, and freezing temperatures above 10,000 feet.
C— showery precipitation covering less than half the area, no turbulence below 18,000 feet, and freezing temperatures above 12,000 feet.

6403. A prognostic chart depicts the conditions

A— existing at the surface during the past 6 hours.
B— which presently exist from the 1,000–millibar through the 700–millibar level.
C— forecast to exist at a specific time in the future.

6404. What information is provided by a Convective Outlook (AC)?

A— It describes areas of probable severe icing and severe or extreme turbulence during the next 24 hours.
B— It provides prospects of both general and severe thunderstorm activity during the following 24 hours.
C— It indicates areas of probable convective turbulence and the extent of instability in the upper atmosphere (above 500 mb).

6405. (Refer to figure 152.) What weather conditions are depicted in the area indicated by arrow A on the Radar Summary Chart?

A— Moderate to strong echoes; echo tops 30,000 feet MSL; line movement toward the northwest.
B— Weak to moderate echoes; average echo bases 30,000 feet MSL; cell movement toward the southeast; rain showers with thunder.
C— Strong to very strong echoes; echo tops 30,000 feet MSL; thunderstorms and rain showers.

6406. (Refer to figure 152.) What weather conditions are depicted in the area indicated by arrow D on the Radar Summary Chart?

A— Echo tops 4,100 feet MSL; strong to very strong echoes within the smallest contour; area movement toward the northeast at 50 knots.
B— Intense to extreme echoes within the smallest contour; echo tops 29,000 feet MSL; cell movement toward the northeast at 50 knots.
C— Strong to very strong echoes within the smallest contour; echo bases 29,000 feet MSL; cell in northeast Nebraska moving northeast at 50 knots.

6407. (Refer to figure 152.) What weather conditions are depicted in the area indicated by arrow C on the Radar Summary Chart?

A— Average echo bases 2,800 feet MSL; thundershowers; intense to extreme echo intensity.
B— Cell movement toward the northwest at 20 knots; intense echoes; echo bases 28,000 feet MSL.
C— Area movement toward the northeast at 20 knots; strong to very strong echoes; echo tops 28,000 feet MSL.

6408. (Refer to figure 152.) What weather conditions are depicted in the area indicated by arrow B on the Radar Summary Chart?

A— Weak echoes; heavy rain showers; area movement toward the southeast.
B— Weak to moderate echoes; rain showers increasing in intensity.
C— Strong echoes; moderate rain showers; no cell movement.

6409. (Refer to figure 154.) What is the height of the 300 MB level at the low–pressure center in Canada?

A— 9,120 meters MSL.
B— 18,000 meters MSL.
C— 11,850 meters MSL.

6410. (Refer to figures 153 through 155.) Interpret the path of the jetstream.

A— Southern California, Nevada, Utah, Nebraska/Kansas, and then southeastward.
B— Oregon, Idaho, Wyoming, Nebraska, Iowa, and across the Great Lakes.
C— The Alaska area, across Canada to Montana, North Dakota, then across the Great Lakes area.

6411. (Refer to figure 153.) What type weather system is approaching the California Coast from the west?

A— LOW.
B— HIGH.
C— Cold front.

6412. (Refer to figures 153 through 155.) What type weather is inferred by the almost vertical extent of the LOW in Canada?

A— A rapid–moving system with little chance of developing cloudiness, precipitation, and adverse flying conditions.
B— A slow–moving storm which may cause extensive and persistent cloudiness, precipitation, and generally adverse flying weather.
C— A rapid–moving storm, leaning to west with altitude, which encourages line squalls ahead of the system with a potential of severe weather.

6413. (Refer to figures 153 through 155.) What is the approximate temperature for a flight from southern California to central Kansas at FL 350?

A— 0916 °C.
B— 0939 °C.
C— 0941 °C.

6414. (Refer to figures 153 through 155.) Determine the approximate wind direction and velocity at FL 240 over the station in central Oklahoma.

A— 280° at 10 knots.
B— 320° at 10 knots.
C— 330° at 13 knots.

6415. (Refer to figures 153 through 155.) What is the relative moisture content of the air mass approaching the California coast?

A— Dry.
B— Moist enough for condensation.
C— Very wet with high potential for clouds and precipitation.

SURFACE AVIATION WEATHER REPORT

AUS SA 1753 40 SCT E250 BKN 8 170 87/69/1911G17/006
BPT SA 1755 30 SCT E250 BKN 7 183/93/74/1704/00
BRO SA 1755 34 SCT 250 –OVC 6H 163/93/74/1415/00
CDS SA 1758 E18 OVC 7RW– 183/73/69/1113/012/RW+E–S BINOVC RB20
CLL SA 1749 30 SCT 250 –BKN 7 181/93/69/2111/ 008
COT SA 1749 40 SCT 200 SCT 10 160/87/70/1310/002
CRP SA 1753 28 SCT E250 BKN 10 169/90/75/1616/003
DAL SA 1755 23 SCT E100 OVC 7 87/71/1605/007/HLYR
DFW SA 1800 35 SCT E120 OVC 10 174/85/68/1707/008
DHT RS 1756 E25 BKN 15 200/72/59/0414/026
DRT SA 1756 20 SCT E100 BKN 250 OVC 10 145/82/72/1212/000
ELP SA 1755 70 SCT 250 SCT 60 131/84/56/0907/015
FTW SA 1750 25 –SCT E100 OVC 7 84/68/1807/008
FTW UA /OV DFW 180005/TM 1803/FL095/TP PA30/SK 036 OVC 060/070 OVC 075/OVC ABV
GGG SA 1745 250 SCT 15 89/69/1508/011
GLS SA 1755 25 SCT E200 BKN 7 92/71
GLS SP 1811 AMOS 86/74/1705/007/000
HOU SA 1752 30 SCT E250 OVC 7 89/70/1506/008
HRL SA 1753 E250 OVC 7 93/72/1415/002/ FEW CU E
IAH SA 1755 33 SCT E250 BKN 10 181/87/73/1207/007
INK SA 1755 E30 BKN 60 BKN 250 BKN 20 144/81/61/0406/088
LBB SA 1750 E12 BKN 30 BKN 100 BKN 12 164/70/67/0413/015/ RB15E40
LBB SP 1818 15 SCT E30 BKN 100 BKN 12 0412/015
LBB UA /OV LBB 045002/TM 1821/FL060/TP B727/SK 045 BKN 053 TOPS RGD
LFK SA 1756 300 –BKN 7 182/91/66/2407/008
MAF SA 1756 M25 BKN 250 OVC 12 142/80/64/0220/009 RWU E–S
MFE SA 1756 250 –BKN 7 151/92/66/1315/998
MRF SA 1752 AMOS E30 BKN 60 125/81/57/0912/000/ PK WND 20
MRF SP 1811 AMOS 83/58/0809/000 PK WND 20

FIGURE 145.—Surface Aviation Weather Report.

SURFACE AVIATION WEATHER REPORT

MWL SA 1756 E11 BKN 50 OVC 10 E169/77/73/1311/006
PSX SA 1755 18 SCT E200 OVC 7 183/87/76/201/0/007
PVW SA 1750 12 SCT E30 OVC 10 185/85/68/0506/011/ BINOVC NE
SAT SA 1756 28 SCT E250 OVC 7 164/85/69/1516/005/ UA OV SAT TM 1739/FL UNKN/TP UNKN/SK TOPS 040
SJT SA 1755 E18 BKN 70 OVC 7 170/77/72/2212/002
SPS RS 1757 9 SCT M25 OVC 6R– 185/75/71/0914/011
SPS SP 1820 M15 OVC 6R– 1010/011
SPS UA /OV SPS/TM 1815/FL090/TP C402/SK OVC 050–060
TPL SA 1751 15 SCT 100 SCT 250 –OVC 15 89/69/1715/007
VCT SA 1755 30 SCT E250 OVC 7 179/88/73/1713/005

AR

ELD SA 1755 250 –BKN 6H 190/88/70/0605/010
FSM SA 1756 E80 BKN 120 OVC 190/86/69/2805/011
FSM UA /OV HRO–FSM/TM 1825/FL290/TP B737/RM SCT TOPS 290
FYV SA 1755 35 SCT E80 BKN 250 OVC 15 202/83/70/1705/016/ RWU SE
HOT SA 1751 40 SCT E150 OVC 15 91/62/3406/010
HRO RS 1755 20 SCT E35 BKN 13T 195/84/72/3007/015/ TB30 S–SSW MOVG E FQT THDR
HRO SP 1825 E15 BKN 35 BKN 13TRW– 1810G20/016/ T SE–W MOVG NE LTGICG RB25
LIT SA 1754 30 SCT E250 BKN 10 182/93/69/0704/007
PBF SA 1753 40 SCT E100 BKN 5H E183/95/68/2907/007
TXK SA 1753 100 SCT 200 –BKN 7 92/66/2503/010

FIGURE 146.—Surface Aviation Weather Report.

TERMINAL FORECAST

TX

ALI FT Ø31515 C12 BKN 6H 1415. 17Z C3Ø BKN 1515G25. 19Z 4Ø SCT C25Ø BKN 1515G25
SLGT CHC C2Ø BKN 3TRW. Ø1Z 15Ø SCT 1315. Ø7Z C8 BKN. Ø9Z IFR CIG. 14Z MVFR CIG..
AMA FT Ø31515 C1Ø BKN 8Ø BKN Ø512 OCNL C1Ø OVC 2TRW+. 18Z C2Ø BKN 8Ø OVC Ø315 OCNL
C1Ø OVC 2TRW+. 14Z C1Ø OVC 5R–F Ø312. Ø9Z IFR CIG RF..
AUS FT COR Ø31615 1545 C25 BKN 1ØØ OVC 171Ø. 18Z C3Ø BKN 1ØØ OVC 171Ø CHC C1Ø OVC 1TRW.
Ø4Z C14 BKN CHC C1Ø OVC 1TRW. Ø9Z MVFR CIG TRW..
BPT FT Ø31515 25 SCT 6H. 17Z 3Ø SCT 25Ø SCT 181Ø. Ø1Z 25Ø SCT. Ø9Z VFR. 1ØZ MVFR FK..
CRP FT Ø31515 2Ø SCT C25Ø BKN 1515G25. 18Z 3Ø SCT C25Ø BKN 1615G25 SLGT CHC TRW. Ø1Z 15Ø
SCT 1615. Ø9Z VFR..
DAL FT Ø31515 3Ø SCT C1ØØ BKN. 17Z C3Ø BKN 1ØØ OVC CHC C1Ø OVC 2TRW. Ø1Z C2Ø BKN CHC
3TRW. Ø9Z MVFR CIG TRW..
DRT FT AMD 1 Ø31515 152ØZ C14 OVC 141Ø. 16Z C2Ø BKN 1ØØ OVC 141Ø CHC 3TRW. 19Z C3Ø
BKN 1ØØ BKN 1412 CHC C2Ø BKN 3TRW. Ø2Z 35 SCT C8Ø BKN CHC C2Ø BKN 3TRW.
Ø9Z MVFR CIG TRW..
ELP FT Ø31515 7Ø SCT 1ØØ SCT. 17Z 7Ø SCT C12Ø BKN Ø812 CHC C5Ø BKN 5TRW G35. Ø8Z C7Ø BKN
Ø712 SLGT CHC RW–. Ø9Z MVFR CIG..
HOU FT Ø31515 2Ø SCT 6H. 17Z 35 SCT 25Ø SCT 181Ø. Ø1. Ø1Z 25Ø SCT. Ø9Z VFR. 11Z MVFR FK..
IAH FT Ø31515 2Ø SCT 6H. 17Z 35 SCT 25Ø SCT 181Ø. Ø1Z 25Ø SCT. Ø9Z VFR. 11Z MVFR FK..
INK FT Ø31515 2Ø SCT 1ØØ SCT 1Ø1Ø. 16Z C25 BKN 8Ø BKN Ø813 CHC C1Ø OVC 1TRW+ A G35.
Ø4Z C2Ø BKN 5Ø OVC Ø513 OCNL C1Ø OVC 1TRW+. Ø9Z MVFR CIG R F. 11Z IFR CIG R F..
LBB FT Ø31515 C1Ø BKN 5Ø OVC Ø612 CHC 1Ø SCT C2Ø OVC 3TRW. 17Z C2Ø BKN 6Ø OVC Ø415
OCNL C1Ø OVC 1TRW+. Ø4Z C1Ø OVC 4R–F OCNL 1TRW+. Ø9Z IFR CIG R F..
SAT FT Ø31515 C16 BKN 6H 171Ø. 17Z C25 BKN 1715. 19Z C3Ø BKN 25Ø OVC 1715 CHC C2Ø BKN
3TRW. Ø5Z C14 BKN CHC 3TRW. Ø9Z MVFR CIG TRW..
SJT FT Ø31515 C15 BKN 5Ø OVC 121Ø OCNL C8 OVC 4TRW. 2ØZ C25 BKN 7Ø OVC Ø912 OCNL C8
OVC 1TRW+. Ø9Z IFR CIG R F..
SPS FT Ø31515 3Ø SCT C8Ø BKN Ø712 CHC C1Ø OVC 2TRW. Ø9Z MVFR CIG TRW..

FIGURE 147.—Terminal Forecast.

CONVECTIVE SIGMET

MKCC WST Ø31755
CONVECTIVE SIGMET 42C
VALID UNTIL 1955Z
TX OK
FROM 5W MLC–PEQ–SJT–5W MLC
AREA SCT EMBDD TSTMS MOVG LTL. TOPS 3ØØ.

CONVECTIVE SIGMET 43C
VALID UNTIL 1955Z
CO KS OK
FROM AKO–OSW–3ØWNW OKC–AKO
AREA SCT TSTMS OCNLY EMBDD MOVG FROM 322Ø. TOPS 38Ø.

CONVECTIVE SIGMET 44C
VALID UNTIL 1955Z
50NE MEM
ISOLD INSTD LVL5 TSTM DIAM 1Ø MOVG FROM 2625. TOP ABV 45Ø.

OUTLOOK VALID UNTIL 2355Z
TSTMS OVR TX AND SE OK WL MOV SEWD 15 KTS.
TSTMS OVER CO, KS, AND N OK WL CONT MOVG SEWD 2Ø KTS.
TSTM OVR TN WL CONT MOVG EWD 25 KTS.

FIGURE 148.—Convective Sigmet.

WINDS AND TEMPERATURES ALOFT FORECASTS

DATA BASED ON 031200Z
VALID 040000Z FOR USE 1800–0300Z. TEMPS NEG ABV 24000

FT	3000	6000	9000	12000	18000	24000	30000	34000	39000
ABI		1306+16	1607+11	1807+06	2108–07	2208–18	240833	250942	300753
ABO			0810+14	0511+08	3415–06	3220–18	312333	312543	302554
AMA		0614	0814+10	0709+05	3210–07	2914–19	281934	282243	292554
ATL	0906	9900+17	9900+12	0205+07	3507–07	3305–19	290534	280543	990054
BNA	9900	9900+17	3205+12	3109+07	3018–07	2918–19	272134	262444	262855
BRO	1510	1614+20	1611+14	1708+08	9900–07	9900–19	990034	990043	990055
DAL	0910	1706+17	2009+11	2011+06	2015–08	2214–19	231333	241342	271153
DEN			9900+09	9900+04	3020–10	3029–21	303636	304145	294756
DSM	3615	3315+07	3118+04	3022+00	2835–12	2748–24	276438	277348	277957
ELP		0610	0614+13	0615+08	0113–05	3614–17	361433	361442	251354
GCK		0611+11	0809+08	9900+03	2817–09	2823–20	273135	273644	284155
HLC		0409+09	0405+07	3106+02	2822–10	2730–21	273936	274545	275256
HOU	0909	1607+19	1606+13	1606+07	1605–08	9900–20	990034	990043	990054
ICT	0516	0613+12	0607+08	9900+04	2718–09	2626–20	263635	264144	274655
IND	3611	3207+12	2912+08	2818+03	2733–09	2643–21	265635	265944	256255
INK		0609+16	0709+12	0608+07	0107–06	3607–18	350833	340842	350855
JAN	3612	3613+18	3611+13	3609+07	0105–08	9900–19	990034	990043	230854
LIT	0310	3608+16	3206+11	2808+06	2517–08	2518–19	252034	252243	262454
LOU	0105	9900+15	2908+10	2913+05	2825–08	2731–20	263834	264143	254454
MEM	0109	0108+17	3408+12	3110+06	2916–07	2717–19	261934	262144	262555
MKC	0316	0211+11	3409+07	3013+03	2728–10	2638–21	265036	265645	276356
MSY	0315	0216+19	0315+13	0414+07	0510–08	0605–20	990034	990043	210854
OKC	0715	0810+14	1106+10	9900+05	2414–08	2419–19	252534	252743	272754
SAT	1107	1713+18	1813+13	1911+07	2006–07	1906–19	180734	170743	990054
SGF	0414	0410+14	3605+09	2908+04	2624–09	2632–20	254135	264444	264655
SHV	0509	9900+18	9900+12	2106+06	2012–08	2109–19	220734	240743	260754
STL	0314	0110+12	3210+08	2915+03	2730–09	2741–21	265435	265744	266055
TUS		0807+23	0814+16	0814+10	0810–05	0505–17	330533	310842	290954

FIGURE 149.—Winds and Temperatures Aloft Forecast.

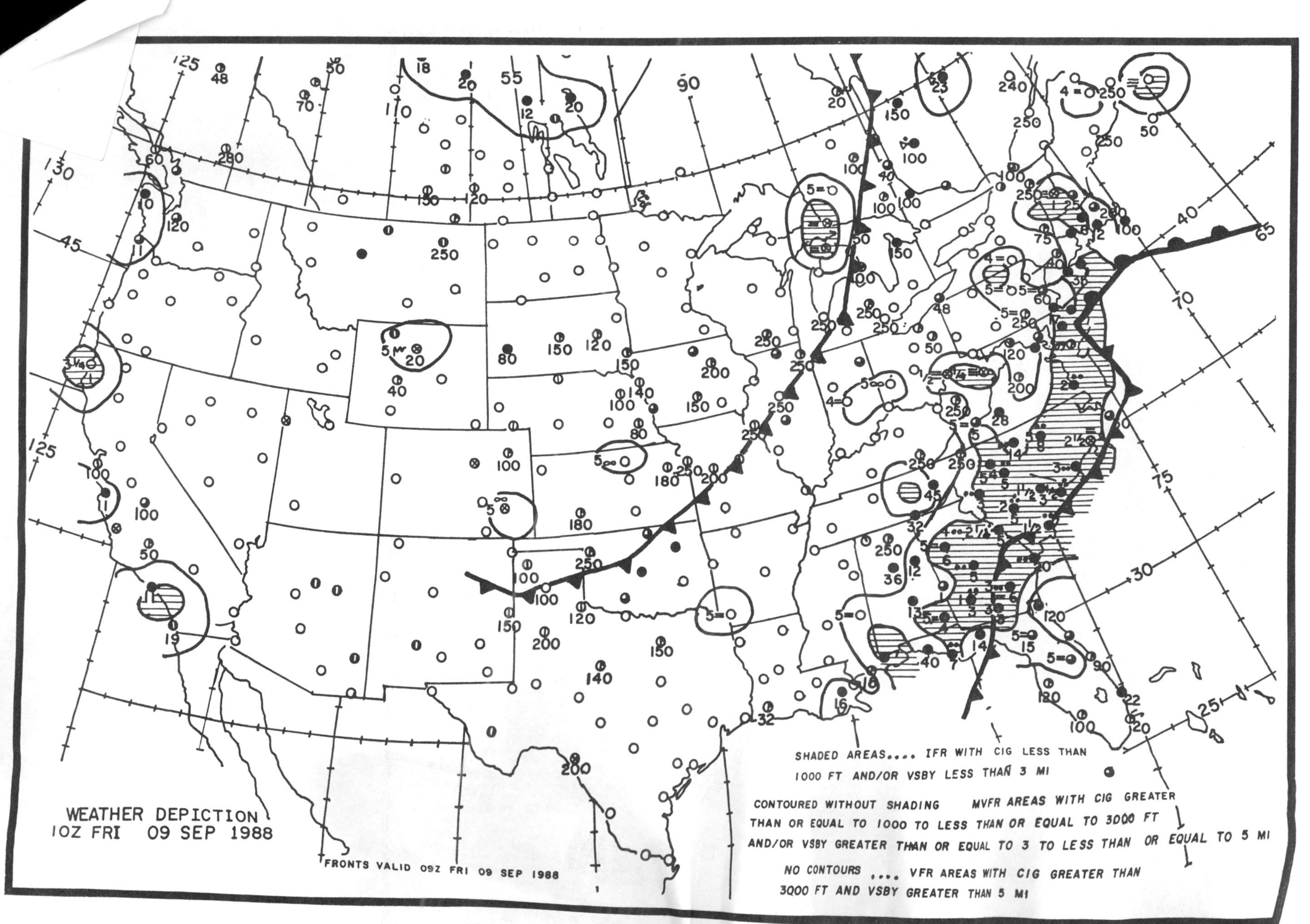

FIGURE 150.—Weather Depiction Chart.

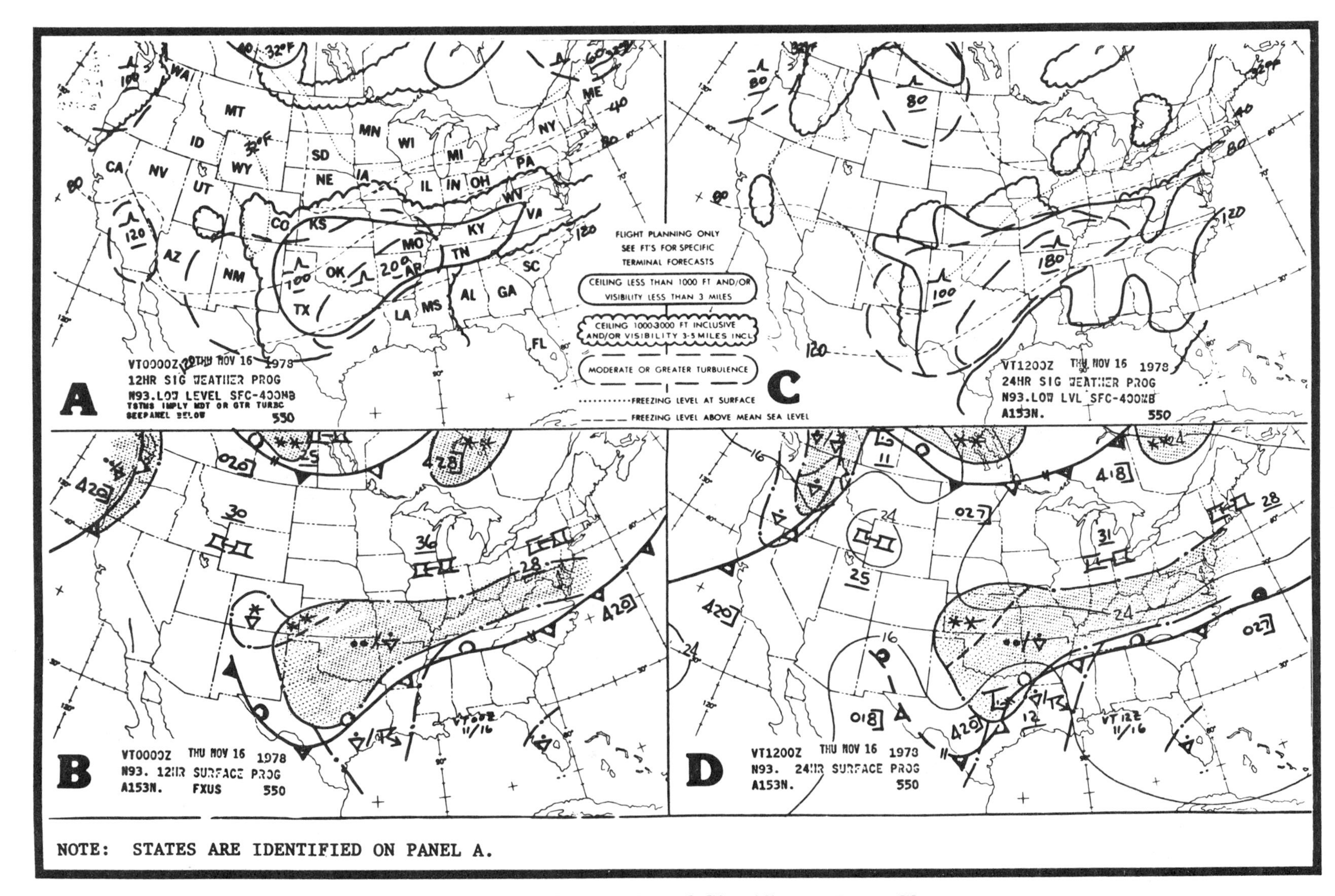

FIGURE 151. U.S. Low-Level Significant Prog Chart.

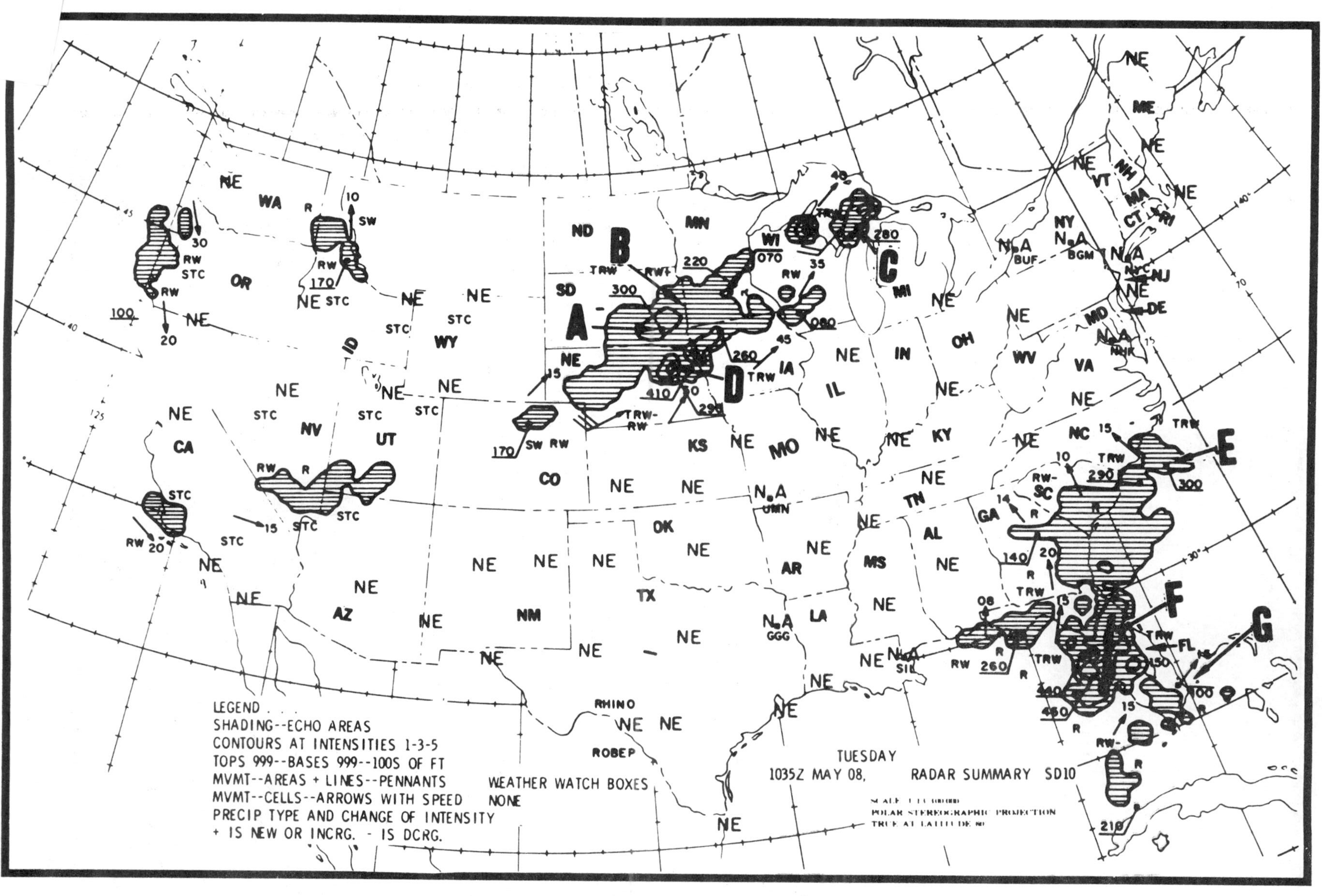

FIGURE 152.—Radar Summary Chart.

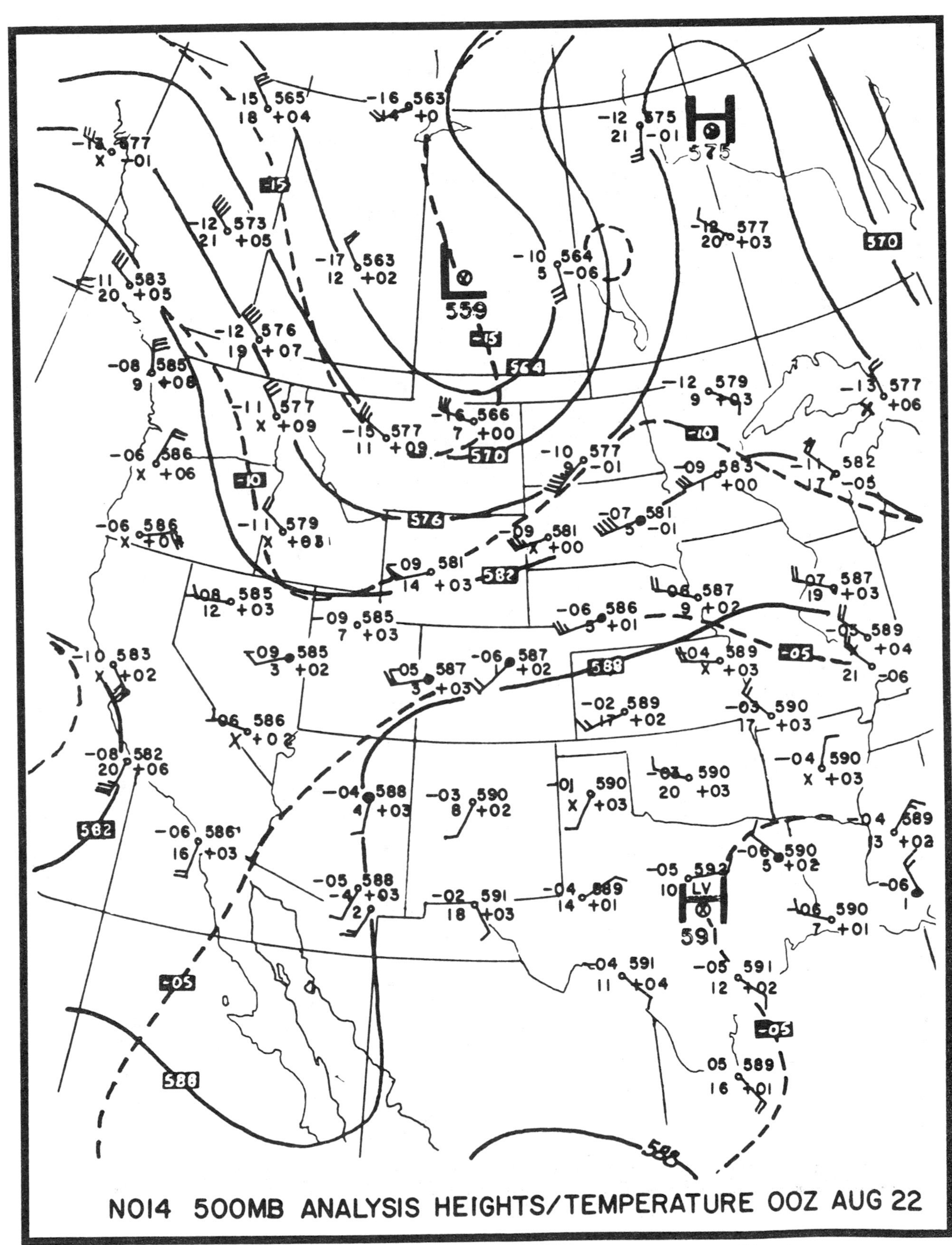

FIGURE 153.—500 MB Analysis Heights / Temperature Chart.

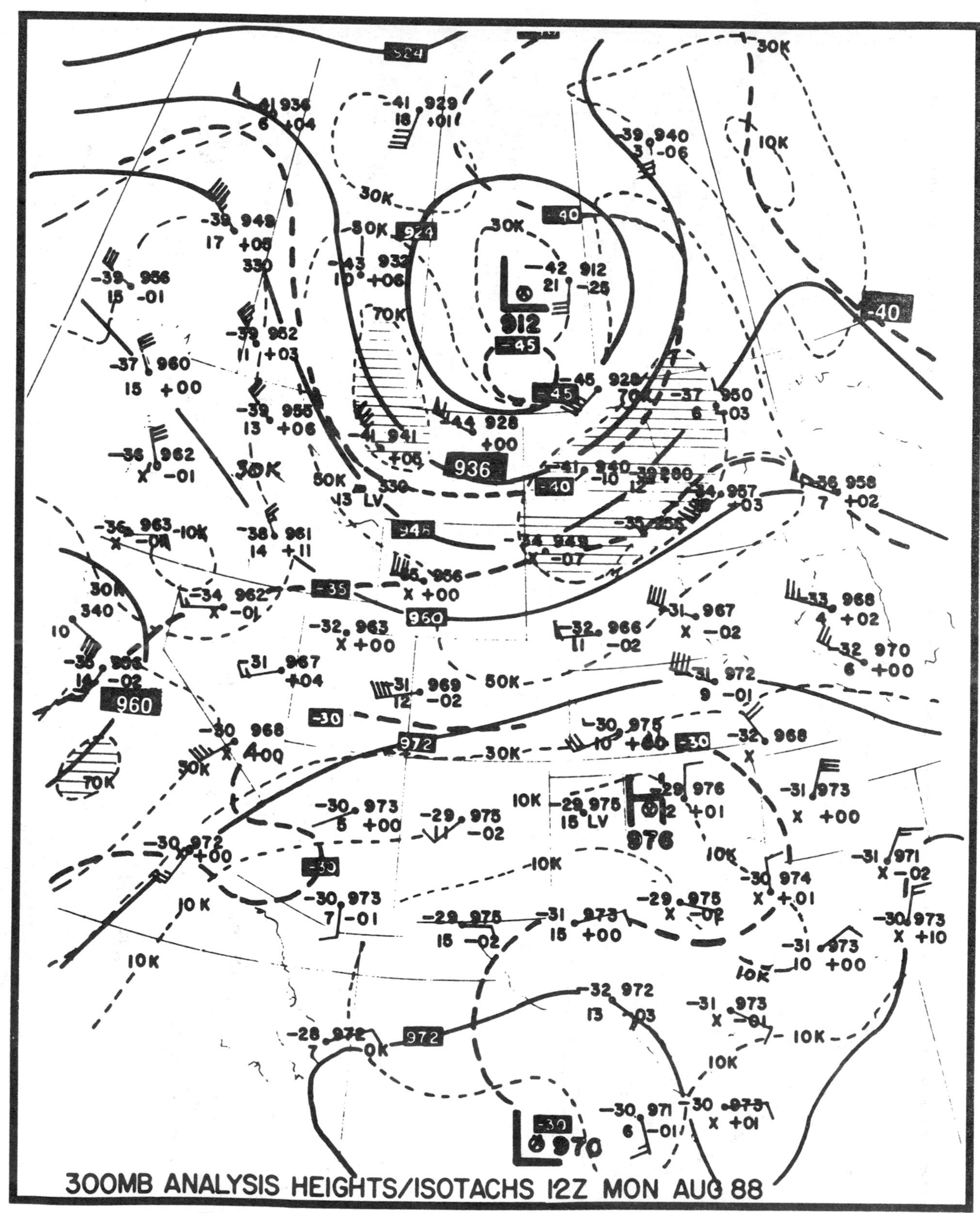

FIGURE 154.—300 MB Analysis Heights / Isotachs Chart.

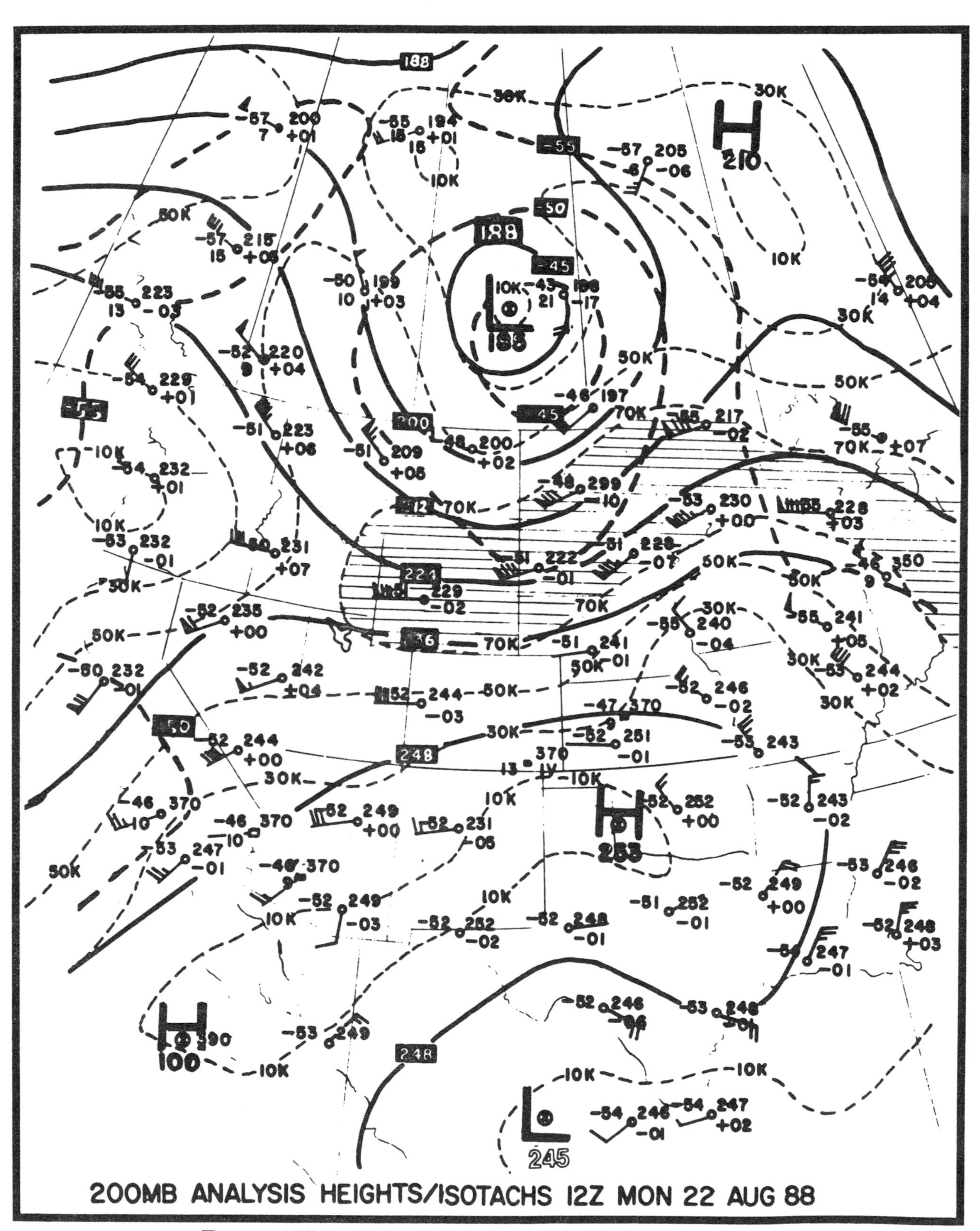

FIGURE 155.—200 MB Analysis Heights / Isotachs Chart.

APPENDIX D

THE RELATIONSHIP BETWEEN SEVERAL ATMOSPHERIC VARIABLES

The following discussion will help in gaining a better understanding of the relationship between temperature, pressure, density, relative humidity, dew point temperature and dew point lapse rate. We will begin by first referring back to our discussion of relative humidity and dew point in Chapter VI. Recall that relative humidity was defined as the ratio (times 100) of the amount (mass) of moisture in the air to the maximum amount of moisture that the air can hold at a given temperature and pressure. Note that if we divide the masses in the numerator and denominator of this ratio by volume, we obtain the water vapor density in the numerator and the maximum water vapor density (capacity) in the denominator. This new ratio (of densities), times 100, gives us another definition for relative humidity; there are other equivalent ways to define relative humidity, but we will not need them.

We know from Chapter II that pressure varies directly as the product of temperature and density. Thus, for a given pressure and temperature, density must be a certain value. The water vapor density will be a part of the total air density, and there will also be a corresponding maximum water vapor density for the volume of air we are considering. Without knowing both temperature and pressure, we could not compute relative humidity.

Suppose we have a balloon, the contents of which will have a certain value of temperature, pressure, and density. If we keep the pressure the same and increase the temperature of the contents of the balloon, then the density of the contents must decrease (recall that pressure varies as the product of temperature and density). Since the mass of the air is trapped inside the balloon, the volume of the balloon must increase in order for the density to decrease. With the larger volume and same mass after the temperature increase, we have a reduced water vapor density and an increased maximum water vapor density (capacity). Thus relative humidity, the ratio of these two densities times 100, will be less. If we now replace a portion of the dry air by water vapor, then the water vapor density goes up and the maximum water vapor density stays the same, and hence the relative humidity rises. Thus for a given pressure, various combinations of changes in temperature and in mass of water vapor cause variable changes in the relative humidity. We stated this fact back in Chapter VI; we could have just a temperature increase or decrease (day or night) without any change in water vapor, a water vapor content change (wind from a water source or from dry land, for example) without a change in temperature, or we could have any combination of these.

Now the dew point is the temperature to which the air must be cooled in order to reach saturation (100% relative humidity), with no change in pressure or water vapor content. We already know that the temperature and dew point are the same value when the air is saturated. For a given pressure and temperature, a dew point less than the air temperature means that the air holds less water vapor per unit volume than it can; therefore, the relative humidity will be less. We now apply these ideas to a comparison of desert and polar climates.

Assume that the temperature and dew point in a desert are 95°F and 45°F, respectively, and 15°F and 12°F in a cold climate. It should be clear that the relative humidity is higher in the cold climate but that the desert air actually holds more water vapor per unit volume since the dew point is much higher in the desert air.

As an air parcel (balloon containing air) rises, its pressure, temperature, and density decrease. Since density is mass per unit volume, the volume of the air parcel must increase, so the original amount of water vapor is now spread out over a larger volume. This means that there is less moisture per unit volume in the air parcel (balloon). We know that this means that the dew point must be less as our balloon rises. Therefore, we must use the difference between the dry adiabatic and dew point lapse rates in order to determine the bases of convective clouds. Incidentally, why does water boil at a lower temperature at a higher elevation (lower pressure)? Because without as strong a "pressure lid", water vapor molecules can escape the surface of boiling water at a slower speed (lower temperature).

APPENDIX E

OBJECTIVES

The reader should demonstrate an understanding of additional/supplemental topics on aviation meteorology to the extent that he/she can answer the following questions (as stated below, or as multiple-choice, etc.):

1. Under what conditions is an alternate not required for the first airport of intended landing on an IFR flight plan?
2. If an alternate is required, state the weather requirements for both a published and a non-published approach procedure.
3. Describe flying weather for interior areas of the Arctic and or near mountain passes in the Arctic.
4. Define "whiteout" and describe its effects on flying.
5. When do thunderstorms typically occur and in which direction do they tend to move in the Arctic?
6. What action should be taken if you are lost when flying during the summer in the Arctic?
7. Why are high pressure belts broken into semi-permanent highs in the Tropics? What determines the weather in the subtropical high pressure belts? Describe the weather over continents, over the open sea, and over islands in the Tropics.
8. How do trade winds blow in the Northern and Southern Hemispheres?
9. How does the "trade wind inversion" affect flying weather?
10. Describe flying conditions over open water, over continents, and over islands in the trade wind belts.
11. What is the effect of convection in the ITCZ? What is weather like under the influence of the ITCZ?
12. Describe how winter and summer monsoons form. What are their effects on flying weather?
13. List the types of migrating tropical weather producers. Describe each and state their effects on weather.
14. Define tropical cyclone and list the international classifications.
15. How and where do tropical cyclones develop? Describe their movement.
16. Describe how a tropical cyclone decays.
17. Describe the weather associated with a tropical depression.
18. Describe the weather in tropical storms and hurricanes.
19. Where is the national hurricane center located? How are storms located?
20. Describe how tropical cyclones affect flying weather.

CHAPTER XVII

ADDITIONAL/SUPPLEMENTAL TOPICS ON AVIATION METEOROLOGY

This chapter includes supplemental information on the meteorology section of the AIM, alternate weather requirements for IFR flight plans, and some information on tropical and arctic weather. The AIM is updated frequently – as often as every six months – consult the latest publication for the most current information.

A. Broadcast of Inflight Weather Advisories – AIM

Inflight weather advisories were described briefly on page 15-7. The specific times of broadcast of such advisories are listed below; further timing for Convective SIGMETs is on page 4-14 of Aviation Weather Services (Appendix A).

1. FSS Broadcasts

FAA FSSs broadcast Severe Weather Forecast Alerts (AWW), Convective SIGMETs, SIGMETs, CWAs, and AIRMETs during their valid period when they pertain to the area within 150 NM of the FSS or a broadcast facility controlled by the FSS as follows:

a. Severe Weather Forecast Alerts (AWW) and Convective SIGMET-Upon receipt and at 15-minute intervals H+00, H+15, H+30, and H+45 for the first hour after issuance.

b. SIGMETs, CWAs, and AIRMETs – Upon receipt and at 30-minute intervals at H+15 and H+45 for the first hour after issuance.

c. Thereafter, a summarized alert notice will be broadcast at H+15 and H+45 during the valid period of the advisories.

d. Pilots, upon hearing the alert notice, if they have not received the advisory or are in doubt, should contact the nearest FSS and ascertain whether the advisory is pertinent to their flights.

2. ARTCC Broadcasts

ARTCCs broadcast a Severe Weather Forecast Alert (AWW), Convective SIGMET, SIGMET, or CWA alert once on all frequencies, except emergency, when any part of the area described is within 150 miles of the airspace under their jurisdiction. These broadcasts contain SIGMET or CWA (identification) and a brief description of the weather activity and general area affected.

NOTE – Terminal control facilities have the option to limit the AWW, Convective SIGMET, SIGMET, or CWA broadcast as follows: local control and approach control positions may opt to broadcast SIGMET or CWA alerts only when any part of the area described is within 50 miles of the airspace under their jurisdiction.

3. HIWAS Broadcasts

This is a continuous broadcast of in-flight weather advisories including summarized AWW, SIGMETs, Convective SIGMETs, CWAs, AIRMETs, and urgent PIREPs. HIWAS has been adopted as a national program and will be implemented throughout the conterminous U.S. as resources permit. In those areas where HIWAS is commissioned, ARTCC, Terminal ATC, and FSS facilities have discontinued the broadcast of in-flight advisories as described in the preceding paragraph. HIWAS is an additional source of hazardous weather information which makes these data available on a continuous basis. It is not, however, a replacement for preflight or in-flight briefings or real-time weather updates from Flight Watch (EFAS). As HIWAS is implemented in individual center areas, the commissioning will be advertised in the Notices to Airmen publication.

a. Where HIWAS has been implemented, a HIWAS alert will be broadcast on all except emergency frequencies once upon receipt by ARTCC and terminal facilities, which will include an alert announcement, frequency instruction, number, and type of advisory updated; e.g., AWW, SIGMET, Convective SIGMET, or CWA.

b. In HIWAS ARTCC areas, FSSs will broadcast a HIWAS update announcement once on all except emergency frequencies upon completion of recording an update to the HIWAS broadcast. Included in the broadcast will be the type of advisory updated; e.g. AWW, SIGMET, Convective SIGMET, CWA, etc.

4. Unscheduled Broadcasts

These broadcasts are made by FSSs on VOR and selected VHF frequencies upon receipt of special weather reports, PIREPs, NOTAMs and other information considered necessary to enhance safety and efficiency of flight. These broadcasts will be made at random times and will begin with the announcement "Aviation Broadcast" followed by identification of the data.

5. Alaskan Scheduled Broadcasts

Selected FSSs in Alaska having voice capability on radio ranges (VOR) or radio beacons (NDB) broadcast weather reports and Notice to Airman information at 15 minutes past each hour from reporting points within approximately 150 miles from the broadcast station.

B. Alternate Weather Requirements for IFR Flight Plans (91.83)

First, an alternate is not required for a Part 97 published standard instrument approach for the first airport of intended landing, if for at least one hour before and one hour after the estimated time of arrival the weather reports or forecasts or any combination of them indicate: (1) the ceiling will be at least 2,000 feet above the airport elevation; and (2) visibility will be at least 3 miles.

Failing to meet the above weather requirements, an IFR alternate airport is required (unless authorized by the Administrator). Current weather forecasts must indicate that, at the estimated time of arrival at the alternate airport, the ceiling and visibility at that airport will be at or above the following alternate airport weather minimums:

(1) If an instrument approach procedure has been published in Part 97 for that airport, the rnate airport minimums specified in that procedure or, if none are so specified, the following ıums:

(i) Precision approach procedure: Ceiling 600 feet and visibility 2 statute miles.

(ii) Nonprecision approach procedure: Ceiling 800 feet and visibility 2 statute miles.

(2) If no instrument approach procedure has been published in Part 97 for that airport, the ceiling and visibility minimums are those allowing descent from the MEA, approach, and landing, under basic VFR.

C. Arctic and Tropical Weather

1. Arctic Weather

If one were to summarize general weather conditions and flight precautions over Alaska, northern Canada, and the Arctic, he would say:

a. Interior areas generally have good flying weather, but coastal areas and Arctic slopes often are plagued by low ceiling, poor visibility, and icing.

b. "Whiteout" conditions over ice and snow covered areas often cause pilot disorientation. "Whiteout" is a visibility restricting phenomenon that occurs in the Arctic when a layer of cloudiness of uniform thickness overlies a snow- or ice-covered surface. Diffused light reflects back and forth between the snow/ice and clouds, eliminating all shadows and resulting in a loss of depth perception. Objects appear to float in the air and the horizon disappears.

c. Flying conditions are usually worse in mountain passes than at reporting stations along the route.

d. Routes through the mountains are subject to strong turbulence, especially in and near passes.

e. Beware of a false mountain pass that may lead to a dead-end.

f. Thundershowers sometimes occur in the interior during May through August. They are usually circumnavigable and generally move from northeast to southwest in the low-level polar north-easterlies, which is opposite the general movement in mid-latitudes. However, winds in the upper levels over polar regions are from the west/southwest as in mid-latitudes, the strongest winds being in winter, and would move the few weather systems that do occur from west to east.

g. Always file a flight plan. Stay on regularly traversed routes, and if downed, stay with your plane.

h. If lost during summer, fly down-drainage, that is, downstream. Most airports are located near rivers, and chances are you can reach a landing strip by flying downstream. If forced down, you will be close to water on which a rescue plane can land. In summer, the tundra is usually too soggy for landing.

i. Weather stations are few and far between. Adverse weather between stations may go undetected unless reported by a pilot in flight. A report confirming good weather between stations is also just as important. Help yourself and your fellow pilot by reporting weather en route.

2. *Tropical Weather*

In chapter 4, we learned that wind blowing out of the subtropical high pressure belts toward the Equator form the northeast and southeast trade winds of the two hemispheres. These trade winds converge in the vicinity of the Equator where air rises. This convergence zone is the "intertropical convergence zone" (ITCZ). In some areas of the world, seasonal temperature differences between land and water areas generate rather large circulation patterns that overpower the trade wind circulation; these areas are "monsoon" regions. Tropical weather discussed here includes the subtropical high pressure belts, the trade wind belts, the intertropical convergence zone, and monsoon regions.

a. Subtropical High Pressure Belts

If the surface under the subtropical high pressure belts were all water of uniform temperature, the high pressure belts would be continuous highs around the globe. The belts would be areas of descending or subsiding air and would be characterized by strong temperature inversions and very little precipitation. However, land surfaces at the latitudes of the high pressure belts are generally warmer throughout the year than are water surfaces. Thus, the high pressure belts are broken into semi-permanent high pressure anticyclones over oceans with troughs or lows over continents. The subtropical highs shift southward during the Northern Hemisphere winter and northward during summer. The seasonal shift, the height and strength of the inversion, and terrain features determine weather in the subtropical high pressure belts.

Over continents, low ceiling and fog often prevent landing at a west coast destination, but a suitable alternate generally is available a few miles inland. Alternate selection may be more critical for an eastern coast destination because of widespread instability and associated hazards. Under a subtropical high over the open sea, cloudiness is scant. The few clouds that do develop have tops from 3,000 to 6,000 feet depending on height of the inversion. Ceiling and visibility are generally quite ample for VFR flight. An island under a subtropical high receives very little rainfall because of the persistent temperature inversion. Surface heating over some larger islands causes light convective showers. Cloud tops are only slightly higher than over open water. Temperatures are mild, showing small seasonal and diurnal changes. A good example is the pleasant, balmy climate of Bermuda.

b. Trade Wind Belts

Figures 17-1 and 17-2 show prevailing winds throughout the tropics for July and January. Note that trade winds blowing out of the subtropical highs over ocean areas are predominantly northeasterly in the Northern Hemisphere and southeasterly in the Southern Hemisphere. The inversion from the subtropical highs is carried into the trade winds and is known as the "trade wind inversion." As in a subtropical high, the inversion is strongest where the trades blow away from the west coast of a continent and weakest where they blow onto an eastern continental shore. Daily variations from these prevailing directions are small except during tropical storms. As a result, weather at any specific location in a trade wind belt varies little from day to day.

In the trade wind belt, skies over open water are about one-half covered by clouds on the average. Tops range from 3,000 to 8,000 feet depending on height of the inversion. Showers, although more common than under a subtropical high, are still light with comparatively little rainfall. Flying weather generally is quite good. Flying weather along continental eastern coasts and mountains is subject to the usual hazards of showers and thunderstorms. Flying over arid regions is good most of the time but can be turbulent in afternoon convective currents; be especially aware of dust devils. Blowing sand or dust sometimes restricts visibility.

The greatest flying hazard near islands is obscured mountain tops. Ceiling and visibility occasionally restrict VFR flight on the windward side in showers. IFR weather is virtually nonexistent on leeward slopes. Islands without mountains have little effect on cloudiness and rainfall. Afternoon surface heating increases convective cloudiness slightly, but shower activity is light. However, any island in either the subtropical high pressure belt or trade wind belt enhances cumulus development even though tops do not reach great heights. Therefore, a cumulus top higher than the average tops of surrounding cumulus usually marks the approximate location of an island. If it becomes necessary to "ditch" in the ocean, look for a tall cumulus. If you see one, head for it. It probably marks a land surface, increasing your chances of survival.

c. The Intertropical Convergence Zone (ITCZ)

Converging winds in the intertropical convergence zone (ITCZ) force air upward. The inversion typical of the subtropical high and trade wind belts disappears. Figures 17-1 and 17-2 show the ITCZ and its seasonal shift. The ITCZ is well marked over tropical oceans but is weak and ill-defined over large continental areas. Convection in the ITCZ carries huge quantities of moisture to great heights. Showers and thunderstorms frequent the ITCZ and tops to 40,000 feet or higher are common as shown in Figure 17-3. Precipitation is copious. Since convection dominates the ITCZ, there is little difference in weather over islands and open sea under the ITCZ. Since the ITCZ is ill-defined over continents, we will not attempt to describe ITCZ continental weather as such. Continental weather ranges from arid to rain forests and is more closely related to the monsoon then to the ITCZ.

d. Monsoon

Asia is covered by an intense high during the winter and a well-developed low during the summer. The same pattern also occurs over Australia and central Africa, although the seasons are reversed in the Southern Hemisphere. The cold, high pressures in winter cause wind to blow from the deep interior outward and offshore. In summer, wind direction reverses and warm moist air is carried far inland into the low pressure area. This large scale seasonal wind shift is the "monsoon." The most notable monsoon is that of southern and southeastern Asia.

During the winter monsoon, excellent flying weather prevails over dry interior regions. Over water, one must pick his/her way around showers and thunderstorms. In the summer monsoon, VFR flight over land is often restricted by low ceilings and heavy rain. IFR flight must cope with the hazards of thunderstorms. The freezing level is quite high in the tropics – 14,000 feet or higher – so icing is restricted to high levels.

e. Transitory Systems

So far, we have concentrated on prevailing circulations. Now we turn to migrating tropical weather producers – the shear line, trough aloft, tropical wave, and tropical cyclone.

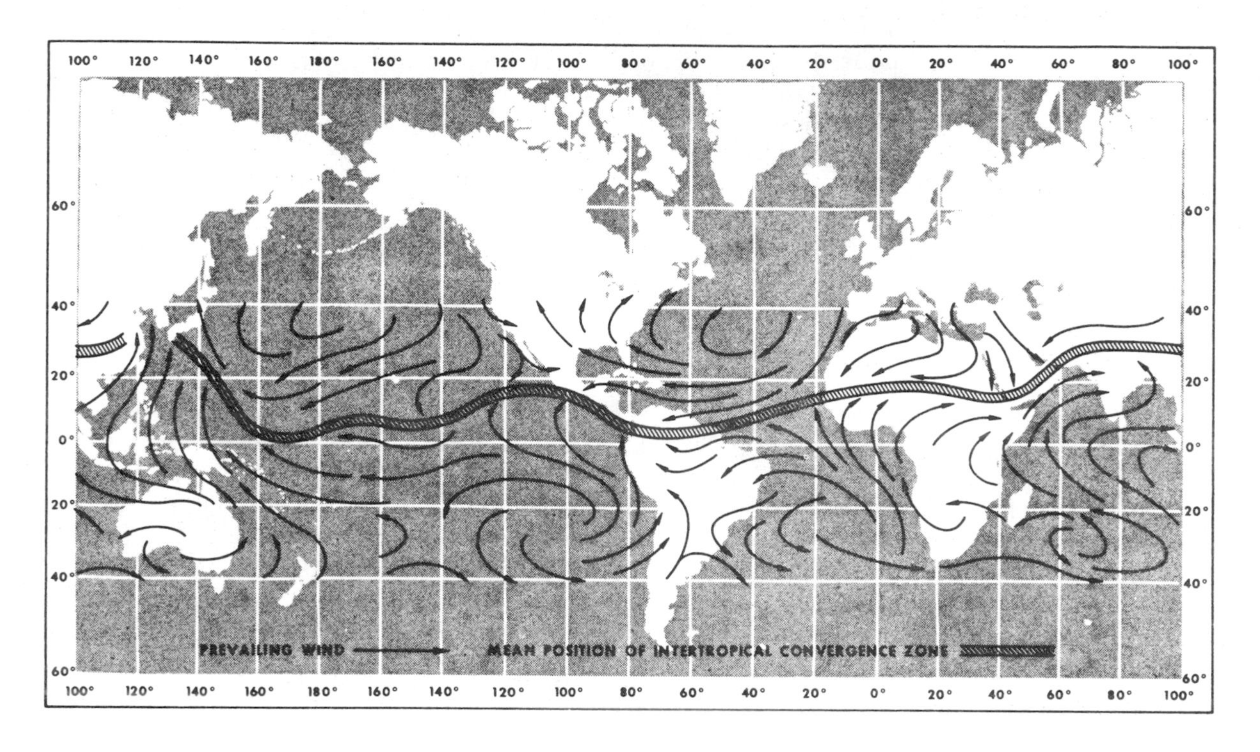

Figure 17-1. Prevailing winds throughout the Tropics in July. Remember that in the Southern Hemisphere, circulation around pressure centers is opposite that in the Northern Hemisphere.

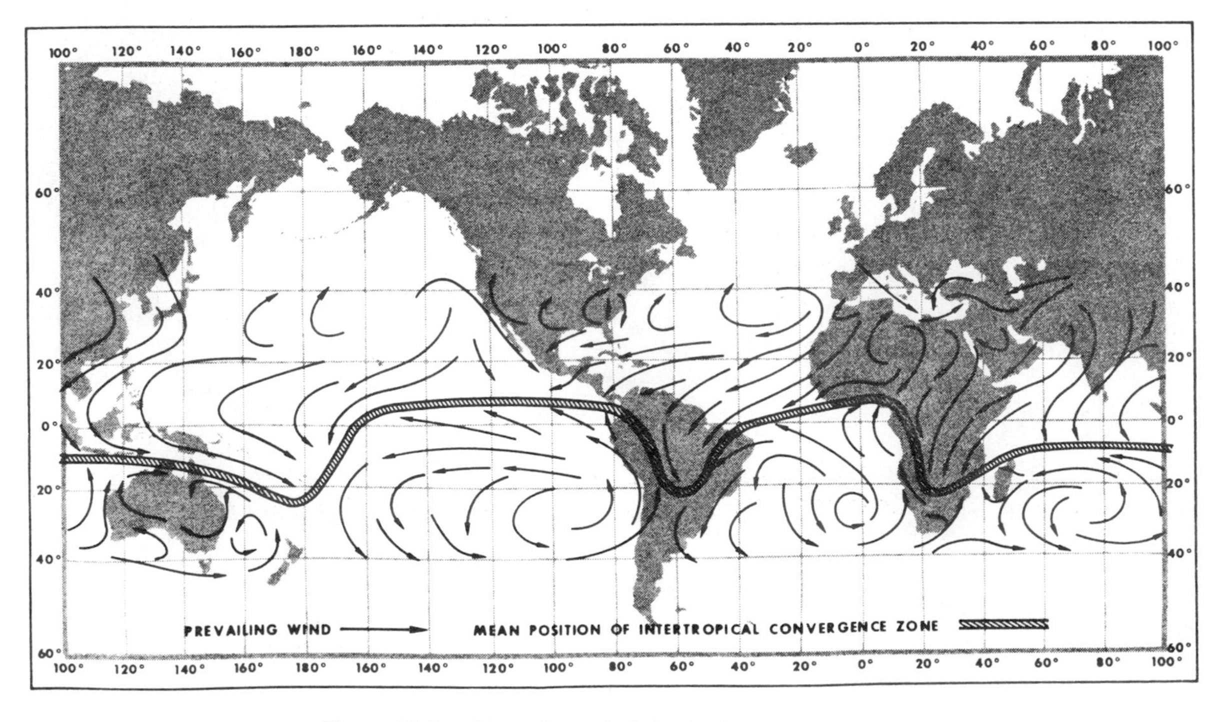

Figure 17-2. Prevailing winds in the Tropics in January.

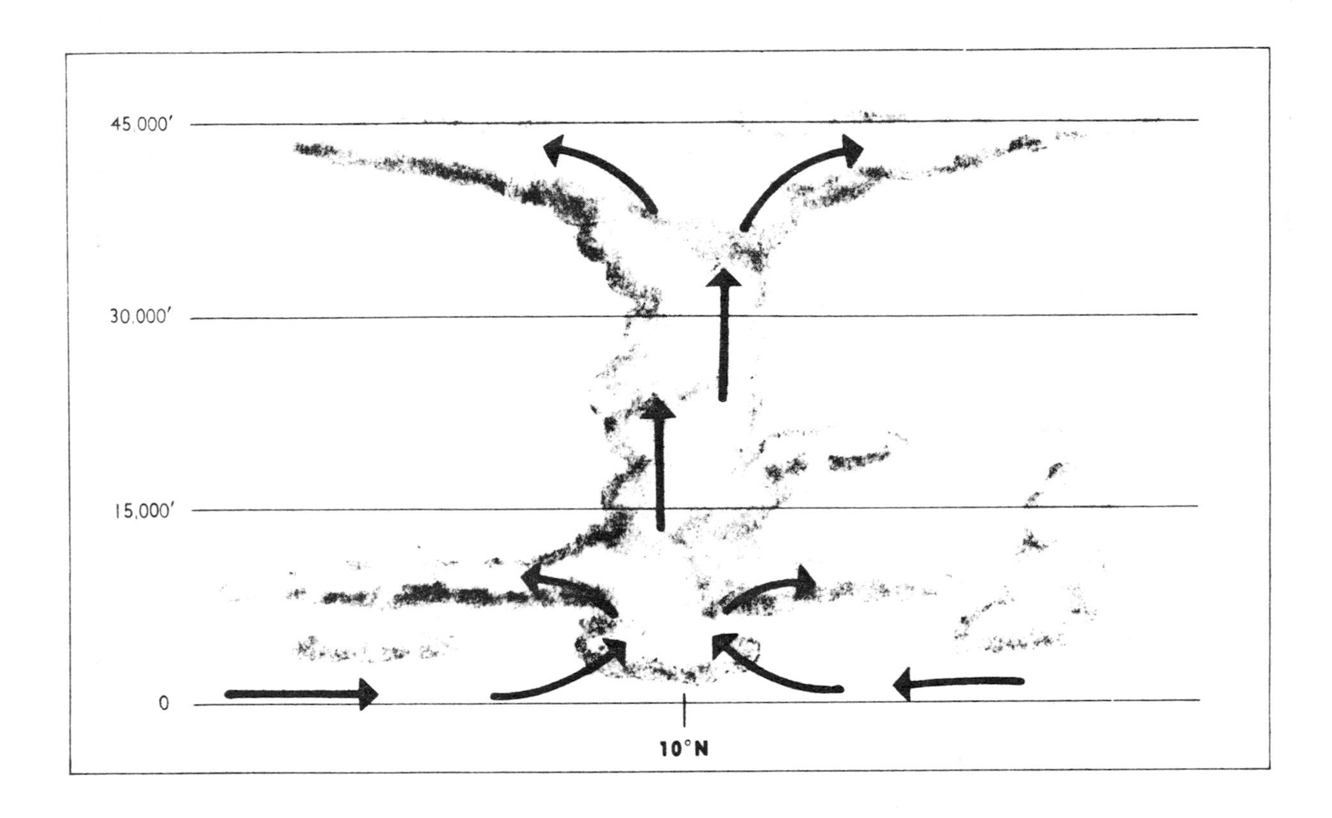

Figure 17-3. Vertical cross section illustrating convection in the Intertropical Convergence Zone.

1. Shear Line

A wind shear line found in the Tropics mainly results from midlatitude influences. In chapter IX we stated that an air mass becomes modified when it flows from its source region. By the time a cold air mass originating in high latitudes reaches the Tropics, temperature and moisture are virtually the same on both sides of the front. A shear line, or wind shift, is all that remains. A shear line also results when a semi-permanent high splits into two cells inducing a trough as shown in Figure 17-4.

These shear lines are zones of convergence creating forced upward motion. Consequently, considerable thunderstorm and rain shower activity occurs along a shear line.

2. Trough Aloft

Troughs in the atmosphere, generally at or above 10,000 feet, move through the Tropics, especially along the poleward fringes. Figure 17-5 shows such a trough across the Hawaiian Island chain. As a trough moves to the southeast or east, it spreads middle and high cloudiness over extensive areas to the east of the trough line. Occasionally, a well-developed trough will extend deep into the Tropics, and a closed low forms at the equatorial end of the trough. The low then may separate from the trough and move westward producing a large amount of cloudiness and precipitation. If this occurs in the vicinity of a strong subtropical jet stream, extensive and sometimes dense cirrus and some convective and clear air turbulence often develop.

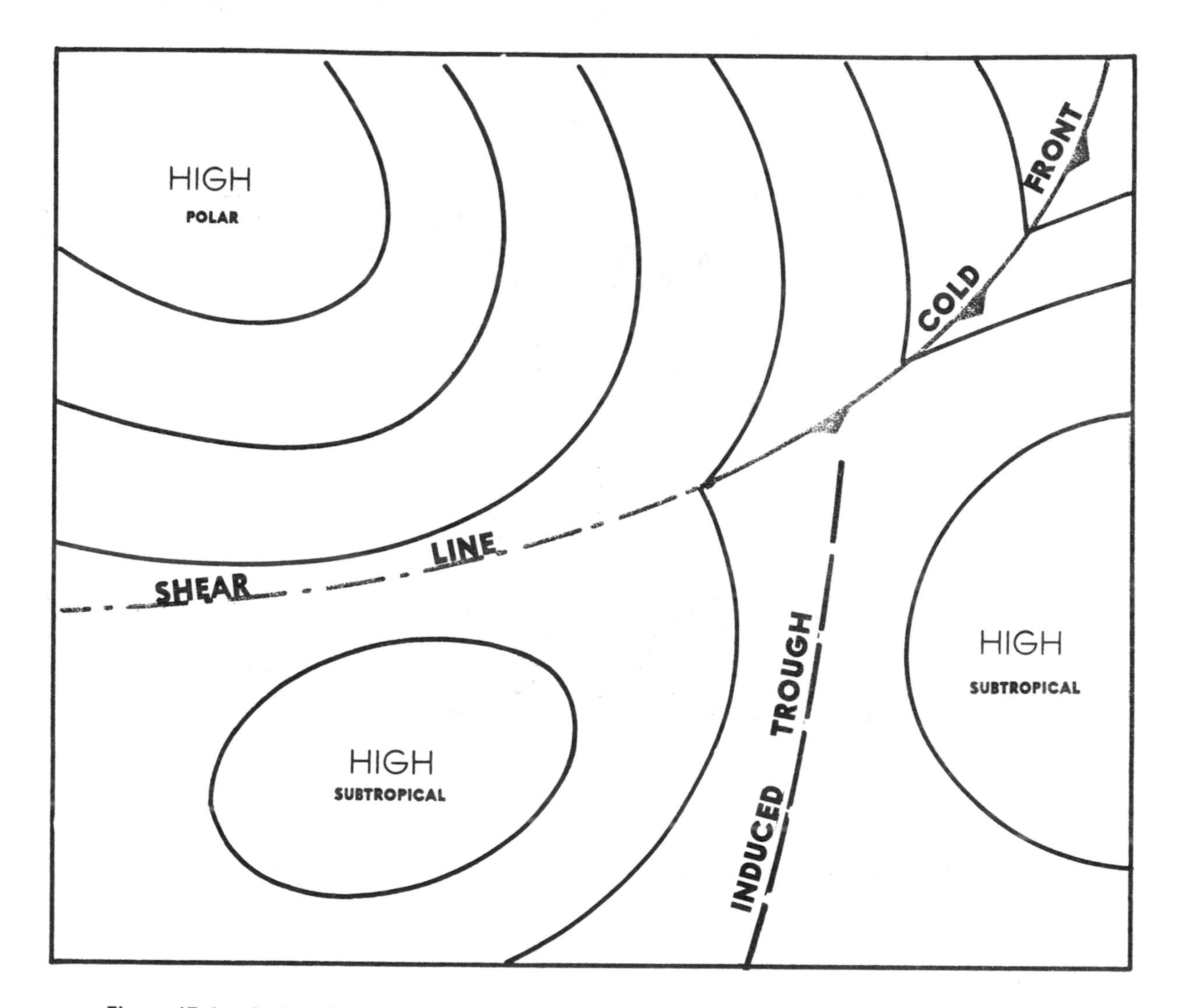

Figure 17-4. A shear line and an induced trough caused by a polar high pushing into the subtropics.

Troughs and lows aloft produce considerable amounts of rainfall in the Tropics, especially over land areas where mountains and surface heating lift air to saturation. Low pressure systems aloft contribute significantly to the record 460 inches average annual rainfall on Mt. Waialeale on Kauai, Hawaii. Other mountainous areas of the Tropics are also among the wettest spots on earth.

3. *Tropical Wave*

Tropical waves (also called easterly waves) are common tropical weather disturbances, normally occurring in the trade wind belt. In the Northern Hemisphere, they usually develop in the southeastern perimeter of the subtropical high pressure systems. They travel from east to west around the southern fringes of these highs in the prevailing easterly circulation of the Tropics. Surface winds in advance of a wave are somewhat more northerly than the usual trade wind direction. As the wave approaches, as shown in Figure 17-6, pressure falls; as it passes, surface wind shifts to the east-southeast or southeast. The typical wave is preceded by very good weather but followed by extensive cloudiness, as shown in Figure 17-7, and often by rain and thunderstorms. The weather activity is roughly in a north-south line.

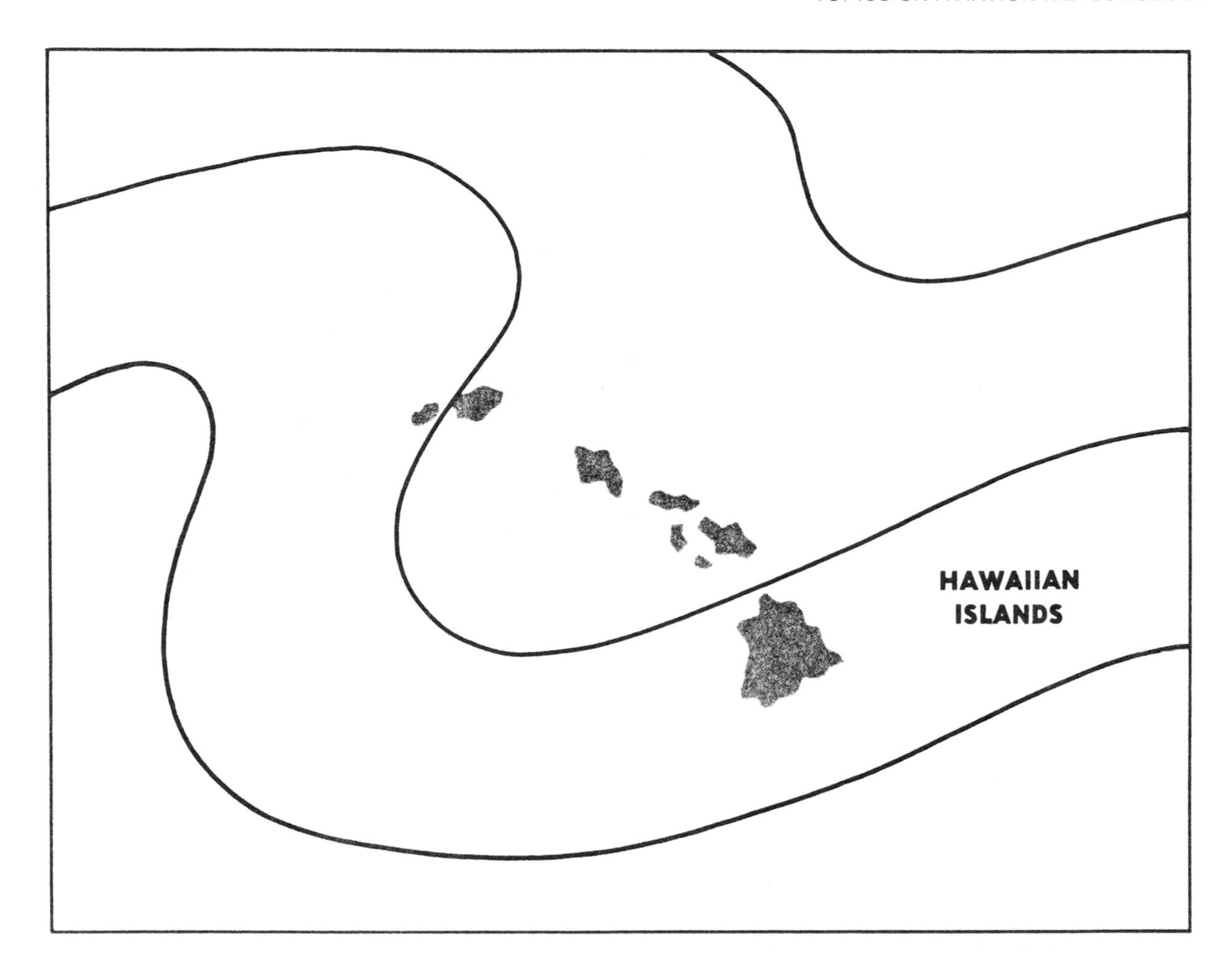

Figure 17-5. A trough aloft across the Hawaiian Islands. Extensive cloudiness develops east of the trough.

Tropical waves occur in all seasons, but are more frequent and stronger during summer and early fall. Pacific waves frequently affect Hawaii; Atlantic waves occasionally move into the Gulf of Mexico, reaching the U.S. coast.

4. Tropical Cyclone

Tropical cyclone is a general term for any low that originates over tropical oceans. Tropical cyclones are classified according to their intensity based on average one-minute wind speeds. Wind gusts in these storms may be as much as 50 percent higher than the average one-minute wind speeds. Tropical cyclone international classifications are:

(1) Tropical Depression – highest sustained winds up to 34 knots (64 km/h),

(2) Tropical Storm – highest sustained winds of 35 through 64 knots (65 to 119 km/h), and

(3) Hurricane or Typhoon – highest sustained winds 65 knots (120 km/h) or more.

Strong tropical cyclones are known by different names in different regions of the world. A tropical cyclone in the Atlantic and eastern Pacific is a "hurricane"; in the western Pacific,

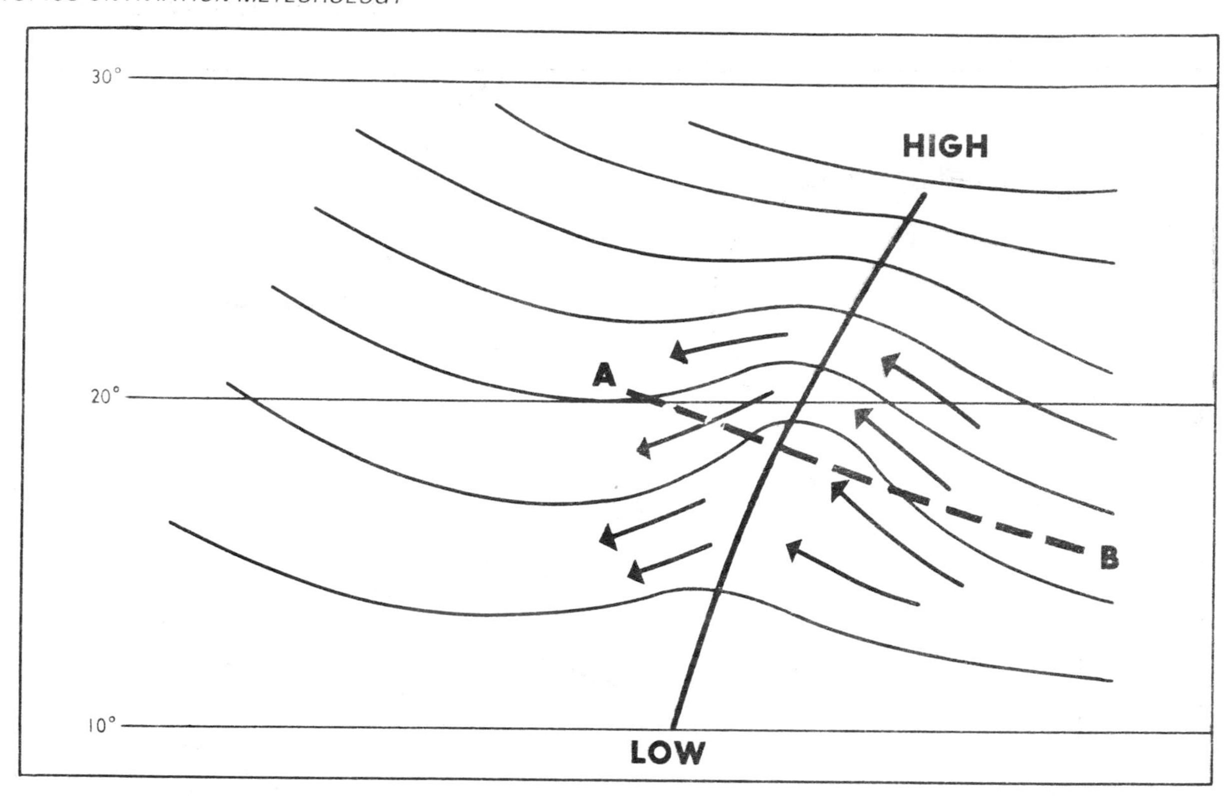

Figure 17-6. A Northern Hemisphere easterly wave. Progressing from (A) to (B), note that winds shift generally from northeasterly to southeasterly. The wave moves toward the west and is often preceded by good weather and followed by extensive cloudiness and precipitation.

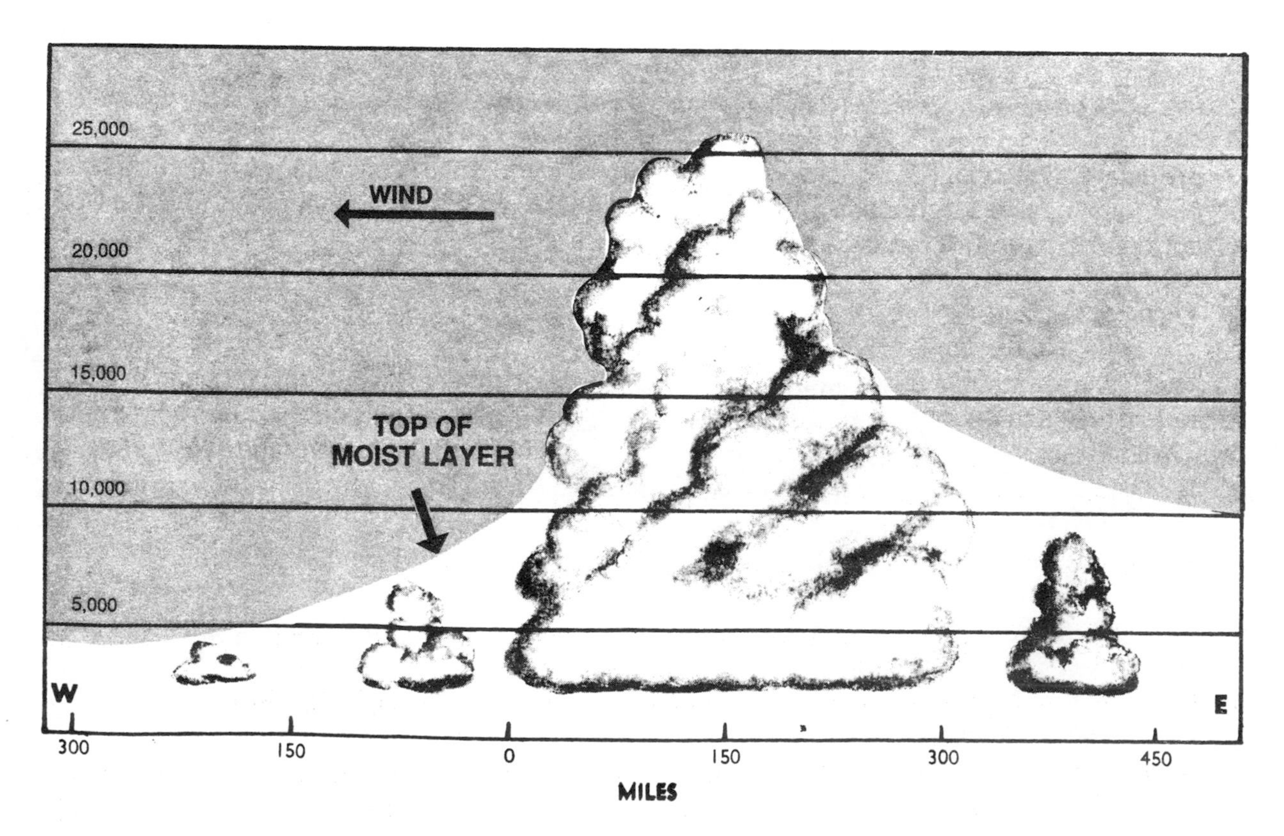

Figure 17-7. Vertical cross section along line A – B in Figure 17-6.

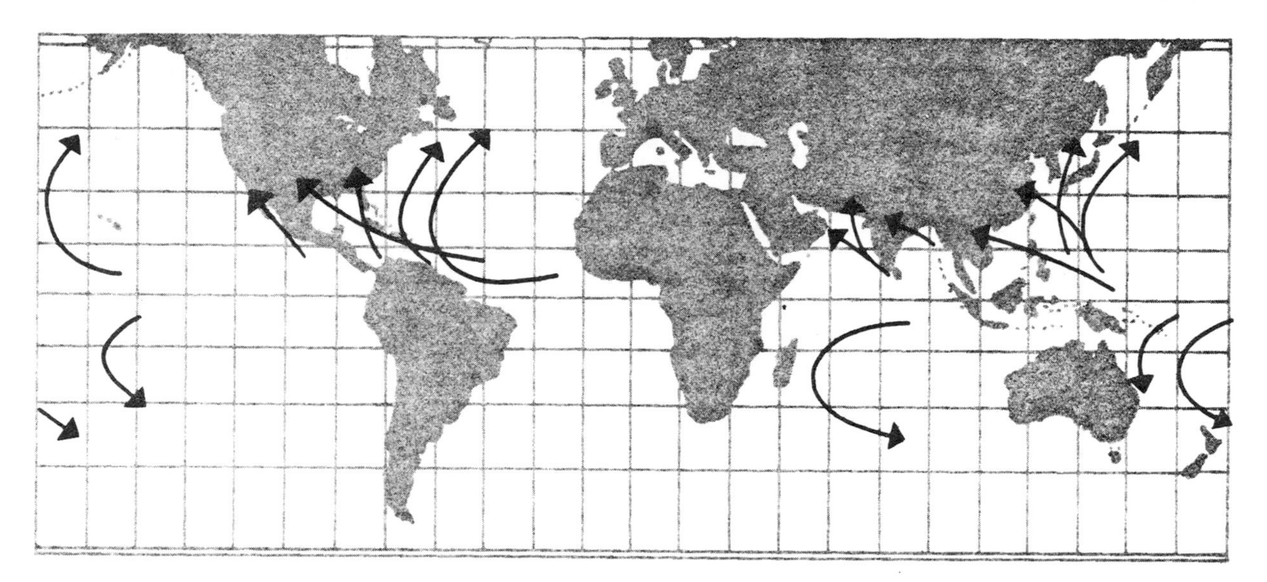

Figure 17-8. Principal regions where tropical cyclones form and their favored directions of movement.

"typhoon"; near Australia, "willy-willy"; and in the Indian Ocean, simply "cyclone." Regardless of the name, these tropical cyclones produce serious aviation hazards. Before we delve into these aspects, let's look at the development, movement, and decay of these cyclones.

(a) Development

Prerequisite to tropical cyclone development are optimum sea surface temperature under weather systems that produce low-level convergence and cyclonic wind shear. Favored breeding grounds are tropical (easterly) waves, troughs aloft, and areas of converging northeast and southeast trade winds along the intertropical convergence zone.

The low level convergence associated with these systems, by itself, will not support development of a tropical cyclone. The system must also have horizontal outflow – divergence – at high tropospheric levels. This combination creates a "chimney," in which air is forced upward causing clouds and precipitation. Condensation releases large quantities of latent heat which raises the temperature of the system and accelerates the upward motion. The rise in temperature lowers the surface pressure which increases low-level convergence. This draws more moisture-laden air into the system. When these chain-reaction events continue, a huge vortex is generated which may culminate in hurricane force winds.

Figure 17-8 shows regions of the world where tropical cyclones frequently develop. Notice that they usually originate between latitudes 5° and 20°. Tropical cyclones are unlikely within 5° of the Equator because the Coriolis force is so small near the Equator that it will not turn the winds enough for them to flow around a low pressure area. Winds flow directly into an equatorial low and rapidly fill it.

(b) Movement

Tropical cyclones in the Northern Hemisphere usually move in a direction between west and northwest while in low latitudes. As these storms move toward the midlatitudes, they come under the influence of the prevailing westerlies. At this time the storms are under the influence of two wind systems, i.e., the trade winds at low levels and prevailing westerlies aloft. Thus a storm may move very erratically and may even reverse course, or circle. Finally, the prevailing westerlies gain control and the storm recurves toward the north, then to the northeast, and finally to the east-northeast. By this time the storm is well into midlatitudes.

(c) Decay

As the storm curves toward the north or east, it usually begins to lose its tropical characteristics and acquires characteristics of lows in middle latitudes. Cooler air flowing into the storm gradually weakens it. If the storm tracks along a coast line or over the open sea, it gives up slowly, carrying its fury to areas far removed from the Tropics. However, if the storm moves well inland, it loses its moisture source and weakens from starvation and increased surface friction, usually after leaving a trail of destruction and flooding.

When a storm takes on middle latitude characteristics, it is said to be "extratropical" meaning "outside the Tropics." Tropical cyclones produce weather conditions that differ somewhat from those produced by their higher latitude cousins and invite our investigation.

(d) Weather in a Tropical Depression

While in its initial developing stage, the cyclone is characterized by a circular area of broken to overcast clouds in multiple layers. Embedded in these clouds are numerous showers and thunderstorms. Rain shower and thunderstorm coverage varies from scattered to almost solid. Diameter of the cloud pattern varies from less than 100 miles in small systems to well over 200 miles in large ones.

(e) Weather in Tropical Storms and Hurricanes

As cyclonic flow increases, the thunderstorms and rain showers form into broken or solid lines paralleling the wind flow that is spiraling into the center of the storm. These lines are the spiral rain bands frequently seen on radar. These rain bands continually change as they rotate around the storm. Rainfall in the rain bands is very heavy, reducing ceiling and visibility to near zero. Winds are usually very strong and gusty and, consequently, generate violent turbulence. Between the rain bands, ceilings and visibilities are somewhat better, and turbulence generally is less intense.

The "eye" usually forms in the tropical storm stage and continues through the hurricane stage. In the eye, skies are free of turbulent cloudiness, and wind is comparatively light. The average diameter of the eye is between 15 and 20 miles, but sometimes is as small as 7 miles and rarely is more than 30 miles in diameter. Surrounding the eye is a wall of cloud that may extend above 50,000 feet. This "wall cloud" contains deluging rain and the strongest winds of the storm. Maximum wind speeds of 175 knots have been recorded in some storms. Figure 17-9 is a radar display of a mature hurricane. Note the spiral rain bands and the circular eye.

(f) Detection and Warning

The National Weather Service has a specialized hurricane forecast and warning service center at Miami, Florida, which maintains constant watch for the formation and development of tropical cyclones. Weather information from land stations, ships at sea, reconnaissance aircraft, long range radars, and weather satellites is fed into the center. The center forecasts the development, movement, and intensity of tropical cyclones. Forecasts and warnings are issued to the public and aviation interests by field offices of the National Weather Service.

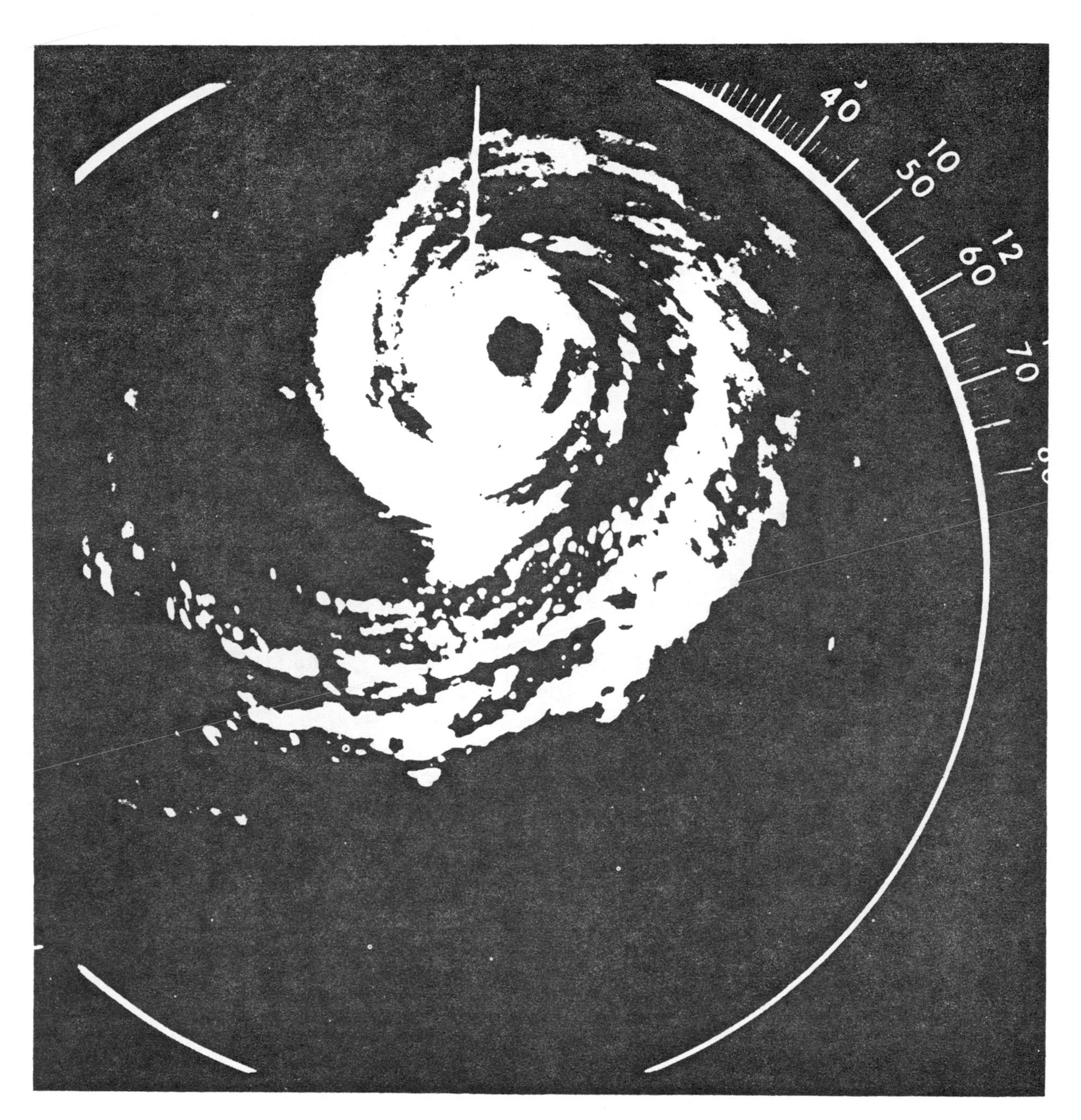

Figure 17-9. Radar photograph of hurricane "Donna" observed at Key West, Florida.

(g) Flying

All pilots except those especially trained to explore tropical storms and hurricanes should AVOID THESE DANGEROUS STORMS. Occasionally, jet aircraft have been able to fly over small and less intense storms, but the experience of weather research aircraft shows hazards at all levels within them.

Tops of thunderstorms associated with tropical cyclones frequently exceed 50,000 feet. Winds in a typical hurricane are strongest at low levels, decreasing with altitude. However, research aircraft have frequently encountered winds in excess of 100 knots at 18,000 feet. Aircraft at low levels are exposed to sustained, pounding turbulence due to the surface friction of the fast-moving air. Turbulence increases in intensity in spiral rain bands and becomes most violent in the wall cloud surrounding the eye.

An additional hazard encountered in hurricanes is erroneous altitude readings from pressure altimeters. These errors are caused by the large pressure difference between the periphery of the storm and its center. One research aircraft lost almost 2,000 feet true altitude traversing a storm while the pressure altimeter indicated a constant altitude of 5,000 feet.

In short, tropical cyclones are very hazardous, so avoid them! To bypass the storm in a minimum of time, fly to the right of the storm to take advantage of the tailwind. If you fly to the left of the storm, you will encounter strong headwinds which may exhaust your fuel supply before you reach a safe landing area.

REVIEW QUESTIONS
(See Appendix C)

Section 2. ALTIMETER SETTING PROCEDURES

7-31. GENERAL

a. The accuracy of aircraft altimeters is subject to the following factors:

1. Nonstandard temperatures of the atmosphere.
2. Nonstandard atmospheric pressure.
3. Aircraft static pressure systems (position error), and
4. Instrument error.

b. EXTREME CAUTION SHOULD BE EXERCISED WHEN FLYING IN PROXIMITY TO OBSTRUCTIONS OR TERRAIN IN LOW TEMPERATURES AND PRESSURES. This is especially true in extremely cold temperatures that cause a large differential between the Standard Day temperature and actual temperature. This circumstance can cause serious errors that result in the aircraft being significantly lower than the indicated altitude.

7-31b NOTE–Standard temperature at sea level is 15 degrees Celsius (59 degrees Fahrenheit). The temperature gradient from sea level is minus 2 degrees Celsius (3.6 degrees Fahrenheit) per 1,000 feet. Pilots should apply corrections for static pressure systems and/or instruments, if appreciable errors exist.

c. The adoption of a standard altimeter setting at the higher altitudes eliminates station barometer errors, some altimeter instrument errors, and errors caused by altimeter settings derived from different geographical sources.

7-32. PROCEDURES

The cruising altitude or flight level of aircraft shall be maintained by reference to an altimeter which shall be set, when operating:

a. Below 18,000 feet MSL:

1. When the barometric pressure is 31.00 inches Hg. or less -- to the current reported altimeter setting of a station along the route and within 100 NM of the aircraft, or if there is no station within this area, the current reported altimeter setting of an appropriate available station. When an aircraft is en route on an instrument flight plan, air traffic controllers will furnish this information to the pilot at least once while the aircraft is in the controllers area of jurisdiction. In the case of an aircraft not equipped with a radio, set to the elevation of the departure airport or use an appropriate altimeter setting available prior to departure.

2. When the barometric pressure exceeds 31.00 inches Hg. -- the following procedures will be placed in effect by NOTAM defining the geographic area affected:

(a) **For all aircraft** - Set 31.00 inches for en route operations below 18,000 feet MSL. Maintain this setting until beyond the affected area or until reaching final approach segment. At the beginning of the final approach segment, the current altimeter setting will be set, if possible. If not possible, 31.00 inches will remain set throughout the approach. Aircraft on departure or missed approach will set 31.00 inches prior to reaching any mandatory/crossing altitude or 1,500 feet AGL, whichever is lower. (Air traffic control will issue actual altimeter settings and advise pilots to set 31.00 inches in their altimeters for en route operations below 18,000 feet MSL in affected areas.)

(b) During preflight, barometric altimeters shall be checked for normal operation to the extent possible.

(c) For aircraft with the capability of setting the current altimeter setting and operating into airports with the capability of measuring the current altimeter setting, no additional restrictions apply.

(d) For aircraft operating VFR, there are no additional restrictions, however, extra diligence in flight planning and in operating in these conditions is essential.

(e) Airports unable to accurately measure barometric pressures above 31.00 inches of Hg. will report the barometric pressure as "missing" or "in excess of 31.00 inches of Hg." Flight operations to and from those airports are restricted to VFR weather conditions.

(f) For aircraft operating IFR and unable to set the current altimeter setting, the following restrictions apply:

(1) To determine the suitability of departure alternate airports, destination airports, and destination alternate airports, increase ceiling requirements by 100 feet and visibility requirements by 1/4 statute mile for each 1/10 of an inch of Hg., or any portion thereof, over 31.00 inches. These adjusted values are then applied in accordance with the requirements of the applicable operating regulations and operations specifications.

EXAMPLE:

Destination altimeter is 31.28 inches, ILS DH 250 feet (200-1/2). When flight planning, add 300-3/4 to the weather requirements which would become 500-1 ¼.

(2) On approach, 31.00 inches will remain set. Decision height or minimum descent altitude shall be deemed to have been reached when the published altitude is displayed on the altimeter.

7-32a1f2 NOTE– Although visibility is normally the limiting factor on an approach, pilots should be aware that when reaching DH the aircraft will be higher than indicated. Using the example above the aircraft would be approximately 300 feet higher.

(3) These restrictions do not apply to authorized Category II and III ILS operations nor do they apply to certificate holders using approved QFE altimetry systems.

(g) The FAA Regional Flight Standards Division Manager of the Affected area is authorized to approve temporary waivers to permit emergency resupply or emergency medical service operation.

(Table 7-32[1])

Altimeter Setting (Current Reported)	Lowest Usable Flight Level
29.92 or higher	180
29.91 to 29.42	185
29.41 to 28.92	190
28.91 to 28.42	195
28.41 to 27.92	200

b. At or above 18,000 feet MSL - to 29.92 inches of mercury (standard setting). The lowest usable flight level is determined by the atmospheric pressure in the area of operation as shown in Table 7-32[1].

(Table 7-32[2])

Altimeter Setting	Correction Factor
29.92 or higher	none
29.91 to 29.42	500 Feet
29.41 to 28.92	1000 Feet
28.91 to 28.42	1500 Feet
28.41 to 27.92	2000 Feet
27.91 to 27.42	2500 Feet

c. Where the minimum altitude, as prescribed in FAR 91.159 and FAR 91.177, is above 18,000 feet MSL, the lowest usable flight level shall be the flight level equivalent of the minimum altitude plus the number of feet specified in Table 7-32[2].

EXAMPLE:

The minimum safe altitude of a route is 19,000 feet MSL and the altimeter setting is reported between 29.92 and 29.42 inches of mercury, the lowest usable flight level will be 195, which is the flight level equivalent of 19,500 feet MSL (minimum altitude plus 500 feet).

7-33. ALTIMETER ERRORS

a. Most pressure altimeters are subject to mechanical, elastic, temperature, and installation errors. (Detailed information regarding the use of pressure altimeters is found in the Instrument Flying Handbook, Chapter IV.) Although manufacturing and installation specifications, as well as the periodic test and inspections required by regulations (Far 43, Appendix E), act to reduce these errors, any scale error may be observed in the following manner:

1. Set the current reported altimeter setting on the altimeter setting scale.

2. Altimeter should now read field elevation if you are located on the same reference level used to establish the altimeter setting.

3. Note the variation between the known field elevation and the altimeter indication. If this variation is in the order of plus or minus 75 feet, the accuracy of the altimeter is questionable and the problem should be referred to an appropriately rated repair station for evaluation and possible correction.

b. Once in flight, it is very important to frequently obtain current altimeter settings en route. If you do not reset your altimeter when flying *from* an area of high pressure or high temperatures *into* an area of low pressures or low temperature, *your aircraft will be closer to the surface than the altimeter indicates*. An inch error on the altimeter equals 1,000 feet of altitude. To quote an old saying: **"GOING FROM A HIGH TO A LOW, LOOK OUT BELOW."**

c. A reverse situation, without resetting the altimeter when going from a low temperature or low pressure area into a high temperature or high pressure area, the aircraft will be higher than the altimeter indicates.

d. The possible results of the above situations is obvious, particularly if operating at the minimum altitude or when conducting an instrument approach. If the altimeter is in error, you may still be on instruments when reaching the minimum altitude (as indicated on the altimeter), whereas you might have been in the clear and able to complete the approach if the altimeter setting was correct.

7-34. HIGH BAROMETRIC PRESSURE

a. Cold, dry air masses may produce barometric pressures in excess of 31.00 inches of Mercury, and many altimeters do not have an accurate means of being adjusted for settings of these levels. As noted in PARA 7-33 *ALTIMETER ERRORS*, when the altimeter cannot be set to the higher pressure setting the aircraft actual altitude will be higher than the altimeter indicates.

b. When the barometric pressure exceeds 31.00 inches, air traffic controllers will issue the actual altimeter setting, and:

1. En Route/Arrivals - Advise pilots to remain set on 31.00 inches until reaching the final approach segment.

2. Departures - Advise pilots to set 31.00 inches prior to reaching any mandatory/crossing altitude or 1,500 feet, whichever is lower.

c. The altimeter error caused by the high pressure will be in the opposite direction to the error caused by the cold temperature.

7-35. LOW BAROMETRIC PRESSURE

When abnormally low barometric pressure conditions occur (below 28.00), flight operations by aircraft unable to set the actual altimeter setting are not recommended.

7-35 Note– The true altitude of the aircraft is lower than the indicated altitude if the pilot is unable to set the actual altimeter setting.

7-36 thru 7-40. RESERVED

Chapter Review Questions — Answer Key

Ch. I.
1. a
2. b
3. a

Ch. II.
1. b
2. b

Ch. III.
1. c
2. b
3. a
4. b
5. d
6. a
7. a
8. b
9. a
10. a
11. d

Ch. IV.
1. b
2. b
3. c
4. a
5. b
6. c
7. d
8. c
9. d
10. b
11. d
12. b
13. a
14. b
15. a

Ch. V.
1. d
2. c
3. a
4. d
5. c
6. c
7. b
8. b
9. c
10. d
11. a
12. a
13. a
14. b
15. b
16. c
17. d
18. a
19. b
20. d
21. a
22. a
23. d
24. b
25. c
26. c
27. b
28. a
29. d
30. a
31. d
32. a
33. c
34. d
35. b
36. b
37. b
38. a
39. a
40. d
41. c
42. c
43. a
44. b
45. a
46. b
47. a

Ch. VI.
1. c
2. c
3. d
4. c
5. c
6. b
7. a
8. a
9. b
10. a
11. b
12. a
13. b

Ch. VII
1. c
2. d
3. b
4. c
5. b

Ch. VIII.
1. c
2. b
3. c
4. a
5. b
6. d
7. a
8. b
9. d
10. a
11. a
12. b
13. b
14. a
15. c
16. d
17. c

Ch. IX
1. d
2. c
3. c

Ch. X
1. a
2. c
3. b
4. c
5. a
6. b
7. b
8. d
9. c
10. d

Ch. XI
1. d
2. c
3. d
4. a
5. c
6. c
7. b
8. a
9. b
10. b
11. b
12. c
13. c
14. b
15. d
16. d
17. d
18. b
19. d
20. d
21. b
22. a
23. d
24. a
25. c
26. c
27. d
28. b
29. a
30. b
31. c
32. d
33. d
34. b
35. a
36. b
37. b
38. b
39. d
40. b
41. a
42. b
43. a
44. c

Ch. XII.
1. c
2. c
3. c
4. d
5. c
6. a
7. c
8. c
9. a
10. c
11. b
12. a
13. a
14. d
15. d
16. b
17. d
18. c
19. c
20. d
21. b
22. c
23. b
24. b
25. b
26. a
27. d
28. b
29. c
30. d
31. a
32. b
33. c
34. c
35. a

Ch. XIII.
1. c
2. d
3. d
4. d
5. a
6. c
7. b
8. c
9. a
10. b
11. c
12. b
13. a
14. d
15. a
16. c
17. b
18. d
19. b
20. b

Ch. XIV.
1. a
2. c
3. a
4. d
5. c
6. d
7. a
8. b
9. b
10. a
11. d
12. a
13. a
14. d
15. a
16. d
17. d
18. a
19. a
20. d
21. d
22. c
23. b
24. c
25. b
26. a
27. c

Ch. XV.
1. b
2. c
3. d
4. b
5. d
6. b
7. a
8. c
9. b
10. c
11. a
12. b
13. c
14. c
15. 4
16. b
17. c
18. a
19. b
20. d
21. a
22. d
23. b
24. c
25. b
26. c
27. c
28. c
29. a
30. b
31. c
32. d
33. a
34. d
35. a
36. b
37. b
38. d
39. d
40. b
41. d
42. d
43. a
44. c
45. d
46. c
47. b
48. a
49. d
50. a
51. b
52. a
53. c
54. a
55. a
56. d
57. d
58. d
59. a
60. a
61. b
62. b
63. d
64. c
65. a

Ch. XVI.
1. c
2. b
3. a
4. d
5. c
6. b
7. d
8. a
9. c
10. a
11. d
12. c
13. a
14. d

GLOSSARY OF WEATHER TERMS

A

absolute instability—A state of a layer within the atmosphere in which the vertical distribution of temperature is such that an air parcel, if given an upward or downward push, will move away from its initial level without further outside force being applied.

absolute temperature scale—*See* Kelvin Temperature Scale.

absolute vorticity—*See* vorticity.

adiabatic process—The process by which fixed relationships are maintained during changes in temperature, volume, and pressure in a body of air without heat being added or removed from the body.

advection—The horizontal transport of air or atmospheric properties. In meteorology, sometimes referred to as the horizontal component of *convection.*

advection fog—Fog resulting from the transport of warm, humid air over a cold surface.

air density—The mass density of the air in terms of weight per unit volume.

air mass—In meteorology, an extensive body of air within which the conditions of temperature and moisture in a horizontal plane are essentially uniform.

air mass classification—A system used to identify and to characterize the different *air masses* according to a basic scheme. The system most commonly used classifies air masses primarily according to the thermal properties of their *source regions:* "tropical" (T); "polar" (P); and "Arctic" or "Antarctic" (A). They are further classified according to moisture characteristics as "continental" (c) or "maritime" (m).

air parcel—*See* parcel.

albedo—The ratio of the amount of electromagnetic *radiation* reflected by a body to the amount incident upon it, commonly expressed in percentage; in meteorology, usually used in reference to *insolation* (solar radiation); i.e., the albedo of wet sand is 9, meaning that about 9% of the incident insolation is reflected; albedoes of other surfaces range upward to 80-85 for fresh snow cover; average albedo for the earth and its atmosphere has been calculated to range from 35 to 43.

altimeter—An instrument which determines the altitude of an object with respect to a fixed level. *See* pressure altimeter.

altimeter setting—The value to which the scale of a *pressure altimeter* is set so as to read true altitude at field elevation.

altimeter setting indicator—A precision *aneroid barometer* calibrated to indicate directly the altimeter setting.

altitude—Height expressed in units of distance above a reference plane, usually above mean sea level or above ground.

(1) **corrected altitude**—Indicated altitude of an aircraft altimeter corrected for the temperature of the column of air below the aircraft, the correction being based on the estimated departure of existing temperature from standard atmospheric temperature; an approximation of true altitude.

(2) **density altitude**—The altitude in the standard atmosphere at which the air has the same density as the air at the point in question. An aircraft will have the same performance characteristics as it would have in a standard atmosphere at this altitude.

(3) **indicated altitude**—The altitude above mean sea level indicated on a *pressure altimeter* set at current local *altimeter setting.*

(4) **pressure altitude**—The altitude in the standard atmosphere at which the pressure is the same as at the point in question. Since an altimeter operates solely on pressure, this is the uncorrected altitude indicated by an altimeter set at standard sea level pressure of 29.92 inches or 1013.2 millibars.

(5) **radar altitude**—The altitude of an aircraft determined by radar-type radio altimeter; thus the actual distance from the nearest terrain or water feature encompassed by the downward directed radar beam. For all practical purposes, it is the "actual" distance above a ground or inland water surface or the true altitude above an ocean surface.

(6) **true altitude**—The exact distance above mean sea level.

altocumulus—White or gray layers or patches of cloud, often with a waved appearance; cloud elements appear as rounded masses or rolls; composed mostly of liquid water droplets which may be supercooled; may contain ice crystals at subfreezing temperatures.

altocumulus castellanus—A species of middle cloud of which at least a fraction of its upper part presents some vertically developed, cumuliform protuberances (some of which are taller than they are wide, as castles) and which give the cloud a crenelated or turreted appearance; especially evident when seen from the side; elements usually have a common base arranged in lines. This cloud indicates instability and turbulence at the altitudes of occurrence.

anemometer—An instrument for measuring *wind speed.*

aneroid barometer—A *barometer* which operates on the principle of having changing atmospheric pressure bend a metallic surface which, in turn, moves a pointer across a scale graduated in units of pressure.

angel—In radar meteorology, an *echo* caused by physical phenomena not discernible to the eye; they have been observed when abnormally strong temperature and/or moisture *gradients* were known to exist; sometimes attributed to insects or birds flying in the radar beam.

anomalous propagation (sometimes called AP)—In radar meteorology, the greater than normal bending of the radar beam such that *echoes* are received from ground *targets* at distances greater than normal *ground clutter.*

anticyclone—An area of high atmospheric pressure which has a closed circulation that is anticyclonic, i.e., as viewed from above, the circulation is clockwise in the Northern Hemisphere, counterclockwise in the Southern Hemisphere, undefined at the Equator.

anvil cloud—Popular name given to the top portion of a *cumulonimbus* cloud having an anvil-like form.

APOB—A *sounding* made by an aircraft.

Arctic air—An air mass with characteristics developed mostly in winter over Arctic surfaces of ice and snow. Arctic air extends to great heights, and the surface temperatures are basically, but not always, lower than those of *polar air.*

Arctic front—The surface of discontinuity between very cold (Arctic) air flowing directly from the Arctic region and another less cold and, consequently, less dense air mass.

astronomical twilight—*See* twilight.

atmosphere—The mass of air surrounding the Earth.

atmospheric pressure (also called barometric pressure)—The pressure exerted by the atmosphere as a consequence of gravitational attraction exerted upon the "column" of air lying directly above the point in question.

atmospherics—Disturbing effects produced in radio receiving apparatus by atmospheric electrical phenomena such as an electrical storm. Static.

aurora—A luminous, radiant emission over middle and high latitudes confined to the thin air of high altitudes and centered over the earth's magnetic poles. Called "aurora borealis" (northern lights) or "aurora australis" according to its occurrence in the Northern or Southern Hemisphere, respectively.

attenuation—In radar meteorology, any process which reduces power density in radar signals.

(1) **precipitation attenuation**—Reduction of power density because of absorption or reflection of energy by precipitation.

(2) **range attenuation**—Reduction of radar power density because of distance from the antenna. It occurs in the outgoing beam at a rate proportional to $1/\text{range}^2$. The return signal is also attenuated at the same rate.

B

backing—Shifting of the wind in a counterclockwise direction with respect to either space or time; opposite of *veering.* Commonly used by meteorologists to refer to a cyclonic shift (counterclockwise in the Northern Hemisphere and clockwise in the Southern Hemisphere).

backscatter—Pertaining to radar, the energy, reflected or scattered by a *target; an echo.*

banner cloud (also called cloud banner)—A banner-like cloud streaming off from a mountain peak.

barograph—A continuous-recording *barometer.*

barometer—An instrument for measuring the pressure of the atmosphere; the two principle types are *mercurial* and *aneroid.*

barometric altimeter—*See* pressure altimeter.

barometric pressure—Same as *atmospheric pressure.*

barometric tendency—The change of barometric pressure within a specified period of time. In aviation weather observations, routinely determined periodically, usually for a 3-hour period.

beam resolution—*See* resolution.

Beaufort scale—A scale of wind speeds.

black blizzard—*Same as* duststorm.

blizzard—A severe weather condition characterized by low temperatures and strong winds bearing a great amount of snow, either falling or picked up from the ground.

blowing dust—A type of *lithometeor* composed of dust particles picked up locally from the surface and blown about in clouds or sheets.

blowing sand—A type of *lithometeor* composed of sand picked up locally from the surface and blown about in clouds or sheets.

blowing snow—A type of *hydrometeor* composed of snow picked up from the surface by the wind and carried to a height of 6 feet or more.

blowing spray—A type of *hydrometeor* composed of water particles picked up by the wind from the surface of a large body of water.

bright band—In radar meteorology, a narrow, intense *echo* on the *range-height indicator* scope resulting from water-covered ice particles of high reflectivity at the melting level.

Buys Ballot's law—If an observer in the Northern Hemisphere stands with his back to the wind, lower pressure is to his left.

C

calm—The absence of wind or of apparent motion of the air.

cap cloud (also called cloud cap)—A standing or stationary cap-like cloud crowning a mountain summit.

ceiling—In meteorology in the U.S., (1) the height above the surface of the base of the lowest layer of clouds or *obscuring phenomena* aloft that is reported as broken or overcast but not thin, or (2) the *vertical visibility* into an *obscuration. See* summation principle.

ceiling balloon—A small balloon used to determine the height of a cloud base or the extent of vertical visibility.

ceiling light—An instrument which projects a vertical light beam onto the base of a cloud or into surface-based obscuring phenomena; used at night in conjunction with a *clinometer* to determine the height of the cloud base or as an aid in estimating the vertical visibility.

ceilometer—A cloud-height measuring system. It projects light on the cloud, detects the reflection by a photo-electric cell, and determines height by triangulation.

Celsius temperature scale (abbreviated C)—A temperature scale with zero degrees as the melting point of pure ice and 100 degrees as the boiling point of pure water at standard sea level atmospheric pressure.

Centigrade temperature scale—Same as *Celsius temperature scale.*

chaff—Pertaining to radar, (1) short, fine strips of metallic foil dropped from aircraft, usually by military forces, specifically for the purpose of jamming radar; (2) applied loosely to *echoes* resulting from chaff.

change of state—In meteorology, the transformation of water from one form, i.e., solid (ice), liquid, or gaseous (water vapor), to any other form. There are six possible transformations designated by the five terms following:

(1) **condensation**—The change of water vapor to liquid water.

(2) **evaporation**—The change of liquid water to water vapor.

(3) **freezing**—The change of liquid water to ice.

(4) **melting**—The change of ice to liquid water.

(5) **sublimation**—The change of (a) ice to water vapor or (b) water vapor to ice. *See* latent heat.

Chinook—A warm, dry *foehn* wind blowing down the eastern slopes of the Rocky Mountains over the adjacent plains in the U.S. and Canada.

cirriform—All species and varieties of *cirrus, cirrocumulus,* and *cirrostratus* clouds; descriptive of clouds composed mostly or entirely of small ice crystals, usually transparent and white; often producing *halo* phenomena not observed with other cloud forms. Average height ranges upward from 20,000 feet in middle latitudes.

cirrocumulus—A *cirriform* cloud appearing as a thin sheet of small white puffs resembling flakes or patches of cotton without shadows; sometimes confused with *altocumulus.*

cirrostratus—A *cirriform* cloud appearing as a whitish veil, usually fibrous, sometimes smooth; often produces *halo* phenomena; may totally cover the sky.

cirrus—A *cirriform* cloud in the form of thin, white featherlike clouds in patches or narrow bands; have a fibrous and/or silky sheen; large ice crystals often trail downward a considerable vertical distance in fibrous, slanted, or irregularly curved wisps called mares' tails.

civil twilight—*See* twilight.

clear air turbulence (abbreviated CAT)—Turbulence encountered in air where no clouds are present; more popularly applied to high level turbulence associated with *wind shear.*

clear icing (or clear ice)—Generally, the formation of a layer or mass of ice which is relatively transparent because of its homogeneous structure and small number and size of air spaces; used commonly as synonymous with *glaze*, particularly with respect to aircraft icing. Compare with *rime icing.* Factors which favor clear icing are large drop size, such as those found in *cumuliform* clouds, rapid accretion of supercooled water, and slow dissipation of *latent heat* of fusion.

climate—The statistical collective of the weather conditions of a point or area during a specified interval of time (usually several decades); may be expressed in a variety of ways.

climatology—The study of *climate.*

clinometer—An instrument used in weather observing for measuring angles of inclination; it is used in conjunction with a *ceiling light* to determine cloud height at night.

cloud bank—Generally, a fairly well-defined mass of cloud observed at a distance; it covers an appreciable portion of the horizon sky, but does not extend overhead.

cloudburst—In popular terminology, any sudden and heavy fall of *rain*, almost always of the *shower* type.

cloud cap—*See* cap cloud.

cloud detection radar—A vertically directed radar to detect cloud bases and tops.

cold front—Any non-occluded *front* which moves in such a way that colder air replaces warmer air.

condensation—*See* change of state.

condensation level—The height at which a rising *parcel* or layer of air would become saturated if lifted adiabatically.

condensation nuclei—Small particles in the air on which water vapor condenses or sublimates.

condensation trail (or contrail) (also called vapor trail)—A cloud-like streamer frequently observed to form behind aircraft flying in clear, cold, humid air.

conditionally unstable air—Unsaturated air that will become unstable on the condition it becomes saturated. *See* instability.

conduction—The transfer of heat by molecular action through a substance or from one substance in contact with another; transfer is always from warmer to colder temperature.

constant pressure chart—A chart of a constant pressure surface; may contain analyses of height, wind, temperature, humidity, and/or other elements.

continental polar air—*See* polar air.

continental tropical air—*See* tropical air.

contour—In meteorology, (1) a line of equal height on a constant pressure chart; analogous to contours on a relief map; (2) in radar meteorology, a line on a radar scope of equal *echo* intensity.

contouring circuit—On weather radar, a circuit which displays multiple contours of *echo* intensity simultaneously on the *plan position indicator* or *range-height indicator* scope. *See* contour (2).

contrail—Contraction for *condensation trail.*

convection—(1) In general, mass motions within a fluid resulting in transport and mixing of the properties of that fluid. (2) In meteorology, atmospheric motions that are predominantly vertical, resulting in vertical transport and mixing of atmospheric properties; distinguished fro *advection.*

convective cloud—*See* cumuliform.

convective condensation level (abbreviated CCL)—The lowest level at which condensation will occur as a result of *convection* due to surface heating. When condensation occurs at this level, the layer between the surface and the CCL will be thoroughly mixed, temperature *lapse rate* will be dry adiabatic, and *mixing ratio* will be constant.

convective instability—The state of an unsaturated layer of air whose *lapse rates* of temperature and moisture are such that when lifted adiabatically until the layer becomes saturated, convection is spontaneous.

convergence—The condition that exists when the distribution of winds within a given area is such that there is a net horizontal inflow of air into the area. In convergence at lower levels, the removal of the resulting excess is accomplished by an upward movement of air; consequently, areas of low-level convergent winds are regions favorable to the occurrence of clouds and precipitation. Compare with *divergence.*

Coriolis force—A deflective force resulting from earth's rotation; it acts to the right of wind direction in the Northern Hemisphere and to the left in the Southern Hemisphere.

corona—A prismatically colored circle or arcs of a circle with the sun or moon at its center; coloration is from blue inside to red outside (opposite that of a *halo*); varies in size (much smaller) as opposed to the fixed diameter of the halo; characteristic of clouds composed of water droplets and valuable in differentiating between middle and cirriform clouds.

corposant—*See* St. Elmo's Fire.

corrected altitude (approximation of true altitude)—*See* altitude.

cumuliform—A term descriptive of all convective clouds exhibiting vertical development in contrast to the horizontally extended *stratiform* types.

cumulonimbus—A cumuliform cloud type; it is heavy and dense, with considerable vertical extent in the form of massive towers; often with tops in the shape of an *anvil* or massive plume; under the base of cumulonimbus, which often is very dark, there frequently exists *virga* precipitation and low ragged clouds (*scud*), either merged with it or not; frequently accompanied by lightning, thunder, and sometimes hail; occasionally produces a tornado or a waterspout; the ultimate manifestation of the growth of a cumulus cloud, occasionally extending well into the stratosphere.

cumulonimbus mamma—A *cumulonimbus* cloud having hanging protuberances, like pouches, festoons, or udders, on the under side of the cloud; usually indicative of severe turbulence.

cumulus—A cloud in the form of individual detached domes or towers which are usually dense and well defined; develops vertically in the form of rising mounds of which the bulging upper part often resembles a cauliflower; the sunlit ~f these clouds are mostly brilliant white; their bases ~ark and nearly horizontal.

nt or strengthening of cyclonic
ıere.

cyclone—(1) An area of low atmospheric pressure which has a closed circulation that is cyclonic, i.e., as viewed from above, the circulation is counterclockwise in the Northern Hemisphere, clockwise in the Southern Hemisphere, undefined at the Equator. Because cyclonic circulation and relatively low atmospheric pressure usually co-exist, in common practice the terms cyclone and low are used interchangeably. Also, because cyclones often are accompanied by inclement (sometimes destructive) weather, they are frequently referred to simply as storms. (2) Frequently misused to denote a *tornado.* (3) In the Indian Ocean, a *tropical cyclone* of hurricane or typhoon force.

D

deepening—A decrease in the central pressure of a pressure system; usually applied to a *low* rather than to a *high*, although technically, it is acceptable in either sense.

density—(1) The ratio of the mass of any substance to the volume it occupies—mass per unit volume. (2) The ratio of any quantity to the volume or area it occupies, i.e., population per unit area, *power density.*

density altitude—*See* altitude.

depression—In meteorology, an area of low pressure; a *low* or *trough*. This is usually applied to a certain stage in the development of a *tropical cyclone*, to migratory lows and troughs, and to upper-level lows and troughs that are only weakly developed.

dew—Water condensed onto grass and other objects near the ground, the temperatures of which have fallen below the initial dewpoint temperature of the surface air, but is still above freezing. Compare with *frost.*

dew point (or dewpoint temperature)—The temperature to which a sample of air must be cooled, while the *mixing ratio* and barometric pressure remain constant, in order to attain saturation with respect to water.

discontinuity—A zone with comparatively rapid transition of one or more meteorological elements.

disturbance—In meteorology, applied rather loosely: (1) any low pressure or cyclone, but usually one that is relatively small in size; (2) an area where weather, wind, pressure, etc., show signs of cyclonic development; (3) any deviation in flow or pressure that is associated with a disturbed state of the weather, i.e., cloudiness and precipitation; and (4) any individual circulatory system within the primary circulation of the atmosphere.

diurnal—Daily, especially pertaining to a cycle completed within a 24-hour period, and which recurs every 24 hours.

divergence—The condition that exists when the distribution of winds within a given area is such that there is a net horizontal flow of air outward from the region. In divergence at lower levels, the resulting deficit is compensated for by subsidence of air from aloft; consequently the air is heated and the relative humidity lowered making divergence a warming and drying process. Low-level divergent regions are areas unfavorable to the occurrence of clouds and precipitation. The opposite of *convergence.*

doldrums—The equatorial belt of calm or light and variable winds between the two tradewind belts. Compare *intertropical convergence zone.*

downdraft—A relative small scale downward current of air; often observed on the lee side of large objects restricting the smooth flow of the air or in precipitation areas in or near *cumuliform* clouds.

drifting snow—A type of *hydrometeor* composed of snow particles picked up from the surface, but carried to a height of less than 6 feet.

drizzle—A form of *precipitation.* Very small water drops that appear to float with the air currents while falling in an irregular path (unlike *rain*, which falls in a comparatively straight path, and unlike *fog* droplets which remain suspended in the air).

dropsonde—A *radiosonde* dropped by parachute from an aircraft to obtain *soundings* (measurements) of the atmosphere below.

dry adiabatic lapse rate—The rate of decrease of temperature with height when unsaturated air is lifted adiabatically (due to expansion as it is lifted to lower pressure). *See* adiabatic process.

dry bulb—A name given to an ordinary thermometer used to determine temperature of the air; also used as a contraction for *dry-bulb temperature.* Compare *wet bulb.*

dry-bulb temperature—The temperature of the air.

dust—A type of *lithometeor* composed of small earthern particles suspended in the atmosphere.

dust devil—A small, vigorous *whirlwind*, usually of short duration, rendered visible by dust, sand, and debris picked up from the ground.

duster—Same as *duststorm.*

duststorm (also called duster, black blizzard)—An unusual, frequently severe weather condition characterized by strong winds and dust-filled air over an extensive area.

D-value—Departure of true altitude from pressure altitude (*see* altitude); obtained by algebraically subtracting true altitude from pressure altitude; thus, it may be plus or minus. On a constant pressure chart, the difference between actual height and *standard atmospheric* height of a constant pressure surface.

E

echo—In radar, (1) the energy reflected or scattered by a *target;* (2) the radar scope presentation of the return from a target.

eddy—A local irregularity of wind in a larger scale wind flow. Small scale eddies produce turbulent conditions.

estimated ceiling—A ceiling classification applied when the ceiling height has been estimated by the observer or has been determined by some other method; because of the specified limits of time, distance, or precipitation conditions, a more descriptive classification cannot be applied.

evaporation—*See* change of state.

extratropical low (sometimes called extratropical cyclone, extratropical storm)—Any *cyclone* that is not a *tropical cyclone*, usually referring to the migratory frontal cyclones of middle and high latitudes.

eye—The roughly circular area of calm or relatively light winds and comparatively fair weather at the center of a well-developed *tropical cyclone.* A *wall cloud* marks the outer boundary of the eye.

F

Fahrenheit temperature scale (abbreviated F)—A temperature scale with 32 degrees as the melting point of pure ice and 212 degrees as the boiling point of pure water at standard sea level atmospheric pressure (29.92 inches or 1013.2 millibars).

Fall wind—A cold wind blowing downslope. Fall wind differs from *foehn* in that the air is initially cold enough to remain relatively cold despite compressional heating during descent.

filling—An increase in the central pressure of a pressure system; opposite of *deepening;* more commonly applied to a low rather than a high.

first gust—The leading edge of the spreading downdraft, *plow wind*, from an approaching thunderstorm.

flow line—A *streamline.*

foehn—A warm, dry downslope wind; the warmness and dryness being due to adiabatic compression upon descent; characteristic of mountainous regions. *See* adiabatic process, Chinook, Santa Ana.

fog—A *hydrometeor* consisting of numerous minute water droplets and based at the surface; droplets are small enough to be suspended in the earth's atmosphere indefinitely. (Unlike *drizzle*, it does not fall to the surface; differs from cloud only in that a cloud is not based at the surface; distinguished from haze by its wetness and gray color.)

fractus—Clouds in the form of irregular shreds, appearing as if torn; has a clearly ragged appearance; applies only to stratus and cumulus, i.e., *cumulus* fractus and *stratus* fractus.

freezing—*See* change of state.

freezing level—A level in the atmosphere at which the temperature is 0°C (32°F).

front—A surface, interface, or transition zone of discontinuity between two adjacent *air masses* of different densities; more simply the boundary between two different air masses. *See* frontal zone.

frontal zone—A *front* or zone with a marked increase in density gradient; used to denote that fronts are not truly a "surface" of discontinuity but rather a "zone" of rapid transition of meteorological elements.

frontogenesis—The initial formation of a *front* or *frontal zone.*

frontolysis—The dissipation of a *front.*

frost (also hoarfrost)—Ice crystal deposits formed by sublimation when temperature and dewpoint are below freezing.

funnel cloud—A *tornado* cloud or *vortex* cloud extending downward from the parent cloud but not reaching the ground.

G

glaze—A coating of ice, generally clear and smooth, formed by freezing of supercooled water on a surface. *See* clear icing.

gradient—In meteorology, a horizontal decrease in value per unit distance of a parameter in the direction of maximum decrease; most commonly used with pressure, temperature, and moisture.

ground clutter—Pertaining to radar, a cluster of *echoes,* generally at short range, reflected from ground *targets.*

ground fog—In the United States, a *fog* that conceals less than 0.6 of the sky and is not contiguous with the base of clouds.

gust—A sudden brief increase in wind; according to U.S. weather observing practice, gusts are reported when the variation in wind speed between peaks and lulls is at least 10 knots.

H

hail—A form of *precipitation* composed of balls or irregular lumps of ice, always produced by convective clouds which are nearly always *cumulonimbus.*

halo—A prismatically colored or whitish circle or arcs of a circle with the sun or moon at its center; coloration, if not white, is from red inside to blue outside (opposite that of a *corona*); fixed in size with an angular diameter of 22° (common) or 46° (rare); characteristic of clouds composed of ice crystals; valuable in differentiating between *cirriform* and forms of lower clouds.

haze—A type of *lithometeor* composed of fine dust or salt particles dispersed through a portion of the atmosphere; particles are so small they cannot be felt or individually seen with the naked eye (as compared with the larger particles of *dust*), but diminish the visibility; distinguished from *fog* by its bluish or yellowish tinge.

high—An area of high barometric pressure, with its attendant system of winds; an *anticyclone.* Also high pressure system.

hoar frost—*See* frost.

humidity—Water vapor content of the air; may be expressed as *specific humidity, relative humidity,* or *mixing ratio.*

hurricane—A *tropical cyclone* in the Western Hemisphere with winds in excess of 65 knots or 120 km/h.

hydrometeor—A general term for particles of liquid water or ice such as rain, fog, frost, etc., formed by modification of water vapor in the atmosphere; also water or ice particles lifted from the earth by the wind such as sea spray or blowing snow.

hygrograph—The record produced by a continuous-recording *hygrometer.*

hygrometer—An instrument for measuring the water vapor content of the air.

I

ice crystals—A type of *precipitation* composed of unbranched crystals in the form of needles, columns, or plates; usually having a very slight downward motion, may fall from a cloudless sky.

ice fog—A type of fog composed of minute suspended particles of ice; occurs at very low temperatures and may cause *halo* phenomena.

ice needles—A form of *ice crystals.*

ice pellets—Small, transparent or translucent, round or irregularly shaped pellets of ice. They may be (1) hard grains that rebound on striking a hard surface or (2) pellets of snow encased in ice.

icing—In general, any deposit of ice forming on an object. *See* clear icing, rime icing, glaze.

indefinite ceiling—A ceiling classification denoting *vertical visibility* into a surface based obscuration.

indicated altitude—*See* altitude.

insolation—Incoming solar *radiation* falling upon the earth and its atmosphere.

instability—A general term to indicate various states of the atmosphere in which spontaneous *convection* will occur when prescribed criteria are met; indicative of turbulence. *See* absolute instability, conditionally unstable air, convective instability.

intertropical convergence zone—The boundary zone between the tradewind system of the Northern and Southern Hemispheres; it is characterized in maritime climates by showery precipitation with cumulonimbus clouds sometimes extending to great heights.

inversion—An increase in temperature with height — a reversal of the normal decrease with height in the *troposphere*; may also be applied to other meteorological properties.

isobar—A line of equal or constant barometric pressure.

iso echo—In radar circuitry, a circuit that reverses signal strength above a specified intensity level, thus causing a void on the scope in the most intense portion of an echo when maximum intensity is greater than the specified level.

isoheight—On a weather chart, a line of equal height; same as *contour* (1).

isoline—A line of equal value of a variable quantity, i.e., an isoline of temperature is an *isotherm*, etc. *See* isobar, isotach, etc.

isoshear—A line of equal *wind shear.*

isotach—A line of equal or constant wind speed.

isotherm—A line of equal or constant temperature.

isothermal—Of equal or constant temperature, with respect to either space or time; more commonly, temperature with height; a zero *lapse rate.*

J

jet stream—A quasi-horizontal stream of winds 50 knots or more concentrated within a narrow band embedded in the westerlies in the high *troposphere.*

K

katabatic wind—Any wind blowing downslope. *See* fall wind, foehn.

Kelvin temperature scale (abbreviated K)—A temperature scale with zero degrees equal to the temperature at which all molecular motion ceases, i.e., absolute zero (0°K = –273°C); the Kelvin degree is identical to the Celsius degree; hence at standard sea level pressure, the melting point is 273°K and the boiling point 373°K.

knot—A unit of speed equal to one nautical mile per hour.

L

land breeze—A coastal breeze blowing from land to sea, caused by temperature difference when the sea surface is warmer than the adjacent land. Therefore, it usually blows at night and alternates with a *sea breeze,* which blows in the opposite direction by day.

lapse rate—The rate of decrease of an atmospheric variable with height; commonly refers to decrease of temperature with height.

latent heat—The amount of heat absorbed during the processes of change of liquid water to water vapor, ice to water vapor, or ice to liquid water; or the amount released during the reverse processes. Four basic classifications are:

(1) **latent heat of condensation**—Heat released during change of water vapor to water.

(2) **latent heat of fusion**—Heat released during change of water to ice or the amount absorbed in change of ice to water.

(3) **latent heat of sublimation**—Heat released during change of water vapor to ice or the amount absorbed in the change of ice to water vapor.

(4) **latent heat of vaporization**—Heat absorbed in the change of water to water vapor; the negative of latent heat of condensation.

layer—In reference to sky cover, clouds or other obscuring phenomena whose bases are approximately at the same level. The layer may be continuous or composed of detached elements. The term "layer" does not imply that a clear space exists between the layers or that the clouds or *obscuring phenomena* composing them are of the same type.

lee wave—Any stationary wave disturbance caused by a barrier in a fluid flow. In the atmosphere when sufficient moisture is present, this wave will be evidenced by *lenticular clouds* to the lee of mountain barriers; also called *mountain wave* or standing wave.

lenticular cloud (or lenticularis)—A species of cloud whose elements have the form of more or less isolated, generally smooth lenses or almonds. These clouds appear most often in formations of orographic origin, the result of *lee waves,* in which case they remain nearly stationary with respect to the terrain (standing cloud), but they also occur in regions without marked orography.

level of free convection (abbreviated LFC)—The level at which a *parcel* of air lifted dry adiabatically until saturated and moist adiabatically thereafter would become warmer than its surroundings in a conditionally unstable atmosphere. *See* conditional instability and adiabatic process.

lifting condensation level (abbreviated LCL)—The level at which a *parcel* of unsaturated air lifted dry adiabatically would become saturated. Compare *level of free convection* and *convective condensation level.*

lightning—Generally, any and all forms of visible electrical discharge produced by a *thunderstorm.*

lithometeor—The general term for dry particles suspended in the atmosphere such as dust, haze, smoke, and sand.

low—An area of low barometric pressure, with its attendant system of winds. Also called a barometric depression or *cyclone.*

M

mammato cumulus—Obsolete. *See cumulonimbus mamma.*

mare's tail—*See* cirrus.

maritime polar air (abbreviated mP)—*See* polar air.

maritime tropical air (abbreviated mT)—*See* tropical air.

maximum wind axis—On a constant pressure chart, a line denoting the axis of maximum wind speeds at that constant pressure surface.

mean sea level—The average height of the surface of the sea for all stages of tide; used as reference for elevations throughout the U.S.

measured ceiling—A ceiling classification applied when the ceiling value has been determined by instruments or the known heights of unobscured portions of objects, other than natural landmarks.

melting—*See* change of state.

mercurial barometer—A *barometer* in which pressure is determined by balancing air pressure against the weight of a column of mercury in an evacuated glass tube.

meteorological visibility—In U.S. observing practice, a main category of *visibility* which includes the subcategories of *prevailing visibility* and *runway visibility.* Meteorological visibility is a measure of horizontal visibility near the earth's surface, based on sighting of objects in the daytime or unfocused lights of moderate intensity at night. Compare *slant visibility, runway visual range, vertical visibility.* *See* surface visibility, tower visibility, and sector visibili

meteorology—The science of the *atmosphere.*

microbarograph—An aneroid *barograph* designed
atmospheric pressure changes of very small ma

millibar (abbreviated mb.)—An internationally used unit of pressure equal to 1,000 dynes per square centimeter. It is convenient for reporting *atmospheric pressure.*

mist—A popular expression for drizzle or heavy fog.

mixing ratio—The ratio by weight of the amount of water vapor in a volume of air to the amount of dry air; usually expressed as grams per kilogram (g/kg).

moist adiabatic lapse rate—*See* saturated-adiabatic lapse rate.

moisture—An all-inclusive term denoting water in any or all of its three states.

monsoon—A wind that in summer blows from sea to a continental interior, bringing copious rain, and in winter blows from the interior to the sea, resulting in sustained dry weather.

mountain wave—A *standing wave* or *lee wave* to the lee of a mountain barrier.

N

nautical twilight—*See* twilight.

negative vorticity—*See* vorticity.

nimbostratus—A principal cloud type, gray colored, often dark, the appearance of which is rendered diffuse by more or less continuously falling rain or snow, which in most cases reaches the ground. It is thick enough throughout to blot out the sun.

noctilucent clouds—Clouds of unknown composition which occur at great heights, probably around 75 to 90 kilometers. They resemble thin *cirrus*, but usually with a bluish or silverish color, although sometimes orange to red, standing out against a dark night sky. Rarely observed.

normal—In meteorology, the value of an element averaged for a given location over a period of years and recognized as a standard.

numerical forecasting—*See* numerical weather prediction.

numerical weather prediction—Forecasting by digital computers solving mathematical equations; used extensively in weather services throughout the world.

O

obscuration—Denotes sky hidden by surface-based *obscuring phenomena* and *vertical visibility* restricted overhead.

[illegible] phenomena—Any *hydrometeor* or *lithometeor* other [illegible] be surface based or aloft.

[illegible] **, also called frontal** [illegible] as a *cold front* over- [illegible] *front.*

[illegible] by mountains as in [illegible] lift, or orographic

y.

to record
nitudes.

ozone—An unstable form of oxygen; heaviest concentrations are in the stratosphere; corrosive to some metals; absorbs most ultraviolet solar radiation.

P

parcel—A small volume of air, small enough to contain a uniform distribution of its meteorological properties, and large enough to remain relatively self-contained and respond to all meteorological processes. No specific dimensions have been defined; however, the order of magnitude of 1 cubic foot has been suggested.

partial obscuration—A designation of sky cover when part of the sky or clouds are hidden by surface based *obscuring phenomena.*

pilot balloon—A small free-lift balloon used to determine the speed and direction of winds in the upper air.

pilot balloon observation (commonly called PIBAL)—A method of winds aloft observation by visually tracking a *pilot balloon.*

plan position indicator (PPI) scope—A radar indicator scope displaying range and azimuth of *targets* in polar coordinates.

plow wind—The spreading downdraft of a *thunderstorm;* a strong, straight-line wind in advance of the storm. *See* first gust.

polar air—An air mass with characteristics developed over high latitudes, especially within the subpolar highs. Continental polar air (cP) has cold surface temperatures, low moisture content, and, especially in its source regions, has great stability in the lower layers. It is shallow in comparison with *Arctic air.* Maritime polar (mP) initially possesses similar properties to those of continental polar air, but in passing over warmer water it becomes unstable with a higher moisture content. Compare *tropical air.*

polar front—The semipermanent, semicontinuous *front* separating air masses of tropical and polar origins.

positive vorticity—*See* vorticity.

power density—In radar meteorology the amount of radiated energy per unit cross sectional area in the radar beam.

precipitation—Any or all forms of water particles, whether liquid or solid, that fall from the atmosphere and reach the surface. It is a major class of *hydrometeor*, distinguished from cloud and *virga*, in that it must reach the surface.

precipitation attenuation—*See* attenuation.

pressure—*See* atmospheric pressure.

pressure altimeter—An *aneroid barometer* with a scale graduated in altitude instead of pressure using *standard atmospheric* pressure-height relationships; shows indicated altitude (not necessarily true altitude); may be set to measure altitude (indicated) from any arbitrarily chosen level. *See* altimeter setting, altitude.

pressure altitude—*See* altitude.

pressure gradient—The rate of decrease of pressure per unit distance at a fixed time.

pressure jump—A sudden, significant increase in *station pressure.*

pressure tendency—*See* barometric tendency.

prevailing easterlies—The broad current or pattern of persistent easterly winds in the Tropics and in polar regions.

prevailing visibility—In the U.S., the greatest horizontal visibility which is equaled or exceeded throughout half of the horizon circle; it need not be a continuous half.

prevailing westerlies—The dominant west-to-east motion of the atmosphere, centered over middle latitudes of both hemispheres.

prevailing wind—Direction from which the wind blows most frequently.

prognostic chart (contracted PROG)—A chart of expected or forecast conditions.

pseudo-adiabatic lapse rate—*See* saturated-adiabatic lapse rate.

psychrometer—An instrument consisting of a *wet-bulb* and a *dry-bulb* thermometer for measuring wet-bulb and dry-bulb temperature; used to determine water vapor content of the air.

pulse—Pertaining to radar, a brief burst of electromagnetic radiation emitted by the radar; of very short time duration. *See* pulse length.

pulse length—Pertaining to radar, the dimension of a radar pulse; may be expressed as the time duration or the length in linear units. Linear dimension is equal to time duration multiplied by the speed of propagation (approximately the speed of light).

Q

quasi-stationary front (commonly called stationary front)—A *front* which is stationary or nearly so; conventionally, a front which is moving at a speed of less than 5 knots is generally considered to be quasi-stationary.

R

RADAR (contraction for radio detection and ranging)—An electronic instrument used for the detection and ranging of distant objects of such composition that they scatter or reflect radio energy. Since *hydrometeors* can scatter radio energy, *weather radars,* operating on certain frequency bands, can detect the presence of precipitation, clouds, or both.

radar altitude—*See* altitude.

radar beam—The focused energy radiated by radar similar to a flashlight or searchlight beam.

radar echo—*See* echo.

radarsonde observation—A *rawinsonde observation* in which winds are determined by radar tracking a balloon-borne target.

radiation—The emission of energy by a medium and transferred, either through free space or another medium, in the form of electromagnetic waves.

radiation fog—*Fog* characteristically resulting when radiational cooling of the earth's surface lowers the air temperature near the ground to or below its initial dewpoint on calm, clear nights.

radiosonde—A balloon-borne instrument for measuring pressure, temperature, and humidity aloft. Radiosonde observation — a *sounding* made by the instrument.

rain—A form of *precipitation*; drops are larger than *drizzle* and fall in relatively straight, althouth not necessarily vertical paths, as compared to drizzle which falls in irregular paths.

rain shower—*See* shower.

range attenuation—*See* attenuation.

range-height indicator (RHI) scope—A radar indicator scope displaying a vertical cross section of *targets* along a selected azimuth.

range resolution—*See* resolution.

RAOB—A *radiosonde* observation.

rawin—A *rawinsonde* observation.

rawinsonde observation—A combined winds aloft and radiosonde observation. Winds are determined by tracking the *radiosonde* by radio direction finder or radar.

refraction—In radar, bending of the *radar beam* by variations in atmospheric density, water vapor content, and temperature.

(1) **normal refraction**—Refraction of the radar beam under normal atmospheric conditions; normal radius of curvature of the beam is about 4 times the radius of curvature of the Earth.

(2) **superrefraction**—More than normal bending of the radar beam resulting from abnormal vertical gradients of temperature and/or water vapor.

(3) **subrefraction**—Less than normal bending of the radar beam resulting from abnormal vertical gradients of temperature and/or water vapor.

relative humidity—The ratio of the existing amount of water vapor in the air at a given temperature to the maximum amount that could exist at that temperature; usually expressed in percent.

relative vorticity—*See* vorticity.

remote scope—In radar meteorology a "slave" scope remoted from weather *radar.*

resolution—Pertaining to radar, the ability of radar to show discrete *targets* separately, i.e., the better the resolutio the closer two targets can be to each other, and stil detected as separate targets.

(1) **beam resolution**—The ability of radar to di between targets at approximately the same r different azimuths.

(2) **range resolution**—The ability of radar to distinguish between targets on the same azimuth but at different ranges.

ridge (also called ridge line)—In meteorology, an elongated area of relatively high atmospheric pressure; usually associated with and most clearly identified as an area of maximum anticyclonic curvature of the wind flow *(isobars, contours,* or *streamlines).*

rime icing (or rime ice)—The formation of a white or milky and opaque granular deposit of ice formed by the rapid freezing of supercooled water droplets as they impinge upon an exposed aircraft.

rocketsonde—A type of *radiosonde* launched by a rocket and making its measurements during a parachute descent; capable of obtaining *soundings* to a much greater height than possible by balloon or aircraft.

roll cloud (sometimes improperly called rotor cloud)—A dense and horizontal roll-shaped accessory cloud located on the lower leading edge of a *cumulonimbus* or less often, a rapidly developing *cumulus*; indicative of turbulence.

rotor cloud (sometimes improperly called roll cloud)—A turbulent cloud formation found in the lee of some large mountain barriers, the air in the cloud rotates around an axis parallel to the range; indicative of possible violent turbulence.

runway temperature—The temperature of the air just above a runway, ideally at engine and/or wing height, used in the determination of density *altitude;* useful at airports when critical values of density altitude prevail.

runway visibility—The *meteorological visibility* along an identified runway determined from a specified point on the runway; may be determined by a *transmissometer* or by an observer.

runway visual range—An instrumentally derived horizontal distance a pilot should see down the runway from the approach end; based on either the sighting of high intensity runway lights or on the visual contrast of other objects, whichever yields the greatest visual range.

S

St. Elmo's Fire (also called corposant)—A luminous brush discharge of electricity from protruding objects, such as masts and yardarms of ships, aircraft, lightning rods, steeples, etc., occurring in stormy weather.

Santa Ana—A hot, dry, *foehn* wind, generally from the northeast or east, occurring west of the Sierra Nevada Mountains ...ially in the pass and river valley near Santa Ana,

...ate of decrease of tempera-
...ed with no gain or
...s with temperature,
...e adiabatic process

...osphere when actual
...m possible at existing

scud—Small detached masses of stratus *fractus* clouds below a layer of higher clouds, usually *nimbostratus.*

sea breeze—A coastal breeze blowing from sea to land, caused by the temperature difference when the land surface is warmer than the sea surface. Compare *land breeze.*

sea fog—A type of *advection fog* formed when air that has been lying over a warm surface is transported over a colder water surface.

sea level pressure—The *atmospheric pressure* at *mean sea level*, either directly measured by stations at sea level or empirically determined from the *station pressure* and temperature by stations not at sea level; used as a common reference for analyses of surface pressure patterns.

sea smoke—Same as *steam fog.*

sector visibility—*Meteorological visibility* within a specified sector of the horizon circle.

sensitivity time control—A radar circuit designed to correct for range *attenuation* so that echo intensity on the scope is proportional to reflectivity of the *target* regardless of range.

shear—*See* wind shear.

shower—*Precipitation* from a *cumuliform* cloud; characterized by the suddenness of beginning and ending, by the rapid change of intensity, and usually by rapid change in the appearance of the sky; showery precipitation may be in the form of rain, ice pellets, or snow.

slant visibility—For an airborne observer, the distance at which he can see and distinguish objects on the ground.

sleet—*See* ice pellets.

smog—A mixture of *smoke* and *fog.*

smoke—A restriction to visibility resulting from combustion.

snow—Precipitation composed of white or translucent ice crystals, chiefly in complex branched hexagonal form.

snow flurry—Popular term for snow *shower*, particularly of a very light and brief nature.

snow grains—*Precipitation* of very small, white opaque grains of ice, similar in structore to *snow* crystals. The grains are fairly flat or elongated, with diameters generally less than 0.04 inch (1 mm).

snow pellet—*Precipitation* consisting of white, opaque, approximately round (sometimes conical) ice particles having a snow-like structure, and about 0.08 to 0.2 inch in diameter; crisp and easily crushed, differing in this respect from *snow grains*; rebound from a hard surface and often break up.

snow shower—*See* shower.

solar radiation—The total electromagnetic *radiation* emitted by the sun. *See* insolation.

sounding—In meteorology, an upper-air observation; a *radiosonde* observation.

source region—An extensive area of the earth's surface characterized by relatively uniform surface conditions where large masses of air remain long enough to take on characteristic temperature and moisture properties imparted by the surface.

specific humidity—The ratio by weight of *water vapor* in a sample of air to the combined weight of water vapor and dry air. Compare *mixing ratio.*

squall—A sudden increase in wind speed by at least 15 knots to a peak of 20 knots or more and lasting for at least one minute. Essential difference between a *gust* and a squall is the duration of the peak speed.

squall line—Any line or narrow band of active *thunderstorms* (with or without *squalls*).

stability—A state of the atmosphere in which the vertical distribution of temperature is such that a *parcel* will resist displacement from its initial level. (*See also* instability.)

standard atmosphere—A hypothetical atmosphere based on climatological averages comprised of numerous physical constants of which the most important are:

(1) A surface *temperature* of 59°F (15°C) and a surface pressure of 29.92 inches of mercury (1013.2 millibars) at sea level;

(2) A *lapse rate* in the troposphere of 6.5°C per kilometer (approximately 2°C per 1,000 feet);

(3) A *tropopause* of 11 kilometers (approximately 36,000 feet) with a temperature of −56.5°C; and

(4) An *isothermal* lapse rate in the stratosphere to an altitude of 24 kilometers (approximately 80,000 feet).

standing cloud (standing lenticular altocumulus)—*See* lenticular cloud.

standing wave—A wave that remains stationary in a moving fluid. In aviation operations it is used most commonly to refer to a *lee wave* or *mountain wave.*

stationary front—Same as *quasi-stationary front.*

station pressure—The actual *atmospheric pressure* at the observing station.

steam fog—Fog formed when cold air moves over relatively warm water or wet ground.

storm detection radar—A weather radar designed to detect *hydrometeors* of precipitation size; used primarily to detect storms with large drops or hailstones as opposed to clouds and light precipitation of small drop size.

stratiform—Descriptive of clouds of extensive horizontal development, as contrasted to vertically developed *cumuliform* clouds; characteristic of stable air and, therefore, composed of small water droplets.

stratocumulus—A low cloud, predominantly *stratiform* in gray and/or whitish patches or layers, may or may not merge; elements are tessellated, rounded, or roll-shaped with relatively flat tops.

stratosphere—The atmospheric layer above the tropopause, average altitude of base and top, 7 and 22 miles, respectively; characterized by a slight average increase of temperature from base to top and is very stable; also characterized by low moisture content and absence of clouds.

stratus—A low, gray cloud layer or sheet with a fairly uniform base; sometimes appears in ragged patches; seldom produces precipitation but may produce *drizzle* or *snow grains.* A *stratiform* cloud.

stratus fractus—*See* fractus.

streamline—In meteorology, a line whose tangent is the wind direction at any point along the line. A flowline.

sublimation—*See* change of state.

subrefraction—See *refraction.*

subsidence—A descending motion of air in the atmosphere over a rather broad area; usually associated with *divergence.*

summation principle—The principle states that the cover assigned to a layer is equal to the summation of the sky cover of the lowest layer plus the additional coverage at all successively higher layers up to and including the layer in question. Thus, no layer can be assigned a sky cover less than a lower layer, and no sky cover can be greater than 1.0 (10/10).

superadiabatic lapse rate—A *lapse rate* greater than the *dry adiabatic lapse rate. See* absolute instability.

supercooled water—Liquid water at temperatures colder than freezing.

superrefraction—*See* refraction.

surface inversion—An *inversion* with its base at the surface, often caused by cooling of the air near the surface as a result of *terrestrial radiation*, especially at night.

surface visibility—Visibility observed from eye-level above the ground.

synoptic chart—A chart, such as the familiar weather map, which depicts the distribution of meteorological conditions over an area at a given time.

T

target—In radar, any of the many types of objects detected by radar.

temperature—In general, the degree of hotness or coldness as measured on some definite temperature scale by means of any of various types of thermometers.

temperature inversion—*See* inversion.

terrestrial radiation—The total infrared *radiation* emitted by the Earth and its atmosphere.

thermograph—A continuous-recording *thermometer.*

thermometer—An instrument for measuring *temperature.*

theodolite—An optical instrument which, in meteorology, is used principally to observe the motion of a *pilot balloon.*

thunderstorm—In general, a local storm invariably produced by a *cumulonimbus* cloud, and always accompanied by lightning and thunder.

tornado (sometimes called cyclone, twister)—A violently rotating column of air, pendant from a cumulonimbus cloud, and nearly always observable as "funnel-shaped." It is the most destructive of all small-scale atmospheric phenomena.

towering cumulus—A rapidly growing *cumulus* in which height exceeds width.

tower visibility—*Prevailing visibility* determined from the control tower.

trade winds—Prevailing, almost continuous winds blowing with an easterly component from the subtropical high pressure belts toward the *intertropical convergenze zone*; northeast in the Northern Hemisphere, southeast in the Southern Hemisphere.

transmissometer—An instrument system which shows the transmissivity of light through the atmosphere. Transmissivity may be translated either automatically or manually into *visibility* and/or *runway visual range.*

tropical air—An air mass with characteristics developed over low latitudes. Maritime tropical air (mT), the principal type, is produced over the tropical and subtropical seas; very warm and humid. Continental tropical (cT) is produced over subtropical arid regions and is hot and very dry. Compare *polar air.*

tropical cyclone—A general term for a *cyclone* that originates over tropical oceans. By international agreement, tropical cyclones have been classified according to their intensity, as follows:

(1) **tropical depression**—Winds up to 34 knots (64 km/h);

(2) **tropical storm**—winds of 35 to 64 knots (65 to 119 km/h);

(3) **hurricane or typhoon**—winds of 65 knots or higher (120 km/h).

tropical depression—*See* tropical cyclone.

tropical storm—*See* tropical cyclone.

tropopause—The transition zone between the *troposphere* and *stratosphere*, usually characterized by an abrupt change of *lapse rate.*

troposphere—That portion of the *atmosphere* from the earth's surface to the *tropopause*; that is, the lowest 10 to 20 kilometers of the atmosphere. The troposphere is characterized by decreasing temperature with height, and by appreciable water vapor.

trough (also called trough line)—In meteorology, an elongated area of relatively low atmospheric pressure; usually associated with and most clearly identified as an area of maximum cyclonic curvature of the wind flow (*isobars, contours,* or *streamlines*); compare with *ridge.*

true altitude—*See* altitude.

true wind direction—The direction, with respect to true north, from which the wind is blowing.

turbulence—In meteorology, any irregular or disturbed flow in the atmosphere.

twilight—The intervals of incomplete darkness following sunset and preceding sunrise. The time at which evening twilight ends or morning twilight begins is determined by arbitrary convention, and several kinds of twilight have been defined and used; most commonly civil, nautical, and astronomical twilight.

(1) **Civil Twilight**—The period of time before sunrise and after sunset when the sun is not more than 6° below the horizon.

(2) **Nautical Twilight**—The period of time before sunrise and after sunset when the sun is not more than 12° below the horizon.

(3) **Astronomical Twilight**—The period of time before sunrise and after sunset when the sun is not more than 18° below the horizon.

twister—In the United States, a colloquial term for *tornado.*

typhoon—A *tropical cyclone* in the Eastern Hemisphere with winds in excess of 65 knots (120 km/h).

U

undercast—A cloud *layer* of ten-tenths (1.0) coverage (to the nearest tenth) as viewed from an observation point above the layer.

unlimited ceiling—A clear sky or a sky cover that does not meet the criteria for a *ceiling.*

unstable—*See* instability.

updraft—A localized upward current of air.

upper front—A *front* aloft not extending to the earth's surface.

upslope fog—Fog formed when air flows upward over rising terrain and is, consequently, adiabatically cooled to or below its initial *dewpoint.*

vapor pressure—In meteorology, the pressure of water vapor in the atmosphere. Vapor pressure is that part of the total atmospheric pressure due to water vapor and is independent of the other atmospheric gases or vapors.

vapor trail—Same as *condensation trail.*

veering—Shifting of the wind in a clockwise direction with respect to either space or time; opposite of backing. Commonly used by meteorologists to refer to an anticyclonic shift (clockwise in the Northern Hemisphere and counterclockwise in the Southern Hemisphere).

vertical visibility—The distance one can see upward into a surface based *obscuration*; or the maximum height from which a pilot in flight can recognize the ground through a surface based obscuration.

virga—Water or ice particles falling from a cloud, usually in wisps or streaks, and evaporating before reaching the ground.

visibility—The greatest distance one can see and identify prominent objects.

visual range—*See* runway visual range.

vortex—In meteorology, any rotary flow in the atmosphere.

vorticity—Turning of the atmosphere. Vorticity may be imbedded in the total flow and not readily identified by a flow pattern.

(a) **absolute vorticity**—the rotation of the Earth imparts vorticity to the atmosphere; absolute vorticity is the combined vorticity due to this rotation and vorticity due to circulation relative to the Earth (relative vorticity).

(b) **negative vorticity**—vorticity caused by anticyclonic turning; it is associated with downward motion of the air.

(c) **positive vorticity**—vorticity caused by cyclonic turning; it is associated with upward motion of the air.

(d) **relative vorticity**—vorticity of the air relative to the Earth, disregarding the component of vorticity resulting from Earth's rotation.

W

wake turbulence—*Turbulence* found to the rear of a solid body in motion relative to a fluid. In aviation terminology, the turbulence caused by a moving aircraft.

wall cloud—The well-defined bank of vertically developed clouds having a wall-like appearance which form the outer boundary of the *eye* of a well-developed *tropical cyclone.*

warm front—Any non-occluded *front* which moves in such a way that warmer air replaces colder air.

warm sector—The area covered by warm air at the surface and bounded by the *warm front* and *cold front* of a *wave cyclone.*

water equivalent—The depth of water that would result from the melting of snow or ice.

waterspout—*See* tornado.

water vapor—Water in the invisible gaseous form.

wave cyclone—A *cyclone* which forms and moves along a front. The circulation about the cyclone center tends to produce a wavelike deformation of the front.

weather—The state of the *atmosphere*, mainly with respect to its effects on life and human activities; refers to instantaneous conditions or short term changes as opposed to *climate.*

weather radar—Radar specifically designed for observing weather. *See* cloud detection radar and storm detection radar.

weather vane—A *wind vane.*

wedge—Same as *ridge.*

wet bulb—Contraction of either *wet-bulb temperature* or *wet-bulb thermometer.*

wet-bulb temperature—The lowest *temperature* that can be obtained on a *wet-bulb thermometer* in any given sample of air, by evaporation of water (or ice) from the muslin wick; used in computing *dewpoint* and *relative humidity.*

wet-bulb thermometer—A thermometer with a muslin-covered bulb used to measure wet-bulb temperature.

whirlwind—A small, rotating column of air; may be visible as a dust devil.

willy-willy—A *tropical cyclone* of hurricane strength near Australia.

wind—Air in motion relative to the surface of the earth; generally used to denote horizontal movement.

wind direction—The direction **from** which wind is blowing.

wind speed—Rate of wind movement in distance per unit time.

wind vane—An instrument to indicate wind direction.

wind velocity—A vector term to include both *wind direction* and *wind speed.*

wind shear—The rate of change of *wind velocity* (direction and/or speed) per unit distance; conventionally expressed as vertical or horizontal wind shear.

X-Y-Z

zonal wind—A west wind; the westerly component of a wind. Conventionally used to describe large-scale flow that is neither cyclonic nor anticyclonic.

INDEX